Novel Approaches to the Structure and Dynamics of Liquids: Experiments, Theories and Simulations

NATO Science Series

A Series presenting the results of scientific meetings supported under the NATO Science Programme.

The Series is published by IOS Press, Amsterdam, and Kluwer Academic Publishers in conjunction with the NATO Scientific Affairs Division

Sub-Series

I. Life and Behavioural Sciences	IOS Press
II. Mathematics, Physics and Chemistry	Kluwer Academic Publishers
III. Computer and Systems Science	IOS Press
IV. Earth and Environmental Sciences	Kluwer Academic Publishers
V. Science and Technology Policy	IOS Press

The NATO Science Series continues the series of books published formerly as the NATO ASI Series.

The NATO Science Programme offers support for collaboration in civil science between scientists of countries of the Euro-Atlantic Partnership Council. The types of scientific meeting generally supported are "Advanced Study Institutes" and "Advanced Research Workshops", although other types of meeting are supported from time to time. The NATO Science Series collects together the results of these meetings. The meetings are co-organized bij scientists from NATO countries and scientists from NATO's Partner countries – countries of the CIS and Central and Eastern Europe.

Advanced Study Institutes are high-level tutorial courses offering in-depth study of latest advances in a field.
Advanced Research Workshops are expert meetings aimed at critical assessment of a field, and identification of directions for future action.

As a consequence of the restructuring of the NATO Science Programme in 1999, the NATO Science Series has been re-organised and there are currently Five Sub-series as noted above. Please consult the following web sites for information on previous volumes published in the Series, as well as details of earlier Sub-series.

http://www.nato.int/science
http://www.wkap.nl
http://www.iospress.nl
http://www.wtv-books.de/nato-pco.htm

Series II: Mathematics, Physics and Chemistry – Vol. 133

Novel Approaches to the Structure and Dynamics of Liquids: Experiments, Theories and Simulations

edited by

Jannis Samios
Department of Chemistry, Physical Chemistry,
University of Athens, Athens, Greece
(Chairman of the NATO-ASI)

and

Vladimir A. Durov
Lomonosov Moscow State University,
Moscow, Russia
(Vice-Chairman of the NATO-ASI)

Kluwer Academic Publishers

Dordrecht / Boston / London

Published in cooperation with NATO Scientific Affairs Division

Proceedings of the NATO Advanced Study Institute on
Novel Approaches to the Structure and Dynamics of Liquids: Experiments, Theories and Simulations
Rhodes, Greece
6–15 September 2002

A C.I.P. Catalogue record for this book is available from the Library of Congress.

ISBN 1-4020-1847-9 (PB)
ISBN 1-4020-1846-0 (HB)
ISBN 1-4020-2384-7 (e-book)

Published by Kluwer Academic Publishers,
P.O. Box 17, 3300 AA Dordrecht, The Netherlands.

Sold and distributed in North, Central and South America
by Kluwer Academic Publishers,
101 Philip Drive, Norwell, MA 02061, U.S.A.

In all other countries, sold and distributed
by Kluwer Academic Publishers,
P.O. Box 322, 3300 AH Dordrecht, The Netherlands.

Printed on acid-free paper

Printed in the Netherlands.

Table of Contents

Preface

The unique behavior of the "liquid state", together with the richness of phenomena that are observed, render liquids particularly interesting for the scientific community. Note that the most important reactions in chemical and biological systems take place in solutions and liquid-like environments. Additionally, liquids are utilized for numerous industrial applications. It is for these reasons that the understanding of their properties at the molecular level is of foremost interest in many fields of science and engineering.

What can be said with certainty is that both the experimental and theoretical studies of the liquid state have a long and rich history, so that one might suppose this to be essentially a solved problem. It should be emphasized, however, that although, for more than a century, the overall scientific effort has led to a considerable progress, our understanding of the properties of the liquid systems is still incomplete and there is still more to be explored. Basic reason for this is the "many body" character of the particle interactions in liquids and the lack of long-range order, which introduce in liquid state theory and existing simulation techniques a number of conceptual and technical problems that require specific approaches. Also, many of the elementary processes that take place in liquids, including molecular translational, rotational and vibrational motions (Trans.-Rot.-Vib. coupling), structural relaxation, energy dissipation and especially chemical changes in reactive systems occur at different and/or extremely short timescales. Therefore, their examination requires the application of a wide range of spectroscopic, diffraction and computational techniques. As a result, it is quite necessary to expand and develop further the existing experimental, theoretical and computer simulation methods in order to understand the properties of liquids more precisely.

The present period is rather favorable for the research of liquids. A large number of sophisticated experimental, theoretical and computer simulation methods is already available, yielding valuable information as far as the bulk-thermodynamic, structural and dynamic properties of liquid systems are concerned. Spectroscopic techniques, such as the X-ray diffraction, synchrotron radiation, EXAFS and small angle neutron scattering (SANS), allow the probing of the intra and intermolecular structure of molecules in a wide range of wave vectors. On the other hand, spectroscopic techniques ranging from conventional dynamical spectroscopy (neutron and light scattering, NMR, dielectric and vibrational spectroscopy) to ultra short laser pulses, transient grating methods, photo acoustic detection, femtosecond spectroscopy etc., the application of which is constantly increasing, have produced novel results and opened up new directions and trends of research in real time dynamics of molecules, energy relaxation, coherent effects, hydrogen bonding and in the time resolution of very rapid chemical phenomena in liquids.

In recent years, the scientific interest has widened, since related subjects concerning confined liquids, liquids in biophysics, in biochemistry and in the earth sciences have also been included. More recently, a sort of refinement and improvement of the liquid state theory (e. g. extended RISM, MF, MMK, DFT, etc.), as well as developments characterized by the extension of the simulation techniques to incorporate quantum effects on various levels (time-independent & dependent quantum simulation methods) have been tested by several investigators in a number of laboratories.

It is obvious that the extent and diversity of the experimental and theoretical efforts required to attain a thorough understanding of the properties of liquids, render their study rather laborious and to some extent discouraging for students and young researchers. At this stage, scientific meetings and advanced study institutes are quite useful since they facilitate, beyond any doubt, the exchange of experience between scientists from theoretical and experimental fields of research in liquid state.

The *Scientific Affairs Division of* NATO has set as one of its aims to promote scientific activities, disseminate advanced knowledge and foster international scientific contacts through high-level teaching courses. Having as guideline previous conferences on the Physical Chemistry of liquid state, organized as NATO Advanced Study Institutes (NATO-ASIs) and encouraged by the "European Molecular Liquids Group" (EMLG), we thought that the above objectives would be met if we managed to attract and bring together experts and students interested in the study of problems related to molecular liquids. In this spirit, a NATO-ASI on the Physical Chemistry of liquid systems, entitled "Novel Approaches to the Structure & Dynamics of Liquids: Experiments, Theories and Simulations", was held at the hotel Esperos Village (Congress Venue) & Esperides, Rhodes, Greece, from September 7 to 15, 2002. The ASI was organized by Assoc. Prof. J. Samios (awarded co-organizer & chair, Athens-Greece), assisted by Prof. Vl. A. Durov (NATO partner country, awarded co-organizer & vice-chair, Moscow-Russia).

The NATO-ASI at Rhodes was planned as an advanced school with seminars and various sessions aiming at graduate students, postdoctoral fellows and researchers, who either enter the field or wish to update their knowledge on the most recent developments in the area of the liquid state. As stated in the title of the meeting, the course of the ASI was organized in such a way as to offer an equally extensive coverage of the three broad areas of interest, namely "*experimental techniques*", "*analytical theory*" and "*computer simulations*".

In general, the primary aim of this ASI was threefold: (a) to offer in a series of lectures, seminars and hands–on computer exercises and workshops a broad overview of the experimental-theoretical and simulation techniques (e.g. training on tools and new software for computer simulations) available today for the study of liquids, (b) to offer extensive and state-of-the-art reviews concerning the progress made in the three aforementioned fields and to discuss problems that need specific experimental and theoretical investigation and (c) to encourage and promote collaborative research projects between groups of researchers who share common problems from all three areas of research, as advocated by the "European Molecular Liquid Group" (EMLG). In this context, morning and early afternoon sessions were devoted to full, one-hour lectures, with the shorter and more specialized ones inserted at the end of the afternoon session.

A third series of lectures and seminars, designed to cover basic aspects and techniques and aiming at graduate students in particular, were delivered before dinner. A number of them were accompanied by hands-on experiments on a computer network, installed for this purpose, in a room very close to the conference hall. The computer seminars gave the students the opportunity to carry out small projects (e.g. quantum mechanical & computer simulations) by using not only the local computers, but also the computing facilities in other countries as well. At this point, we should have to acknowledge the valuable help of Dr D. Dellis (Athens-Greece) who prepared the educational computer projects on the simulation techniques and Assist. Prof. A. Koutselos for his assistance on several aspects of the meeting. Also, a poster session was organized with more than 100 entries and with topics that covered a broad variety of experimental techniques, as well as theoretical and computer simulation methods. Prominent scientists and researchers in the field, having put, in first place, an emphasis on the educational aspect of their lectures, presented their impressive work before an international audience. In this manner, it was possible for young participants to interact extensively with scientific leaders in the field.

At this ASI we had the opportunity to come in contact and discuss new advances in the study of liquid state phenomena. The topics of the presentations were devoted to novel approaches of problems related to the structural and dynamic properties of systems ranging from simple molecular liquids to complex systems, such as extremely anisotropic molecules, aqueous and non-aqueous solutions, ionic fluids and solutions, liquids in porous materials, polymers, polymer mixtures and solutions as well as membranes and proteins in polar solvents. The advances made in complex liquids have been stimulated by the fact that they play an important role in various chemical and technological processes and in biological systems. In general, considerable attention was devoted to the study of the effects the molecular interactions have on various simple and collective processes in liquid systems. A particularly interesting aspect of this ASI was the contribution, in terms of experience and ideas, of researchers working on classical and quantum molecular simulation techniques. Special attention was also given on the most recent developments in the Statistical-Mechanical analytical techniques and in their implementation in the study of complex fluids and critical phenomena. A considerable part of the scientific program was devoted to lectures with main emphasis in recent developments and advances of experimental techniques. Studies concerning effects of the solute-solvent interactions on translational, rotational and vibrational molecular motions and on the inter- intra-molecule structural changes in specific solutions were also presented and critically discussed.

It is generally recognized that the aims of this summer school have been fulfilled. The present NATO-ASI book contains all main lectures, as well as some selected shorter contributions that were presented in the meeting. The contents of this volume address the results of this ASI since they cover recent developments in the three areas of current research in liquid state. The Editors would like to thank all contributing authors for the systematic and careful preparation of their articles. Finally, it is a great pleasure to thank all the participants of the meeting who have significantly contributed to the success of this NATO-ASI at Rhodes.

Jannis Samios (Chair)
(Athens-Greece)

Vladimir A. Durov (vice-Chair)
(Moscow-Russia)

Acknowledgments

At this point, I would like to express our gratefulness to a number of people, who have contributed to the development and success of this Summer School. It is a great pleasure to thank all the sponsors of this ASI and foremost the *Scientific Affairs Division* of NATO for the generous financial support that made this meeting possible. In particular, special thanks go to Dr. F. Pedrazzini, NATO Program Director, PST, and his Secretary Mrs A. Trapp for their valuable assistance provided in numerous aspects during the organization of the meeting. We are also thankful to the IUPAC, to the Ministry of Development of Greece and to the University of Athens, Greece for financial support. I would like to mention here the fruitful cooperation and assistance, especially on the scientific part of the meeting, of Prof. Philippe A. Bopp (Univ. Bordeaux France) and Prof. Vladimir A. Durov (Lomonosov Univ. Moscow Russia). Finally, I would like to thank my students in Athens for their help in editing this book.

J. Samios

INTERMOLECULAR INTERACTIONS AND COOPERATIVE EFFECTS FROM ELECTRONIC STRUCTURE CALCULATIONS: AN EFFECTIVE MEANS FOR DEVELOPING INTERACTION POTENTIALS FOR CONDENSED PHASE SIMULATIONS

SOTIRIS S. XANTHEAS
Chemical Sciences Division, Pacific Northwest National Laboratory,906 Battelle Boulevard, PO Box 999, MS K1-83, Richland, WA 99352 U.S.A.

1. INTRODUCTION

The modeling of the macroscopic properties of homogeneous and inhomogeneous systems via atomistic simulations such as molecular dynamics (MD) or Monte Carlo (MC) techniques is based on the accurate description of the relevant solvent-solute and solvent-solvent intermolecular interactions. The total energy (U) of an n-body molecular system can be formally written as [1,2,3]

$$U = U_{1-body} + U_{2-body} + U_{3-body} + ... + U_{k-body} + ... + U_{n-body}. \quad (1)$$

For simplicity, we will furthermore assume that U is evaluated as the energy difference with respect to the energies of the isolated molecular constituents (or "bodies") at infinite distances between each other, which can be conveniently chosen as the zero of the energy scale. Then, U_{1-body} represents the energy penalty (or deformation) due to the change of the individual intramolecular geometries of all bodies from their gas phase values due to their mutual interaction. Because of the fact that this term can be relatively small, especially for systems exhibiting weak intermolecular (solvent-solvent and solute-solvent) interactions, it is often not taken into account [4]. However, for systems like negative ion-water clusters it can exceed 5 kcal/mol [4,5]. Even for molecular systems exhibiting weak intermolecular interactions, this term does describe a physical effect that has a distinct spectroscopic signature manifested experimentally by the shift of the individual infrared (IR) bands of the interacting molecular constituents from their isolated gas phase values. As a typical example, in the water dimer the deformation energy is ~0.04 kcal/mol as a result of typical changes of ~0.07 Å in the bond lengths and 0.4° in the bond angles [6] from the gas phase monomer values. These internal geometry changes induce a measured red shift of 155 cm^{-1} for the hydrogen bonded OH stretching frequency [7]. The deformation energy is larger for liquid water and ice as a result of the larger changes in water's internal geometry (0.972/1.008 Å for the bond lengths and 106.6/109.5° for the bond angles for liquid water[8]/ice[9], respectively) from the experimental gas phase monomer values [10] of 0.957Å and 104.52°, a fact that in turn induces larger red shifts for the OH stretches [11].

J. Samios and V.A. Durov (eds.), Novel Approaches to the Structure and Dynamics of Liquids: Experiments, Theories and Simulations, 1–15.

The second term in the many-body expansion (1) is the energy of all the combinations of the constituent dimers,

$$U_{2\text{-}body} = \sum_{i=1}^{n-1}\sum_{j>i}^{n} \Delta^2 E(ij) \tag{2}$$

where

$$\Delta^2 E(ij) = E(ij) - \{E(i) + E(j)\}, \tag{2a}$$

the third term is the energy of all the trimers

$$U_{3\text{-}body} = \sum_{i=1}^{n=2}\sum_{j>i}^{n-1}\sum_{k>j}^{n} \Delta^3 E(ijk) \tag{3}$$

where

$$\Delta^3 E(ijk) = E(ijk) - \{E(i) + E(j) + E(k)\} - \{\Delta^2 E(ij) + \Delta^2 E(ik) + \Delta^2 E(jk)\} \tag{3a}$$

and so on. The 1- and 2-body terms constitute the additive, whereas the 3-body and higher terms constitute the non-additive components of the many body expansion.

Given the form of the many-body expansion (1) the question naturally arises of ways to evaluate the energy U for each nuclear configuration. This is usually done via

(i) *a brute force approach* during which the total energy U is obtained from electronic structure calculations with (semi-empirical) or without (*ab-initio*) adjustable parameters or

(ii) *a step-by-step approach* during which the individual pieces of the many-body expansion (1) are evaluated from a chosen function that describes the interactions between the various constituents.

The first approach offers the advantage of including the electronic degrees of freedom (different states) and can therefore describe reactions during which chemical bonds are broken and new ones are formed, it is in principle more accurate provided that the chosen level of electronic structure theory is adequate for the problem at hand but suffers from the large computational expense which currently makes it impractical for simulations of macroscopic systems with realistic choices of simulation times and large periodically repeated unit cells. The second approach is computationally more tractable but is usually less accurate with respect to the first one.

Important issues that are associated with the previous discussion consist of the accuracy of the electronic structure methods, the order of the convergence of the many-body expansion and the choice of the functional forms that are used to describe the intermolecular interactions. In the following we will present brief overviews of these issues and use the interactions in water as an example to further discuss their significance.

2. SYSTEMATIC CONVERGENCE OF BINDING ENERGIES FROM ELECTRONIC STRUCTURE CALCULATIONS

The binding energy of an assembly of n bodies is defined as

$$\Delta E = E(ijk...n) - \{E(i) - E(j) - E(k) - ... - E(n)\} \quad (4)$$

where $E(ijk...n)$ is the energy of the system at its equilibrium position and $E(i)$ the corresponding energy of fragment i at its isolated equilibrium geometry, obtained as solutions of the non-relativistic Schrödinger equation. It has been previously shown [2,6] that the Hartree-Fock (HF) level of theory underestimates the binding energies of weakly bound (~ 5 kcal/mol or less) hydrogen bonded systems and produces intramolecular distances that are too large with respect to experiment. Among the various methods that incorporate electron correlation [12] and are based on the HF wavefunction, the Møller-Plesset perturbation [13] theory (MP2, MP3, MP4) and coupled-cluster [14] based [CCSD, CCSD(T)] methods have been previously reported to yield accurate binding energies for small hydrogen bonded clusters [15]. A recent study [16] of the binding energies of the first few water clusters n=2-6 furthermore suggested that the MP2 binding energies are within <0.5% from the ones obtained at the CCSD(T) level of theory. Although the difference between MP2 and CCSD(T) can be larger for other hydrogen bonded systems (such as 2% for the HF dimmer [15,17]), the suggestion of initial calculations at the MP2 level with subsequent ones at the CCSD(T) level of theory seems to provide a realistic approach towards obtaining accurate estimates of the binding energies of hydrogen bonded systems.

Coupled to the expansion of the level of theory to better describe electron correlation is the necessary expansion of the one-electron orbital basis set to better describe the radial and angular parts of the electronic wavefunction. Among the numerous basis sets that have been developed based on different philosophies [18], the family of correlation consistent basis sets of Dunning and co-workers [19] clearly provides a systematic avenue towards reaching this goal. These sets were constructed by grouping together all basis functions that contribute roughly equal amounts to the correlation energy of the atomic ground states. In this approach functions are added to the basis sets in shells. The sets approach the complete basis set (CBS) limit, for each succeeding set in the series provides an ever more accurate description of both the atomic radial and angular spaces. It has been previously found that, for hydrogen bonded dimmers [17,20], the family of correlation consistent basis sets that is augmented with additional diffuse functions in order to better describe the atomic negative ions (aug-cc-pVxZ, x=D, T, Q, 5, 6) provides a faster convergence to the CBS limit when compared to the plain (non-augmented) cc-pVxZ sets. A caveat of the interplay between the level of electronic structure theory and the size of the orbital basis set towards obtaining the "correct answer for the right reason" has been recently emphasized [21].

In the calculation of the binding energies of weakly bound complexes it has been argued [22] that due to the fact that the size of the basis sets used in practice are finite, the basis functions on each fragment help lower the energy of a neighboring fragment and vice versa. Correction for this effect, which is termed Basis Set Superposition Error (BSSE), is usually estimated via the function counterpoise (fCP) method [23] of Boys and Bernardi. In this method, the (BSSE-uncorrected) binding energy ΔE of a dimer consisting of two fragments A and B is

$$\Delta E = E_{AB}^{\alpha\cup\beta}(AB) - E_{A}^{\alpha}(A) - E_{B}^{\beta}(B) \quad (5)$$

where $E_G^{\sigma}(M)$ corresponds to the energy of a molecular system M at geometry G computed with basis set σ. The BSSE-corrected binding energy $\Delta E(BSSE)$ is [4]

$$\Delta E(BSSE) = E_{AB}^{\alpha\cup\beta}(AB) - E_{AB}^{\alpha\cup\beta}(A) - E_{AB}^{\alpha\cup\beta}(B) + E_{rel}^{\alpha}(A) + E_{rel}^{\beta}(B) \quad (6)$$

where

$$E_{rel}^{\alpha}(A) = E_{AB}^{\alpha}(A) - E_{A}^{\alpha}(A) \quad (7a)$$

$$E_{rel}^{\beta}(B) = E_{AB}^{\beta}(B) - E_{B}^{\beta}(B) \quad (7b)$$

are the fragment deformation energies, i.e. the energy penalty associated with the distortion of the molecular constituents from their isolated geometries to the ones they assume in the cluster. Equation (6) can be recast with the help of equations (5) and (7a)-(7b) as

$$\Delta E(BSSE) = \Delta E - \{E_{AB}^{\alpha\cup\beta}(A) - E_{AB}^{\alpha}(A)\} - \{E_{AB}^{\alpha\cup\beta}(B) - E_{AB}^{\beta}(B)\}, \quad (8)$$

suggesting that ΔE and $\Delta E(BSSE)$ do converge to the same value at the CBS limit since, at this limit, the term in the brackets is zero. The problems arising from the omission of the fragment deformation energies from the calculation of $\Delta E(BSSE)$ especially when large basis sets are used has been previously discussed [4]. The difference between the uncorrected and BSSE-corrected binding energies therefore provides an estimate of the basis set incompleteness. The variation of the binding energy of the NH_3-H_2O dimer at the MP2 level of theory with the aug-cc-pV*x*Z sets (*x*=D, T, Q, 5) is shown in Figure 1. It is seen that the difference between the uncorrected and BSSE-corrected binding energies is approximately halved upon increasing the cardinal number *x* of the basis set, i.e., going from *x* to *x*+1. Even with the aug-cc-pV5Z set the correction amounts to 0.12 kcal/mol at the MP2 level.

The smooth variation of the binding energies with the cardinal number *x* allows for a *heuristic* extrapolation of the binding energy with *x*. It should be noted that this extrapolation, for this case performed using the polynomial [24]

$$\Delta E = \Delta E_{CBS} + \gamma/(\ell_{max} + 1)^4 + \delta/(\ell_{max} + 1)^5 \quad (9)$$

where ℓ_{max} is the value of the highest angular momentum function in the basis set, γ, δ, and ΔE_{CBS} (the CBS limit) are constants determined by a least-mean-squares fit, only results in an improvement of <0.1 kcal/mol with respect to the best computed values with the largest aug-cc-pV5Z. Indeed with this basis set the uncorrected and BSSE-corrected binding energies are –6.56 and –6.44 kcal/mol, respectively, whereas the CBS limit is –6.51 kcal/mol.

It should be once again emphasized that the extrapolation process is a *heuristic* approach that is based on the graphical inspection of the results rather than a formal mathematical formulation and as such has spurred several suggestions as to the choice of the functional form used in the extrapolation process (see Reference [25] for a recent account).

Figure 1 nevertheless illustrates the importance of performing accurate electronic structure calculation in order to quantify the two-body interaction. For instance, the use of the aug-cc-pVDZ set overestimates the two-body ammonia-water interaction by 7%.

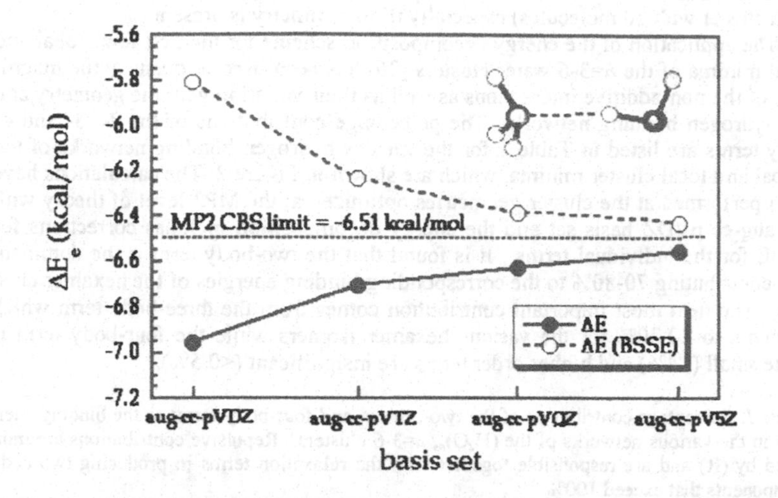

Figure 1. Variation of the MP2 binding energy of the NH_3-H_2O dimer with the family of augmented correlation consistent basis sets of double through quintuple zeta quality.

3. MAGNITUDE OF MANY-BODY, NON-ADDITIVE INTERACTIONS

Provided that the many-body interaction terms can be quantitatively determined from high level electronic structure calculations, it is desirable to assess the order of convergence of the many-body expansion [eq. (1)], i.e., the importance of the 3-body, 4-body and higher terms. This is achieved by examining the energetics of the larger (n>2) clusters and performing the decomposition of their binding energies according to eq. (1). The effect of BSSE in the evaluation of the many body terms is straightforward [2].

The enumeration of the many-body terms corresponds to a combinatorial problem as for a cluster of size n the energies of all monomers, dimers, trimers etc need to be computed. Since the number of ways of choosing m molecules from a collection of n is

$$\binom{n}{m} = \binom{n}{n-m} = \frac{n!}{m!(n-m)!} \qquad (10)$$

and furthermore

$$\sum_{m=1}^{n} \binom{n}{m} = 2^n - 1 \qquad (11)$$

one needs to perform (2^n-1) calculations in order to evaluate the 1-, 2-, ..., n-body terms of a cluster of n molecules. This process can be quite tedious (>1,000 terms for a cluster with 10 molecules) especially if no symmetry is present.

The application of the energy decomposition scheme for the first few global and local minima of the n=3-6 water clusters [26] has been used to quantify the magnitude of the non-additive interactions as well as their variation with the geometry and the hydrogen bonding network. The percentage contributions of the 2-, 3- and 4-body terms are listed in Table 1 for the various hydrogen bonding networks of the global and local cluster minima, which are shown in Figure 2. The calculations have been performed at the cluster geometries optimized at the MP2 level of theory with the aug-cc-pVDZ basis set and the energy decomposition includes corrections for BSSE for the individual terms. It is found that the two-body term is the dominant one contributing 70-80% to the corresponding binding energies of the hexamer clusters. The next most important contribution comes from the three-body term which amounts to 20-30% for the various hexamer isomers while the four-body term is quite small (<4%) and higher order terms are insignificant (<0.5%).

Table 1. Percentage contribution of the two-, three- and four-body terms to the binding energies in the various networks of the $(H_2O)_n$, n=3-6 clusters. Repulsive contributions are indicated by (R) and are responsible together with the relaxation terms in producing two-body components that exceed 100%.

Cluster	Network	Two-body	Three-body	Four-body
Trimer	*(da,da,da)*	85.1	17.7	
	(a,dd,a)	106.7	5.6 (R)	
	(d,aa,d)	104.6	3.9 (R)	
Tetramer	*(da,da,da,da)*	76.3	25.6	2.2
	Cage	87.3	16.5	0.6 (R)
	(aa,dd,aa,dd)	109.2	9.6 (R)	0.7
	(aa,da,dd,da)	106.8	4.5 (R)	
Pentamer	*(da,da,da,da,da)*	71.8	28.6	3.7
	Cage	80.4	23.1	1.3
Hexamer	Prism	80.3	22.8	1.5
	Cage	79.9	23.2	1.2
	Book	75.6	26.1	2.5
	S_6 *(da,da,da,da,da,da)*	69.5	29.7	4.4

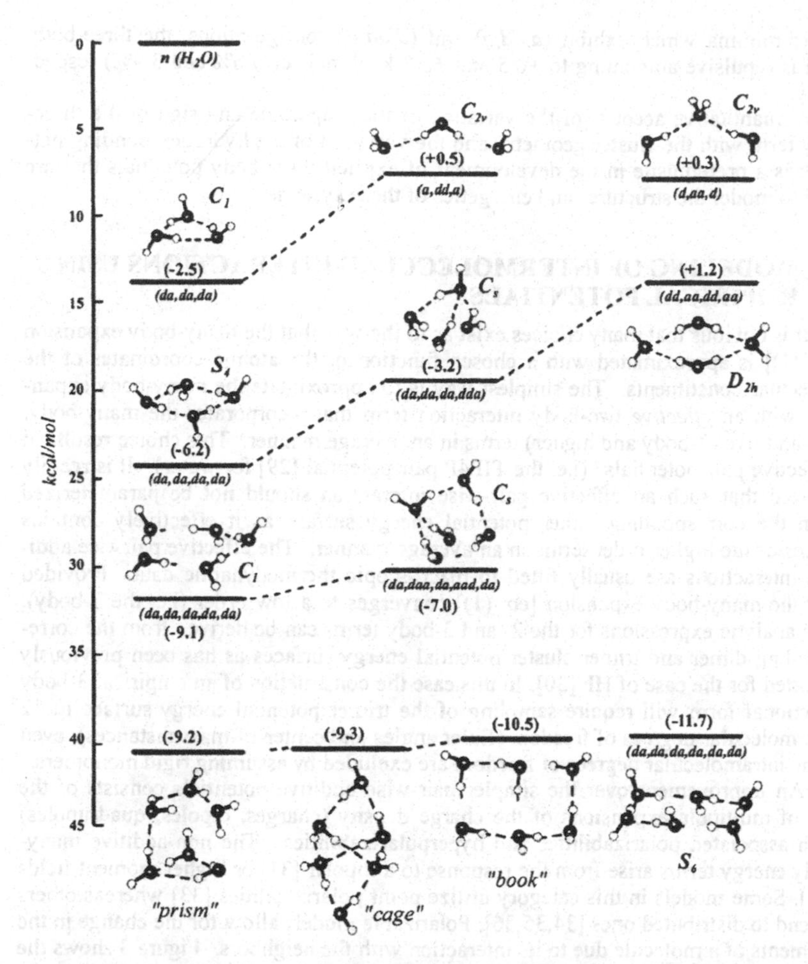

Figure 2. **Geometries of the various water cluster minima $(H_2O)_n$, n=3-6. The hydrogen bonding network is denoted as (*d*)=donor, (*a*)=acceptor to neighbors and their combinations. The magnitude of the 3-body term (in kcal/mol) is listed in parentheses.**

These results furthermore suggest that purely 2-body interaction potentials derived from the water dimer potential energy surface will result in errors exceeding 20% for clusters whereas 2- plus 3-body empirical potentials that are parameterized from the dimer and trimer potential energy surfaces will be more accurate, being in error of just 4% or less for the minimum energy configurations of these clusters. The results of Table 1 and Figure 2 furthermore indicate that the magnitude of the 3-body term depends both on the cluster geometry [27] as well as the nature of the hydrogen bonding network [28]. For instance, for the global "ring" minimum of the water trimer in which all 3 water molecules act as simultaneous donors-acceptors (*da*) to neighbors, the 3-body term is attractive and amounts to -2.5 kcal/mol or 17.7% of the total cluster binding energy. However, for the higher lying "open"

trimer minima which exhibit (*a,dd,a*) and (*d,aa,d*) configurations, the three-body term is repulsive amounting to +0.5 and +0.3 kcal/mol (or 5.6% and 3.9%) respectively.

A quantitative account of the variation of the magnitude and sign of the three-body term with the cluster geometry and the variation of the hydrogen bonding network is a prerequisite in the development of explicit three-body potentials that are used to model the structures and energetics of these systems.

4. MODELING OF INTERMOLECULAR INTERACTIONS USING EMPIRICAL POTENTIALS

It is obvious that many choices exist as to the way that the many-body expansion [eq. (1)] is approximated with a chosen function of the atomic coordinates of the molecular constituents. The simplest way is to approximate the many-body expansion with an *effective* two-body interaction term that incorporates the many-body, non-additive (3-body and higher) terms in an average manner. This choice results in "effective pair potentials" (i.e. the TIP4P pair potential [29] for water). It is readily realized that such an effective pair-wise interaction should not be parameterized from the corresponding dimer potential energy surface as it effectively contains pieces of the higher order terms in an average manner. The effective pairwise additive interactions are usually fitted to macroscopic thermodynamic data. Provided that the many-body expansion [eq. (1)] converges to a low order (i.e. the 3-body), then analytic expressions for the 2- and 3-body terms can be derived from the corresponding dimer and trimer cluster potential energy surfaces as has been previously reported for the case of HF [30]. In this case the construction of an empirical 3-body functional form will require sampling of the trimer potential energy surface in 12 intermolecular degrees of freedom (Euler angles and center-of-mass distances), even if the intramolecular degrees of freedom are excluded by assuming rigid monomers.

An improvement over the simpler pair-wise additive potentials consists of the use of multipole expansions of the charge density (charges, dipoles, quadrupoles) with associated polarizabilities and hyperpolarizabilities. The non-additive many-body energy terms arise from the response to a dipolar [31] or higher moment fields [32]. Some models in this category utilize point polarizabilities [33] whereas others extend to distributed ones [34,35,36]. Polarizable models allow for the change in the moments of a molecule due to its interaction with the neighbors. Figure 3 shows the variation with cluster size of the total (permanent + induced) molecular dipole moment in water clusters n=2-20 with the TTM model [36]. It should be emphasized that the *individual* molecular dipole moment in a cluster is not an observable, only the total dipole moment of the cluster is. It has been shown that different schemes based on partitioning the charge density obtained from electronic structure calculations can lead to widely different values for the molecular multipoles in a cluster, indicating how arbitrary their definition is [37]. However, within a polarizable empirical potential framework, these are used to account for the non-additive components of the intermolecular interactions. The fact that the pairwise additive potentials can be thought of describing implicitly the non-additive components is manifested by the fact that they are associated with a permanent constant dipole moment whose value is larger than the isolated molecule value (i.e., 2.18 Debye for the TIP4P model, compared to 1.855 Debye for the isolated molecule).

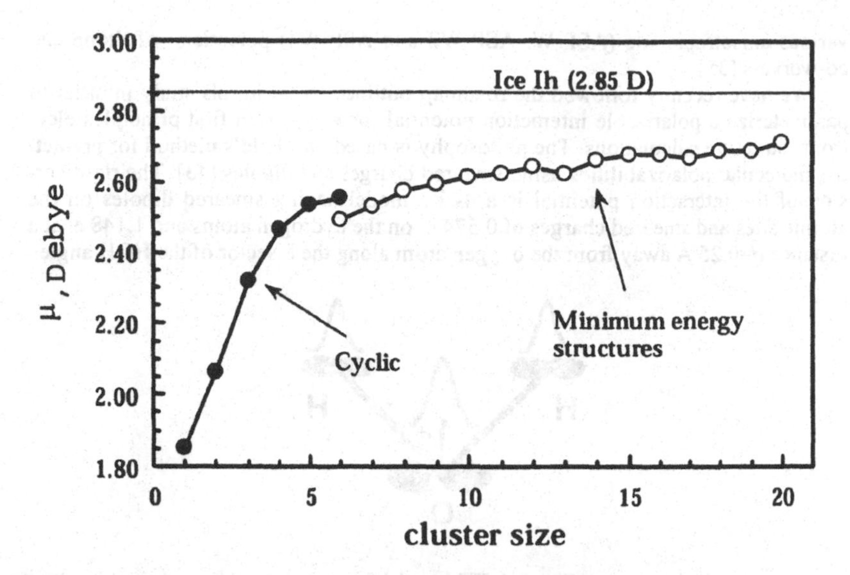

Figure 3. Variation of the total (permanent + induced) dipole moment (in Debye) of individual water molecules in water clustes n=2-20 with the TTM model. The value for ice Ih is also indicated.

Some of the most sophisticated emprirical models by Stone and co-workers [35] describe the interaction between water molecules as the sum of electrostatic, induction, charge-transfer, dispersion and short-range repulsion terms. The long range electrostatic energy is expressed in terms of multipoles distributed on the oxygen (charge, dipole and quadrupole) and hydrogen (charge, dipole) atoms, respectively [38]. The induction interaction is modeled using one-site polarizabilities up to quadrupole-quadrupole on the oxygen atom. Charge transfer terms are included via an atom-atom functional form involving the four oxygen-hydrogen pairs and evaluated from intermolecular perturbation theory calculations on the dimer. The short-range repulsive energy is described using an exponential site-site function (both isotropic and anisotropic terms). Finally the dispersion energy is accounted either using Tang-Toennies damping functions or with a site-site model. This potential encompasses over 70 parameters and has not yet been used for liquid simulations. A recent review [39] provides a useful overview of the developed empirical potentials for water.

5. USE OF THE RESULTS OF AB-INITIO CALCULATIONS IN THE PARAMETRIZATION OF EMPIRICAL POTENTIALS

Many intermolecular potentials for water are built empirically (i.e., their parameters are fitted to macroscopic thermodynamic data) although there have been previous attempts to use the results of electronic structure calculations in order to fit empirical interaction potentials for water. Most notable effort is probably the development of polarizable models by Clementi and co-workers (NCC [40] and NCC-vib [41] potentials), with other parallel efforts such as the NEMO potential [42] and the various anisotropic site (ASP-W, ASP-W2 and ASP-W4) potentials of Stone and co-workers [35].

We have recently followed the roadmap outlined so far in this study in order to parameterize a polarizable interaction potential for water from first principles electronic structure calculations. The philosophy is based on Thole's method for predicting molecular polarizabilities usind smeared charges and dipoles [43]. The rigid version of the interaction potential is a 4-site model having smeared dipoles on the atomic sites and smeared charges of 0.574 e on the hydrogen atoms and 1.148 e^- at a distance d=0.25 Å away from the oxygen atom along the bisector of the HOH angle.

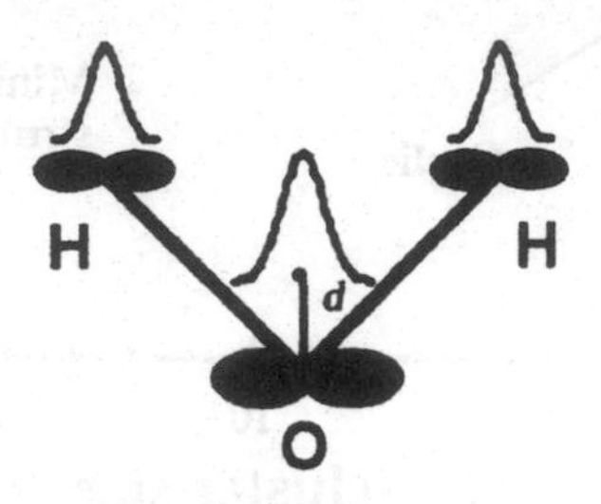

Figure 4. TTM model for water

The OH bond lengths are fixed at 0.9572 Å and the HOH angle at 104.52°. The gas phase molecular permanent dipole is 1.853 Debye. The only intramolecular contributions arise from the atomic dipole-dipole interactions. We used Thole's method for expressing the dipole tensor in terms of the "reduced distance" $(r^{red})_{ij} = r_{ij}/(\alpha_i\alpha_j)^{1/6}$, where α_i, α_j are the corresponding polarizabilities. Among the many choices proposed by Thole for the charge density we used

$$\rho(r) = \frac{1}{A^3}\frac{3a}{4\pi}\exp\left(-a\left(\frac{r}{A}\right)^3\right), \tag{12}$$

where $A=(\alpha_i\alpha_j)^{1/6}$, and α_i, α_j are the atomic polarizabilities of atoms i and j, respectively, and a=0.572 is the dimensionless width parameter parametrized for this density ($a^{CC} = a^{CD} = a^{DD} = 0.572$). The total interaction is written as

$$U_{tot} = U_{pair} + U_{elec} + U_{pol} \tag{13}$$

where

$$U_{pair} = \sum_{i=1}^{N}\sum_{j=1}^{N}\left(\frac{C_1}{r^{12}} + \frac{C_2}{r^{10}} + \frac{C_3}{r^6}\right) \tag{14}$$

$$U_{elec} = U_{charge-charge} + U_{charge-dipole} + U_{dipole-dipole} \tag{15}$$

$$U_{pol} = \sum_{i=1}^{N}\frac{D_i D_i}{2\alpha_i} \tag{16}$$

are the pair, electrostatic and polarization components, respectively, D_i is the dipole and α_i the polarizability of atomic site i, C_1, C_2 and C_3 are constants and the sums are over N atoms in the system.

The initial parametrization (TTM) of the model [36] was based on about 15 points of the water dimer potential energy surface which were obtained at the MP2 level of theory with the aug-cc-pVQZ basis set. The contributions from $U_{elec} + U_{pol}$ in eq. (13) are determined by the partial charges and atomic polarizabilities. We determined U_{pair} by assuming that the only other contribution to the energy comes from a pair-wise additive oxygen-oxygen interaction and fitting to the MP2 data for the dimer, i.e., $U_{pair} = U_{MP2} - (U_{elec} + U_{pol})$ using the functional form eq. (14) for U_{pair}. Recognizing the fact that the 2-body interaction is the most difficult to converge [44] we subsequently refitted [45] the pair interaction to *scaled* MP2/CBS-quality energy differences along symmetry-constrained minimum energy paths (MEPs) of the water dimer potential energy surface as a function of the O-O separation corresponding to the approach of the two water molecules along C_s, C_i and C_{2v} symmetries as shown in Figure 5 [46]. The resulting new version of the model is denoted as TTM2-R (R for rigid).

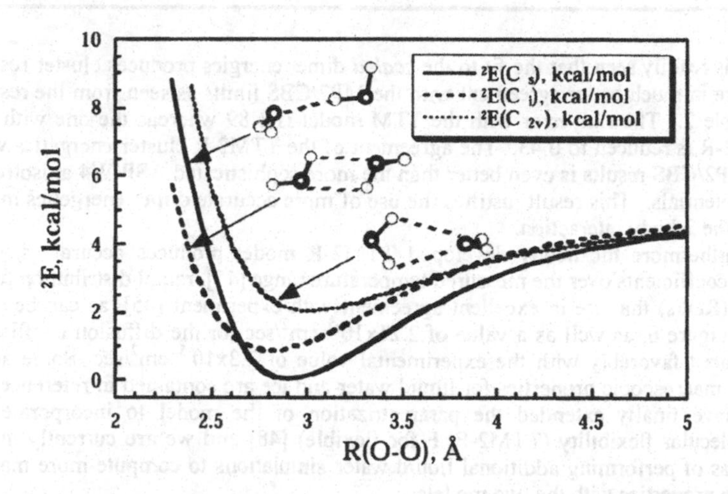

Figure 5. Symmetry constrained MEPs on the water dimer PES.

At the same time, using the methodology outlined in Section 2 we have established [16] for the first time accurate binding energies for the water trimer through pentamer global ring minima and four low-lying hexamer isomers (cage, prism, book, cyclic S_6) from first principles electronic structure calculations. These are -15.8 kcal/mol (trimer), -27.6 kcal/mol (tetramer), -36.3 kcal/mol (pentamer), -45.9 kcal/mol (prism hexamer), -45.8 kcal/mol (cage hexamer), -45.6 kcal/mol (book hexamer) and –44.8 kcal/mol (ring S_6 hexamer). Effects of higher correlation, estimated at the coupled cluster plus single and double with a perturbative estimate of the triple excitations [CCSD(T)] level of theory, as well as inclusion of estimates for core-valence correlation suggest that these estimates are accurate to within 0.2

kcal/mol. This task was deemed necessary in order to be able to assess the accuracy of the developed empirical model in the absence of experiemental measurements for the cluster energetics.

Table 2. Binding energies of water clusters n=2-6 with the various models and comparison with theMP2/CBS results.

N	RWK2	DC	ASP-W	ASP-VRT	ASP-W4	TTM	TTM2-R	MP2/CBS
2	-6.14	-4.69	-4.68	-4.91	-4.99	-5.33	-4.98	-4.98
3 cyclic	-15.23	-13.34	-14.81	-15.65	-15.49	-16.68	-15.59	-15.8
4 cyclic	-27.84	-23.95	-24.32	-25.93	-26.97	-28.57	-27.03	-27.6
5 cyclic	-36.22	-31.80	-30.88	-33.52	-35.09	-37.91	-36.05	-36.3
5 cage	-36.30	-31.06	-32.65	-34.77	-35.23		-34.75	-35.1
6 prism	-46.89	-40.97	-42.54	-45.90	-47.06	-48.54	-45.11	-45.9
6 cage	-47.28	-40.76	-43.31	-45.00	-45.86	-48.91	-45.65	-45.8
6 book	-46.39	-40.38	-40.48	-43.44	-45.00	-47.73	-45.14	-45.6
6 cyclic	-43.86	-39.34	-37.58	-41.09	-43.36	-46.48	-44.28	-44.8
RMS Error	0.93	4.25	3.99	1.18	0.80	1.89	0.43	

It is readily seen that the fit to the *scaled* dimer energies produces cluster results that are in much better agreement with the MP2/CBS limits as seen from the results of Table 2. The rms error with the TTM model is 1.89 whereas the one with the TTM2-R is reduced to 0.43. The agreement of the TTM2-R cluster energetics with the MP2/CBS results is even better than the more sophisticated ASP-W4 anisotropic site potentials. This result justifies the use of more accurate dimer energetics in the fit of the 2-body interaction.

Furthermore the newly developed TTM2-R model produces accurate second virial coefficients over the measured temperature range [47], radial distribution functions (RDFs) that are in excellent agreement with experiment [45], as can be seen from Figure 6, as well as a value of 2.21×10^{-5} cm^2/sec for the diffusion coefficient compared favorably with the experimental value of 2.3×10^{-5} cm^2/sec. Some additional macroscopic properties for liquid water and ice are contained in reference 45. We have finally extended the parametrization of the model to incorporate intramolecular flexibility (TTM2-F, F for flexible) [48] and we are currently in the process of performing additional liquid water simulations to compute more macroscopic properties with the two models.

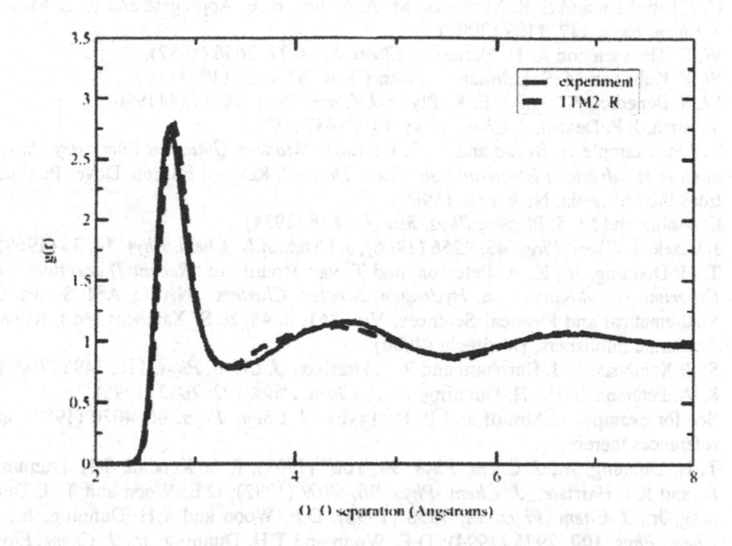

Figure 6. Liquid water O-O Radial Distribution Function with the TTM2-R model.

6. CONCLUSIONS

The results presented in this study justify the use of high level electronic structure calculations in the development of empirical models used to describe intermolecular interactions. The use of systematically improvable methodological approaches together with new developments in software and hardware provide a route to the calculation of accurate PESs of molecular systems. In the absence of experimental data quantifying the molecular interactions, this approach provides indispensable and currently irreplaceable information needed for the development of accurate interaction potentials. To this end, theory can provide a useful and in most cases necessary complement to experiment.

Acknowledgements: This research effort has greatly benefited from various helpful discussions with Dr. C. J. Burnham. This work was supported by the Division of Chemical Sciences, Office of Basic Energy Sciences, US Department of Energy. Battelle operates the Pacific Northwest National Laboratory for the Department of Energy.

REFERENCES

1. F. H. Stillinger, *J. Chem. Phys.* **57**, 1780 (1972).
2. S. S. Xantheas, *J. Chem. Phys.* **100**, 7523 (1994).
3. G. Kaplan, R. Santamaria and O. Novaro, *Mol. Phys.* **84**, 105 (1995).
4. S. S. Xantheas, *J. Chem. Phys.* **104**, 8821 (1996).
5. S. S. Xantheas, *J. Phys. Chem.* **98**, 13489 (1994); S. S. Xantheas, *J. Amer. Chem. Soc.* **117**, 10373 (1995).

6. S. S. Xantheas and T. H. Dunning, Jr., *J. Chem. Phys.* **99**, 8774 (1993).
7. C. J. Burnham and S. S. Xantheas, M. A. Miller, B. E. Applegate and R. E. Miller, *J. Chem. Phys.* **117**, 1109 (2002).
8. W. E. Thiessen and A. H. Narten, *J. Chem. Phys.* **77**, 2656 (1982).
9. W. F. Kuhs and M. S. Lehman, *J. Phys. Chem.* **87**, 4312 (1983).
10. W. S. Benedict, N. Gailar, E. K. Plyler *J. Chem. Phys.* **24**, 1139 (1956).
11. V. Buch, J. P. Devlin, *J. Chem. Phys.* **110**, 3437 (1999).
12. See for example A. Szabo and N. S. Ostlund, "*Modern Quantum Chemistry: Introduction to Advanced Electronic Structure Theory*", Revised Edition, Dover Publications Inc., Mineola, New York (1989).
13. C. Møller and M. S. Plesset, *Phys. Rev.* **46**, 618 (1934).
14. J. Cizek, *J. Chem. Phys.* **45**, 4256 (1966); J. Cizek, *Adv. Chem. Phys.* **14**, 35 (1969).
15. T. H. Dunning, Jr., K. A. Peterson, and T. van Mourik, in "*Recent Theoretical and Experimental Advances in Hydrogen Bonded Clusters*" NATO ASI Series C: Mathematical and Physical Sciences, Vol. **561**, p. 45, S. S. Xantheas (ed.), Kluver Academic Publishers, Dordrecht (2000).
16. S. S. Xantheas, C. J. Burnham and R. J. Harrison, *J. Chem. Phys.* 116, 1493 (2002).
17. K. A. Peterson and T. H. Dunning, Jr., *J. Chem. Phys.* 102, 2032 (1995).
18. See for example J. Almlöf and P. R. Taylor, *J. Chem. Phys.* 86, 4070 (1987) and references therein.
19. T. H. Dunning, Jr., *J. Chem. Phys.* **90**, 1007 (1989); R.A. Kendall, T.H. Dunning, Jr. and R.J. Harrison, *J. Chem. Phys.* **96**, 6769 (1992); D.E. Woon and T.H. Dunning, Jr., *J. Chem. Phys.* **98**, 1358 (1993); D.E. Woon and T.H. Dunning, Jr., *J. Chem. Phys.* **100**, 2975 (1994); D.E. Woon and T.H. Dunning, Jr., *J. Chem. Phys.* **103**, 4572 (1995); A. K. Wilson, T. van Mourik and T. H. Dunning, Jr., *J. Molec. Struct. (Theochem)* **388**, 339 (1996); T. H. Dunning, Jr., K. A. Peterson, A. K. Wilson, *J. Chem. Phys.* **114**, 9244 (2001); K. A. Peterson, T. H. Dunning, Jr., *J. Chem. Phys.* **117**, 10548 (2002).
20. D. Feller, *J. Chem. Phys.* **96**, 6104 (1992).
21. T. H. Dunning, Jr., *J. Phys. Chem. A* **104**, 9062 (2000).
22. B. Liu and A. D. McLean, *J. Chem. Phys.* **59**, 4557 (1973).
23. S. F. Boys and F. Bernardi, *Mol. Phys.* **19**, 553 (1970).
24. C. F. Bunge, *Theor. Chim. Acta* **16**, 126 (1970); V. Termath, W. Klopper and W. Kutzelnigg, *J.Chem. Phys.* **94**, 2002 (1991); W. Klopper, *J. Chem. Phys.* **102**, 6168 (1995).
25. S. Huh, J. S. Lee, *J. Chem. Phys.* **118**, 3035 (2003).
26. S. S. Xantheas, *Chem. Phys.* **258**, 225 (2000).
27. G. Chalasinski, M. M. Szczesniak, C. Cieplak and S. Scheiner, *J. Chem. Phys.* **94**, 2873 (1991).
28. S. S. Xantheas, *Phil. Mag. B* **73**, 107 (1996).
29. W. L. Jorgensen, J. Chandrasekhar, J. D. Madura, R. W. Impey and M. L. Klein, *J. Chem. Phys.* **79**, 926 (1983).
30. M. Quack, J. Stohner, M. A. Shum, *J. Mol. Struct.* **294**, 33 (1993); W. Klopper, M. Quack, M. A. Shum, *Mol. Phys.* **94**, 105 (1998); M. Quack, J. Stohner, M. A. Suhm, *J. Mol. Struct.* **599**, 381 (2001).
31. L. Dang and T.-M.Chang, *J. Chem. Phys.*, **106**, 8149 (1997).
32. E. R. Batista, S. S. Xantheas, H. Jonsson, *J. Chem. Phys.* **109**, 4546 (1998).
33. P. Ahlström, A. Wallqvist, S. Engström and B. Jöhnsson, *Mol. Phys.* **68**, 563 (1989); M. Sprik, M. L. Klein, *J. Chem. Phys.* **89**, 7556 (1988).
34. D. N. Bernardo, Y. Ding, K Krogh-Jespersen and R. M. Levy, *J. Phys. Chem.* **98**, 4180 (1994).
35. C. Millot and A. J. Stone, *Mol. Phys.* **77**, 439 (1992); C. Millot, J-C Soetens, M. T. C. Martins Costa, M. P. Hodges, A. J. Stone, *J. Phys. Chem.* A, **102**, 754 (1998).

36. C. J. Burnham, J. C. Li, S. S. Xantheas, and M. Leslie, *J. Chem. Phys.* **110**, 4566 (1999).
37. E. R. Batista, S. S. Xantheas, H. Jonsson, *J. Chem. Phys.* **111**, 6011 (1999).
38. A. J. Stone and M. Alderton, *Mol. Phys.* **56**, 1047 (1985).
39. B. Guillot, *J. Mol. Liq.* **101**, 219 (2002).
40. U. Niesar, G. Corongiu, E. Clementi,G. R. Kneller, D. K. Bhattacharya, *J. Phys. Chem.* **94**, 7949 (1990).
41. G. Corongiu, *Int. J. Quant. Chem.* **42**, 1209 (1992).
42. P. O. Åstrand, P. Linse, G. Kalström, *Chem. Phys.* **191**, 195 (1995); P. O. Åstrand, A. Wallqvist, G. Kalström, *J. Chem. Phys.* **100**, 1262 (1994).
43. B. T. Thole, *Chem. Phys.* **59**, 341 (1981).
44. J. M. Pedulla and K. D. Jordan in "*Recent Theoretical and Experimental Advances in Hydrogen Bonded Clusters*" NATO ASI Series C: Mathematical and Physical Sciences, Vol. **561**, p. 35, S. S. Xantheas (ed.), Kluver Academic Publishers, Dordrecht (2000).
45. C. J. Burnham and S. S. Xantheas, *J. Chem. Phys.* **116**, 1500 (2002).
46. C. J. Burnham and S. S. Xantheas, *J. Chem. Phys.* **116**, 1479 (2002).
47. G. K. Schenter, *J. Chem. Phys.* **117**, 6573 (2002).
48. C. J. Burnham and S. S. Xantheas, *J. Chem. Phys.* **116**, 5115 (2002).

MODELS IN THEORY OF MOLECULAR LIQUID MIXTURES: STRUCTURE, DYNAMICS, AND PHYSICOCHEMICAL PROPERTIES

V. A. DUROV

Department of Physical Chemistry, Faculty of Chemistry, Lomonosov Moscow State University, Moscow, RU-119899, Russia; E-mail: durov@phys.chem.msu.ru

Abstract: Models of structure and properties of liquid mixtures have been outlined. Main attention is given to the extended quasichemical approach for modeling supramolecular ordering in mixtures, selforganized by H-bonds and unified description of their physicochemical properties. Models of polyvariable supramolecular species as regard to structure and composition, taking in account the cooperativity on H-bonding, as well as the methods for describing their structure, composition, electric (dipole moment), and optic (polarisability) properties are developed. Interrelations between thermodynamic functions (Gibbs energy, enthalpy, entropy), dielectric (permittivity), and optic (refractive index and its fluctuation derivatives, determining Rayleigh light scattering) properties of mixtures, and microscopic characteristics of aggregates are analyzed. The methods for obtaining both the integral and differential parameters of aggregation are developed, applicable for structural study of the long range molecular correlations, including supramolecular aggregates of nanosizes. Models of the media with internal parameters of different nature and tensor dimension have been outlined. Backgrounds for their application to study dynamic processes of the supramolecular reorganization, intramolecular transitions, and energy transfer, as well as fluctuation and relaxation phenomena are considered. Fluctuation and relaxation contributions of internal parameters to both equilibrium and kinetic properties and Rayleigh ratio of mixtures were established. The applications to individual liquids and mixtures are illustrated. New data on the structure of aggregates and thermodynamics of their formation were obtained. Supramolecular assemblies in liquids with the long range molecular correlations were revealed. Macroscopic manifestations of the supramolecular organization in physicochemical properties of liquids are discussed.

Key words: Liquids; Mixtures; H-Bonds; Aggregates; Extended Quasichemical Models; Thermodynamics; Permittivity; Fluctuations; Rayleigh Light Scattering; Internal Variables; Molecular Thermal Motion; Kinetic Phenomena; Supramolecular Organisation; Long Range Molecular Correlations

J. Samios and V.A. Durov (eds.), Novel Approaches to the Structure and Dynamics of Liquids: Experiments, Theories and Simulations, 17–40.

1. INTRODUCTION

Development of the theory of liquids is based upon investigations of their structure, molecular interactions, and dynamic processes proceeding during the molecular thermal motion. An increasing role of investigations of liquids by combining the methods, as well as the molecular design of liquid materials require approaches for analyzing and predicting macroscopic properties on the basis of a general concept. The main goal of the liquid theory is connection between microscopic properties of molecules and their interactions to structure and macroscopic properties of liquids.

First, we are interested is the intermolecular potential. In fact potentials being applied are the model ones. In quantum-mechanical calculations the result depends on the level of the calculation, e.g., Hartree-Fock approach, many configuration approach, perturbation theory, and then the results obtained are further simplified, e.g., an atom-atom representation of the intermolecular potential. [1].

The main subject of the paper is the models applied both to describing structure and properties of liquids and dynamic processes proceeding in them. Last years there is an growing interest in studying thermodynamics and kinetics of aggregation due to noncovalent intermolecular forces and its manifestations in macroscopic properties of matter. The aggregation phenomena are important in phase transitions, glass formation, nucleation phenomena etc., and in the interdisciplinary fields of science like theory of selforganisation of a matter and synergetics [2-4]. This problem is related to constructing devices based on the microscopic properties of aggregates. The main progress has been achieved for crystal solids where structure and properties of aggregates may be characterized quite definitely [5].

The situation for liquids is much more complicated due to the temporal mobility of the structure during the molecular thermal motion. Various labile aggregates may be realized in liquids, and their structure and properties are being changed under variation of the thermodynamic conditions. Main attention is given to the local ordering inside the nearest coordination shells, and thus in general the role of the long range molecular correlations for structure and properties of liquids is not quite recognized [6-13].

Molecular liquids represent the promising object for studying structure and molecular thermal motion. The structural, thermodynamic and kinetic features of processes occurring in the molecular liquids represent the elementary stages to more complex systems, hierarchically e.g., liquid crystals, polymers, micelles and microemulsions, colloids, and biological systems. Mixtures with hydrogen bonds, which are the attractional directed molecular interactions intermediate between the chemical bonds and the van der Waals forces, present the attractive object to study supramolecular ordering, because aggregates polyvariable both in structure and composition may be expected in them.

2. ON DESCRIPTION OF LIQUID STRUCTURE

As to the principal direction for mixture modelling first statistical-mechanical theory and the simulation methods may be mentioned, where the physical language, based on the correlation functions F_s $(s = 2,3,...,N)$, determining the probabilities of the spatial configurations of particles has been applied [14-16]. A binary function $F_2(R) = g(R)$ plays the central role because, first, the thermodynamic functions are expressed in terms of it, and second, it can be studied by the diffraction techniques.

Many approaches have been proposed for calculation of radial function e.g., Born-Bogolubov-Green-Kirkwood-Ivon chain of the integrodifferential equations, Ornstein-Zernicke integral equation (Percus-Yevick's and hyperchain equations corresponding to different closures of Ornstein-Zernicke equation), and then perturbation theory, *etc.* [14,15]. The site-site versions of these approaches have been developed. In computer simulation such as Molecular Dynamics and Monte-Carlo methods the correlation functions are applied both for presenting and understanding of the simulation data [16].

The radial function $g(R)$ gives a one-dimensional picture of liquid structure and includes averaging on translation and orientation of molecules. Diffraction techniques allow study of this function for nearest molecules, reflecting just the local ordering of molecules in a liquid, and thus the long range molecular correlations are staying in shadow. In computer simulation both due to periodical boundary condition and limited time of calculation, the long range molecular correlations may be cut off artificially. Thus really the role of supramolecular ordering in liquids is not understood in contrast with crystals [10-13].

Different approaches are being applied to complex liquids, and especially for those with attractional directed forces like H-bonds. They may be subdivided to physical, lattice, and chemical models [6-13,17]. First ones are statistical-mechanical models, e.g., SAFT (Statistical Associated Fluid Theory) [18,19], based on Wertheim's perturbation theory, second, the lattice models, based on Guggenheim-Barker theory [20,21], and its further hole modifications, taking into account the results on the lattice gas statistics, e.g. LFHB (Lattice Fluid Hydrogen Bond) [22], and the last ones are the chemical models, combining the extension of the hard sphere approach for anisotropic molecules with chemical equilibria equation [17].

Main area for application of these approaches [17-22] is the equation of state and thermodynamic functions of mixing. Thus these models at present firstly cannot describe other equilibrium properties of mixtures such as dielectric and optic ones and, secondly they are not extended to decribing kinetic properties of mixtures and processes occurring in them during the molecular thermal motion.

3. ON EXTENDED QUASICHEMICAL MODELS

In chemistry the concept of concentrations of particle and chemical reactions between them is applied to describe the state of matter. In the framework of the quasichemical models the processes are represented by the equations of chemical reaction:

$$\sum_i \nu_{i\alpha} M_i \underset{k'_\alpha}{\overset{k_\alpha}{\Leftrightarrow}} \sum_i \nu'_{i\alpha} M_i \, , \ (\alpha = 1,2,...,r), \tag{1}$$

$$d\xi_\alpha = \frac{dn_{i\alpha}}{\Delta\nu_{i\alpha}}, \quad \Delta\nu_{i\alpha} = \nu'_{i\alpha} - \nu_{i\alpha}, \quad A_\alpha = -\sum_i \mu_i \Delta\nu_{i\alpha}, \tag{2}$$

where M_i designates the reagent, $\nu_{i\alpha}$ and $\nu'_{i\alpha}$ denote its stoichiometric coefficients in the reaction α. The internal processes are characterised by the extent (degree of advancement) ξ_α and the affinity A_α, where μ_i is the chemical potential of the

i-th reagent, and constants of rate k_α, k'_α. For nonequilibrium states the extents are independent parameters and corresponding affinities are not equal to zero.

Quasichemical models combine both macroscopic and molecular theories. One can treat this approach as an extension of chemical thermodynamics and kinetics methods to internal processes occurring in matter, such as spatial reorganisations of the intermolecular structure, and in particular, the association processes, energy transfer between molecular degrees of freedom, conformation transitions of molecules, and related fluctuation phenomena. Molecular-statistical models are used both to specify the nature of processes represented by Eqns. (1), (2) and to obtain equations for physicochemical properties. Thus quasichemical models represent the language for both investigating equilibrium and nonequilibrium properties of liquids and phenomena occurring in them during the molecular thermal motion, and presenting results in terms common for chemical systems [7-13].

The extents $\{\xi_\alpha\}$ (2) represent the scalar internal variables. The examples of the vector internal variable are fluctuating local velocity and polarization of liquid. Tensor variables are entered for describing anisotropic states of system. Mechanic tensor variables are represented by the tensors of deformation $\{u_{ik}\}$ and stress $\{\sigma_{ik}\}$. Internal tensor variables $\{\eta_{\beta.ik}\}$ characterize anisotropic states of matter, e.g., local polarisability tensor, and those are not generally reduced to mechanic ones. The tensor dimension of the internal variables is determined by the property of a system or the phenomena considered. The internal variables were introduced for describing chemical reactions, viscoelasticity and relaxation phenomena, light scattering, phase transitions and critical phenomena [1,7-13,23-45]. The concept of internal variables is especially important for theory of the irreversible processes.

The term "quasichemical models" [7-13] underlines the description on the chemical language of the wide spectra of the processes occurring in matter, which in general are not accompanied by the change of the chemical nature of the reagents, and thus allow to consider both different properties and phenomena under the unified approach. These models are named sometimes as association or chemical ones when association processes are considered [17], though their applicability is much wider. Then, these models should not be mixed with the "quasichemical approach" introduced by Guggenheim [20] at the estimation of the partition function of the lattice model, proposed for thermodynamic functions of the mixtures. Thus, it has only the formal analogy with the real processes, occurring in a matter during molecular thermal motion.

Quasichemical models of association processes were applied usually to thermodynamic and spectroscopic properties of liquid systems [6-13,46-49]. Recently, along with the elaboration of association models and thermodynamics of associated solutions in general [6-13,50-53], this approach has been extended to dielectric [6-13,54-58], optic [6-13,59-63], and kinetic [7-13,64,65] properties of pure liquids as well as mixtures. The properties considered are thermodynamic functions, Gibbs energy G, enthalpy H, entropy S, and corresponding excess functions for mixtures, G^E, H^E, S^E, activity coefficients of the components γ_A, γ_B, dielectric properties, permittivity ε_S, optical properties - Rayleigh ratios for isotropic and anisotropic light scattering R_{is}, and R_{an}, and kinetic ones, related to relaxation phenomena. It is important to underline that the properties of mixtures considered are determined by different molecular parameters.

The results were applied to study molecular interactions, thermodynamics of aggregation, and the structure of aggregates in liquid mixtures [53,66-85] (see for the review [6-12]). New data on the structural and thermodynamic parameters of aggregates and the molecular nature of the macroscopic properties of mixtures were obtained.

Next, we have to solve the interrelated problems of description of supramolecular ordering and properties of liquids in the quasichemical approach [6-12]. Firstly, we have to describe the structure, composition, electric, and optic properties of aggregates. Secondly, we have to develop the thermodynamic model of the aggregate mixture. Thirdly, we have to derive the equations for macroscopic properties of liquid mixtures. The topological features of the aggregates, determined mainly by the number of specific bonds, which can form molecule with neighboring ones, should be taken into account.

The model of associated species of arbitrary composition with the different ways of H-bonding of molecule in the aggregates was introduced (see for the review [6-12]). These aggregates are polyvariable in composition and structure. Actually, firstly, on each site of an aggregate may be located any molecule from the mixture components, and, secondly, the different ways for bonding lead to the structural polyvariability of aggregates. Really, starting from the "root" molecule of an aggregate the various ways for further aggregate growth are possible. Thus the structural and compositional multiformity of aggregates leads to the new types of those structures. For instance even for homogeneous on composition chain-like aggregates, the structures differing in molecular correlations were established and, thus, the new models of chain-like aggregates were introduced.

The model developed is taking into account the cooperative (collective) character of H-bonding. In common sense it means the dependence of the energy of consecutive H-bond formation on the size of aggregates. This simplest type of H-bond cooperativity was accounted for on the thermodynamic level by introducing different equilibrium constants for dimerisation and that for formation of larger aggregates. New types of the H-bonds cooperativity were allowed due to considering of the more long range molecular correlations in aggregates.

For a binary solution of components A and B the aggregate comprising n molecules is described by the sets of subscript $\{n, j_1, j_2, \ldots j_n, l_1, l_2, \ldots l_{n-1}\} \equiv n, \{j_k\}, \{l_i\}$, where $j_k = a,b$ $(k = 1,2,\ldots n)$ indicates the type of the k-th molecule, and $l_i = 1,2,\ldots,q$ $(i=1,2,\ldots n-1)$ denotes the state of the i-th bond in it.

The properties of these aggregates, e.g., composition, structure, dipole moment, polarizability tensor, differ, as do the macroscopic properties of the respective liquids. Therefore, this required the approaches for describing both microscopic properties of aggregates and macroscopic properties of liquids. The matrix technique was developed to consider aggregation, and both properties of aggregates and mixtures [50,51,54,55,57,60].

Structural and compositional polyvariability of the aggregates leads to the possibility for realizing systems, self-organised by noncovalent directed interactions like H-bonds, conserving information in double or more multivalent forms. Thus controlling both compositional and structural reorganisations of the aggregates is promising both for molecular information processing in general and especially for that in living nature.

4. QUASICHEMICAL MODELS IN THERMODYNAMICS OF MIXTURES

Thermodynamic model for mixture of aggregates is the first step for relating macroscopic properties of solution to microscopic properties of the aggregates. The approach for molecular interaction contributions to thermodynamic functions (Gibbs energy, enthalpy, and entropy) of both liquids and mixtures was developed and then the quasichemical non-ideal associated solution (QCNAS) model has been constructed [6-12,50,51]. Repulsive interactions are treated within the framework of Flory approach, dipolar interactions are described using an extended Onsager model, dispersion interactions are allowed for using the continuous approach, and short range specific ones are dealt with by the association model.

The molecular force contributions to Gibbs energy, enthalpy, and entropy of pure liquids and mixtures, self-organised by H-bonds, taking into account their supramolecular organisation have been calculated. For moderately polar substances such as monohydric alcohols the major contributions come from the dispersion interaction and H-bonding, whereas the long range constituents of dipole force are of only minor importance. As to the strongly polar N-amides the three contributions mentioned are comparable. The heat of evaporation of liquids is reproduced quite well [6-11].

Activity coefficients of components γ_A, γ_B and excess functions of mixtures, e.g., excess enthalpy H^E, include three interrelated contributions, namely, first, associative one describing the deviations from ideality due to repulsion forces and association, second one due to dispersion interactions, and third one due to dipolar forces:

$$\gamma_{A,B} = \gamma^{ass}_{A,B} \cdot \gamma^{dis}_{A,B} \cdot \gamma^{dip}_{A,B}, \quad H^E = H^{E,ass} + H^{E,dis} + H^{E,dip}. \tag{3}$$

The first terms in Eqn. (3) are:

$$\gamma^{ass}_{A,B} = \frac{C_{a,b}}{\overline{C}_{a,b} x^o_{A,B}} \exp\left(\frac{1}{\overline{n}_{A,B}} - \frac{\varphi^o_{A,B}}{x^o_{A,B} n_{AB}} \right). \tag{4}$$

In Eqn. (4) $C_{a,b}$ are the molar concentrations of the monomers a and b, $x^o_{A,B}$ and $\varphi^o_{A,B}$ are the mole and volume fractions of the component A and B, and n_A, n_B, n_{AB} are the average numbers of association in the pure liquids A, B, and in the mixture, respectively.

The successive formation of the aggregates is described by the equilibrium constants $\left\{K^T_{j_1 j_2, l_1}\right\}$ for dimerisation and those $\left\{K^{T'}_{j_1 j_2, l_1 l_2}\right\}$ for larger ($n \geq 3$) aggregates expressed in terms of the thermodynamic activities of the aggregates. These constants depend on the molecules, participating in bonding ($j_1 j_2 = a, b$), and on the states of the bond, formed at the given association step l_2, and at the preceding one l_1. Thus the cooperative (collective) character of H-bonding is taken into account.

The equilibrium constants are expressed in the following form

$$K^T = K \cdot K^{dis}_{\gamma} \cdot K^{dip}_{\gamma}, \tag{5}$$

where the terms K_{γ}^{dis}, K_{γ}^{dip} are expressed via the dispersion and dipolar forces contributions to the activity coefficients of aggregates $\{\gamma_i^{dis}\}$, $\{\gamma_i^{dip}\}$. The concentration "equilibrium constants" K are expressed by the molar (volume) concentrations of aggregates $\{C_i\}$ ($\{\varphi_i\}$) and they include the non-ideality of the aggregates mixture due to the difference in their "size", characterizing the repulsive part of the intermolecular potential, which are described by the respective activity coefficients $\{\gamma_i^{ass}\}$. The last ones represent formally the transforming from the molar fractions $\{x_i\}$ to the molar (volume) concentrations $\{C_i\}$ ($\{\varphi_i\}$) of aggregates.

The values n_{AB}, $\overline{n}_A$, $\overline{n}_B$ (4) represent the integral parameters of the aggregation. The differential parameters of that are expressed e.g., by molar f_n and weight w_n distribution functions, which give the fraction of the aggregates, consisting of n molecules, and the fraction of molecules, included to them, respectively.

Estimation of both dispersion $\gamma_{A,B}^{dis}$ and dipolar $\gamma_{A,B}^{dip}$ contributions [6-10,50-52] to the activity coefficients is important for extracting the aggregation contributions $\gamma_{A,B}^{ass}$ (3).

The approach developed relates the thermodynamic nonideality of the mixture to molecular properties, rather than employing commonly the empirical parameters, and allows going beyond the approach of just short range interactions. The thermodynamic nonideality of the aggregates mixture leads to the dependence of the concentration "equilibrium constants" K, which are applied to calculate the aggregates concentration, from the solution composition. These dependencies are described within the QCNAS model. Thermodynamic equilibrium constants K^T (5) are independent on the mixture composition and thus they may be applied for describing the same process in different systems. The representation of the thermodynamic equilibria constant K^T (5) seems promising for describing chemical equilibria in liquid mixtures in general.

The multivariance of the aggregates both on composition and structure results to various dependencies of excess functions on solution composition, but the influence of them on thermodynamic functions may be relatively essential only if there is a difference in the stoichiometry of an association. Thermodynamic functions do not include the structural features of the aggregates and thus, for instance the models with the same stoichiometry and different structure of the aggregates e.g., linear and cyclic cannot be discriminated by thermodynamic study.

In general, the results obtained allow to describe thermodynamics of the internal processes in non-ideal mixtures, especially supramolecular ordering phenomena, from the molecular based model.

5. QUASICHEMICAL MODELS IN DIELECTROMETRY OF MIXTURES

Applications of dielectric constant data to study supramolecular organisation of liquids are based both on models for mixture of aggregates and the theory of dielectric properties. For the permittivity ε_S of mixture Eqn. (6) was derived [6-13,54-56]:

$$\frac{9k_B T V_m}{4\pi N_A \varepsilon_S} \sum_{j=1}^{k} \left(\frac{\varepsilon_s - \varepsilon_{\infty j}}{2\varepsilon_s + \varepsilon_{\infty j}} \right) \varphi_j^o = \sum_{j=1}^{k} \left(\frac{\varepsilon_{\infty j} + 2}{2\varepsilon_s + \varepsilon_{\infty j}} \right)^2 x_j^o \mu_j^2 g_j^d = g_s^d \sum_{j=1}^{k} \left(\frac{\varepsilon_{\infty j} + 2}{2\varepsilon_s + \varepsilon_{\infty j}} \right)^2 x_j^o \mu_j^2 \quad (6)$$

where k_B, N_A are the Boltzmann and Avogadro constants, $\varepsilon_{\infty.j}$, μ_j, and g_j^d = $<\mu_j\mu_j^*>/\mu_j^2$ (j = 1,2,...,k) are dipole moment, the deformation permittivity, and the dipole correlation factor of the j-th component, where μ_i^* is the dipole moment of a sphere in dielectric at the fixed position and orientation of one (central) molecule j in it, and g_s^d is the dipole factor of the solution. The deviations of dipole factor g_s^d, g_j^d from the unity are due to the molecular spatial correlations. The dipole factors g_j^d describe the correlations of both similar and different components of a mixture. Equations relating the dipole factors g_j^d, g_s^d to the structure of aggregates and thermodynamics of their formation were established [6-10,54-56].

The applications to aliphatic [66-68], alicyclic [69], aromatic [70-72], unsaturated [80] alcohols, and N-amides [73,74] (see for the review [6-12]) demonstrated the possibility of a fairly reliable discrimination of the aggregation models based on the analysis of permittivity and dipole correlation factor. Models with the same stoichiometry and different structures of aggregates (e.g., chain-like and cyclic) may be distinguished even qualitatively. For most of the pure alcohols the dominating aggregates are the chain-like ones in agreement with the computer simulation [86]. The new structures of the chain-like aggregates were revealed.

The thermodynamic parameters of the aggregation, e.g. equilibrium constants K, the enthalpy ΔH and the entropy ΔS of H-bonding were found, integral and differential parameters of aggregation were determined, and the cooperativity of H-bonding has been revealed.

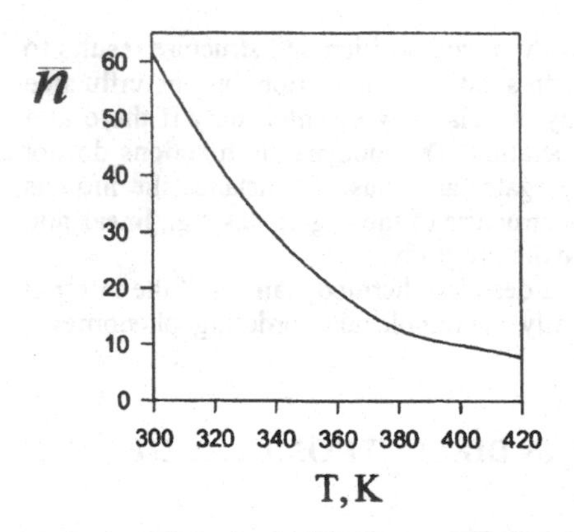

Figure 1. The average numbers of association in liquid methanol.

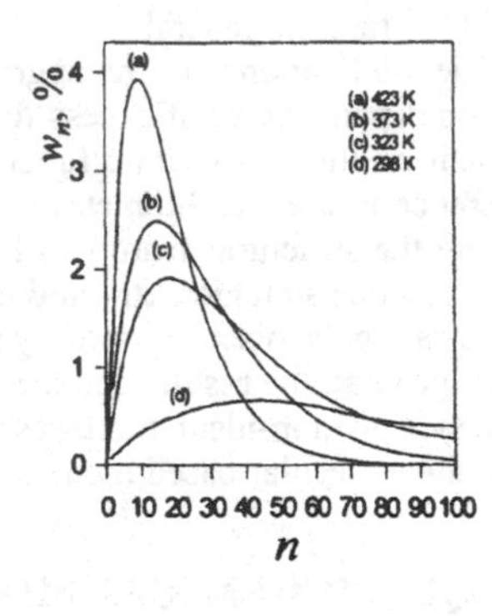

Figure 2. Weight distribution function w_n. of methanol aggregates

For the first time found the long range correlations by H-bonding should be underlined. The average numbers of association $\overline{n}$ in alkanols are equal to ten and above at room temperature and increase with decreasing temperature (Fig. 1). The distribution functions of aggregates on size are quite wide, and thus the evidence for larger aggregates was found (Fig. 2). Thus the methods developed allow to study the long range molecular correlations in liquids outside of the nearest coordination shells and obtain data on the supramolecular aggregates of nanosizes. The structural sensitivity of the approach developed is limited by the saturation like dependence of dipole correlation factor on the numbers of association $\overline{n}$ at high values of those. Thus data for the high association numbers may be less exact in comparison with the low ones.

6. EXTENDED QUASICHEMICAL MODELS IN THEORY OF FLUCTUATIONS AND RAYLEIGH LIGHT SCATTERING

Rayleigh light scattering is subdivided to isotropic and anisotropic ones, corresponding to fluctuations of isotropic and symmetric components of the permittivity tensor (the refractive index $\varepsilon = n^2$) [87], and respectively for the Rayleigh ratio $R = R_{is} + R_{an}$. Rayleigh ratio for isotropic scattering R_{is} is usually expressed as a sum of the terms, interpreted as due to independent fluctuations of temperature, density, and concentration [88-90]:

$$R_{is} \sim <\Delta\varepsilon^2>_{is} = <\Delta\varepsilon^2>_T + <\Delta\varepsilon^2>_\rho + <\Delta\varepsilon^2>_c . \qquad (7)$$

Firstly, this common consideration of isotropic fluctuations of the permittivity tensor and Rayleigh ratio in mixture (7) is not quite correct. Actually the fluctuation of temperature are independent both from density and concentration fluctuations, and thus in reality the last two terms in Eqn. (7) are determined as a joint result of the correlated fluctuations of density and concentration. According to another interpretation of Eqn. (7), the first two terms are due to correlated fluctuations of temperature and pressure, and the last one is due to the independent on those concentration fluctuations [6-13,45,63].

Next, the expressions for fluctuations of density $<\Delta\rho^2>$, concentration $<\Delta x_B^{02}>$ and their correlation $<\Delta\rho\Delta x_B^0>$ contain in a denominator $(\partial\mu_B / \partial x_B^0)_{T,p}$. Therefore approaching to the critical point of demixing of solution the increasing fluctuations of concentration are accompanied by growing both fluctuations of density and correlations of fluctuations of density and concentration. At $T \rightarrow T_{crit}$ their correlation coefficient $r_{\rho x_B^0} \rightarrow 1$, and that means the linear dependence of density and concentration fluctuations. Thus, the common conclusion on the dominating role of the concentration scattering near critical temperature of demixing of solution [40,41,91] also requires updating [6-13,63].

Scattering due to concentration fluctuations is related to the activity coefficient of the components γ_A, γ_B (3) by Einstein-Smoluchovsky relationship. Therefore investigation of the molecular nature of both the concentration fluctuations and scattering is possible; the applications for that of the QCNAS model were performed [8-12,77,83].

Secondly, in the traditional approach fluctuations of internal variables such as the extents of quasichemical process $\{\xi_\alpha\}$ are not taken into account. Thus, Eqn. (7) is derived under the assumption that the external thermodynamic variables e.g., *{T, ρ, and c}*, determine uniquely of the fluctuation state of a system. Therefore it allows for the fluctuations which do not disturb the equilibrium of a system with respect to the internal processes. In general the nonequilibrium states are determined by both external and internal variables of different nature and tensor dimension [8-13].

Thermodynamic potentials describing the non-equilibrium states of a system, taking into account both scalar $\{\xi_\alpha\}$, vector, and second rank tensor $\{\eta_{\beta,ik}\}$ internal variables were constructed, starting from statistical-mechanical ensembles. On the basis of this results the fluctuation theory was developed [7-13]. Fluctuations of scalar $\{\xi_\alpha\}$ and tensor $\{\eta_{\beta,ik}\}$ internal variables in isotropic medium are independent and thus may be considered separately.

It was shown that for fluctuations non-equilibrium as to the scalar internal parameters ($A_\alpha \neq 0$) the formulas for fluctuations of the extensive variables, e.g., energy, volume, coincide with those obtained by the Gibbs ensemble method [91], or by the theory of fluctuations equilibrium as regards to internal parameters ($A_\alpha = 0$) [39,45], $<\Delta E^2> = kT^2 C_V^o$, $<\Delta V^2> = kTV\beta_T^o$. In contrast, the fluctuations of intensive variable, e.g., temperature, pressure, are determined by the instant ($\xi = const$) values of the parameters, e.g., $<\Delta T^2> = kT^2 / C_V^\infty = <T^2>_A + kT^2(1/C_V^\infty - 1/C_V^0)$, where C_V^0 and C_V^∞ are the equilibrium ($A = 0$) and instant ($\xi = const$) isochoric heat capacity, and the second term expresses contribution of the extent fluctuations to temperature ones.

The contributions from fluctuation of the extent to thermodynamic properties, e.g., isobaric heat capacity C_p and isothermal compressibility β_T, are given by the formulas:

$$C_P^o - C_P^\infty = \frac{1}{kT^2}\left(\frac{\partial H}{\partial \xi}\right)_{T,P}^2 \langle \xi^2 \rangle,\ \beta_T^0 - \beta_T^\infty = \frac{1}{kTV}\left(\frac{\partial V}{\partial \xi}\right)_{T,P}^2 \langle \xi^2 \rangle \quad (8)$$

where $(\partial H/\partial \xi)_{T,P} = \Delta H$, $(\partial V/\partial \xi)_{T,P} = \Delta V$ are the enthalpy and the volume variations due to the quasichemical process.

Because for liquids the equilibrium shear modulus $\mu^0 = 0$, the quadratic form of the fluctuating Gibbs energy, responding to tensor internal parameters, is degenerated and thus, it describes states of so-called indifferent equilibrium. It means the linear relationships of the internal tensor variables $\{\eta_{\beta,ik}\}$ and the tensor of deformation $\{u_{ik}\}$, and therefore only determining linear combinations of them is possible. The fluctuations of tensor parameters determine the instant $(\eta = const)$ stress modulus μ^∞ of liquid [13].

Fluctuations of the internal variables make an additional contribution to scattering. Isotropic fluctuations of the permittivity and respective Rayleigh ratio in an individual liquid are given by the formula [7-13]:

$$R_{is} \sim \langle \Delta\varepsilon^2 \rangle_{is} = \frac{kT^2}{C_V^o}\left(\frac{\partial \varepsilon}{\partial T}\right)_{V,A}^2 + kTV\beta_T^o\left(\frac{\partial \varepsilon}{\partial V}\right)_{T,A}^2 - kT\left(\frac{\partial \varepsilon}{\partial \xi}\right)_{S,V}^2\left(\frac{\partial A}{\partial \xi}\right)_{S,V}^{-1} \quad (9)$$

in which the first two terms correspond to scattering by temperature and density fluctuations [Eqn. (7)], and the last one due to affinity A (extent ξ) fluctuations. For manifesting a process in isotropic light scattering, the reagents must be different in the enthalpy and the scalar polarizability. It follows from the experimental data that fluctuations of the internal variable should not be ignored in light scattering, because experimental values of Rayleigh ratios R^{exp} are systematically greater than calculated ones R^{calc}. Therefore it may be a source of the new data on the internal processes. Extraction of these terms requires both the precise measurements of the intensity of scattered light and the accurate data on the first two terms in Eqn. (9). On the other hand, additional terms in Eqn. (9) are not universal and estimation of them requires the specification of the internal processes. Therefore it may be a source of an unrecognized error in the scattering methods.

Rayleigh ratio for anisotropic scattering R_{an} is determined by the average molecular anisotropy $<\gamma^2>$ [6-13,92]:

$$R_{an} \sim <\gamma^2> = g^{opt}\gamma^2 = <\tilde{\gamma}\cdot\gamma> = \frac{1}{2N} < \sum_{p,q=1}^{N} \alpha_{p\sigma}\alpha_{q\tau}(3\cos^2\theta_{p\sigma,q\tau} - 1) >, \quad (10)$$

where $\alpha_{p\sigma}$ are the principal values of the molecular polarizability tensor, $\theta_{p\sigma,q\tau}$ are the angles between its principal axes, γ^2 is the molecular anisotropy, and $g^{opt} = <\gamma^2>/\gamma^2$ is the optical correlation factor of liquid. The relationships of optical factors with structure of aggregates and thermodynamics of their formation both for liquids and mixtures were established [6-12,60].

The dipole factors g^d (6) are determined by the mutual orientations of dipole moment of molecules and the optical ones g^{opt} (10) by those of molecular polarizability. Thus it allows complimentary information on the structure of supramolecular species to be obtained, and moreover there is a possibility to calculate dielectric properties from optical ones and *vice versa*. A lower dependence of optical factors on the number of association favors using dielectric data for strongly associated liquids.

Thus methods developed allow to obtain new data on the internal processes, occurring in liquids, especially on supramolecular ordering, based on the interrelations of them with thermodynamic, dielectric, and optic properties of mixtures. They give detailed data on the structure and properties of aggregates. Secondly, they allow to obtain both the thermodynamic parameters and integral and differential parameters of the aggregation, and to study the supramolecular aggregates with the long range molecular correlations. It seems promising to study the ordering of the supramolecular aggregates in liquid mixtures similar to that in liquid crystals.

7. EXTENDED QUASICHEMICAL MODELS IN INVESTIGATIONS OF RELAXATION PHENOMENA AND DYNAMIC PROCESSES IN LIQUIDS

Macroscopic description of irreversible processes is based on the nonequilibrium thermodynamics methods [1,23-26]. The local balance equations for extensive variables in combination with the relationships of fluxes of mass, momentum, energy, etc., to thermodynamic forces (gradients of velocity, concentration, temperature, etc.) form the set of equations to describe the spatial-temporal evolution of nonequilibrium systems. Onsager kinetic coefficients $L_{\alpha\beta}$, relating thermodynamic fluxes and forces, are connected to the rate constants of processes k, diffusion coefficient D, and bulk η_v and shear η_s viscosity, etc., and they should be determined either from experiment or using molecular-kinetic theory.

Study of dynamic processes occurring in liquids during the molecular thermal motion (Fig. 3) both by statistical mechanics and computer simulation methods present difficulties nowadays not only because of computational problems, but due to the necessity to construct the rather complex initial molecular models. Experimental methods are of considerable importance for investigating dynamic processes and models interpret their results. Thermodynamic models of the systems with internal parameters hold a prominent position among them [7-13,27-34,36-38]. Models provide the molecular interpretation of the internal variables, for instance they are the extents of corresponding quasichemical process both for conformation transitions and excitation or deactivation processes of molecules, and they are scalar, vector, or tensor variables, describing spatial supramolecular ordering for structural and dipole relaxation, etc.

Therefore, the importance of the extended quasichemical models to study dynamic processes in liquids may be stated. First one is the theory of the kinetic coefficients of nonequilibrium thermodynamics, second one is the theory of relaxation (acoustic, dielectric, etc.) spectra, and third one is the applications to studying dynamic processes occurring in liquids at the molecular thermal motion (Fig. 3). Extending this approach to non-linear irreversible phenomena seems promising. Molecular thermal motion is responsible for local equilibrium in a matter, and that is the basic point for thermodynamic approach to the irreversible phenomena. The mechanisms of the molecular thermal motion in liquids represent the elementary stages both for more complex systems and chemical reactions.

Processes of molecular thermal motion contribute both to equilibrium properties of liquids, e.g., thermodynamic potentials, excess functions of mixtures (3), isothermal β_T^0 (8) and adiabatic β_S^0 (13) compressibility, isochoric C_V^0 and isobaric C_p^0 (8) heat capacity, coefficient of thermal expansion θ^0, static permittivity ε_S (6), integral Rayleigh ratio R (9), (10) and kinetic properties, e.g., rate constants of the chemical reactions k, k' and those for internal processes k_α, k_α' (12), bulk η_v (11) and shear η_s (14) viscosity, diffusion coefficient D, dynamic adiabatic compressibility $\beta_S^*(\omega) = \beta_S'(\omega) + i\omega\eta_v(\omega)$, (11), shear module $\mu^*(\omega) = \mu(\omega) + i\omega\eta_s(\omega)$ (15), permittivity $\varepsilon^*(\omega) = \varepsilon'(\omega) - i\varepsilon''(\omega)$, and Rayleigh ratio $R(\omega)$, which depend on the frequency ω (Fig. 3). Thus the relationships of the thermodynamics and kinetics of the processes, occurring in a matter are exhibited.

Let us consider the scalar internal variables $\{\xi_\alpha\}$ (2), and examples from acoustic spectroscopy. The complex velocity of the longitudinal acoustic waves is determined by the expression $V_l^{*2} = [K_S^*(\omega) + 4\mu^*(\omega)/3]/\rho$, where $K_S^*(\omega) = 1/\beta_S^*(\omega)$ and $\mu^*(\omega)$ are the complex adiabatic bulk and the shear modules. The complex adiabatic compressibility $\beta_S^*(\omega)$ is given by the equation [28,30]:

$$\beta_S^*(\omega) = \beta_S'(\omega) + i\omega\eta_v(\omega) = \beta_S^\infty(\omega) + \sum_{\alpha=1}^{r} \frac{\delta_\alpha \beta_S}{1 + \omega^2 \tau_{pS,\alpha}^2} + i\sum_{\alpha=1}^{r} \frac{\delta_\alpha \beta_S \omega \tau_{pS,\alpha}}{1 + \omega^2 \tau_{pS}^2}. \quad (11)$$

In Eqn. (11) $\delta_\alpha \beta_S = \beta_{S,\alpha} - \beta_{S,\alpha}^\infty$ are the contributions of normal reaction to the equilibrium adiabatic compressibility β_S^0; $\eta_v(\omega)$ is the dynamic bulk viscosity; and $\tau_{pS,\alpha}$ are the isobaric-adiabatic relaxation times of normal reaction, determined by the eigenvalue problem:

$$\frac{1}{\tau} x = LHx. \quad (12)$$

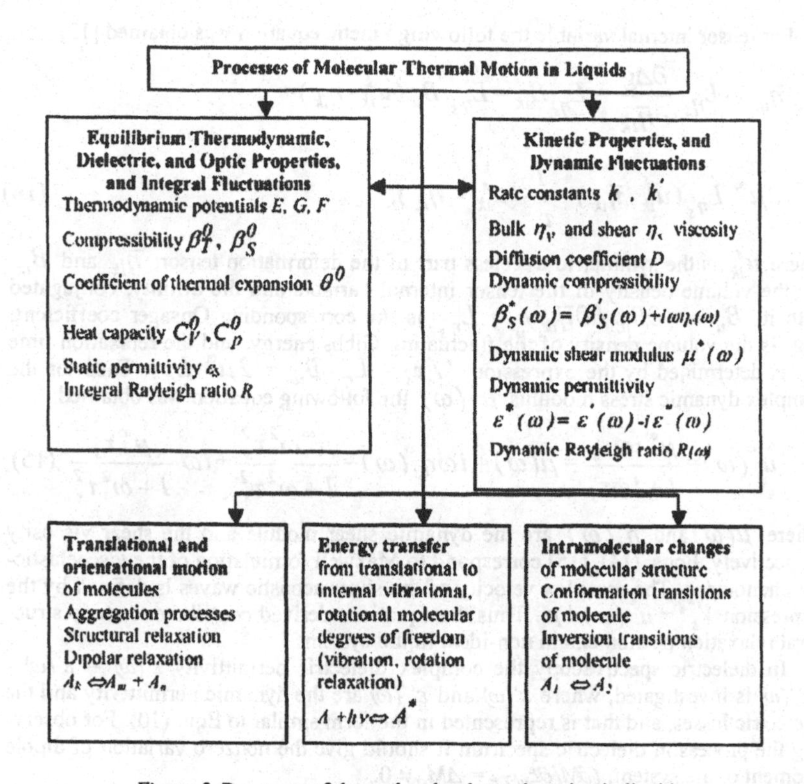

Figure 3. Processes of the molecular thermal motion in liquids

In Eqn. (12) L is the diagonal matrix, and the elements of that are proportional both to rate constants k_α (or k'_α) for direct (or reverse) process α (1), and activities (concentrations) of the reagent according to the mass action law. The elements of the symmetric matrix H are the second derivatives of enthalpy with respect to the extents of natural reaction $\{\xi\}$. The eigenvectors x_α form the matrix X of the linear transformation $\xi = X\zeta$ from extents of natural reaction ξ to those of normal reaction ζ. The relaxation strengths b_α are determined by the expressions:

$$b_\alpha = \frac{\delta_\alpha \beta_S}{\beta_S^0} = \frac{1}{V\beta_S^0}\left[\tilde{x}_\alpha \cdot (\partial V/\partial\xi)_{pS}\right]^2 \tau_{pS,\alpha}, \qquad (13)$$

where $(\partial V/\partial\xi_\alpha)_{pS} = \Delta V_\alpha - V_m\theta^\infty \Delta H_\alpha / C_p^\infty$. For observing the process in acoustic spectrum it should possess nonzero enthalpy ΔH_α, or volume ΔV_α, variations. Eqns. (11)-(13) were derived without using specific solution model; the simplest model of the ideal solution (ideal associated solution) is applied usually [27-32].

For tensor internal variable the following kinetic equation was obtained [13]:

$$\dot{\overline{\eta}}_{ik} = -L_{\eta_s} \frac{\partial \Delta g}{\partial \overline{\eta}_{ik}} = L_{\eta_s} B_{ik} = L_{\eta_s} B_u (u^s_{ik} - \overline{\eta}_{ik}) =$$

$$2\mu^{\infty} L_{\eta_s} (u^s_{ik} - \overline{\eta}_{ik}) = \frac{1}{\tau_s} (u^s_{ik} - \overline{\eta}_{ik}), \quad (14)$$

where u^s_{ik} is the symmetric traceless part of the deformation tensor; $\overline{\eta}_{ik}$ and B_{ik} are the volume density of the tensor internal variable and the affinity, conjugated with it, $B_u = (\partial B_{ik} / \partial \overline{\eta}_{ik})_{u^s}$; L_{η_s} is the corresponding Onsager coefficient; Δg is the volume density of the fluctuating Gibbs energy, and the relaxation time τ_s is determined by the expression $1/\tau_s = L_{\eta_s} B_u = 2\mu^{\infty} L_{\eta_s}$. Thus for the complex dynamic stress modulus $\mu^*(\omega)$ the following equation was obtained:

$$\mu^*(\omega) = \frac{\mu^{\infty} i\omega\tau_s}{1 + i\omega\tau_s} = \mu(\omega) + i\omega\eta_s(\omega) = \frac{\mu^{\infty}\omega^2\tau_s^2}{1+\omega^2\tau_s^2} + i\omega \frac{\mu^{\infty}\tau_s}{1+\omega^2\tau_s^2}, \quad (15)$$

where $\mu(\omega)$ and $\eta_s(\omega)$ are the dynamic shear module and the shear viscosity respectively. Eqns. (14), (15) correspond to Maxwell formulation of the viscoelasticity phenomena. The complex velocity of the shear acoustic waves is defined by the expression $V_s^{*2} = \mu^*(\omega)/\rho$. Thus the equations derived contribute to study structural relaxation phenomena in non-ideal liquid systems.

In dielectric spectroscopy the complex dielectric permittivity $\varepsilon^*(\omega) = \varepsilon'(\omega) - i\varepsilon''(\omega)$ is investigated, where $\varepsilon'(\omega)$ and $\varepsilon''(\omega)$ are the dynamic permittivity and the dielectric losses, and that is represented in the form similar to Eqn. (10). For observing the process in dielectric spectrum it should give the nonzero variation of dipole moment of the system, $(\partial M/\partial \xi_\alpha)_{T,P} = \Delta M_\alpha \neq 0$.

The equations defining the relaxation times τ_α (12) and strengths b_α (13) for nonideal solutions (nonideal aggregate mixtures), based on the QCNAS model were obtained. The structural and thermodynamic data on internal processes obtained from analysis of the equilibrium properties are useful for these purposes. The thermodynamic non-ideality of the solution leads to the essential corrections to thermodynamic and kinetic parameters of the internal processes, determined from relaxation spectra [7-13].

The Markov character of the processes is implied in Eqn. (1). Taking into account the spatial-temporal correlation of processes may change the sense of Eqn. (1) with the respective modifying the dispersion relationships (11), (15). This is the most important for both critical points and phase transition phenomena [40-42] and vitrification of liquid [93]. The spatial correlation of the dynamic processes is taken into account in the model of collective reactions [93-96]. It leads to both essential deviations of the kinetic properties from Arrhenius equation, and widening the relaxation spectrum of supercooled liquids, and thus allows explanation of these important features of the glass formation. Studying the interrelations of the models of collective reactions and collective coupling modes [97-99] seems promising.

Of the special interest is to further extend this approach for vector and tensor internal variables. Among more particular problems models with continuous variation in the internal parameters could be mentioned. It seems interesting to investigate interrelations of dynamic processes, different on nature, e.g. structural relaxation and conformation transitions of molecules, etc.

The applications allow obtaining new data both on the thermodynamics and kinetics of supramolecular ordering, conformation transitions of molecules, and energy transfer processes, as well as on the spatial-temporal hierarchy of the processes, occurring in liquids during the molecular thermal motion [7-13,100-105].

In general the results obtained extend the quasichemical concept to dynamic fluctuation and relaxation phenomena and kinetic properties of liquid systems.

8. APPLICATIONS FOR BINARY MIXTURES

Let us characterize briefly the applications to binary solutions, which rely on the analysis of their thermodynamic, dielectric and optic properties (see for the review [6-12]). The binary mixtures were subdivided according to thermodynamic properties to those with negative and positive deviation from ideal solution behavior. On the basis of general results the specific equations for thermodynamic functions of mixing, permittivity, and Rayleigh ratio were derived. The properties of the mixture were considered over as wide concentration and temperature range as possible.

8.1 Binary Mixtures with Negative Deviation from Ideal Solution Behavior

Propanone-trichloromethane [77] and dimethylsulfoxide-trichloromethane [75] mixtures were considered as the examples of solutions with negative deviation from ideality. The propanone-trichloromethane mixture was described by the model $A + B + AB + AB_2$, and for dimethylsulfoxide-trichloromethane one it was found necessary also to include dimerisation of dimethylsulfoxide (model $A + B + AB + AB_2 + A_2$), where symbol A means propanone or dimethylsulfoxide, and B denotes trichloromethane.

Models proposed describe the activity coefficients of the components, excess Gibbs energy, enthalpy, and entropy, as well as permittivity, and Rayleigh ratio of the mixtures investigated in the whole concentration range. The major contributions to the excess thermodynamic functions are made by specific interactions, and the role of universal interactions is reduced to partial compensation of the association contribution. Universal interactions lead to the dependence of the "concentration" equilibrium constants on the mixture composition, which is varied up to 30 %, and this is highly essential for studying thermodynamics of aggregation, and in particularly at the evaluating concentration of aggregates.

Mixtures with negative deviation from ideality are characterizing by relatively small numbers of association, for instance in propanone-trichloromethane mixture the maximum number of association n_{AB} = 1.25 (T = 313 K). Thus the molecular correlations extend to no more than first coordination shell.

8.2 Binary Mixtures with Positive Deviation from Ideal Solution Behavior

Monohydric alcohol solutions comprise the binary mixtures demonstrating generally positive deviation from ideality. We can recognize at least four types of mixtures of alcohol [6-12]. First one is the mixture of nonpolar "inert" solvent-associating solute, e.g., cyclohexane-cyclohexanol [76], cyclohexane-ethanol [106]. Second one is the mixture of nonpolar "solvating" solvent-associating solute, e.g., tetrachloromethane-methanol [83], tetrachloromethane-unsaturated alcohol [80], and

1,4-dioxane-methanol [107]. Third one is the mixture polar "solvating" solvent-associating solute, e.g., ketone-alcohol [78,79,84], and trichloromethane-alcohol [81]. Fourth one is the mixture of associating solvent-associating solute, e.g., methanol-butanol, and alcohol-N-amide.

The mixtures similar cyclohexane-ethanol were described by the model of infinite chain-like alcohol association, added by the cyclic alcohol aggregate, $A_1 + B_1 + B_2 + B_3 + ... + B_k + ... + B_{l,cycl}$, hereinafter A denotes the molecule of the solvent, and B means alcohol. The mixtures of the second and third type were treated using the model $A_1 + B_1 + B_2 + B_3 + ... + B_k + ... + B_{l,cycl} + AB_1 + AB_2 + AB_3 + ... + AB_m + ...$. For the mixtures of the fourth type more general model $A_1 + B_1 + B_{l,cycl} + A_mB_n$ $(m,n \geq 1)$ was applied. Though the supramolecular structure of the solutions belonging to both second and third types was described by the same stoichiometric model the polar nature of the solvent in the ketone-alcohol, and trichloromethetane-alcohol mixtures required a detailed description of the structure of complexes solvent - solute.

For establishing the most probable structure of the cyclic aggregates of alcohol the data on Monte-Carlo simulation of methanol containing 500 molecules in the main box [86] were applied. The detailed analysis of the topological structure of aggregates containing up to 62 molecules was performed. The most probable cyclic aggregate of methanol consists of four molecules [10-12,86].

The numbers of association being very high in pure methanol ($\bar{n}_B \cong 60$, $T = 298$ K), decrease dramatically upon dilution (Figs. 1, 4). The solvation increases the number of association of mixture n_{AB}, but greatly decreases of that for alcohol n_B due to competition between association and solvation of alcohol. The number of association in methanol has been found much greater than in Monte-Carlo simulation [86] ($\bar{n}_B \cong 12$, T= 298 K), probably due to insufficient size of the simulation main box [9-12]. Cyclic aggregates play important role only in the mixtures poor of alcohol, e.g., $w_{cycl} = 24\%$ at $x_B^o = 0.1$ m. f. (Fig. 5), according to the higher thermodynamic stability of non-polar aggregates in a less polar medium. In pure alcohol the role of cyclic aggregates is less evident because the chain-like aggregates are dominating, $w_{cycl} \cong 2\%$. Thus the model of chain-like association seems a good approach for pure methanol. It follows from the results obtained that the numerous models of the alcohol mixtures, taking into account one or few types of the aggregates, are not adequate.

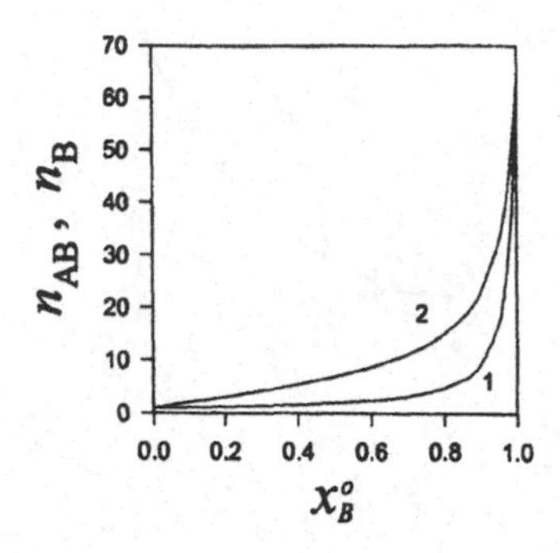

Figure 4. The average numbers of association of mixture n_{AB} (1) and alcohol n_B (2) in the mixture tetrachloromethane-methanol, T = 298 K; x_B^o is m.f. of methanol.

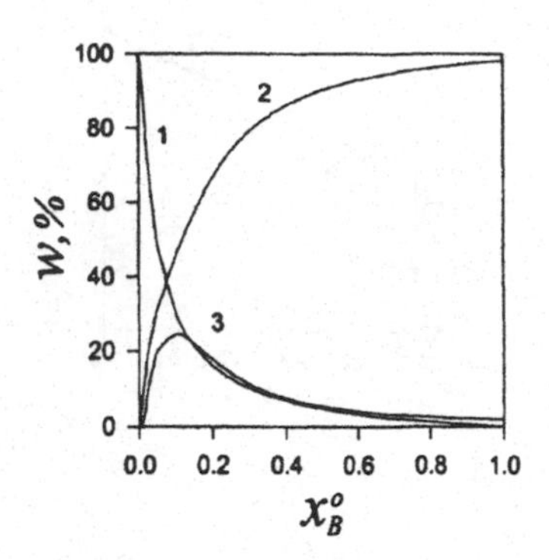

Figure 5. The weight fractions of alcohol monomers (1), linear aggregates(2) and cyclic tetramers (3) in the mixture tetrachloromethane-methanol, T = 298 K; x_B^o is m.f. of methanol.

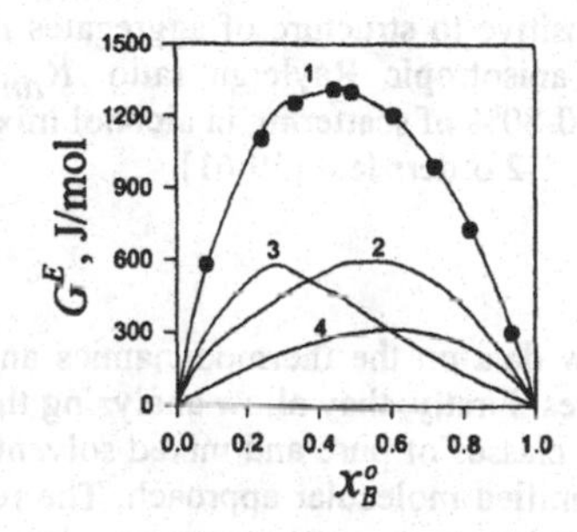

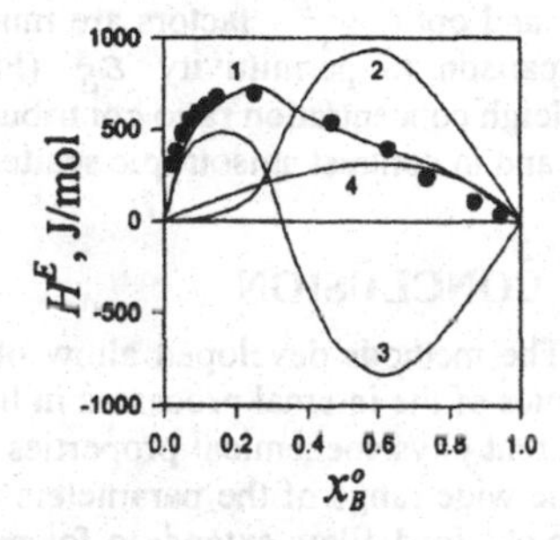

Figure 6. Excess Gibbs energy G^E, T = 293 K (a) and enthalpy H^E , T = 323 K (b) of the tetrachloromethane-methanol mixture. Points, experimental data; lines, calculated values, (1) total values, (2) association contribution, (3) dipole contribution, (4) dispersion contribution; x_B^o is m.f. of methanol.

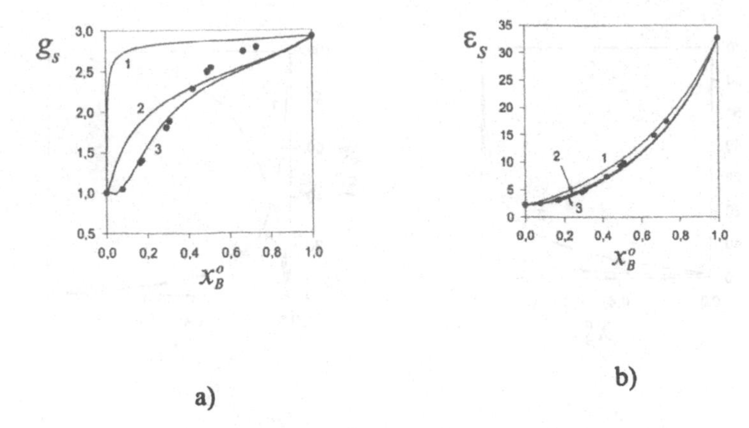

a) b)

Figure 7. Dipole correlation factor g_s^d (a) and permittivity ε_S (b) of tetrachloromethane-methanol mixture, T = 298 K. Points, experiment, lines, calculated values without allowance for solvation and cyclic tetramers (1), with allowance for solvation (2), and with allowance for solvation and cyclic tetramers (3); x_B^o is m.f. of methanol.

In general the models of alcohol solutions developed give a good description of thermodynamic functions, permittivity, and Rayleigh light scattering. Universal interaction contributions to the excess thermodynamic functions [Fig. (6)] are comparable with the association one [83] and do not reduce to their partial counterbalancing as e.g., for the propanone-trichloromethane mixture [77]. Therefore again the universal contributions are necessary for the proper study of the association. Dipole g_s^d and optic g_s^{opt} factors are much more sensitive to structure of aggregates in comparison to permittivity ε_S (Fig. 7) and anisotropic Rayleigh ratio R_{an}. Rayleigh concentration ratio contributes up to ~70-80% of scattering in alcohol mixture and in contrast anisotropic scattering is about 1-2 orders less [59-61].

9. CONCLUSION

The methods developed allow obtaining new data on the thermodynamics and kinetics of the internal processes in liquid mixtures. Firstly, they allow analyzing the different physicochemical properties of the wide classes of pure and mixed solvents in the wide range of the parameters within the unified molecular approach. The results obtained allow extending for multicomponent mixtures, and thus properties of ternary mixtures from the data obtained for binary ones may be predicted.

Secondly, they allow studying structure, composition, and properties of the aggregates, as well as thermodynamics and kinetics of aggregation. The differential and integral characteristics of aggregation in mixtures can be investigated. Thus the detailed description of the supramolecular organisation of mixtures can be obtained.

Thirdly, the methods developed allow structural investigations of liquids, and especially the long range supramolecular ordering. These data are difficult or impossible to obtain by other methods, e.g., spectroscopy or diffraction ones.

Fourthly, the models developed describe properties of mixture in the whole concentration range, and thus include pure solvents, diluted and concentrated solutions. They enable both description and prediction of properties over a wide temperature range.

Fifthly, the equations derived determine the thermodynamic potentials of the non-equilibrium states, fluctuations of both external and internal variables, and contributions of the internal processes to macroscopic properties. Thus it extends the quasichemical concept to kinetic and relaxation properties of the nonideal mixtures.

Supramolecular ordering reveals itself in the macroscopic properties of mixture. The different supramolecular organisation is characterizing of mixtures with negative and positive deviations from ideal solution behavior. In mixtures with the negative deviation from ideality the main role belongs to complexation between the dissimilar molecules, and in contrast in mixtures with the positive deviation the association of homogeneous molecules plays usually the main role.

Thermodynamic properties are relatively sensitive to the stoichiometry of the aggregation, but not to the structure of aggregates. Both dielectric and optical properties are related directly to the spatial ordering of molecules. Therefore joint application of the methods developed to different properties of mixtures allows obtaining the complementary information on the structure of aggregates, impossible by other methods especially as to the long range molecular correlations.

Selforganisation in mixtures by the attractional directed molecular interactions like H-bonds leads to the formation of the supramolecular aggregates consisted of tens of molecule. The supramolecular aggregates revealed, having the nanosizes, are important for new understanding of the long range molecular correlations for structure and properties of liquid state of matter.

It may be concluded, that the backgrounds of the unified approach, based on the concept of supramolecular assemblies, to thermodynamic, dielectric, optic, and kinetic properties of liquids have been elaborated. The results presented contribute to the supramolecular thermodynamics and kinetics as the background for analyzing properties of liquids, related to different molecular parameters and studying the long range supramolecular ordering phenomena. It allows prediction of macroscopic properties and thus contributes to molecular design of liquid materials. It could be proposed that the models developed for describing microscopic stages of supramolecular ordering in molecular liquids reflect the general features of the selforganisation phenomena in systems with larger spatial-temporal scales of the processes.

REFERENCES

1. Hobza, P. and Zagradnik, R. (1989) *Intermolecular Complexes,* Academia, Prague.
2. Kondepudi, D. and Prigogine, I. (1999) *Modern Thermodynamics.* Wiley, New York.
3. Haken, H. (1978) *Synergetics. An Introduction to Nonequilibrium Phase Transitions and Self-Organisation in Physics, Chemistry and Biology,* Springer, Berlin.
4. Haken, H. (1988) *Information and Self-organisation*, Springer, Berlin.
5. Lehn, J.-M. (1995) *Supramolecular Chemistry,* VCH, Weinheim.
6. Durov, V.A. (1989) Models of Associative Interactions in Physical Chemistry of Liquid Non-Electrolytes, in G.A. Krestov (ed.), *Solutions of Non-Electrolyte in Liquids* (monograph in Russian), pp. 36-102, Science, Moscow.
7. Durov, V.A. (1993) Quasichemical Models of Liquid Non-Electrolytes, *Zh. Fiz. Khim.* 67, 290-304 (*Russ. J. Phys. Chem.* (1993) 67, 264-278).

8. Durov, V.A. (1998) Modeling of Supramolecular Ordering in Molecular Liquids: Structure, Physicochemical Properties, and Macroscopic Manifestations, *J. Mol. Liq.* 78 51-82.
9. Durov, V.A. (2000) Molecular Modelling of Thermodynamic and Related Properties of Mixtures, *J. Therm. Anal. Cal.* 62, 15-27.
10. Durov, V.A. (2002) Models of Liquid Mixtures: Supramolecular Organisation and Physicochemical Properties, in A.M. Kutepov (ed.), *Concentrated and Saturated Mixtures* (monograph in Russian), pp. 170-254, Science, Moscow.
11. Durov, V.A. (2003) Modeling of Supramolecular Ordering in Mixtures: Structure, Dynamics and Properties, *J. Mol. Liq.* 103-104, 41-82.
12. Durov, V.A. (2003) Models of Liquid Mixtures: Structure, Dynamics, and Properties, *Pure and Applied Chem.*, accepted.
13. Durov, V.A. (2003) Thermodynamic Models of the System with Internal Variables: Fluctuation and Relaxation Phenomena, *J. Mol. Liq.*, accepted.
14. Croxton, C.A. (1974) *Liquid State Physics*, Cambridge University Press, Cambridge.
15. Chapman, W.G., Gubbins, K.E., Toslin, C.G., and Gray, C.G. (1987) Mixtures of Polar and Associating Molecules, *Pure and Appl. Chem.* 59, 53-60.
16. Allen, M.P. and Tildesley, D.J. (1987) *Computer Simulation of Liquids.* Clarendon Press, Oxford.
17. Economou, G. and Donohue, M.D. (1991) Chemical, Quasi-Chemical and Perturbation Theories for Associating Fluids, AICHEJ 37, 1875-1894.
18. Chapman, W.G., Gubbins, K.E., Jackson, J., and Radosz, M. (1989) SAFT: Equation-of-State Solution Model for Associating Fluids, *Fluid Phase Equil.* 52, 31-38.
19. Adidharma, H. and Radosz, M. (1999) A Study of Square-well Statistical Associating Fluid Theory Approximations, *Fluid Phase Equil.* 161, 1-20
20. Guggenheim, E.A. (1952) *Mixtures*, Clarendon Press, London.
21. Barker, J.A. (1963) *Lattice Theories of Liquid State*, Pergamon Press, Oxford.
22. Panayiotou, C. and Sanchez, I.C. (1991) Hydrogen Bonding in Fluids: An Equation of State Approach, *J. Phys. Chem.* 95, 90-97.
23. Prigogine, I. and Defay, P. (1954) *Chemical Thermodynamics*, Longmans Green, London.
24. De-Groot, S.R. and Mazur, P. (1962) *Non-Equilibrium Thermodynamics*, North Holland, Amsterdam.
25. Gyarmati, I. (1970) *Non-Equilibrium Thermodynamics. Field Theory and Variational Principles*, Springer, Berlin.
26. Keizer, J. (1987) *Statistical Thermodynamics of Irreversible Processes*, Springer, New York.
27. Mikhailov, I.G., Soloviev, V.A., and Syurnikov Yu.P. (1964) *The Foundations of Molecular Acoustics* (monograph in Russian), Science, Moscow.
28. Bauer, U. (1965) Phenomenological Theory of Relaxation Phenomena in Gases, in W.P. Mason (ed.), *Physical acoustics*, Academic Press, New York, 2, Part A, pp. 61-154.
29. Litovitz, T. and Davis, T. (1965) Structural and shear relaxation in liquids, in W.P. Mason (ed.), *Physical acoustics*, Academic Press, New York, 2, Part A, pp. 298-370.
30. Eigen, M. and De-Meyer, L. (1974) Theoretical Backgrounds of Relaxation Spectroscopy, in G. Hammes (ed.), *Investigation of Rates and Mechanisms of Fast Reactions*, pp. 79-129, Wiley, New York.
31. Wyn-Jones, E. (ed.), (1975) *Chemical and Biological Applications of Relaxation Spectrometry*, Reidel, Dordrecht.
32. Gettins, E. and Wyn-Jones, E. (eds.), (1979) *Techniques and Applications of Fast Reactions in Solutions*, Reidel, Dodrecht.
33. Mountain, R.D. (1966) Spectral Distribution of Scattered Light in a Simple Liquid, *Rev. Mod. Phys.* 58, 204- 214.

34. Leontovich, M. (1941) Relaxation in liquids and scattering of light, *J. Physics (USSR)* 4, 499-514.
35. Kluitenberg, G.A. (1962) Thermodynamical Theory of Elasticity and Plasticity, *Physica* 28, 217-232; A Note on the Thermodynamics of Maxwell Bodies, Kelvin Bodies (Voigt Bodies), and Fluids, 28, 561-568.
36. Rytov, S.M. (1970) Relaxation theory of Rayleigh light scattering, *Zh. Exper. and Theor. Phys.* 58, 2155-2170.
37. Volterra, V. (1969) Theory of light scattering from shear waves in liquids, *Phys. Rev.* 180, 156-166.
38. Wang, C.H. (1980) Depolarised Rayleigh-Brillouin scattering of shear waves and molecular reorientation in a viscoelastic liquid, *Mol. Phys.* 41, 541-565.
39. Landau, L.D. and Lifshitz, E.M. (1976) *Statistical Physics*, 3rd ed., Part 1, Chapters XII, XIV, Pergamon Press, New York.
40. Stanley, E.U. (1971) *Introduction to Phase Transition and Critical Phenomena*, Clarendon Press, Oxford.
41. Shang-keng Ma, (1976) *Modern Theory of Critical Phenomena*, Benjamin, London.
42. Toledano, J-Cl. and Toledano, P. (1987) *The Landau Theory of Phase Transitions*, World Scientific, Singapore.
43. Landau, L.D. and Lifshitz, E.M. (1987) *Theory of Elasticity* (monograph in Russian), 4th ed., Science, Moscow.
44. Leontovich, M.A. (1983) *Introduction to Thermodynamics. Statistical Physics* (monograph in Russian), Science, Moscow.
45. Durov, V.A. and Ageev, E.P. (2003) *Thermodynamic Theory of Solutions* (monograph in Russian), 2nd ed., Chapters 4,7, URSS Editorial, Moscow.
46. Acree, W.E. (1984) *Thermodynamic Properties of Nonelectrolyte Solutions*. Academic Press, Orlando.
47. March, K. and Kohler, F. (1985) Thermodynamic Properties of Associated Solutions. *J. Mol. Liq.*, 30, 13-55.
48. Prausnitz, J.M., Lichtenthaler, R.N., and de Azevedo, R.N. *Molecular Thermodynamics of Fluid-Phase Equilibria*, 2nd ed., Prentice-Hall, Englewood Cliffs, New York.
49. Schuster, P., Zundel, G., and Sandorfy, S. (eds.) (1976) *The Hydrogen Bond: Recent Development in Theory and Experiment*, 1-3, North Holland, Amsterdam.
50. Durov, V.A. (1989) Contributions from Electrostatic Intermolecular Interactions to Thermodynamic Properties of Liquid Non-Electrolytes, I. One-component Liquids, *Zh. Fiz. Khim.* 63, 1750-1758; II. Solutions, 63, 1759-1766; III. Thermodynamics of Associative Equilibria, 63, 2033-2040 (*Russ. J. Phys. Chem.* (1989) 63, 967-971; 971-974; 1121-1124).
51. Durov, V.A. (1991) Thermodynamics of Non-Ideal Aggregates Mixture and Excess Functions of Non-Electrolytes Solution, *Zh. Fiz. Khim.* 65, 1766-1777 (*Russ. J. Phys. Chem.* (1991) 65, 939-945).
52. Durov, V.A. and Shilov, I.Yu. (1994) On Thermodynamics of Solutions of Polar Substances, *Zh. Fiz. Khim.* 68, 184-186 (*Russ. J. Phys. Chem.* (1994) 68, 165-166).
53. Durov, V.A. and Shilov, I.Yu. (1996) On Thermodynamics of Nonelectrolyte Solutions, *Zh. Fiz. Khim.* 70, 659-663 (*Russ. J. Phys. Chem.* (1996) 70, 600-604).
54. Durov, V.A. (1981) Structure and Dielectric Properties of One-Component Liquids, *Zh. Fiz. Khim.* 55, 2833-2841 (*Russ. J. Phys. Chem.* (1981) 55, 1612-1616).
55. Durov, V.A. (1982) Theory of Dielectric Properties of Liquid Associated Mixtures, *Zh. Fiz. Khim.* 56, 1950-1956 (*Russ. J. Phys. Chem.* (1982) 56, 1191-1194).
56. Durov, V.A. (1989) Theory of Static Dielectric Constant of Liquid Systems, *Zh. Fiz. Khim.* 63, 1587-1594 (*Russ. J. Phys. Chem.* (1989) 63, 875-878).
57. Durov, V.A. (1986) Theory of Static Permittivity of Associated Liquids and New Possibilities for Studying their Molecular Structure, in M.I. Shackhparonov and L.P. Filippov

(eds.), *Studies on the Structure, Thermal Motion and Properties of Liquids* (monograph in Russian), Moscow University Press, Moscow, pp. 35-67.
58. Durov, V.A. (1999) Dielectric Materials, in T.Letcher (ed.), *Chemical Thermodynamics. A 'Chemistry for the 21st Century'*, Blackwell Science, London, pp. 327-334.
59. Durov, V.A. (1976) On the Theory of Rayleigh Light Scattering in Liquids, Containing Chain-Like Aggregates, in M.I. Shakhparonov and L.P. Filippov (eds.), *Physics and Physical Chemistry of Liquids* (monograph in Russian), 3rd Issue, Moscow University Press, Moscow, pp. 125-137.
60. Durov, V.A. (1981) Calculation of Molecular Anisotropy of Chain-Like Aggregates. III. Statistically Averaged Molecular Anisotropy, *Zh. Fiz. Khim.* 55, 882-889 (*Russ. J. Phys. Chem.* (1981) 55).
61. Durov, V.A., Zhuravlev, V.I., and Romm, Th.A. (1985) Theoretical Study of the Molecular Structure of Liquids by Rayleigh Spectroscopy. I. Monohydric Alcohols. *Zh. Fiz. Khim.* 59, 96-101; II. Structure and Optical Properties of Liquid Methanol, 59, 102-106.
62. Durov, V.A. (1995) Liquid Systems Supramolecular Organization: Modelling of Properties, Phenomena, and Molecular Design, *Hung. J. Ind. Chem.* 23, 195-200.
63. Durov, V.A. (1987) *Vestn. Mosk. Univ., Ser. Khim.* 28, 54-61 (*Proc. Moscow State University, Chem.* 28, 45-50).
64. Durov, V.A. (1987) On Problems of Studying Molecular Thermal Motion in Liquids and Structure of Those by Relaxation Spectroscopy, in G.A. Krestov (ed.) *Theoretical Methods on the Description of Solution Properties* (monograph in Russian), Academic Press, Ivanovo, pp. 57-63.
65. Durov, V.A. (1994) Models of Liquid Systems with Internal Variables: Thermodynamics, Supramolecular Organisation, and Kinetic Phenomena, in M. Zeidler (ed.), *Ultrafast Phenomena in Liquids and Glasses. Vibrational and Electronic Dynamics*, Zakopane, Poland, p. 61.
66. Durov, V.A. (1982) Theory of the Static Dielectric Constant of Liquid Monohydric Alcohols, *Zh. Fiz. Khim.* 56, 384-390 (*Russ. J. Phys. Chem.* (1982) 56, 232-236).
67. Durov, V.A. and Usacheva, T.M. (1982) Dielectric Properties and Molecular Structure of Liquid n-Alkanols, *Zh. Fiz. Khim.* 56, 648-652 (*Russ. J. Phys. Chem.* (1982) 56, 394-396).
68. Durov, V.A., Bursulaya, B.Dj., and Ivanova, N.A. (1990) Molecular Interactions and Thermodynamic Functions of Liquid n-Alkanols, *Zh. Fiz. Khim.* 64, 34-39 (*Russ. J. Phys. Chem.* (1990) 64, 19-22).
69. Durov, V.A., Bursulaya, B.Dj., Artykov, A., and Ivanova, N.A. (1989) Molecular Interactions and Thermodynamic Functions of Liquid Alicyclic Alcohols, *Zh. Fiz. Khim.* 63, 3192-3198.
70. Durov, V.A., Pukhala, Ch., and Lifanova, N.V. (1982) Structure and Dielectric Properties of Aromatic Alcohols. Phenylmethanol and 2-Phenylethanol, *Vestn. Mosk. Univ. Ser. Khim.* 23, 33-37 (*Proc. Moscow State University*, Chem. (1982), 23).
71. Durov, V.A. and Pukhala, Ch. (1984) Molecular Structure of Liquid Aromatic Alcohols. 1-Phenylethanol and Phenyl-t-Butanol, *Zh. Fiz. Khim.* 58, 391-395 (*Russ. J. Phys. Chem.* (1984) 58, 232-235).
72. Durov, V.A., Bursulaya, B.Dj., and Ivanova, N.A. (1990) Molecular Interactions, Thermodynamic Functions and Structure of Liquid Aromatic Alcohols, *Zh. Fiz. Khim.* 64, 620-626 (*Russ. J. Phys. Chem.* (1990) 64, 331-333).
73. Durov, V.A. (1981) Structure and Dielectric Properties of Liquid N-Monosubstituted Amides, *Zh. Fiz. Khim.* 55, 2842-2848 (*Russ. J. Phys. Chem.* (1981), 55, 1616-1620).
74. Durov, V.A. and Bursulaya, B.Dj. (1991) Thermodynamics of Molecular Interactions in Liquid N-Monosubstituted Amides, *Zh. Fiz. Khim.* 65, 2066-2071 (*Russ. J. Phys. Chem.* (1991) 65, 1098-1100).

75. Durov, V.A. and Shilov, I.Yu. (1996) Supramolecular Organization and Physico-chemical Properties of Dimethylsulfoxide-Trichloromethane Solutions, *Zh. Fiz. Khim.* 70, 818-824 (*Russ. J. Phys. Chem.* (1996) 70, 757-763).
76. Durov, V.A. and Shilov, I.Yu. (1996) Supramolecular Organization and Physicochemical Properties of Cyclohexane-Cyclohexanol Solutions, *Zh. Fiz. Khim.* 70, 1224-1229 (*Russ. J. Phys. Chem.* (1996) 70, 1138-1143).
77. Durov, V.A. and Shilov, I.Yu. (1996) Molecular Structure and Physicochemical Properties of Acetone-Chloroform Mixture, *J. Chem. Soc. Faraday Trans.* 92, 3559-3564.
78. Durov, V.A. and Shilov, I.Yu. (1996) Supramolecular Organization and Physicochemical Properties of Cyclohexanone-Cyclohexanol Solutions, *Zh. Fiz. Khim.* 70, 2180-2186 (1996) (*Russ. J. Phys. Chem.* (1996) 70, 2016-2022).
79. Durov, V.A. and Shilov, I.Yu. (1997) Molecular Structure and Dielectric Properties of Cyclohexanone-4-Methylcyclohexanol Mixtures, *Zh. Fiz. Khim.* 71, 450-454 (*Russ. J. Phys. Chem.* (1997) 71, 381-385).
80. Durov, V.A. and Shilov, I.Yu. (1998) Dielectric Properties and Structure of Liquid Unsaturated Monohydric Alcohols and Their Solutions in Tetrachloromethane, *Zh. Fiz. Khim.* 72, 1245-1250 (*Russ. J. Phys. Chem.* (1998) 72, 1114-1119).
81. Durov, V.A., Tereshin, O.G., and Shilov, I.Yu. (2001) Supramolecular Structure and Physicochemical Properties of Chloroform-Methanol Mixture, *Zh. Fiz. Khim.* 75, 1618-1627 (*Russ. J. Phys. Chem.* (2001) 75, 1593-1602).
82. Durov, V.A., Tereshin, O.G., and Shilov, I.Yu. (2001) Supramolecular Structure and Physicochemical Properties of Chloroform-Ethanol Mixture, *Zh. Fiz. Khim.* 75, 1927-1934 (*Russ. J. Phys. Chem.* (2001) 75, 1832-1839).
83. Durov, V.A. and Shilov, I.Yu. (2001) Supramolecular Structure and Physicochemical Properties of the Mixture Tetrachloromethane - Methanol, *J. Mol. Liq.* 92, 165-184.
84. Durov, V.A. and Tereshin, O.G. (2003) Supramolecular Structure and Physicochemical Properties of Acetone-Methanol Mixture, *Struct. Chem.*, accepted.
85. Durov, V.A. and Tereshin, O.G. (2003) Models of Halogenated Hydrocarbon-Organic Solvent Mixtures: Molecular Interactions, Structure and Physicochemical Properties, *Fluid Phase Equil.*, accepted.
86. Shilov, I.Yu., Rode, B.M., and Durov V.A. (1999) Long Range Molecular Correlations and Hydrogen Bonding in Liquid Methanol. A Monte-Carlo Simulation. *Chem. Phys.* 241, 75-82.
87. Landau, L.D. and Lifshitz, E.M. (1982) *Electrodynamics of Continuous Media* (monograph in Russian), 2nd. ed., Science, Moscow.
88. Berne, B.J. and Pecora, R. (1976). *Dynamic Light Scattering with Applications to Chemistry, Biology, and Physics*. Wiley, New York.
89. Flygare, W.H. (1977) Light Scattering in Pure Liquids and Solutions, *Chem. Soc. Rev.* 6, 109-138.
90. Kivelson, D. and Madden, P.A. (1980) Light Scattering Studies of Molecular Liquids, *Ann. Rev. Phys. Chem.* 31, 523-558.
91. Hill, T. (1956) *Statistical Mechanics*, Mc-Graw-Hill, New York.
92. Keilich, S. (1981) *Nonlinear Molecular Optics* (monograph in Russian), Mir, Moscow.
93. Shakhparonov, M.I. and Durov, V.A. (1980). *Mechanisms of Fast Reactions in Liquids* (monograph in Russian), pp. 307-340, Vyushaya Shkola, Moscow.
94. Shakhparonov, M.I. and Durov, V.A. (1979) Theory of Collective Reactions in Liquids. I. Formulation of the Basic Concept, *Zh. Fiz. Khim.* 53, 1401-1406; V. Collective Reactions and Vitrification, 53, 2451-2455 (*Russ. J. Phys. Chem.* (1979), 53, 792-795; 1401-1403).
95. Durov, V.A. and Shakhparonov, M.I. (1979) Theory of Collective Reactions in Liquids. II. Properties of Correlation Functions for Reaction Events, *Zh. Fiz. Khim.* 53, 1833-1834; III. Simple Collective Reaction, 53, 2251-2255; IV. Rate of the Collective Reaction, 53, 2256-2260; VI. Williams-Landell-Ferry Equation, 53, 2456-2460 (*Russ. J. Phys. Chem.* (1979) 53, 1041; 1282-1285; 1285-1287; 1404-1406).

96. Durov, V.A. (1979) Theory of Collective Reactions in Liquids. VII. System with Two Collective Reactions, *Zh. Fiz. Khim.*, 53, 3086-3091; 8. System with Multiple Collective Reactions, 53, 3092-3096 (*Russ. J. Phys. Chem.* (1979) 53, 1772-1774; 1775-1777).
97. Gotze, W. (1999) Recent Tests of the Mode-Coupling Theory for Glassy Dynamics. *J. Phys.-Condens. Matter* 11, A1-A45.
98. Theis, C. and Schilling, R. (1999) Neutron-Scattering and Molecular Correlations in a Supercooled Liquid, *Phys. Rev. E.* 60, 740-750.
99. Lunkenheimer, P., Schneider, U., Brand, R., and Loidl, A. (2000) Glassy Dynamics, *Contemporary Phys.* 41, 15-36.
100. Durov, V.A., Rabitchev, E.O., and Shakhparonov M.I. (1980) Acoustic Spectroscopy of Glycerol and Its Mixtures with Butanol, in Ya.I. Gerasimov, P.A. Akishin, and M.I. Shackhparonov (eds.), *Modern Problems in Physical Chemistry* (monograph in Russian), 12th Issue, Moscow University Press, Moscow, pp. 180-218.
101. Durov, V.A. (1986) Acoustic Spectroscopy of the Conformation Transitions of Molecule. I. Three States Model, *Zh. Fiz. Khim.* 60, 618-623; II. Three States Model: Analysis of the General Solution, 60, 624-630; III. Molecules with Six-Membered Rings, 60, 1754-1761; 4. Alkanes. Basic Problems. n-Pentane, 60, 1762-1767; VI. Spectrum of Liquid n-Pentane, 60, 2826-2833 (*Russ. J. Phys. Chem.* (1986), 60, 367-370; 370-374; 1050-1054; 1054-1058; 1826-1833).
102. Durov, V.A. and Artykov, A. (1988) Kinetics and Mechanisms of Acoustic Relaxation in Liquid Six-Membered Ketones, *Zh. Fiz. Khim.* 62, 461-471 (*Russ. J. Phys. Chem.* (1988), 62, 210-215).
103. Durov, V.A. and Artykov, A. (1988) Kinetics and Mechanisms of Molecular Thermal Motion in Liquid Cyclohexane, *Zh. Fiz. Khim.* 62, 2477-2483 (Russ. J. Phys. Chem. (1988), 62).
104. Durov, V.A. and Ziyaev, G.M. (1988) Kinetics and Mechanisms of Acoustic Relaxation in 2,3-Dimethylbutane-2,3-diol and Its Binary Mixtures with Water and 1-Butanol, *Zh. Fiz. Khim.* 62, 450-460 (*Russ. J. Phys. Chem.* (1988), 62, 205-210).
105. Durov, V.A. and Ziyaev, G.M. (1989) Kinetics and Mechanisms of Acoustic Relaxation in Butane-2,3-diol, and Its Binary Mixtures with Water and 1-Butanol, *Zh. Obsh. Khim.* 59, 204-210.
106. Durov, V.A. and Tereshin O.G. (2004) Supramolecular Structure and Physicochemical Properties of Cyclohexane-Ethanol Mixture. On the Role of the Cyclic Aggregates, *Zh.Fiz.Khim.* 75 (*Russ. J. Phys. Chem.* (2004), 75), accepted.
107. Durov, V.A. and Tereshin, O.G. (2003) Supramolecular Structure and Physicochemical Properties of 1,4-Dioxane-Methanol Mixture. *Zh.Fiz.Khim.* 74 (*Russ. J. Phys. Chem.* (2003), 74), accepted.

CLASSICAL VERSUS QUANTUM MECHANICAL SIMULATIONS: THE ACCURACY OF COMPUTER EXPERIMENTS IN SOLUTION CHEMISTRY

B. M. RODE, C.F. SCHWENK AND B.R. RANDOLF
Department of Theoretical Chemistry, Institute of General, Inorganic and Theoretical Chemistry, University of Innsbruck, Innrain 52a, A-6020 Innsbruck, Austria

Abstract: Sophisticated simulation techniques in combination with high-speed computing provide a very powerful tool for the elucidation of structural data and dynamics of solutions, which in several aspects can be superior to any experimental technique. A careful analysis and comparison of simulation results achieved at different levels of accuracy shows that classical simulations, even including 3-body corrections, do not supply sufficiently precise data for all structural details and dynamical processes. As simulation techniques based on small clusters and simple density functionals also fail in the prediction of ion solvate structures, mixed quantum mechanical / molecular mechanical (QM/MM) simulations at Hartree-Fock level with medium-sized basis sets appear as the only viable method within today's computational affordability to achieve the necessary accuracy for a theoretical approach to the details of microspecies structures and their dynamics in electrolyte solutions. Results of QM/MM-MD simulations for numerous main group and transition metal cations presented here exemplify the capability of this method and clearly show the limits not only of classical simulation techniques, but also of the models being used for the interpretation of experimental measurements.

1. INTRODUCTION

Monte Carlo (MC) and Molecular Dynamics (MD) simulation techniques have become a well-established tool for the study of liquid systems. The basic approach common to both methods is a separation of the interactions between all particles present in the system into contributions of pairs, triples, quadruples ... a.s.o. of particles:

$$V_{total} = \sum V(i,j) + \sum V(i,j,k) + \ldots + \sum V(i,j,k,\ldots,n) \quad (1)$$

In most simulations, only the term describing pair interactions is taken into account, assuming pairwise additiivity of the energies / forces in the system, although it has been shown quite early that higher terms are by no means negligible, even for

J. Samios and V.A. Durov (eds.), Novel Approaches to the Structure and Dynamics of Liquids: Experiments, Theories and Simulations, 41–52.

weakly interacting solvent molecules [11]. In particular, 3 -body terms describing the mutual polarisation of interacting species can decisively influence the structure of microspecies formed in solution, and this influence increases with the interaction energy between species, *e.g.* going from single-charged to double-charged ions interacting with a solvent.

A good method to estimate the importance of the higher terms in eq. 1 is the performance of *ab initio* calculations of small clusters of molecules, revealing the differences in binding energy per molecule with increasing number of interacting particles. Having recognised once the importance of such terms, there are several ways to account for them. The simplest approach is the use of empirical pair potentials considering mutual polarisation effects in an averaged way and fitted to reproduce some experimentally known properties of the system studied. A more systematic approach was the evaluation of pair potentials from quantum chemical calculations of species dimers in a fixed geometry with a further species, as realised in the Nearest-Neigbour-Ligand Correction formalism for simulations of hydratedions [33].

2. *AB INITIO* GENERATION OF INTERACTION POTENTIALS

The methodically most correct and controllable way to develop interaction potentials is based on quantum mechanical *ab initio* calculations of the corresponding energy surfaces for pair, 3-body, and higher interaction terms, with a sufficiently accurate level of theory and basis set. This procedure does not pose too much problems for pair potentials, where a few thousand configurations may represent the energy surface with sufficient accuracy and completeness, and where the interaction potential can be fitted to an analytical function of the type

$$\Delta E_{fit} = \sum_{i=1}^{3} \frac{A_{ic}}{r_{ic}^{a}} + \frac{B_{ic}}{r_{ic}^{b}} + \frac{C_{ic}}{r_{ic}^{c}} + \frac{D_{ic}}{r_{ic}^{d}} + E_{ic}\exp(-F_{ic}r_{ic}) + \frac{q_i q_c}{r_{ic}} \quad (2)$$

For the often very essential 3-body terms, the same procedure can be followed in principle, constructing the energy surface from a representative number of 3-particle configurations calculated by the same *ab initio* formalism employed for the evaluation of the pair function. Due to the larger number of degrees of freedom, however, several ten thousands of these configurations are needed, and the fitting of the energy contributions to an analytical function becomes considerably more difficult [37,50,39]. 4-body and higher terms make this procedure almost unfeasible, due to the large number of needed energy surface points and the problems to represent the surface by a suitable function.

3. *ab Initio* MC AND MD SIMULATIONS

A quantum mechanical evaluation of the energy and all forces of the system would be the "natural" solution to all of these problems. However, even today's fastest available parallelizing computers do not allow to deal this way with the number of molecules needed to represent a liquid system, for which the simulation box should contain at least several hundred particles within the framework of periodic boundary conditions/minimal image convention, in order to avoid artificial symmetry effects and to obtain reliable structural data up to at least 10 Å distance from a given center.

For these reasons, a compromise was sought by dividing a given chemical system into a "sensitive", chemically most relevant region, where quantum mechanical methods would be applied, and an "outer" region, where traditional molecular mechanics appear to be sufficiently accurate. This separation, usually referred to as "QM/MM" approach has been considered for the treatment of chemically active sites of biomolecules [4,14] and in its first approaches considered semiempirical MO methods as the quantum mechanical tool to be employed [12]. Later, this separation was applied to solutions of ions [20], where - most conveniently - the separation of the QM subsystem does not involve any breaking of covalent bonds in the transition region to the MM subsystem. However, the enormous computational effort related to the evaluation of energies for several million configurations (MC) or of forces in tens of thousands of time steps (MD) set up technical limits for this method, restricting it to Hartree-Fock level of theory and to moderate basis sets. Unfortunately, semiempirical MO methods as MNDO or AMI as well as *ab initio* calculations with simple basis sets as STO-3G proved not suitable for this approach [20, 40]. Investigations with several basis sets have shown that double-zeta plus polarisation quality is the minimum required for a good description of the QM region [20].

Another attempt to avoid the classical w-body problem was the introduction of the Car-Parinello method [10], where small clusters of 30 — 60 molecules are treated by an approximate quantum mechanical formalism, based on the Becke-LYP density functional [16,6]. Although sometimes also quoted as *"ab initio"* dynamics, the use of this functional rather classifies the method as a semiempirical procedure, and the still necessary restriction to small clusters seems a serious obstacle to obtain reliable structural data.

4. DO N-BODY EFFECTS PLAY AN IMPORTANT ROLE IN SOLUTION CHEMISTRY?

In order to assess the need for high-level, *i.e.* quantum mechanics - based simulation techniques, the actual importance of higher w-body effects has to be clarified. The comparison of already available results from *ab initio* QM/MM simulations with classical pair and 2 + 3-body simulations allows to obtain a quite good insight into the relevance of the w-body terms. The effects of these terms will be summarized in the following as "quantum effects", as they are of non-classical nature and only accountable through quantum mechanical evaluation methods allowing for mutual polarisation and charge transfer.

For ions interacting rather weakly with the solvent, one would expect a negligible influence of such effects. However, it has been found that the structure-breaking behaviour of the K(I) ion in water is only revealed by a quantum mechanical treatment of the first hydration shell [41]. This finding proved of essential importance for the interpretation of the functionality of the potassium-specific ion channels in cell membranes [5, 41]. For Na(I) a very similar hydration shell structure is obtained by classical and QM/MM MD simulations, whereas only the QM/MM formalism elaborates the details of the experimentally known coordination number of ~ 4 for Li (I) [21, 42]

For alkaline earth metal ions, quantum effects are even more important. Ca(II) is one of the most thoroughly investigated ions, due to its biological importance. Due to the Ca...O bond length of ~ 2.5 Å, which leads to a partial occlusion of the related peak in diffraction studies by the broad O...O peak, experimental determination of the hydration structure of this ion is difficult, and the simulation results were quite ambiguous until recently, when QM/MM simulations were performed at HF and B3LYP-DFT level with 499 water molecules [37], leading to an 8-fold coordinated, square-antiprism shaped structure which is in agreement with the most reliable experimental data for dilute solutions of Ca(II) [29,31]. At the same time, a Car-Parinello simulation of Ca(II) with 50 water molecules resulted in a coordination number of 6 [35], indicating the tendency of this method to underestimate coordination numbers (an even more convincing example for this will be discussed below).

When ion solvation in mixed solvents is studied, the differences between classical and quantum mechanical simulations become even more striking: QM/MM MD simulations of Li(I), Na(I), Mg(II) and Ca(II) [43,44,45,47] in aqueous ammonia show striking differences for the composition of the first shell and even more deviations in size and composition of the second solvation shells, where average coordination numbers of 18-20 are reduced to values of 10-12 by the quantum mechanical treatment. Solvation in mixed solvents usually leads to a number of microspecies simultaneously present, which often do not correspond to the average composition of the solvate accessible through experimental methods. In such cases the theoretical approach is clearly superior to predict the chemical reactivity of solvate complexes [48].

Another famous example, where classical simulations, even 3-body corrected, will fail to reflect well-known chemical behaviour is the Jahn-Teller effect of hy-drated Cu(II) ion: The typical square-pyramidal distortion observed experimentally both in the solid and the liquid state [8, 13], could only be reproduced by the first *ab initio* QM/MM MC simulation of hydrated Cu(II) ion [27].

Summarizing these widespread observations of a crucial influence of w-body effects in most ion solvates, the conclusion that only QM/MM simulations are capable of delivering accurate structural data for such systems (without fitting potentials empirically to experimental data), seems fully justified. As a correct structure evidently is the basis for correct dynamical data, the same conclusion will be valid for them, *i.e.* for molecular rotations/vibrations/translations and for ligand exchange processes.

In the following, therefore, only data obtained by *ab initio* QM/MM MD simulations will be discussed, with further comparisons to classical simulations for the same systems being available in the references.

5. SIMULATION PROTOCOL

All QM/MM MD simulations reported here have been performed for one metal ion immersed in 499 water molecules positioned in a cubic box of the experimental density of pure water. The QM region included the full first hydration sphere, and water molecules were allowed to enter/leave the QM region through a transition region of 0.2 Å width, ensuring a smooth transition between quantum mechanical and molecular mechanical forces. Periodic boundary conditions were applied, and long-range interactions were handled by the reaction field method. The temperature of the *NVT* ensembles (298 K) was controlled by the Berendsen algorithm [7] with a relaxation time of 0.1 ps. For the MM part of the systems, *ab initio* generated pair and 3-body potential functions for ion-water interactions were used, whose details are given in the corresponding references. For water, the flexible BJH-CF2 model [38,9] was used, which allows explicit hydrogen movements, requesting thus a time step of 0.2 fs for the simulations. QM/MM simulations were started from classically equilibrated configurations, sampling a 15-30 ps interval after equilibration had been achieved. Further simulation details are available in the references [19,21,50,37] Typical CPU times required for QM/MM simulations are in the range of 3 to 9 months on a parallel computer cluster with 6-8 1 Ghz processors, the QM part being executed with the parallelized version of the TURBOMOL program [49,2,1]

6. STRUCTURE OF HYDRATED TRANSITION METAL IONS AND HYDRATION ENERGIES

Tables 1 and 2 list the characteristic structural data for the first and second hydration shells of a series of main group and transition metal ions investigated by means of QM/MM MD simulations and comparé these values to available experimental data and some results of classical simulations.

It has to be mentioned that, except of EXAFS measurements, most experiments have been carried out with solutions of rather high concentrations (1-5 M), in which counterion effects, ion pairing and even lack of sufficient solvent molecules to form full hydration shells can strongy influence the actual coordination numbers. The simulation data are valid for dilute solutions and hence also correspond to concentrations relevant for the activity of these ions in biological systems. It can be seen, however, that wherever experimental data are available for dilute solutions, an excellent agreement is found between QM/MM simulation and experiment.

Hydration energies computed from the simulations can be compared to experimentally evaluated enthalpies of hydration. Data for single-ion hydration enthalpies can only be evaluated on the basis of some assumptions, attributing parts of the enthalpies measured for salts to separate anion and cation contributions. Literature values show considerable variations and therefore the data from an evaluation based on the same TATB extrathermo-dynamic assumptions [26, 25, 24] have been taken for a comparison, which is displayed in Table 3.

It is interesting to observe that experimental and simulation values converge with increasing charge of the cations. Whereas for mono- and divalent ions the energies obtained by the simulation are in average 10 – 20% above the experimental enthalpies, the differences become almost negligible for tervalent ions. As there are no methodical differences in the simulations, a different validity of the experimental assumptions for the ions under consideration may be held responsible.

TABLE 1. Structural data obtained by classical and *ab ini-tio* QM/MM MD simulations, and comparison with experimental values and classical simulations: alkaline and alkaline earth metal ions, Ag(I) and Hg(II)

Ion	Method	Ri[A]	Ni	R2[A]	N2	Ref
Li+	QM	1.95	4.2	4	—	[21]
	Class	2.05	4	4.5	-	[42]
	Exp	2.25	4	4.44	9.5	[30]
Na+	QM	2.33	5.6	-	-	[41]
	Class	2.36	6.5	-	-	[41]
	Exp	2.41	6	-	-	[30]
K+	QM	2.81	8.3	-	-	[41]
	Class	2.78	7.8	-	-	[41]
	Exp	2.7	6	-	-	[30]
Mg2+	QM	2.03	6	4.12	18.3	[46]
	Class	2.13	8	4.23	26.6	[46]
	Exp	2.0	6	4.1	12	[30]
Ca2+	QM	2.46	7.6	4.78	19.1	[37]
	Class	2.5	7.1	5	14.8	[37]
	Exp	2.39	6.9	-	-	[30]
Ag+	QM	2.5	5	5	25	[34]
	Class	2.6	5	4.8	25	[34]
	Exp	2.42	4	4.29	17.3	[30]
Hg2+	QM	2.42	6.2	4.6	22	[34]
	Class	2.46	6	4.8	33	[34]
	Exp	2.41	6	4.1	18.4	[30]

TABLE 2. **Structural data obtained by classical and *ab initio* QM/MM MD simulations, and comparison with experimental values and classical simulations: first-row transition metal ions**

Ion	Method	Ri[A]	Ni	R_2[A]	N_2	Ref
V^{2+}	QM	2.23	6	4.4	15.8	[23]
	Class	2.31	7	4.85	20.12	[50]
	Exp	-	-	-	-	
Mn^{2+}	QM	2.25	6	4.4	15.9	[22]
	Class	2.35	6.08	4.77	21.25	[50]
	Exp	2.2	6	4.34	10.7	[30]
Fe^{2+}	QM	2.1	6	4.5	12.4	[34]
	Class	2.15	6	4.6	12.9	[34]
	Exp	2.09	6	4.3	12	[30]
Co^{2+}	QM	2.17	6	4.6	15.9	[34]
	Class	2.27	5.94	4.6	22.73	[34]
	Exp	2.1	6	4.28	14.8	[30]
Ni^{2+}	QM	2.14	6	4.5	13	[19]
	Class	2.25	6	4.68	16	[19]
	Exp	2.06	6	4.33	12	[30]
Cu^{2+}	QM	2.07/2.17	6	4.62	11.7	[34]
	Class	2.12	6	4.72	12.4	[34]
	Exp	1.94	6	3.95	11	[30]
Ti^{3+}	QM	2.03	6	4.2	11	[34]
	Class	2.08	6	4.8	11.2	[34]
	Exp	2.0/2.08	-	-	-	[3]
Fe^{3+}	QM	2.02	6	4.3	13.4	[34]
	Class	2.05	6	4.3	15.1	[34]
	Exp	2.01	6	4.8	11.2	[30]
Co^{3+}	QM	1.97	6	4.3	15.2	[34]
	Class	2.03	6	4.4	18.8	[34]
	Exp	-	-	-	-	

TABLE 3. **Hydration energies for transition metal ions obtained from QM/MM MD simulations in comparison to experimentally estimated ionic hydration enthalpies**

Ion	Calc [kcal.mol^{-1}]	Exp [kcal.mol^{-1}] [26]
Ag^{+}	-152	-118
V^{2+}	-560	-404
Mn^{2+}	-550	-447
Fe^{2+}	-500	-465
Co^{2+}	-547	-487
Cu^{2+}	-530	-507
Hg^{2+}	-553	-443
Ti^{3+}	-1086	-1038
Fe^{3+}	-1100	-1060
Co^{3+}	-1144	-1122

7. DYNAMICS OF SOLVATION AND EXCHANGE PROCESSES

As the dynamics of ligand exchange reactions at metal ions are one of the essential factors determining their chemical and biological (re)activity, they have been a prominent research topic in the past decades, mostly investigated by experimental techniques [30,15]. The time scale for the exchange of ligands in pure water spans from picoseconds to days, and the lower end is hardly accessible to spectroscopic techniques (a promising way though is femtosecond laser pulse spectroscopy in the future). Whenever the first hydration shell is very stable, second shell exchange processes become of interest, again occurring in the ps range. Further, some investigations have shown that heteroligands bound to the cations may labilize an otherwise rather stable hydration shell, pushing the exchange rates to a much faster scale by several orders of magnitude [28].

Slow exchange processes are beyond the capability of today's accessible simulation timescale, but the fast processes in the picosecond range, which are so hard to estimate by experimental means, are the ideal topic for investigations by MD methods. As an accurate description of the molecular structure of the complexes and the acting forces in the system are *a conditio sine qua non* for the determination of exchange processes, classical MD simulations cannot be regarded as suitable tool, and only the inclusion of quantum effects promises reliable data for the dynamics of these processes.

Several exchange processes have been investigated, therefore, by means of *ab initio* QM/MM MD simulations recently [36,18], and the data are compiled in Table 4. Besides the evaluation of vibrational spectra, exchange rates and mean residence times of ligands, the simulations also allow a very detailed visualization of the dynamics of hydrated ions. A new visualisation tool has been developed for this purpose, and example trajectory files are available to observe several characteristic processes at http://www.molvision.com.

A most significant example of the need for quantum effects is the simulation of hydrated Cu(II). The well-known Jahn-Teller effect cannot be reproduced, as mentioned previously, by a classical simulation, but it determines in a most crucial way the speed of ligand exchange, increasing it by a factor of almost 1000 compared to neighboring ions as Ni(II).

The details of the dynamics of hydrated Cu(II) also show that the distortions of the basically octahedral arrangement of water ligands change much faster than estimated from experimental data [15], and that it is not always 2 ligands in trans-position which are located at "elongated" positions, although this configuration dominates in the case of Cu(II) - in contrast to Ti(III), where an equilibrium of several distorted species is present [34]. The mean life time of a specific Jahn-Teller distorted configuration lies between 30 and 200 femtoseconds, according to the simulation results. Figure 1 shows some snapshots of hydrated Cu(II) with differently distorted configurations. The mean residence time for a second shell water ligand is not much shorter than for Fe(II) or Co(II), indicating that the Jahn-Teller effect has not much influence on the second shell of this ion. This seems to be different in the case of the also Jahn-Teller distorted hydrated Ti(III) ion, where the mean residence time of second shell ligands is considerably shorter than that for Co(III) and Fe(III). At this time it should be mentioned that a Car-Parinello MD simulation of hydrated Cu(II) has predicted a 5-coordinated copper ion [32], in clear contrast to almost all experimental evidence and *ab initio* QM/MM MD simulations, thus indicating once more the tendency of the CP method to underestimate hydration numbers.

TABLE 4. Mean residence times of water ligands in the first and/or second hydration shell of transition metal ions (in ps) using r* = 2ps[17].

Ion	1st shell [ps]	2nd shell [ps]
Li^+	11	
Ag^+	1440	10
Ca^{2+}		17
V^{2+}	—	18
Mn^{2+}	–	24
Fe^{2+}	–	24
Co^{2+}	–	26
Cu^{2+}	–	23
Hg^{2+}	42	1526
Ti^{3+}		
Fe^{3+}	—	48
Co^{3+}	-	55

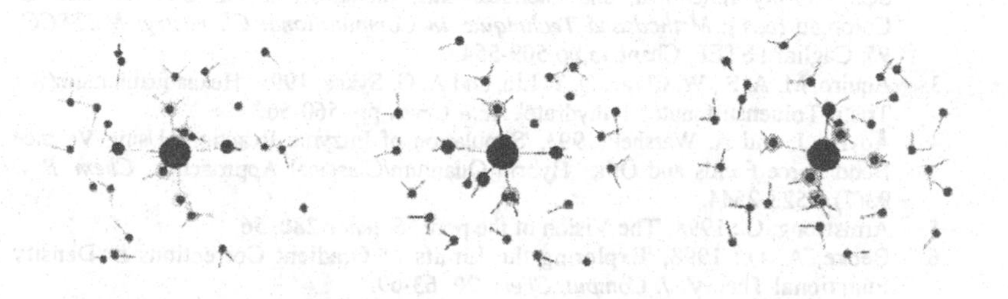

Figure 1. Snapshots of the dynamical Jahn-Teller effect in hydrated Cu(II), obtained by QM/MM MD simulation and visualised by MOLVISION®. (An animated color picture can be downloaded from www.molvision.com, where also a trajectory file for the full visualisation is available) 1st shell igands: dark grey, the larger and smaller size corresponds to ligands at shorter (2.05 A) and larger (2.25 A) distances; 2nd shell ligands: small, light grey bulk: only visualised as line models

8. LIMITATIONS OF THE METHOD AND OUTLOOK

The very brief and exemplary presentation of results achieved by *ab initio* QM/MM MD simulations given here should have allowed a preliminary appreciation of the capability of this method to perform "computer experiments" of remarkable accuracy in the elucidation of ion solvation phenomena. At the same time the present limitations in terms of size of the QM region and methodical accuracy (level of theory, basis sets) as well as length of the accessible time span have become visible. Even within these limitations, which are being quickly shifted to new frontiers by every improvement of computer speed and capacity, the simulation methods including quantum effects have been able to position themselves as one of the most powerful tools in solution chemistry, often providing more detailed information about species formation, structure and dynamics in solution than most presently available experimental techniques. The "computer experiments" thus can be regarded as most instrumental for the interpretation of spectroscopic data, for the elucidation of reaction mechanisms in solution, and for the stimulation of new experimental approaches in solution chemistry.

9. ACKNOWLEDGEMENT

This work was supported by the Austrian Science Foundation (FWF) (project P13644-TPH).

REFERENCES

1. Ahlrichs, R., M. *Ear,* M. Haser, H. Horn, and C. Kolmel: 1989, 'Electronic Structure Calculations on Workstation Computers: The Program System TURBOMOLE'. *Chem. Phys.Lett.* **162**(3), 165-169.
2. Ahlrichs, R. and M. von Arnim: 1995, 'TURBOMOLE, parallel implementation of SCF, density functional, and chemical shift modules.'. In: E. Clementi and G. Corongiu (eds.): *Methodsand Techniques in Computational Chemistry: METECC-95.* Cagliari: STEF, Chapt.13,pp.509-554.
3. Aquiro, M. A. S., W. Clegg, Q. T. Liu, and A. G. Sykes: 1995, 'Hexaaquatitanium(III) Tris(p-Toluensulfonate) Trihydrate'. *Acta Cryst.* pp. 560-562.
4. Åqvist, J. and A. Warshel: 1993, 'Simulation of Enzyme Reactions Using Valence Bond Force Fields and Other Hybrid Quantum/Classical Approaches'. *Chem. Rev.* **93**(7), 2523-2544.
5. Armstrong, C.: 1998, 'The Vision of the pore. *Science* **280**, 56.
6. Becke, A. D.: 1998, 'Exploring the Limits of Gradient Corrections in Density Functional Theory'. *J. Comput. Chem.* **20**, 63-69.
7. Berendsen, H. J. C., J. P. M. Postma, W. F. van Gunsteren, A. DiNola, and J. R. Haak: 1984, 'Molecular Dynamics with coupling to an external bath'. *J. Phys. Chem.* **81**, 3684-3690.
8. Bersuker, I. B.: 2001, 'Modern Aspects of the Jahn-Teller Effect Theory and Applications To Molecular Problems'. *Chem. Rev.* **101**(4), 1067-1114.
9. Bopp, P., G. Jansco, and K. Heinzinger: 1983, 'AN IMPROVED POTENTIAL FOR NON-RIGID WATER MOLECULES IN THE LIQUID PHASE'. *Chem. Phys. Lett.* **98**(2), 129-133.
10. Car, R. and M. Parinello: 1985, 'Unified Approach for Molecular-Dynamics and Density Functional Theory'. *Phys. Rev. Lett.* **55**(22), 2471-2474.

11. Clementi, E., H. Kistenmacher, W. Kolos, and S. Romano: 1980, 'Non-Additivity in Water-Ion-Water Interactions'. *Theor. Chim. Acta* **55**, 257-266.
12. Cummins, P. L. and J. E. Gready: 1997, 'Coupled Semiempirical Molecular Orbital and Molecular Mechanics Model (QM/MM) for Organic Molecules in Aqueous Solution'.*J.Comput. Chem.* **18**(12), 1496-1512.
13. Curtiss, L., J. W. Halley, and X. R. Wang: 1992, 'Jahn Teller effect in Liquids'. *Phys. Rev.Lett.* **69**(16), 2435-2438.
14. Gao, J.: 1996, 'Hybrid Quantum and Molecular Mechanical Simulations: An Alternative Avenue to Solvent Effects in Organics Chemistry'. *Ace. Chem. Res.* **29**(6), 298-305.
15. Helm, L. and A. E. Merbach: 1999, 'Water exchange on metal ions: experiments and simulations'. *Coord. Chem. Rev.* **187**, 151-181.
16. Hertwig, R. H. and W. Koch: 1997, 'On the parameterization of the local correlation functional. What is Becke-3-LYP?'. *Chem. Phys. Lett.* **268**(5-6), 345-351.
17. Impey, R. W., P. A. Madden, and I. R. McDonald: 1983, 'Hydration and Mobility of Ions in Solution'. *J. Phys. Chem.* **87**(25), 5071-5083.
18. Inada, Y., H. H. Loeffler, and B. M. Rode: 2002a, 'Librational, vibrational, and exchange motions of water molecules in aqueous Ni(II) solution: Classical and QM/MM molecular dynamics simulations'. *Chem. Phys. Lett.* **358**, 449^4-58.
19. Inada, Y., A. M. Mohammed, H. H. Loeffler, and B. M. Rode: 2002b, 'Hydration Structure and Water Exchange Reaction of Nickel(II) Ion: Classical and QM/MM Simulations'. *J. Phys. Chem. A* **106**(29), 6783-6791.
20. Kerdcharoen, T., K. R. Liedl, and B. M. Rode: 1996, 'A QM/MM simulation method applied to the solution of Li^+ in liquid ammonia'. *Chem. Phys.* **211**, 313-323.
21. Loeffler, H. H. and B. M. Rode: 2002, 'The hydration structure of the lithium ion'. *J. Chem.Phys.* **117**(1), 110-117.
22. Loeffler, H. H., J. I. Yague, and B. M. Rode: 2002a, 'Many-Body Effects in Combined Quantum Mechanical/Molecular Mechanical Simulations of Hydrated Manganous Ion'. *J.Phys.Chem. A* **106**, 9529-9532.
23. Loeffler, H. H., J. I. Yagtie, and B. M. Rode: 2002b, 'QM/MM-MD Simulation of Hydrated Vanadium(H) Ion'. *Chem. Phys. Lett.* **363**, 367-371.
24. Marcus, Y: 1987a, 'Thermodynamics of ion hydration and its interpretation in terms of acommon model'. *Pure & Appl. Cham.* **59**(9), 1093-1101.
25. Marcus, Y.: 1987b, 'The Thermodynamics of Solcation of Ions'. *J. Chem. Soc., FaradayTrans.* **83**, 339-349.
26. Marcus, Y.: 1991, 'Thermodynamics of Solvation of Ions. Part 5.—Gibbs Free Energy of Hydration at 298.15 K'. *J. Chem. Soc., Faraday Trans.* **87**(17), 2995-2999.
27. Marini, G. W., K. R. Liedl, and B. M. Rode: 1999, 'Investigations of Cu^{2+} Hydration and the Jahn-Teller Effect in Solution by QM/MM Monte Carlo Simulations'. *J. Phys. Chem.A* **103**(51), 11387-11393.
28. Nagypal, I. and F. Debreczeni: 1984, 'NMR Relaxation Studies in Solution of Transition Metal Complexes. XL Dynamics of Equilibria in Aqueous Solutions of the Copper(II)-Ammonia System'. *Inorg. Chim. Acta* **81**, 69-74.
29. Neilson, G. W. and J. E. Enderby: 1989, 'The Coordination of Metal Aquaions'. In: *Advancesin Inorganic Chemistry,* Vol. **34**. Academic Press, Inc., pp. 195-218.
30. Ohtaki, H. and T. Radnai: 1993, 'Structure and Dynamics of Hydrated Ions'. *Chem. Rev.* **93**(3), 1157-1204.
31. Palinkas, G. and K. Heinzinger: 1986, 'HYDRATION SHELL STRUCTURE OF THE CALCIUM ION'. *Chem. Phys. Lett.* **126**, 251-254.
32. Pasquarello, A., I. Petri, P. S. Salmon, O. Parisel, R. Car, E. Toth, D. H. Powell, H. E. Fischer, L. Helm, and A. Merbach: 2001, 'First Solvation Shell of the Cu(II) Aqua Ion: Evidence for Fivefold Coordination'. *Science* **291**, 856-859.

33. Rode, B. M. and S. M. Islam: 1990, 'Monte Carlo Simulations with an Improved Potential Function for Cu(II)-Water Including Neighbour Ligand Corrections'. **46**, 357-362.
34. Rode, B. M., C. F. Schwenk, R. Armunanto, T. Remsungnen, and C. Kritayakornupong, 'unpublished results'.
35. Samios, J. (ed.): 2002, *Novel Approaches to the Dynamics of Liquids: Experiments, Theories and Simulations*. Advanced Study Institute.
36. Schwenk, C. F, H. H. Loeffler, and B. M. Rode: 2001a, 'Dynamics of the solvation process of Ca^{2+} in water'. *Chem. Phys. Lett.* **349**(1-2), 99-103.
37. Schwenk, C. F., H. H. Loeffler, and B. M. Rode: 2001b, 'Molecular dynamics simulations of Ca^{+} in water: Comparison of a classical simulation including three-body corrections and Born-Oppenheimer ab initio and density functional theory quantum mechanical/molecular mechanics simulations'. *J. Chem. Phys.* **115**(23), 10808-10813.
38. Stillinger, F. H. and A. Rahman: 1978, 'Revised central force potentials for water'. *J. Chem. Phys.* **68**(2), 666-670.
39. Texler, N. R. and B. M. Rode: 1997, 'Monte Carlo simulations of copper chloride solutions at various concentrations including full 3-body corrections terms'. *Chem. Phys.* **222**, 281-288.
40. Tongraar, A., K. R. Liedl, and B. M. Rode: 1997, 'Solvation of Ca^{2+} in Water Studied by Born-Oppenheimer ab Initio QM/MM Dynamics'. *J. Phys. Chem. A* **101**(35), 6299-6309.
41. Tongraar, A., K. R. Liedl, and B. M. Rode: 1998a, 'Born-Oppenheimer ab Initio QM/MM Dynamics Simulations of Na^{+} and K^{+} in Water: From Structure Making to Structure Breaking Effects'. *J. Phys. Chem. A* **102**(50), 10340-10347.
42. Tongraar, A., K. R. Liedl, and B. M. Rode: 1998b, 'The hydration shell structure of Li+ investigated by Born-Oppenheimer ab initio QM/MM dynamics'. *Chem. Phys. Lett.* **286**, 56-64.
43. Tongraar, A. and B. M. Rode: 1999, 'Preferential Solvation of Li+ in 18.45 % Aqueous Ammonia: A Born-Oppenheimer ab Initio Quantum Mechanics/Molecular Mechanics MD Simulation'. *J. Phys. Chem. A* **103**(42), 8524-8527.
44. Tongraar, A. and B. M. Rode: 2001, 'A Born-Oppenheimer Ab Initio Quantum Mechanical/Molecular Mechanical Molecular Dynamics Simulation of Preferential Solvation of Na^{+} in Aqueous Ammonia Solution'. *J.Phys.Chem. A* **105**(2), 506-510.
45. Tongraar, A., K. Sagarik, and B. M. Rode: 2001a, 'Effects of Many-Body Interactions on the Preferential Solvation of Mg^{2}+ in Aqueous Ammonia Solution: A Born-Oppenheimer ab Initio QM/MM Dynamics Study'. *J. Phys. Chem. B* **105**(54), 10559-10564.
46. Tongraar, A., K. Sagarik, and B. M. Rode: 2001b, 'Non-additive contributions on the Hydration Shell structure of Mg^{2+} studied by Born-Oppenheimer ab Initio Qnauntum Mechanical/Molecular Mechanical Molecular Dynamics Simulation'. *Chem. Phys. Lett.* **346**, 485-491.
47. Tongraar, A., K. Sagarik, and B. M. Rode: 2002, 'Preferential solvation of Ca^{2+} in aqueous ammonia solution: Classical and combined ab initio quantum mechanical/molecular mechanical molecular dynamics simulations'. **4**, 628-634.
48. Vizoso, S., M. G. Heinzle, and B. M. Rode: 1994, 'Hydroxylamine-water: Intermolecular Potential Functionand Simulation of hydrated NH_2OH'. *J. Chem. Soc., Faraday Trans.* **90**(16), 2377-2344.
49. von Arnim, M. and R. Ahlrichs: 1998, 'Performance of Parallel TURBOMOLE for Density Functional Calculations'. *J. Comput. Chem.* **19**(15), 1746-1757.
50. Yagüe, J. I., A. M. Mohammed, H. Loeffler, and B. M. Rode: 2001, 'Classical and Mixed Quantum Mechanical/Molecular Mechanical Simulation of Hydrated Manganous Ion'. *J. Phys. Chem. A* **105**(32), 7646-7650.

BASIC CONCEPTS AND TRENDS IN *AB INITIO* MOLECULAR DYNAMICS

MARK E. TUCKERMAN
Dept. of Chemistry and Courant Institute of Mathematical Sciences, New York University, New York, NY 10003

Abstract: The field of *ab initio* molecular dynamics, in which finite temperature molecular dynamics trajectories are generated using forces obtained from electronic structure calculations performed "on the fly", is a rapidly evolving and growing technology that allows chemical processes in condensed phases to be studied in an accurate and unbiased way. This article is intended to present the basics of the *ab initio* molecular dynamics method and to highlight some recent trends. Beginning with a derivation of the method from the Born-Oppenheimer approximation, issues including the density functional representation of electronic structure, basis sets, calculation of observables, and the Car-Parrinello extended Lagrangian algorithm and extensions of the latter are discussed.

1. Introduction

Modern theoretical methodology, aided by the advent of high speed computing, has advanced to a level that the microscopic details of chemical processes in condensed phases can now be treated on a relatively routine basis. One of the most commonly used theoretical approaches for such studies is the molecular dynamics (MD) method, in which the classical Newtonian equations of motion for a system are solved numerically starting from a prespecified initial state and subject to a set of boundary conditions appropriate to the problem. MD methodology allows both equilibrium thermodynamic and dynamical properties of a system at finite temperature to be computed, while simultaneously providing a "window" onto the microscopic motion of individual atoms in the system. One of the most challenging aspects of an MD calculation is the specification of the forces. In many applications, these are computed from an empirical model or

J. Samios and V.A. Durov (eds.), Novel Approaches to the Structure and Dynamics of Liquids: Experiments, Theories and Simulations, 53–91.

force field, in which simple mathematical forms are posited to describe bond, bend, and dihedral angle motion as well as van der Waals and electrostatic interactions between atoms, and the model is parameterized by fitting to experimental data or high level *ab initio* calculations on small clusters or fragments. This approach has enjoyed tremendous success in the treatment of systems ranging from simple liquids and solids to polymers and biological systems such as proteins and nucleic acids.

Despite their success, force fields have a number of serious limitations. First, charges appear as static parameters in the force field, and therefore, electronic polarization effects are not included. This limitation has long been recognized, and attempts to rectify the problem have been proposed in the form of so called polarizable models (see Ref. [1] for a review), in which charges and/or induced dipoles are allowed to fluctuate in response to a changing environment. While these models have also enjoyed considerable success, they also have a number of serious limitations, including a lack of transferability and standardization. Second, force fields generally suffer from an inability to describe chemical bond breaking and forming events. The latter problem can be treated in an approximate manner using techniques such as the empirical valence bond method [2] or other semi-empirical approaches. However, such methods are also not transferable and, therefore, need to be reparameterized for each type of reaction and may end up biasing the reaction path in undesirable ways.

Because of the limitations of force field based approaches, one of the most important recent developments in MD, which addresses these problems, is the so called *ab initio* molecular dynamics (AIMD) method [3–11], which combines finite temperature dynamics with forces obtained from electronic structure calculations performed "on the fly" as the MD simulation proceeds (see, in particular, the recent excellent reviews by Marx and Hutter [10] and by Tuckerman [12] for detailed discussions). Because the electronic structure is treated explicitly in AIMD calculations, many-body forces, electronic polarization, and bond-breaking and forming events are described to within the accuracy of the electronic structure representation. Moreover, the AIMD methodology can be easily extended to incorporate nuclear quantum effects via the Feynman path integral approach [13, 14], leading to the *ab initio* path integral (AIPI) technique [15–18].

The AIMD and AIPI methods have been used to study a wide variety of chemically interesting and important problems in areas such as liquid structure, acid-base chemistry, industrial and biological catalysis, and materials. Applications include (but are certainly not limited to) calculations of the structure and dynamics of water [19–31] and other hydrogen-bonded liquids [32–36], structure and dynamics in acidic [37–42] and basic solutions [43, 44], proton transport in aqueous [45–49] and other environments [34, 35, 50–53] and in clusters [54–58], structure, proton order/disorder and dynamical properties of ice [28,59–67], structure of liquid silicates and glasses [68–76], mechanisms of Ziegler-Natta industrial catalysis [77–79] and other surface catalytic processes [66,80–84] and polymer knotting [85–88], to give just a sampling of the application areas. More recently, AIMD methods have started to impact the biological sciences and have been applied in calculations of NMR chemical shifts in drug-enzyme complexes [89], structure of nucleic acids [90], exploration of the design of possible biomimetics [91,92], and structure, dynamics and binding mechanisms in myoglobin [93,94]. In many of these applications, new physical phenomena have been revealed, which could not have been uncovered using empirical models, often leading to new interpretations of experimental data and even suggesting new experiments to perform.

Not unexpectedly, the power and flexibility of the AIMD (and AIPI) methodology come at the price of a significant increase in computational overhead compared to force field based approaches. Whereas the latter can currently be applied routinely to systems consisting of 10^4-10^6 atoms and access time scales on the order of tens of nanoseconds or longer, AIMD calculations can currently be applied routinely to systems of just a few tens or hundred of atoms and access time scales on the order of tens of picoseconds. Moreover, whereas force fields can be finely tuned for specific situations, the accuracy of AIMD calculations is limited by the accuracy of the electronic structure method employed. Currently, the most commonly used electronic structure theory in AIMD is the Kohn-Sham formulation of the density functional theory (DFT) [95–97], in which the electronic orbitals are expanded in a plane-wave basis set. This protocol provides a reasonably accurate description of the electronic structure for many types of chemical environments while maintaining an acceptable computational overhead and constitutes the original Car-Parrinello (CP) formulation of the method [3]. Clearly, then, AIMD calculations are limited by the accuracy of currently available density functionals. However, it is important to note that AIMD is a general approach that can be used with any electronic structure method, and a number of examples exist in the literature which employ more accurate or more empirical electronic structure representations [98,99] as well as different basis sets [100–102].

The organization of this article is as follows. In Sec. 2, we shall begin with a derivation of the AIMD method starting from the Born-Oppenheimer approximation and arriving at the classical equations of motion on the exact ground electronic surface. We shall then discuss, in Sec. 3, the use of the DFT representation of the electronic structure and various approximation schemes. Next, in Sec. 4, we will discuss different basis set expansions of the electronic orbitals and indicate the advantages and disadvantages of different choices. Following this, in Sec. 5, we will discuss the adiabatic dynamics method and its relation to the Car-Parrinello (CP) algorithm [3] for AIMD calculations. In Sec. 6, we will then discuss the calculation of observable properties. In particular, we will focus on properties for which having direct access to the electronic structure, a unique feature of AIMD calculations, is important. These include IR and Raman spectra and NMR chemical shifts.

2. The Born-Oppenheimer approximation and *ab initio* molecular dynamics

We begin our discussion of AIMD by considering a system of N nuclei described by coordinates, $\mathbf{R}_1,...,\mathbf{R}_N \equiv \mathbf{R}$, momenta, $\mathbf{P}_1,...,\mathbf{P}_N \equiv \mathbf{P}$, and masses $M_1,...,M_N$, and N_e electrons described by coordinates, $\mathbf{r}_1,...,\mathbf{r}_{N_e} \equiv \mathbf{r}$, momenta, $\mathbf{p}_1,...,\mathbf{p}_{N_e} \equiv \mathbf{p}$, and spin variables, $s_1,...,s_{N_e} \equiv s$. Throughout the discussion, we shall employ atomic units. The Hamiltonian of the system is given by

$$\begin{aligned} H &= \sum_{I=1}^{N} \frac{\mathbf{P}_I^2}{2M_I} + \sum_{i=1}^{N_e} \frac{\mathbf{p}_i^2}{2} + \sum_{i>j} \frac{1}{|\mathbf{r}_i - \mathbf{r}_j|} + \sum_{I>J} \frac{Z_I Z_J}{|\mathbf{R}_I - \mathbf{R}_J|} - \sum_{i,I} \frac{Z_I}{|\mathbf{R}_I - \mathbf{r}_i|} \\ &\equiv T_{\mathrm{N}} + T_{\mathrm{e}} + V_{\mathrm{ee}}(\mathbf{r}) + V_{\mathrm{NN}}(\mathbf{R}) + V_{\mathrm{eN}}(\mathbf{r},\mathbf{R}) \end{aligned} \tag{1}$$

where Z_I is the charge on the Ith nucleus. In the second line, T_{N}, T_{e}, V_{ee}, V_{NN}, and V_{eN} represent the nuclear and electron kinetic energy operators and electron-electron, electron-nuclear, and nuclear-nuclear interaction potential operators, respectively. In order to solve the complete quantum mechanical problem, we start by seeking the eigenfunctions and eigenvalues of this Hamiltonian, which will be given by solution of the time-independent Schrödinger equation

$$[T_{\mathrm{N}} + T_{\mathrm{e}} + V_{\mathrm{ee}}(\mathbf{r}) + V_{\mathrm{NN}}(\mathbf{R}) + V_{\mathrm{eN}}(\mathbf{r},\mathbf{R})]\Psi(\mathbf{x},\mathbf{R}) = E\Psi(\mathbf{x},\mathbf{R}) \tag{2}$$

where $\mathbf{x} \equiv (\mathbf{r}, s)$ denotes the full collection of electron position and spin variables, and $\Psi(\mathbf{x},\mathbf{R})$ is an eigenfunction of H with eigenvalue E. Clearly, an exact solution of Eq. (2) is not possible and approximations must be made. We first invoke the Born-Oppenheimer approximation by recognizing that, in a dynamical sense, there is a strong separation of time scales between the electronic and nuclear motion, since the electrons are lighter than the nuclei by three orders of magnitude. In terms of Eq. (2), this can be exploited by assuming a quasi-separable ansatz of the form

$$\Psi(\mathbf{x},\mathbf{R}) = \phi(\mathbf{x},\mathbf{R})\chi(\mathbf{R}) \tag{3}$$

where $\chi(\mathbf{R})$ is a nuclear wave function and $\phi(\mathbf{x},\mathbf{R})$ is an electronic wave function that depends parametrically on the nuclear positions. We note, at this point, that an alternative derivation using a fully separable ansatz to the time-dependent Schrödinger equation was presented by Marx and Hutter in Ref. [10]. Substitution of Eq. (3) into Eq. (2) and recognizing that the nuclear wave function $\chi(\mathbf{R})$

is more localized than the electronic wave function, i.e. $\nabla_I\chi(\mathbf{R}) \gg \nabla_I\phi(\mathbf{x},\mathbf{R})$, yields

$$\frac{[T_e + V_{ee}(\mathbf{r}) + V_{eN}(\mathbf{r},\mathbf{R})]\,\phi(\mathbf{x},\mathbf{R})}{\phi(\mathbf{x},\mathbf{R})} = E - \frac{[T_N + V_{NN}(\mathbf{R})]\,\chi(\mathbf{R})}{\chi(\mathbf{R})} \tag{4}$$

From the above, it is clear that the left side can only be a function of $\mathbf{R}$ alone. Let this function be denoted, $\varepsilon(\mathbf{R})$. Thus,

$$\frac{[T_e + V_{ee}(\mathbf{r}) + V_{eN}(\mathbf{r},\mathbf{R})]\,\phi(\mathbf{x},\mathbf{R})}{\phi(\mathbf{x},\mathbf{R})} = \varepsilon(\mathbf{R})$$

$$[T_e + V_{ee}(\mathbf{r}) + V_{eN}(\mathbf{r},\mathbf{R})]\,\phi(\mathbf{x},\mathbf{R}) = \varepsilon(\mathbf{R})\phi(\mathbf{x},\mathbf{R}) \tag{5}$$

Eq. (5) is an electronic eigenvalue equation for an electronic Hamiltonian, $H_e(\mathbf{R}) = T_e + V_{ee}(\mathbf{r}) + V_{eN}(\mathbf{r},\mathbf{R})$ which will yield a set of normalized eigenfunctions, $\phi_n(\mathbf{x},\mathbf{R})$ and eigenvalues, $\varepsilon_n(\mathbf{R})$, which depend parametrically on the nuclear positions, $\mathbf{R}$. For each solution, there will be a nuclear eigenvalue equation:

$$[T_N + V_{NN}(\mathbf{R}) + \varepsilon_n(\mathbf{R})]\,\chi(\mathbf{R}) = E\chi(\mathbf{R}) \tag{6}$$

Moreover, each electronic eigenvalue, $\varepsilon_n(\mathbf{R})$, will give rise to an electronic surface on which the nuclear dynamics is described by a time-dependent Schrödinger equation for the time-dependent nuclear wave function $X(\mathbf{R},t)$:

$$[T_N + V_{NN}(\mathbf{R}) + \varepsilon_n(\mathbf{R})]X(\mathbf{R},t) = i\frac{\partial}{\partial t}X(\mathbf{R},t) \tag{7}$$

will evolve. The physical interpretation of Eq. (7) is that the electrons respond instantaneously to the nuclear motion, therefore, it is sufficient to obtain a set of instantaneous electronic eigenvalues and eigenfunctions at each nuclear configuration, $\mathbf{R}$ (hence the parametric dependence of $\phi_n(\mathbf{x},\mathbf{R})$ and $\varepsilon_n(\mathbf{R})$ on $\mathbf{R}$). The eigenvalues, in turn, give a family of (uncoupled) potential surfaces on which the nuclear wave function can evolve. Of course, these surfaces can (and often do) become coupled by so called non-adiabatic effects, contained in the terms that have been neglected in the above derivation.

In many cases, non-adiabatic effects can be neglected, and we may consider motion *only* on the ground electronic surface described by:

$$[T_e + V_{ee}(\mathbf{r}) + V_{eN}(\mathbf{r},\mathbf{R})]\,\phi_0(\mathbf{x},\mathbf{R}) = \varepsilon_0(\mathbf{R})\phi_0(\mathbf{x},\mathbf{R})$$

$$[T_N + \varepsilon_0(\mathbf{R}) + V_{NN}(\mathbf{R})]X(\mathbf{R},t) = i\frac{\partial}{\partial t}X(\mathbf{R},t) \tag{8}$$

Moreover, if nuclear quantum effects can be neglected, then we may arrive at classical nuclear evolution by assuming $X(\mathbf{R},t)$ is of the form

$$X(\mathbf{R},t) = A(\mathbf{R},t)e^{iS(\mathbf{R},t)} \tag{9}$$

and neglecting all but the lowest order term, which yields an approximate equation for $S(\mathbf{R},t)$:

$$H_N(\nabla_1 S, ..., \nabla_N S, \mathbf{R}_1, ..., \mathbf{R}_N) + \frac{\partial S}{\partial t} = 0 \tag{10}$$

which is just the classical Hamiltonian-Jacobi equation with

$$H_N(\mathbf{P}_1, ..., \mathbf{P}_N, \mathbf{R}_1, ..., \mathbf{R}_N) = \sum_{I=1}^{N} \frac{\mathbf{P}_I^2}{2M_I} + V_{NN}(\mathbf{R}) + \varepsilon_0(\mathbf{R}) \tag{11}$$

denoting the classical nuclear Hamiltonian. The Hamilton-Jacobi equation is equivalent to classical motion on the ground-state surface, $E_0(\mathbf{R}) = \varepsilon_0(\mathbf{R}) + V_{NN}(\mathbf{R})$ given by

$$\begin{aligned} \dot{\mathbf{R}}_I &= \frac{\mathbf{P}_I}{M_I} \\ \dot{\mathbf{P}}_I &= -\nabla_I E_0(\mathbf{R}) \end{aligned} \tag{12}$$

Note that the force $-\nabla_I E_0(\mathbf{R})$ contains a term from the nuclear-nuclear repulsion and a term from the derivative of the electronic eigenvalue, $\varepsilon_0(\mathbf{R})$. Because of the Hellman-Feynman theorem, the latter can be expressed as

$$\nabla_I \varepsilon_0(\mathbf{R}) = \langle \phi_0(\mathbf{R}) | \nabla_I H_e(\mathbf{R}) | \phi_0(\mathbf{R}) \rangle \tag{13}$$

Equations (12) and Eq. (13) form the theoretical basis of the AIMD approach. The practical implementation of the AIMD method requires an algorithm for the numerical solution of Eq. (12) with forces obtained from Eq. (13) at each step of the calculation. Moreover, since an exact solution for the ground state electronic wave function, $|\phi_0(\mathbf{R})\rangle$ and eigenvalue, $\varepsilon_0(\mathbf{R})$ are not available, in general, it is necessary to introduce an approximation scheme for obtaining these quantities. This is the topic of the next section.

3. Representation of the electronic structure

At this point, a simple form for $E_0(\mathbf{R})$ could be introduced, giving rise to a force field based approach. Such a form would necessarily be specific to a particular system and, therefore, not be transferable to other situations. If, on the other hand, one derives forces directly from very accurate electronic structure calculations, the computational overhead associated with the method will be enormous. It is clear, therefore, that the practical utility of the AIMD approach relies on a compromise between accuracy and efficiency of the electronic structure representation based on available computing resources. One approach that has proved particularly successful in this regard is the density functional theory (DFT). The DFT is based on the Hohenberg-Kohn theorem [103], which states that a one-to-one mapping exists between ground state electronic densities and external potentials. The ground state density, $n_0(\mathbf{r})$, is given in terms of the ground state wave function by

$$n_0(\mathbf{r}) = \sum_{s,s_2,\ldots,s_{N_e}} \int d\mathbf{r}_2 \cdots d\mathbf{r}_{N_e}\, |\phi_0(\mathbf{r},s,\mathbf{r}_2,s_2,\ldots,\mathbf{r}_{N_e},s_{N_e})|^2 \tag{14}$$

(Here, $\mathbf{r}$ and s represent a single position and spin variable, respectively.) A consequence of the Hohenberg-Kohn theorem is that the exact ground state energy, $\varepsilon_0(\mathbf{R})$, can be obtained by minimizing a certain functional, $\varepsilon[n]$ over all electronic densities, $n(\mathbf{r})$ that can be associated with an antisymmetric ground state wavefunction, $|\phi_0\rangle$ of a Hamiltonian H_e for some potential V_{eN} (the so called v-representability conditions) subject to the condition that $\int d\mathbf{r}\, n(\mathbf{r}) = N_e$. The theorem can also be extended to so called N-representable densities (obtained from any antisymmetric wavefunction) via the Levy prescription [96, 97]. The functional $\varepsilon[n]$ is given as a sum $T[n] + W[n] + V[n]$, where $T[n]$ and $W[n]$ represent the kinetic energy and Coulomb repulsion energies, respectively, and $V[n] = \int d\mathbf{r}\, V_{eN}(\mathbf{r})n(\mathbf{r})$. Although the functional $T[n] + W[n]$ is a universal for all systems of N_e electrons, its form is not known. Thus, in order that DFT be of practical utility, Kohn and Sham (KS) introduced the idea of a non-interacting reference system with a potential $V_{KS}(\mathbf{r},\mathbf{R})$ such that the ground state energy and density of the non-interacting system equal those of the true interacting system [95]. Within the KS formulation of DFT, a set of single-particle orbitals, $\psi_i(\mathbf{r})$, $i = 1,\ldots,n_{occ}$ with occupation numbers f_i, where $\sum_{i=1}^{n_{occ}} f_i = N_e$ is introduced. These are known as the KS orbitals. In terms of the KS orbitals, the density is given by

$$n(\mathbf{r}) = \sum_{i=1}^{n_{occ}} f_i |\psi_i(\mathbf{r})|^2 \tag{15}$$

and the functional takes the form

$$\begin{aligned}
\varepsilon[\{\psi_i\}] &= -\frac{1}{2}\sum_{i=1}^{n_{occ}} f_i \langle\psi_i|\nabla^2|\psi_i\rangle + \frac{e^2}{2}\int d\mathbf{r}\, d\mathbf{r}'\, \frac{n(\mathbf{r})n(\mathbf{r}')}{|\mathbf{r}-\mathbf{r}'|} \\
&+ \varepsilon_{xc}[n] + \int d\mathbf{r}\, n(\mathbf{r}) V_{eN}(\mathbf{r},\mathbf{R}) \\
&\equiv T_{\mathrm{nonint}}[\{\psi\}] + J[n] + \varepsilon_{xc}[n] + V[n]
\end{aligned} \tag{16}$$

The first term in the functional represents the quantum kinetic energy, the second is the direct Coulomb term from Hartree-Fock theory, the third term is the exact exchange-correlation energy, whose form is unknown, and the fourth term is the interaction of the electron density with the external potential due to the nuclei. Thus, the KS potential is given by

$$V_{KS}(\mathbf{r},\mathbf{R}) = \frac{e^2}{2}\int d\mathbf{r}' \, \frac{n(\mathbf{r}')}{|\mathbf{r}-\mathbf{r}'|} + \frac{\delta\varepsilon_{xc}}{\delta n(\mathbf{r})} + V_{eN}(\mathbf{r},\mathbf{R}) \tag{17}$$

and the Hamiltonian of the non-interacting system is, therefore,

$$H_{KS} = -\frac{1}{2}\nabla^2 + V_{KS}(\mathbf{r},\mathbf{R}) \tag{18}$$

The KS orbitals will be the solutions of a set of self-consistent equations known as the *Kohn-Sham equations* [96, 97]:

$$H_{KS}\psi_i(\mathbf{r}) = \varepsilon_i\psi_i(\mathbf{r}) \tag{19}$$

where ε_i are the KS energies. The preceding discussion makes clear the fact that DFT is, in principle, an exact theory for the ground state of a system. However, because the exchange-correlation functional, defined to be $\varepsilon_{xc}[n] = T[n] - T_{\text{nonint}}[\{\psi\}] + W[n] - J[n]$, is unknown, in practice, approximations must be made. One of the most successful approximations is the so called *local density (LDA)* approximation, in which the functional is taken to be the spatial integral over a local function that depends only on the density:

$$\varepsilon_{xc}[n] \approx \int d\mathbf{r} \, f_{LDA}(n(\mathbf{r})) \tag{20}$$

The LDA is physically motivated by the notion that the interaction between the electrons and the nuclei creates only weak inhomogeneities in the electron density. Therefore, the form of f_{LDA} is obtained by evaluating the exact expressions for the exchange and correlation energies of a homogeneous electron gas of uniform density n at the inhomogeneous density $n(\mathbf{r})$. The LDA has been successfully used in numerous applications of importance in solid state physics, including studies of semiconductors and metals. In many instances of importance in chemistry, however, the electron density possesses sufficient inhomogeneities that the LDA breaks down. This is particularly true in hydrogen-bonded systems, for example. In such cases, the LDA can be improved by adding an additional dependence on the lowest order gradients of the density:

$$\varepsilon_{xc}[n] \approx \int d\mathbf{r} \, f_{GGA}(n(\mathbf{r}), |\nabla n(\mathbf{r})|, \nabla^2 n(\mathbf{r})) \tag{21}$$

which is known as the *generalized gradient (GGA)* approximation. Among the most widely used GGAs are those of Becke [104], Lee, Yang and Parr [105], Perdew and Wang [106], Perdew, Burke and Erznerhof [107] and Cohen and Handy [108–110]. Typically, these can be calibrated to reproduce some subset of the known properties satisfied by the exact exchange-correlation functional. GGAs such as these have been used successfully in nearly all of the application areas discussed in Sec. 1. However, GGAs are also known to underestimate transition state barriers and cannot adequately treat dispersion. Attempts to incorporate dispersion interactions in an empirical way have recently been proposed [111]. In order to improve reaction barriers, new approximation schemes such as Becke's 1992 functional [112], which incorporates exact exchange, and the so called *meta-GGA* functionals [113–116], which include an additional dependence on the electron kinetic energy density

$$\tau(\mathbf{r}) = \sum_{i=1}^{n_{occ}} f_i |\nabla\psi_i(\mathbf{r})|^2 \tag{22}$$

have been proposed with reasonable success. However, the problem of designing accurate approximate exchange-correlation functionals remains one of the greatest challenges in DFT.

Finally, in order to overcome the limitations of DFT in the context of AIMD, it is, of course, possible to employ a more accurate electronic structure method, and approaches using full configuration-interaction representations have been proposed [98]. Typically, these have a higher computational overhead and, therefore, can only be used to study much smaller systems such as very small clusters. However, as computing platforms become more powerful and new algorithms are developed, it is conceivable that approaches other electronic structure methods will be used more routinely in AIMD studies.

4. Basis set expansions

4.1. PLANE WAVE BASIS SETS

In MD calculations, the most commonly employed boundary conditions are periodic boundary conditions, in which the system is replicated infinitely in space. This is clearly a natural choice for solids and is particularly convenient for liquids. In an infinite periodic system, the KS orbitals become Bloch functions of the form

$$\psi_{i\mathbf{k}}(\mathbf{r}) = e^{i\mathbf{k}\cdot\mathbf{r}} u_{i\mathbf{k}}(\mathbf{r}) \tag{23}$$

where $\mathbf{k}$ is a vector in the first Brioullin zone and $u_{i\mathbf{k}}(\mathbf{r})$ is a periodic function. A natural basis set for expanding a periodic function is the Fourier or plane wave basis set, in which $u_{i\mathbf{k}}(\mathbf{r})$ is expanded according to

$$u_{i\mathbf{k}}(\mathbf{r}) = \frac{1}{\sqrt{\Omega}} \sum_{\mathbf{g}} c_{i,\mathbf{g}}^{\mathbf{k}} e^{i\mathbf{g}\cdot\mathbf{r}} \tag{24}$$

where Ω is the volume of the cell, $\mathbf{g} = 2\pi\mathbf{h}^{-1}\hat{\mathbf{g}}$ is a reciprocal lattice vector, $\mathbf{h}$ is the cell matrix, whose columns are the cell vectors ($\Omega = \det(\mathbf{h})$), $\hat{\mathbf{g}}$ is a vector of integers, and $\{c^{\mathbf{k}}_{i,\mathbf{g}}\}$ are the expansion coefficients. An advantage of plane waves is that the sums needed to go back and forth between reciprocal space and real space can be performed efficiently using fast Fourier transforms (FFTs). In general, the properties of a periodic system are only correctly described if a sufficient number of **k**-vectors are sampled from the Brioullin zone. However, for the applications to be considered herein, which are largely concerned with nonmetallic systems, it is generally sufficient to consider a single **k** point, $\mathbf{k} = (0,0,0)$, known as the Γ-point ($\mathbf{k} = (0,0,0)$), so that the plane wave expansion reduces to

$$\psi_i(\mathbf{r}) = \frac{1}{\sqrt{\Omega}} \sum_{\mathbf{g}} c_{i,\mathbf{g}} e^{i\mathbf{g}\cdot\mathbf{r}} \tag{25}$$

At the Γ-point, the orbitals can always be chosen to be real functions. Therefore, the plane-wave expansion coefficients satisfy the following property

$$c^*_{i,\mathbf{g}} = c_{i,-\mathbf{g}} \tag{26}$$

which requires keeping only half of the full set of plane-wave expansion coefficients. In actual applications, plane waves up to a given cutoff, $|\mathbf{g}|^2/2 < E_{\text{cut}}$, only are kept. Similarly, the density $n(\mathbf{r})$ given by Eq. (15) can also be expanded in a plane wave basis:

$$n(\mathbf{r}) = \frac{1}{\Omega} \sum_{\mathbf{g}} n_{\mathbf{g}} e^{i\mathbf{g}\cdot\mathbf{r}} \tag{27}$$

However, since $n(\mathbf{r})$ is obtained as a square of the KS orbitals, the cutoff needed for this expansion is $4E_{\text{cut}}$ for consistency with the orbital expansion.

Using Eqs. (25) and (27) and the orthogonality of the plane waves, it is straightforward to compute the various energy terms. The kinetic energy can be easily shown to be

$$\varepsilon_{\text{KE}} = -\frac{1}{2} \sum_i \int d\mathbf{r}\, \psi_i^*(\mathbf{r}) \nabla^2 \psi_i(\mathbf{r}) = \frac{1}{2} \sum_i \sum_{\mathbf{g}} g^2 |c^i_{\mathbf{g}}|^2 \tag{28}$$

where $g = |\mathbf{g}|$. Similarly, the Hartree energy becomes

$$\varepsilon_{\text{H}} = \frac{1}{2} \int d\mathbf{r}\, d\mathbf{r}' \frac{n(\mathbf{r})n(\mathbf{r}')}{|\mathbf{r}-\mathbf{r}'|} = \frac{1}{\Omega} \sum_{\mathbf{g}}{}' \frac{4\pi}{g^2} |n_{\mathbf{g}}|^2 \tag{29}$$

where the summation excludes the $\mathbf{g} = (0,0,0)$ term.

The exchange and correlation energy, $\varepsilon_{\text{xc}}[n]$ in the LDA or GGA, is evaluated on the real-space FFT grid so that it can be expressed as

$$\varepsilon_{\text{xc}}[n] = \frac{\Omega}{N_{\text{grid}}} \sum_{\mathbf{r}} f_{\text{GGA}}(n(\mathbf{r}), |\nabla n(\mathbf{r})|, \nabla^2 n(\mathbf{r})) \tag{30}$$

where $N_{\rm grid}$ is the number of real-space grid points. As was shown by White and Bird [117], the use of the grid eliminates the complexity of functional differentiation by allowing the contribution to the KS potential from $\varepsilon_{\rm xc}$ be computed from

$$\begin{aligned}\frac{d\varepsilon_{\rm xc}}{dn(\mathbf{r})} &= \frac{\Omega}{N_{\rm grid}}\frac{\partial f_{\rm GGA}}{\partial n(\mathbf{r})} \\ &+ \frac{\Omega}{N_{\rm grid}}\sum_{\mathbf{r}'}\left[\frac{\partial f_{\rm GGA}}{\partial|\nabla n(\mathbf{r}')|}\frac{\partial|\nabla n(\mathbf{r}')|}{\partial n(\mathbf{r})}+\frac{\partial f_{\rm GGA}}{\partial\nabla^2 n(\mathbf{r}')}\frac{\partial\nabla^2 n(\mathbf{r}')}{\partial n(\mathbf{r})}\right]\end{aligned} \tag{31}$$

The gradient and (if needed) the Laplacian of the density can be computed efficiently using FFTs:

$$\begin{aligned}\nabla n(\mathbf{r}) &= \sum_{\mathbf{g}} i\mathbf{g}e^{i\mathbf{g}\cdot\mathbf{r}}\sum_{\mathbf{r}'} n(\mathbf{r}')e^{-i\mathbf{g}\cdot\mathbf{r}'} \\ \nabla^2 n(\mathbf{r}) &= -\sum_{\mathbf{g}} g^2 e^{i\mathbf{g}\cdot\mathbf{r}}\sum_{\mathbf{r}'} n(\mathbf{r}')e^{-i\mathbf{g}\cdot\mathbf{r}'}\end{aligned} \tag{32}$$

Equation (32) also shows how the derivatives needed in Eq. (31) can be easily computed using FFTs.

The external energy is made somewhat complicated by the fact that, in a plane wave basis, very large basis sets are needed to treat the rapid spatial fluctuations of core electrons. Therefore, core electrons are often replaced by atomic pseudopotentials [118–120] or augmented plane wave techniques [121]. Here, we shall discuss the former. In the atomic pseudopotential scheme, the nucleus plus the core electrons are treated in a frozen core type approximation as an "ion" carrying only the valence charge. In order to make this approximation, the valence orbitals, which, in principle must be orthogonal to the core orbitals, must see a different pseudopotential for each angular momentum component in the core, which means that the pseudopotential must generally be nonlocal. In order to see this, we consider a potential operator of the form

$$\hat{V}_{\rm pseud} = \sum_{l=0}^{\infty}\sum_{m=-l}^{l} v_l(r)|lm\rangle\langle lm| \tag{33}$$

where r is the distance from the ion, and $|lm\rangle\langle lm|$ is a projection operator onto each angular momentum component. In order to truncate the infinite sum over l in Eq. (33), we assume that for some $l \geq \bar{l}$, $v_l(r) = v_{\bar{l}}(r)$ and add and subtract the function $v_{\bar{l}}(r)$ in Eq. (33):

$$\hat{V}_{\rm pseud} = \sum_{l=0}^{\infty}\sum_{m=-l}^{l}(v_l(r)-v_{\bar{l}}(r))|lm\rangle\langle lm| + v_{\bar{l}}(r)\sum_{l=0}^{\infty}\sum_{m=-l}^{l}|lm\rangle\langle lm|$$

$$= \sum_{l=0}^{\infty}\sum_{m=-l}^{l}(v_l(r)-v_{\bar{l}}(r))|lm\rangle\langle lm| + v_{\bar{l}}(r)$$

$$\approx \sum_{l=0}^{\bar{l}-1}\sum_{m=-l}^{l}\Delta v_l(r)|lm\rangle\langle lm| + v_{\bar{l}}(r) \tag{34}$$

where the second line follows from the fact that the sum of the projection operators is unity, $\Delta v_l(r) = v_l(r) - v_{\bar{l}}(r)$, and the sum in the third line is truncated before $\Delta v_{\bar{l}}(r) = 0$. The complete pseudopotential operator will be

$$\hat{V}_{\text{pseud}}(r;\mathbf{R}_1,\ldots,\mathbf{R}_N) = \sum_{I=1}^{N}\left[v_{\text{loc}}(|\mathbf{r}-\mathbf{R}_I|) + \sum_{l=0}^{\bar{l}-1}\Delta v_l(|\mathbf{r}-\mathbf{R}_I|)|lm\rangle\langle lm|\right] \tag{35}$$

where $v_{\text{loc}}(r) \equiv v_{\bar{l}}(r)$ is known as the local part of the pseudopotential (having no projection operator attached to it). Now, the external energy, being derived from the ground-state expectation value of a one-body operator, will be given by

$$\varepsilon_{\text{ext}} = \sum_i f_i\langle\psi_i|\hat{V}_{\text{pseud}}|\psi_i\rangle \tag{36}$$

The first (local) term gives simply a local energy of the form

$$\varepsilon_{\text{loc}} = \sum_{I=1}^{N}\int d\mathbf{r}\, n(\mathbf{r})v_{\text{loc}}(|\mathbf{r}-\mathbf{R}_I|) \tag{37}$$

which can be evaluated in reciprocal space as

$$\varepsilon_{\text{loc}} = \frac{1}{\Omega}\sum_{I=1}^{N}\sum_{\mathbf{g}} n^*_{\mathbf{g}}\tilde{v}_{\text{loc}}(\mathbf{g})e^{-i\mathbf{g}\cdot\mathbf{R}_I} \tag{38}$$

where $\tilde{V}_{\text{loc}}(\mathbf{g})$ is the Fourier transform of the local potential. Note that at $\mathbf{g} = (0,0,0)$, only the nonsingular part of $\tilde{v}_{\text{loc}}(\mathbf{g})$ contributes. In the evaluation of the local term, it is often convenient to add and subtract a long-range term of the form $Z_I\text{erf}(\alpha_I r)/r$, where $\text{erf}(x)$ is the error function, each ion in order to obtain the nonsingular part explicitly and a residual short-range function $\bar{v}_{\text{loc}}(|\mathbf{r}-\mathbf{R}_I|) = v_{\text{loc}}(|\mathbf{r}-\mathbf{R}_I|) - Z_I\text{erf}(\alpha_I|\mathbf{r}-\mathbf{R}_I|)/|\mathbf{r}-\mathbf{R}_I|$ for each ionic core. For the nonlocal contribution, Eq. (25) is substituted into Eq. (35), an expansion of the plane waves in terms of spherical Bessel functions and spherical harmonics is made, and, after some algebra, one obtains

$$\varepsilon_{\text{NL}} = \sum_i f_i\sum_I\sum_{\mathbf{g},\mathbf{g}'} e^{-i\mathbf{g}\cdot\mathbf{R}_I}c^*_{i,\mathbf{g}}v_{\text{NL}}(\mathbf{g},\mathbf{g}')c_{i,\mathbf{g}'}e^{i\mathbf{g}'\cdot\mathbf{R}_I} \tag{39}$$

where

$$v_{\rm NL}(\mathbf{g},\mathbf{g}') = (4\pi)^2 \sum_{l=0}^{\bar{l}-1} \sum_{m=-l}^{l} \int dr\, r^2\, j_l(gr) j_l(g'r) \Delta v_l(\mathbf{r}) \times Y_{lm}(\theta_{\mathbf{g}},\phi_{\mathbf{g}}) Y^*_{lm}(\theta_{\mathbf{g}'},\phi_{\mathbf{g}'}) \quad (40)$$

where $\theta_{\mathbf{g}}(\theta_{\mathbf{g}'})$ and $\phi_{\mathbf{g}}(\phi_{\mathbf{g}'})$ are the spherical polar angles associated with the vector $\mathbf{g}(\mathbf{g}')$, Y_{lm} are the spherical harmonics, and $j_l(x)$ is a spherical Bessel function. Eq. (40), which is known as the *semi-local* form, shows that the evaluation of the nonlocal energy can be quite computationally expensive. It also shows, however, that the matrix element is *almost* separable in g and g′ dependent terms. A fully separable approximation can be obtained by writing

$$v_{\rm NL}(\mathbf{g},\mathbf{g}') = (4\pi)^2 \sum_{l=0}^{\bar{l}-1} \sum_{m=-l}^{l} \int dr\, r^2 \int dr'\, r'^2 j_l(gr) j_l(g'r') \Delta v_l(\mathbf{r}) \frac{\delta(r-r')}{rr'} \times Y_{lm}(\theta_{\mathbf{g}},\phi_{\mathbf{g}}) Y^*_{lm}(\theta_{\mathbf{g}'},\phi_{\mathbf{g}'}) \quad (41)$$

where a radial δ-function has been introduced. Now, the δ-function is expanded in terms of a set of radial eigenfunctions (usually taken to be those of the Hamiltonian from which the pseudopotential is obtained) for each angular momentum channel

$$\frac{\delta(r-r')}{rr'} = \sum_{n=0}^{\infty} \phi^*_{nl}(r)\phi_{nl}(r') \quad (42)$$

If this expansion is now substituted into Eq. (41), the result is

$$v_{\rm NL}(\mathbf{g},\mathbf{g}') = (4\pi)^2 \sum_{n=0}^{\infty} \sum_{l=0}^{\bar{l}-1} \sum_{m=-l}^{l} \left[\int dr\, r^2\, j_l(gr) \Delta v_l(\mathbf{r}) \phi^*_{nl}(r) Y_{lm}(\theta_{\mathbf{g}},\phi_{\mathbf{g}})\right] \times \left[\int dr'\, r'^2 j_l(g'r') Y^*_{lm}(\theta_{\mathbf{g}'},\phi_{\mathbf{g}'}) \phi_{nl}(r')\right] \quad (43)$$

which is now fully separable at the expense of another infinite sum that needs to be truncated. The sum over n can be truncated after a finite number of terms, although some care is required in order to effect the truncation, the so called *Kleinman-Bylander approximation* [122] is the result of truncating it at just a single term. The result of this truncation can be shown to yield the approximate form:

$$v_{\rm NL}(\mathbf{g},\mathbf{g}') \approx (4\pi)^2 \sum_{l=0}^{\bar{l}-1} \sum_{m=-l}^{l} N^{-1}_{lm} \left[\int dr\, r^2\, j_l(gr) \Delta v_l(\mathbf{r}) \phi^*_l(r) Y_{lm}(\theta_{\mathbf{g}},\phi_{\mathbf{g}})\right] \times \left[\int dr'\, r'^2 \Delta v_l(r') j_l(g'r') Y^*_{lm}(\theta_{\mathbf{g}'},\phi_{\mathbf{g}'}) \phi_l(r')\right] \quad (44)$$

where

$$N_{lm} = \int dr\, r^2\, \phi_l^*(r)\Delta v_l(r)\phi_l(r) \qquad (45)$$

and $\phi_l(r) \equiv \phi_{0l}(r)$. Finally, substituting Eq. (44) into Eq. (39) gives the nonlocal energy as

$$\varepsilon_{\rm NL} = \sum_{i=1}^{N_e}\sum_{I=1}^{N}\sum_{l=0}^{\bar{l}-1}\sum_{m=-l}^{l} Z^*_{iIlm} Z_{iIlm} \qquad (46)$$

where

$$Z_{iIlm} = \sum_{\mathbf{g}} c^i_{\mathbf{g}} e^{i\mathbf{g}\cdot\mathbf{R}_I} \tilde{F}_{lm}(\mathbf{g}) \qquad (47)$$

and

$$\tilde{F}_{lm}(\mathbf{g}) = 4\pi N_{lm}^{-1/2} \int dr\, r^2 j_l(gr)\Delta v_l(ur)\phi_l(r) Y_{lm}(\theta_{\mathbf{g}},\phi_{\mathbf{g}}) \qquad (48)$$

For certain elements, it has been shown that the simple Kleinman-Bylander form can lead to spurious or unphysical bound states known as *ghost levels*. Refs. [123, 124] contains analyses and techniques for treating spurious ghost states. Alternatively, ghosts can be eliminated by taking more terms that just the first in Eq. (43) [125] or working directly with the semilocal form.

The last issue we shall discuss is that of boundary conditions within the plane wave description. Plane waves naturally describe fully periodic systems, such as solids, or systems that can be treated with periodic boundary conditions, such as liquids. What if we wish to treat a systems, such as cluster, surface, or wire, in which one or more boundaries is *not* periodic? It turns out that such situations can be treated rigorously within the plane-wave formalism. One approach is based on a direct solution to the Poisson equation in a box containing the cluster [126, 127]. Here, we shall discuss a simpler and more direct approach developed by Martyna and Tuckerman [128–130], which involves the use of a screening function in the long-range energy terms, i.e. the Hartree and local pseudopotential terms. The idea is to use the so called first image form of the average energy in order to form an approximation to a cluster, wire, or surface system, whose error can be controlled by the dimensions of the simulation cell. Thus, given any density, $n(\mathbf{r})$, and any interaction potential, $\phi(\mathbf{r}-\mathbf{r}')$, the average potential energy in this approximation is given by

$$\langle\phi\rangle^{(1)} = \frac{1}{2\Omega}\sum_{\mathbf{g}} |n_{\mathbf{g}}|^2 \bar{\phi}(-\mathbf{g}) \qquad (49)$$

where $\bar{\phi}(\mathbf{g})$ is a Fourier expansion coefficient of the potential given by

$$\bar{\phi}(\mathbf{g}) = \int_{-L_c/2}^{L_c/2} dz \int_{-L_b/2}^{L_b/2} dy \int_{-L_a/2}^{L_a/2} dx\, \phi(\mathbf{r}) e^{-i\mathbf{g}\cdot\mathbf{r}} \qquad \text{(Cluster)} \quad (50)$$

$$\bar{\phi}(\mathbf{g}) = \int_{-L_c/2}^{L_c/2} dz \int_{-L_b/2}^{L_b/2} dy \int_{-\infty}^{\infty} dx\, \phi(\mathbf{r}) e^{-i\mathbf{g}\cdot\mathbf{r}} \qquad \text{(Wire)} \qquad (51)$$

$$\bar{\phi}(\mathbf{g}) = \int_{-L_c/2}^{L_c/2} dz \int_{-\infty}^{\infty} dy \int_{-\infty}^{\infty} dx\, \phi(\mathbf{r}) e^{-i\mathbf{g}\cdot\mathbf{r}} \qquad \text{(Surface)} \qquad (52)$$

Here, L_a, L_b, and L_c are the dimensions of the simulation cell (assumed to be orthorhombic for simplicity) in the x, y, and z directions ($\mathbf{h} = \text{diag}(L_a, L_b, L_c)$). Note that, unlike Eq. (29), the $\mathbf{g} = (0,0,0)$ term is not excluded. In order to have an expression that is easily computed within the plane wave description, consider two functions $\phi^{(\text{long})}(\mathbf{r})$ and $\phi^{(\text{short})}(\mathbf{r})$, which are assumed to the long and short range contributions to the total potential, i.e.

$$\begin{aligned} \phi(\mathbf{r}) &= \phi^{(\text{long})}(\mathbf{r}) + \phi^{(\text{short})}(\mathbf{r}) \\ \bar{\phi}(\mathbf{g}) &= \bar{\phi}^{(\text{long})}(\mathbf{g}) + \bar{\phi}^{(\text{short})}(\mathbf{g}). \end{aligned} \qquad (53)$$

We require that $\phi^{(\text{short})}(\mathbf{r})$ vanish exponentially quickly at large distances from the center of the parallelepiped and that $\phi^{(\text{long})}(\mathbf{r})$ contain the long range dependence of the full potential, $\phi(\mathbf{r})$. With these two requirements, it is possible to write

$$\begin{aligned} \bar{\phi}^{(\text{short})}(\mathbf{g}) &= \int_{D(\Omega)} d\mathbf{r}\, \exp(-i\mathbf{g}\cdot\mathbf{r}) \phi^{(\text{short})}(\mathbf{r}) \\ &= \int_{\text{all space}} d\mathbf{r}\, \exp(-i\mathbf{g}\cdot\mathbf{r}) \phi^{(\text{short})}(\mathbf{r}) + \varepsilon(\mathbf{g}) \\ &= \tilde{\phi}^{(\text{short})}(\mathbf{g}) + \varepsilon(\mathbf{g}) \end{aligned} \qquad (54)$$

with exponentially small error, $\varepsilon(\mathbf{g})$, provided the range of $\phi^{(\text{short})}(\mathbf{r})$ is small compared size of the parallelepiped. In order to ensure that Eq. (54) is satisfied, a convergence parameter, α, is introduced which can be used to adjust the range of $\phi^{(\text{short})}(\mathbf{r})$ such that $\varepsilon(\mathbf{g}) \sim 0$ and the error, $\varepsilon(\mathbf{g})$, will be neglected in the following.

The function, $\tilde{\phi}^{(\text{short})}(\mathbf{g})$, is the Fourier transform of $\phi^{(\text{short})}(\mathbf{r})$. Therefore,

$$\begin{aligned} \bar{\phi}(\mathbf{g}) &= \bar{\phi}^{(\text{long})}(\mathbf{g}) + \tilde{\phi}^{(\text{short})}(\mathbf{g}) \\ &= \bar{\phi}^{(\text{long})}(\mathbf{g}) - \tilde{\phi}^{(\text{long})}(\mathbf{g}) + \tilde{\phi}^{(\text{short})}(\mathbf{g}) + \tilde{\phi}^{(\text{long})}(\mathbf{g}) \\ &= \hat{\phi}^{(\text{screen})}(\mathbf{g}) + \tilde{\phi}(\mathbf{g}) \end{aligned} \qquad (55)$$

where $\tilde{\phi}(\mathbf{g}) = \tilde{\phi}^{(\text{short})}(\mathbf{g}) + \tilde{\phi}^{(\text{long})}(\mathbf{g})$ is the Fourier transform of the full potential, $\phi(\mathbf{r}) = \phi^{(\text{short})}(\mathbf{r}) + \phi^{(\text{long})}(\mathbf{r})$ and

$$\hat{\phi}^{(\text{screen})}(\mathbf{g}) = \bar{\phi}^{(\text{long})}(\mathbf{g}) - \tilde{\phi}^{(\text{long})}(\mathbf{g}). \qquad (56)$$

Thus, Eq. (56) becomes leads to

$$\langle \phi \rangle = \frac{1}{2\Omega} \sum_{\hat{\mathbf{g}}} |\bar{n}(\mathbf{g})|^2 \left[\tilde{\phi}(-\mathbf{g}) + \hat{\phi}^{(\text{screen})}(-\mathbf{g}) \right] \tag{57}$$

The new function appearing in the average potential energy, Eq. (57), is the difference between the Fourier series and Fourier transform form of the long range part of the potential energy and will be referred to as the screening function because it is constructed to "screen" the interaction of the system with an infinite array of periodic images. The specific case of the Coulomb potential,

$$\phi(\mathbf{r}) = \frac{1}{r} \tag{58}$$

can be separated into short and long range components via

$$\frac{1}{r} = \frac{\text{erf}(\alpha r)}{r} + \frac{\text{erfc}(\alpha r)}{r} \tag{59}$$

where the first term is long range. The screening function for the cluster case is easily computed by introducing an FFT grid and performing the integration numerically [128]. For the wire [130] and surface [129] cases, analytical expressions can be worked out and are given by

$$\begin{aligned}
\bar{\phi}^{(\text{screen})}(\mathbf{g}) = & -\frac{4\pi}{g^2} \left\{ \cos\left(\frac{g_c L_c}{2}\right) \right. \\
& \times \left[\exp\left(-\frac{g_s L_c}{2}\right) - \frac{1}{2} \exp\left(-\frac{g_s L_c}{2}\right) \text{erfc}\left(\frac{\alpha^2 L_c - g_s}{2\alpha}\right) \right. \\
& \left. - \frac{1}{2} \exp\left(\frac{g_s L_c}{2}\right) \text{erfc}\left(\frac{\alpha^2 L_c + g_s}{2\alpha}\right) \right] \\
& \left. + \exp\left(-\frac{g^2}{4\alpha^2}\right) Re \left[\text{erfc}\left(\frac{\alpha^2 L_c + i g_c}{2\alpha}\right) \right] \right\} \qquad \text{(Surface)}
\end{aligned} \tag{60}$$

$$\begin{aligned}
\bar{\phi}^{(\text{screen})}(\mathbf{g}) = & \frac{4\pi}{g^2} \left[\exp(-g^2/4\alpha^2) E(\alpha, L_b, g_b) E(\alpha, L_c, g_c) \right. \\
& + \cos\left(\frac{g_b L_b}{2}\right) \frac{4\sqrt{\pi}}{\alpha L_b} \exp(-g_c^2/4\alpha^2) I(\alpha, L_b, L_c, g_c) \\
& \left. + \cos\left(\frac{g_c L_c}{2}\right) \frac{4\sqrt{\pi}}{\alpha L_c} \exp(-g_b^2/4\alpha^2) I(\alpha, L_c, L_b, g_b) \right] \\
& - \frac{4\pi}{g^2} e^{-g^2/4\alpha^2} \qquad \text{(Wire)}
\end{aligned} \tag{61}$$

where

$$I(\alpha,L_1,L_2,g) = \int_0^{\alpha L_1/2} dx\, xe^{-g_a^2 L_1^2/16x^2} e^{-x^2} E\left(\frac{2x}{L_1},L_2,g\right) \tag{62}$$

and

$$E(\lambda,L,g) = \text{erf}\left(\frac{\lambda^2 L + ig}{2\lambda}\right) \tag{63}$$

where $\mathbf{g} = (g_a, g_b, g_c)$ and $g_s = \sqrt{g_a^2 + g_b^2}$. The one-dimensional integrals in Eq. (62) are well suited to be performed by Gaussian quadrature techniques. It should be noted that a simplified expression for the surface screening function can be obtained in the limit $\alpha \rightarrow \infty$ [10, 129, 131]:

$$\bar{\phi}^{(\text{screen})}(\mathbf{g}) = -\frac{4\pi}{g^2}\left[\cos\left(\frac{g_c L_c}{2}\right) e^{-g_s/L_c/2}\right] \tag{64}$$

However, as is discussed Ref. [129], some care is needed for $g_s = 0$. Moreover, the wire screening function also simplifies in the limit $\alpha \rightarrow \infty$ to

$$\bar{\phi}^{\text{screen}} \longrightarrow \frac{16}{g^2}\left[\cos\left(\frac{g_b L_b}{2}\right)\frac{J(g_c, g_a, L_c, L_b)}{L_b} + \cos\left(\frac{g_c L_c}{2}\right)\frac{J(g_b, g_a, L_b, L_c)}{L_c}\right] - \frac{4\pi}{g^2} \tag{65}$$

where

$$J(g_1, g_2, , L_1, L_2) = \int_0^{L_1/2} dx\, e^{ig_1 x}\sqrt{\theta(x, g_2, L_2)}K_1\left(\sqrt{4\theta(x, g_2, L_2)}\right) \tag{66}$$

$$\theta(x,g,L) = \frac{g^2 L^2/16}{1 + 4x^2/L^2} \tag{67}$$

and $K_1(z)$ is a modified Bessel function.

4.2. GAUSSIAN BASIS SETS

There is a great advantage to be gained by the use of localized basis sets over a delocalized basis like plane waves. In particular, the computations scale better for well localized orbitals. One of the most widely used localized basis sets is the Gaussian basis. In a Gaussian basis, the KS orbitals are expanded according to

$$\psi_i(\mathbf{r}) = \sum_{\alpha,\beta,\gamma} C^i_{\alpha\beta\gamma} G_{\alpha\beta\gamma}(\mathbf{r};\mathbf{R}) \tag{68}$$

where the basis function, $G_{\alpha\beta\gamma}(\mathbf{r};\mathbf{R})$ are centered on atoms and, therefore, are dependent on the positions of the atoms. The basis functions generally take the form

$$G_{\alpha\beta\gamma}(\mathbf{r};\mathbf{R}_I) = N_{\alpha\beta\gamma}x^{\alpha}y^{\beta}z^{\gamma}\exp\left[-|\mathbf{r}-\mathbf{R}_I|^2/2\sigma^2_{\alpha\beta\gamma}\right] \tag{69}$$

with integer α, β, and γ for Gaussian centered on atom I. The advantage of Gaussians is that many of the integrals appearing in the DFT functional (and in other electronic structure methods) can be done analytically. The main disadvantage, however, is that Gaussian basis sets are nonorthogonal and, hence, the overlap matrix between basis functions needs to be included in the various energy terms. Moreover, being dependent on atomic positions, the derivatives of the basis functions with respect to positions need to be computed. This leads to a considerable degree of complication for molecular dynamics. Finally, with Gaussian bases, it is difficult to reproduce the correct asymptotic behavior of the density, and one must always be aware of the effects of basis set superposition errors. Many of these problems can be eliminated by choosing to work with a simpler localized orthonormal basis set to be discussed in the next subsection.

4.3. DISCRETE VARIABLE REPRESENTATIONS

While the plane wave basis set has the advantage of simplicity and the lack of spatial bias inherent in Gaussian basis sets, they lead to $O(N^2M)$ scaling for DFT calculations, where N is the number of electronic states and M is the number of basis functions. Moreover, because they are spatially delocalized, they are not optimal for use on emerging massively parallel computing architectures because of their high communication overhead. As noted above, the problem of delocalization is largely eliminated with Gaussian basis sets at the expense of considerably increased complexity due to the nonorthogonality of the basis functions and the introduction of basis set superposition error. A significant advantage might be gained if the localized character of Gaussian basis sets could be achieved using simple, orthonormal basis functions that are not centered on atomic positions, thereby avoiding the complexity of Pulay forces. In fact, such a basis set is possible in the form of the discrete variable representation (DVR) [132–136], and implementation of AIMD with DVR basis sets was recently introduced [102]. A one-dimensional DVR is composed of a set of N functions, $u_i(x)$, $i = 1,...,N$ and a set of N grid point, x_i, such that the basis functions satisfy a Kroenecker δ property:

$$u_i(x_j) = \frac{\delta_{ij}}{a_i} \tag{70}$$

is satisfied on the grid points. Here, a_i is a generally complex number such that $w_i = |a_i|^2$ defines a set of quadrature weights. Therefore, the DVR functions behave as coordinate eigenfunctions on a particular grid. In this sense, they are the real-space analogs of plane waves, which are momentum eigenfunctions. Moreover, they can be constructed in such a way that each DVR function is well localized about a point on the grid. DVRs are commonly used in accurate bound-state and scattering calculations [132–136] and have been recently adapted for use in DFT based electronic structure calculations [102]. The basis functions then satisfy orthogonality and completeness relations of the form

$$\begin{aligned} \sum_{k=1}^{N} w_k u_i^*(x_k) u_j(x_k) &= \delta_{ij} \\ \sum_{k=1}^{N} w_k u_k^*(x_i) u_k(x_j) &= \delta_{ij} \end{aligned} \tag{71}$$

An important property of a DVR is that any position-dependent operator $\Omega(x)$ is approximately diagonal DVR basis set. By defining a projection operator, $P_N = \sum_{i=1}^{N} |u_i\rangle\langle u_j|$, it can be shown that

$$\begin{aligned} \langle u_i | P_N \Omega(x) P_N | u_j \rangle &= \Omega(x_i)\delta_{ij} \\ \langle u_i | \Omega(x) | u_j \rangle &\approx \Omega(x_i)\delta_{ij} \\ \lim_{N\to\infty} \langle u_i | \Omega(x) | u_j \rangle &= \Omega(x_i)\delta_{ij} \end{aligned} \tag{72}$$

so that as the basis set size approaches infinite, position-dependent operators become exactly diagonal.

DVR functions that satisfy Eq. (70) can be constructed from simpler basis functions according to the boundary conditions of the problem. For example, a DVR appropriate for periodic boundary conditions on a one-dimensional grid of N points can be constructed from a sum of cosine functions according to

$$u_l^{(\text{per})}(x) = \sqrt{\frac{1}{NL}} \sum_{\lambda=1}^{N} \cos[k_\lambda (x - x_l)] \tag{73}$$

where $k_\lambda = 2\pi(\lambda - N' - 1)/L$, $\lambda, l = 1, \ldots, N$ and $N' = (N-1)/2$, while a fixed-node DVR appropriate for a one-dimensional cluster system can be constructed from sine functions according to

$$u_l^{(\text{clus})}(x) = \frac{2}{\sqrt{(N+1)L}} \sum_{\lambda=1}^{N} \sin k_\lambda x \sin k_\lambda x_l \tag{74}$$

where $k_\lambda = \pi\lambda/L$, $\lambda = 1, \ldots, N$. Finally, a useful basis set for expansion of the KS orbitals can be constructed from a direct product of these one-dimensional DVRs according to

$$\Phi_{lmn}(\mathbf{r}) = u_l^{(a)}(x) v_m^{(b)}(y) q_n^{(c)}(z) \tag{75}$$

where a, b, and c indicate the type of function used, i.e. periodic (per) or cluster (clus). Thus, the basis functions can be tailored for different types of boundary conditions, e.g. a periodic DVR in one dimension and cluster DVRs in the other two could yield a basis set appropriate for surface calculations, etc. Once a direct product basis set is chosen, the KS orbitals are expanded according to

$$\psi_i(\mathbf{r}) = \sum_{l,m,n} C^i_{lmn} \Phi_{lmn}(\mathbf{r}) \tag{76}$$

where C^i_{lmn} is a set of expansion coefficients.

Specific terms in the energy functional follow directly from the form of the basis functions. Thus, the kinetic energy can be expressed analytically as

$$\varepsilon_{\mathrm{KE}} = \sum_i \sum_{l,l'} \sum_{m,m'} \sum_{n,n'} C^{i*}_{lmn} T_{ll',mm',nn'} C^i_{l'm'n'} \tag{77}$$

where the kinetic energy matrix has the highly sparse form

$$T_{ll',mm',nn'} = t^{(a)}_{ll'} \delta_{mm'} \delta_{nn'} + \delta_{ll'} t^{(b)}_{mm'} \delta_{nn'} + \delta_{ll'} \delta_{mm'} t^{(c)}_{nn'} \tag{78}$$

where the one-dimensional matrices are given by

$$t^{(\mathrm{per})}_{nn'} = \begin{cases} -(\frac{2\pi}{L})^2 \frac{N'}{3}(N'+1) & n = n' \\ \frac{-(\frac{2\pi}{L})^2 (-1)^{n-n'} \cos[\frac{\pi(n-n')}{N}]}{2\sin^2[\frac{\pi(n-n')}{N}]} & n \neq n' \end{cases} \tag{79}$$

for periodic and

$$t^{(\mathrm{clus})}_{nn'} = \begin{cases} -\frac{1}{L^2}\frac{\pi^2}{2}[\frac{2(N+1)^2+1}{3} - \frac{1}{\sin^2(\frac{\pi n}{N+1})}] & n = n' \\ -\frac{(-1)^{n-n'}}{L^2}\frac{\pi^2}{2}[\frac{1}{\sin^2(\frac{\pi(n-n')}{2(N+1)})} - \frac{1}{\sin^2(\frac{\pi(n+n')}{2(N+1)})}] & n \neq n' \end{cases} \tag{80}$$

for fixed-node DVRs, respectively. Terms in the functional involving the electron density can be computed by substituting Eq. (76) into Eq. (15):

$$n(\mathbf{r}) = \sum_{i=1}^{n_{\mathrm{occ}}} \left[\sum_{l,m,n} C^i_{lmn} \Phi_{lmn}(\mathbf{r}) \right]^2 \tag{81}$$

The resulting expression simplifies considerably when the density is subsequently evaluated at the points of the DVR grid, and the latter are all that is required for computing energies and forces. The gradient of the density, which is needed for GGA functionals, is computed by differentiating Eq. (81) and evaluating the result at the DVR grid points. Resulting derivatives of the DVR functions are computed once in the beginning and stored on one-dimensional grids. A simplifying approximation to this approach can also be introduced by postulating a similar DVR expansion for the density

$$n(\mathbf{r}) = \sum_{l,m,n,} N_{lmn}\Phi_{lmn}(\mathbf{r}) = \sum_{l,m,n} N_{lmn}u_l(x)v_m(y)q_n(z) \tag{82}$$

This allows the gradient of the density to be computed according to

$$\nabla n(x_l,y_m,z_n) = \sum_{l'} N_{l'mn}u'_{l'}(x_l) + \sum_{m'} N_{lm'n}v'_{m'}(y_m) + \sum_{n'} N_{lmn'}q'_{n'}(z_n) \tag{83}$$

Here, again, the derivatives of the DVR functions can be computed once at the beginning of a simulation and the values stored on one-dimensional grids. Our experience, thus far, is that the approximation in Eq. (83) tends to converge the total energy with the same number of grid points as the exact method, however, the errors in the exact method are smaller for DVR grids with fewer points. The nonlocal part of the pseudopotential is also evaluated in real space using the expansion coefficients of the KS orbitals. In principle, the Hartree energy could also be evaluated in real space using fast multipole moment (FMM) techniques. However, for simplicity, we choose to employ a hybrid approach in which the Hartree energy and long range part of the local pseudopotential energy are evaluated in reciprocal space using the screening function methodology described in Sec. 4.1. The short range part of the local pseudopotential is still evaluated in real space. This involves the evaluation of only two FFTs, one to obtain the density in reciprocal space and the other to transform the long-range contributions to the KS potential back to real space.

5. The Car-Parrinello Algorithm

5.1. THE BASIC ALGORITHM

In order to obtain the ground state energy and forces, the KS functional must be minimized over the set of single-particle orbitals subject to an orthonormality condition

$$\langle\psi_i|\psi_j\rangle = \delta_{ij} \tag{84}$$

Moreover, in order to combine this minimization with the nuclear dynamics of Eq. (12), it is necessary to carry out the minimization at each nuclear configuration. Thus, if Eq. (12) is integrated in a MD calculation using a numerical integrator, then the minimization would need to be carried out at each step of the MD simulation and the forces computed using the orbitals thus obtained.

Although a step-by-step minimization is an acceptable method for performing AIMD simulations, its use can have a high computational overhead if accurate energy conservation is desired [10]. In 1985, Car and Parrinello (CP) showed that this coupling between nuclear time evolution and electronic minimization could be treated efficiently via an implicit adiabatic dynamics approach [3]. In their scheme, a fictitious dynamics for the electronic orbitals is invented which, given orbitals initially at the minimum for an initial nuclear configuration, allows them to follow the nuclear motion adiabatically, and thus, be automatically at the approximately minimized configuration at each step of the MD evolution.

In order to understand how the CP scheme works, consider a simple model system with two degrees of freedom, x and y, described by a Hamiltonian $H = p_x^2/2m_x + p_y^2/2m_y + V(x,y)$ with $m_x \ll m_y$. Suppose, further, that x and y are maintained at separate temperatures T_x and T_y, via a thermostat coupling, such that $T_x \ll T_y$. The resulting dynamics under these conditions can be analyzed in a rigorous manner [18]. The essence of the analysis is that the y motion can be shown to be driven by a time-averaged force over the x motion. That is, if Δt is a time interval characteristic of the y motion, then a numerical evolution step for y appears as [137]:

$$y(\Delta t) = y(0) + \Delta t v_y(0) + \frac{\Delta t}{m_y}\int_0^{\Delta t/2} dt\, F_y\left[x_{\rm adb}(x(0), v_x(0), y(0)); t\right] \tag{85}$$

where $x_{\rm adb}(x(0), v_x(0), y(0); t)$ denotes the adiabatic evolution of x up to a time $t < \Delta t/2$ starting from initial conditions $x(0)$, $v_x(0)$ at fixed $y = y(0)$. If the system is assumed to be ergodic, then the time average in Eq. (85) can be replaced by a phase space average at fixed y [137],

$$\frac{2}{\Delta t}\int_0^{\Delta t/2} dt\, F_y\left[x_{\rm adb}(x(0), v_x(0), y(0)); t\right] = \frac{\int dx\, F_y(x,y) e^{-\beta_x V(x,y)}}{\int dx\, e^{-\beta_x V(x,y)}} \tag{86}$$

The averaged force is derivable from $\langle F_y(x,y)\rangle = (\partial/\partial y)(1/\beta_x)\ln Z_x(y;\beta_x)$ where

$$Z_x(y;\beta_x) = \int dx\, e^{-\beta_x V(x,y)} \tag{87}$$

from which it follows that the probability distribution of y in configuration space is $P(y) \propto (\int dx \exp[-\beta_x V(x,y)])^{\beta_y/\beta_x}$, where $\beta_x = 1/kT_x$ and $\beta_y = 1/kT_y$. In the limit, $T_x \to 0$, $\beta_x \to \infty$, the distribution can be shown to reduce to $P(y) \propto \exp[-\beta_y \min_x V(x,y)]$, where the potential is minimized with respect to x at fixed y. This means that y can be described by an effective Hamiltonian $H_y = p_y^2/2m_y + \min_x V(x,y)$, which is the desired form for AIMD. That is, for $T_x \ll T_y$, only very fluctuations about $\min_x V(x,y)$ will contribute to the distribution, and y will move on a very good approximation to the correctly minimized surface. As an example illustrating good vs. poor adiabatic following for this simple x-y problem, the reader should see Fig. 10 of Ref. [137]. The dynamics of x will be a fictitious *adiabatic* dynamics in this case that only serves to generate the approximately minimized potential at each value of y.

This idea can be realized in AIMD is controlled by introducing a fictitious dynamics for the electrons (analogous to x) via a set of orbital "velocities" $\{\dot{\psi}_i(\mathbf{r})\}$ and a fictitious electronic "kinetic energy" (not to be confused with the true quantum kinetic energy) given by

$$K_{\text{fict}} = \mu \sum_i \langle \dot{\psi}_i | \dot{\psi}_i \rangle \tag{88}$$

where μ is a fictitious mass parameter (having units of energy$\times$time^2) that controls the time scale on which the electrons "evolve" with the condition that $T_{\text{elec}} \ll T_{\text{ion}}$. The orbitals are incorporated into an extended dynamical system described by a Lagrangian [138] of the form [3]:

$$L = \mu \sum_i \langle \dot{\psi}_i | \dot{\psi}_i \rangle + \frac{1}{2} \sum_{I=1}^{N} M_I \dot{\mathbf{R}}_I^2 - E[\{\psi\},\mathbf{R}] + \sum_{i,j} [\Lambda_{ij} (\langle \psi_i | \psi_j \rangle - \delta_{ij})] \tag{89}$$

where $E[\{\psi\},\mathbf{R}] = \varepsilon[\{\psi\},\mathbf{R}] + V_{\text{NN}}(\mathbf{R})$. Here $V_{\text{NN}}(\mathbf{R})$ is now the electrostatic ion-ion repulsion. The matrix Λ_{ij} is a set of Lagrange multipliers introduced in order to ensure that the condition $\langle \psi_i | \psi_j \rangle = \delta_{ij}$ is satisfied dynamically as a constraint. This Lagrangian specifies the true dynamics of the ions and a fictitious adiabatic dynamics for the electrons that generates the instantaneous forces on the ions from the approximately minimized electronic configuration. for the electrons that The equations of motion are obtained from the usual Euler-Lagrange equations:

$$\begin{aligned}
\frac{d}{dt}\left(\frac{\delta L}{\delta \dot{\psi}_i^*(\mathbf{r})}\right) - \frac{\delta L}{\delta \psi_i^*(\mathbf{r})} &= 0 \\
\frac{d}{dt}\left(\frac{\partial L}{\partial \dot{\mathbf{R}}_I}\right) - \frac{\partial L}{\partial \mathbf{R}_I} &= 0
\end{aligned} \tag{90}$$

which gives the following coupled dynamical equations of motion:

$$\begin{aligned}
M_I \ddot{\mathbf{R}}_I &= -\nabla_I E[\{\psi\},\mathbf{R}] \\
\mu \ddot{\psi}_i(\mathbf{r}) &= -\frac{\delta}{\delta \psi_i^*(\mathbf{r})} E[\{\psi\},\mathbf{R}] + \sum_j \Lambda_j \psi_j(\mathbf{r})
\end{aligned} \tag{91}$$

These are known as the Car-Parrinello (CP) equations. The electronic equation can also be written in an abstract bra-ket form as:

$$\mu |\ddot{\psi}_i\rangle = -\frac{\partial E}{\partial \langle \psi_i |} + \sum_j \Lambda_{ij} |\psi_j\rangle \tag{92}$$

Note that $\partial E/\partial\langle\psi_i|$ can be represented as $-f_i H_{KS}|\psi_i\rangle$, where H_{KS} is the Kohn-Sham Hamiltonian discussed above, which, in the present case is given by

$$\begin{aligned} H_{KS} &= -\frac{1}{2}\nabla^2 + \int d\mathbf{r}' \frac{n(\mathbf{r}')}{|\mathbf{r}-\mathbf{r}'|} \\ &+ \frac{\delta\varepsilon_{xc}}{\delta n(\mathbf{r})} + V_{\text{ext}}(\mathbf{r},\mathbf{R}) + \hat{V}_{\text{NL}}(\mathbf{R}) \end{aligned} \tag{93}$$

In reality, Eq. (92) formally represents an equation of motion for the expansion coefficients of the orbitals. Thus, for plane wave and DVR basis sets, the equation of motion reads:

$$\begin{aligned} \mu\ddot{c}_{i,\mathbf{g}} &= -\frac{\partial E}{\partial c^*_{i,\mathbf{g}}} + \sum_j \Lambda_{ij} c_{j,\mathbf{g}} \\ \mu\ddot{C}^i_{lmn} &= -\frac{\partial E}{\partial C^{i*}_{lmn}} + \sum_j \Lambda_{ij} C^j_{lmn} \end{aligned} \tag{94}$$

respectively. For Gaussian basis sets, the equations of motion are somewhat more complicated by the position dependence of the basis functions and their nonorthogonality and, therefore, will not be discussed here. Below, an algorithm [139] for integrating the CP equations subject to the orthonormality constraint will be presented.

Beginning with an initially minimized set of Kohn-Sham orbitals, $\{|\psi_i(0)\rangle\}$ corresponding to an initial nuclear configuration, $\mathbf{R}(0)$ and initial velocities, $\{\dot{\psi}_i(0)\rangle\}$, $\dot{\mathbf{R}}(0)$, the first step is a velocity update:

$$\begin{aligned} |\dot{\psi}_i^{(1)}(0)\rangle &= |\dot{\psi}_i(0)\rangle + \frac{\Delta t}{2\mu}|\varphi_i(0)\rangle \qquad i=1,\ldots,n_{\text{occ}} \\ \dot{\mathbf{R}}_I(\Delta t/2) &= \dot{\mathbf{R}}_I(0) + \frac{\Delta t}{2M_I}\mathbf{F}_I(0) \qquad I=1,\ldots,N \end{aligned} \tag{95}$$

followed by a position/orbital update:

$$\begin{aligned} |\tilde{\psi}_i\rangle &= |\psi_i(0)\rangle + \Delta t|\dot{\psi}_i^{(1)}\rangle \qquad i=1,\ldots,n_{\text{occ}} \\ \mathbf{R}_I(\Delta t) &= \mathbf{R}_I(0) + \Delta t\dot{\mathbf{R}}_I(\Delta t/2) \qquad I=1,\ldots,N \end{aligned} \tag{96}$$

where $|\varphi_i(0)\rangle = (\partial E/\partial\langle\psi_i|)|_{t=0}$ is the initial force on the orbital, $|\psi_i\rangle$. At this point, we do not yet have the orbitals at $t=\Delta t$ or orbital velocities at $t=\Delta t/2$ because the constraint force $\Lambda_{ij}|\psi_j\rangle$ needs to be applied to both the orbitals and orbital velocities via the Lagrange multiplier matrix. This is accomplished by enforcing the orthogonality constraint on the orbitals at $t=\Delta t$:

$$\langle\psi_i(\Delta t)|\psi_j(\Delta t)\rangle = \delta_{ij} \tag{97}$$

where

$$|\psi_i(\Delta t)\rangle = |\tilde{\psi}_i\rangle + \sum_j X_{ij}|\psi_j(0)\rangle \tag{98}$$

where $X_{ij} = (\Delta t^2/2\mu)\Lambda_{ij}$. The multipliers, X_{ij}, are determined such that Eq. (97) is satisfied. Substituting Eq. (98) into Eq. (97) yields a matrix equation for the Lagrange multipliers:

$$\mathrm{XX}^\dagger + \mathrm{XB} + \mathrm{B}^\dagger\mathrm{X}^\dagger + \mathrm{A} = \mathrm{I} \tag{99}$$

where $A_{ij} = \langle\tilde{\psi}_i|\tilde{\psi}_j\rangle$ and $B_{ij} = \langle\psi_i(0)|\tilde{\psi}_j\rangle$. Noting that $\mathrm{A} = I + O(\Delta t^2)$ and $\mathrm{B} = I + O(\Delta t)$, the matrix equation can be solved iteratively via

$$\mathrm{X}_{n+1} = \frac{1}{2}\left[\mathrm{I} - \mathrm{A} + \mathrm{X}_n(\mathrm{I} - \mathrm{B}) + (I - \mathrm{B}^\dagger)X_n^\dagger - \mathrm{X}_n^2\right] \tag{100}$$

starting from an initial guess

$$\mathrm{X}_0 = \frac{1}{2}(\mathrm{I} - \mathrm{A}) \tag{101}$$

Once the matrix X_{ij} is obtained, the orbitals are updated using Eq. (98) and an orbital velocity update

$$|\dot{\psi}_i^{(2)}\rangle = |\dot{\psi}_i^{(1)}\rangle + \frac{1}{\Delta t}\sum_j X_{ij}|\psi_j(0)\rangle \tag{102}$$

is performed.

At this point, the new orbital and nuclear forces, $|\varphi_i(\Delta t)\rangle$ and $\mathbf{F}_I(\Delta t)$, are calculated, and a velocity update of the form

$$\begin{aligned}
|\dot{\psi}_i^{(3)}\rangle &= |\dot{\psi}_i^{(2)}\rangle + \frac{\Delta t}{2\mu}|\varphi_i(\Delta t)\rangle \qquad i = 1, \ldots, n_{\mathrm{occ}} \\
\dot{\mathbf{R}}_I(\Delta t) &= \dot{\mathbf{R}}_I(\Delta t/2) + \frac{\Delta t}{2M_I}\mathbf{F}_I(\Delta t)
\end{aligned} \tag{103}$$

is performed. Again, we do not have the final orbital velocities until an appropriate constraint force is applied. For the velocities, the appropriate force is the first time derivative of the orthogonality constraint:

$$\langle\psi_i(\Delta t)|\dot{\psi}_j(\Delta t)\rangle + \langle\dot{\psi}_i(\Delta t)|\psi_j(\Delta t)\rangle = 0 \tag{104}$$

where

$$|\dot{\psi}_i(\Delta t)\rangle = |\dot{\psi}_i^{(3)}\rangle + \sum_j Y_{ij}|\psi_i(\Delta t)\rangle \tag{105}$$

and Y_{ij} are a new set of Lagrange multipliers for enforcing the condition Eq. (104). Substituting Eq. (105) into Eq. (104) gives a simple solution for Y_{ij}:

$$\mathrm{Y} = -\frac{1}{2}(\mathrm{C}+\mathrm{C}^{\dagger}) \tag{106}$$

where $C_{ij} = \langle\psi_i(\Delta t)|\dot{\psi}_i^{(3)}\rangle$. Given the matrix, Y_{ij}, the final orbital velocities are obtained via Eq. (105).

Under certain circumstances, the orthonormality constraint can become dependent on atomic positions, in which case, the basic CP algorithm acquires another level of complexity. The position dependence can come about, for example, when Gaussian basis sets are used. It can also occur when using the so called ultrasoft pseudopotential scheme [120, 140, 141], in which the standard orthogonality constraint is replaced by a more general constraint of the form

$$\langle\psi_i|\hat{B}(\mathbf{R})|\psi_j\rangle = \delta_{ij} \tag{107}$$

resulting from a relaxation of the norm-conservation condition. Here, $\hat{B}$ is a position-dependent operator required in the pseudopotential formulation. For details of this scheme, the interested reader is referred to Ref. [120, 140, 141]. Defining the general overlap matrix as $S_{ij}(\mathbf{R}) = \langle\psi_i|\hat{B}(\mathbf{R})|\psi_j\rangle$, which reduces to the ordinary overlap when $\hat{B} = I$, it is clear that whenever S is position dependent, the CP equations need to be modified to read [140–142]

$$\begin{aligned}
\mu|\ddot{\psi}_i\rangle &= -\frac{\partial E}{\partial\langle\psi_i|} + \sum_{k,j}\Lambda_{kj}\frac{\partial S_{kj}(\mathbf{R})}{\partial\langle\psi_i|} = -\frac{\partial E}{\partial\langle\psi_i|} + \sum_j \Lambda_{ij}\hat{B}(\mathbf{R})|\psi_j\rangle \\
M_I\ddot{\mathbf{R}}_I &= -\frac{\partial E}{\partial\mathbf{R}_I} + \sum_{i,j}\Lambda_{ij}\frac{\partial S_{ij}(\mathbf{R})}{\partial\mathbf{R}_I} = -\frac{\partial E}{\partial\mathbf{R}_I} + \sum_{i,j}\Lambda_{ij}\langle\psi_i|\nabla_I\hat{B}(\mathbf{R})|\psi_j\rangle
\end{aligned} \tag{108}$$

which includes a contribution to the nuclear equation of motion from the constraint. Because the constraint is now coupled to both the electronic and nuclear equations of motion, it is necessary to iterate the constraint procedure through the nuclear update [141, 142], which can have a high computational overhead. One way to overcome this problem is to employ nonorthogonal orbitals as described in the next section.

5.2. NONORTHOGONAL ORBITALS

As alluded to above, the reformulation of the electronic structure problem in terms of a set of nonorthogonal orbitals has a number of advantages in AIMD, in particular, it can simplify the problem of a position-dependent overlap matrix. It can also aid in the control of adiabaticity as will be discussed in Sec. 5.3. The standard orthogonal orbitals $|\psi_i\rangle$ may be transformed to a set of nonorthogonal orbitals $|\phi_i\rangle$ via a transformation of the form

$$|\psi_i\rangle = \sum_j |\phi_j\rangle T_{ji} \tag{109}$$

where the matrix T is defined to be

$$\mathrm{T} = \mathrm{O}^{-1/2} \tag{110}$$

where O is the overlap matrix with respect to the nonorthogonal orbitals:

$$O_{ij} = \langle\phi_i|\hat{B}(\mathbf{R})|\phi_j\rangle \tag{111}$$

Note that if $\hat{\mathrm{B}} = \mathrm{I}$, O is just the standard overlap of the nonorthogonal orbitals. It is easily verified that the this transformation preserves the generalized orthogonality of the original orbitals.

For use in the CP equations of motion, we begin with an extended Lagrangian of the form [142]

$$L = \mu\sum_i \langle\dot{\phi}_i|\dot{\phi}_i\rangle + \frac{1}{2}\sum_I M_I \dot{\mathbf{R}}_I^2 - f\sum_{i,j} M_{ji}\langle\phi_i|H_{\mathrm{KS}}|\phi_j\rangle + \sum_\alpha \lambda_\alpha\sigma_\alpha[\{\phi\}] \tag{112}$$

where $\mathrm{M} = \mathrm{O}^{-1}$ and uniform occupation number, $f_i = f$ have been assumed. The case of unequal occupation numbers is more complicated but can, nevertheless, be formulated as a nonorthogonal orbital scheme. For details, see Ref. [143]. In Eq. (112), the fictitious kinetic energy is expressed directly in terms of the nonorthogonal orbitals, and the energy functional is expressed in terms of the nonorthogonal orbitals via the transformation in Eq. (109). The last term in Eq. (112) involves an arbitrary set of constraint, $\sigma_\alpha[\{\phi\}]$, $\alpha = 1,\ldots,N_c$ enforced by a set of Lagrange multipliers, $\{\lambda_\alpha\}$. The introduction of a constraint into the CP dynamics is done as a means of preventing the orbitals from becoming linearly dependent as the dynamics proceeds. The precise form of the constraint is not particularly important and, therefore, can be chosen to be simpler than the generalized orthogonality constraint obeyed by the original orbitals. This method is known as the constrained nonorthogonal orbital (CNO) approach [142]. Three possibilities are a simple orthogonality constraint:

$$\langle\phi_i|\phi_j\rangle = \delta_{ij} \tag{113}$$

which involves $n_{\mathrm{occ}} \times n_{\mathrm{occ}}$ constraints, a simple norm constraint:

$$\langle\phi_i|\phi_i\rangle = 1 \tag{114}$$

which involves just n_{occ} constraint, or a single constraint of the form

$$\sum_i \langle\phi_i|\phi_i\rangle = n_{\mathrm{occ}} \tag{115}$$

Finally, it is easily verified that the energy derivatives required for the CP equations of motion are

$$\begin{aligned}
\frac{\partial E}{\partial\langle\phi_i|} &= f\sum_j\left(\frac{\partial E}{\partial\langle\psi_i|} - \sum_k \hat{B}|\psi_k\rangle\langle\psi_k|\hat{H}_{\rm KS}|\psi_j\rangle\right)T_{ji} \\
\frac{\partial E}{\partial \mathbf{R}_I} &= f\sum_i\left(\langle\psi_i|\nabla_I\hat{H}_{\rm KS}|\psi_i\rangle - \sum_j\langle\psi_j|\nabla_I\hat{B}|\psi_j\rangle\langle\psi_j|\hat{H}_{\rm KS}|\psi_i\rangle\right)
\end{aligned} \quad (116)$$

expressed in term of matrix elements and derivatives involving the original orbitals. The use of the CNO method with a simple constraint such as the norm constraint condition can yield a non-negligible savings in computational overhead in both the ultrasoft pseudopotential and standard (norm-conserving) pseudopotential ($\hat{B} = I$) schemes [142].

5.3. ADIABATICITY CONTROL THROUGH ISOKINETIC CONSTRAINTS

The CP technique relies heavily on the assumption that an adiabatic separation between the fictitious electron dynamics and the nuclear dynamics can be maintained. In general, this will only be true if the separation between the ground and first excited electronic surfaces, i.e. the band gap, is large compared to kT for all nuclear configurations. While this is generally true for insulators and semi-conductors, it is not true for metals and can be problematic in numerous chemical reactions where the two surfaces approach each other at a transition state. In fact, for metals, the CP dynamics is not correct because the motion does not occur on the ground state surface, although it can, nevertheless, yield some useful information. In such cases, the fictitious thermal energy in the electronic subsystem can lead to excitations that destroy adiabaticity and lead to a rapid exchange of energy between the nuclear and electronic subsystems. This will cause the nuclei to cool and the electrons to heat, and the CP dynamics will cease to be meaningful.

One way to ameliorate this problem is to employ one of the widely used thermostatting methods on the electronic subsystem, such as the Nosé-Hoover [144] or Nose-Hoover chain [145] methods, and several such approaches have been reported [139, 146]. However, these may not be robust enough in certain systems or may allow fluctuations large enough that adiabaticity is lost despite the action of the thermostat. Here, we describe an alternative, highly robust, approach based on the Gaussian isokinetic ensemble method. The adaptation of this method for CP dynamics was recently described by Minary, *et al* [147]. The isokinetic ensemble method employs a nonholonomic constraint to keep the fictitious electronic kinetic energy in the CP Lagrangian fixed:

$$\mu\sum_i\langle\dot{\psi}_i|\dot{\psi}_i\rangle = K_e \quad (117)$$

In order to impose this constraint, an additional Lagrange multiplier, α, is introduced into the CP equations. Thus, in terms of orthogonal orbitals, the new electronic equation of motion reads

$$\mu|\ddot{\psi}_i\rangle = |\varphi_i\rangle + \sum_j \Lambda_{ij}|\psi_j\rangle - \alpha|\dot{\psi}_i\rangle \tag{118}$$

where $|\varphi_i\rangle = \partial E/\partial\langle\psi_i|$. Applying Gauss' principle of least constraint, analytical expressions for the Lagrange multipliers Λ_{ij} and α can be determined analytically by differentiating the orthogonality constraint twice and the isokinetic constraint once yielding

$$\begin{aligned} \langle\psi_i|\ddot{\psi}_j\rangle + 2\langle\dot{\psi}_i|\dot{\psi}_j\rangle + \langle\ddot{\psi}_i|\psi_j\rangle &= 0 \\ \sum_i [\langle\dot{\psi}_i|\ddot{\psi}_i\rangle + \langle\ddot{\psi}_i|\dot{\psi}_i\rangle] &= 0 \end{aligned} \tag{119}$$

Then, using the equation of motion to substitute in for $|\ddot{\psi}_i\rangle$ and its complex conjugate for $\langle\ddot{\psi}_i|$ and solving for the two multipliers, one obtains

$$\begin{aligned} \lambda_{ij} &= \frac{f_i + f_j}{2}\langle\psi_i|\hat{H}_{\mathrm{KS}}|\psi_j\rangle - \mu\langle\dot{\psi}_i|\dot{\psi}_j\rangle \\ \alpha &= \frac{1}{2K_e}\sum_i [\langle\dot{\psi}_i|\hat{H}_{\mathrm{KS}}|\psi_i\rangle + \langle\psi_i|\hat{H}_{\mathrm{KS}}|\dot{\psi}_i\rangle] \end{aligned} \tag{120}$$

Remarkably, each of these expressions is what would be obtained for each multiplier in the absence of the other constraint, showing that the two constraints are completely uncoupled. Substituting Eqs. (120) into the electronic equation of motion yields:

$$\begin{aligned} \mu|\ddot{\psi}_i\rangle &= -f_i\hat{H}_{\mathrm{KS}}|\psi_i\rangle + \frac{f_i + f_j}{2}\sum_j |\psi_i\rangle\langle\psi_i|\hat{H}_{\mathrm{KS}}|\psi_i\rangle \\ &- \mu\sum_j |\psi_j\rangle\langle\dot{\psi}_j|\dot{\psi}_i\rangle + \frac{1}{K_e}|\dot{\psi}_i\rangle\sum_j f_j\,[\langle\dot{\psi}_j|\hat{H}_{\mathrm{KS}}|\psi_j\rangle + \langle\psi_j|\hat{H}_{\mathrm{KS}}|\dot{\psi}_j\rangle] \end{aligned} \tag{121}$$

In principle, Eq. (121) could be solved using a Liouville operator based approach as discussed in Ref. [139]. Alternatively, the analytical expression for α could be used while retaining a numerical approach for Λ_{ij}, however, as was shown in Ref. [147], this requires an iterative procedure. Therefore, the simplest approach is to combine the isokinetic method with the CNO approach outlined in Sec. 5.2. This scheme would be described by an equation of motion of the form

$$\mu|\ddot{\phi}_i\rangle = |\varphi_i^{\mathrm{CNO}}\rangle + \sum_\alpha \lambda_\alpha \frac{\partial\sigma_\alpha}{\partial\langle\phi_i|} + \frac{f}{2K_e}|\dot{\phi}_i\rangle\sum_j [\langle\dot{\phi}_j|\hat{H}_{\mathrm{KS}}|\phi_j\rangle + \langle\phi_j|\hat{H}_{\mathrm{KS}}|\dot{\phi}_j\rangle] \tag{122}$$

where $|\varphi_i^{(\mathrm{CNO})}\rangle$ is the CNO force in Eq. (116). An algorithm for integrating Eq. (122) was recently presented by Minary, *et al* [147].

6. Calculating Observables

Up to now, we have discussed a wide variety of simulation techniques including basis sets, orbital choices, and adiabaticity control methods. All of this methodology would, of course, be useless if one could not compute experimentally measurable observables. In this regard, AIMD simulations have some distinct advantages over force field based MD in that the former permit direct access to the electronic structure and, hence, any observable that can be derived directly from it. Thus, greatly widens the range of observables that can be computed from MD simulations.

In MD calculations, observables are computed by performing averages of appropriate functions, $O(\mathbf{P},\mathbf{R})$ of the momenta and coordinates of the particles in the system. The procedure relies on the ergodic hypothesis which states that given an infinite amount of time, a system will visit all of its accessible phase space so that ensemble averages of $O(\mathbf{P},\mathbf{R})$ can be directly related to time averages of the MD trajectory:

$$\langle O(\mathbf{P},\mathbf{R})\rangle = \lim_{T\to\infty}\frac{1}{T}\int_0^T dt\, O(\mathbf{P}(t),\mathbf{R}(t)) \tag{123}$$

Equation (123) will, therefore, yield any equilibrium average for the system. The ensemble average in Eq. (123) could refer to any pertinent ensemble. For example, a microcanonical (NVE) ensemble average would be given by

$$\langle O(\mathbf{P},\mathbf{R})\rangle = \frac{1}{N!h^{3N}\Omega(N,V,E)}\int d^N\mathbf{P}\, d^N\mathbf{R}\, O(\mathbf{P},\mathbf{R})\delta(H_{\mathrm{N}}(\mathbf{P},\mathbf{R})-E) \tag{124}$$

where $H_{\mathrm{N}}(\mathbf{P},\mathbf{R})$ is the classical nuclear Hamiltonian of Eq. (11) and $\Omega(N,V,E)$ is the microcanonical ensemble partition function. An average in the canonical (NVT) ensemble is given by

$$\langle O(\mathbf{P},\mathbf{R})\rangle = \frac{1}{N!h^{3N}Q(N,V,T)}\int d^N\mathbf{P}\, d^N\mathbf{R}\, O(\mathbf{P},\mathbf{R})e^{-\beta H_{\mathrm{N}}(\mathbf{P},\mathbf{R})} \tag{125}$$

As a concrete examples, note that a radial distribution function is given as an average:

$$\begin{aligned} g(r) &= \frac{1}{4\pi r^2\rho N^2 Q(N,V,T)}\int d^N\mathbf{P}\, d^N\mathbf{R}\sum_{I\neq J}\delta(|\mathbf{R}_I-\mathbf{R}_J|-r)e^{-\beta H_{\mathrm{N}}(\mathbf{P},\mathbf{R})} \\ &= \frac{1}{4\pi\rho r^2}\left\langle \frac{1}{N^2}\sum_{I\neq J}\delta(|\mathbf{R}_I-\mathbf{R}_J|-r)\right\rangle \end{aligned} \tag{126}$$

Similarly, the elastic neutron scattering structure factor can be computed by Fourier transforming the radial distribution function or by directly performing an ensemble (or trajectory) average over $S(\mathbf{k}) = (1/N)|\sum_{I=1}^{N}\exp(i\mathbf{k}\cdot\mathbf{R}_I)|^2$.

Dynamical properties such as spectra and transport coefficients can be obtained within classical linear response theory from time correlation functions. The time correlation function between two observables, $A(\mathbf{P},\mathbf{R})$ and $B(\mathbf{P},\mathbf{R})$ is given by

$$\langle A(0)B(t)\rangle = \frac{1}{Q(N,V,T)}\int d^N\mathbf{P}\, d^N\mathbf{R}\, A(\mathbf{P},\mathbf{R})B(\mathbf{P}_t(\mathbf{P},\mathbf{R}),\mathbf{R}_t(\mathbf{P},\mathbf{R})) \times e^{-\beta H_N(\mathbf{P},\mathbf{R})} \tag{127}$$

where $(\mathbf{P}_t(\mathbf{P},\mathbf{R}),\mathbf{R}_t(\mathbf{P},\mathbf{R}))$ designates the phase space trajectory obtained from the initial condition $(\mathbf{P},\mathbf{R}))$. For $A = B$, Eq. (127) becomes an autocorrelation function. For example, the diffusion coefficient can be computed from the velocity-velocity autocorrelation function:

$$D = \frac{1}{3}\int_0^{\infty} dt\, \frac{1}{N}\sum_{I=1}^{N}\langle \mathbf{V}_I(0)\cdot\mathbf{V}_I(t)\rangle \tag{128}$$

The velocity autocorrelation function can also be used to obtain a frequency spectrum for the system known as the *power spectrum* by Fourier transformation:

$$I(\omega) = \int_{-\infty}^{\infty} dt\, e^{i\omega t}\frac{1}{N}\sum_{i=1}^{N}\langle \mathbf{V}_I(0)\cdot\mathbf{V}_I(t)\rangle \tag{129}$$

Although the power spectrum, itself, it not a directly measurable quantity, the frequencies at which the peaks in the function occur can be compared to other spectroscopic measurements. While this can be a useful comparison, it is obviously preferable to compute, for example, the infrared (IR) or Raman spectrum, which can can be directly measured. For such quantities, having access to the electronic structure is a significant advantage of the AIMD method. Below, we discuss the calculation of several types of spectra based on the use of the electronic structure.

6.1. INFRARED SPECTRUM

In linear response theory, the infrared absorption coefficient, $\alpha(\omega)$, is given by the Fourier transform of the electric dipole moment correlation function

$$\alpha(\omega) = \frac{4\pi\omega\tanh(\beta\hbar\omega/2)}{3\hbar n(\omega)cV}\int_{-\infty}^{\infty} dt\, e^{-i\omega t}\left\langle \frac{1}{2}\sum_{k=x,y,z}[\hat{M}_k(0),\hat{M}_k(t)]_+\right\rangle \tag{130}$$

where $n(\omega)$ is the index of refraction of the medium, V is the volume, c is the speed of light, $\hat{M}_k(t)$ is the kth component of the electric dipole moment operator, and $[\hat{M}_k(0), \hat{M}_k(t)]_+$ is the anticommutator between the operators, $[A,B]_+ = AB + BA$. Of the two terms in the anti-commutator, $\langle \hat{M}_k(t)\hat{M}_k(0)\rangle$ corresponds to an absorption process while $\langle \hat{M}_k(0)\hat{M}_k(t)\rangle$ corresponds to an emission process. In the approximation of classical nuclei, the dipole moment operator is replaced by the classical dipole moment function. However, in order to retain some of quantum information, the factor $\tanh(\beta\hbar\omega/2)$ is retained so that the IR absorption is expressed as

$$\alpha(\omega) = \frac{4\pi\omega\tanh(\beta\hbar\omega/2)}{3\hbar n(\omega)cV} \int_{-\infty}^{\infty} dt\, e^{-i\omega t} \left\langle \sum_{k=x,y,z} \hat{M}_k(0)\hat{M}_k(t) \right\rangle \tag{131}$$

Note that more sophisticated quantum corrections, for example, Egelstaff type corrections [148], can also be applied. The total dipole moment can be decomposed into ionic and electronic contributions according to

$$\mathbf{M} = \mathbf{M}^{(\mathrm{ion})} + \mathbf{M}^{(\mathrm{elec})} \tag{132}$$

Since the ions are treated as classical point particles, the ionic contribution can be computed straightforwardly from the particle positions. The electronic contribution is more subtle. For a non-periodic or cluster system, the expression

$$\begin{aligned} \mathbf{M}^{(\mathrm{elec})} &= -e \int d\mathbf{x}_1 \cdots d\mathbf{x}_{N_e} \phi_0^*(\mathbf{x}_1,\ldots,\mathbf{x}_{N_e}) \left[\sum_{i=1}^{N_e} \mathbf{r}_i\right] \phi_0(\mathbf{x}_1,\ldots,\mathbf{x}_{N_e}) \\ &= -e \int d\mathbf{r}\, n_0(\mathbf{r})\mathbf{r} \end{aligned} \tag{133}$$

where ϕ_0 is the exact ground state wavefunction, $n_0(\mathbf{r})$ is the ground state density, and $\mathbf{r}_i$ is the ith electron position operator, can be evaluated easily because the dipole moment operator is a one-body operator. For periodic systems, however, Eq. (133) is not translationally invariant. For periodic systems, the proper generalization of the dipole moment expression is based on the so called Berry phase approach [149–152] and takes the form

$$\begin{aligned} M_k^{(\mathrm{elec})} &= -e\mathrm{Im}\ \ln \int d\mathbf{x}_1 \cdots d\mathbf{x}_{N_e} \\ &\times\ \phi_0^*(\mathbf{x}_1,\ldots,\mathbf{x}_{N_e}) \left[e^{2\pi i \sum_{i=1}^{N_e} r_{i,k}/L_k}\right] \phi_0(\mathbf{x}_1,\ldots,\mathbf{x}_{N_e}) \end{aligned} \tag{134}$$

where $M_k^{(\mathrm{elec})}$ is the kth component of the electronic contribution to the dipole moment, $r_{i,k}$ is the kth component of the ith electron position operator, and L_k is the length of the supercell in the kth direction (assuming an orthorhombic cell – see Refs. [153, 154] for generalizations to arbitrary cell shapes), which is assumed to be cubic in this case. Clearly, Eq. (134) possesses the correct translational symmetry. However, it introduces an additional complication because the operator $\exp(2\pi i \sum_i r_{i,k}/L)$ is now a many-body operator [151] so that the expectation value in Eq. (134) cannot be expressed simply in terms of the electron density. As was shown in Refs. [149, 150, 155], in order to compute the electronic component of the dipole moment within KS theory, it is necessary to compute the matrix elements

$$R_{ij,k} = \langle \psi_i | e^{2\pi i r_k / L_k} | \psi_j \rangle \tag{135}$$

using the KS orbitals (or their periodic parts when **k**-points other than the Γ-point are used) and compute the dipole moment contribution from

$$M_k^{(\text{elec})} = -\frac{eL_k}{\pi} \text{Im} \ln \det \left(R_{ij,k} \right) \tag{136}$$

It should be noted that the Berry phase approach is also the starting point for methods to find unitary transformations among the orbitals that lead to a maximally set of orbitals known as *Wannier functions* [153–155], which have been shown to be of great utility in studying other electronic properties of a system, for example, local dipole moments [25].

6.2. OTHER SPECTRA

Recently, a general formalism within DFT was introduced to compute the response of a system to a small applied external perturbation [156], which is often what is measured in experiments. The resulting variational DFT perturbation theory, which is in the same spirit as, though conceptually different from, more standard formulations [157–159], leads to a general scheme for computing the second derivative of the energy with respect to an applied field, and, therefore, incorporates observables such as anharmonic Raman spectra [67] and nuclear magnetic resonance (NMR) chemical shifts [160].

In the case of Raman spectroscopy, it is necessary to compute the full polarizability tensor [67, 161]

$$\alpha_{\mu\nu} = -\frac{\partial \mathbf{P}_\mu}{\partial \mathbf{E}_\nu} = \frac{\partial^2 E}{\partial \mathbf{E}_\mu \mathbf{E}_\nu} \qquad \mu, \nu = x, y, z \tag{137}$$

where **P** is the induced polarization vector due to an externally applied electric field, **E**, and E is the total energy. Within linear response theory, the Raman scattering cross section can then be related to the autocorrelation functions of the polarizability tensor. For cubic systems, the full tensor can be expressed as [67]

$$\alpha_{\mu\nu}(t) = \lambda(t)\delta_{\mu\nu} + \beta_{\mu\nu}(t) \tag{138}$$

where $\beta_{\mu\nu}(t)$ is traceless. This subdivision will lead to isotropic and anisotropic spectra given by [67]

$$
\begin{aligned}
I_{\mathrm{iso}}(\omega) &= \frac{N}{2\pi}\int dt\, e^{-i\omega t}\langle\lambda(0)\lambda(t)\rangle \\
I_{\mathrm{aniso}}(\omega) &= \frac{N}{2\pi}\int dt\, e^{-i\omega t}\frac{1}{10}\langle \mathrm{Tr}[\beta(0)\beta(t)]\rangle
\end{aligned}
\tag{139}
$$

Again, the Berry phase formalism can be used to compute the polarization of the system [67].

Finally, it is important to mention that the DFT perturbation theory has also been employed to derive a new approach to the calculation of NMR chemical shifts in periodic systems [160]. NMR spectra are among the most important tools used in chemistry to characterize a chemical environment. Indeed, a variety of novel approaches have been introduced for computing the chemical shifts [28, 160, 162, 163]. Unfortunately, the complexities of this methodology are sufficient that they will not be dealt with here in any amount of detail, but the interested reader is referred to the above mentioned literature. However, in order to give a feel for the problem, we note that the chemical shielding tensor, like the polarizability density, is expressible as the derivative of an induced effect due to an external field. In particular, it is the derivative of the induced local magnetic field due to an externally applied magnetic field, **B**:

$$
\sigma_{\mu\nu}(\mathbf{r}) = \frac{\partial \mathbf{B}_{\mu}^{(\mathrm{ind})}(\mathbf{r})}{\partial \mathbf{B}_{\nu}} \tag{140}
$$

From $\sigma_{\mu\nu}$, the chemical shift tensor $\delta_{\mu\nu}$ can easily be computed. From the eigenvalues of $\delta_{\mu\nu}$, for example, the isotropic and MAS chemical shifts can be determined. The induced field is determined by the total electronic current $\mathbf{j}(\mathbf{r})$ via

$$
\mathbf{B}^{(\mathrm{ind})}(\mathbf{r}) = \frac{\mu_0}{4\pi}\int d\mathbf{r}' \frac{(\mathbf{r}-\mathbf{r}')}{|\mathbf{r}-\mathbf{r}'|} \times \mathbf{j}(\mathbf{r}) \tag{141}
$$

where μ_0 is the permeability of free space. For a Hamiltonian based formalism, the applied magnetic field, **B**, is represented in terms of a vector potential, $\mathbf{A}(\mathbf{r})$ with $\mathbf{B} = \nabla \times \mathbf{A}(\mathbf{r})$. Thus, a gauge choice must be made for $\mathbf{A}(\mathbf{r})$, and a typical form is [160]

$$
\mathbf{A}(\mathbf{r}) = -\frac{1}{2}(\mathbf{r}-\mathbf{R}) \times \mathbf{B} \tag{142}
$$

where **R** is known as the gauge origin. Thus, it can be seen that the position operator problem in periodic systems is prevalent here as well. In order to solve the problem, Mauri, *et. al* proposed modulating **B** by a periodic function [28, 164–168], whereas Sebastiani and Parrinello have applied the Berry phase approach [160]. Another technical difficulty that arises in a plane-wave basis stems from the fact that the chemical shifts are very sensitive to the shape of the wavefunction in the core region, which is pseudized. This problem has been addressed by Gregor, *et al* [162], who have developed a set of additive constant corrections that accurately reproduce the all-electron magnetic shieldings.

7. Summary

Basic concepts and several extensions of the Car-Parrinello *ab initio* molecular dynamics algorithm have been discussed. In particular, the algorithm has been derived starting from the Born-Oppenheimer approximation, the use of density functional theory has been discussed as well as basis set choices, and propagation algorithms. Finally, the problem of computing observables, including different types of spectra has been overviewed. Other extensions of the algorithm not discussed here include the combination of Car-Parrinello with path integral molecular dynamics [15–18], excited states [169, 170], and linear scaling methods [171–174]. For further details, the reader is referred to recent review [10, 12]. We hope the present discussion serves as a useful starting point for those interested in becoming practitioners and developers in the field of *ab initio* molecular dynamics.

8. Acknowledgements

The author gratefully acknowledges his colleagues throughout the world for many longstanding and fruitful collaborations, in particular, Michele Parrinello, ETH Zürich and Swiss Center for Scientific Computing, Dominik Marx, Ruhr-Universität, Bochum, Jürg Hutter, University of Zürich, Magali Benoit, Université Montpellier, and Glenn Martyna, IBM, Yorktown Heights, NY. The author also acknowledges the members of his research group at New York University, Yi Liu, Zhongwei Zhu, Lula Rosso, Peter Minary, and Joseph A. Morrone for their hard work and dedication. Much of the original work presented herein was supported by NSF CHE-9875824, NSF CHE-0121375, the Research Corporation RI0218, an NYU Whitehead Fellowship in Biomedical and Biological Sciences, and the Camille and Henry Dreyfus Foundation, Inc. TC-02-012.

References

1. S. W. Rick and S. J. Stuart, Rev. Comp. Chem. **18**, (2002).
2. A. Warshel and R. M. Weiss, J. Am. Chem. Soc. **102**, 6218 (1980).
3. R. Car and M. Parrinello, Phys. Rev. Lett. **55**, 2471 (1985).
4. D. K. Remler and P. A. Madden, Mol. Phys. **70**, 921 (1990).
5. M. C. Payne, M. P. Teter, D. C. Allan, T. A. Arias, and J. D. Joannopoulos, Rev. Mod. Phys. **64**, 1045 (1992).

6. G. Galli and M. Parrinello, Computer simulation in chemical physics, NATO ASI Series C **397**, 261 (1993).
7. M. E. Tuckerman, P. J. Ungar, T. von Rosenvinge, and M. L. Klein, J. Phys. Chem. **100**, 12878 (1996).
8. M. J. Gillan, Contemp. Phys. **38**, 115 (1997).
9. M. Parrinello, Solid State Commun. **102**, 107 (1997).
10. D. Marx and J. Hutter, In *Modern Methods and Algorithms of Quantum Chemistry* J. Grotendorst, ed. (PUBLISHER, Forschungszentrum, Juelich, NIC Series Vol. 1, 2000), pp. 301–449.
11. R. Car, Quant. Struct. Act. Rel. **21**, 97 (2002).
12. M. E. Tuckerman, J. Phys. Condens. Matter **14** (2002).
13. R. P. Feynman and A. R. Hibbs, *Quantum Mechanics and Path Integrals* (McGraw-Hill, New York, 1965).
14. R. Feynman, *Statistical Mechanics*. (Benjamin, Reading, (1972)).
15. D. Marx and M. Parrinello, Z. Phys. B **95**, 143 (1994).
16. D. Marx and M. Parrinello, J. Chem. Phys. **104**, 4077 (1996).
17. M. Tuckerman, D. Marx, M. L. Klein, and M. Parrinello, J. Chem. Phys. **104**, 5579 (1996).
18. D. Marx, M. E. Tuckerman, and G. J. Martyna, Comp. Phys. Comm. **118**, 166 (1999).
19. K. Laasonen, M. Sprik, M. Parrinello, and R. Car, J. Chem. Phys. **99**, 9080 (1993).
20. E. S. Fois, M. Sprik, and M. Parrinello, Chem. Phys. Lett. **223**, 411 (1994).
21. M. Sprik, J. Hutter, and M. Parrinello, J. Chem. Phys. **105**, 1142 (1996).
22. P. L. Silvestrelli, M. Bernasconi, and M. Parrinello, Chem. Phys. Lett. **277**, 478 (1997).
23. B. L. Trout and M. Parrinello, Chem. Phys. Lett. **288**, 343 (1998).
24. L. D. Site, A. Alavi, and R. M. Lynden-Bell, Mol. Phys. **96**, 1683 (1999).
25. P. L. Silvestrelli and M. Parrinello, Phys. Rev. Lett. **82**, 3308 (1999).
26. M. Sprik, Chem. Phys. **258**, 139 (2000).
27. M. Krack and M. Parrinello, Phys. Chem. Chem. Phys. **2**, 2105 (2000).
28. B. G. Pfrommer, F. Mauri, and S. G. Louie, J. Am. Chem. Soc. **122**, 123 (2001).
29. P. L. Geissler, C. Dellago, D. Chandler, J. Hutter, and M. Parrinello, Science **291**, 2121 (2001).
30. E. Schwegler, G. Galli, F. Gygi, and R. Q. Hood, Phys. Rev. Lett. **87**, 265501 (2001).
31. S. Izvekov and G. A. Voth, J. Chem. Phys. **116**, 10372 (2002).
32. M. Diraison, G. J. Martyna, and M. E. Tuckerman, J. Chem. Phys. **111**, 1096 (1999).
33. E. Tsuchida, Y. Kanada, and M. Tsukada, Chem. Phys. Lett. **311**, 236 (1999).
34. Y. Liu and M. E. Tuckerman, J. Phys. Chem. B **105**, 6598 (2001).
35. J. A. Morrone and M. E. Tuckerman, J. Chem. Phys. **117**, 4403 (2002).
36. J. A. Morrone and M. E. Tuckerman, Chem. Phys. Lett. (submitted).
37. K. Laasonen and M. L. Klein, J. Am. Chem. Soc. **116**, 11620 (1994).
38. K. Laasonen and M. L. Klein, Mol. Phys. **88**, 135 (1996).
39. M. Sprik, J. Phys. Condensed Matter **8**, 9405 (1996).
40. K. Laasonen and M. L. Klein, J. Phys. Chem. A **101**, 98 (1997).
41. E. J. Meijer and M. Sprik, J. Am. Chem. Soc. **120**, 6345 (1998).
42. D. Kim and M. L. Klein, J. Phys. Chem. B **104**, 10074 (2000).
43. Z. Zhu and M. E. Tuckerman, J. Phys. Chem. B **106**, 8009 (2002).
44. B. Chen, J. M. Park, I. Ivanov, G. Tabacchi, M. L. Klein, and M. Parrinello, J. Am. Chem. Soc. **124**, 8534 (2002).
45. M. E. Tuckerman, K. Laasonen, M. Sprik, and M. Parrinello, J. Phys. Chem. **99**, 5749 (1995).
46. M. E. Tuckerman, K. Laasonen, M. Sprik, and M. Parrinello, J. Chem. Phys. **103**, 150 (1995).
47. D. Marx, M. E. Tuckerman, J. Hutter, and M. Parrinello, Nature **367**, 601 (1999).
48. D. Marx, M. E. Tuckerman, and M. Parrinello, J. Phys. Condens. Matt. **12**, A153 (2000).
49. M. E. Tuckerman, D. Marx, and M. Parrinello, Nature **417**, 925 (2002).
50. D. E. Sagnella, K. Laasonen, and M. L. Klein, Biophys. J. **71**, 1172 (1996).
51. H. S. Mei, M. E. Tuckerman, D. E. Sagnella, and M. L. Klein, J. Phys. Chem. B **102**,

10446 (1998).
52. M. Pavese, D. R. Berard, and G. A. Voth, Chem. Phys. Lett. **300**, 93 (1999).
53. L. Rosso and M. E. Tuckerman, J. Am. Chem. Soc. (submitted).
54. K. Laasonen, M. Parrinello, R. Car, C. Y. Lee, and D. Vanderbilt, Chem. Phys. Lett. **207**, 208 (1993).
55. K. Laasonen and M. L. Klein, J. Phys. Chem. **98**, 10079 (1994).
56. M. E. Tuckerman, D. Marx, M. L. Klein, and M. Parrinello, Science **275**, 817 (1997).
57. H. Arstila, K. Laasonen, and A. Laaksonen, J. Chem. Phys. **108**, 1031 (1998).
58. P. L. Geissler, C. Dellago, D. Chandler, J. Hutter, and M. Parrinello, Chem. Phys. Lett. **321**, 225 (2000).
59. C. Y. Lee, D. Vanderbilt, K. Laasonen, R. Car, and M. Parrinello, Phys. Rev. Lett. **69**, 462 (1992).
60. C. Y. Lee, D. Vanderbilt, K. Laasonen, R. Car, and M. Parrinello, Phys. Rev. B **47**, 4863 (1993).
61. M. Benoit, M. Bernasconi, and M. Parrinello, Phys. Rev. Lett. **76**, 2934 (1996).
62. M. Bernasconi, M. Benoit, M. Parrinello, G. L. Chiarotti, P. Focher, and E. Tosatti, Physica Scripta A **166**, 98 (1996).
63. M. Bernasconi, P. L. Silvestrelli, and M. Parrinello, Phys. Rev. Lett. **81**, 1235 (1998).
64. M. Benoit, D. Marx, and M. Parrinello, Nature **392**, 258 (1998).
65. M. Benoit, D. Marx, and M. Parrinello, Solid State Ionics **125**, 23 (1999).
66. Z. F. Liu, C. K. Siu, and J. S. Tse, Chem. Phys. Lett. **309**, 335 (1999).
67. A. Putrino and M. Parrinello, Phys. Rev. Lett. **88**, 176401 (2002).
68. J. Sarnthein, A. Pasquarello, and R. Car, Science **275**, 1925 (1997).
69. A. Pasquarello and R. Car, Phys. Rev. Lett. **79**, 1766 (1997).
70. M. Boero, A. Pasquarello, J. Sarnthein, and R. Car, Phys. Rev. Lett. **78**, 887 (1997).
71. A. Pasquarello, J. Sarnthein, and R. Car, Phys. Rev. B **57**, 14133 (1998).
72. A. Pasquarello and R. Car, Phys. Rev. Lett. **80**, 5145 (1998).
73. C. Massobrio, A. Pasquarello, and R. Car, J. Am. Chem. Soc. **121**, 2943 (1999).
74. F. Mauri, A. Pasquarello, B. G. Pfrommer, Y. G. Yoon, and S. G. Louie, Phys. Rev. B **62**, R4786 (2000).
75. M. Benoit, S. Ispas, and M. E. Tuckerman, Phys. Rev. B **64**, 224205 (2001).
76. C. J. Pickard and F. Mauri, Phys. Rev. Lett. **88**, 086403 (2002).
77. M. Boero, M. Parrinello, and K. Terakura, J. Am. Chem. Soc. **120**, 2746 (1998).
78. M. Boero, M. Parrinello, S. Hueffer, and H. Weiss, J. Am. Chem. Soc. **122**, 501 (2000).
79. M. Boero, M. Parrinello, H. Weiss, and S. Hueffer, J. Phys. Chem. A **105**, 5096 (2001).
80. K. C. Haas, W. F. Schneider, A. Curioni, and W. Andreoni, Science **282**, 265 (1998).
81. C. Stampfl and M. Scheffler, Surf. Sci. **435**, 119 (2000).
82. K. C. Haas, W. F. Schneider, A. Curioni, and W. Andreoni, J. Phys. Chem. B **104**, 5527 (2000).
83. C. Stampfl, M. V. Ganduglia-Pirovano, K. Reuter, and M. Scheffler, Surf. Sci. **500**, 368 (2002).
84. G. J. Kroes, A. Gross, E. J. Baerends, M. Scheffler, and D. A. McCormack, Acc. Chem. Res. **35**, 193 (2002).
85. M. Saitta and M. L. Klein, Nature **399**, 46 (1999).
86. M. Saitta and M. L. Klein, J. Chem. Phys. **111**, 9434 (1999).
87. M. Saitta and M. L. Klein, J. Am. Chem. Soc. **121**, 11827 (1999).
88. M. Saitta and M. L. Klein, J. Phys. Chem. B **105**, 6495 (2001).
89. S. Piana, D. Sebastiani, P. Carloni, and M. Parrinello, J. Am. Chem. Soc. **123**, 8730 (2001).
90. J. Hutter, P. Carloni, and M. Parrinello, J. Am. Chem. Soc. **118**, 871 (1996).
91. U. Roethlisberger and P. Carloni, Intl. J. Quant. Chem. **73**, 209 (1999).
92. W. Andreoni, A. Curioni, and T. Mordasini, IBM J. Res. and Development **45**, 397 (2001).
93. C. Rovira and M. Parrinello, Intl. J. Quant. Chem. **80**, 1172 (2000).
94. C. Rovira, B. Schulze, M. Eichinger, J. D. Evanseck, and M. Parrinello, Biophys. J. **81**, 435 (2001).
95. W. Kohn and L. J. Sham, Phys. Rev. **140**, A1133 (1965).

96. R. G. Parr and W. Yang, *Density Functional Theory of atoms and molecules* (Oxford University Press, Oxford, 1989).
97. R. M. Dreizler and E. K. U. Gross, *Density Functional Theory* (Springer-Verlag, Berlin/Heidelberg, 1990).
98. Z. H. Liu, L. E. Carter, and E. A. Carter, J. Phys. Chem. **99**, 4355 (1995).
99. B. D. Martino, M. Celino, and V. Rosato, Comp. Phys. Comm. **120**, 255 (1999).
100. R. A. Friesner, Chem. Phys. Lett. **116**, 39 (1985).
101. G. Lippert, J. Hutter, and M. Parrinello, Mol. Phys. **92**, 477 (1997).
102. Y. Liu and M. E. Tuckerman, Phys. Rev. Lett. (submitted).
103. P. Hohenberg and W. Kohn, Phys. Rev. B **136**, 864 (1964).
104. A. D. Becke, Phys. Rev. A **38**, 3098 (1988).
105. W. Y. C. Lee and R. C. Parr, Phys. Rev. B **37**, 785 (1988).
106. J. P. Perdew and Y. Wang, Phys. Rev. B **45**, 13244 (1992).
107. J. P. Perdew, K. Burke, and M. Ernzerhof, Phys. Rev. Lett. **77**, 3865 (1996).
108. N. C. Handy and A. J. Cohen, Mol. Phys. **99**, 403 (2001).
109. A. J. Cohen and N. C. Handy, Mol. Phys. **99**, 607 (2001).
110. N. C. Handy and A. J. Cohen, J. Chem. Phys. **116**, 5411 (2002).
111. Q. Wu and W. Yang, J. Chem. Phys. **116**, 515 (2002).
112. A. D. Becke, J. Chem. Phys. **96**, 2155 (1992).
113. A. D. Becke and M. R. Roussel, Phys. Rev. A **39**, 3761 (1989).
114. A. D. Becke, J. Chem. Phys. **112**, 4020 (2000).
115. E. Proynov, H. Chermette, and D. R. Salahub, J. Chem. Phys. **113**, 10013 (2000).
116. M. Ernzerhof, S. N. Maximoff, and G. E. Scuseria, J. Chem. Phys. **116**, 3980 (2002).
117. J. A. White and D. M. Bird, Phys. Rev. B **50**, 4954 (1994).
118. G. Bachelet, D. Hamann, and M. Schluter, Phys. Rev. B **26**, 4199 ((1982)).
119. N. Troullier and J. L. Martins, Phys. Rev. B **43**, 1993 (1991).
120. D. Vanderbilt, Phys. Rev. B **41**, 7892 (1990).
121. P. E. Bloechl, Phys. Rev. B **50**, 17953 (1994).
122. L. Kleinman and D. M. Bylander, Phys. Rev. Lett. **48**, 1425 (1982).
123. X. Gonze, P. Kaeckell, and M. Scheffler, Phys. Rev. B **41**, 12264 (1990).
124. X. Gonze, R. Stumpf, and M. Scheffler, Phys. Rev. B **44**, 1991 (1991).
125. M. E. Tuckerman and G. J. Martyna, (To be submitted).
126. R. W. Hockney, Phys. Rev. B **48**, 2081 (1993).
127. R. N. Barnett and U. Landmann, Methods Comput. Phys. **9**, 136 (1978).
128. G. Martyna and M. Tuckerman, J. Chem. Phys. **110**, 2810 (1999).
129. P. Minary, M. E. Tuckerman, K. A. Pihakari, and G. J. Martyna, J. Chem. Phys. **116**, 5351 (2002).
130. M. E. Tuckerman, P. Minary, K. A. Pihakari, and G. J. Martyna, In *Computational Methods for Macromolecules: Challenges and Applications* T. Schlick and H. H. Gan, eds. (PUBLISHER, Springer, Berlin, 2002), p. 381.
131. J. J. Mortensen and M. Parrinello, J. Phys. Chem. B **104**, 2901 (2000).
132. J. C. Light, I. P. Hamilton, and J. V. Lill, J. Chem. Phys. **82**, 1400 (1985).
133. J. T. Muckerman, Chem. Phys. Lett. **173**, 200 (1990).
134. D. T. Colbert and W. H. Miller, J. Chem. Phys. **96**, 1982 (1992).
135. R. G. Littlejohn, M. Cargo, T. Carrington, K. A. Mitchell, and B. Poirier, J. Chem. Phys. **116**, 8691 (2002).
136. S. Guerin and H. R. Jauslin, Comp. Phys. Comm. **121-122**, 496 (1999).
137. L. Rosso, P. Minary, Z. Zhu, and M. E. Tuckerman, J. Chem. Phys. **116**, 4389 (2002).
138. Although the problem could just as well be formulated in terms of an extended Hamiltonian as in the simple *x-y* model, we prefer to use the Lagrangian formulation as in the original CP paper [3].
139. M. E. Tuckerman and M. Parrinello, J. Chem. Phys. **101**, 1301 (1994).
140. K. Laasonen, R. Car, C. Lee, and D. Vanderbilt, Phys. Rev. B **43**, 6796 (1991).
141. K. Laasonen, A. Pasquarello, R. Car, C. Lee, and D. Vanderbilt, Phys. Rev. B **47**, 10142 (1993).

142. J. Hutter, M. E. Tuckerman, and M. Parrinello, J. Chem. Phys. **102**, 859 (1995).
143. M. Tuckerman and G. J. Martuna (to be submitted).
144. W. Hoover, Phys. Rev. A **31**, 1695 (1985).
145. G. Martyna, M. Klein, and M. Tuckerman, J. Chem. Phys. **97**, 2635 (1992).
146. P. Blochl and M. Parrinello, Phys. Rev. B **45**, 9413 (1991).
147. P. Minary, G. J. Martyna, and M. E. Tuckerman, J. Chem. Phys. (submitted).
148. P. A. Egelstaff, Adv. Phys. **11**, 203 (1962).
149. R. D. King-Smith and D. Vanderbilt, Phys. Rev. B **47**, 1651 (1993).
150. R. Resta, Rev. Mod. Phys. **66**, 899 (1994).
151. R. Resta, Phys. Rev. Lett. **80**, 1800 (1998).
152. R. Resta, J. Phys. Condens. Matter **14**, R625 (2002).
153. P. L. Silvestrelli, Phys. Rev. B **59**, 9703 (1999).
154. G. Berghold, C. J. Mundy, A. H. Romero, J. Hutter, and M. Parrinello, Phys. Rev. B **61**, 10040 (2000).
155. N. Marzari and D. Vanderbilt, Phys. Rev. B **56**, 12847 (1997).
156. A. Putrino, D. Sebastiani, and M. Parrinello, J. Chem. Phys. **113**, 7102 (2000).
157. S. Baroni, P. Gianozzi, and A. Testa, Phys. Rev. Lett. **58**, 1861 (1985).
158. X. Gonze and J. P. Vigneron, Phys. Rev. B **39**, 13120 (1989).
159. X. Gonze, Phys. Rev. A **52**, 1096 (1995).
160. D. Sebastiani and M. Parrinello, J. Phys. Chem. A **105**, 1951 (2001).
161. B. J. Berne and R. Pecora, *Dynamic Light Scattering* (John Wiley and Sons, Inc., New York, 1976).
162. T. Gregor, F. Mauri, and R. Car, J. Chem. Phys. **111**, 1815 (1999).
163. C. J. Pickard and F. Mauri, Phys. Rev. B **63**, 245101 (2001).
164. F. Mauri and S. Louie, Phys. Rev. Lett. **76**, 4246 (1996).
165. F. Mauri, B. Pfrommer, and S. Louie, Phys. Rev. Lett. **77**, 5300 (1996).
166. F. Mauri, B. Pfrommer, and S. Louie, Phys. Rev. Lett. **79**, 2340 (1997).
167. Y. Yoon, B. Pfrommer, F. Mauri, and S. Louie, Phys. Rev. Lett. **80**, 3388 (1998).
168. F. Mauri, B. Pfrommer, and S. Louie, Phys. Rev. B **60**, 2941 (1999).
169. A. Alavi, J. Kohanoff, M. Parrinello, and D. Frenkel, Phys. Rev. Lett. **73**, 2599 (1994).
170. N. L. Doltsinis and D. Marx, Phys. Rev. Lett. **88**, 166402 (2002).
171. G. Galli and M. Parrinello, Phys. Rev. Lett. **69**, 3547 (1992).
172. X. P. Li, R. W. Nunes, and D. Vanderbilt, Phys. Rev. B **48**, 14646 (1993).
173. G. Galli and F. Mauri, Phys. Rev. B **50**, 4316 (1994).
174. D. R. Bowler, T. Miyazaki, and M. J. Gillan, J. Phys. Condens. Matter **14**, 2781 (2002).

CONCEPTS OF IONIC SOLVATION

HARTMUT KRIENKE
Institut für Physikalische und Theoretische Chemie der Universität Regensburg, D-93040 Regensburg, Germany

Abstract: The general features of ion solvation, of the solvent structure around solutes, and their influence on solution properties, are discussed in the framework of simple theoretical approaches, starting from classical molecular models. Ionic solvation is studied with integral equation methods. Analytical solutions for pair correlations and thermodynamics are discussed for the case of solution of charged hard spheres in polarizable dipolar hard spheres in the framework of the MSA and LIN approximations.

1. INTRODUCTION

Polar polarizable solvents and the solution of salts in these systems are the subject of this research. Calculations on the molecular Born - Oppenheimer (BO) level permits to study in detail the influence of the molecular structure of the solvent on solvation phenomena. Thermodynamic and dielectric properties of the solvents are derived on this level. The solvents studied are representatives of different classes, belonging to three types of solvents [3,4] : (i) protic solvents, (ii) dipolar aprotic solvents and (iii) low polarity and inert solvents.

We consider anisotropic molecules and ions with multipolar interactions. Suitable models for the calculations in the framework of classical statistical mechanics are interaction site models (ISM). The potential energy $U(1,\ldots,N)$ is decomposed into a sum of intermolecular pair interactions $U_{\alpha\beta}(12)$.

$$U(1,\ldots,N) = \sum_{\alpha\beta} U_{\alpha\beta}(12) \tag{1}$$

Generally, two-particle configurations are characterized by the intercenter vector $\vec{r}_{12}$ and sets $\vec{\Omega}_1$ and $\vec{\Omega}_2$ of the three Eulerian angles $(\alpha_i, \beta_i, \gamma_i)$ $(i = 1,2)$. In the ISM the interaction between two molecules $U_{\alpha\beta}(12) = U_{\alpha\beta}(\vec{r}_{12}, \vec{\Omega}_1, \vec{\Omega}_2)$, is decomposed into a sum of spherically symmetric interactions between interaction centers (sites) of the two particles.

$$U(12) = \sum_{ij} u_{ij}(r) \tag{2}$$

Interactions of partial charges, repulsion and Van der Waals (VdW) dispersion interactions are described by Lennard-Jones - Coulomb (LJC) potentials between the sites.

J. Samios and V.A. Durov (eds.), Novel Approaches to the Structure and Dynamics of Liquids: Experiments, Theories and Simulations, 93–110.

$$u_{ij}(r) = 4\varepsilon_{ij}\left[\left(\frac{\sigma_{ij}}{r}\right)^{12} - \left(\frac{\sigma_{ij}}{r}\right)^{6}\right] + \frac{z_i z_j e^2}{4\pi\varepsilon_0 r} \tag{3}$$

Molecular geometries, partial charges are available from experiments and from quantummechanical calculations. The pair potential parameters $\sigma_{ij}, \varepsilon_{ij}$ of the ISM may be estimated either from quantummechanical supermolecule calculations or from experimental fits. Lorentz - Berthelot rules are often used to combine interactions of different sites:

$$\sigma_{ij} = \frac{1}{2}(\sigma_{ii} + \sigma_{jj}) \quad ; \quad \varepsilon_{ij} = \sqrt{\varepsilon_{ii}\varepsilon_{jj}} \tag{4}$$

For analytical calculations simpler potential models are used such as hard spheres or Lennard - Jones particles with or without point multipoles on it.

2. MOLECULAR PAIR CORRELATION FUNCTIONS

2.1 Definitions

From the interaction potentials molecular pair correlation functions are calculated. These functions are defined as integrals of the Boltzmann factor of the potential energy U_N over $N-2$ sets of molecular coordinates

$$g(12) = \frac{V^2\Omega^2}{Q_N}\int\ldots\int \exp\left[-\frac{U_N(1,\ldots,N)}{kT}\right] d(3)\ldots d(N) \tag{5}$$

Q_N is the configuration integral of the molecular system,

$$Q_N = \int\ldots\int \exp\left[-\frac{U_N(1,\ldots,N)}{kT}\right] d(1)\ldots d(N) \tag{6}$$

with $\Omega = 4\pi$ for linear molecules and $\Omega = 8\pi^2$ for nonlinear molecules. The molecular pair correlation function is at least a function of 4 variables. The different possibilities of representation are (i) expansion of the correlation functions into rotational invariants [9] and representation of the r-dependent expansion coefficients

$$g(12) = \sum_{\mu\nu}\sum_{mnl} g_{\mu\nu}^{mnl}(r_{12})\Phi_{\mu\nu}^{mnl}(12) \tag{7}$$

(ii) presentation of site - site correlation functions $g_{ij}(r)$ calculated from the molecular pair correlation functions by

$$g_{ij}(r) = \frac{1}{(8\pi^2)^2}\int\int g(12)\delta[r_{ij}(12) - r]d(1)d(2) \tag{8}$$

Results for the pair correlation functions $g(12)$ in polar solvents and solutions based on molecular integral equation theories and on simulations are discussed in several papers [24,14,25,31,32,26].

2.2 Correlation functions and screening

In classical many-particle systems the intermolecular pair potential $u_{\alpha\beta}(12)$ may be conceived as a sum of two parts

$$u_{\alpha\beta}(12) = u_{\alpha\beta}^{\mathrm{SR}}(12) + u_{\alpha\beta}^{\mathrm{LR}}(12) \tag{9}$$

where $u_{\alpha\beta}^{\mathrm{SR}}(12)$ is the short-range and $u_{\alpha\beta}^{\mathrm{LR}}(12)$ is the long-range contribution to the interaction of two particles of species α and β, respectively. We assume that the long-range interactions are of electrostatic character. The short-range interactions make up the reference system describing the mutual impenetrability of classical particles as well as the short-range attractive van der Waals forces. Long- and short-range interactions exert different effects on the properties of the system demanding different approaches.

The calculation of the pair correlation function is performed by integral equation or by simulation methods [4]. We use the Ornstein - Zernike formalism introducing the concept of the direct correlation function $c_{\alpha\beta}(12)$.

The long-range forces are introduced by a perturbation theory [12,19] with renormalized chain sums with hypervertices into which the static structure factor of the reference system is introduced [16,2,36]. The corresponding OZ equations and closure relations for the correlation functions of the system are

$$h_{\alpha\beta}(12) = c_{\alpha\beta}(12) + \sum_{\gamma} \rho_{\gamma} \int c_{\alpha\gamma}(13) h_{\gamma\beta}(32)\, d(3) \tag{10}$$

$$g_{\alpha\beta}(12) = \exp[-\beta u_{\alpha\beta}(12) + h_{\alpha\beta}(12) - c_{\alpha\beta}(12) + E_{\alpha\beta}(12)] \tag{11}$$

$E_{\alpha\beta}(12)$ is the sum of bridge or elementary graphs.

The subdivision of the potential in Equation (9) entails the subdivision of the direct and total correlation functions into parts of the reference system, $c_{\alpha\beta}^{0}(12)$ and $h_{\alpha\beta}^{0}(12)$, depending only from the short range interactions $u_{\alpha\beta}^{\mathrm{SR}}(12)$, and into remaining parts.

It is assumed that $c_{\alpha\beta}^{0}(12)$ and $h_{\alpha\beta}^{0}(12)$ are known in their entire definition range. For simple approximations of the reference interactions (e.g. hard spheres or LJ particles) there are analytical or simple numerical solutions for these functions (some of the solution methods will be discussed in the next subsection). In that case one also assumes that information about the sum of bridge graphs of the reference system $E_{\alpha\beta}^{0}(12)$ is available.

The remaining parts consist of the long-range contributions to the direct correlation function (given by the long- range parts of the interactions $-\beta u_{\alpha\beta}^{\mathrm{LR}}(12)$, $\beta = 1/k_B T$), and screened potentials $G_{\alpha\beta}(12)$, and of the correction terms $\delta c_{\alpha\beta}(12)$ and $\delta h_{\alpha\beta}(12)$

$$c_{\alpha\beta}(12) = c_{\alpha\beta}^{0}(12) - \beta u_{\alpha\beta}^{\mathrm{LR}}(12) + \delta c_{\alpha\beta}(12)$$

$$h_{\alpha\beta}(12) = h_{\alpha\beta}^{0}(12) + G_{\alpha\beta}(12) + \delta h_{\alpha\beta}(12) \tag{12}$$

An integral equation for the screened potential is given by [18,12,24]

$$h^0_{\alpha\beta}(12)+G_{\alpha\beta}(12)=c^0_{\alpha\beta}(12)-\beta u^{\mathrm{LR}}_{\alpha\beta}(12)+$$

$$\sum_\gamma \rho_\gamma \int [c^0_{\alpha\gamma}(13)-\beta u^{\mathrm{LR}}_{\alpha\gamma}(13)][h^0_{\gamma\beta}(32)+G_{\gamma\beta}(32)]d(3) \tag{13}$$

The screened potential $G_{\alpha\beta}(12)$ is also called the chain sum [16,2,1]. If the reference system is a mixture of hard spheres with contact distances $\sigma_{\alpha\beta}=(\sigma_\alpha+\sigma_\beta)/2$

$$u^{\mathrm{SR}}_{\alpha\beta}(r)=\infty \quad \text{for} \quad r\le\sigma_{\alpha\beta} \quad ; \quad u^{\mathrm{SR}}_{\alpha\beta}(r)=0 \quad \text{for} \quad r>\sigma_{\alpha\beta} \tag{14}$$

an optimizing procedure is possible, leading to a closure relation of

$$G_{\alpha\beta}(12)=0 \quad \text{for} \quad r<\sigma_{\alpha\beta} \tag{15}$$

for Equation (13). This means that the long-range perturbation potential $u^{\mathrm{LR}}_{\alpha\beta}(12)$ is assumed to be state-dependent for distances $r<\sigma_{\alpha\beta}$. The direct correlation functions of the reference system are calculated in the Percus - Yevick approximation for a mixture of hard spheres, and the closure relations are

$$h^0_{\alpha\beta}(r)=-1 \quad \text{for} \quad r<\sigma_{\alpha\beta} \quad ; \quad c^0_{\alpha\beta}(r)=0 \quad \text{for} \quad r>\sigma_{\alpha\beta} \tag{16}$$

The optimized RPA, Equation (13), with the closure (15) is called the Mean Spherical Approximation (MSA) for the correlation functions of a multicomponent fluid with long-range interactions. For point charges and point dipoles on hard spheres very simple forms of the pair correlation functions and of thermodynamic functions result in this approximation, derived by simple analytical procedures. These solutions give a basic feeling of the structure, of thermodynamic and of dielectric properties of ionic and polar fluids and of the phenomena of solvation and association in these fluids.

After the calculation of a screened potential $G_{\alpha\beta}(12)$, cluster expansions for the correction terms can be derived. They start with

$$g^{(1)}_{\alpha\beta}(12)=g^0_{\alpha\beta}(12)\exp[G_{\alpha\beta}(12)+\delta W^{(3)}_{\alpha\beta}(12)] \tag{17}$$

The cluster coefficient $\delta W^{(3)}_{\alpha\beta}(12)$ is given by

$$\delta W^{(3)}_{\alpha\beta}(12)=\sum_\gamma \rho_\gamma \int [S_{\alpha\gamma}(13)S_{\gamma\beta}(32)+S_{\alpha\gamma}(13)H_{\gamma\beta}(32)+H_{\alpha\gamma}(13)S_{\gamma\beta}(32)]d(3) \tag{18}$$

with

$$S_{\alpha\beta}(12)=g^0_{\alpha\beta}(12)(\exp[G_{\alpha\beta}(12)]-1)-G_{\alpha\beta}(12) \tag{19}$$

and

$$H_{\alpha\beta}(12)=h^0_{\alpha\beta}(12)+G_{\alpha\beta}(12) \tag{20}$$

Neglecting the coefficient $\delta W^{(3)}_{\alpha\beta}(12)]$ and linearizing the exponent leads to the so called LIN - approximation for the pair correlation function.

$$g_{\alpha\beta}^{(\mathrm{LIN})}(12) = g_{\alpha\beta}^{0}(12)[1+G_{\alpha\beta}(12)] \tag{21}$$

This approximation has been studied for dipolar fluids [18,4] , and it will be shown also useful in the case of solvation of hard sphere ions by dipolar hard spheres. Integral equation approximations are obtained as

$$g_{\alpha\beta}^{\mathrm{RHNC}}(12) = g_{\alpha\beta}^{0}(12)\exp[G_{\alpha\beta}(12)+\delta h_{\alpha\beta}(12)-\delta c_{\alpha\beta}(12)] \tag{22}$$

which together with the OZ - Equation (10) yields the so-called Reference Hypernetted Chain (RHNC) equation [28]. A further linearization of Equation (22) with respect to $(\delta h - \delta c)$ leads to the Reference Percus-Yevick-Allnatt (RPYA) closure relation

$$g_{\alpha\beta}^{\mathrm{RPYA}}(12) = g_{\alpha\beta}^{0}(12)\exp[G_{\alpha\beta}(12)][1+\delta h_{\alpha\beta}(12)-\delta c_{\alpha\beta}(12)] \tag{23}$$

The integral equations of the MSA-, RHNC-, or RPYA- type are used to calculate the screened potentials and the correction terms of the pair-correlation functions for given interaction potentials $u_{\alpha\beta}(12)$ on different levels of approximation.

It should be stressed that the general theory developed here is especially useful when long-range perturbations (*e.g.* multipolar interactions) are added to the properties of an already known reference system with short-range interactions. This concept is helpful when the pecularities resulting from additional electrostatic interactions are the matter of discussion. If there is no division of the molecular interaction (e.g. in the case of ISM for pure polar fluids) these equations turn into the usual Molecular Ornstein - Zernike (MOZ) equations with the HNC - or PY - closures.

2.3 MSA of charged hard spheres

2.3.1 Screening by long range forces

This approach is attractive for an examination of the excess thermodynamic properties of electrolyte solutions because it gives rather simple analytical expressions in terms of a single screening parameter, Γ, and it also satisfies Onsagers high-charge, high-density limits [30,34,35].

The ionic MSA for charged hard spheres was first considered by Waisman and Lebowitz [38]. It consists in the solution of the integral equation, eq. (13), written as

$$h_{\alpha\beta}^{0}(12)+G_{\alpha\beta}(12) = h_{\alpha\beta}(r) = c_{\alpha\beta}(r)+\sum_{\gamma}\rho_{\gamma}\int c_{\alpha\gamma}(|\vec{r}-\vec{r}'|)h_{\gamma\beta}(\vec{r}')d\vec{r}' \tag{24}$$

with the closure relation

$$h_{\alpha\beta}(r) = -1 \quad \text{for} \quad r \le \sigma_{\alpha\beta} \quad ; \quad c_{\alpha\beta}(r) = -\beta U_{\alpha\beta}^{\mathrm{C}}(r) \quad \text{for} \quad r \ge \sigma_{\alpha\beta} \tag{25}$$

with $U_{\alpha\beta}^{\mathrm{C}}(r)$ the Coulomb potential part of eq. (3). The classical solution methods for these problems are integral transform methods [37,39] and the Wiener Hopf factorization [5,6,7] which are currently used in textbooks on liquid state theory [4,13,15,16,29,11]. They make use of complex algebra for the evaluation of simple results, thus avoiding the general use of Percus - Yevick and MSA solutions of charged hard spheres. In contrast, a very simple derivation of the MSA results for charged hard spheres of equal diameter is possible by direct differentiation of the Ornstein-Zernike equation, as exemplified in reference [20] for the PY solution of hard spheres, and in reference [33] for the MSA solution for the restricted primitive model of electrolytes. The total correlation functions of the restricted primitive model (RPM) with $\sigma_{\alpha\beta} = \sigma$ and $|z_\alpha| = |z_\beta|$ are divided in a hard-sphere part $h^0(r)$ and in a Coulomb part $h^C(r)$

$$h^0(r) = \frac{1}{2}[h_{++}(r) + h_{+-}(r)] \quad ; \quad h^C(r) = \frac{1}{2}[h_{++}(r) - h_{+-}(r)] \tag{26}$$

Corresponding expressions hold for the direct correlation functions $c^0(r)$ and $c^C(r)$.

2.3.2 PY -approximation for hard spheres

For $h^0(r)$ follows from Equations (24) and (25) the PY -approximation for hard spheres.

$$h^0(r) = c^0(r) + \rho \int c^0(r') h^0(|\vec{r} - \vec{r}'|) d\vec{r}' \quad ; \quad \rho = \rho_+ + \rho_- \tag{27}$$

The use of bipolar coordinates leads to the integrated form of Equation (1)

$$h^0(r) = -1 \quad \text{if} \quad r < \sigma \quad ; \quad c^0(r) = 0 \quad \text{if} \quad r > \sigma \tag{28}$$

$$H(r) = C(r) + 2\pi\rho \int_0^\infty C(s) ds \int_{|r-s|}^{r+s} H(t) dt \tag{29}$$

where the functions

$$C(r) = rc^0(r) \quad ; \quad H(r) = rh^0(r) \tag{30}$$

were introduced. Differentiation of eq. (3) leads to

$$H'(r) = C'(r) + 2\pi\rho \left(\int_0^\infty C(s) H(r+s) ds - \int_0 C(s) H(r-s) ds + \int^\infty C(s) H(s-r) ds \right) \tag{31}$$

The last two terms can be collected into one by expansion of the total correlation function $h(r)$ to negative r values. $h(r)$ must be an even function of r

$$h(-r) = h(r) \tag{32}$$

This leads to the following prescription for the continuation of $H(r)$

$$H(-r) = -H(r) \tag{33}$$

and for Eq. (5) follows

$$H'(r) = C'(r) + 2\pi\rho \int_0^\infty C(s)[H(r+s) - H(r-s)] ds \tag{34}$$

With the conditions $H(r) = -r$ in $r < \sigma$ and $C(s) = 0$ in $s > \sigma$ one has for $r < \sigma$:

$$-1 = C'(r) + 2\pi\rho\left(\int_0^\sigma C(s)H(r+s)ds + r\int_0^\sigma C(s)ds - \int_0^\sigma C(s)s\,ds\right) \tag{35}$$

Then

$$C'(0) = -1 + 4\pi\rho\int_0^\sigma C(s)s\,ds \tag{36}$$

To get the second derivative, eq. (5) is transformed in $r < \sigma$ to

$$H'(r) = C'(r) + 2\pi\rho\int_0^\infty [C(s) - H(s)][H(r+s) + H(r-s)]ds$$

$$+4\pi\rho r\int_0^\sigma C(s)ds - 4\pi\rho\int_0^\sigma sC(s)ds + \frac{\pi}{3}\rho r^3 \tag{37}$$

Then the second derivative in $r < \sigma$ is :

$$0 = C''(r) + 4\pi^2\rho^2\int_0^\sigma C(t)J(r,t)dt + 4\pi\rho\int_0^\sigma C(s)ds + \pi\rho r^2 \tag{38}$$

The kernel $J(r,t)$

$$J(r,t) = \int_0^\infty [H(r+s) - H(r-s)][H(s+t) - H(s-t)]ds \tag{39}$$

is after a transformation given by

$$J(r,t) = -\int_{-t} H(u)H(u-r+t)du + \int H(u)H(u-r-t)du$$

$$= -\int_{-t} (-u)(-u+r-t)du + \int (-u)(-u+r-t)du = tr^2 + \frac{t^3}{3} \tag{40}$$

The resulting equation in $r < \sigma$ is

$$0 = C''(r) + \frac{(2\pi\rho)^2}{3}\int_0^\sigma C(s)s^3ds - (2\pi\rho r)^2\int_0^\sigma C(s)s\,ds + 4\pi\rho\int_0^\sigma C(s)ds + \pi\rho r^2 \tag{41}$$

An expression for $C''(0)$ is given as follows:

$$C''(0) = -4\pi\rho\int_0^\sigma C(s)\left(1 + \frac{\pi\rho s^3}{3}\right)ds \tag{42}$$

Further derivatives in the region $r < \sigma$ simply yield

$$C^{(3)}(r) = 2\pi\rho r\left(-1 + 4\pi\rho\int_0^\sigma C(s)s\,ds\right) \tag{43}$$

Thus

$$C^{(3)}(0) = 0 \tag{44}$$

and

$$C^{(4)}(r) = 2\pi\rho(-1 + 4\pi\rho \int_0^\sigma C(s)s\,ds) = C^{(4)}(0) = 2\pi\rho C'(0) \tag{45}$$

lead to the series development

$$C(r) = rC^{'}(0) + \frac{1}{2}r^2C^{''}(0) + \frac{1}{6}r^3C^{(3)}(0) + \frac{1}{24}r^4C^{(4)}(0) \quad ; \quad r < \sigma \tag{46}$$

with the coefficients

$$C^{'}(0) = -1 + 4\pi\rho \int_0^\sigma C^0(s)s\,ds \quad ; \quad C^{''}(0) = -4\pi\rho \int_0^\sigma C^0(s)\left(1 + \frac{\pi\rho s^3}{3}\right)ds \tag{47}$$

$$C^{(3)}(0) = 0 \quad ; \quad C^{(4)}(0) = 2\pi n\left[1 + 4\pi\rho \int_0^\sigma C^0(s)s\,ds\right] \tag{48}$$

All higher $C^{(i)}(0) = 0$. Therefore the direct correlation function may be written in the form, $x = r/\sigma$,

$$c^0(x) = a_1 + a_2x + a_3x^3 \quad \text{for} \quad x \le 1 \quad ; \quad c^0(x) = 0 \quad \text{for} \quad x > 1 \tag{49}$$

where

$$a_1 = -\frac{(1+2\eta)^2}{(1-\eta)^4} \quad ; \quad a_2 = 6\eta\frac{(1+0.5\eta)^2}{(1-\eta)^4} \quad ; \quad a_3 = \frac{\eta}{2}a_1 \tag{50}$$

$\eta = (\pi\rho\sigma^3)/6$ is the space filling factor.

2.3.3 The Coulomb contribution

The screened potential $G_{\alpha\beta}(12)$ is in the case of charged hard spheres of equal diameter σ given by the Coulomb contribution $h^C(r)$. This contribution or, equivalently, the function $c^C(r)$, are derived from the equation

$$h^C(r) = c^C(r) + \rho \int c^C(r')h^C(|\vec{r} - \vec{r}'|)d\vec{r}' \tag{51}$$

and from the closure relations

$$h^C(r) = 0 \quad \text{for} \quad r < \sigma \quad ; \quad c^C(r) = -\frac{b\sigma}{r} \quad \text{for} \quad r > \sigma \tag{52}$$

with the Bjerrum parameter

$$b = \frac{\beta(ze)^2}{4\pi\varepsilon_0\sigma} \tag{53}$$

Blum used for the calculation of $c^C(r)$ the method of the Wiener - Hopf factorization [7,8].

Recently Rickayzen also succeeded to solve this problem by direct differentiation of the OZ equation [33]. For this purpose $c^C(r)$ is divided into two parts

$$c^C(r)=c_s^C(r)+c_l^C(r) \quad ; \quad c_s^C(r)=0 \quad \text{for} \quad r>\sigma \quad ; \quad c_l^C(r)=-\frac{b\sigma}{r} \quad \text{for all r} \tag{54}$$

Then Equation (25) leads to

$$h^C(r)=c^C(r)+\rho\int c_s^C(r')h^C(|\vec{r}-\vec{r}'|)d\vec{r}'+F(r) \tag{55}$$

with

$$F(r)=\rho\int c_l^C(r')h^C(|\vec{r}-\vec{r}'|)d\vec{r}' \tag{56}$$

Rickayzen finds from the Green function structure of $c_l^C(r)$ that $F(r)$ is a constant, $F(0)=A$. Introducing

$$rc^C(r)=C^C(r)=C_s^C(r)-b\sigma \tag{57}$$

and $H^C(r)=rh^C(r)$, Equation (29), in bipolar coordinates reads

$$H^C(r)=C_s^C(r)-b\sigma+2\pi\rho\int_0^\infty C_s^C(s)ds\int_{|r-s|}^{r+s}H^C(t)dt+Ar \tag{58}$$

Subsequent differentiation of this equation shows for $r<\sigma$

$$0=[C_s^C]'(r)+A+\pi\rho\int_0^\sigma C_s^C(s)H^C(r+s)ds \tag{59}$$

$$[C_s^C]''(r)=-2\pi A\rho\int_0^\sigma H^C(s)ds \quad ; \quad [C_s^C]^{(3)}(r)=0 \tag{60}$$

so that $C_s^C(r)$ is a quadratic polynom in r of the form

$$C_s^C(r)=C_s^C(0)+[C_s^C]'(0)r+\frac{1}{2}[C_s^C]''(0)r^2 \quad ; \quad r<\sigma \tag{61}$$

with

$$C_s^C(0)=\sigma b \quad ; \quad [C_s^C]'(0)=-A \quad ; \quad [C_s^C]''(0)=\frac{A^2}{2\sigma b} \tag{62}$$

where

$$A=-4\pi\sigma b\rho\int_0^\infty H^C(s)ds \tag{63}$$

is connected to the internal excess energy of the charged hard sphere system. This leads immidiately to the well known solution for $c(r)$

$$c^C(x) = -b(2B - B^2 x) \quad \text{for} \quad x = \frac{r}{\sigma} \le 1 \;\; ; \;\; c^C(x) = -\frac{b}{x} \quad \text{for} \quad x > 1 \tag{64}$$

with

$$\frac{A}{2b} = B = \frac{\Gamma\sigma}{1+\Gamma\sigma} \;\; ; \;\; \Gamma\sigma = \frac{1}{2}[\sqrt{1+2\kappa\sigma} - 1] \tag{65}$$

Γ is Blum's MSA screening parameter for charges [7]. The reduced internal excess energy is simply

$$\frac{E^{ex,C}}{NkT} = -bB \tag{66}$$

Charged hard spheres are very often used to describe the subsystem of charged particles in a polar solvent. The equilibrium structural properties and thermodynamic excess functions of this subsystem are well represented by this approximation in the case of small Bjerrum parameters b . Together with the concept of ion association in the framework of a chemical picture also excess properties of systems with high Bjerrum parameters are well reproduced [21,22,23]. A simplified derivation of the basic formulae of the MSA for charged hard spheres therefore seemed to be useful at this place.

2.4 Dipolar screening - MSA and LIN approximations for dipolar hard spheres

2.4.1 Nonpolarizable models

Taking into account a very simplified molecular picture of the polar solvent, it can be modelled as a system of hard spheres with point dipoles in the center. The MSA solution for the pair correlation functions can be traced back to the solution of hard sphere problem in Percus - Yevick approximation, as Wertheim has shown [40]. As a linear combination of the contributions to the pair correlation functions from the short range reference interactions and from the long range electrostatic contributions the MSA solution is of limited use. This has been shown in numerous comparisons with computer simulation results (see e.g. [4] and the references cited there). The LIN approximation, eq. (21) for dipolar hard spheres, which is also based on a knowledge of hard sphere distributions in Percus - Yevick approximation, has been shown to give much better agree ment with the simulation results. We start from the equation (24) of section 2.2 and calculate the screened potential $G(1,2)$ (omitting the indicees, because we treat a one component system now) for hard spheres of diameter σ with point dipoles in the center.

The center to center distance of two point dipoles is $\vec{r} = r\vec{r}^{\,0}$, the dipole moment vectors are $\vec{m}_i = \mu\vec{m}_i^0$; $i = 1,2$ (the superscript $\vec{r}^{\,0}$ means a unit vector). The potential is then

$$\beta U^D = (1,2) = \infty \quad \text{for} \quad x \le 1 \;\; ; \;\; \beta U^D(1,2) = \frac{(\mu^*)^2}{x^3} D(1,2) \quad \text{for} \quad x > 1 \tag{67}$$

$D(1,2)$ and $\Delta(1,2)$ are the corresponding angular functions for the dipole problem:

$$D(1,2)=3(\vec{m}_1^0\vec{r}^0)(\vec{m}_2^0\vec{r}^0)-\Delta(1,2) \quad ; \quad \Delta(1,2)=(\vec{m}_1^0\vec{m}_2^0) \tag{68}$$

Reduced units are used for distances and dipole moments:

$$x=\frac{r}{\sigma} \quad ; \quad \mu^*=\frac{\mu}{\sqrt{4\pi\varepsilon_0 k_B T\sigma^3}} \tag{69}$$

The MSA problem is defined by the OZ equation

$$h(1,2)=h^0(\rho,r)+G(1,2)=c(1,2)-\rho\int c(1,3)h(3,2)d(3) \tag{70}$$

and the closure relation

$$h(1,2)=-1 \quad \text{for} \quad r\le\sigma \quad ; \quad c(1,2)=-\beta U^D(1,2) \quad \text{for} \quad r>\sigma \tag{71}$$

$h^0(\rho,r)$ is, as before ,the corresponding solution of the PY problem for hard spheres of diameter σ and number density ρ. The convolution in Eq. (4) now contains integrations in the angular space where

$$d(3)=d\vec{r}_3 d\vec{\Omega}_3 \quad ; \quad \int d\vec{\Omega}_3=\int_0^\pi \sin(\theta)d\theta\int_0^{2\pi} d\pi=4\pi \tag{72}$$

for linear molecules. The angular dependence of the correlation function $G(1,2)$ is given by an expansion in the $\Delta(1,2)$ - and $D(1,2)$ - terms with r - dependent coefficients

$$G(1,2)=h_\Delta(r)\Delta(1,2)+h_D(r)D(1,2) \tag{73}$$

The solution of the dipolar problem in MSA can be traced back to the hard sphere problem in PY approximation with number densities $\rho_+=2K\rho$ and $\rho_-=-K\rho$ [40] as

$$h_\Delta(r)=2K[h^0(\rho_+,r)-h^0(\rho_-,r)] \quad \text{for} \quad r\ge\sigma \tag{74}$$

and

$$\hat{h}_D(r)=K[2h^0(\rho_+,r)+h^0(\rho_-,r)] \quad \text{for} \quad r\ge\sigma \tag{75}$$

and

$$h_D(r)=\hat{h}_D(r)-\frac{3}{r^3}\int_0^r \hat{h}_D(r')r'^2dr' \tag{76}$$

The factor K is connected with the dipolar excess energy in MSA and given by

$$\frac{E^{ex,MSA}}{Nk_BT}=-\frac{4\pi\rho\mu^2}{3k_BT}\int_\sigma^\infty\frac{h_D(r)}{r}dr=-3yK \tag{77}$$

where y is the dimensionless coupling parameter for dipolar systems.

$$y = \frac{\rho\mu^2}{9\varepsilon_0 k_B T} \tag{78}$$

The static permittivity ε follows from the Kirkwood formulae [16]

$$\frac{(\varepsilon-1)(2\varepsilon+1)}{9\varepsilon} = y g_K \tag{79}$$

The Kirkwood factor g_K describes the correlations between dipoles in the system and is calculated from the mean square of the total dipole moment $M = \sum_{i=1}^{N} \vec{m}_i$ of the system, given by

$$g_K^{MSA} = 1 + \frac{4\pi\rho}{3} \int_0^\infty h_\Delta(r) r^2 dr \tag{80}$$

In the present approximation this is

$$g_K^{MSA} = 1 + \frac{8K\pi\rho}{3} \int_0^\infty [h^0(\rho_+, r) - h^0(\rho_-, r)] r^2 dr \tag{81}$$

A very simple expression results then for the dielectric constant results in the MSA, namely

$$\varepsilon^{MSA} = \frac{(1+4K\eta)^2 (1+K\eta)^2}{(1-2K\eta)^6} \tag{82}$$

In the LIN approximation the pair distribution function g(1,2) reads

$$g(1,2) = g^0(r)[1 + h_\Delta(r)\Delta(1,2) + h_D(r)D(1,2)] \tag{83}$$

which is equivalent to

$$h_\Delta^{LIN}(r) = g^0(r) h_\Delta(r) \quad ; \quad h_D^{LIN}(r) = g^0(r) h_D(r) \tag{84}$$

Eq. (18) together with Eq. (14) leads to a corrected Kirkwood factor g_K^{LIN}, ($x = r/R$)

$$g_K^{LIN} = g_K^{MSA} + \delta g_K \tag{85}$$

with

$$\delta g_K = 16K\eta \int_0^\infty h^0(\rho, x)[h^0(\rho_+, x) - h^0(\rho_-, x)] x^2 dx \tag{86}$$

The correction term can simply be calculated by using the Fourier transform of the PY hard sphere direct correlation function [18,4].

With the corrected Kirkwood - factor g_K^{LIN} a new approximation for the static permitivity ε is calculated according to Eq. (13).

$$\frac{(\varepsilon^{LIN}-1)(2\varepsilon^{LIN}+1)}{9\varepsilon^{LIN}} = yg_K^{LIN} \tag{87}$$

A LIN approximation for the internal excess energy is derived as

$$\frac{E^{ex,LIN}}{Nk_BT} = -3yK^L \tag{88}$$

with K^L calculated from

$$K^L = \int_\sigma^\infty \frac{g^0(r)h_D(r)}{r}dr \tag{89}$$

Some results of these calculations are compared in Table 1 with the results of RHNC calculations derived according to eq. (22).

Table 1 Thermodynamic functions for dipolar hard sphere model systems [4]:

$(\mu^*)^2$	MSA	LIN	RHNC
Static permittivity ε			
1.0	7.8	9.26	9.82
2.0	20.0	27.0	32.0
Reduced internal excess energy $-E^{ex}/NkT$			
1.0	0.689	0.746	0.972
2.0	1.99	2.19	2.613

For simplified models of polar fluids these approximations give a first impression on the dielectric constants of real systems. We consider the polar liquids water, acetone (AC), chloroform (CF), methylen chloride (MC), tetrahydrofurane (THF), and dimethylformamide (DMF). Taking into account experimental number densities ρ, dipole moments μ and effective values of the particle diameter σ, which are estimated in prescribing the packing fraction of the molecules a fixed value $\eta = .45$, the following Table 2 results

Table 2 Static dielectric permittivity ε of polar hard spheres at $298.15\,K$ - MSA - and LIN approximations.

Liquid	ρ	μ	σ	ε^{MSA}	ε^{LIN}	α	ε_α^{LIN}	ε^{exp}
	$[10^{28}\,m^{-3}]$	$[10^{-30}\,Cm]$	$[10^{-10}\,m]$			$[10^{30}\,m^{-3}]$		
Water	3.33	6.21	2.88	45.80	67.96	0.4 (1.44)	80.2	78.4
AC	0.81	9.51	4.61	19.99	26.97	-	-	21.0
CF	0.743	3.84	4.75	2.34	2.40	6.5 (8.53)	4.86	4.8
MC	0.934	3.80	4.40	2.75	2.87	13.0 (6.82)	8.81	8.9
THF	0.738	5.84	4.76	4.99	5.57	4.0 (6.84)	7.51	7.5
DMF	0.719	12.87	4.80	40.89	59.96	-	-	37.4

2.4.2 Polarizability of Solvent Molecules

If polar solvent molecules have a finite polarizability, calculations in the liquid phase have to be fulfilled with an effective dipole moment which is the sum of the permanent dipole moment and a contribution induced by the local electrostatic field of the surrounding molecules. An approximate value of this quantity is given in terms of a scalar polarizability α from a mean field calculation [41,17]. By defining a reduced polarizability $\alpha^* = \alpha/\sigma^3$ it follows that

$$\mu^*_{eff} = \frac{\mu^*}{(1-16K\eta\alpha^*)}\sqrt{1+\frac{3\alpha^*}{(\mu^*)^2}(1-16K\eta\alpha^*)} \tag{90}$$

The value of μ^*_{eff} calculated from Eq. (24) is now introduced into the theory instead of μ^*. Some calculations illuminate this approach, the results of which are also depicted in Table 2. If the polarizability α is used as an adjustable parameter instead of the experimental value (which is given in brackets and calculated as one third of the trace of the corresponding polarizability tensor), then the experimental values of the dielectric constant of the liquids studied are fitted.

3. SOLVATION OF CHARGED HARD SPHERES IN A DIPOLAR HARD SPHERE SOLVENT

3.1 MSA and LIN approximation

The ideas underlying the concept of ion solvation may be illustrated here for the simple case of charged hard spheres representing the ions in a solvent made up of dipolar hard spheres. We consider the simplified situation that all particles have the same diameter σ. The center to center vector between the point dipole and the ion is $\vec{r} = r\vec{r}^{\,0}$. The ion-dipole interactions read for an ion with a charge $e_i = z_i e$ and a solvent molecule with a dipole moment $\vec{m} = \mu\vec{m}^{\,0}$

$$U_{is}(1,2) = \infty \quad ; \quad r \le \sigma \,, \qquad U_{is}(1,2) = -\frac{ze\mu}{4\pi\varepsilon_0 r^2}E(1,2) \quad ; \quad r > \sigma \tag{91}$$

where the angular function $E(1,2)$ is given by the scalar product

$$E(1,2) = \vec{m}^{\,0}\vec{r}^{\,0} \tag{92}$$

It may be represented with the help of the angular function $\vec{\Omega} = (\theta,\phi)$ yielding for $E(\Omega)$ the relations

$$\int E(\vec{\Omega})d\vec{\Omega} = 0 \quad ; \quad \int E(\vec{\Omega})^2 d\vec{\Omega} = \frac{1}{3} \tag{93}$$

The calculation of solvation energy starts from the energy equation for ion - dipole mixtures at vanishing ion concentration. At the limit of infinite dilution the sums vanish, and the number density $\rho = N/V$ is that of the pure solvent. For the ion - dipole pair distribution function in the infinite dilution case results an expansion according to eq. (21)

$$g_{is}(1,2) = g^0(r)(1 + z_i h^E(r)E(\vec{\Omega}) + \ldots) \tag{94}$$

Insertion of Eqs. (1) and (4) into the expression for the internal energy yields the Born energy of solvation E^B [10,27,18]

$$E^B = \rho \int_0^\infty r^2 dr \int \left[\frac{-ze\mu E(\vec{\Omega})}{\varepsilon_0 r^2} \right] [g^0(r)(1 + zh^E(r)E(\vec{\Omega}))] d\vec{\Omega}$$

$$= -\frac{\rho z^2 e\mu}{\varepsilon_0} \int_0^\infty h^E(r) dr \int E(\vec{\Omega})^2 d\vec{\Omega} = -\frac{\rho z^2 e\mu}{3\varepsilon_0} \int_0^\infty g^0(r) h^E(r) dr \tag{95}$$

The integrand of eq.(5) $g^0(r)h^E(r)$ is related to the polarization density $P(r)$ around the solvated ion.

At large distances ($r \to \infty$) the polarization vector $\vec{P}(\vec{r}) = P(r)\vec{r}/r$ has its macroscopic value obtained in continuum theory

$$\vec{P}_i(r) = \varepsilon_o(\varepsilon - 1)\vec{E}_i(\vec{r}) \quad ; \quad \vec{E}_i(\vec{r}) = \frac{z_i e}{4\pi\varepsilon_0 \varepsilon r^3} \vec{r} \tag{96}$$

where the polarization density $P_i(r)$ is given by

$$P_i^{macro}(r) = \frac{(\varepsilon - 1) z_i e}{4\pi\varepsilon r^2} \tag{97}$$

The microscopic expression $P_i^{micro}(r)$ for the polarization density follows from the angular average of the orientational correlations of the dipole molecules at distance r from the central ion

$$P_i^{micro}(r) = \rho\mu \int g_{is}(\vec{r}, \vec{\Omega}) E(\vec{\Omega}) d\vec{\Omega} = \frac{\rho\mu z_i}{3} g^0(r) h^E(r) \tag{98}$$

which at large distances must be equal to its macroscopic value $P_i^{macro}(r)$, Eq. (7). Therefore

$$\lim_{r\to\infty} h^E(r) = \frac{3(\varepsilon - 1)e}{4\pi\rho\mu\varepsilon r^2} = h^{E,\infty}(r) \tag{99}$$

The solution of the MSA equations for the ion dipole mixture in the infinite dilution case leads to an integral equation for a function $F(r)$ which is related to $h^E(r)$ in the following way:

$$\frac{h^E(r)}{h^{E,\infty}(r)} = 1 + F(r) - r\frac{dF(r)}{dr} \quad ; \quad r > \sigma_{is} \tag{100}$$

The integral equation for $F(r)$ is expressed as [10]

$$F(x) = \frac{x}{\left(1 - \frac{3\xi}{1+4\xi}\right)} \quad ; \quad x \le 1$$

$$F(x)=24\xi\int_0^1 F(x-y)\left[\frac{(1+4\xi)}{2(1-2\xi)^2}(y^2-1)-\frac{3\xi}{(1-2\xi)^2}(y-1)\right]dy \quad ; \quad x>1 \tag{101}$$

In Eq. (101)

$$\xi=\frac{K}{6}\pi\rho\sigma^3 \tag{102}$$

where K is the energy constant of the pure dipolar system, defined by Eq. (77) The solution of Eq. (101) has to be done numerically.

The Born solvation energy E_i^B in the MSA is from Eq.(5) with $g^0(r)=1$

$$(E_i^B)^{MSA}=-\frac{z_i e^2}{4\pi\varepsilon_0\sigma}\frac{\left(1-\frac{1}{\varepsilon}\right)}{\left(1-\frac{3\xi}{1+4\xi}\right)}=-b_i\frac{\varepsilon^{MSA}-1}{\left(1-\frac{3\xi}{1+4\xi}\right)} \tag{103}$$

Here b_i is the Bjerrum parameter of the ion i,

$$b_i=\frac{z_i e^2}{4\pi\varepsilon_0\varepsilon^{MSA}\sigma} \tag{104}$$

and ε^{MSA} is the dielectric constant of the pure dipolar system, calculated in the MSA.

In the LIN approximation is the relation between microscopic and macroscopic polarisation density given by

$$\frac{P_i^{micro}(r)}{P_i^{macro}(r)}=\frac{g^0(r)h^E(r)}{h^{E,\infty}(r)}=g^0(r)\left(1+F(r)-r\frac{dF(r)}{dr}\right) \quad ; \quad r>\sigma \tag{105}$$

The Born solvation energy in LIN approximation is

$$(E_i^B)^{LIN}=(E_i^B)^{MSA}(1+\delta E_i^B) \tag{106}$$

The additional contribution is calculated as

$$\delta E_i^B=\left(1-\frac{3\xi}{1+4\xi}\right)\int_\sigma^\infty \frac{h^0(r)}{r^2}\left(1+F(r)-r\frac{dF(r)}{dr}\right)dr \tag{107}$$

4. CONCLUSIONS

To estimate structural and thermodynamic properties of solvation shells of ions on the Born - Oppenheimer level, it is advantageous to use simplified models of ions and solvent molecules as a starting point. Complementary to more refined statistical mechanical methods, such as simulation methods and numerical solutions of integral equations, systems like hard sphere ions in dipolar hard spheres can be treated on the MSA or LIN level by simple analytical methods of real analysis, avoiding complex mathematical methods like numerical Fourier transforms or complex analysis. Properties like internal energies, pressures and dielectric constants can be estimated in a first approximation and the general behaviour of these quantities is described in a correct way if the input parameters of the calculations are altered.

REFERENCES

1. H.C. Andersen and D. Chandler, *J.Chem.Phys.* , 547 (1970).
2. H.C. Andersen and D. Chandler, *J.Chem.Phys.* , 1918 (1972).
3. J. Barthel, R. Neueder et al., *Electrolyte Data Collection* in G. Kreysa (ed.): *DECHEMA Chemistry Data Series, Vol. XII, Alcohols I (1992), II (1992), III (1994), Aprotic protophobic solvents I (1996), II (1999), III (2000), IV (2000), Protophilic H - bond donor and aprotic solvents (2001)*, DECHEMA, Frankfurt/M.
4. J. Barthel, H. Krienke and W. Kunz, *Physical Chemistry of Electrolyte Solutions - Modern Aspects*, Steinkopff, Darmstadt, and Springer, New York (1998)
5. R.J. Baxter, *Phys.Rev.* , 170 (1967).
6. R.J. Baxter, in *Physical Chemistry - An Advanced Treatise*, Vol.8A,H *Eyring,D.Henderson,andW.Jost,eds.,AcademicPress,NewYork*(1971),*p.267*.
7. L. Blum, *Mol.Phys.* , 1529 (1975).
8. L. Blum, in *Theoretical Chemistry, Advances and Perspectives*, Vol. 5, Academic Press, New York (1980), p.1.
9. L. Blum and A.J. Toruella, *J.Chem.Phys.* , 303 (1972)
10. D.Y.C. Chan, D.J. Mitchell, and B.W. Ninham, *J.Chem.Phys.* , 2946 (1979).
11. H.T. Davis, *Statistical Mechanics of Phases, Interfaces, and Thin Films*, Verlag Chemie, Weinheim (1996).
12. M.F. Golovko and H. Krienke, *Mol.Phys.* , 967 (1989).
13. C.G. Gray and K.E. Gubbins, *Theory of Molecular Fluids*, Clarendon Press, Oxford (1984).
14. R. Fischer, P.H.Fries, J. Richardi and H. Krienke, *J.Chem.Phys.* , 8467 (2002).
15. H.L. Friedman, *A Course in Statistical Mechanics*, Prentice-Hall, Englewood Cliffs (1985).
16. J.-P. Hansen and I.R. McDonald, *Theory of Simple Liquids*, Academic Press, London (1986).
17. J.S. Høye, G. Stell *J.Chem.Phys.* , , 461 (1980).
18. H. Krienke and G. Weigl, *Ann.Phys.(Leipzig)* , 313 (1988).
19. H. Krienke and R. Thamm, *Mol.Phys.* , 757 (1992).
20. H. Krienke, *J.Mol.Liq.* , 263 (1998).
21. H. Krienke and J. Barthel, *Z.Phys.Chem.* , 71 (1998).
22. H.Krienke and J. Barthel, *J.Mol.Liq.* , 123, (1998).
23. H. Krienke, J. Barthel, M. Holovko, I. Protsykevich, and Yu. Kalyushnyi, *J.Mol.Liq.* ., 191, (2000).
24. H. Krienke and J.Barthel, in: *Equations of state for fluids and fluid mixtures, Ch.16: Ionic fluids* (Eds.: J.V. Sengers et al.), pp. 751-804, Elsevier, Amsterdam (2000)
25. H. Krienke, R. Fischer and J. Barthel, *J.Mol.Liq.* **-99**, 329 (2002) .
26. H. Krienke, *Pure Appl.Chem.* in press (2003) .
27. D. Levesque, J.J. Weis, and G.N. Patey, *J.Chem.Phys.* , 1887 (1980).
28. F. Lado, *Phys.Rev.A* , 2548 (1973).
29. L.L. Lee, *Molecular Thermodynamics of Nonideal Fluids*, Butterworths, Boston (1988).
30. L. Onsager, *J.Phys.Chem.* , 189 (1939).
31. J. Richardi, P.H. Fries and H. Krienke, *J.Phys.Chem.B* , 5196 (1998).
32. J. Richardi, P.H. Fries and H. Krienke, *J.Chem.Phys.* , 4079 (1998).
33. G. Rickayzen, *Mol.Phys.* , 721 (1999).
34. Y. Rosenfeld and L. Blum, *J.Phys.Chem.* , 5149 (1986).
35. Y. Rosenfeld and L. Blum, *J.Chem.Phys.* , 1556 (1986).
36. G. Stell, in *Phase Transitions and Critical Phenomena*, Vol.V, C. Domb and M.S. Green, eds., Academic Press, London (1976), p.205.

37. E. Thiele, *J.Chem.Phys.* , 474 (1963).
38. E. Waisman and J.L. Lebowitz, *J.Chem.Phys.* , 4307 (1970).
39. M. Wertheim, *Phys.Rev.Lett.* , 321 (1963).
40. M. Wertheim, *J.Chem.Phys.* , 4291 (1971).
41. M. Wertheim *Mol.Phys.* , 1425 (1973).

REAL TIME VISUALIZATION OF ATOMIC MOTIONS IN DENSE PHASES

S. BRATOS, J-CL.LEICKNAM, F.MIRLOUP, R. VUILLEUMIER
Laboratoire de Physique Théorique des Liquides,
Université Pierre et Marie Curie, Case Courrier 121,
4 Place Jussieu, 75252 Paris Cedex 05, France

G. GALLOT
Laboratoire d'Optique et Biosciences, Ecole Polytechnique,
Route de Saclay, 91128 Palaiseau Cedex, France

M. WULFF AND A. PLECH
European Synchrotron Radiation Facility, BP 220,
6 rue Jules Horowitz, Grenoble Cedex 38043, France

AND

S. POMMERET
CEA/Saclay, DSM/DRECAM/SCM/URA 331 CNRS,
91191 Gif-sur-Yvette,France

1. INTRODUCTION.

It has always been a dream of physicists and chemists to follow temporal variations of molecular geometry during a chemical reaction in real time, to "film" them in a way similar as in the everyday life. Unfortunately, chemical events take place on tiny time scales comprised between 10 fs and 100 ps, approximately. Visualizing atomic motions thus remained a dream over two centuries. This is no longer true today, consequence of an immense instrumental development the last decades. Two methods are particularly important. The first of them is ultrafast optical spectroscopy employing the recently developed laser technology. In his breakthrough work A. Zewail was able to show how can this method be used to follow the photoelectric dissociation of gaseous ICN in real time[1,2]. It has later been applied to several other problems, and particularly so to visualize OH..O motions in liquid water[3,4]. Unfortunately, visible light interacts predominantly with outer shell rather than with deeper lying core

J. Samios and V.A. Durov (eds.), Novel Approaches to the Structure and Dynamics of Liquids: Experiments, Theories and Simulations, 111–128.

electrons that most directly indicate molecular geometry. It is thus difficult to convert spectral data into data on molecular geometry. The second method refers to time resolved x-ray diffraction and absorption. As x-rays interact predominantly with deeply lying core electrons which are tightly bonded to the nuclei, converting x-ray data into data relative to molecular geometry is, in principle at least, much more straightforward than in optical spectroscopy. Unfortunately, pulsed x-rays techniques are technically very demanding. Heavy constraints are imposed on pulse duration, brilliance, and photon flux. A number of instruments are presently available, but none of them meets all the requirements needed for a real time probing of molecular motions. An intense effort is actually being done to improve them. Nevertheless, in spite of these difficulties, important results have already been reached following this route[5-8].

The purpose of the present paper is to discuss these two methods of visualization of molecular motions in dense phases. Time-resolved electron diffraction will not be discussed: electrons having a short penetration depth, this technique is more useful for studying gases and surfaces than for analyzing condensed matter. Time scales considered range from the nanosecond to the femtosecond. The methodology employed as well as the information accumulated in this way will be illustrated on a few examples, hopefully representative of the current state-of-the-art. The two subsequent sections refer to time-resolved optical spectroscopy, and to time-resolved x-ray diffraction, respectively. A last section will summarize main conclusions.

2. NONLINEAR OPTICAL SPECTROSCOPY

2.1. GENERALITIES

There exists a large number of nonlinear optical techniques [9,10]. In some of them the applied fields are stationary, and in the others they have the form of short pulses. The techniques of the first class are called frequency-domain techniques; they are particularly useful to detect level positions, transition dipole moments, etc. They are currently employed in studying simple systems like low density atomic vapour, small molecules in supersonic beams, etc. The techniques of the second class are designated as time-domain techniques; they are carried out on systems involving many degrees of freedom and permit to study different relaxation processes, to monitor molecular motions in real time, etc. As a rule, frequency-domain techniques are more appropriate for small and simple systems, whereas time-domain techniques are more adequate for complex and large systems.

Another division concerns time-domain techniques. What is the pulse duration which is required? According to the basic principles of physics, it must

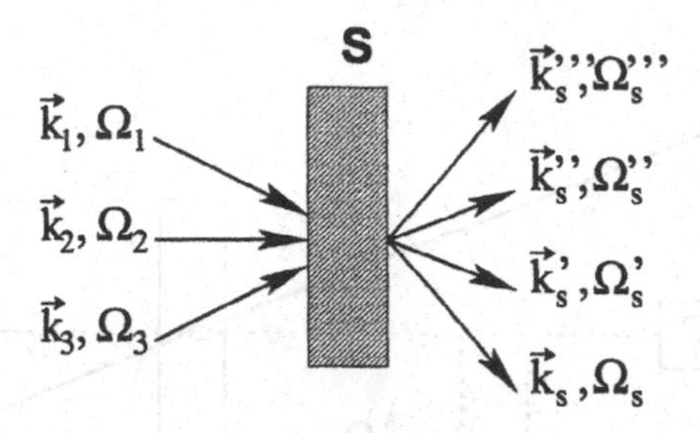

Figure 1. **Schematic representation of four-wave mixing experiments: $\mathbf{k}_1$, $\mathbf{k}_2$, $\mathbf{k}_3$ are wave vectors of the three incident pulses, and $\mathbf{k}_s$ that of the emerging coherent signal; the corresponding frequencies are Ω_1, Ω_2, Ω_3 and Ω_s. Different choices of wave vectors and frequencies lead to different kinds of optical spectroscopy. In a pump-probe experiment $\mathbf{k}_1$=$\mathbf{k}_2$, $\mathbf{k}_s$=$\mathbf{k}_3$ and Ω_1=Ω_2, Ω_s=Ω_3; in CARS $\mathbf{k}_1$=$\mathbf{k}_3$, $\mathbf{k}_s$=2$\mathbf{k}_1$-$\mathbf{k}_2$and Ω_1=Ω_3, Ω_s=2Ω_1-Ω_2; and in a 3-pulse photon echo experiment $\mathbf{k}_s$=$\mathbf{k}_3$+$\mathbf{k}_2$-$\mathbf{k}_1$ and Ω_s=Ω_3+Ω_2-Ω_1.**

correspond to time scale of the process under investigation. How many incident pulses are employed? In some techniques only one pulse is generated, and its scattering by the system under investigation is examined. Between these techniques one can cite time-resolved fluorescence and time-resolved Raman. In four-wave mixing techniques three laser fields are employed with wave vectors $\mathbf{k}_1$, $\mathbf{k}_2$, $\mathbf{k}_3$ and frequencies ω_1, ω_2 ω_3 respectively. The fourth field is that of the emerging coherent signal where the wave vector $\mathbf{k}_s$ and the frequency ω_s are:

$$\begin{aligned} \mathbf{k}_s &= \pm\mathbf{k}_1 \pm \mathbf{k}_2 \pm \mathbf{k}_3 \\ \omega_s &= \pm\omega_1 \pm \omega_2 \pm \omega_3 \end{aligned} \quad (1)$$

Various processes then differ in the choice of $\mathbf{k}_s$ and ω_s (Fig. 1). The most familiar examples are pump-probe spectroscopy where $\mathbf{k}_1$=$\mathbf{k}_2$ and $\mathbf{k}_s$=$\mathbf{k}_3$; Coherent Antistokes Raman Scattering (CARS) where $\mathbf{k}_1$=$\mathbf{k}_3$ and $\mathbf{k}_S$=2$\mathbf{k}_1$-$\mathbf{k}_2$; or else 3-pulse photon echo where $\mathbf{k}_s$=$\mathbf{k}_3$+$\mathbf{k}_2$-$\mathbf{k}_1$. In these processes the signal field is in a direction $\mathbf{k}_s$, generally different from that of the incoming wave vectors $\mathbf{k}_i$. It is possible to interpret this signal in terms of a grating formed by two beams and a third beam that undergoes a Bragg diffraction from the grating. The pump-probe spectroscopy is the simplest of these processes.

A last division of nonlinear optical techniques is into resonant and off-resonant techniques. In the case of a resonant technique, the incident field frequency, or a combination of them, is equal to one of characteristic frequencies of the matter. The pump-probe spectroscopy is an example of such techniques. They provide a direct probe for specific eigenstates and their dynamical behaviour. They are also sensitive to relaxation processes including spontaneous

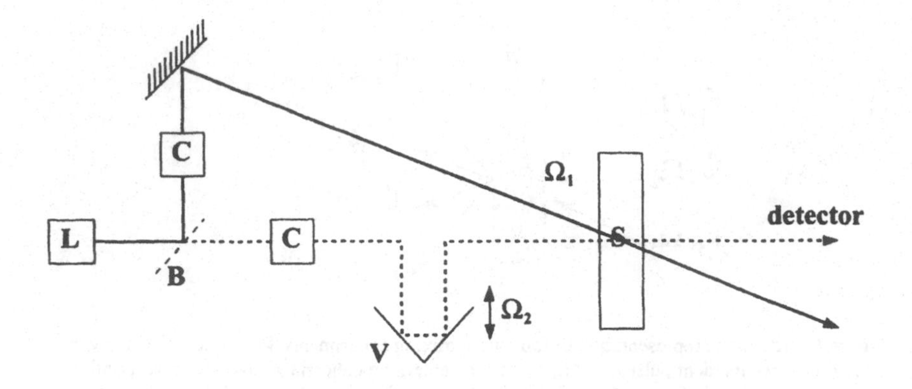

Figure 2. Schematic representation of a pump-probe experiment.

emission. In the case of an off-resonant technique, the incident field frequency is far detuned from any natural frequency of the system under consideration. Time-resolved Raman scattering is a representative of techniques of this class. Off-resonant techniques permit to avoid absorptive losses and other competing processes. Resonant and off-resonant techniques are both of common use.

2.2. PUMP-PROBE SPECTROSCOPY

A pump-probe spectroscopic experiment is pictured in Fig. 2. A powerful laser L generates short, pico- or subpicosecond pulses with energy in the 10μJ range. Each of these pulses is split by a beam splitter B into two pulses of unequal intensity, an intense pump pulse P_1 and a weak probe pulse P_2. Their frequencies Ω_1, Ω_2 are obtainable from the initial laser frequency Ω_L by using a particular optical device C called wave length tuner. The time delay τ between them is controlled by another optical device V consisting of elements that elongate optical path at will. The two pulses meet in the sample cell S where they mix and interact with matter. The pump pulse brings the molecules in an excited state. Prepared in this way, the system is unstable and relaxes back to the equilibrium. This return to the equilibrium is then monitored by measuring the probe absorption for different probe frequencies Ω_2 and for different delay times τ.

The measured quantity is the signal $S(\Omega_1, \Omega_2, \tau)$ defined as time integrated probe absorption $W(\Omega_1, \Omega_2, \tau)$ in presence of the pump, minus probe absorption $W_0(\Omega_2)$ in absence of the pump. The signal may be positive or negative, according to whether $W < W_0$ or $W > W_0$; these two processes are called bleaching and induced absorption, respectively. It results from this definition that the signal S is a differential quantity. Experimental data are generally presented in one of the following two ways. The signal S may be plotted for a fixed pump frequency Ω_1 and a given delay time τ, as a function of the probe frequency Ω_2. This plot represents a frequency-resolved spectrum. Alternatively, it can also be recorded for fixed pump and probe frequencies Ω_1, Ω_2, as a function of the delay time τ. This forms a time-resolved spectrum. Of course, a complete collection of frequency-resolved spectra contains exactly the same information as a full collection of time-resolved spectra. However, one never has at one's disposal all of them, and it is generally found useful to consider them both.

2.3. HYDROGEN BONDING AND ITS PROPERTIES

The material chosen to show how atomic motions can be visualized is liquid water. This fluid has a huge importance for all living bodies, and thus merits a special concern. For convenience, the experiments have most often been realized in diluted solutions HDO/D_2O rather than in normal water. In these solutions the dominant force is hydrogen bonding, a weak chemical bond between the two water molecules. The OH grouping of one of them acts as proton donor, and the oxygen atom of the other acts as proton acceptor. The length r of an OH..O group fluctuates in wide limits, somewhere between 2.75A and 3.00A; liquid water may thus be viewed as a mixture of hydrogen bonds of different length. Which are molecular dynamics to be monitored? The motions to be visualized are the stretching motions of the OH..O group, and the rotations of HDO molecules captured in it.

Two important characteristics of hydrogen bonding must be mentioned. (i) The OH stretching vibrations are strongly affected by this sort of interaction. The stronger the hydrogen bond, the softer the OH link and the lower is its frequency Ω: the covalent OH bond energy is lent to the OH..O bond and reinforces the latter. Although the connection between Ω and r is not strictly one-to-one, a number of extremely useful relationships linking them were published, following the initial proposal of Rundle and Parasol[11,12]. The one frequently adopted is the relationship due to Mikenda[13]. It was obtained by compilation of spectroscopic and structural data of 61 solid hydrates containing more than 250 hydrogen bonds (Fig. 3a). As in HDO/D_2O the mean value of Ω is $\Omega_0 = 3420$ cm^{-1}, the equilibrium OH..O bond length $r_0 = 2.86$A is predicted by the Mikenda relationship. (ii) Not only the OH vibrations, but also the HDO rotations are influenced noticeably by hydrogen bonding. This is due to steric forces that hinder the HDO rotations. As they are stronger in short than in long hydrogen bonds, these rotations are slower in the first case than in the second. Although this effect was only recently discovered[14], its existence can hardly be contested. With this information in mind, the experiments can be described as follows.

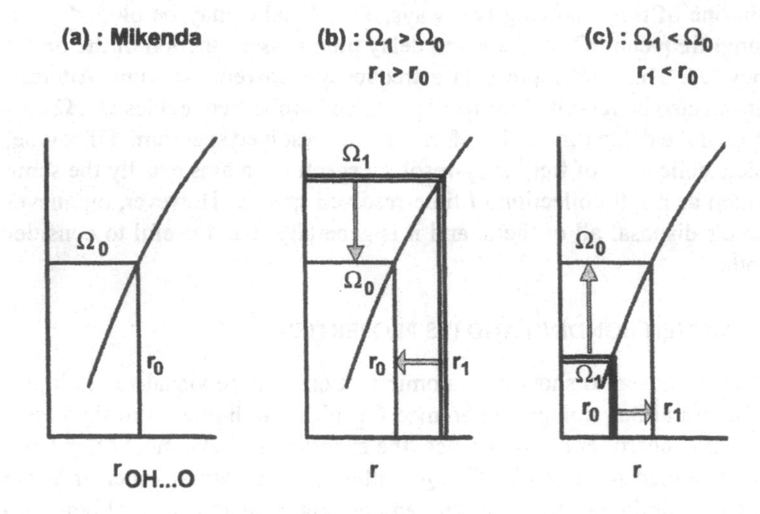

Figure 3. (a) Mikenda relation connecting the OH stretching frequency Ω and the OH..O bond length r: Ω_0 and r_0 denote the equilibrium values of Ω and r in a diluted HDO/D_2O solution. (b, c) Principle of the experiment: Ω_1 and r_1 represent the mean values of Ω and r in a coherently excited wave packet.

2.4. PRINCIPLE OF THE EXPERIMENTS

The pump-probe experiment set-up to visualize the OH..O stretching motions in water is as follows. An ultrafast pulse of frequency Ω_1, belonging to the conventional OH stretching band of HDO, is used to excite OH vibrations; this excitation results in selecting OH..O bonds of a given length r_1. Prepared in this way, the system relaxes back to the equilibrium and is monitored by a probe pulse; the OH band is recorded for different time delays τ. Two choices exist for the pump frequency Ω_1. First, let Ω_1 be larger than Ω_0, the mean frequency of HDO in D_2O (Fig. 3b); the laser selected OH..O bonds, initially longer than in equilibrium where their length is equal to r_0, contract with time, and a low frequency shift of the OH band from Ω_1 to Ω_0 may be anticipated. The opposite behaviour is expected if Ω_1 is smaller than Ω_0 (Fig. 3c); the OH..O bonds, initially too short, expand on relaxing and a high frequency shift is expected. No band shift should be observed if Ω_1 is equal to Ω_0. Conversely, knowing the peak frequency $\Omega(\tau)$ of the signal at time τ and using the Mikenda relation between OH frequency and OH..O bond length permits determination of this distance at time τ. The OH..O stretching motions in water can be visualized by proceeding in this way.

It is also possible to monitor molecular rotations in water. However, a polarization resolved pump-probe experiment is required for that purpose. The method consists in selecting hydrogen bonds of a specified length r_1 by pumping the system with an appropriate frequency Ω_1, as described above. Molecular rotations in the sub-set of hydrogen bonds obtained in this way are analyzed next by measuring the so called rotational anisotropy $R = (S_{\parallel} - S_{\perp})/(S_{\parallel} + 2S_{\perp})$ for different time delays τ; $S_{\parallel}$ is the pump-probe signal measured with pump and probe electric fields parallel to each other, and $S_{\perp}$ that obtained when they are perpendicular. Rotational anisotropy is an important indicator of molecular rotations in liquid systems. In fact, it was shown [15,16] that under certain conditions

$$R(\tau) = (2/5) < P_2(cos(\theta(\tau))) > = (2/5)\exp(-3/2 < \theta^2(\tau) >) \quad (2)$$

where P_2 is the second-order Legendre polynomial and $\theta(\tau)$ the angle between the transition moment vectors in times 0 and τ. It is also important to notice that $R(\tau)$ as given above is independent of Ω_1 and Ω_2. The main message contained in this formula is that the square averaged rotation angle $\sqrt{<\theta^2(\tau)>}$ is deducible from the rotational anisotropy $R(\tau)$. This sort of experiment thus has the intrinsic power of visualizing molecular rotations.

2.5. THEORY OF BAND SHAPES

Modern theories of band shapes all employ the density matrix approach of statistical mechanics. This is true for both linear and nonlinear spectroscopy. Noted by $\rho(t)$, the density matrix of a system submitted to electromagnetic radiation may be calculated treating the von Neumann equation for $\rho(t)$ in the frame of the perturbation theory. The energy of interaction between the system and the radiation field is considered as a perturbation. If this interaction energy is small enough, only linear terms need to be retained. This is the case for all standard versions of absorption spectroscopy, nuclear magnetic resonance, dielectric absorption and relaxation, etc. If, on the contrary, the energy deposited by a laser is large enough, the perturbation series can no longer be truncated after its linear terms, and higher-order terms must be considered. For example, terms up to the third order in perturbation energy are needed in pump-probe spectroscopy where [17,18]

$$\begin{aligned} S(\Omega_1,\Omega_2,\tau) &= (2/\hbar)Im(\textstyle\int_{-\infty}^{\infty}dt \int_0^{\infty}d\tau_1 \int_0^{\infty}d\tau_2 \int_0^{\infty}d\tau_3 \\ &\times < \dot{E}_{2i}(\mathbf{r},\mathbf{t})\mathbf{E_j}(\mathbf{r},\mathbf{t}-\tau_3)\mathbf{E_k}(\mathbf{r},\mathbf{t}-\tau_3-\tau_2)\mathbf{E_l}(\mathbf{r},\mathbf{t}-\tau_3-\tau_2-\tau_1) >_{\mathbf{E}} \\ &\times < M_l(0)[M_k(\tau_1),[M_j(\tau_1+\tau_2),M_i(\tau_1+\tau_2+\tau_3)]] >_S \end{aligned} \quad (3)$$

This expression involves two kinds of four-time correlation function, the correlation functions of the total and the probe electric fields $\mathbf{E}$ and $\mathbf{E}_2$, and that of electric dipole moment **M** of the system. The indices i,j,k,l denote Cartesian components of these vectors; the Einstein convention is employed indicating a summation over doubled indices. The average $< >_S$ is over the states of the non-perturbed liquid system, and the average $< >_E$ is over all possible realizations of the incident electric fields. The symbol [,] denotes a commutator and the dot a time derivative. Choosing the electric fields $\mathbf{E}_1$, $\mathbf{E}_2$ of appropriate form, all possible cases of polarization may be treated. The above expression for $S(\Omega_1, \Omega_2, \tau)$ represents an exact third-order perturbation theory result.

Applying it to the study the OH..O stretching motions in water is straightforward. On the contrary, a difficulty arises in examining HDO rotations in water. The problem is that the experiment envisaged to visualize rotational dynamics of HDO in D_2O is based on Eq. (2), derived under a number of restrictive conditions. It was supposed that (i) the coupling between molecular rotations and other degrees of freedom of the system is absent, and (ii) the time delay between the pump and probe pulses is long enough to avoid their overlap. Are they satisfied in the case of water? The condition (ii) certainly fails at small τ's where the pump and probe pulses overlap, whatever the material under consideration. Unfortunately, the condition (i) is not satisfied neither for water, although it is in many other liquids: HDO rotations depend on the OH..O bond length, and are thus correlated with OH..O stretching motions. Can Eq. (2) still be applied in spite of this difficulty? To answer this question an important theoretical effort based on Eq. (3) was necessary, and led to the following conclusions [19]: The first is that correlations between rotations and OH..O motions transform an initially unique curve $R = R(\tau)$ into a family of curves $R = R(\Omega_1, \Omega_2, \tau)$. The second effect is the interference of the pump and probe pulses at short times where they overlap. Rotational anisotropy no longer remains a useful indicator of rotational dynamics at these time scales. Finally, if this short time domain is eliminated, $R(\Omega_1, \Omega_2, \tau)$ still permits to visualize molecular rotations, even if correlations are present. A mono-colour experiment, where $\Omega_1 = \Omega_2$, is required to monitor molecular rotations in hydrogen bonds of a fixed length; and a two-colour experiment, where $\Omega_1 \neq \Omega_2$, is needed to follow these motions in contracting or expanding hydrogen bonds. If necessary precautions are taken, ultrafast laser spectroscopy permits to visualize molecular rotations in water!

2.6. "FILMING" MOLECULAR MOTIONS IN LIQUID WATER

Hydrogen bond motions discussed in this Section were all probed with 150 fs resolution. The data presented hereafter concern OH..O stretching and HDO rotational dynamics. The first of them [3,4] were monitored at two excitation frequencies, at $\Omega_1 = 3510$ and $3340\ cm^{-1}$. If the excitation is at $3510\ cm^{-1}$, the length of the pump selected hydrogen bonds is $r_1 = 2.99$A, longer than $r_0 =$

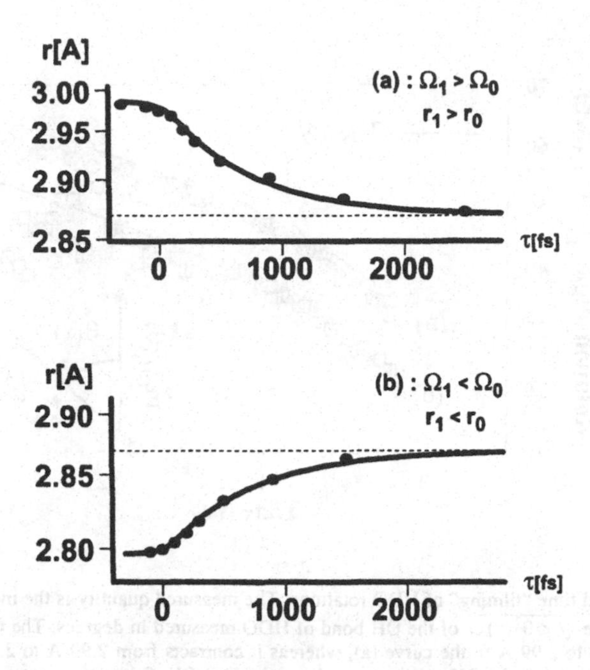

Figure 4. Observing OH..O motions in real time: The length $r_{OH..O}$ bond is expressed as a function of the time τ. The initial OH..O bond length is 2.99 A in (a) and 2.80 A in (b). The points represent experimental data. The full curves are given by Eq. (5) of the text.

2.86A: Fig. (4a) thus illustrates the contraction of initially elongated hydrogen bonds. If the excitation is at 3340 cm^{-1}, the length of the pump selected hydrogen bonds is $r_1 = 2.80$A, smaller than $r_0 = 2.86$A: Fig. 4b thus pictures the extension of initially compressed hydrogen bonds. Each point represents an experimentally measured bond length. The OH..O dynamics were thus studied in real time, and the bonds were "filmed" during their evolution in the liquid sample. It is interesting to notice that these motions are all monotonic and that no oscillations appear. They are strongly damped as expected; though, it is fascinating to "see" them directly!

This Section may be concluded by presenting two "films" showing HDO rotations in liquid D_2O [19-21]. The mean squared rotational angle $\sqrt{<\theta^2(\tau)>}$ of the OH bond of HDO is illustrated as a function of time τ (Fig. 5). The OH..O bond length is kept constant and equal to 2.99A in curve (a), whereas it contracts from 2.99A to 2.86A in curve (b). One notices that rotational angles of the order of 35° are attained in times of the order of 700 fs. As bending of this magnitude leads to a breaking of hydrogen bonds, one

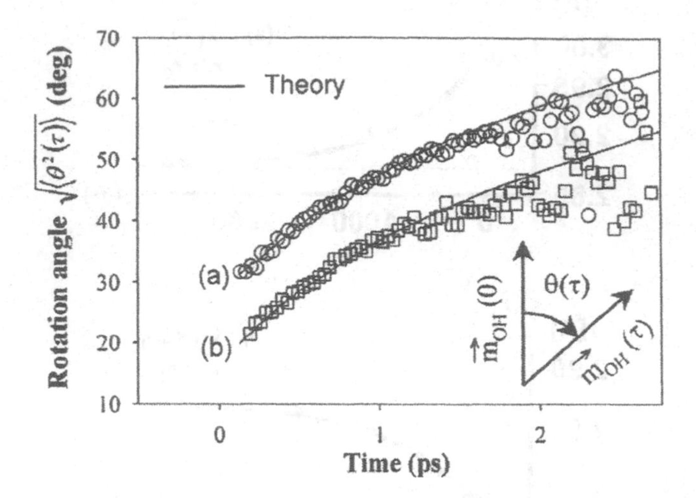

Figure 5. Real time "filming" of HDO rotations: The measured quantity is the mean squared rotational angle $\sqrt{<\theta^2(\tau)>}$ of the OH bond of HDO measured in degrees. The OH..O bond length is equal to 2.99 A in the curve (a), whereas it contracts from 2.99 A to 2.86 A in the curve (b). As expected, the OH rotations are slower in short OH..O bonds.

concludes that their lifetime in water is rotation-limited. Unfortunately, our camera is blind at very short times where the pump and probe pulses interact coherently: no visualization is possible at these time scales. Note however that $\sqrt{<\theta^2(0)>}$ is larger than zero for any finite pump-probe pulse duration. Nonlinear optical spectroscopy thus brings new contributions to the science of hydrogen bonding.

3. TIME RESOLVED X-RAY DIFFRACTION

3.1. GENERALITIES

A number of excellent text-books exist describing x-ray diffraction; see e.g. Ref. [22]. In a time-resolved x-ray diffraction experiment, the system is first submitted to an intense optical pump pulse, triggering a chemical reaction. Its time evolution is monitored next by the help of a x-ray probe pulse, emitted later with a well defined time delay. Changing it then permits to visualize the reaction products in real time. Unfortunately, this experiment, although conceptually simple, is difficult to realize in practice. Two major techniques are actually in use. (i) In the first of them, a synchrotron is employed to generate

(a) Synchrotron source

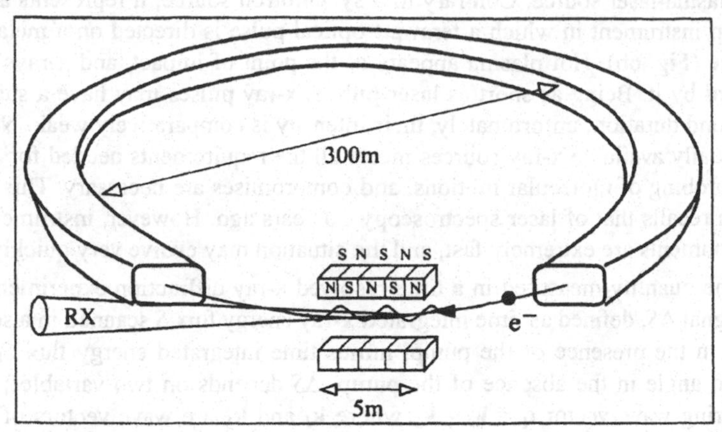

(b) Laser plasma source

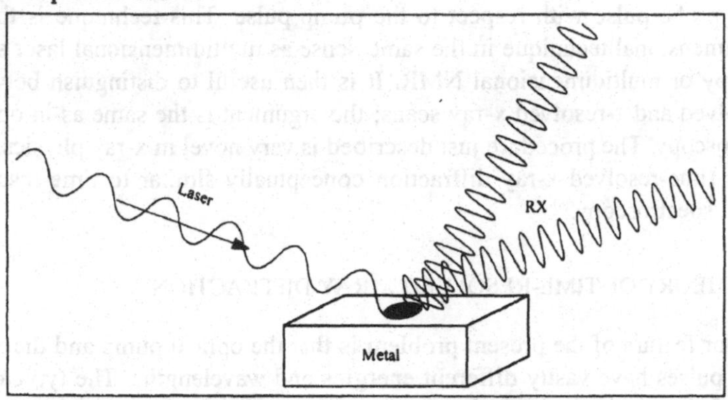

Figure 6. **Sources of pulsed x-ray radiation: (a) In a synchrotron source fast electrons circulate in a storage ring. Extracted from it, they cross an undulator which forces them to oscillate and to generate high intensity x-rays. Electronic devices are used to transform continuous x-ray radiation into pulses. (b) In a laser plasma source, a terawatt laser beam is focused on a metallic surface. The high temperature at the point of impact produces hot plasma emitting pulsed x-rays.**

fast electrons circulating in a storage ring; the dimensions of the latter are of the order of 300m (Fig. 6a). This ring is designed with straight sections, where an insertion device is placed. Called undulator, this device consists of a lattice of magnets which force the electrons to execute small oscillations; intense beams of radiation are produced in this way. If the electron current inside the storage ring is not constant, but is formed of bunches, pulsed x-rays are created. This source may generate high flux beams with pulses between 50 and 200 ps and a well defined structure and polarization. (ii) The second source is a plasma-laser source. Contrary to a synchrotron source, it represents a table top instrument in which a terawatt optical pulse is directed on a metallic surface (Fig. 6b). Hot plasma appears at the point of impact, and x-rays are emitted by it. Being as short as laser pulses, x-ray pulses may have a subpicosecond duration; unfortunately, their intensity is comparatively weak. None of actually available x-ray sources meets all the requirements needed for real time probing of molecular motions, and compromises are necessary. This situation recalls that of laser spectroscopy 30 years ago. However, instrumental developments are extremely fast, and the situation may evolve very quickly.

The quantity measured in a time-resolved x-ray diffraction experiment is the signal ΔS, defined as time integrated x-ray energy flux S scanned in a solid angle in the presence of the pump, minus time integrated energy flux S_0 in a solid angle in the absence of the pump. ΔS depends on two variables, the scattering wave vector $\mathbf{q} = \mathbf{k}_I - \mathbf{k}_S$ where $\mathbf{k}_I$ and $\mathbf{k}_S$ are wave vectors of the incident and scattered x-ray radiation, respectively, and on the time delay τ of the probe pulse with respect to the pump pulse. This technique is thus a two-dimensional technique in the same sense as multidimensional laser spectroscopy or multidimensional NMR. It is then useful to distinguish between $\mathbf{q}$-resolved and τ-resolved x-ray scans; the argument is the same as in optical spectroscopy. The procedure just described is very novel in x-ray physics, and makes time-resolved x-ray diffraction conceptually similar to time-resolved optical spectroscopy.

3.2. THEORY OF TIME-RESOLVED X-RAY DIFFRACTION

A major feature of the present problem is that the optical pump and the x-ray probe pulses have vastly different energies and wavelengths. The typical energy of photons in the optical pulse is on the order of a few eV, whereas that of photons in the x-ray pulse is in the keV range. This duality strongly complicates the experimental work but simplifies, rather than complicates, theoretical calculations. It makes a separate study of x-ray probing and of optical pumping possible, which reduces the complexity of the problem considerably. The following route may thus be chosen to construct a theory of time-resolved x-ray diffraction. The first step consists in developing a Maxwell-type theory of x-ray scattering by an optically excited system of charges and currents, and the second step in presenting a statistical description of pump-induced changes in the electron density. The following expression for the differential signal $\Delta S(\mathbf{q}, \tau)$ was reached in this way [28]:

$$\Delta S(\mathbf{q}, \tau) = \int_{-\infty}^{\infty} dt\, I_X(t-\tau)\, \Delta S_{inst}(\mathbf{q}, \mathbf{t}), \tag{4a}$$

$$\Delta S_{inst}(\mathbf{q},\mathbf{t}) = -(\frac{\mathbf{e}^2}{\mathbf{mc}^2\hbar})^2\mathbf{P}\int_0^\infty \mathbf{d}\tau_1 \int_0^\infty \mathbf{d}\tau_2$$
$$\times < E_i(\mathbf{r},\mathbf{t}-\tau_1)\mathbf{E}_j(\mathbf{r},\mathbf{t}-\tau_1-\tau_2) >_\mathbf{E}$$
$$\times < [[f(\tau_1+\tau_2)f^*(\tau_1+\tau_2), M_i(\tau_2)], M_j(0)] >_S \qquad (4b)$$

In this equation $I_X(\mathrm{t})$ is the x-ray beam profile, P is a factor depending on the polarization of the x-ray electric field, E_i, E_j are cartesian components of the pump electric field $\mathbf{E}$, M_i, M_j are cartesian components of the dipole moment vector $\mathbf{M}$ of the system and $f(t) = \int d\mathbf{r}\,\mathbf{n}(\mathbf{r},\mathbf{t}) \exp(-\mathbf{iqr})$ is the time-dependent x-ray form factor where $n(\mathbf{r},\mathbf{t})$ is the time-dependent electron density. The Einstein convention is employed all along indicating a summation over doubled indices. As in optical spectroscopy, the description of the experiment requires two correlation functions. The first of them involves the pump electric field $\mathbf{E}$, and the second the form factor f as well as the electric dipole moment $\mathbf{M}$ of the system. The symbol $< >_E$ indicates an averaging over different realizations of the incident optical field, and the symbol $< >_S$ that over the states of the non-perturbed system. Eq. (4) is the basic equation of time-resolved x-ray diffraction; it is of a very recent date. The similarity of Eqs. (3) and (4) should be emphasized.

3.3. PRINCIPLE OF THE EXPERIMENT

This technique will now be applied to re-examine an old problem of chemistry, the recombination of photo-dissociated iodine in solution. Being considered as a prototype of a "simple" chemical reaction, this process has been extensively studied in the past. It can be realized in two different ways. The solvent can trap the dissociated atoms in a solvent cage and force them to recombine; this is what is called geminate recombination. Alternatively, the atoms can also escape the cage and recombine with another partner; this is non-geminate recombination. Two points emerged from a long lasting research effort: (i) The iodine molecule possesses a large number of valence shell electrons, and thus has a large number of electronic states [23] (Fig. 7). Ten of them correlate with the atomic states $^2P_{3/2} + ^2P_{3/2}$ where the two iodine atoms are in the ground state $^2P_{3/2}$; and eleven other molecular states correlate with atomic states $^2P_{3/2} + ^2P_{1/2}$ and $^2P_{1/2} + ^2P_{1/2}$ where one iodine atom at least is in its first excited state $^2P_{1/2}$. These states are all repulsive, with exception of the states X,A,A' and B. (ii) As the electronic structure of molecular iodine is complex, the I_2 recombination reaction can not be simple, contrary to the initial intuition. Powerful techniques of time-resolved spectroscopy [24,25] and of computer simulation [26] were needed to elucidate essential features of these complex dynamics. Time scales involved are comparatively long and range in the nanosecond time domain.

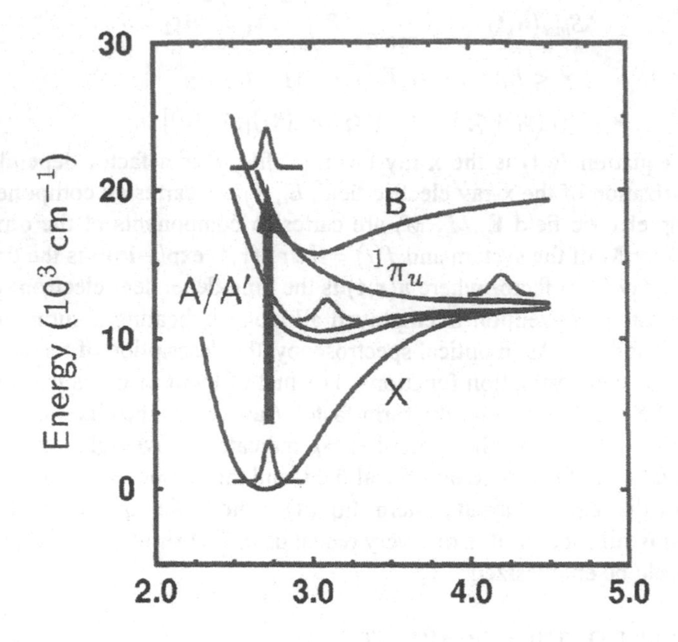

Figure 7. Electronic energy surfaces of I_2. Only four of ten electronic states correlating with the ground state X of I_2 are illustrated. The states A/A' and B are attractive and all the others are repulsive. The interatomic distance R_{eq} of I_2 is 2.68 A, and the interatomic distance R_0 of the hot molecule I_2^* is 4 A.

We shall now re-examine this problem by employing time-resolved x-ray diffraction techniques [27]. The principle of our experiment is as follows. One starts by exciting optically at 520 nm a dilute I_2/CCl_4 solution; this brings the iodine molecule into the electronic states B and $^1\pi_u$. These states dissociate in times of the order of 1 ps into the atomic state $^2P_{3/2}$ + $^2P_{3/2}$. A hot iodine molecule I^*_2 results, having a length R_0 of the order of 4A. Subsequent transformations $I^*_2 \Rightarrow 2I$ and $I^*_2 \Rightarrow I_2$ are just those we wish to study. Once triggered, the geminate and non-geminate recombinations are monitored using time-delayed x-ray diffraction. The x-ray signal $\Delta S(\mathbf{q}, \tau)$ of the solution was then measured as a function of both wave vector $\mathbf{q}$ and time τ. The optical pulse duration was of the order of 100 fs, and that of the x-ray pulse was of the order of 100 ps. Unfortunately, the x-ray source of the European Synchrotron Radiation Facility in Grenoble does not permit a better adjustment of these two pulses.

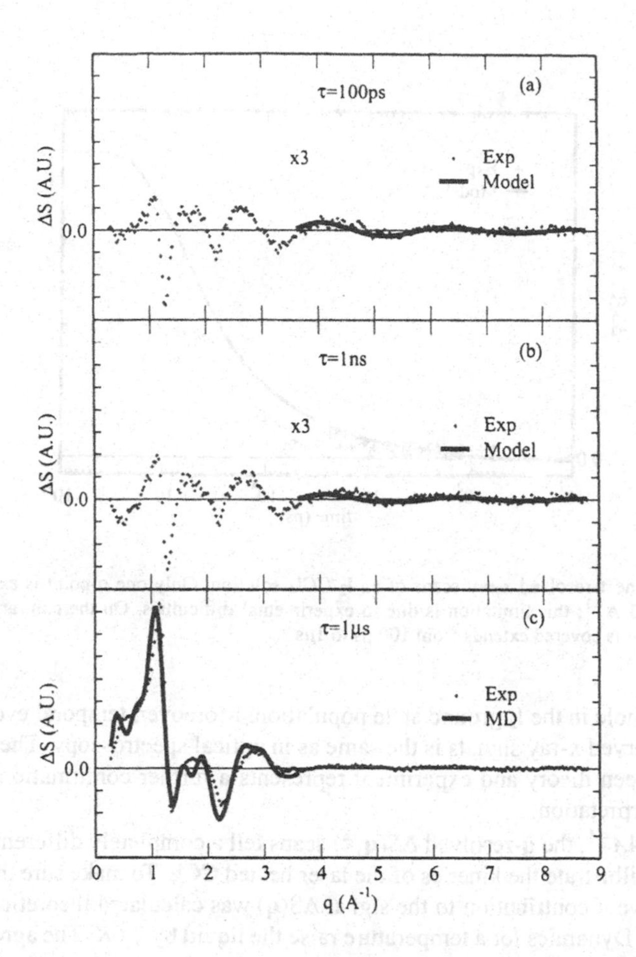

Figure 8. The q-resolved x-ray scans of an I_2/CCl_4 solution: The region which is explored extends from 0.5 A^{-1} to 9 A^{-1}. Three time points are presented where τ=100 ps (a), 1 ns (b) and 1μs (c). The low q-region where $q < 4A^{-1}$ indicates the CCl_4 heating, whereas the high q-region where $q > 4A^{-1}$ is governed by kinetics of photo-excited I_2.

3.4. MONITORING IODINE RECOMBINATION IN REAL TIME

The following conclusions were reached from our study. It was first noted that the **q**-resolved $\Delta S(\mathbf{q},\tau)$ scans permit to explore different chemical processes in different domains of **q** (Fig. 8). If $q > 4A^{-1}$ the x-ray scans mainly reflect the I_2 kinetics. The maxima of ΔS(q,τ) coincide nearly completely with the minima of S(q), which is expected if x-ray diffraction is due predominantly to the laser

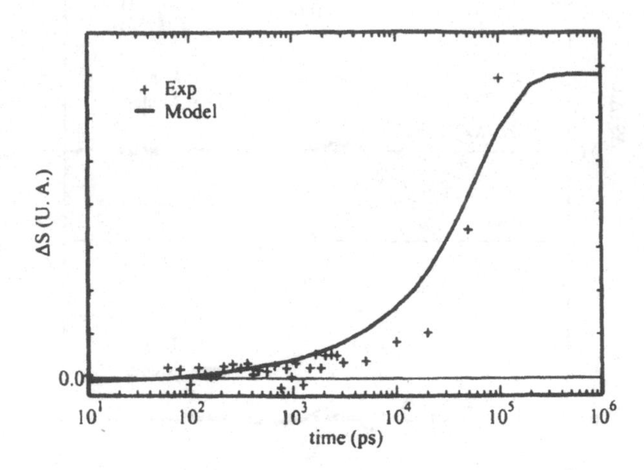

Figure 9. The τ-resolved x-ray scans of an I_2/CCl_4 solution: Only one q-point is examined where q=1.15 A^{-1}; this limitation is due to experimental difficulties. On the contrary, the τ domain which is covered extends from 100 ps to 1μs.

generated hole in the I_2 ground state population. Moreover, temporal evolution of the observed x-ray signals is the same as in optical spectroscopy. The agreement between theory and experiment represents a further confirmation of the above interpretation.

If $q < 4A^{-1}$, the **q**-resolved $\Delta S(\mathbf{q}, \tau)$ scans tell a completely different story. They then illustrate the kinetics of the laser heated CCl_4. To make sure that it is so, the solvent contribution to the signal $\Delta S(\mathbf{q})$ was calculated theoretically by Molecular Dynamics for a temperature raise the liquid by $2.6K$. The agreement between experiment and theory at long times, where the contribution of the iodine disappears, is perfect. All major peaks are properly located. Although the calculation is less accurate at short times, the agreement still remains good enough to confirm the present analysis. The I_2 and CCl_4 kinetics are coupled to each other through their common relaxation times: it should then be possible to study the kinetics of I_2 by "looking" on CCl_4!

Finally, the τ-resolved $\Delta S(\mathbf{q}, \tau)$ scans still remain to be discussed. Measuring them is more difficult than recording **q**-resolved scans: only the most intense diffraction peak at q=1.15 A^{-1} was examined for that reason. Its temporal evolution is illustrated in Fig. 9. As in this case $q < 4A^{-1}$, this peak mainly describes the expansion of the solvent due to its heating by recombination of I_2. The agreement between theory and experiment is, here again, satisfactory. Combining data provided by **q**-resolved and τ-resolved scans thus increases considerably our information about this prototype reaction.

4. CONCLUSIONS

It results from the present review that observing temporal variations of molecular geometry, as chemical reaction goes on, is no longer an unrealizable dream. Time-resolved optical spectroscopy, time-resolved x-ray diffraction and absorption are the main techniques permitting to attain this goal. Pulsed radiolysis or time-resolved electron diffraction may also be employed. As far as the theory is concerned, its main tools are statistical mechanics of non-linear optical processes and large-scale computer simulations. It should be strongly emphasized that basic theoretical concepts are all the same, in spite of huge differences in the experimental technology when going from optical spectroscopy to x-ray diffraction. As a whole, the field is still far from reaching its maturity. Nevertheless, a rapid progress may be anticipated for the next future

References

1. Dantus, M., Rosker, M., and Zewail, A.H. (1987) Real-time femtosecond probing of "transition states" in chemical reactions, *J. Chem. Phys.* **87**, 2395-2397.
2. Bernstein, R.B., and Zewail, A.H. (1989) Femtosecond real-time probing of reactions. III. Inversion to the potential from femtosecond transition-state spectroscopy experiments, *J. Chem. Phys.* **90**, 829-842.
3. Gale, G., Gallot, G., Hache, F., Lascoux, N., Bratos, S., and Leicknam, J-Cl. (1999), Femtosecond dynamics of hydrogen bonds in liquid water: A real time study, *Phys. Rev. Lett.* **82**, 1068-1071.
4. Bratos, S., Gale, G., Gallot, G., Hache, F., Lascoux, N., and Leicknam, J-Cl. (2000), Motion of hydrogen bonds in diluted HDO/D_2O solutions: direct probing with 150 fs resolution, *Phys. Rev. E* **61**, 5211-5217.
5. Rose-Petruck, C, Jimenez, R., Guo, T., Cavalleri, A., Siders, C.W., Ráksi, F., Squier, J.A., Walker, B.C., and Wilson, K.R. (1999), Picosecond-milliangström lattice dynamics measured by ultrafast x-ray diffraction, *Nature* **398**, 310-312.
6. Rousse, A., Rischel, C., Fourmaux, S., Uschmann, I., Sebban, S., Grillon, G., Balcou, Ph., Frster, E., Geindre, J.P., Audebert, P., Gauthier, J.C., and Hulin, D. (2001), Non-thermal melting in semiconductors measured at femtosecond resolution, *Nature* **410**, 65-68.
7. Srajer, V., Teng, T., Ursby, T., Pradervand, C., Ren, Z., Adachi, S., Schildkamp, W., and Bourgeois, D. (1996), Photolysis of the Carbon Monoxide Complex of Myoglobin: Nanosecond Time-Resolved Crystallography, *Science* **274**, 1726-1729.
8. Srajer, V., Ren, Z., Teng, T.Y., Schmidt, M., Ursby, T., Bourgeois, D., Pradervand, C., Schildkamp, W., Wulff, M., and Moffat, K., Protein Conformational Relaxation and Ligand Migration in Myoglobin: A Nanosecond to Millisecond Molecular Movie from Time-Resolved Laue x-ray Diffraction, *Biochemistry* **40**, 13802-13815.
9. Shen, Y.R. (1984), The Principles of Nonlinear Optics *Wiley*, New York.
10. Mukamel, S. (1995), Principles of Nonlinear Optical Spectroscopy, *Oxford University Press*, New York.
11. Rundle, R.E., and Parasol, M. (1952), O-H stretching frequencies in very short and possibly symmetrical hydrogen bond, *J. Chem. Phys.* **20**, 1487-1488.

12. Novak, A. (1974), Hydrogen bonding in solids. Correlation of spectroscopic and crystallographic data. *Structure and Bonding* **18**, 177-216.
13. Mikenda, W. (1986), Stretching frequency versus bond distance correlation of O-D(H)...Y (Y=N, O, S, Se, Cl, Br, I) hydrogen bonds in solid hydrates, *J. Mol. Struct.* **147**, 1-15.
14. Woutersen, S., Emmerichs, U., and Bakker, H.J. (1997), Femtosecond mid-ir pump-probe spectroscopy of liquid water: evidence for a two-component structure, *Science* **278**, 658-660.
15. Tao, T. (1969), Time dependent fluorescence depolarization and Brownian rotational diffusion coefficients of macromolecules, *Biopolymers* **8**, 609-632.
16. Fleming, G.R., Morris, J.M., and Robinson, G.W. (1976), Direct observation of rotational diffusion by picosecond spectroscopy, *Chem. Phys.* **17**, 91-100.
17. Bratos, S., and Leicknam, J-Cl. (1994), Ultrafast infrared pump-probe spectroscopy of water: a theoretical description, *J. Chem. Phys.* **101**, 4536-4546.
18. Bratos, S., and Leicknam, J-Cl. (1998), Anisotropy of pump-probe absorption of the hydrated electron. A statistical model, *J. Chem. Phys.* **109**, 9950-9957.
19. Gallot, G., Bratos, S., Pommeret, S., Lascoux, N., Leicknam, J-Cl., Kozinski, M., Amir, W., and Gale, G.M. (2002), Coupling between molecular rotations and OH..O motions in liquid water: theory and experiment, *J. Chem. Phys.* **117**, 11301-11309.
20. Nienhuys, H-K., Van Santen, R.A., and Bakker, H.J. (2000), Orientational relaxation of liquid water molecules as an activated process, *J. Chem. Phys.* **112**, 8487-8494.
21. Bakker, H.J., Woutersen, S., and Nienhuys, H-K. (2000), Reorientational motion and hydrogen-bond stretching dynamics in liquid water, *Chem. Phys.* **258**, 233-245.
22. Als-Nielsen, S., Morrow, D.MC. (2001), Elements of modern x-ray physics, *Wiley*, New York.
23. Mulliken, R. (1971), Iodine revisited, *J. Chem. Phys.* **55**, 288-309.
24. Chuang, T.J., Hoffman, G.W., and Eisenthal, K.B. (1974), Picosecond studies of the cage effect and collision induced predissociation of iodine in liquids, *Chem. Phys. Lett.* **25**, 201-205.
25. Harris, A.L., Brown, J.K., and Harris, C.B. (1988), The nature of simple photodissociation reactions in liquids on ultrafast time scale, *Ann. Rev. Phys. Chem.* **39**, 341-366.
26. Bergsma, J. P., Coladonato, M. H., Edelsten, P. M., Kahn, J. D., Wilson, K. R. and Fredkin, D. R. (1986), Transient x-ray scattering calculated from molecular dynamics, *J. Chem. Phys.* **84**, 6151-6160.
27. Neutze, R., Wouts, R., Techert, S., Davidsson, J., Kocsis, M., Kirrander, A., Schotte, F., and Wulff, M. (2001), Visualizing photochemical dynamics in solution through picosecond x-ray scattering, *P. R. L.* **87**, 195508.
28. Bratos, S., Mirloup, F., Vuilleumier, R., and Wulff, M. (2002) Time-resolved x-ray diffraction: Statistical theory and its application to the photo-physics of molecular iodine, *J. Chem. Phys.* **116**, 10615-10625.

REVERSE MONTE CARLO ANALYSES OF DIFFRACTION DATA ON MOLECULAR LIQUIDS

L. PUSZTAI
Research Institute for Solid State Physics and Optics, Hungarian Academy of Sciences, Budapest, P.O.Box 49., H-1525, Hungary

Abstract: The way Reverse Monte Carlo (RMC) modelling facilitates the interpretation of diffraction data taken on molecular liquids is described. It is suggested that the subtraction of the *intra*-molecular contributions, which is prone to numerical errors, can successfully be replaced by modelling the full structure factor using flexible molecular units in the RMC simulation. Moreover, details of the molecular structure in the liquid state may also be obtained in such a way, as will be demonstrated by the example of molten tungsten-hexachloride. It is shown that in many cases, one single *total* structure factor can provide information on the *partial* pair correlations. The required scattering vector range for a successful experiment is also discussed: the indications are that in a number of instances, measuring the structure factor up to about 10 Å^{-1} may be sufficient for capturing the most important features of the microscopic structure.

1. INTRODUCTION

The microscopic (atomic level) *structure* of a material is arguably its most basic property; the knowledge of it is essential for understanding (and/or evaluating) other properties, as well. The experimental technique that can provide direct information on the structure of a system in the condensed (solid and liquid) phases is (X-ray or neutron) *diffraction*.

For crystalline materials, in general, it is possible to determine the coordinates of each particle (atom or ion) in a 'building block' (unit cell) of the structure. For disordered (liquid and amorphous solid) systems, however, this is not possible, for the lack of such structural units; the desription of the structure is only feasible in terms of *correlation functions* (see, e.g., Ref. [1]). As a further principal restriction, diffraction data, i.e. the most important experimental source of information, can uniquely characterise *pair correlations* only. In practice, diffraction data are always subject to limitations (most importantly, in terms of the scattering vector (Q-) range available) and errors (systematic and statistical); due to these, even the pair correlation function is burdened with (sometimes, significant) uncertainties (see, e.g., Refs. [2,3]).

Our current subjects, molecular liquids have been the subjects of structural studies for decades (for an early study, see, e.g., Ref. [4]). Although the number of experiments carried out is really impressive, in most of the cases no in-depth knowledge of the atomic level structure could be gained. Most frequently, it is only rather

J. Samios and V.A. Durov (eds.), Novel Approaches to the Structure and Dynamics of Liquids: Experiments, Theories and Simulations, 129–142.

vague, qualitative statements that could be made on, for instance, the mutual orientations of neighbouring molecules (even when coupled with highly sophisticated theory, see, e.g., Refs. [5,6]).

The reasons for this deficiency of experiment-based knowledge are manyfold (multicomponent systems; unfortunate weighting factors; materials contain hydrogen which is bad for both X-rays and neutrons; complicated internal structure; etc...) but if one wants to name the origin of them then it can possibly be said that we simply want to derive far too many details on the grounds of measured quantities that are far too much of averaged in nature. The main consequence of all this is that there are many (sometimes, very) different local arrangements (orientations) of neighbouring molecules that will result in the same two-particle correlation function - and we are able to measure only this latter (or rather, something related to this latter) directly. There is not much to improve on this sad situation (because we have no experimental means for that). However, one always has to bear this inherent feature in mind while dealing with structural features of molecular systems (and not to over-interpret results, either from experiment or modelling).

2. EVALUATING DIFFRACTION DATA: THE TRADITIONAL WAY

This contribution focuses on making the point that *modelling must be an essential constituent of the interpretation of diffraction data* from molecular liquids. After considering general issues, some particular details of the interpretation procedure will be dealt with. It may be helpful if the traditional way of evaluating diffraction data (on molecular liquids) is described shortly, marking the points which we wish to address later.

(i) *Standard correction procedures, for calculating the structure factor from the measured angle dependent scattering intensity curve.* Although 'inverse' methods, such as the MCGR programme [2], can help during the correction stage by, for instance, subtracting a quadratic background (and thus mimicking inelastic corrections [7]), these possibilities are not considered here in detail.

(ii) *Subtraction of the intra-molecular contribution from the structure factor.* This stage is executed with the purpose of 'cleaning' the signal from contributions that are characteristic to the structure of the molecule only (for examples, see Refs. [8,9]). The way it is done is that one exponential expression is assigned to each (kind of) intra-molecular distance (the two parameters of the expressions are the distance itself and its variance, the Debye-Waller factor). The sum of these exponentials, the 'intra-molecular structure factor' is then fitted to the 'high Q part' of the measured (total) structure factor (tsf). Finally, by making use of the parameters of the best fit, it is possible to extrapolate the 'intra-molecular structure factor' to lower Q values and then, to subtract it from the total structure factor. While this is a perfectly legitimate procedure in principle, in practice one never knows exactly when the 'high Q part' of the structure factor, where inter-molecular contributions can be neglected, begins. Since very small uncertainties at the high Q bit can cause large discrepancies in terms of the values of the parameters of the exponential terms, subtraction at low Q may lead to large errors (in terms of the inter-molecular structure factor). Additionally, by definition, the inter-molecular part is considered as correlations between (atoms of) *rigid* molecules, which may not be justified in many cases. It is suggested here that this step should be omitted and the full structure factor, containing intra-

and inter-molecular contributions, should be considered throughout the entire evaluation process. As it will be shown by the example of tungsten-hexachloride, proper handling of flexible molecules can provide information on the molecular structure, as well.

(iii) *Separation of the partial structure factors (in the case of multi-component systems).* Traditionally, this step is made by inverting the matrix consisting of the coefficients of the partial structure factors in the different total structure factors (see, e.g., Ref. [10]). However, as many independent total structure factors are required for performing this step as many partials are present in the system - for a two component liquid, like water, three measurements (using different techniques, or more frequently, different isotopic compositions) are mandatory. The availability of suitable techniques/isotopes is far from guaranteed and therefore most frequently, carrying out the necessary number of experiments is impossible, making the determination of the partials impossible. By the example of carbon-tetrachloride, it will be shown that under favourable conditions, only one total structure factor is able to provide information on more than one partial.

(iv) *Fourier transformation to real space: obtaining (partial) pair correlation functions.* In order to minimise truncation errors thay may occur during this step, measurements should be made covering the widest Q-range, at least up to 20 $Å^{-1}$, that is available only at the best instruments. It will be shown here that by making use of inverse methods (like Reverse Monte Carlo modelling), that do *not* perform the Fourier transform from Q-space to r-space but only the inverse of this, the above requirement on the Q-range covered will be much less strict. (Note that similar statement can be made concerning point (ii), as well.)

(v) *Interpretation: calculating coordination numbers, constructing model orientational correlation fucntions, etc...* Based only on the (partial) pair correlation function(s), possibilities are rather limited. Carrying these post-evaluation steps out is much easier if structural models (with particle coordinates) are present. Here, only a limited exploitation will be demonstrated, by showing bond angle distribution functions for some liquids. Note, however, that anything that is calculable from particle coordinates can be evaluated if structural models are available.

3. WHY MODELLING ?

By 'models', we mean large collections of particles (atoms, ions) in a simulation box, represented by Cartesian coordinates; the number of particles nowadays is - routinely - about 10000. There are (at least) two major reasons why constructing models is thought to be unavoidable when dealing with (the structure of molecular) liquids:

(1) **Extending available experimental information**. We have to know (as much as possible about) the structure of (molecular) liquids; however, solely from experiments, it is only the pair correlation function, which is a kind of '(weighted, according to the relative scattering power of particles) interparticle distance spectrum', that can be accessed. A three dimensional model that represents the experimentally measured structure factor(/pair correlation function) will provide additional information, like details of the local symmetries around particles, that can be calculated from the particle coordinates.

(2) **Confirming the quality/reliability of measured data.** As it was already mentioned, (experiment-related) difficulties are encountered while obtaining struc-

ture factors and pair correlation functions from diffraction measurements. Suitable structural models can provide more confidence in experimental data, by setting up a *consistent system of measurement, the pair correlation function and three dimensional models.*

As an example of what ways proper modelling can provide valuable help, Figure 1 shows results of two different neutron diffraction measurements of liquid D_2O at ambient pressure and temperature. The two structure factors (one from a pulsed [11] and the other from a steady state [12] neutron source) differ noticably, particularly at higher scattering vector values. Since it is the same material and (at least, very nearly) the same technique, they should not: only *one* of the two measurements may be reliable. Reverse Monte Carlo modelling studies (see below) have been carried out for both sets of data and it turned out that when fitting one of the structure factors (from Ref. [12]), the fit actually resembles better to the *other* one (Ref. [11]). That is, modelling could actually tell more and less reliable sets of data apart.

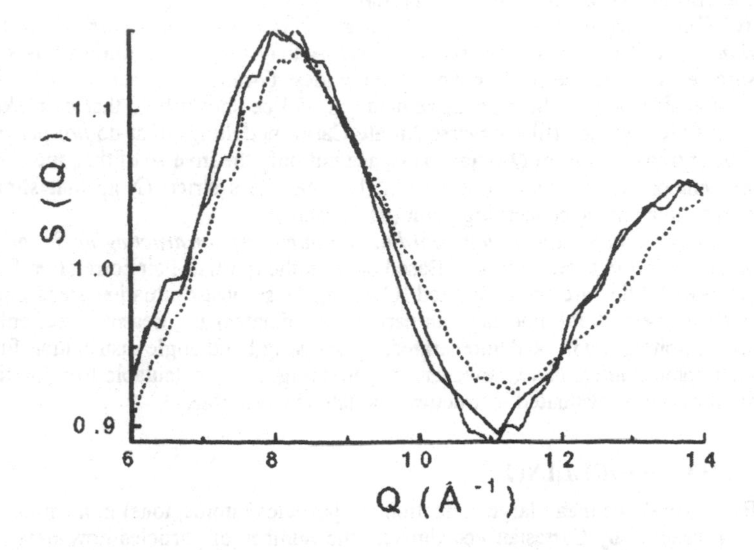

Figure 1. Total neutron structure factors for D2O. Solid line: from Ref. [11]; dotted line: from Ref. [12]; dash-dotted line: RMC-fit to data from Ref. [12].

As a consequence of the above, for learning the most that can be learnt about the structure of (molecular) liquids on the basis of diffraction experiments, we need large structural models that are *consistent with diffraction data* within their uncertainties. Unfortunately, more standard (molecular dynamics, Metropolis Monte Carlo) simulation techniques [13] are not, in general, able to provide such models, for the lack of suitable interatomic potential functions. (Also note that their aim is different, which is the understanding of the nature of interparticle interactions; modelling simply aims at the interpretation of diffraction results. It would be extremely desirable if the two went together - unfortunately, at the time of writing, this is practically impossible.)

In the following Section, one possible way of making use of molecular modelling will be shown.

4. REVERSE MONTE CARLO MODELLING OF MOLECULAR SYSTEMS

Here, we would first like to describe shortly what RMC is and then, to emphasize those properties which make it useful for interpreting diffraction results, particularly on molecular liquids. For more detailed descriptions of the RMC method, see Refs. [14,15].

Reverse Monte Carlo is a simple tool for constructing large, 3D structural models that are consistent (within the estimated level of their errors) with tsf's obtained from diffraction experiments. Via random movements of particles, the difference (calculated similarly to the χ^2-statistics) between experimental and model total structure factors is minimised. From the particle configurations, the partial pair correlation functions, as well as other structural characteristics (neighbour distributions, cosine distribution of bond angles) can be calculated.

For a molecular system, the choice of 'particles' to be moved is of particular importance. As it has already been mentioned, if the intra-molecular part is subtracted from the measured structure factor then it is rigid molecules that must be moved - this is because *no* information on the molecular structure remains in the intermolecular structure factor. However, as it was pointed out in, for instance, Ref. [16], the concept of rigid molecules, strictly speaking, is not consistent with measured diffraction data which are limited in terms of the available scattering vector range and contain errors. The use of flexible molecules is therefore being advocated, which is realised via atomic movements. The proper molecular structure is maintained via a kind of coordination constraints, coined 'fixed neighbour constraints' (fnc) [17,18]. Fnc's keep specified (by serial number) neighbour atoms within specified distance ranges of central atoms - that is, fnc's are essentially a (rather purpose-built) neighbour list. Although fnc's work fairly well for small molecules (see below), it has to be noted that they cannot provide the ideal solution for molecular liquids since they cannot cope with (or 'excite') rotational degrees of freedom directly - which are essential features of molecular systems. Rotation of small molecules can be brought about by atomic movements in practice, but even then, rod-like and flat molecules (or parts of molecules) cannot be handled properly. For this reason, another RMC code has been written very recently (and is being tested right now) which allow rotations and 'vibrations' (atomic displacements) alike [19].

The Reverse Monte Carlo algorithm for molecular systems, containing *flexible molecules*, may be given as follows:

1. Start with an initial configuration with periodic boundary conditions. The positions of the N(>4000) atoms should be consistent with the molecular structure already at the start. Assume some tolerance for the intramolecular (bonded and non-bonded) distances.

2. Calculate the partial pair distribution functions for this configuration

$$g_{ij}^{C,o}(r)=\frac{n_j^{C,o}(r)}{4\pi r^2\Delta r\rho_j} \tag{1}$$

where $n_j^{C,o}$ is the number of atoms of type j at a distance between r and $r+\Delta r$ from a central atom of type i, averaged over all atoms of type i as centres. The superscripts C and o refer to 'calculated' and 'old', respectively. ρ_j is the number density of particles type j.

3. Compose the total pair distribution function(s) from the partials, weighted according to the concentrations and scattering powers of atoms:

$$G^{C,o}(r) = \sum_i \sum_j c_i c_j \overline{b_i} \overline{b_j} \left(g_{ij}(r) - 1\right) \tag{2}$$

where c's are molar fractions and b's are the (neutron) scattering lengths of the components. (Note that for X-rays, a more complicated way has to be followed since the X-ray scattering power of atoms depends on the scattering vector; for simplicity, throughout this contribution, the formalism of neutron scattering will be applied.) Transform to the total structure factor

$$S^{C,o}(Q) - 1 = \frac{4\pi\rho}{Q} \int_0^\infty r G^{C,o}(r) \sin Qr dr \tag{3}$$

It is worthwhile remembering that in the RMC procedure, only transforms from r to Q space occur, which makes it easier to handle truncation errors, since the extent of the available r-range can be increased by using larger models.

4. Calculate the difference between the experimental total structure factor $A^E(Q)$ and that calculated from the configuration $A^C(Q)$

$$\chi_o^2 = \sum_k \frac{\left\{S^{C,o}(Q_k) - S^E(Q_k)\right\}^2}{\sigma^2} \tag{4}$$

where the sum is over all the experimental points and σ is the - estimated - experimental error, which functions as a 'control parameter' for the simulation. (By prescribing how close a fit to experiental data is required, σ controls the ratio of the numbers of accepted/attempted moves).

5. Move one atom at random.

6. Check if all the atoms are within the prescribed tolerance values of the intramolecular distances (i.e., if the molecules are still held together) after the attempted move. Try another move if not.

7. Calculate the new pair distribution function, $G^{C,n}(r)$ and total structure factor, $S^{C,n}(Q)$, and the new difference between model and experiment, χ^2_n.

8. If $\chi^2_n < \chi^2_o$ the move is accepted and the new configuration becomes the old configuration. If $\chi^2_n > \chi^2_o$ then the move is accepted with probability $\exp\left\{-\left(\chi_n^2 - \chi_o^2\right)/2\right\}$.

9. Repeat from step 5.

In a 'normal' case (reliable data; meaningful input parameters), χ^2 will decrease until it reaches its minimum value, which minimum is linked to the given value of σ. After reaching its minimum (the state 'structural equilibrium'), χ^2 will oscillate so that the ratio of the numbers of accepted/attempted moves will not decrease. If it does (and the run eventually would 'freeze in') then it is an indication that the structure formed is not the kind of 'global' minimum the algorithm is seeking. For a demonstration of what an acceptable agreement between experimental and model structure factors is, in Figure 2, the structure factor from an RMC model of liquid D_2O is compared to the experimental data of Ref. [11].

Over the years, RMC modelling helped us to understand a number of issues concerning structural disorder in general. Some of them are really obvious - so obvious that many times, we would be inclined to overlook them. It would be impossible to provide full background to every one of them here, so let us just list some of these points, with - at least - references to more detailed descriptions.

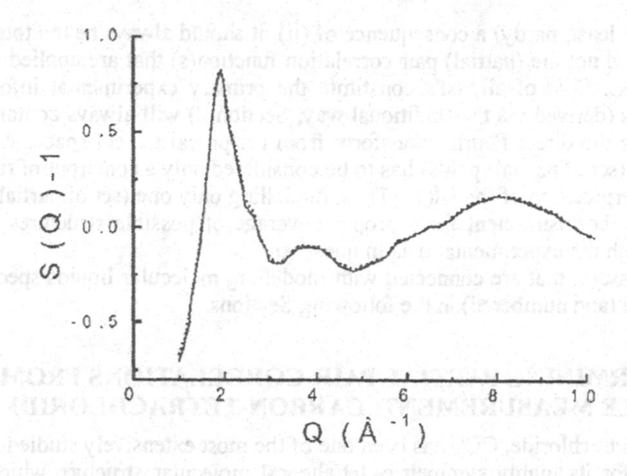

Figure 2. Total neutron structure factors for D_2O. Solid line: experimental result from Ref. [11]; dotted line: RMC-fit.

(i) *There is always more than one structural model that would fit a given (set of) data -- and this has nothing to do with RMC!* RMC has just happened to be the technique which started to produce different models for a given data (see e.g. Refs. [20,21]), but it has to be recognised that it is the diffraction data that allow for diversity of models. Therefore, one must explore, as fully as (technically) possible, the 'configuration space' available for models connected to a (set of) measurement(s).

(ii) *In practice, there is always more than one (set of partial) pair correlation function(s) corresponding to a given (set of) tsf's.* This fact, demonstrated, for instance, in Refs. [3,22,23], is even less appreciated than the one under (i). There is no magic about it: this is just the consequence of the fact that there is no 'perfect data': your data will never span an infinite scattering vector range (which would be necessary for a perfect Fourier-transform) and your data will never be error-free, either (which would be a pre-requisite for a unique separation of the partials). That is, one has to consider more than one solution in terms of the pair correlation functions, as well.

(iii) One way of attempting to meet the above requirements is *the introduction of geometrical* (i.e., ones that can be formulated on the basis of the particle coordinates) *constraints*. These constraints can, for instance, represent specific ideas about the microscopic structure of the material in question; in this way, one will be able to tell if a given idea is consistent with existing measurements or not. That is, RMC can help us to select ideas/models which are (or are not) acceptable, on a very strict basis.

(iv) Modelling the structure factor with RMC will help to spot if there are problems (systematic errors) with that tsf -- a clear demonstration of this feature can be seen in Figure [1] (see also in Ref. [23]). Experience shows that if sensibly executed Reverse Monte Carlo runs do not lead to a satisfactory agreement with tsf's then it is quite likely that it is the experimental structure factor that should be reconsidered (re-analysed; a little more theoretical approach to this problem can be found in Ref. [24]).

(v) As (at least, partly) a consequence of (ii), it should always be the total *structure factor* and not the (partial) pair correlation function(s) that are applied as input data for RMC. First of all, tsf's constitute the primary experimental information whereas pcf's (derived via the traditional way, Section 2) will always contain errors related to the the direct Fourier transform from reciprocal to real space. Additionally, a given (set of partial) pcf('s) has to be considered only as one (out of the many possible) interpretation of the tsf('s). Thus, modelling only one (set of partial) pcf('s) will certainly be insufficient for a proper coverage of possible structures that are consistent with the experimental data in question.

Further issues, that are connected with modelling molecular liquids specifically, are discussed (and numbered) in the following Sections.

5. DETERMINING PARTIAL PAIR CORRELATIONS FROM A SINGLE MEASUREMENT: CARBON-TETRACHLORIDE

Carbon-tetrachloride, CCl_4 has been one of the most extensively studied molecular liquids, for its highly symmetric, tetrahedral molecular structure which is the prototype for many other systems (for references and more detailed description, see Refs. [16,18]). In Figure 3, the structure factor for the material is shown, as measured at two neutron wavelengths on the PSD diffractometer [25] at the Budapest Reserch Reactor (Hungary). Both results can be modelled easily by RMC and they provide equal structural characteristics [26].

The main challenge about the microscopic structure of this (and also, many other) liquid is the nature of the *orientational correlations*, the kind of mutual orientations of neighbour molecules. This question has been resectly been discussed in detail [18]; in short, it was found that the 'corner-to-face' type docking ('Apollo-model' [8]) of CCl_4 neighbours, that had been considered at face value for at least three decades, cannot be the dominant in carbon-tetrachloride. (The dominance of Apollo-connections could be excluded by putting Apollo-pairs in the RMC models directly; as a result, the goodness of fit has deteriorated noticeably [18].) Instead, a (admittedly, much less plausible - but at least, possibly correct) 'corner-to-corner' connection is suggested. Here, these particulars are not discussed further; instead, a couple of more general points are to be made.

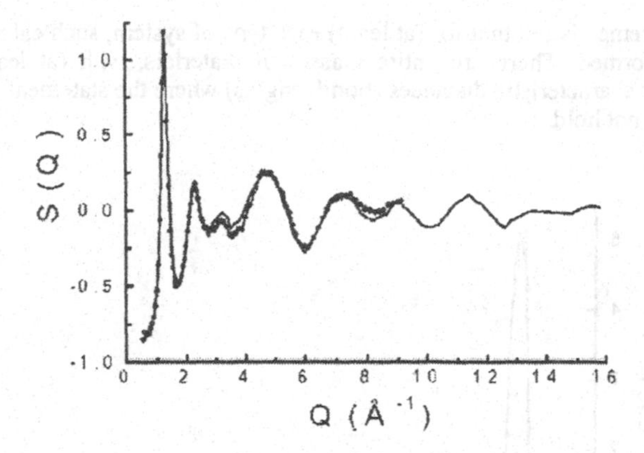

Figure 3. Total neutron structure factors for liquid CCl4. Solid line: λ=0.66 Å; dotted line: λ=1.06 Å.

(vi) *In quite a few cases, partial pair correlation functions can be determined with high accuracy from only one experiment, by means of RMC modelling.* For this to be done, molecules have to have a centre which is surrounded by a shell of its ligands; in CCl_4, the centre is the carbon atom which is buried in the shell of the four chlorines. The molecular geometry has to be known well and it should not be too floppy. Furthermore, it is a great advantage if the shell-forming atoms scatter the most. If all these requirements are met then the ligand-ligand pair correlations will dominate the diffraction signal. Since the position of the ligands determine the position of the central atom, from a 3D structural model that is consistent with the (single) experimental structure factor, all the three partials can be calculated. This is how for carbon-tetrachloride, the three partial pcf's (and structure factors) could be determined by RMC modelling the first time [16].

(vii) *In quite a few cases, relatively short structure factors, measured up to about 10 $Å^{-1}$, will be sufficient for a satisfactory description of the structure.* This statement can be checked by RMC modelling, for instance, the two structure factors of Figure 3 independently and then comparing (partial) pair correlation functions and other characteristics, like bond angle distributions, for the two models. In Figure 4 the comparison of the partial Cl-Cl pair correlation functions is shown. Differences arc minute, although the *Q*-range was nearly twice as wide for the short wavelength measurement - which indicates that if proper methods are applied then even features that are thought to be hidden at higher scattering vector values can be retrieved with excellent accuracy. (If cosine distributions of Cl-C-Cl bong angles are compared then differences become invisible, see Ref. [18].)

The above simple demonstration is, obviously, insufficient for making generalised statements - although it does indicate that not even for molecular liquids is always mandatory to measure the structure factor over very wide *Q*-ranges. Additionally, we note that for many other types of (model) systems, much more detailed investigations have been carried out - with essentially the same results. For amorphous Si, the necessary upper *Q*-value was found to be 6 $Å^{-1}$, for liquid Ga, about 4 $Å^{-1}$ and for an amorphous Ni-Nb alloy, about 8 $Å^{-1}$ [28]. However impressive these findings are, it has to be remembered that for (at least) each type of system, such calculations have to be performed. There are entire classes of materials, with (at least two) slightly different characteristic distances (bond lengths) where the statement of point (vii) will clearly not hold.

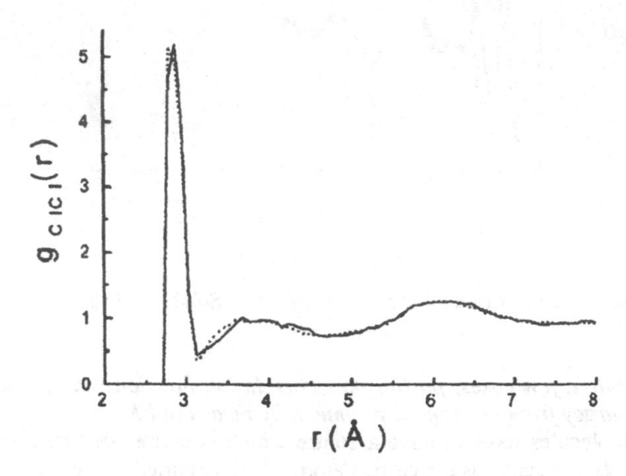

Figure 4. Cl-Cl partial pair correlation functions for liquid CCl4, calculated from RMC models. Solid line: λ=0.66 Å; dotted line: λ=1.06 Å.

6. DETERMINATION OF THE INTRAMOLECULAR STRUCTURE: LIQUID TUNGSTEN-HEXACHLORIDE

Having understood (thought-to-be) well known systems, it is logical to move towards the totally unknown. In this spirit, we have carried out neutron diffraction measurements on liquid tungsten-hexachloride, WCl_6. WCl_6 is a solid at room temperature, melting at 557 K and even that was unknown if it became a molten salt (with free ions) or a molecular liquid. For this reason, the subtraction of the intramolecular part was not just unnecessary but also, impossible this time. The measured structure factor, together with RMC modelling results, is shown in Figure 5 (the measurement was carried out at the Studsvik Neutron Research Laboratory, in Sweden [29]).

In order to find out about the species in the liquid, several RMC calculations with different constraints have been performed. As it is obvious from Figure 5, flexible molecules (held together by fnc's) can be used without difficulties. What is even more interesting is that if do a hard sphere Monte Carlo (HSMC) calculation using the same species (flexible molecules, with the same fnc - but no data to fit!) as in the RMC then the molecular geometries for the two models differ enormously. This is demonstrated in Figure 6, by comparing the cosine distribution for Cl-W-Cl bond angles. The HSMC geometry seems more or less random, within the (quite strong) steric limitations imposed by the fnc's. In contrast, the RMC geometry, brought about by the experimental data, shows up as a very well defined octahedron, with Cl-W-Cl bond angles found only at 90 and 180°. This is a remarkable result, also found for liquid $SbCl_5$ [29] (with different actual values).

From other structural characteristics calculated from the particle coordinates, it appeared that the intermolecular correlations are much weaker than found in carbon-tetrachloride. Looking now at the structure factors (Figure 5), it is clear that the intensity of the first ('sharp diffraction') peak (or 'pre'-peak) must depend on the *intra*molecular structure, as well. Without pressing this point any further, we note that this finding is in sharp contrast with the conventional ('textbook') wisdom that first sharp diffraction peaks are chracteristic to intermediate range order exclusively.

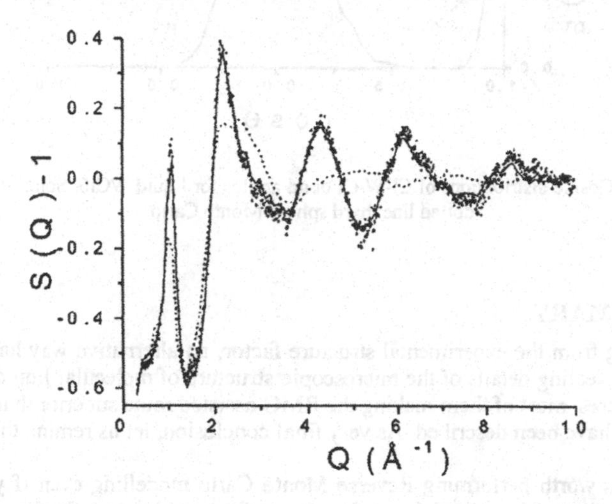

Figure 5. Total structure factors for liquid WCl_6, calculated from RMC models. Markers: experiment [29]; solid line: RMC fit, using flexible molecules (fnc); dotted line: hard sphere Monte Carlo with flexible molecules (same molecules as in RMC; see text).

Now it is possible to formulate a few more points concerning structural studies of molecular liquids by RMC:

(viii) In some cases, the relatively *short structure factor* allows us to *determine details of the intramolecular structure*. This is a clear indication for that the subtraction of the intramolecular contribution is unnecessary.

(ix) It has proved very helpful (also, for liquid CCl_4 [16]) to set up a 'reference system' for molecular liquids, by imposing the same molecular structure on hard sphere Monte Carlo calculations as on RMC. The HSMC runs will then produce particle configurations that reflect the effects of all constraints (density, cut-offs, molecular structure). Comparing corresponding RMC and HSMC structures allow us to separate features that can be ascribed to the experimental structure factor (that is, the physical nature of the system under study).

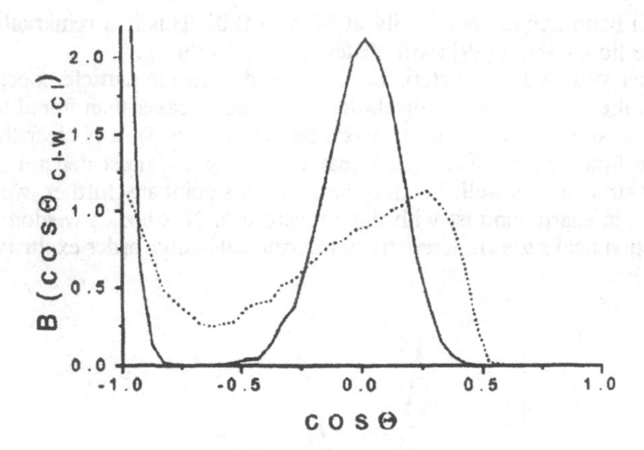

Figure 6. Cosine distributions of Cl-W-Cl bond angles for liquid WCl6. Solid line: RMC; dotted line: hard sphere Monte Carlo.

7. SUMMARY

Starting from the experimental structure factor, an alternative way has been devised for revealing details of the microscopic structure of molecular liquids. A number of features, most of them making the RMC-assisted route superior than the traditional one, have been described. As very final conclusion, let us remind the reader of two things:

(1) It is worth performing Reverse Monte Carlo modelling even if you do not want to use the resulting particle configurations for anything (c.f. Section 2).

(2) If you do wish to use the models, however, then producing one model is not sufficient - by the application of (geometrical) constraints, the range of structures that are consistent with the experiment has to be explored.

Acknowledgment

We thank the Hungarian Basic Research Fund (OTKA), under Grant No. T 32308, for financial support.

REFERENCES

1. Hansen, J.-P., McDonald, I.R. (1986) *The Theory of Simple Liquids,* Academic Press, London.
2. Pusztai, L., McGreevy, R.L. (1999) MCGR: an Inverse Method for Deriving the Pair Correlation Function, *J. Neutron Research* **8**, 17-35.
3. Pusztai, L., McGreevy, R.L. (1998) The structure of molten CuBr, *J. Phys. : Cond. Matter* **10**, 525-532.
4. Narten, A.H., Danford, M.H., Levy, H.A. (1967) Structure and Intermolecular Potential of Liquid Carbon Tetrachloride Derived from X-ray Diffraction Data *J. Chem. Phys.* **46**, 4875-4880.

5. Montague, D.G., Chowdhury, M.R., Dore, J.C., Reed, J. (1983) A RISM analysis of structural data for tetrahedral molecular systems, *Mol. Phys.* **50**, 1-23.
6. Bermejo, F.J., Enciso, E., Alonso, J., Garcia, N., Howells, W.S. (1988) How well do we know the structure of simple molecular liquids ? CCl_4 revisited, *Mol. Phys.* **64**, 1169-1184.
7. Pusztai, L., McGreevy, R.L. (1997) MCGR: an inverse method for deriving the pair correlation function from the structure factor, *Physica B* **234-236**, 357-358.
8. Egelstaff, P.A., Page, D.I., Powles, J.G. (1971) Orientational correlations in molecular liquids by neutron scattering. Carbon tetrachloride and germanium tetrabromide, *Mol. Phys.* **20**, 881-894.
9. Clarke, J.H., Dore, J.C., Gibson, I.P., Granada, J.R., Stanton, G.W. (1978) Neutron Diffraction Studies of Tetrachloride Liquids, *Faraday Disc. Chem. Soc.* **66**, 277-286.
10. Enderby J.E., North D.N., Egelstaff P.A. (1966) The partial structure factors of liquid Cu-Sn, *Phil. Mag.* **14**, 961-970.
11. Soper A.K., Bruni F., Ricci M.A. (1997) Site-site correlation functions of water from 25 to 400 °C: Revised analysis of new and old diffraction data, *J. Chem. Phys.* **106**, 247-254.
12. Bellisent-Funel, M.-C., Teixeira, J., Bosio, L. (1987) Structure of high density amorphous water. II. Neutron scattering study, *J. Chem. Phys.* **87**, 2231-2235.
13. Allen, M.P., Tildesley, D.J (1987) *Computer Simulation of Liquids*, Clarendon Press, Oxford.
14. McGreevy, R.L., Pusztai, L. (1988) Reverse Monte Carlo Simulation: A New Technique for the Determination of Disordered Structures, *Molec. Simul.* **1**, 359- 367.
15. Pusztai, L. (1998 Structural modelling using the reverse Monte Carlo technique: Application to amorphous semiconductors *J. Non-Cryst. Sol.* **227-230**, 88-95; McGreevy, R.L. (2001) Reverse Monte Carlo modelling, *J. Phys. : Cond. Matter* **13**, R877-R913.
16. Pusztai, L., McGreevy, R.L. (1997) The structure of liquid CCl_4, *Mol. Phys.* **90**, 533-540.
17. Pusztai, L., McGreevy, R.L. (1997) RMC: introduction of a new type of constraint for molecular systems and network glasses, Studsvik NFL Annual Report for 1996, OTH :21.
18. Jóvári, P., Mészáros, Gy., Pusztai, L., Sváb, E. (2001) The structure of liquid tetrachlorides CCl_4, $SiCl_4$, $GeCl_4$, $TiCl_4$, VCl_4 and $SnCl_4$, *J. Chem. Phys.* **114**, 8082-8090.
19. Evrard, G. (2003) The RMC++ software (private communication)
20. Howe, M.A., McGreevy, R.L. (1991) Determination of three body correlations in liquids by RMC modelling of diffraction data. I. Theoretical tests, *Phys. Chem. Liq.* **24**, 1-12.
21. Gereben, O. and L Pusztai, L. (1994) Structure of amorphous semiconductors: Reverse Monte Carlo studies on a-C, a-Si and a-Ge', *Phys. Rev. B*, **50**, 14136- 14143.
22. Pusztai, L. (1999) On the partial pair correlation functions of liquid water, *Phys. Rev. B* **60**, 11851-11854.
23. Pusztai, L. (2000) On the structure of high and low density amorphous ice, *Phys. Rev. B* **61**, 28-31.
24. Pusztai, L., Gereben, O., Baranyai, A. (1994) Some remarks on the measured structure factor, *Physica Scripta*, **T57**, 69-71.
25. Sváb, E., Mészáros, Gy., Deák, F. (1996) Neutron powder diffractometer at the Budapest reseach reactor, *Materials Science Forum* **228-231**, 247-252.
26. Jóvári, P., Mészáros, Gy., Pusztai, L., Sváb, E. (2000) Neutron diffraction studies on liquid CCl_4 and C_2Cl_4. *Phyica B* **276-278**, 491-492.

27. Gereben, O., Pusztai, L. (1995) Determination of the microscopic structure of disordered materials on the basis of limited Q-space information, *Phys. Rev. B* **51**, 5768-5772.
28. Pusztai, L., McGreevy, R.L. (2003) The structure of simple molecular liquids $SbCl_5$, WCl_6 and CS_2 (in preparation)

STRUCTURAL CHANGE AND NUCLEATION CHARACTERISTICS OF WATER/ICE IN CONFINED GEOMETRY

JOHN DORE, BEAU WEBBER
Physics Lab, University of Kent, Canterbury CT2 7NR, UK

DAN MONTAGUE
ex-Willamette University, Oregon, USA

THOMAS HANSEN
Institut Laue-Langevin, F 38042 Grenoble Cedex 9, France

Contact author: J.C.Dore@ukc.ac.uk

1. INTRODUCTION

The confinement of liquids in the restricted space of a mesoporous solid leads to significant changes in their properties. The main effect is a reduction in the nucleation temperature that, according to the Gibbs-Thomson formalism is inversely proportional to the pore size. Other effects on the liquid structure and transport properties also occur and depend on the detailed characteristics of the mesoporous material. Confined liquids occur in various situations, both in the natural environment and also in industrial processes, so that there is a growing interest in understanding the fundamental principles that govern the behaviour and modified characteristics.

For scientific study it is important to have a well-characterised system in which the solid matrix has a large pore volume, controlled narrow pore size distribution function and high surface area. The most convenient material for this purpose is silica and there is a wide range of pore sizes available that can be produced by the sol-gel process. More recently, other forms of mesoporous silica such as MCM and SBA-types have been fabricated with an ordered arrangement of cylindrically-shaped pores made by a template process. The internal silica surface is normally smooth and hydrophilic so that surface wetting is not usually a problem and the water is easily adsorbed into the pore volume. In contrast, the activated carbons, which are also extremely important in industrial use, have a much more complex pore and surface structure which makes a detailed interpretation of the experimental measurements more difficult. Consequently, the main interest has focused on silicas as the ideal substrate and the initial investigation of liquids in confined geometry has been centred on these materials.

J. Samios and V.A. Durov (eds.), Novel Approaches to the Structure and Dynamics of Liquids: Experiments, Theories and Simulations, 143–156.

2. THEORETICAL BACKGROUND

The formalism for neutron scattering by molecular liquids has been presented elsewhere [1,2] and only a brief summary will be given here. The measured diffraction pattern can be converted into a molecular structure factor, $S_M(Q)$, which may be divided into intra- and inter-molecular contributions, i.e.

$$SM(Q)=fl(Q)+DM(Q) \tag{2.1}$$

where $f_l(Q)$ is the molecular form factor and $D_M(Q)$ defines the spatial relationship between the molecules. The real-space distribution function $d_L(r)$ can be obtained from the Fourier-Bessel transform:-

$$d_L(r) = 4\pi r\, \rho_M\, [g_L^{-}(r) - 1] = \frac{2}{\pi} \int_0^{\infty} QD_M(Q)\sin Qr\, dQ \tag{2.2}$$

where ρ_M is the molecular number density and $g_L^{-}(r)$ is a composite pair correlation function, which for D_2O is :-

$$g_L^{-}(r) = 0.092\, g_{OO}(r) + 0.423\, g_{OD}(r) + 0.485\, g_{DD}(r) \tag{2.3}$$

The measurement for liquids confined in the voids of a solid porous substrate requires the subtraction of the substrate scattering from the total scattering by the 'liquid + substrate'. This method needs careful adjustment of the relative intensities to allow for the variable attenuation factors. The direct analysis also assumes that the interference contributions resulting from the liquid-substrate cross-term are negligible and this factor depends on the pore size, which influences the relative contributions. This approximation is satisfactory for pore sizes of >50 Å but is inadequate at lower values. The dispersed nature of the liquid also leads to an effective broadening of the diffraction profile if the pore dimensions are small. This phenomenon is similar to that occurring in the diffraction broadening of Bragg peaks for small crystallite sizes and is a result of the limitations in the volume integral which leads to an effective convolution of the full pattern with a function M(Q), that is dependent on the pore size distribution [3]. In the case of liquids, the overall shape of the diffraction pattern does not have sharp features and the diffraction broadening effect is significant only for small pores (<20Å).

Another method of analysis is to use a first-order difference function to analyse the structural changes with temperature. This technique has a number of advantages, particularly for studies in confined geometry, since it eliminates the substrate and intra-molecular scattering in a systematic manner. The temperature difference function is defined by

$$\Delta D_M(Q, \Delta T) = S_M(Q,T) - S_M(Q,T_o) \tag{2.4}$$

in relation to a chosen reference temperature, T_o with $\Delta T = |T-T_0|$ and the corresponding transform yields the real-space function

$$\Delta d_L(r, \Delta T) = \frac{2}{\pi} \int_0^{\infty} Q\, \Delta D_M(Q, \Delta T) \sin Qr\ dQ, \tag{2.5}$$

which defines the structural re-arrangement. The low r-value behaviour of $d_L(r)$, where g(r) is zero, gives the microscopic density through the relation

$$d_L(r) = -(4\pi\, \rho_M)\, r \tag{2.6a}$$

and the corresponding expression for $\Delta d_L(r)$ is

$$\Delta d_L(r) = -(4\pi\, \Delta\rho_M)\, r, \tag{2.6b}$$

which depends on the change in density, $\Delta\rho_M$ (ΔT).

3. REVIEW OF EARLIER RESULTS

a) Super-cooled water in sol-gel silicas

The first neutron diffraction studies of water in a series of commercial sol-gel silicas [3,4] were reported in the early 1980's and were followed by further studies covering different pore sizes which have been reviewed comprehensively [5] for work done up to 2000. Normally, super-cooled water would be metastable at these temperatures but the effect of the confinement means that the liquid phase is in an equilibrium state and forms readily under these conditions. The neutron results confirmed the depression of nucleation point, which had been observed by other techniques, and provided information on the changing structure of the water network. The main features are illustrated in Figs 1 and 2, which show the variation in the diffraction pattern for D_2O water with temperature and the shift of the main diffraction peak position as a function of temperature, $Q_0(T)$, for several mesoporous silicas of varying pore size. The peak profile in Fig 1 changes systematically and indicates the development of a shoulder on the high-Q side as the temperature is reduced. The variation of the peak position shown as a solid line in Fig 2a is monotonic across the whole temperature range but shows a systematic increase in slope below the point of maximum density, which is 11°C for D_2O. The trend is towards a value of 1.7 $Å^{-1}$ at a temperature of –40°C, which corresponds to the limiting case of homogeneous nucleation of D_2O at ambient pressure [6]. The confinement of water in pores gives a displacement of the peak position which may be attributed to enhanced hydrogen-bonding and is shown as single points for a range of silicas with different pores sizes.

Most molecular liquids would have a very small change in the Q_0 value arising simply from density changes but water exhibits a much larger effect that can be linked to the extended formation of the hydrogen-bonded network. The studies of deeply super-cooled water in the bulk phase have demonstrated the enhancement of the intermediate-range correlations as the temperature is reduced, as shown in Fig 2b. The picture to emerge from these studies is of an evolution towards the extended continuous random network of tetrahedral hydrogen-bonds that characterises the structure of low-density amorphous ice [7]. The overall behaviour of the confined water is therefore similar to that of bulk water, except that the hydrogen-bonded network appears to be more developed relative to that of bulk water at the same temperature, as shown by the data in Fig 2.

b) Water and ice in MCM silicas

MCM silicas consist of two main types with cylindrical pores of 30-35 Å diameter; MCM41 is based on straight parallel channels arranged in an approximately hexagonal lattice, whereas MCM48 is formed from a cubic liquid crystal phase which has branched channels of a similar diameter. Neutron diffraction measurements of D_2O water in both of these materials have been made and show that deep super-cooling occurs in which the liquid phase may exist at 45°C below the normal freezing point [5°C for D_2O]. The diffraction pattern undergoes a systematic change as the temperature is lowered, as shown in Fig 3a. The temperature difference function, $\Delta D_M(Q,\Delta T)$, can be evaluated from these datasets and transformed to give the $\Delta d_L(r,\Delta T)$ function shown in Fig 3b.

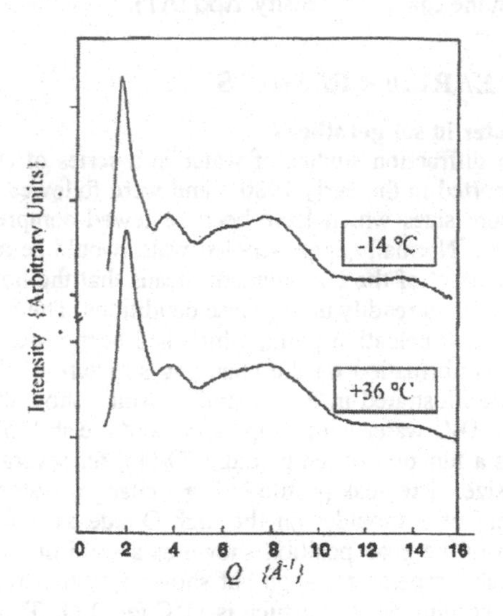

Figure 1. The change in the diffraction pattern for D_2O with temperature.

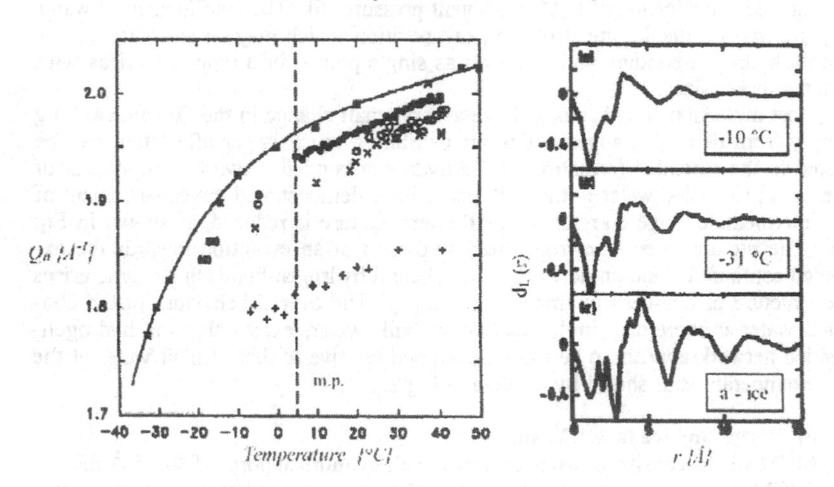

Figure 2. Super-cooled water a) variation of $Q_0(T)$ for bulk and confined water in various sol-gel silicas, b) change of $d_l(r)$, with temperature [-10 °C, -31°C] and amorphous ice.

It is clear that these results are similar to those seen in the bulk phase behaviour but extend to much lower temperatures and show the continued displacement of the peak $Q_0(T)$ towards the position for hexagonal ice; the profile also sharpens due to the presence of longer-range correlations. The spatial distribution function also shows an increase in the correlations over the intermediate range (5Å < r < 15Å) corresponding to a growth of the hydrogen-bond network, as expected. Another unexpected finding was that the transition was completely reversible and showed no evidence of the hysteresis effects that usually occur for nucleation and melting in confined geometry; this feature is discussed further in Sec 3e..

The reasons for the formation of the defective cubic ice are still not understood but it seems to be reproduced in all studies involving the formation of small ice crystallites that are restricted in growth.

c) Over-filled samples of water and ice in MCM silicas

The neutron investigation is sensitive to the total amount of water in the sample so that excess water above the amount needed to fill the pore volume will reside on the outer surfaces of the particles or grains. On cooling, this external water will freeze before that inside the pores and, as shown in Fig 3a for an over-filled MCM48 sample, there is some super-cooling before nucleation to hexagonal ice at 249K, which is indicated by the triplet profile of the first diffraction peak. If the temperature is further reduced, the pore water freezes as cubic ice and the central peak of the triplet grows. However, the peak on the low-Q side also increases in intensity showing that the defective form of cubic ice is again created, so there is no growth of the external ice with hexagonal form into the pore volume. Another feature shown in these curves is that even after nucleation, the diffraction pattern changes slightly as the temperature is further reduced to 206K, indicating that there are further structural changes after the main nucleation event has occurred. The additional Bragg peaks at higher Q-values also mimic these changes as shown in the difference functions. Similar results are shown for MCM41 measurements with an over-filled water sample.

A more detailed picture can be obtained by transforming the ΔD_M (Q, ΔT) function with a reference temperature of T_0 =236K, to give a representation of the changes in real space after the hexagonal ice has formed and relative to the deeply super-cooled liquid at ~45K below its normal bulk freezing point. Fig.4a shows three sets of diffraction data for over-filled MCM48 silicas at temperatures of 226, 216 and 206K; the corresponding Δd_L (r, ΔT) curves are shown in Fig 4b. The absence of a peak at 1.0Å indicates that the molecular terms of the $f_1(Q)$ form-factor have been accurately eliminated in the difference function analysis. The presence of sharp peaks at 1.8 and 2.1Å arises from local hydrogen-bonds in the ice lattice and suggests that there is a sharpening of the spatial correlations as the cubic ice forms. This feature presumably means that the H-bond angle variation becomes more restricted and consequently affects the longer-range correlations, which show a similar behaviour out to 20Å. It therefore becomes clear that the deeply super-cooled water has considerable disorder relative to the defective form of cubic ice. Furthermore, there are additional changes in the 20K region below the main nucleation event, as observed previously but less clearly in earlier studies. The development of deep minima at 6.5, 10.3 and 14Å is also a characteristic feature of the d(r) function for amorphous ice and seems to be a signature of long-range order in H-bonded networks. However, there are also interesting features in the shape of the curves in the 4-10Å region that have not yet been fully interpreted.

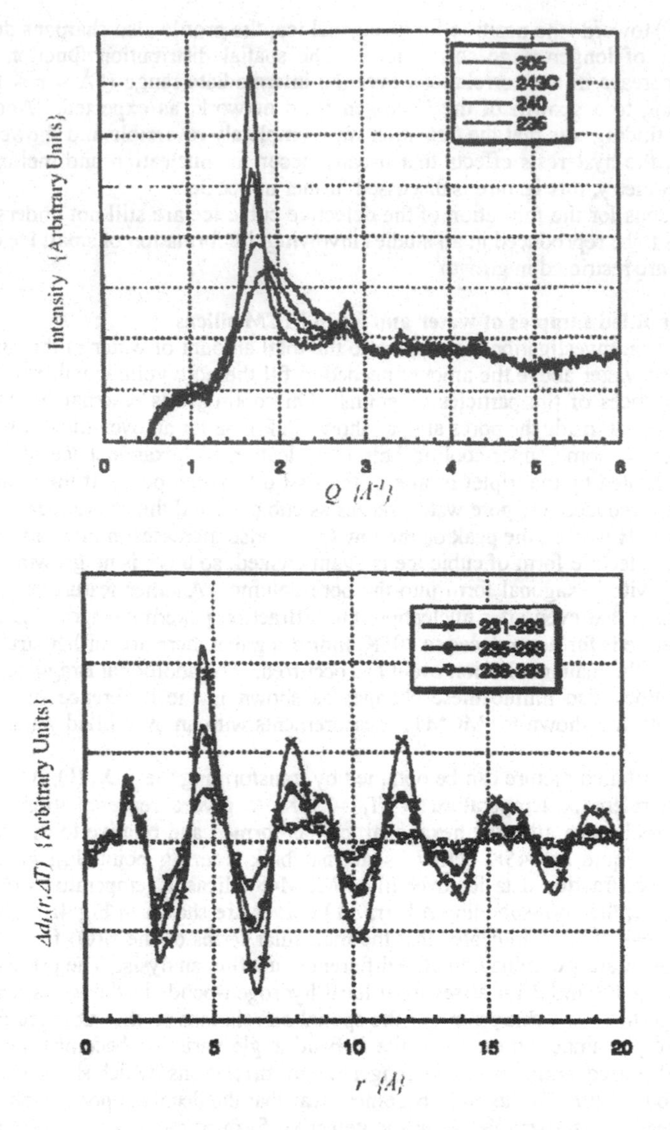

Figure 3. Neutron diffraction studies of a sample of D_2O in MCM41 silica measured on the D4 diffractometer: a) scattering intensity from water/ice as a function of temperature, b) the structural change, $\Delta d_L(r, \Delta T)$, for the same datasets.

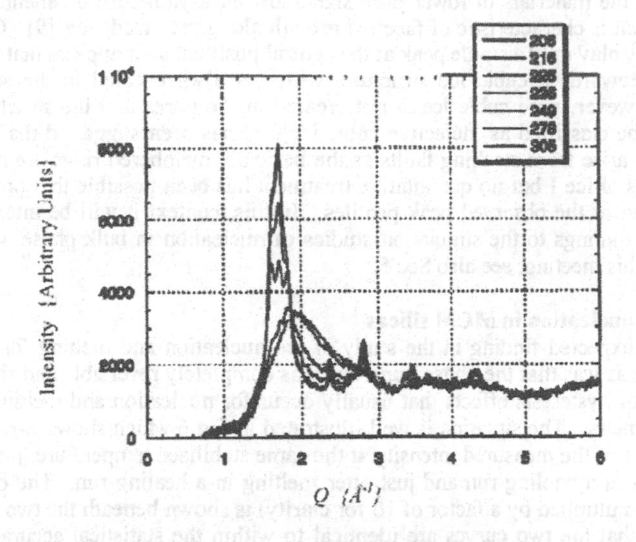

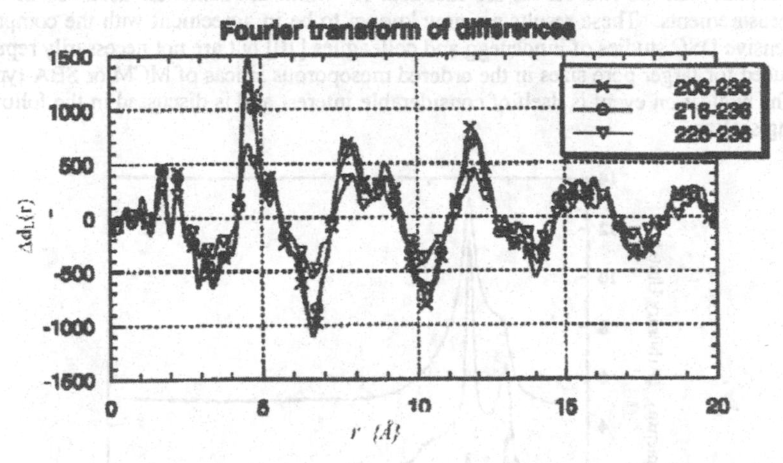

Figure 4. Neutron diffraction of D_2O in an overfilled sample of MCM48 silica taken on the D20 diffractometer: a) scattering intensity from water/ice as a function of temperature, b) the structural change, $\Delta d_L(r, \Delta T)$, for the same datasets with a reference temperature of 236K.

d) Ice nucleation in various sol-gel silicas

A neutron diffraction study was made of nucleation in a range of sol-gel silicas of different sizes [8] where the effective water thickness was maintained at approximately 20Å by partial filling of the larger pore materials. The comparison of the primary ice peaks is shown in Fig 5. For the large pore size, the triplet that characterises hexagonal ice I_c is clearly shown. However, the peak profiles for the ice formed in the materials of lower pore size show an asymmetric broadening of the peaks which is characteristic of facetted growth along preferred axes [9]. Cubic ice I_c would display only a single peak at the central position, so it appears that there is a tendency towards a cubic ice structure, which is also revealed in the secondary peaks. However, pure cubic ice is not created in the pores and the structure must therefore be classified as 'defective cubic ice'. It has been suggested that this feature could arise from stacking faults as the basic six-membered rings are present in both forms of ice I but no quantitative treatment has been possible that provides an explanation of the observed peak profiles. In this, context it will be interesting to relate the findings to the simulation studies of nucleation in bulk phase water presented at this meeting; see also Sec 5.

e) Ice nucleation in MCM silicas

An unexpected finding in the study of ice nucleation and melting for water in MCM silicas was that the phase transition was completely reversible and showed no evidence of hysteresis effects that usually occur for nucleation and melting in confined geometry. The situation is well illustrated in Fig 6.which shows two superimposed plots of the measured intensity at the same stabilised temperature, just prior to nucleation in a cooling run and just after melting in a heating run. The difference function (multiplied by a factor of 10 for clarity) is shown beneath the two plots and indicates that the two curves are identical to within the statistical accuracy of the measurements. These results are now known to be in agreement with the comprehensive DSC studies of Findenegg and colleagues [10] but are not necessarily reproduced for larger pore sizes in the ordered mesoporous silicas of MCM or SBA-type. The nucleation event is itself of considerable interest and is discussed in the following section.

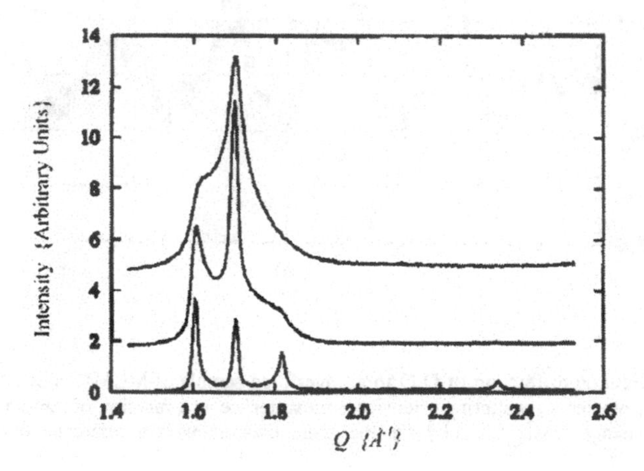

Figure 5. The first diffraction peak for ice nucleation in sol-gel silicas with different pore size; see the text for details.

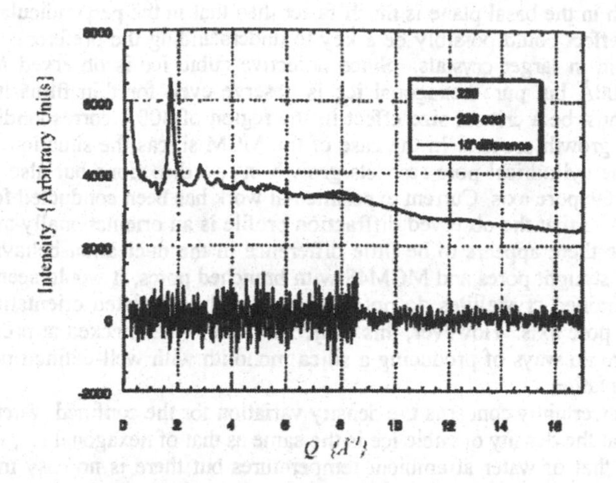

Figure 6. **Two superimposed datasets for cooling and heating runs at a temperature of 236K taken with the two multidetectors on the D4 diffractometer, showing the complete reversibility of the transition after the nucleation of cubic ice; the difference between the two datasets is shown below multiplied by a factor of x10.**

f) The cubic ice phase

The cubic ice phase has a similar local structure to that of hexagonal ice, consisting of six-membered rings of hydrogen-bonded molecules. The structure factor has a single peak at 1.70 Å^{-1}, which is at the central peak position for the hexagonal ice triplet. High resolution studies of nucleation in several sol-gel silicas have revealed important changes in the peak profile. The hexagonal ice triplet has Bragg peaks at Q-values of 1.55, 1.70 and 1.85Å^{-1} corresponding to (1,0,0), (0,0,2) and (1,0,1) reflections. The results indicate that a defective form of cubic ice is formed in the smaller pores. There is also a secondary peak at 2.33 Å^{-1} that is present for ice I_c and not for ice I_h so that the two phases are easily distinguished, in principle. However, all the diffraction observations give a broadened first peak with a shoulder on the low-Q side, or even a separated peak, if the Q-resolution is sufficient (Fig.5). It therefore seems that pure cubic ice is never formed even though there are several different routes to its production from either the liquid or the amorphous ice phase and also when produced by slow vapour-deposition at temperatures just above 145K. The reasons for this behaviour are still not understood but it seems to be reproduced in all studies involving the formation of small ice crystallites that are restricted in growth.

The most likely explanation of this phenomenon rests with the initial formation of a small hydrogen-bonded assembly that will act as a nucleus for crystal growth. When the proto-crystallite is small the crystal axes will not be well defined and the subsequent initial growth will contain a number of stacking faults. As subsequent layers are added during the growth process, it seems that these defects become 'frozen in' but when the crystallite has grown to a certain size, the change in surface-to-volume factor causes a modification in molecular orientations and the crystal readily grows into the hexagonal form. Furthermore, it is well known that, for hexagonal ice, the growth in the basal plane is much faster than that in the perpendicular direction and this effect could possibly be a key to understanding the preference for the hexagonal form in larger crystals. Since defective cubic ice is observed for pore sizes up to 200Å but pure hexagonal ice is observe even for thin films in 500Å pores, there must be a critical size effect in the region of 300Å corresponding to a change in the growth process. In the case of the MCM silicas the situation is more complex as the cylindrical pore restricts growth in two directions but also permits growth along the pore axis. Current experimental work has been conducted for powdered samples so that the observed diffraction profile is an orientationally-averaged pattern. Since there appears to be little difference in the nucleation behaviour for MCM41 with straight pores and MCM48 with branched pores, it would seem likely that the constrained crystallites do not have a strongly-correlated orientation with respect to the pore axis. However, this conjecture cannot be checked at present because there are no ways of producing a silica monolith with well-defined pore size and axial orientation.

Another uncertainty concerns the density variation for the confined water/ice. It is assumed that the density of cubic ice is the same as that of hexagonal ice, which is 8% less than that of water at ambient temperatures but there is no easy means of determining the real variation in the density of the super-cooled water in the mesopores. It is convenient to assume that the changes correspond to situation for bulk water with a continuing decrease of the density towards that of ice I as the temperature is reduced towards the limiting value of –40°C but no confirmation of this behaviour has yet been made. It may be possible to check this feature by temperature-dependent SANS measurements but this experiment is difficult due to the possibility of systematic errors and has not yet been attempted.

4. RECENT STUDIES

A further series of measurements has been made using a continuous temperature variation rather than a set of measurements at stabilised temperatures. Since the Q-range for significant changes in the $\Delta D_M(Q,\Delta T)$ term are restricted to Q-values below ~10 $Å^{-1}$, it is also more convenient to use a diffractometer with a wavelength in the region of 1.3 Å and a high neutron flux. This approach is conveniently achieved by using the rebuilt D20 diffractometer at ILL [11], which is fitted with a 140° position-sensitive detector. The sample was contained in a 5 mm diameter vanadium cell, and the temperature was controlled using a ramp with a speed of either 1 or 2K/min. Measurements of the whole diffraction pattern up to a Q-value of 6.5 $Å^{-1}$, were made with a typical readout every two minutes. The advantage of this technique is that the datasets are continuously recorded and no time is wasted on temperature stabilisation but there is a minor disadvantage due to the temperature gradient that exists in the sample material throughout the run. Since the changes are systematic with temperature, even near the phase transition points, the $\Delta D_M(Q,\Delta T)$ function is well described but the actual mean temperature of the sample is less well defined. A further advantage is that the whole of the diffraction pattern is measured simultaneously so that any change in the relative intensities of ice peaks near a phase transition can be readily observed. Two examples of the new measurements are presented below but there is further analysis to be made on these and other datasets.

a) Deeply super-cooled water in MCM silicas

The results for D_2O in MCM41 silica are shown in the form of stacked $\Delta D_M(Q,\Delta T)$ plots in Fig 7 for a cooling run to 220K with a reference temperature of 292K. Several adjacent runs have been added together to give improved statistics. The curves have a constant profile and increase in intensity as the temperature is reduced, confirming the results obtained in the earlier studies. It is also possible to display the data as a contour plot, as given in a preliminary presentation of the measurements at a earlier conference [12]. The full analysis of the data in real-space should improve the accuracy of the previous study and give a more detailed understanding of the way in which the hydrogen-bonded network builds up in this region of deep super-cooling. It should provide a very stringent test of current models of liquid water based on various types of interaction potential as the spatial correlations appear to be very sensitive to the temperature and also extend over a wider r-range than is normally considered in computer simulation studies. It seems probable that a full quantum calculation will be required rather than the use of empirical potentials and some possible approaches have already been displayed in other papers presented at this meeting [13].

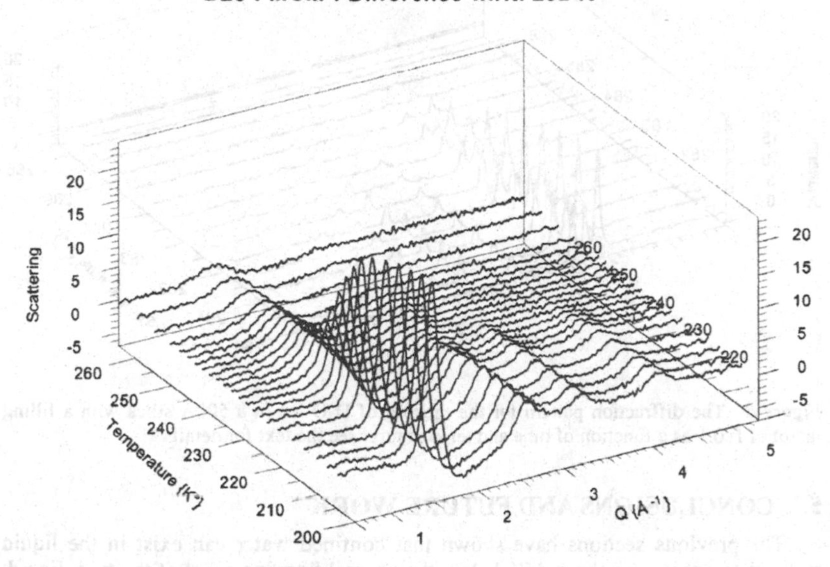

Figure 7. The difference function ΔD_M (Q, ΔT) for D_2O water in MCM41 silica taken on the D20 diffractometer as a function of temperature in continuous read-out mode.

b) Ice melting in larger-pore sol-gel silicas

Measurements were also made for a sol-gel silica with a 500 Å pore size. In this case, hexagonal ice is formed in the pores even for a partial filling factor of only 0.10. The good resolution of the D20 instrument and the reduced diffraction broadening effect in the larger pores means that the individual ice triplet is well observed and it is interesting to study the way the profile changes as the temperature approaches the melting temperature on a heating cycle. The results are shown as stacked intensity plots in Fig 8 A close examination of the relative intensities indicates that the central peak reduces more rapidly than the other two peaks as the transition temperature is approached, indicating some pre-melting effects within 3-5K of the phase change. This effect was not expected but presumably indicates that the breakdown of the ordered hydrogen-bond structure of the crystal proceeds by increased vibrational or librational motion with specific characteristics. This phenomenon would be impossible to study in the bulk phase due to the rapid annealing of crystallites and the growth of single domains. It is only because the individual crystallites are isolated within the pore matrix that this behaviour can be identified. Clearly, it will be interesting to make a more detailed investigation using a slower heating rate or for set temperatures as the phenomenon is probably influenced by both time- and temperature-dependent effects.

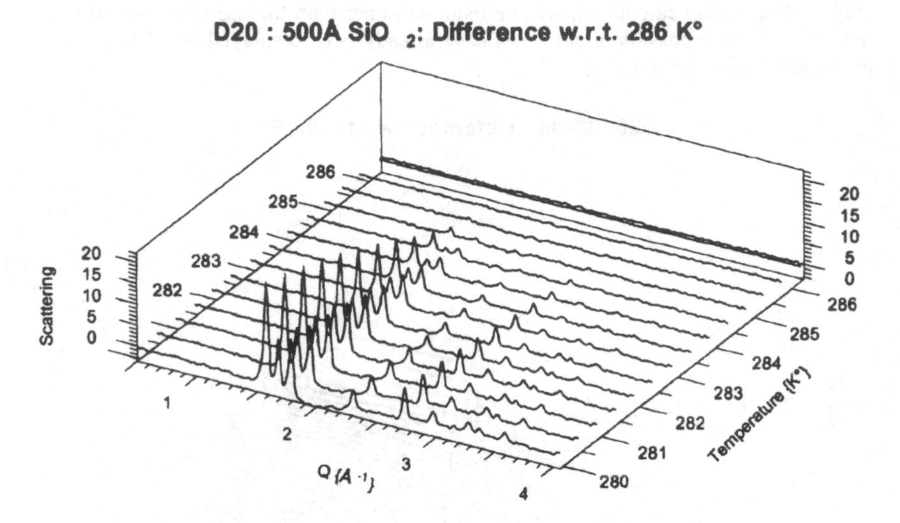

Figure 8. The diffraction pattern for the melting of D_2O ice in a 500Å silica with a filling factor of f=0.1 as a function of time and temperature; see the text for details

5. CONCLUSIONS AND FUTURE WORK

The previous sections have shown that confined water can exist in the liquid state, at temperatures about 45K below the normal freezing point of the bulk liquid. The structure of the deeply super-cooled liquid follows the trends already observed for the temperature variation of the bulk phase, in which there is a continuous evolution towards the open network structure of low-density amorphous ice. Eventually the system undergoes a phase transition to give a solid phase but the crystalline form is strongly dependent on the geometrical constraints. For small pores, a defective form of cubic ice is created and there is a complete reversibility across the transition. In the larger pores, hexagonal ice is formed even if the liquid film is quite thin and hysteresis effects are observed.

The picture that emerges from the new datasets [Figs 3a and 3b.] confirms earlier conjectures that the onset of nucleation involves a local fluctuation giving rise to an ice-like nucleus based on tetrahedrally-bonded rings, comprising six molecules. This proto-nucleus is only partially ordered and does not adopt a specific stacking pattern that relates directly to either hexagonal or cubic ice. The diffraction pattern is substantially broadened in the narrow pores of the MCM silicas but probably mimics the pattern seen in the earlier data for the sol-gel silicas with <100Å pore size. The profile shown in Fig 8 has not yet been fully explained but presumably arises from crystallites with a high proportion of correlated stacking faults. There is no physical explanation of why this particular form should be the stable (or metastable) state under these conditions but obviously there is a nucleation and growth process that stabilises the crystallites with a finite size determined by the morphology of the silica matrix.

For increased pore sizes, the crystallites can grow larger and it would seem that when a critical size has been reached the subsequent growth is essentially of hexagonal form. The present data suggest that ice formed under normal ambient conditions begins as a defective cubic ice nucleus. This observation is in qualitative agreement with the two-phase process seen in the MD computations of Ohmine and colleagues for nucleation of bulk water and also presented at this meeting [14] but a detailed analysis of the developing network structures will need to be compared with the experimental results. Another experimental technique, reported at this meeting [15] uses Mie scattering of laser light to study nucleation in a levitated droplet of 30μ diameter.

The importance of water/ice in various environments with varying pore morphology and different substrate materials will ensure a continuation of this fascinating subject. Plans have already been made to investigate the behaviour of water/ice in the SBA silicas, which cover an interesting range of 60-100Å in pore diameter; this range is comparable with many situations involving water channels in biomembrane systems. Modification of the silica interface can also be made through the grafting of radicals onto the pore walls but it will probably be necessary to check that a high coverage factor is achieved to ensure that unambiguous data are obtained. Another exciting possibility is to investigate the behaviour of water/ice in the very small diameter [5-10 Å] pores of carbon nanotubes. Water readily enters into open-ended nanotubes and the vapour-adsorption measurements of Kaneko and colleagues [16] on carbon nanohorns provide the necessary control conditions for conducting this difficult experiment [17]. The diffraction broadening effects will be large and there will be a major contribution from the carbon-water cross terms, so simulation studies will probably be needed to interpret the observations. A new form of ice has already been predicted by Koga et al [18] for the low-temperature solid phase in carbon nanotubes, using MD simulation techniques.

In summary, a clear and consistent qualitative picture of the liquid phase is beginning to emerge from these neutron studies. The temperature difference function is consistent for all the systems studied so far and corresponds to enhanced long-range correlations of the hydrogen-bond network as the temperature is reduced. On nucleation, a defective proto-crystallite is formed with a predominantly cubic ice structure but as the crystal size increases, the hydrogen bond network adopts the familiar pattern of hexagonal ice. Although the experimental findings seem incontrovertible, there is still uncertainty about the basic origin of these features and the understanding of water/ice characteristics still presents us with a substantial scientific challenge.

The use of the small pore diameters of the MCM silicas enables the liquid to be routinely cooled to much lower temperatures than previously attained and therefore gives a unique possibility of studying deeply 'super-cooled' water and cubic ice in a stable environment. The use of mesoporous solids therefore provides a convenient 'laboratory' for conducting experimental investigations in conditions that would normally be regarded as highly metastable. The techniques developed here for water/ice are, of course, capable of use for any other molecular liquid and have already been applied to cyclohexane [19].

REFERENCES

1. J.G.Powles, Advances in Physics 22, 1 (1973)
2. I.P.Gibson and J.C.Dore, Mol Phys. 48 1019 (1983)
3. D.C.Steytler, J.C.Dore and C.J.Wright, Mol Phys 48 1031 (1983)
4. D.C.Steytler, J.C.Dore and C.J.Wright, Mol Phys 56 1001 (1985)
5. J.C.Dore, Chem Phys 258 327 (2000)
6. M.C.Bellissent-Funel et al, Europhys Letts 2 241
7. D.Blakey, PhD Thesis, University of Kent (1994) and M.R.Choudhury, J.C.Dore and J.T.Wenzel, J.Non.Cryst Solids 53 247 (1982)
8. J.M.Baker, J.C.Dore and P.Behrens, J.Phys.Chem B101 6226 (1997)
9. The theoretical formalism was originally described in the interpretation of diffraction data for carbon blacks and is often referred to as 'the Warren profile'; B.E.Warren Phys. Rev. 59 693 (1941).
10. The application to the cubic ice profiles is discussed in J.M.Baker, PhD
11. G.Findenegg, private communication and Phys. Chem. Chem Phys., 3, 1185-1195 (2001).
12. The new D20 diffractometer is described in the instrument section of the ILL website; www.ill.fr
13. J.Dore, B.Webber, M.Hartl, P.Behrens and T.Hansen, Physica A314 501 (2002)
14. S.S.Xantheas, this volume
15. M.Matsumoto, S.Saito and I.Ohmine, Nature 416 409 (2002) and also I. Ohmine, this volume
16. P.Stoeckel, J.Klein, I. Weidinger, H.Baumgartel and T.Leisner, see also H.Baumgartel in this volume
17. K.Kaneko, private communication and E.Bekyrova et al, Chem Phys Letts 366 463 (2002).
18. An experimental study of water/ice in carbon nanohorns by neutron diffraction is feasible and has been proposed; J.C.Dore, A.Burian and K.Kaneko; currently under consideration (2002-03)
19. K.Koga, G.T.Gao,H. Tanaka and X.C.Zeng, Nature 412 802 (2001)
20. H.Farman, J.C.Dore and J.B.W.Webber, J.Mol.Liq. 96 357 (2002)

SOLVATION STRUCTURE OF CHLORIDE AND IODIDE IONS STUDIED BY MEANS OF EXAFS USING A COMPACT SYNCHROTRON RADIATION SOURCE

K. OZUTSUMI and H. OHTAKI
Department of Applied Chemistry, Faculty of Science and Engineering, Ritsumeikan University, 1-1-1 Noji-Higashi, Kusatsu 525-8577, Japan

Abstract: An EXAFS (extended X-ray absorption fine structure) beamline set in a compact synchrotron radiation (SR) source at the SR Center of Ritsumeikan University consists of three windows, a slit, a double-crystal monochromator and two detectors. By using Si(220), Ge(220), and InSb(111) monochromator crystals, the EXAFS spectrometer covers the K-edge absorption energies of elements from silicon to zinc. Structural investigation of solvated ions with relatively low atomic numbers became possible under an atmospheric pressure. The EXAFS data of chloride and iodide ions in water, methanol and ethanol measured by the EXAFS equipment are reported. This is the first case to study the solvation structure of chloride ions by EXAFS.

1. INTRODUCTION

Ionic solvation is one of the most fundamental subjects of solution chemistry. Knowledge on solvation structure of ions such as ion-solvent bond length and solvation number is essential to elucidate static and dynamic properties of electrolyte solutions. Bond length between an ion and a solvating molecule is obtainable by the X-ray (XD) and neutron diffraction (ND) and extended X-ray absorption fine structure (EXAFS) methods. On the other hand, solvation number can be determined by the diffraction (XD and ND) and various spectroscopic methods (EXAFS, nuclear magnetic resonance, Raman, *etc.*). Of these, the diffraction and EXAFS methods give solvation numbers of cations and anions independently without an extrathermodynamic assumption.

A huge number of structural data on ionic hydration have been accumulated. The data are compiled in reviews recently published [1-3]. Determination of solvation structure of metal ions in nonaqueous solutions has been conducted in these decades by using mainly XD and EXAFS methods. The review [3] contains a large number of structural data in nonaqueous solutions. The feature of EXAFS is its high selectivity for atoms and high sensitivity. Hence, the EXAFS method is widely used for dilute systems to which the XD and ND methods can hardly be applied. However, structural information detected by EXAFS is usually limited to the first solvation sphere and the second solvation structure can hardly be discussed by this method. Nevertheless, the EXAFS technique is a powerful means for electrolytes in

J. Samios and V.A. Durov (eds.), Novel Approaches to the Structure and Dynamics of Liquids: Experiments, Theories and Simulations, 157-165.

aqueous and nonaqueous media, where solubility of electrolytes is usually more restricted in nonaqueous solvents than in aqueous solution. Synchrotron radiation (SR) X–rays are the most powerful source for EXAFS measurements. A compact superconducting SR source at Ritsumeikan University [4] supplies intense soft X–rays as shown in Fig. 1 and is very suitable for EXAFS investigations in soft X–ray regions. Although the photon flux from the ring steeply decreases above 5 keV, the X–ray intensity from the ring is much stronger than that from a conventional X–ray source. The distribution of X–ray intensities from the SR source is quite clean while white X–rays emitted from a conventional X–ray generator are usually contaminated by characteristic lines from a filament material. Thus, the SR ring is still a good source for elements with their absorption–edge energies above 5 keV. An EXAFS spectrometer attached to the compact SR source of Ritsumeikan University is equipped with Si(220) and Ge(220) crystals, and an InSb(111) crystal is newly introduced last year. Then, the EXAFS spectrometer makes it possible to measure K–edge EXAFS spectra of elements from silicon (atomic number 14) to zinc (30) under an atmospheric pressure.

In contrast to a large number of investigations of cationic solvation structure in aqueous and nonaqueous solvents, structural studies of anions in solutions are much less. Since anionic solvation is usually weaker than cationic solvation and solvated solvent molecules are rapidly exchanged with those in the bulk, the NMR method provides a quite small values of solvation number in the first solvation shell of anions (0 – 1) in spite of their large surface areas. When one applies the EXAFS method to structural studies on anionic solvation, weaker EXAFS oscillations than those by cations are usually obtained. Hence, high–precision EXAFS measurements are necessary for the determination of anionic solvation structure. Solvation structure of bromide and iodide ions in various solvents has been investigated by the EXAFS method [5–7], while that of chloride ion has never been examined. This is because the energy of the chlorine K–edge absorption is 2.82 keV.

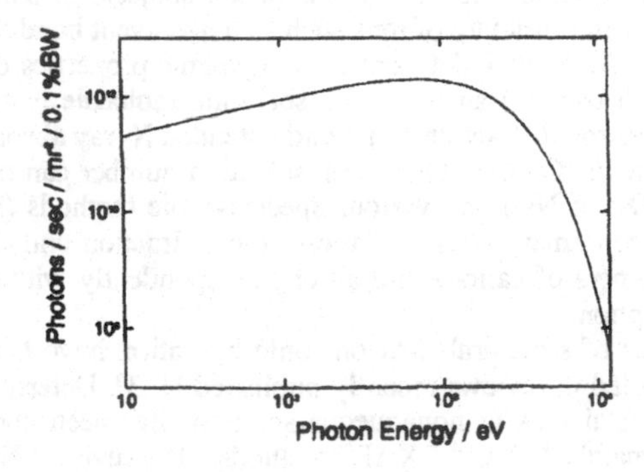

Figure 1. Spectrum of photons emitted from the compact SR source of Ritsumeikan University.

For example, soft X–rays with energies of 2 and 3 keV lose 99.9 and 88 % of their intensities, respectively, while passing through air with thickness of 10 cm. Thus, it has been commonly understood that EXAFS spectra in soft X–ray regions should be measured in vacuum. Also, owing to strong absorption of soft X–rays by samples, it is thought that a transmission EXAFS method is difficult to be applied. However, since the measurement with a transmission mode is the most convenient for obtaining reliable EXAFS spectra, it should be very useful if an EXAFS measurement with a transmission mode under an atmospheric pressure could be established for studies on solvation structure of ions with low atomic numbers. In this report K–edge EXAFS spectra of chloride ions in solutions measured under an atmospheric pressure are presented. This is the first case that EXAFS spectra of the chloride ion in solutions are reported. K–Edge EXAFS spectra of iodide ions in solutions have been obtained in various solvents [7]. However, measurements of iodine K–edge (33.163 keV) spectra are not very suitable by the EXAFS method, because in very high X–ray energy region changes in absorption coefficients above and below the edge are small and thus the amplitude of EXAFS oscillations becomes weak as compared with those of chloride and bromide ions. Instead, iodine L–edges are in soft X–ray regions and a difference in absorption coefficients between above and below the L_{III}–edge is large. Therefore, EXAFS spectra of a good quality can be easily obtained when absorption spectra at the L_{III}–edge are measured. However, measurements of L_{III}–edge EXAFS spectra of iodide ions in solutions have never been examined so far. In the present report solvation structure of chloride and iodide ions in water, methanol and ethanol studied by means of soft X–ray EXAFS is presented.

2. EXAFS SPECTROMETER AT SR CENTER OF RITSUMEIKAN UNIVERSITY

The outline of the EXAFS beamline employed is depicted in Fig. 2, which consists of three windows, a slit, a double–crystal monochromator and two detectors. Broad band radiation from the SR ring passes through the first Be window (a), which eliminates low energy lights and prevents monochromator crystals from heating caused by irradiation. A desired size of the incident beam is adjustable by the slit (b). Then, monochromatic X–rays are obtained by a Ge(220), Si(220), or InSb(111) double–crystal monochromator (c). This part of the beamline is under the environment of high vacuum ($1 - 4 \times 10^{-6}$ Pa). The monochromatic X–rays enter a low vacuum part (2 – 3 Pa) and the second Be window (d) separates the low and high vacuum regions. By using such an arrangement of a monochromator and windows, EXAFS measurements could be performed under an atmospheric pressure. The third window (e) separates the vacuum of the beamline from the atmosphere and a Be foil or a Capton film was used as a separator. The incident monochromatic X–ray intensity I_0 and transmitted intensity I after passing through a sample (g) are simultaneously measured by two ionization chambers with a path length of 4.5 and 31 cm (f and h), respectively (S–1329A and S–1196B, OKEN, Fussa, Japan). He(100%), N_2(100%), N_2(85%)+Ar(15%), N_2(50%)+Ar(50%), and Ar(100%) gases are used for the detection. Suitable gases for two ionization chambers could be selected based on the detection efficiency of the gases in the K–edge absorption energy of an element to be measured. The output currents are amplified by two current amplifiers (428, Keithley, Cleveland, USA) and changed to frequency by two 1 MHz V/F converters (NVF–02, Tsuji Denshi, Chiyoda, Japan). The signals are then fed to a scaler (974, ORTEC, Oak Ridge, USA) and finally stored in an computer (PC–9801FA, NEC, Tokyo, Japan). For measurements in the vicinity of the K–

edges of cobalt, nickel, copper, and zinc, where intensities of X–rays emitted from the SR ring steeply decrease (see Fig. 1), a scintillation counter with 2 inches in diameter (SC–50, Rigaku, Tokyo, Japan) is also employed as an I detector.

The double–crystal monochromator employed is of Golovchenko type [13]. The mechanical movement of the monochromator is in a range from 15° to 75°. The photon flux passing through a 2 mm × 9 mm slit is estimated at the sample position from the ion current of the ionization chamber. The evaluated values at various X–ray energies are summarized in Table 1. The InSb(111) crystal supplies monochromatic X–rays from 1.7 to 6.4 keV, which covers the K–edge absorption energies of Si, P, S, Cl, Ar, K, Ca, Sc, Ti, and V. However, EXAFS measurements for K, Ca, and Sc are difficult by this monochromator because L–edge absorption of In and Sb appears in a region of 3.7 – 4.7 keV. The X–ray energy ranges provided by using Si(220) and Ge(220) are 3.4 – 12.4 keV and 3.2 – 12.0 keV, respectively. However, the compact SR source does not supply sufficient X–ray intensities above 11 keV as seen from Table 1. A practical use of Si(220) and Ge(220) crystals is thus limited to the elements K, Ca, Sc, Ti, V, Cr, Mn, Fe, Co, Ni, Cu, and Zn.

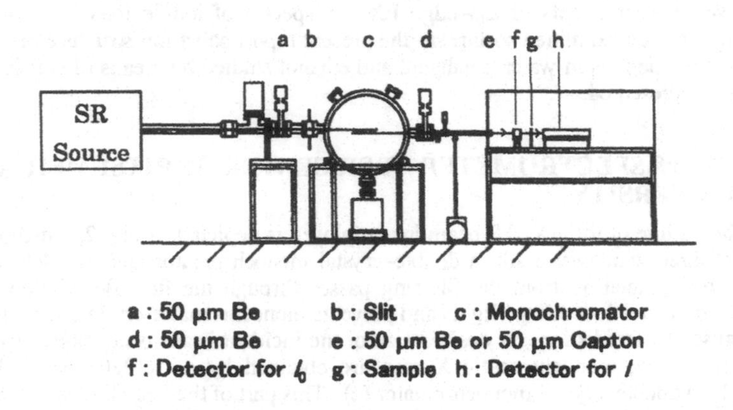

Figure 2. The outline of EXAFS spectrometer at the SR Center of Ritsumeikan University. The two detectors (f and h) almost contact with each other and sample (g) is attached to f at the measurements of K–edge EXAFS spectra of chlorine.

The quality of EXAFS spectra is much affected by higher–order harmonics. The steep decrease in X–ray intensities above 5 keV is characteristic of the intensity distribution of the compact SR source. The number of photons at 10 keV is a few orders of magnitude smaller than that at 5 keV as is seen in Table 1 and, hence, the contamination of higher–order harmonics is negligible above 5 keV. Therefore, measurements of EXAFS spectra of a good quality become easy in the energy range of 5.0 – 10.5 keV by using the spectrometer.

The InSb(111) monochromator is employed for EXAFS measurements in soft X––ray regions. The X–ray intensities reflected by the InSb crystal are related to the atomic scattering factors f_{In} and f_{Sb} of In and Sb, respectively. The crystal structure of InSb is of the zinc blend type. The X–ray intensity scattered by the InSb(222) plane is proportional to $(f_{In} - f_{Sb})^2$ for this structure type [9]. The atomic numbers of

In and Sb differ only by 2 and the amount of the second–order harmonics involved in the incident beam must be negligibly small. The X–rays diffracted by the InSb(333) face may significantly contaminate the incident beam because the intensity is related to $(f_{In}^2 + f_{Sb}^2)$ [9]. However, helium gas is used for the detection of the incident and transmitted soft X–rays and the detection efficiency is low enough for the third–order harmonic X–rays with energies of 5 – 8 keV.

Table 1. Photon flux at the sample position at various incident X–ray energies by using Si(220), Ge(220) and InSb(111) double crystal.[a]

X–ray energy (keV)	Photon flux Si(220)	Ge(220)	InSb(111)
2.0			8×10^9
2.5			1×10^{10}
3.0			8×10^9
3.5	1×10^9	2×10^9	3×10^9
4.0	1×10^9	2×10^9	1×10^9
5.0	9×10^8	9×10^8	3×10^8
6.0	4×10^8	5×10^8	1×10^8
7.0	2×10^8	2×10^8	
8.0	6×10^7	6×10^7	
9.0	3×10^7	2×10^7	
10.0	7×10^6	5×10^6	
11.0	3×10^6	1×10^6	

[a] The beam size is 2mm × 9 mm and the photon flux is normalized to 300 mA storage.

3. SOLVATION STRUCTURE OF CHLORIDE AND IODIDE IONS IN METHANOL AND ETHANOL

Figure 3 shows the chlorine K–edge EXAFS spectrum of aqueous 1 mol dm^{-3} tetraethylammoium chloride solution and the iodine L_{III}–edge EXAFS spectrum of aqueous 0.5 mol dm^{-3} NaI solution measured by using InSb(111) and Ge(220) monochromators, respectively. A sample solution was absorbed on a piece of paper (CLEAN WIPE–P, Asahi Kasei, Tokyo, Japan) and sealed in a bag made of a thin Mylar film. In order to avoid intensity loss due to absorption by air at the measurement of the chloride solutions, the I chamber is located just behind the I_0 one and the separation distance is about 5 mm.

The extracted EXAFS oscillations and the Fourier transformed structure functions for 1 mol dm^{-3} $(C_2H_5)_4NCl$ in water, methanol, and ethanol are depicted in Fig. 4. The broad and large peaks around 230 pm in the $|F(r)|$ functions, which were uncorrected for the phase–shifts, are mainly due to the Cl^-–O interactions (Fig. 4b). Analyses of these peaks by the Fourier–filtering method were performed assuming that the chloride ion in water binds with six water molecules and the Cl^-–O distance is 320 pm [2]. In the fitting procedure, the theoretical backscattering amplitude and phase functions reported by McKale, *et al.* [10] were used. The $k^3\chi(k)$ curves in methanol and ethanol solutions showed shoulders around 3×10^{-2} pm^{-1}, which was not seen in the aqueous solution. As a first attempt, only one kind of scatterer, *i.e.*, oxygen atom, was considered, and the solvation numbers in methanol and ethanol were found to be 3.4 and 3.8, respectively.

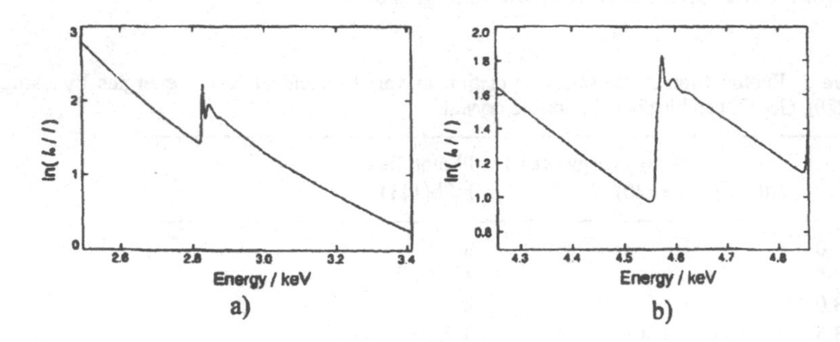

Figure 3. (a) The K–edge EXAFS spectrum of aqueous 1 mol dm^{-3} $(C_2H_5)_4NCl$ solution with an InSb(111) monochromatoe, (b) the L_{III}–edge EXAFS spectrum of aqueous 0.5 mol dm^{-3} NaI solution with a Ge(220) monochromator.

These values were significantly smaller than six found in water. Furthermore, the Hamilton *R*–factors for methanol and ethanol are appreciably larger than that for water (see Table 2). Then, contribution of two scatterers to the oscillations was taken into consideration. The other scatterer is carbon atom in a methyl group of methanol and a methyl or a methylene group in ethanol. The least–squares calculations assuming two scatterers, in which the numbers of the Cl^-–O and Cl^-...C interactions were treated as the same, gave much smaller *R*–factors both for the methanol and ethanol systems than those under the single–scatterer assumption. The total numbers of atoms participating in the chloride solvation are 6.4 and 7.8 for methanol and ethanol solutions, respectively. The chloride ion in the alcohol solutions interacts with oxygen atoms as well as methyl or methylene groups. The interaction of methyl groups with halide ions in DMSO has been reported [11,12].

When the numbers of the Cl^-–O and Cl^-...C interactions were optimized as independent parameters, the *n* value for the Cl^-...C pair did not well converge. Thus, it could not be definitely concluded from the present study that all the alcohol molecules in the solvation sphere bind with the chloride ion through both oxygen and carbon atoms or some of the alcohol molecules solvate the chloride ion with only hydroxyl oxygen atom.

The $k^3\chi(k)$ and $|F(r)|$ curves for 0.5 mol dm^{-3} NaI in water and 0.5 mol dm^{-3} $(n\text{–}C_4H_9)_4NI$ in methanol and ethanol shown in Fig. 5 were analyzed in a similar manner.

Looking at various literature values for the hydration number of iodide ion, the value of eight was adopted as reference [2]. The I^-–O distance was set to be 360 pm in water [2]. The results are given in Table 3.

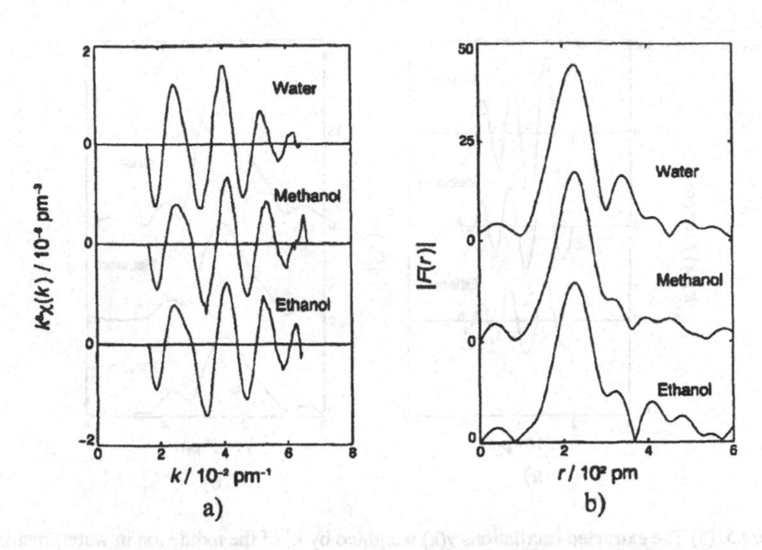

Figure 4. (a) The extracted oscillations $\chi(k)$ weighted by k^3 of the chloride ion in water, methanol, and ethanol, (b) the Fourier transforms of the $k^3\chi(k)$ curves, uncorrected for the phase shift.

Table 2. The number of interaction n, the bond length r, the Debye–Waller factor s, and the Hamilton R–factor for the solvated chloride ion in 1 mol dm–3 water, methanol and ethanol solutions of $(C_2H_5)_4NCl$ at 25 °C.

Solvent	Interaction	n	r/pm	σ/pm	R
Water	Cl^-–O	6[a]	320[a]	20.7 ± 0.1	0.028
	Cl^-–O	3.4 ± 0.3	316 ± 1	15.9 ± 0.6	0.153
Methanol	Cl^-–O	3.2 ± 0.2	317 ± 1	16.6 ± 0.2	0.018
	Cl^-...C	3.2 ± 0.2	389 ± 1	19.7 ± 0.1	
	Cl^-–O	3.8 ± 0.4	319 ± 1	16.8 ± 0.7	0.183
Ethanol	Cl^-–O	3.9 ± 0.5	320 ± 1	18.1 ± 0.7	0.056
	Cl^-...C	3.9 ± 0.5	391 ± 1	19.6 ± 0.4	

[a] Fixed during the calculation.

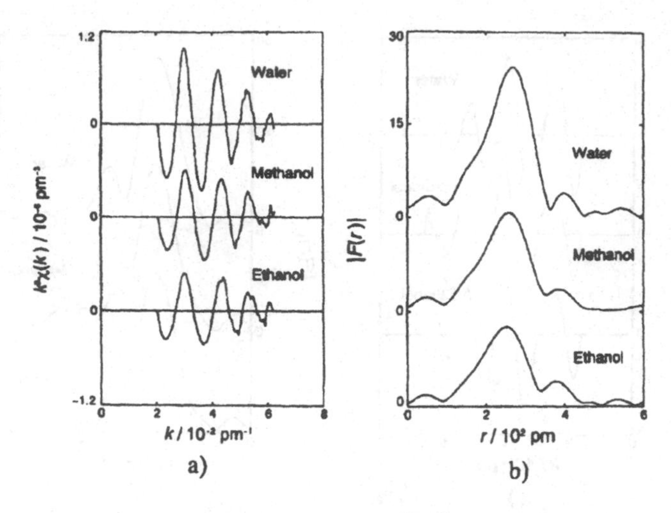

Figure 5. (a) The extracted oscillations $\chi(k)$ weighted by k^3 of the iodide ion in water, methanol, and ethanol, (b) the Fourier transforms of the $k^3\chi(k)$ curves, uncorrected for the phase shift.

Table 3. The number of interaction n, the bond length r, the Debye–Waller factor s, and the Hamilton R–factor for the solvated iodide ion in 0.5 mol dm^{-3} NaI in water, and 0.5 mol dm^{-3} $(n\text{–}C_4H_9)_4NI$ in methanol and ethanol at 25 °C.

Solvent	Interaction	n	r/pm	σ/pm	R
Water	I^-–O	8[a]	360[a]	20.1 ± 0.1	0.044
	I^-–O	3.9 ± 0.3	353 ± 1	20.0 ± 0.7	0.190
Methanol	I^-–O	3 ± 1	361 ± 1	17 ± 3	0.046
	I^-...C	3 ± 1	405 ± 1	20 ± 3	
	I^-–O	3.5 ± 0.4	352 ± 1	17.7 ± 0.7	0.171
Ethanol	I^-–O	3 ± 1	358 ± 1	17 ± 2	0.059
	I^-...C	3 ± 1	404 ± 1	20 ± 5	

[a] Fixed during the calculation.

The one–scatterer calculation led to an unrealistically small solvation number of I^- in the alcohols compared with that in water. The two–scatterers calculation including oxygen and carbon atoms gave the total number of interactions of 6 ± 2 for methanol and ethanol (Table 3). The differences between the I^-–O and I^-...C interaction distances are 44 and 46 pm in methanol and ethanol, respectively. The values are much smaller than those for the Cl^-–O and Cl^-...C interactions (72 and 71 pm for methanol and ethanol, respectively), which suggest easier accommodation of alcohol molecules around the iodide ion than the chloride ion. Thus, the smaller solvation number in methanol and ethanol than that in water may be caused by ligand–ligand repulsive interactions in the solvation shell of the anion due to the larger volumes of the alcohol molecules than water. The errors involved in the numbers of the I^-–O and I^-...C interactions are rather large, which may in part arise from a large fluctuation in the position of considering alcohol molecules around the iodide ion.

REFERENCES

1. Marcus, Y. (1988) Ionic radii in aqueous solutions, *Chem. Rev.* **88**, 1475–1498.
2. Ohtaki, H. and Radnai, T. (1993) Structure and dynamics of hydrated ions, *Chem. Rev.* **93**, 1157–1204.
3. Ohtaki, H. (2001) Ionic solvation in aqueous and nonaqueous solutions, *Monatsh. Chem.* **132**, 1237–1268.
4. Iwasaki, H., Nakayama, Y., Ozutsumi, K., Yamamoto, Y., Tokunaga, Y., Saisho, H., Matsubara, T., and Ikeda, S. (1998) Compact superconducting ring at Ritsumeikan University, *J. Synchrotron Rad.*, **5**, 1162–1165.
5. Tanida, H., Sakane, H., and Watanabe, I. (1994) Solvation structure for bromide ion in various solvents by extended X–ray absorption fine structure, *J. Chem. Soc., Dalton Trans.* 2321–2326.
6. Sawa, Y., Miyanaga, T., Taida, H., and Watanabe, I. (1995) Temperature dependence of EXAFS for bromide ions in solution, *J. Chem.. Soc., Faraday Trans.* 91, 4389–4393.
7. Tnida, H. and Watanabe, I. (2000) Dependence of EXAFS (extended X–ray absorption fine structure) parameters of iodide anions in various solvents upon a solvent parameter, *Bull. Chem. Soc. Jpn.* 73, 2747–2752.
8. Golovchenko, J.A., Levesque, R.A., and Cowan, P.L. (1981) X–ray monochromator system for use with synchrotron radiation source, *Rev. Sci. Instrum.* **52**, 509–516.
9. Warren, B. E. (1990) *X–Ray Diffraction*, Dover, New York.
10. McKale, A.D., Veal, B.W., Paulikas, A.P., Chan, S.–K., and Knapp, G.S. (1988) Improved ab initio calculations of amplitude and phase functions for extended X–ray absorption fine structure spectroscopy, *J. Am. Chem. Soc.* **110**, 3763–3768.
11. Wakabayashi, K., Maeda, Y., Ozutsumi, K., Ohtaki, H. (in press) The structure of solvated halide ions in dimethyl sulfoxide studied by Raman spectroscopy and X–ray diffraction, *J. Mol. Liquids*.
12. Onthong, U., Bako, I., Radnai, T., Hermansson, K., Probst, H. (2002) Ab–initio study of the interaction of dimethyl sulfoxide with the ions Li^+ and I^-, *Abstract of "Novel Approaches to the Structure and Dynamics of Liquids; Experiments, Theory and Simulations"*, 161, Rhodes, Greece, September 7–15 (2002).

ULTRASONICALLY INDUCED BIREFRINGENCE IN LIQUIDS AND SOLUTIONS

H. NOMURA[a], T. MATSUOKA[b] AND S. KODA[b]
[a]*Laboratory of Chemistry, Department of Natural Science, School of Science and Technology, Tokyo Denki University, Hatoyama, Hiki-Gun, Saitama, 350-0394 Japan*
[b]*Department of Molecular Design and Engineering, Graduate School of Engineering, Nagoya University, Furo-cho, Chikusa-ku, Nagoya, 464-8603, Japan*

1. INTRODUCTION

The birefringence is induced in liquids and solutions containing certain amount of nonspherical molecules and particles as a result of the orientation of the molecules or particles due to longitudinal ultrasonic waves [1-23]. This phenomenon was called as the ultrasonically induced birefringence. Early theoretical studies of the birefringence have been reviewed by Hilyard and Jerrard [1].

For neat liquids consisting of the small anisotropic molecules, the orientational relaxation time is smaller than the period of ultrasound so that the velocity gradient caused by ultrasound can directly induce the sinusoidal orientation. This causes the sinusoidal birefringence and it is proportional to the ultrasonic amplitude, that is, the square root of the ultrasonic intensity $\sqrt{W_U}$ [1-7]. On the other hand, for large anisotropic particles such as colloidal particles, the orientational relaxation time is much larger than the period of the ultrasonic wave so that the orientational motion cannot follow the sinusoidal velocity gradient. However, the radiation pressure, which is one of the typical quadratic acoustic effects, produces the stationary torque on the particle that induces the uniform and stationary orientation of the particles in the solutions [8-11]. In this case, the stationary birefringence is induced. The induced birefringence is proportional to the ultrasonic intensity W_U.

Ultrasonically induced birefringence will be a useful method for the investigation of orientational behavior in liquids and solutions since it is applicable to wide variety of the solution systems because it does not require the dipole moment or magnetic anisotropy. In addition, it is suitable to investigation of the orientational motion of ionic system because it is free from the presence of ions or electric double layer of ionic colloids. We have been studied ultrasonically induced birefringence in liquid crystals in isotropic phase, colloidal polymer solutions. In this review, we will summarize our recent theoretical and experimental studies.

In the first part of this review, we will describe our experimental system for the ultrasonically induced birefringence [10,20].

J. Samios and V.A. Durov (eds.), Novel Approaches to the Structure and Dynamics of Liquids: Experiments, Theories and Simulations, 167–192.

In the second part, we will briefly introduce the theory of ultrasonically induced birefringence for neat liquids and show the experimental results in liquid crystals in isotropic phase [7,20].

In the third part, we will briefly review our extension of the modified Oka theory for to rod-like particles the show experimental results. The frequency dependence of birefringence will be discussed on the basis of the modified Oka theory for rod-like particles. In addition, orientational relaxation times of rod-like particles have also been evaluated from the trace of the transient birefringence.

It is well known that at the concentration above critical micelle concentration (CMC), surfactants molecules aggregate to spherical micelles. At higher concentration and/or in addition of salts, it changes to rod-like micelles. The birefringence has been observed in such a rod-like micelles systems. We also discuss anomalous transient curve observed in entangled rod-like micelles [11,12].

Finally, we would like to review the ultrasonically induced birefringence in polymer solutions. The mechanism of the ultrasonically induced birefringence in polymer solutions is of strong interest since a polymer molecule is a long flexible chain consisting of the small monomer unit. We will show our recent results [20-24].

2. EXPERIMENTAL SYSTEM OF ULTRASONICALLY INDUCED BIREFRINGENCE

To measure ultrasonically induced birefringence, two different detecting techniques were used [20]. One is the "non-biased" measurement and the other is the "biased" measurement. Figure 1 show a block diagram of a measurement system. The direction of the incident light is perpendicular to the direction of the sound wave propagation. The light from a He-Ne laser (632.8 nm, 5 mW) passes through a polarizer with an angle of polarization at 45°, a sample cell, and an analyzer. For the non-biased detection, the analyzer is set at extinction angle. For the biased detection, a $\lambda/4$ plate is inserted and the analyzer is set with a small offset angle β from its extinction position.

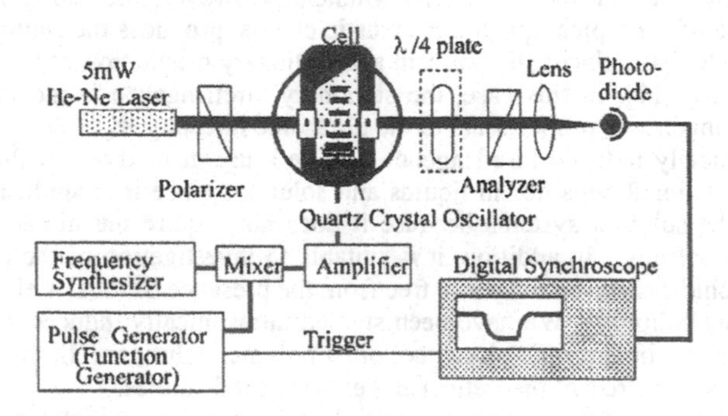

Figure 1. . Block diagram of ultrasonically induced birefringence measurement system

If the acoustic field is applied to solutions, the optical phase retardation δ is produced and the intensity of light passed through the analyzer increases. For $|\beta|<<1$ and $|\delta|<<1$, the intensities of light passed through the analyzer are given by

$$I = I_0\left((\delta/2)^2 + \beta\delta + \beta^2\right) + I_b \qquad (1)$$

where I_0 and I_b are the light intensities in the absence of the sound wave with the polarizer and the analyzer being parallel and perpendicular, respectively. For the non-biased detection, $\beta = 0$, the phase retardation is given as, $\delta = 2\sqrt{(I - I_b)/I_0}$. While for the biased detection, the ph ase retardation is given as, $\delta = (I_+ - I_-)/(2I_0\beta)$, where the light intensities with offset angles $\pm\beta$ are described by I_+ and I_-, respectively. The birefringence Δn is obtained from the phase retardation δ as follows.

$$\Delta n = \lambda\delta/(2\pi d) \qquad (2)$$

where λ is the wavelength of the laser light and d is the optical path length.

As indicated in Eq.(1), the phase retardation obtained by the non-biased measurements is the root mean square one and thus the sign of birefringence cannot be obtained. The merit of this method is that both the sinusoidal and stationary term can be detected. The Δn as much as 10^{-8} can be detected. In the biased measurement, the sinusoidal birefringence is averaged and only the stationary component is observed. In this case, the sign of the birefringence can be obtained and high signal to noise (S/N) ratio can be realized if values of β are properly set as $|\delta|<<|\beta|<<1$. The Δn in the order of 10^{-11} can be measured and we can distinguish the weak stationary birefringence against the sinusoidal one.

In the case of stationary birefringence, application of burst ultrasound enables us to measure the reorientational relaxation time τ by analyzing the extinction curve after burst wave is turned off [8,10]. If the reorientational motion is expressed in terms of a single relaxation process, the extinction curve of the birefringence is given as

$$\Delta n(t) = \Delta n_{st}\exp(-t/\tau) \qquad (3)$$

where Δn_{st} the stationary value of $\Delta n(t)$.

Figure 2 shows the typical trace of ultrasonically induced birefringence signal for the α-hematite (Fe_2O_3) sol using the biased measurement technique and the waveform of the ultrasonic pulse used. The sign of the birefringence is negative.

To discuss the frequency dependence of the birefringence, the ultrasonic intensity must be measured precisely. To do so, we measured the ultrasonic intensity on the basis of the light diffracted by the propagating sound waves. The Raman-Nath parameter v_R is defined as,

$$v_R = (2\pi d/\lambda)\cdot(\partial n/\partial\rho)\cdot\delta\rho \qquad (4)$$

where n is the refractive index and $\delta\rho$ is the density perturbation caused by ultrasound. The Raman-Nath parameter can be estimated from the diffracted light intensity using the numerical calculation given by Klein and Cook [20,25]. The Raman-Nath parameter is related to the ultrasonic intensity W_U as,

$$W_U = v_R^{\,2}\left(c_0^{\,3}/2\rho\right)\cdot(\lambda/2\pi d)^2\cdot(\partial\rho/\partial n)^2 \qquad (5)$$

where c_0 is the sound velocity. Detailed method to obtain ultrasonic intensity was given in the literature [7,20].

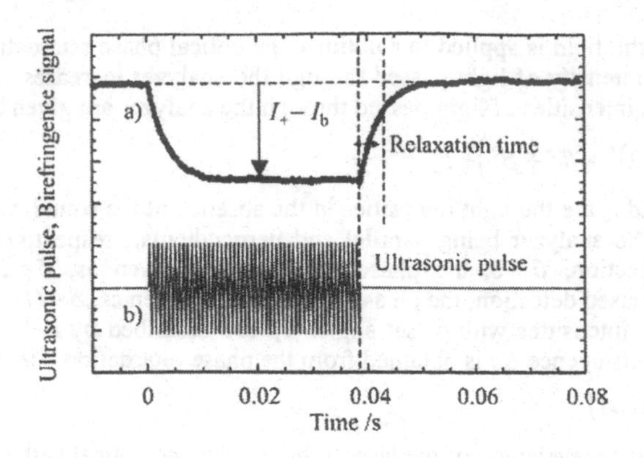

Figure 2 (a) Trace of transient ultrasonically induced birefringence of α-Fe_2O_3 sols at volume fraction 4.7×10^{-6} at 25°C (25 MHz, 0.005 W·cm^{-2}). (b) Applied ultrasonic pulse.

3. ULTRASONICALLY INDUCED BIREFRINGENCE IN NEAT LIQUIDS

3.1 INTRODUCTION

For neat liquids, a sinusoidal velocity gradient can directly produce sinusoidal orientational order. The reorientational order produced is not uniform but propagates as waves and the induced birefringence should be proportional to the ultrasonic amplitude, that is, the square root of the ultrasonic intensity. In the neat liquids, we cannot obtain the transient decay curve after rapid cessation of ultrasonic irradiation. If the relaxation frequency for reorientational motion is close to the applied ultrasound frequency, the birefringence per amplitude of the applied ultrasonic wave should be affected by the reorientational relaxation processes of the molecules.

The ultrasonically induced birefringence due to the velocity gradient is treated in terms of the coupling of the reorientational mode with the translational mode. Problems with regard to the coupling phenomena have been investigated mainly by flow birefringence and the VH depolarized light scattering [26-28]. Theories for the flow birefringence and the VH depolarized light scattering were constructed independently on the basis of the de Gennes' phenomenological theory [26] and irreversible statistical mechanics [27,28]. In both theories, the coupling of reorientational mode with the shear mode was taken into account. The expressions derived from both theories are consistent with each other [29]. In the case of ultrasonically induced birefringence, where coupling of the reorientational mode with the longitudinal mode should be taken into account, the formula for the birefringence from the irreversible statistical treatment has already been derived by Lipeles and Kivelson. We derived a theoretical expression of the ultrasonically induced birefringence in terms of the de Gennes' phenomenological theory [7,26] and here, we brief review the derivation and compare it with that obtained by Lipeles and Kivelson [30].

The liquid crystal in isotropic phase is a most appropriate sample to examine the expression obtained here, because the relaxation frequencies of the orientational motions are the same order magnitude as the ultrasonic frequencies usually used for ultrasonic relaxation studies. Besides, the flow birefringence for isotropic phase of liquid crystal has been already reported [31] by Martinoty *et al.* The values of the flow birefringence and the relaxation frequencies are required to reproduce the frequency dependence of the ultrasonically induced birefringence.

3.2 PHENOMENOLOGICAL THEORY OF ULTRASONICALLY INDUCED BIREFRINGENCE IN NEAT LIQUIDS [7,24]

Expression of ultrasonically induced birefringence in neat liquids can be obtained from the extension of the de Gennes' phenomenological equations [26] to more general ones [7,24]. The basic equations are transport equations of the mass and momentum conservation as,

$$\delta\dot{\rho} + \rho\partial_k v_k = 0 \tag{6}$$

$$\rho\dot{v}_\alpha = \partial_\beta\left(\sigma_{\alpha\beta} - \delta p\,\delta_{\alpha\beta}\right) \tag{7}$$

where ρ is the equilibrium density, $\delta\rho$ and δp are the mass and pressure perturbations, respectively, $\sigma_{\alpha\beta}$ is the stress tensor and v_α is the velocity vector and repeated indices are summed. We neglect here the effects of temperature fluctuations. The stress tensor and strain rate tensor, $\sigma_{\alpha\beta}$ and $\dot{\gamma}_{\alpha\beta}$ can be written as the sum of isotropic and anisotropic parts: $\sigma_{\alpha\beta} = \sigma_{\alpha\beta}{}^{i} + \sigma_{\alpha\beta}{}^{a}$ and $\dot{\gamma}_{\alpha\beta} = \dot{\gamma}_{\alpha\beta}{}^{i} + \dot{\gamma}_{\alpha\beta}{}^{a}$. The isotropic and anisotropic parts of the strain rate tensor are respectively written as follows, $\dot{\gamma}_{\alpha\beta}{}^{i} = \partial_k v_k \delta_{\alpha\beta}/3$ and $\dot{\gamma}_{\alpha\beta}{}^{a} = (\partial_\alpha v_\beta + \partial_\beta v_\alpha)/2 - \dot{\gamma}_{\alpha\beta}{}^{i}$. To consider the coupling of translational mode with orientational mode, we will take the tensor order parameter of molecular orientation $Q_{\alpha\beta}$. and its conjugated stress $\phi_{\alpha\beta}$ as internal variables [26]. Since the tensor order parameter is anisotropic (i. e. traceless), the anisotropic part of the $\sigma_{\alpha\beta}$ and $\dot{\gamma}_{\alpha\beta}$ can couple with the orientational mode. The relation between "force" and "flow" can be thus written by the expression of non-equilibrium thermodynamics as follows:

$$\sigma_{\alpha\beta}{}^{i} = 3\eta_v\dot{\gamma}_{\alpha\beta}{}^{i} \tag{8}$$

$$\sigma_{\alpha\beta}{}^{a} = 2\eta_s\dot{\gamma}_{\alpha\beta}{}^{a} + 2\mu\dot{Q}_{\alpha\beta} \qquad \phi_{\alpha\beta} = 2\mu\dot{\gamma}_{\alpha\beta}{}^{a} + \nu\dot{Q}_{\alpha\beta} \tag{9}$$

where η_v is the volume viscosity, η_s is the shear viscosity, and μ and ν are viscosity parameters introduced by de Gennes: μ is a measure of coupling between $Q_{\alpha\beta}$ and anisotropic part of the flow field, while ν is related to the relaxation of $Q_{\alpha\beta}$. If fluctuation of $Q_{\alpha\beta}$ is small, the relation of $Q_{\alpha\beta}$ and $\phi_{\alpha\beta}$ is written as, $\phi_{\alpha\beta} = -A\,Q_{\alpha\beta}$, where A is the quadratic expansion coefficient of free energy in $Q_{\alpha\beta}$. We also need the thermodynamic relation, $\delta\rho = (\partial\rho/\partial p)_s\delta p = c_0^{-2}\delta p$, where c_0 is the sound velocity in the low frequency limit.

Using the Eq.(6)-(9), we discuss problems of the longitudinal modes. Gradient of the displacement is taken in the x direction. The linearized differential equations of motion are expressed respectively as follows,

$$\begin{pmatrix} \delta\dot{\rho} \\ \dot{v}_x \\ \dot{Q}_{xx} \end{pmatrix} = \begin{pmatrix} 0 & -\rho\partial_x & 0 \\ -\dfrac{c_0{}^2\partial_x}{\rho} & \dfrac{1}{\rho}\left(\eta_v + \dfrac{4}{3}\eta_s(1-R)\right)\partial_x{}^2 & -\dfrac{2\mu}{\rho}\Gamma\partial_x \\ 0 & -\dfrac{4}{3}\dfrac{\mu}{\nu}\partial_x & -\Gamma \end{pmatrix} \begin{pmatrix} \delta\rho \\ v_x \\ Q_{xx} \end{pmatrix} \tag{10}$$

where Γ is the reorientational relaxation rate defined as $\Gamma = A/\nu$ and R is the translational-reorientational coupling parameter and it is related to the de Gennes viscosity coefficient as $R=(2\mu^2)/(\eta_s\nu)$ [26]. For longitudinal case, Q_{yy} and Q_{zz} are obtained as $Q_{yy}= Q_{zz}= -Q_{xx}/2$.

From the Eq. (9), we can derive the expression of the ultrasonically induced birefringence. The tensor order parameter $Q_{\alpha\beta}$ is related to the local dielectric tensor as, $Q_{\alpha\beta}= 3\left(\varepsilon_{\alpha\beta}- \varepsilon,^{-}\ \delta_{\alpha\beta}\right) / \left(2\Delta\varepsilon\right)$, where $\varepsilon_{\alpha\beta}$ is the dielectric tensor, $\varepsilon,^{-}$ is the mean dielectric constant in the absence of external perturbations and $\Delta\varepsilon$ is the anisotropy in the dielectric constant if all molecules are perfectly aligned in one direction. We choose z axis to be the direction of the optical beam whose polarization lies in the xy plane.

For ultrasonically induced birefringence measurement, the direction of the polarization of incident light is at 45° from the y axis. The relation of Q_{xx} to the birefringence can be obtained as follows [7,24],

$$\Delta n = \{\Delta\varepsilon/(2\bar{n})\}Q_{xx} \tag{11}$$

Applying of the Fourier transform to Eq.(10) and using the definition of the ultrasonic intensity, $W_{\mathrm{U}} = \delta p^2/(2\rho c_0)$, the root mean square of the ultrasonically induced birefringence Δn_{rms} is obtained as,

$$\Delta n_{\mathrm{rms}} = \frac{\Delta\varepsilon}{2\bar{n}}\cdot\frac{|Q_{xx}|}{\sqrt{2}} = \frac{2\Delta\varepsilon}{3\bar{n}}\sqrt{\frac{R\eta_S}{2A}}\sqrt{\frac{W_{\mathrm{U}}}{\rho c_0{}^3}}\frac{\omega}{\sqrt{1+(\omega/\Gamma)^2}} \tag{12}$$

Expression for flow birefringence is obtained from the same framework as [7,24,26],

$$\Delta n_{\mathrm{f}} = -\{2\bar{\dot{\gamma}}\Delta\varepsilon/(3\bar{n})\}\sqrt{R\eta_S/(2A)} \tag{13}$$

From Eqs. (12) and (13), the ultrasonically induced birefringence is related to the flow birefringence $\left|\Delta n_{\mathrm{f}} / \dot{}, ^{-}, \gamma\right|$ by,

$$\Delta n_{\mathrm{rms}} = \left|\frac{\Delta n_{\mathrm{f}}}{\bar{\dot{\gamma}}}\right|\sqrt{\frac{W_{\mathrm{U}}}{\rho c_0{}^3}}\frac{\omega}{\sqrt{1+(\omega/\Gamma)^2}} \tag{14}$$

Expressions for flow [32] and ultrasonically induced birefringence [30] were obtained by Kivelson and coworkers based on an irreversible statistical treatment as,

$$\left|\frac{\Delta n_{\mathrm{f}}}{\bar{\dot{\gamma}}}\right|^{\mathrm{Kivelson}} = \frac{1}{\bar{n}}\left(\nu_{\mathrm{K}}\cdot\frac{R_{\mathrm{Total}}\Delta_\mu}{1+\Delta_\mu}\cdot\frac{\eta_S}{k_{\mathrm{B}}T}\cdot\frac{R}{\Gamma}\right)^{1/2}\frac{\lambda_0{}^2}{\pi} \tag{15}$$

$$\Delta n_{\mathrm{rms}}^{\mathrm{Kivelson}} = \frac{1}{2}\left|\frac{\Delta n_{\mathrm{f}}}{\bar{\dot{\gamma}}}\right|^{\mathrm{Kivelson}}\sqrt{\frac{W_{\mathrm{U}}}{\rho c_0{}^3}}\frac{\omega}{\sqrt{1+(\omega/\Gamma)^2}} \tag{16}$$

where R_{Total} is the total light scattering per unit volume, Δ_μ is the depolarization ratio, ν_{K} is the ratio of the central Lorenztian HH Rayleigh line relative to the total depolarized light scattering intensity, k_{B} is the Boltzmann constant, T is the temperature. It should be noted, however, the proportionality constant in Eq.(16) differs by a factor of 2. This difference comes from the choice of coupled valuable in longitudinal mode. Details of the problem in the coupling between longitudinal and orientational mode were discussed in the literature [7,24].

3.3 EXPERIMENTAL RESULTS FOR ISOTROPIC PHASE IN LIQUID CRYSTALS [7]

As shown in the preceding section, the theoretical expressions for the ultrasonically induced birefringence of liquids and solutions are obtained in the framework of the de Gennes' phenomenological theory.

The liquid crystal in isotropic phase is a most appropriate sample to examine the expression obtained, because the relaxation frequencies of the orientational motions are in the same order of ultrasonic frequencies usually used. Besides, the flow birefringence for isotropic phase of liquid crystals, *p*-*n*-pentyl *p*'-cyanobiphenyl (5CB), has been already reported by Martinoty *et al* [4]. The values of the flow birefringence and the relaxation frequencies are required to reproduce the frequency dependence of the ultrasonically induced birefringence.

To confirm our expression quantitatively, we measured the ultrasonically induced birefringence of 5CB in the isotropic phase as a function of frequency (frequencies at 3,5,7,11, and 13 MHz) at temperature of 50 and 55 °C.

Figure 3 shows the ultrasonic intensity dependence of the birefringence in 5CB at 50 °C. The slope of the solid line for this logarithmic plot is 0.5, indicating the birefringence is proportional to the square root of the ultrasonic intensity. This result is in good agreement with Eq. (14). Figure 4 shows frequency dependence of $\Delta n_{rms} \cdot \sqrt{W_U}^{-1}$. The value of $\Delta n_{rms} \cdot \sqrt{W_U}^{-1}$ increased but the slope $\partial\left(\Delta n_{rms} \cdot \sqrt{W_U}^{-1}\right) / \partial f$ decreased with increasing frequency. The values of $\left|\Delta n_f / \dot{,} \bar{,} \gamma\right|$ from flow birefringence measurements [31] and Γ from optical beating light scattering experiments [33]. The values of $\left|\Delta n_f / \dot{,} \bar{,} \gamma\right|$ are 1.86 and 1.26 ns and those of $\Gamma/2\pi$ are 10.0 and 15.3 MHz at 50 and 55°C, respectively. The solid curves in Figure 4 were the calculated ones obtained by substituting those values into Eq. (14). The observed values of $\Delta n_{rms} \cdot \sqrt{W_U}^{-1}$ in this study were satisfactorily reproduced by Eq. (14).

Martinoty and Bader measured the ultrasonically induced birefringence for the isotropic phase of liquid crystals [4,5]. In their experiment, however, only the relative values of the ultrasonic intensity were measured and they examined Eq.(14) in terms of the temperature dependence of the birefringence at three different frequencies. We directly measured the ultrasonically induced birefringence as a function of frequency and ultrasonic intensity. Our results were reproduced completely by the equation derived here. Moreover, the absolute values of ultrasonically induced birefringence obtained in this work were in good agreement with those calculated using the flow birefringence and reorientational relaxation frequency obtained by light scattering. This indicates clearly that our theoretical treatment is valid and the method of ultrasonic intensity measurement presented here is very usefull.

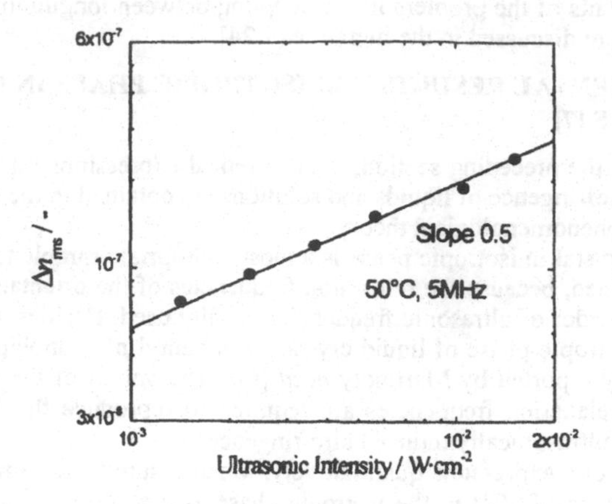

Figure 3. Ultrasonic intensity dependence of the root mean square of the induced birefringence of 5CB (Closed circle) at 50 °C. The solid line indicates the slope of 0.5

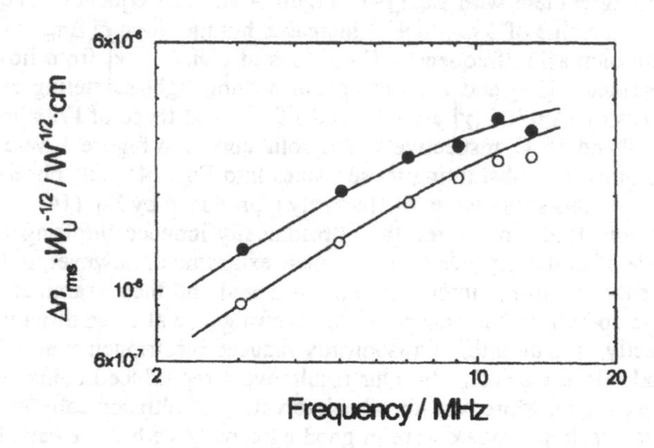

Figure 4. Frequency dependence the birefringence per the square root of the ultrasonic intensity at 50°C (closed) and 55°C (open). Solid curves indicate the estimation from Eq.(20). The detailed explanations are given in the text

4. ULTRASONICALLY INDUCED BIREFRINGENCE IN COLLOIDAL SOLUTIONS

4.1 INTRODUCTION

The theory developed by Oka [15,16] indicated that large disc-like rigid particles align by hydrodynamics torque which is produced by the radiation pressure due to the passage of the ultrasonic waves. The normal of disk-like particles is parallel to the ultrasonic field, and the sign of the birefringence of disk particles is negative. The measurements of the birefringence of large rigid disc-like particles were carried out on bentonite [2] and gold sols [3,8,18]. These experimental results show that the birefringence is proportional to the ultrasonic intensity as predicted by the Oka theory. Ou-Yang *et al.* [8] made use of the Raman-Nath diffraction effect to measure the ultrasonic intensity at the frequency range from 1 to 19 MHz and investigated the ultrasonically induced birefringence of gold sols as a function of solvent viscosity, particle size and ultrasonic intensity and frequency. The birefringence increased with increasing ultrasonic frequency and decreasing viscosity and the sign of the birefringence was positive. They modified the Oka theory to explain the sign of the birefringence and the dependence of the birefringence on the frequency and viscosity.

For large rigid rod-like particles, no theoretical study of the ultrasonically induced birefringence has been reported. Experimental investigations on the birefringence of V_2O_5 sols [9,13,14,17,19] have been carried out so far. The ultrasonically induced birefringence of V_2O_5 sols was proportional to the ultrasonic intensity and the volume fraction as predicted by the Oka theory [15,16]. Petralia also indicated that the birefringence depends on the viscosity and the sign of the birefringence is positive [17]. We carried out the investigation of the frequency dependence of the birefringence of rod-like rigid particles and discussed the results in terms of the modified Oka theory.

Some surfactant molecules form rod-like micelles in aqueous solution [34,35]. In the dilute concentration range, the solutions of these surfactants behave like sol [36]. At higher concentrations, the rod-like micelles form the entanglement networks and the solutions show the pronounced viscoelasticity. The rod-like micelles are anisotropic aggregates and the solutions become birefringent by application of external fields. The dynamics of the entanglement networks of the rod-like micelles has been investigated by birefringence measurements such as electric birefringence [37] and flow birefringence [38-40]. The results of studies showed some anomalous behaviors characteristic of the entanglement networks.

Compared with other birefringence methods, the advantages of the ultrasonically induced birefringence one is that it enables us to investigate the motion of rod-like micelles under the fast mechanical perturbation. In addition, it is free from the presence of ions or electric double layer of rod-like micelle. We will here introduce the investigation of the dynamics of the entanglement networks of the rod-like micelles hexadecyldimethylamine oxide (CDAO) and mixtures of hexadecyltrimethylammonium bromide (CTAB) and sodium salicylate (NaSal) by ultrasonically induced birefringence measurements. The elastic property of the entanglement networks of the rod-like micelles will be discussed from the transient spectrum of ultrasonically induced birefringence.

4.2 MODIFIED OKA THEORY FOR ROD-LIKE PARTICLES

Here we briefly review the modified Oka theory for the rod-like particles [10].

Application of external fields such as electric, magnetic and ultrasonic fields will induce the birefringence Δn, which is given by

$$\Delta n = 2\pi \Delta G \Theta \phi / n_s \tag{17}$$

where Θ is the mean orientation of the particles, ΔG is the optical anisotropy, ϕ is the volume fraction of particles and n_s is the refractive index of the solution. The mean orientation can be calculated from the distribution function of particles which is obtained by considering the potential energy due to the torque caused by the applied field. For ultrasonic field, Oka [15,16] assumed that the passage of acoustic waves sets up the radiation pressure so that large rigid disc-like particles are subject to the turning torque that will align them. The ultrasonic torque is estimated by considering the Rayleigh disc [41,42] problem. We extended the modified Oka theory for disc-like particles in view of application to prolate particles. The time averaged torque $|M|$ for prolate particles is written as,

$$|M| = X \rho_0 a b^2 (v-u)^2 \sin 2\theta \tag{18}$$

where v is the fluid velocity, u is the translational velocity of particle, ρ_0 is the densities of solvent, the a and b are the radii of major and minor axes of the particle, respectively and θ is an angle between the major axis of prolate spheroids and the direction of ultrasonic propagation. Symbol X is given as,

where

$$X = \frac{2\pi \varepsilon_s^{\,3} \left[2\varepsilon_s^{\,3}/3 + \ln\{(1+\varepsilon_s)/(1-\varepsilon_s)\} - \varepsilon_s \right]}{Y\left(2\varepsilon_s^{\,3} + Y\right)}, \quad Y = \frac{1-\varepsilon_s^{\,2}}{2} \ln\left[\frac{1+\varepsilon_s}{1-\varepsilon_s}\right] - \varepsilon_s \tag{19}$$

with $\varepsilon_s = \sqrt{a^2 - b^2}/a$. If the ultrasonic energy is small compared with the thermal disturbance, the birefringence is expressed for the prolate particles as

$$\Delta n = 4X\pi a b^2 \Delta G \phi W_U |F|^2 / (15 k_B T c_s n_s) \tag{20}$$

where k_B is the Boltzmann constant, T is the temperature, c_s is the sound velocity of solvent, and the factor F is the non-dimensional factor related to the relative velocity between the fluid and particles. The optical anisotropy of the prolate particles can be obtained by an extension of the theory of Peterlin and Stuart [44].

In the Oka theory, the factor F was calculated in an ideal fluid, one without viscosity, it is found that for the disk for the prolate spheroid [10] as,

$$F = (\rho - \rho_0) / \{\rho + \rho_0 \alpha_p / (2 - \alpha_p)\} \tag{21}$$

where ρ is the densities of the particle and the parameter α_p is given as,

$$\alpha_p = ab^2 \int_0^\infty \left(a^2 + x\right)^{-1/2} \left(b^2 + x\right)^{-2} dx \tag{22}$$

In the Oka theory, the factor F is independent of the viscosity and the frequency so that the birefringence is also independent of them.

Ou-Yang *et al.* [8] indicated that the suspended particles which execute translational oscillations in fluid are subject to the drag force that is shown in terms of the librational Reynolds number $R_L = a^2 \omega \rho_0 / (2\eta_S)$ [45]. They supposed that the translational motion is much faster than the orientational motion and the equation of translational motion for particles in dilute region is expressed as

$$F = (\rho - \rho_0) / \left[\rho + i(\zeta / V\omega)\left\{\left(1 + \sqrt{R_L}\right) - i\left(\sqrt{R_L} + \gamma R_L\right)\right\}\right] \tag{23}$$

where V is the particle volume, ζ is the Stokes' drag coefficient and γ is a constant. When the rod-like particles are translating in the direction perpendicularly to major axis, the ζ is given [46] as

$$\zeta = 16\pi\eta_S ab^2 / \left(\chi + a^2\alpha_p\right), \quad \chi = ab^2 \int_0^\infty \left(a^2 + x\right)^{-1/2} \left(b^2 + x\right)^{-1} dx \tag{24}$$

In the high frequency limit, $R_L \to \infty$, the parameter F in viscous fluids is reduced to that in ideal fluid. Comparing Eqs.(23) and (24) with Eq.(21) leads to the relation, $\gamma = \alpha_p (\chi + b^2\alpha_p) / \{(6a^2)\cdot(2-\alpha_p)\}$. The frequency dependence of the birefringence for the rod-like can be expressed in terms of F.

4.3 ULTRASONICALLY INDUCED BIREFRINGENCE IN ROD-LIKE SOL SOLUTIONS [10]

In order to elucidate the frequency dependence of the ultrasonically induced birefringence and the orientational relaxation time of rod-like particles, we used the hematite (α-Fe_2O; axial ratio=5.5, the volume fraction was varied from 9.4×10^{-8} to 2.4×10^{-5}) sol of rod-like particles and poly(tetrafluoroethylene) (PTFE; the average length of major axis of 0.55μm and the average axial ratio of 1.9. The volume fraction was varied from 1.3×10^{-7} to 8.2×10^{-4}) latex of prolate particles. We will show here the experimental results and discuss the frequency dependence of the stationary birefringence and the relaxation time of them.

As described in subsection 4.2, the Oka theory [15,16] predicted that the birefringence is proportional to the ultrasonic intensity and volume fraction. In addition the birefringence does not depend on the ultrasonic frequency nor the solvent viscosity.

Figure 5 shows the intensity dependence of the birefringence of α-Fe_2O_3 and PTFE sols. The birefringence obtained here is proportional to the ultrasonic intensity. This is in accordance with the Oka theory. Figure 6 shows the frequency dependence of the birefringence multiplied by the temperature to compensate for the $1/(k_BT)$ factor associated with thermal randomization. The birefringence depends on the frequency. This means that the birefringence of rod-like particles is not fully interpreted by the Oka theory.

From the theoretical expression in the above subsection, the theoretical estimation of the frequency dependence of the stationary birefringence can be obtained. The calculated birefringence increased with frequency. This tendency is qualitatively in accordance with the experimental results. However, for α-Fe_2O_3 sols, the experimental values were about 20 times larger than the theoretical value while, for PTFE latex, the experimental values are about 100 times as large as the theoretical values [10].

The large deviations from the theoretical values cannot be explained in terms of the modified Oka theory for rod-like particles, even if one takes into account the experimental and estimation errors in the refractive index and the size of the particle.

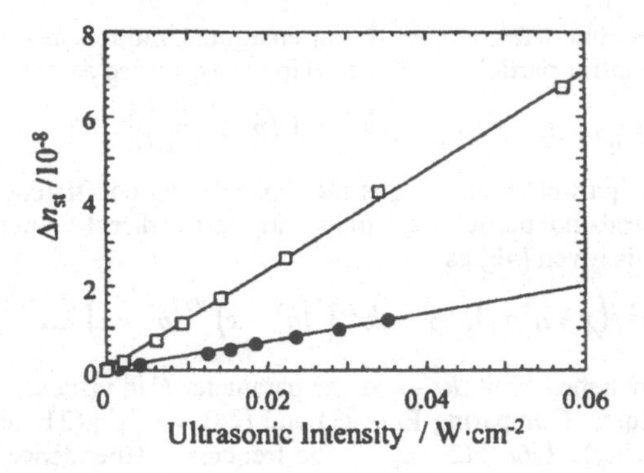

Figure 5.. Birefringence against ultrasonic intensity at 25°C (25 MHz): (●) α-Fe_2O_3 sols at volume fraction 4.7×10^{-6}; (□) PTFE latex at volume fraction 7.6×10^{-5}. Birefringence against ultrasonic intensity at 25°C (25 MHz): (●) α-Fe_2O_3 sols at volume fraction 4.7×10^{-6}; (□) PTFE latex at volume fraction 7.6×10^{-5}.

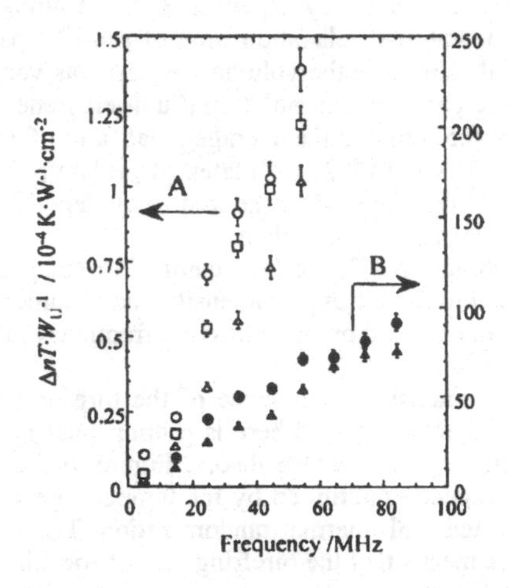

Figure 6.. Plots of $\Delta nT/W_U$ against the ultrasonic frequency. A. α-Fe_2O_3 sols of (volume fraction=4.7 x 10^{-6}). Temperature: (O) 50 °C; (□) 25 °C; (△) 10°C. B. PTFE latex (volume fraction=8.2 x 10^{-4}). Temperature: (●) 50°C; (▲) 10°C.

In the modified Oka theory, the parameter F in the Oka theory was replaced by the one derived from the Navier-Stokes equation for viscous fluids. However, the turning torque also should be modified on the basis of the Navier-Stokes equation. In addition, under oscillatory motion, the flow around the rod-like particles is complicated and the flow may be in eddy or stagnation. A new theory of the ultrasonically induced birefringence should be required for the case of rod-like particles after taking into account of turning torque in viscous fluids and the boundary condition of fluids around the rod-like particles.

The birefringence decays after ultrasonic field disappears as shown in Figure 2 in the experimental section. This decay reflects the orientational relaxation process of the particles. The orientational relaxation time of the particles is obtained by fitting the extinction process of the birefringence to Eq.(3). The orientational relaxation times of α-Fe_2O_3 sols and PTFE latex are plotted against $\eta_S/(k_BT)$ in Figure 7. As is shown in Figure 7, the orientational relaxation times of both α-Fe_2O_3 and PTFE particles could be expressed by the Debye-Einstein equation. The orientational relaxation time τ of α-Fe_2O_3 and PTFE particles in dilute solution followed the Debye-Einstein equation, $\tau = \eta_s V^* / (k_BT)$, where V^* is the effective volume of a particle. The effective volume of α-Fe_2O_3 and PTFE particle estimated from the slopes in Figure 7 were 1.5×10^{-20} and 7.6×10^{-20} m^3, respectively. The effective volume V^* of a prolate particle was given as [47,48]

$$V^* = \frac{8\pi a^3}{9}\left(\frac{1-p^4}{(2-p^2)G(p)-1}\right),\ G(p) = \frac{1}{(1-p^2)^{1/2}}\ln\left[\frac{1+(1-p^2)^{1/2}}{p}\right] \qquad (25)$$

where p is the axial ratio b/a. For PTFE latex, $2a$ calculated from Eq. (32) was 0.80 μm with p=1/1.9. The size of PTFE particles obtained from the electron microscopy was within 0.4 to 1.0 μm with an average value of 0.55 μm and it was smaller than the size estimated from the birefringence measurements. For α-Fe_2O_3 particles, the value of $2a$ was 0.55 μm with p=1/5.5 and was consistent with the value reported by Ozaki [49].

This fact means that the orientational motion of these particles is not affected by the interparticle interaction. In other words, information of the orientational motion of isolated particles with large anisotropy is obtained by the ultrasonically induced birefringence, since the measurements in very dilute solution can be carried out.

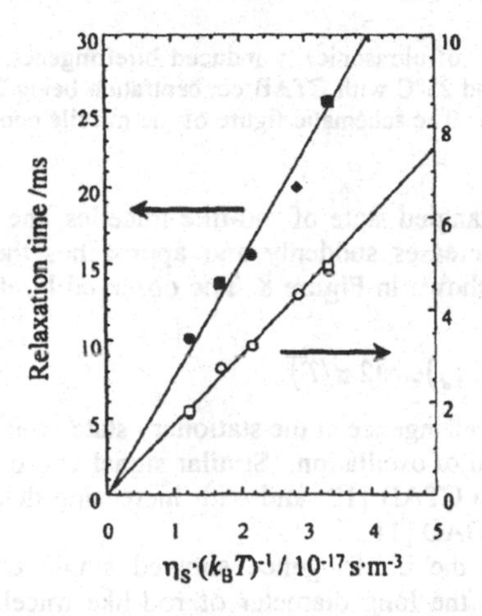

Figure 7. Orientational relaxation time for α-Fe_2O_3 sols and PTFE latex are plotted against η_S/k_BT. A. α-Fe_2O_3 sols (volume fraction (□)2.4×10^{-5}; (○)4.7×10^{-6}). B. PTFE latex(volume fraction (■)4.1×10^{-3}; (■)7.6×10^{-5}; (◆)2.6×10^{-6}.

4.4 ULTRASONICALLY INDUCED BIREFRINGENCE OF ROD-LIKE MICELLES IN ENTANGLEMENT NETWORKS [11,12]

Figure 8 gives transient trace of the ultrasonically induced birefringence measurements observed for aqueous solutions of CTAB-NaSal at 55, 35 and 25°C with CTAB concentration being 20mM and molar ratio of NaSal to CTAB being 0.6 [12]. From viscosity and rheological measurements, spherical micelles were formed at 55°C and at 35°C, it changed to rod-like but not in entanglements states. However, at 25°C the entanglement network was recognized by rheological measurements. In spherical micelles at 55°C, a birefringence signal was not observed. In the region of rod-like micelles at 35°C, the negative birefringence with single exponential decay was observed.

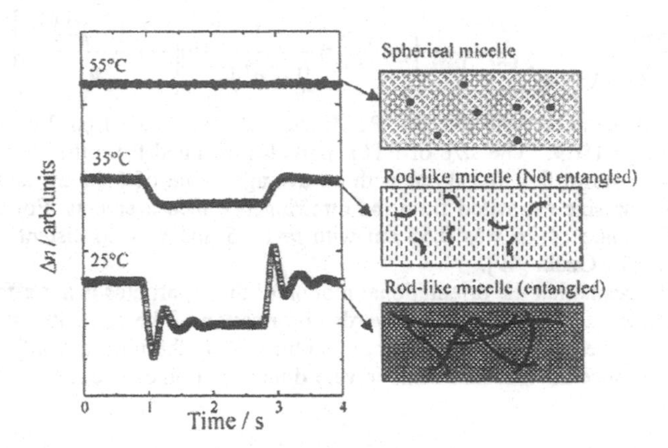

Figure 8. Transient trace of ultrasonically induced birefringence for aqueous solutions of CTAB-NaSal at 55, 35 and 25°C with CTAB concentration being 20mM and molar ratio of NaSal to CTAB being 0.6. The schematic figure of the micelle condition is illustrated in the right side.

At 25°C, in the entangled state of rod-like micelles, the birefringence after the onset of ultrasound increases suddenly and approaches the steady value after a damped oscillation as shown in Figure 8. The observed birefringence signal can be expressed by as

$$\Delta n(t) = \Delta n_{st} \exp(-t/\tau_d)\cos(2\pi t/T) \qquad (26)$$

where Δn_{st} is the birefringence at the stationary state, τ_d is the damping time constant and T is the period of oscillation. Similar signal was observed with increasing molar ratio of NaSal to CTAB [12] and with increasing detergent concentration in aqueous solutions of CDAO [11].

In the region that the birefringence showed single exponential behavior in CTAB-NaSal solution, the long diameter of rod-like micelles could be estimated from the Debye-Einstein plot and it was about 2 μm. However, the mean distance between the rod-like micelles was 0.3 μm. The concentration thus was not in dilute region so that application of the Debye-Einstein relation was not valid. From the rheological measurements, the complex shear modulus behaved like the Rouse like polymer. The effective viscosity should be taken into account in the Debye-Einstein relation as was done in the polymer solution. The experimental results for rod-like micelles when the entanglement was formed were summarized in the following [11,12],

1. The transient birefringence signal showed the damped oscillation.
2. The stationary values of the birefringence was proportional to the ultrasonic intensity .
3. The $\Delta n_{st} / W_U$ value increased with increasing frequency.
4. The period of the oscillation decreased with increasing molar ratio of the added salt to micelle, increasing concentration of micelles and decreasing the temperature.

The ultrasonic intensity dependence of the birefringence of the entanglement networks indicated that the radiation pressure induced the birefringence. The frequency dependence of the birefringence resulted from the translational velocity difference between the entanglement networks and the solvent. With increasing ultrasonic frequency, the translational motion of entanglement networks with respect to solvent became more hindered. As the radiation pressure on the rod-like micelles increased with the translational velocity difference, the birefringence increased with the ultrasonic frequency.

No theory has been proposed yet which can explain the damped oscillation behavior of transient ultrasonically induced birefringence of the rod-like micelles. The viscoelastic properties [50-53] of the entanglement networks of CDAO and the mixtures composed of CTAB and NaSal have been described by the Maxwell model which has only one stress relaxation time, τ_s, and the entanglement networks are regarded as elastic body with plateau modulus, G_0, at frequencies much higher than $1/\tau_s$. The shear stress on the entanglement networks of the mixtures of CTAB and NaSal exhibited the damped oscillation according as the start of shear flows of high rate above 0.24 s^{-1} [54]. We considered the ultrasonically induced birefringence of rod-like micelles as follows. The entanglement networks suffer the stress due to the radiation pressure of the ultrasonic wave and shrink in the direction of sound propagation with damped elastic oscillation. In the shrunk of entanglement network, the major axis of the rod-like micelles is elongated and orientated perpendicularly to the direction of the ultrasonic wave propagation. After the cessation of ultrasonic fields, the entanglement networks relax from a shrunk state with damped elastic oscillation. Since the fact that G_0 of the entanglement networks was approximately proportional to the square of the concentration of surfactant molecules is essentially equal to that observed in entangling polymer systems. Using simple model of the damped oscillation, period T, is inversely proportional to the square root of G_0. We have measured the birefringence and the complex shear modulus in the aqueous solutions of CTAB and NaSal [12]. Figure 9 shows the plot of the period T against G_0. The solid line in the figure indicates the slope $-1/2$ and the data almost follow the slope. This indicates that the damped oscillation of the ultrasonically induced birefringence observed in entangled micelles can be described by the above discussion.

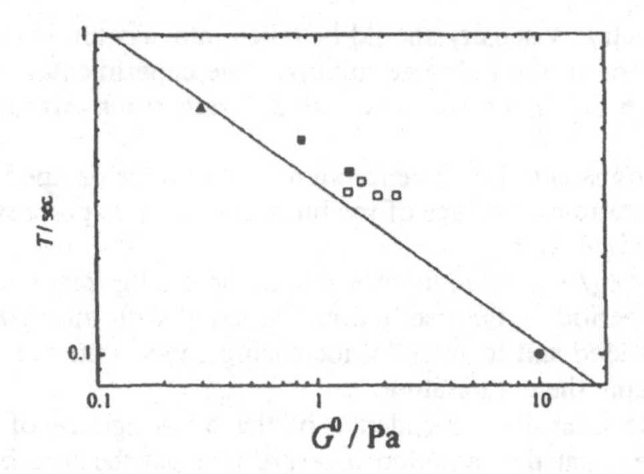

Figure 9. Plot of the period of the damped oscillation against plateau shear modulus G_0 for aqueous solutions of CTAB-NaSal. CTAB concentration id 20mM. ((■):with different temperatures at molar ratio of 0.6). (□): with different molar ratio of NaSal to CTAB at 25°C), (●) 10mM CTAB at molar ratio of 0.9, (▲) CDAO 100mM.

5. ULTRASONICALLY INDUCED BIREFRINGENCE IN POLYMER SOLUTIONS

5.1 INTRODUCTION

As shown above sections, the mechanism of the ultrasonically induced birefringence has been classified into two. The mechanism of the ultrasonic induced birefringence in polymer solutions is of strong interest since a polymer molecule is a long chain consisting of the small monomer unit. In 1950, Peterlin has expected that the ultrasonically induced birefringence should be observed in polymer solutions and proposed a theory for the birefringence [1,55]. The theory predicted that the birefringence is proportional to the square root of the ultrasonic intensity.

In 1964, Jerrard measured first the ultrasonically induced birefringence of polystyrene (PS)-toluene and polyisobutylene (PIB)-cyclohexane solution as a function of ultrasonic intensity and frequency (1-5 MHz). He has reported that ultrasonically induced birefringence of polymer solutions is proportional to the square root of the ultrasonic intensity. The results were in agreement with Peterlin's theory.

Recently we observed that not only the root mean square of the birefringence Δn_{rms} by the "non-biased" technique but also the stationary birefringence Δn_{st} by "biased" technique in PS-toluene solutions [20]. The former was proportional to the square root of the ultrasonic intensity, $\sqrt{W_U}$ but the latter was proportional to W_U. The mechanism of the ultrasonically induced birefringence should be investigated more in details. In this section, we show our recent result for polymer solutions.

5.2 ULTRASONIC INTENSITY DEPENDENCE OF BIREFRINGENCE IN POLYMER SOLUTIONS [20-24]

Peterlin considered that polymer solutions will behave more as liquids than colloidal solutions, because of the flexibility of the polymer chain being permeable spheres in which the solvent was partially in immobilized [1,55]. The optical behavior was calculated by finding an expression for the mean polarizabilities in two directions and the method of Langevin was used to connect the anisotropy of the polarizability with the birefringence. He obtained the following expression for ultrasonically induced birefringence of polymer solutions.

$$\Delta n_{\mathrm{rms}} = \frac{4\pi}{45} N_0 \frac{(n^2+2)^2}{n} (\alpha_1 - \alpha_2) \sqrt{\frac{W_{\mathrm{U}}}{\rho c_0^{\,3}}} \frac{\omega\tau}{\sqrt{1+\omega^2\tau^2}} \tag{27}$$

where α_1 and α_2 are polarizabilities along and perpendicular to bond axis, respectively, N_0 is the number of molecules per unit volume and the relaxation time τ is a function of both rotational and internal diffusion constant and ω is the angular frequency of ultrasonic wave. His theory predicted the birefringence is proportional to $\sqrt{W_{\mathrm{U}}}$.

Recently we measured the ultrasonically induced birefringence of PS-toluene solutions using not only the non-biased measurement which was used in Jerrard's experiment but also the biased measurement [20]. Figure 10 shows the ultrasonic intensity dependence of Δn_{rms} and Δn_{st}. As shown in Figure 10, Δn_{rms} was proportional to the square the root of the ultrasonic intensity, but Δn_{st} are proportional to the ultrasonic intensity. The experimental results showed that Δn_{rms} was proportional to the square root of the ultrasonic intensity were in agreement with those obtained by Jerrard for PS-toluene solutions.

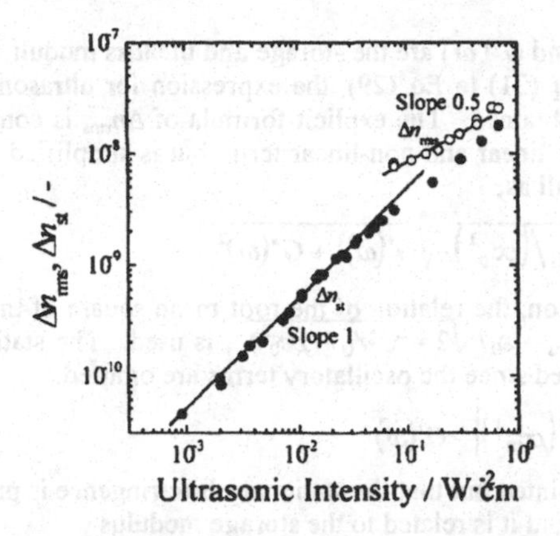

Figure 10. Ultrasonic intensity dependence of the root mean square birefringence Δn_{rms} and the stationary birefringence Δn_{st} in toluene solution of polystyrene (10 wt%, 25MHz, 25°C) [20]. The solid and dashed lines indicate the slopes of 1 and 0.5, respectively.

In polymer solutions, the normal stress difference can be observed and it is well known the normal stress difference is a typical non-linear effect and it can be defected by flow birefringence measurement [56,57]. The theory of flow birefringence in polymer solutions can be extended to the ultrasonically induced birefringence. Although details are given elsewhere [24], we describe its derivation here briefly.

The birefringence caused by the longitudinal deformation is written as,

$$\Delta n = C\left(\sigma_{xx}^{\,*} - \sigma_{yy}^{\,*}\right) = C\left(\sigma_{xx} - \sigma_{yy}\right) \tag{28}$$

where C is the stress optical coefficient. Using the Lodge's phenomenological equations [56-58], the response of stress difference is given as,

$$\sigma_{xx} - \sigma_{yy} = \int_{-\infty}^{t} \frac{\partial}{\partial t'} G(t-t')\left[B_{xx}(t,t') - B_{yy}(t,t')\right] dt' \tag{29}$$

where $B_{\alpha\beta}(t, t')$ is the Finger strain tensor [56-58] at time t referred to the state time t' and $G(t)$ is a shear relaxation modulus. The Finger strain tensor for longitudinal deformation is written as,

$$\mathbf{B}(t,t') = \begin{pmatrix} \lambda(t,t')^2 & 0 & 0 \\ 0 & 1 & 0 \\ 0 & 0 & 1 \end{pmatrix} \tag{30}$$

where $\lambda(t, t')$ is a expansion factor in the x direction and is related to the Hencky strain [56,57], $\varepsilon(t, t')$ as $\lambda(t, t') = \exp(\varepsilon(t, t'))$. Under the condition of the steady oscillatory strain, when the exponential factor is expanded to the second order, the stress difference is written as,

$$\begin{aligned} \sigma_{xx} - \sigma_{yy} &= 2\varepsilon_0 \left(G'(\omega)\cos\omega t - G''(\omega)\sin\omega t\right) + \\ &+ 2\varepsilon_0{}^2 \left\{G'(\omega) + \left(G'(\omega) - G'(2\omega)/2\right)\cos(2\omega t) - \left(G''(\omega) - G''(2\omega)/2\right)\sin(2\omega t)\right\} \end{aligned} \tag{31}$$

where $G'(\omega)$ and $G''(\omega)$ are the storage and the loss moduli, respectively.

Substituting Eq. (31) to Eq. (29), the expression for ultrasonically induced birefringence can be obtained. The explicit formula of Δn_{rms} is complicated because of the coupling of the linear and non-linear term, but is simplified when the non-linear contribution is small as,

$$\Delta n_{\mathrm{rms}} = C\sqrt{W_{\mathrm{U}} / \left(\rho c_0{}^3\right)} \cdot \sqrt{G'(\omega)^2 + G''(\omega)^2} \tag{32}$$

In the calculation, the relation of the root mean square of the strain to the ultrasonic intensity, $\varepsilon_{\mathrm{rms}} = \varepsilon_0 / \sqrt{2} = \sqrt{W_{\mathrm{U}} / (\rho c_0{}^3)}$, is used. The stationary birefringence Δn_{st} is easily derived since the oscillatory terms are omitted:

$$\Delta n_{\mathrm{st}} = C\left[W_{\mathrm{U}} / \left(\rho c_0{}^3\right)\right] \cdot G'(\omega) \tag{33}$$

It should be pointed out that the stationary birefringence is proportional to the ultrasonic intensity and it is related to the storage modulus

In the microscopic theory of viscoelasticity of Rouse-Zimm, the storage and the loss moduli are respectively written as [58],

$$G'(\omega) = N_0 k_B T \sum_{P=1}^{P_{\max}} \frac{(\omega\tau_P)^2}{1 + (\omega\tau_P)^2}, \quad G''(\omega) = N_0 k_B T \sum_{P=1}^{P_{\max}} \frac{\omega\tau_P}{1 + (\omega\tau_P)^2} \tag{34}$$

where $P_{\max}$ is the maximum mode number, and τ_P is the relaxation time of Pth normal mode. In case of the Gaussian chain, the stress optical coefficient is written as,

$$C = 4\pi\left(n^2 + 2\right)^2 (\alpha_1 - \alpha_2) / (45 k_B T \cdot n) \tag{35}$$

Substituting Eqs. (34) and (35) to Eq. (32), Eq. (32) is the same form as the Peterlin's result (Eq. (27)) if the only first normal mode is taken into account.

The relaxation time of the Pth mode is written in terms of the longest relaxation time τ_1 and the scaling constant related to the dimension of the polymer chain ν as,

$$\tau_P = \tau_1 / P^{3\nu} \qquad \tau_1 = [\eta] K_1 M \eta_s / (N_A k_B T) \tag{36}$$

η_s is the solvent viscosity, $[\eta]$ is the intrinsic viscosity, N_A is the Avogadro's number and K_1 is first order relaxation time factor determined by theory [59].

We compared our experimental results with our expression mentioned above [21]. The results are listed in Table I. The calculated value of Δn_{rms}, which was proportional to $\sqrt{W_U}$, was in the same order of magnitude as the experimental one. However, the calculated result for Δn_{st} was about 10^5 times smaller than that obtained by the experiments. In our expression, Δn_{st} resulted from the quadratic term in the Finger strain tensor in the Lodge equation. In the framework of the Lodge equation, the non-linearity is a result of *frame invariance* and the stress-strain relation itself belongs to the linear region [56,57]. To estimate the Δn_{st} more correctly, nonlinear terms of viscoelasticity and acoustic field should be taken into account.

Table I. Comparison of the calculation with experiment for the ultrasonically induced birefringence for PS-toluene solutions [21].

	$\Delta n_{rms} \cdot \sqrt{W_U}^{-1}$ / $cm \cdot W^{-0.5}$	$\Delta n_{st} \cdot W_U^{-1}$ / $cm^2 \cdot W^{-1}$
Experiment	3.0×10^{-8}	4.8×10^{-8}
Calculation	1.5×10^{-8}	6.6×10^{-13}

In order to elucidate the detailed mechanisms of ultrasonically induced birefringence in polymer solutions, it is very important to know the interactions between the segments in a chain. For the reason, it is better to measure the birefringence in lower concentration ranges. In our study, we focused on the experimental studies about the stationary birefringence because the sign of the birefringence is obtained and the high signal to noise ratio is realized [20]. We have started the systematic measurement of ultrasonically induced birefringence in polymer solutions for the molecular structure of segment, concentration, molecular weight and frequency dependence and so on [22].

5.3 EFFECT OF MOLECULAR STRUCTURE AND MOLECULAR WEIGHT ON THE STATIONARY ULTRASONICALLY INDUCED BIREFRINGENCE IN POLYMER SOLUTIONS [21-23]

Figure 11 (a) shows typical traces of the transient signal of ultrasonically induced birefringence of PS -toluene, polycarbonate (PC)-chloroform and polybutadiene (PBD)-toluene solutions. The sign of the birefringence was positive for PS-toluene and negative for the PC-chloroform and PBD-toluene solutions. The rise and decay time of the trace of the signal reflects the orientational relaxation times of molecular motions. The rise time of the trace of the birefringence was about 4μs, which was almost the same as the rise time of the ultrasonic pulse. This means that the orientational relaxation time contributing to the birefringence is smaller than 4μs.

The difference in the sign of birefringence is reflected by the local configuration of polymer chain in the ultrasonic field. It is reasonable to consider that for the flexible polymer chain, the orientation of the segment units of a main chain of the polymer will be perpendicular to the sound propagation. The configuration of the phenyl group will be essentially related to the sign of the birefringence signal. The difference in the sign reflects the difference in the way of bonding of the phenyl ring to the main chain structure as shown schematically in Figure 11 (b).

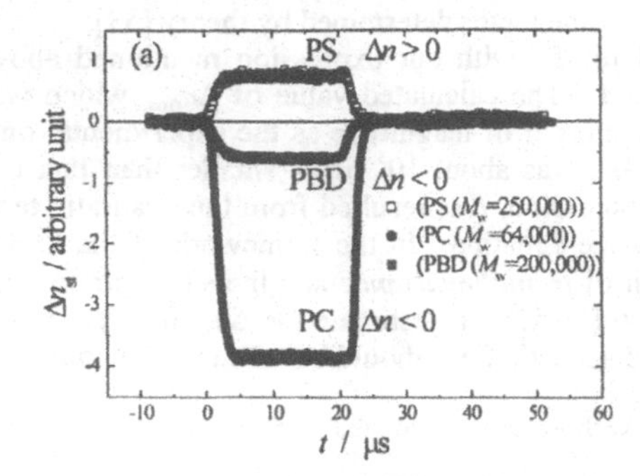

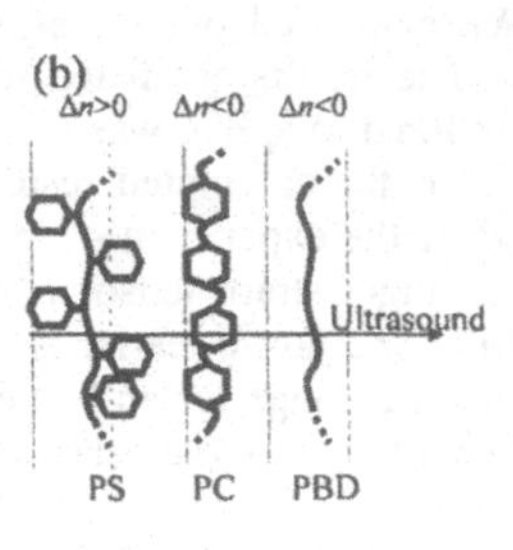

Figure 11. (a) Traces of the birefringence signal for PS-toluene, PC-chloroform and PBD-toluene solutions. (b) Schematic illustration of the local orientation of PS, PC and PBD under the ultrasound.

The molecular weight dependence of the $\Delta n_{st} \cdot W_U^{-1}$ value is shown for PS-toluene solutions at constant concentration of 1.28 mol·dm^{-3} in Figure 12. At the molecular weight lower than 1×10^4, the birefringence increased as molecular weight increased. However, molecular weight dependence was not recognized when the molecular weight was higher than 1×10^4.

It is well known that polymer solutions show the relaxation processes in the frequency range of MHz which have been observed by ultrasonic relaxation [60,61] and dielectric dispersion spectroscopy [60]. These relaxation processes arise from the local segmental motions and the relaxation time does not depend on the molecular weight of the polymer chain in solutions in the region of molecular weight above 10^3 to 10^4. In PS-toluene solutions, no molecular weight dependence of the ultrasonic relaxation spectra was observed above 1×10^4 [60,61]. This *critical* molecular weight of the ultrasonic relaxation was the same as that of the ultrasonically induced birefringence. This suggests that the Δn_{st} is mainly related to the local orientational segmental motions in polymer solutions.

In order to confirm the local segmental model for Δn_{st} of polymer solutions, we define the intrinsic values of the stationary ultrasonic birefringence and compared Δn_{st} with the segmental anisotropy in polarizability, $\Delta \alpha_s$. Since extrapolated values of $\Delta n_{st} \cdot W_U^{-1} / C$ took finite ones, we defined the intrinsic value of Δn_{st} as follows,

$$\left[\Delta n_{st} \cdot W_U^{-1}\right] \equiv \lim_{C \to 0} \Delta n_{st} \cdot W_U^{-1} / C \tag{37}$$

The segmental anisotropy in polarizability, $\Delta \alpha_s$, was estimated from the flow, magnetic and electric birefringence [62-64]. To compare the birefringence with the anisotropy in polarizability on the basis of the segment unit, the intrinsic values of the birefringence multiplied by the number of monomer unit in a segment, n_s,

$[\Delta n_{st} \cdot W_U^{-1}] \cdot n_s$ was calculated. The value of n_s was estimated from the molecular weight of monomer unit, M_0 and the molecular weight of a segment, M_s, which was determined by limiting rigidity of the Rouse mode obtained by the viscoelastic measurements at high frequencies [65,66]. Figure 13 shows the plots the intrinsic values of the stationary birefringence per a segment against the segmental anisotropy in polarizability. The linear relationship is obtained. This means that the stationary ultrasonically induced birefringence of polymer solutions is related to the segmental anisotropy in polarizability of polymer chains.

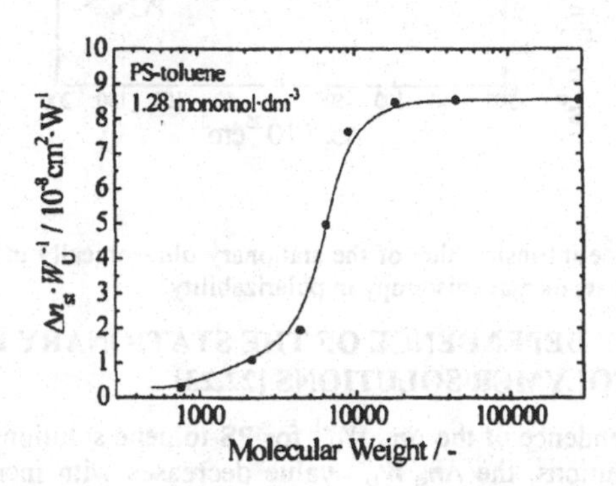

Figure 12. Molecular weight dependence of $\Delta n_{st} \cdot W_U^{-1}$ for PS-toluene solution at the concentration of 1.28 mol·dm^{-3}.

The segmental anisotropy in polarizability, $\Delta\alpha_s$, is related to the monomer anisotropy in polarizability, $\Delta\alpha_m$, as,

$$\Delta\alpha_s = Z\Delta\alpha_m \tag{38}$$

where Z is the parameter which reflect the stiffness of the polymer chain, the restricted rotation and the steric effects between monomer units [67]. In aqueous solutions of polyelectrolyte, we can change the stiffness of polymer chain drastically with addition of the salt. We measured the ultrasonically induced birefringence in aqueous solutions of sodium polystyrenesulfonate (NaPSS) and tetramethylammonium polystyrenesulfonate (TMAPSS) with addition of the salt [22]. The birefringence deceased with ionic strength. It is well known that the decrease of the persistence length with addition of the salts decreases the viscosity of aqueous solutions of polyelectrolyte. Since the persistence length is directly correlated to the chain stiffness, the segmental anisotropy in polarizability decreases with the addition of salt. Therefore the decrease of the $\Delta n_{st} \cdot W_U^{-1}$ values with addition of the salt is caused by the decrease in the stiffness. The $\Delta n_{st} \cdot W_U^{-1}$ values of NaPSS and TMAPSS solutions approached that of PS-toluene solution at the same molarity at higher ionic strength. This result is quite reasonable because the polyelectrolyte behaves flexible polymer in the presence of excess salt and the fact also supports our assumption that the stationary birefringence of polymer solutions is directly related to the segmental anisotropy in polarizability of polymer chains.

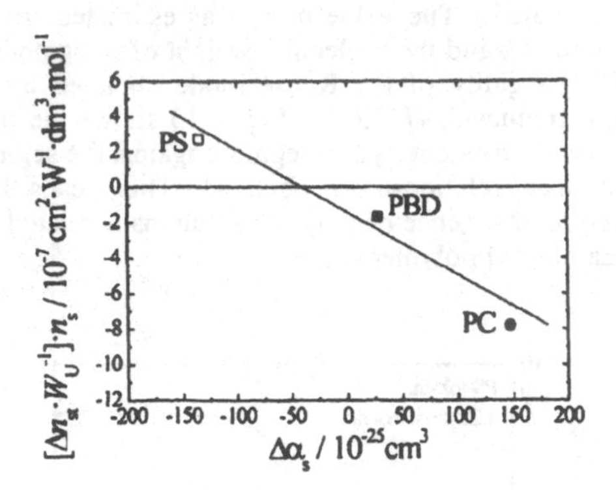

Figure 13. Plots of the intrinsic value of the stationary ultrasonically induced birefringence per a segment against segmental anisotropy in polarizability..

5.4 FREQUENCY DEPENDENCE OF THE STATIONARY BIREFRINGENCE IN POLYMER SOLUTIONS [22,23]

Frequency dependence of the $\Delta n_{st} \cdot W_U^{-1}$ for PS-toluene solutions is shown in Figure 14. For all solutions, the $\Delta n_{st} \cdot W_U^{-1}$ value decreases with increasing frequency likely to "relaxation phenomena" in the frequency range investigated. The difference in $\Delta n_{st} \cdot W_U^{-1}$ values between at lower and higher frequency region increased with increasing the molecular weight. For PC-chloroform and PBD-toluene solutions, the $\Delta n_{st} \cdot W_U^{-1}$ values also decreased with increasing frequency[22].

For the root mean square of birefringence, Δn_{rms}, in polymer solutions, Jerrard observed that it was proportional to $\sqrt{W_U}$ and the value of $\Delta n_{rms} \cdot \sqrt{W_U^{-1}}$ increased with increasing frequency in the frequency range from 1 to 5MHz [2]. It can then be concluded that the sinusoidal birefringence, i.e. the linear birefringence, in polymer solutions is caused by the sinusoidal velocity gradient generated by the ultrasound.

For colloidal solutions, the stationary birefringence increased with increasing frequency of the ultrasound as shown in the section 4. Since the velocity difference between the fluid and particles increases with an increase in frequency in a non-ideal fluid, the $\Delta n_{st} \cdot W_U^{-1}$ values increased with frequency.

In subsection 5.2, we derived the expression of the stationary birefringence. However, the estimated value for the PS-toluene solution was 10^5 times smaller than the experimental value. In addition, the predicted frequency dependence indicated by Eqs.(33) and (34) was not consistent with the experimental results.

Frequency dependence of the stationary birefringence in polymer solution cannot be explained by the mechanism of for the ultrasonically birefringence discussed so far. Ultrasonic intensity used in the birefringence measurement ranged from 0.001 to 1 $W \cdot cm^{-2}$. The corresponding pressure amplitude of ultrasound estimated as from 0.054 to 1.7 atm using values of the density and sound velocity of $10^3\ kg \cdot m^{-3}$ and $1500\ m \cdot s^{-1}$, respectively. Since the pressure amplitude is not so small, the linearization of the fluid dynamics equation is not adequate for our situation. This suggests strongly that quadratic acoustic effect including the streaming will occur in solutions and it causes the birefringence of polymer chains in solutions. Furthermore, in polymer solutions, it is considered that many nonlinear effects, arising from the coupling between local segmental motion and the solvent flow and so on, exist. To clarify the mechanism the stationary birefringence, the further theoretical treatments including the non-linearity of fluid dynamics and the polymer dynamics is required.

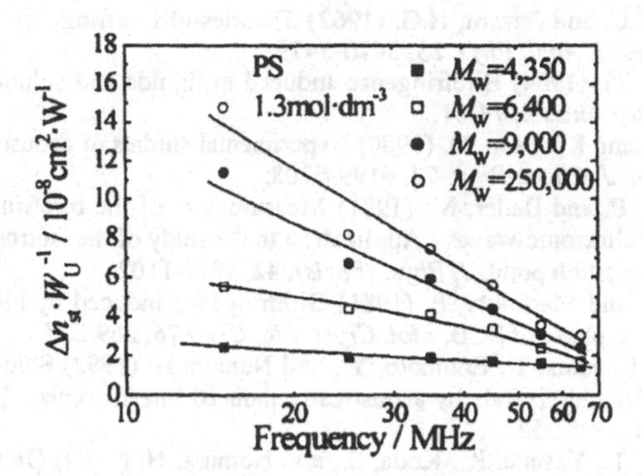

Figure 14. Frequency dependence of the stationary birefringence per ultrasonic intensity for PS-toluene solutions [22].

6. SUMMARY

In this review, we showed our experimental and theoretical work on the ultrasonically induced birefringence for in neat liquids, colloidal solutions and polymer solutions.

In section 2, the experimental setup for ultrasonically induced birefringence was introduced. The feature of the "non-biased" and "biased" measurement was described.

In section 3, we described the theoretical treatment of the ultrasonically induced birefringence in neat liquids on the basis of the framework of de Gennes treatment. The experimental results in isotropic phase liquid crystal were in agreement with the theory.

In section 4, we reviewed the results for the ultrasonically induced birefringence for colloidal solutions. In this system, the radiation pressure causes the stationary birefringence which is proportional to the ultrasonic intensity. The modified Oka theory for rod-like particle could quantitatively explain the frequency dependence but agreement was inadequate. Transient trace of the birefringence for the entangled rod-like micelles showed anomalous damped oscillation.

In section 5, our resent results for polymer solution were explored. In polymer solutions the both sinusoidal and stationary birefringence was observed. From the molecular weight and the configuration dependence studies showed the stationary birefringence was to be related to the local motion of the polymer segment not to the Rouse-Zimm mode.

As noted in the introduction the ultrasonically induced birefringence is a one of the useful methods to investigate the orientational behavior and the coupling between the translational and orientational motion of the molecules and particles in liquids and solutions. However, there still exist many problems to be solved. Our work of the ultrasonically induced birefringence is to be continued.

REFERENCES

1. Hilyard, N.C. and Jerrard, H.G. (1962) Theories of birefringence induced by ultrasonic waves, *J. Appl. Phys.* **33**, 3470-3479.
2. Jerrard, H. G. (1964) Birefringence induced in liquids and solutions by ultrasonic waves, *Ultrasonics* **2**, 74-81.
3. Lipeles, R.and Kivelson, D. (1980) Experimental studies of acoustically induced birefringence, *J. Chem. Phys.* **72**, 6199-6208.
4. Martinoty, P. and Bader, M. (1981) Measurement of the birefringence induced in liquids by ultrasonic-waves - Application to the study of the isotropic-phase of PAA near the transition point, *J. Phys. (Paris)*, **42**, 1097-1102.
5. Bader, M. and Martinoty, P. (1981) Birefringence induced by ultrasonic-waves in the isotropic-phase of PCB, *Mol. Cryst. Liq. Cryst.***76**, 269-277.
6. Koda, S., Koyama, T., Enomoto, Y., and Nomura H. (1992) Study on orientational motion of liquid-crystals by acoustically induced birefringence, *Jpn. J. Appl. Phys.* **31** Suppl.31-1, 51-53.
7. Matsuoka, T., Yasuda, K., Koda, S., and Nomura, H. (1999) On the frequency dependence of ultrasonically induced birefringence in isotropic phase of liquid crystal: 5CB (p-n-pentyl p'-cyanobiphenyl), *J. Chem. Phys.* **11**, 1580-1586.
8. Ou-Yang, H.D., MacPhail, R.A., and Kivelson D. (1986) Nonlinear ultrasonically induced birefringence in gold sols: Frequency-dependent diffusion, *Phys. Rev.* **A33**, 611-619.
9. Yasuda, K., Matsuoka, T., Koda, S., and Nomura, H. (1994) Dynamics of V_2O_5 sol by measurement of ultrasonically induced birefringence, *Jpn. J. Appl. Phys.* **33**, 2901-2904.
10. Yasuda, K., Matsuoka, T., Koda, S., and Nomura, H. (1996) Frequency dependence of ultrasonically induced birefringence of rodlike particles, *J. Phys. Chem.* **100**, 5892-5897.
11. Yasuda, K., Matsuoka, T., Koda, S., and Nomura, H. (1996) Dynamics of entanglement networks of rodlike micelles studied by measurements of ultrasonically induced birefringence, *J. Phys. Chem. B* **101**, 1138-1141.
12. Matsuoka, T., Yamamoto, K., Koda, S., and H. Nomura, in preparation.
13. Kawamura, H. (1937) Chouonpa no ba no hikari ni taisuru ichi kouka, *Kagaku* (Tokyo) **7**, 6-7, 54-55 [in Japanese]
14. Kawamura, H. (1937) Chouonpaba ni okeru hikari no fukukussetsu, *Kagaku* (Tokyo) **7**, 139 [in Japanese].
15. Oka S. (1939) Zur Theorie Doppelbrechunf bei nicht-Kugelförmingen Kolloiden in Ultraschallfelde *Kolloid Z.* **87**, 37-43 [in German].
16. Oka S. (1940) Zur Theorie de akustischen Doppelbechung von Kollidalen Lösungen, *Z. Physik.* **116**, 632-656 [in German].
17. Petralia, S. (1940) Sopra la birifrangenza provocata nei liquidi da ultrasuoni, *Nuovo Cimento*, **17**, 378-389 [in Italian].
18. Yasunaga, T., Tatsumoto, N., and Inoue, H. (1969) Birefringence induced in gold sol by ultrasonic wave, *J. Colloid & Interface Sci.* **29**, 178-180.
19. Watanabe, T., Ikeda, Y., Hibino, M., Kudo, T., Hosoda, M., Miyayama, M., Sakai, K. (2002) Ultrasonic and light scattering characterization of anisotropic colloidal particles in sol, *Jpn. J. Appl. Phys.* **41**, 3157-3158.
20. Matsuoka, T., Koda, S., and Nomura, H. (2000) Linear and nonlinear ultrasonically induced birefringence in polymer solutions, *Jpn. J. Appl. Phys.* **39**, 2902-2905.

21. Nomura, H., Ando, S., Matsuoka, T., and Koda, S. (2003) Effect of chain structure and molecular weight on ultrasonically induced birefringence, J. Mol. Liq., **103-104,** 111-119.
22. Nomura, H., Ando, S., Matsuoka, T., and Koda, S., Relationship between segmental anisotropy in polarizability and stationary ultrasonically induced birefringence in polymer solutions, *J. Mol. Liq.*, submitted.
23. Nomura, H., Matsuoka, T., and Koda, S. Ultrasonically induced birefringence in polymer solutions, *Pure and Appl. Chem.*, submitted.
24. Nomura, H., Matsuoka, T., and Koda, S. (2002) Translational-orientational coupling motion of molecules in liquids and solutions, *J. Mol. Liq.* **96-97**, 135-151.
25. Klein, W.R. and Cook, B. D. (1967) Unified Approach to Ultrasonic Light Diffraction, *IEEE Trans. Sonics. Ultrason.*, **SU-14**, 123-134.
26. De Gennes, P.G. and Prost, J. (1993) *The Physics of Liquid Crystals 2nd ed. Chap. 2*, Clarendon, Oxford and references therein. Su
27. Berne, B. and Pecora R., (1976) *Dynamic Light Scattering*, Wiley, New York..
28. Kivelson, D. and Madden P.A. (1980) Light-scattering-studies of molecular liquids, *Ann. Rev. Phys. Chem.* **31**, 523-558 and references therein.
29. Alms, G.R., Gierke, T. D., and Patterson G.D. (1977) Observation and analysis of depolarized Rayleigh doublet in isotropic MBBA and measurement of de Gennes viscosity coefficients, *J. Chem. Phys.*, **67**, 5779-5787.
30. Lipeles, R.and Kivelson, D. (1977) Theory of ultrasonically induced birefringence, *J. Chem. Phys.* **67**, 4564-4570.
31. Martinoty., P., Kiry, F., Nagai. S., Candau, S., and Debeauvais, F. (1977) Viscosity coefficients in isotropic phase of a nematic liquid-crystal, *J. Phys. (Paris)* **38** 159-162.
32. Kivelson, D., Keyes, T., Champion, J. (1976) Theory of molecular-reorientation rates, flow birefringence, and depolarized light-scattering, *Mol. Phys.* **31**, 221-232 .
33. Shibata, T., Matsuoka, T., Koda, S., and Nomura, H. (1998) Depolarized light scattering in the isotropic phase of liquid crystals, *J. Chem. Phys.* **109**, 2038-2042 .
34. Hoffmann, H., Oetter, G., and Schwandner, B. (1987) The aggregation behavior of tretradecyldimetylaminoxide, *Prog. Colloid Polym. Sci.*, **73**, 95-106.
35. Pilsl, H., Hoffmann, H., Hofmann, S., Kalus, J., Kencono, A.W., Lindner, P., and Ulbricht W. (1993) Shape investigation of mixed micelles by small-angle neutron-scattering , *J. Phys. Chem.* **97**, 2745-2754.
36. Rehage, H. and Hoffmann, H. (1988) Rheological properties of viscoelastic surfactant systems, *J. Phys. Chem.* **92**, 4712-4719.
37. Hoffmann, H., Krämer, U., and Thurn H. (1990) Anomalous behavior of micellar solutions in electric birefringence measurements, *J. Phys. Chem.*, **94**, 2027-2033.
38. Shikata, T.,. Dahman, S.J., and Pearson, D.S. (1994) Rheooptical behavior of wormlike micelles, *Langmuir* **10**, 3470-3476.
39. Hofmann, S., Rauscher, A., and Hoffmann, H. (1991) Shear induced micellar structures, *Ber. Bunsenges. Phys. Chem.* **95**, 153-164.
40. Hu Y.T., Wang, S.Q., and Jamieson, A.M. (1993) Kinetic-studies of a shear thickening micellar solution, *J. Colloid Interface Sci.* **156**, 31-37.
41. King, L.V. (1935) On the acoustic radiation pressure on circular discs; inertia and diffraction corrections, *Proc Roy. Soc. London* **A153**, 1-16.
42. King, L.V. (1935) On the theory of the inertia and diffraction correction for Rayleigh disc **A153**, 17-40.
43. Rayleigh, J.W.S. (1945) *Theory of Sound 2nd ed.*, Dover, New York.
44. Peterlin, A. and Stuart, H.A. (1939) Über die Bestimmung der Größe und Form, sowie der elektrischen , optischen und magnetischen Anisotropie von submikroskopischen Teilchen mit Hilfe der künstlichen Doppelbrechung und der inneren Reibung, *Z. Physik* **112**, 129-147 [in German].
45. Landau L.D., Lifshitz P.M. (1959) *Fluid Mechanics*, Pergamon Press, Oxford.

46. Lamb, S.H. (1932) *Hydrodynamics 6th ed.*, Cambridge University Press, Cambridge.
47. Perrin, F. (1934) Mouvement brownien d'un ellipsoide (I):dispersion dielectrique pour des molecules ellipsoidailes, *J. Phys. Rad. (Paris)* **5**, 497-511 [in French].
48. Perrin, F. (1936) Mouvement brownien d'un ellipsoide (II) : rotation libre et depolarisation des fluorescences. Tranclation et diffusion de molecules ellipsoidles, *J. Phys. Rad. (Paris)* **7**, 1-11 [in French].
49. Ozaki, M., Kratohvil, S., and Matijevic, E. (1984) Formation of monodispersed spindle-type hematite particles, *J. Colloid Interface Sci.* **102**, 146-151.
50. Platz, G., Thunig, C., and Hoffmann H. (1990) Iridescent phases in aminoxide surfactant solutions, *Prog. Colloid Polym. Sci.* **83**, 167-175.
51. Hashimoto, K. and Imae, T. (1991) Rheological properties of aqueous-solutions of alkyldimethylamine and oleyldimethylamine oxides - spinnability and viscoelasticity, *Langmuir* **7**, 1734-1741.
52. Shikata, T. and Kotaka, T. (1991) Entanglement network of thread-like micelles of a cationic detergent, *J. Non-Crystalline Solids* **131-133**, 831-835.
53. Wheeler, E.K., Izu, P., and Fuller, G.G. (1996) Structure and rheology of wormlike micelles, *Rheol. Acta* **35**, 139-149.
54. Shikata, T., Hirata, H., Takatori E., and Osaki, K. (1988) Nonlinear viscoelastic behavior of aqueous detergent solutions, *Non-Newtonian Fluid Mech.* **28**, 171-182.
55. Peterlin, A. (1950) La biréfringence acoustique des solutions Macromoléculaires, *Rec. Trav. Chim.* **69**, 14-21 [in French].
56. Larson, R.G. (1988) *Constitutive Equations for Polymer Melts and Solutions*, Butterworth, London.
57. Larson, R.G. (1999) *The structure and Rheology of Complex Fluids*, Oxford Univ. Press, New York.
58. Doi M. and Edwards, S.F. (1986) *The Theory of polymer Dynamics*, Clarendon Press, Oxford.
59. Tanaka, H., Sakanishi, A. and Kaneko, M. (1966) Dynamic viscoelastic properties of dilute polymer solutions, *J. Polym. Sci.* **C15**, 317-330.
60. Bailey, R.T., North, A.M., and Pethrick, R.A. (1981) *Molecular Motion in High Polymers*, Clarendon, Oxford.
61. Nomura, H., Kato, S., and Miyahara, Y. (1975) Ultrasonic absorption in polymer solutions, *Mem. Fac. Eng. Nagoya Univ.* **27** 73-125.
62. Tsvetkov, V.N. (1965) Anisotropy of the segment and monomer units selected polymer molecules in Brandrup and E. H. Immergut (eds.), *Polymer Handbook*, Wiley, New York, pp.V-75-77.
63. Champion, J.V., Desson R.A. and Meeten, G. H. (1974) Conformation of polycarbonate by flow and magnetic birefringence, *Polymer* **15**, 301-305.
64. Champion, J.V., Meeten, G.H., and Southwell, G.W. (1976) Electro-optic Kerr birefringence of polystyrenes in dilute solutions, *Polymer* **17**, 651-655.
65. Inoue, T. and Osaki, K. (1996) Role of polymer chain flexibility on the viscoelasticity of amorphous polymers around the glass transition zone, *Macromolecules* **29**, 1595-1599.
66. Inoue, T. and Osaki, K. (1996) Dynamic birefringence of vinyl polymers, *Macromolecules* **29**, 6240-6245.
67. Stein , R.S. and Tobolsky, A.V. (1952) Determination of the statisitical segment size of polymer chains from stress-birefringence studies, *J. Poly. Sci.* **11**, 285-288.

NOVEL APPROACHES IN SPECTROSCOPY OF INTERPARTICLE INTER-ACTIONS. VIBRATIONAL LINE PROFILES AND ANOMALOUS NON-COINCIDENCE EFFECTS.

S.A. KIRILLOV

Institute for Sorption and Problems of Endoecology, Ukrainian National Academy of Sciences, Gen. Naumov St. 3, 03142 Kyiv - 142, Ukraine, and Institute for Technological and Information Innovations, P.O. Box 263, 03134 Kyiv - 134, Ukraine. Electronic address: kir@i.kiev.ua

Abstract: **This chapter deals with the theories of vibrational line profiles and anomalous (negative) frequency non-coincidence effects in condensed media. First, a novel approach to the line profile analysis is described. It is based on a new, flexible time-correlation function (TCF), which has an analytical counterpart in the frequency domain. Using this function one can fit vibrational line profiles obtaining dynamical information at the same time. Numerous applications of this TCF are considered, including analyses of both line profiles and dynamics. Second, direct methods of estimation of repulsion contributions to frequency shifts and non-coincidence effects are described, and model calculations enabling one to separate the contributions of repulsion and attraction forces resulting in frequency non-coincidences are presented.**

1. INTRODUCTION

Studies of condensed matter by means of vibrational spectroscopy can be considered on two levels. On the first level, one finds the number of lines and bands in infrared (IR) and Raman spectra and their degrees of polarization, and quantitatively estimates if IR and Raman spectra coincide. This information is sufficient for conclusions regarding the geometrical structure of molecules studied, and sometimes allows one to make a judgement about *intra*molecular interactions. The description of properties of molecules in terms of their vibrational transition frequencies and band parameters (widths, intensities, spectral moments, polarization coefficients) is the scope of *molecular spectroscopy* in its usual meaning.

An important feature of the spectra of molecules in condensed media is their sensitivity to phase transformations or dissolution, showing up in essential changes of spectral lines and bands. Interactions between complex particles significantly perturb their potential functions and affect the position of vibrational levels, leading to vibrational spectral shifts relating to the gas-phase. In different liquids, solids and glasses, rotational mobility and the rate of energy exchange between molecules sig-

J. Samios and V.A. Durov (eds.), Novel Approaches to the Structure and Dynamics of Liquids: Experiments, Theories and Simulations, 193–227.

nificantly vary. In some cases, interactions cause the distortion of the probe particle making forbidden lines observable. Therefore on the second level of spectroscopic studies of condensed matter one deals with changes in vibrational transition frequencies and band parameters (frequency shifts, the appearance of forbidden lines due to interaction-induced phenomena, frequency non-coincidence effects, line broadening, etc.) caused by phase transformations or dissolution. Proper interpretation of such changes enabling one to draw conclusions regarding interactions of the probe particle with its surrounding in the system studied is the scope of *spectroscopy of interparticle interactions*. The methods of spectroscopy of interparticle interactions are well illuminated in several books [1-4] and numerous review articles (see, e.g., Refs [5-17]).

The aim of this chapter is to put under focus recent advances in spectroscopy of interparticle interactions. First, the theories of vibrational line profiles in condensed media will be outlined, and a novel approach to the line profile analysis will be described. It is based on a new, flexible time-correlation function (TCF), which has an analytical counterpart in the frequency domain. Using this function one can fit vibrational line profiles obtaining dynamical information at the same time. Numerous applications of this TCF will be considered, including analyses of both line profiles and dynamics.

Second, we discuss the systems with spectroscopic "anomalies" (negative, blue frequency shifts and, especially, negative frequency non-coincidence effects), and try to cast some light on the obscure problem of repulsion forces in vibrational spectroscopy. Direct methods of calculation of repulsion contributions to frequency shifts and non-coincidence effects will be described, and a model enabling one to separate the contributions of repulsion and attraction forces resulting in frequency non-coincidences will be presented.

2. VIBRATIONAL LINE PROFILES IN CONDENSED MEDIA: MODEL DESCRIPTION

2.1 INTRODUCTORY REMARKS

Vibrational line shape analysis is known as a standard tool for the studies of vibrational and rotational relaxation of molecules in liquids: its theory is well established, and significant advances in the field are explicitly documented in numerous monographs and reviews [2-16]. Experimentally, the simplest approach to studies of vibrational and rotational relaxation has been proposed in the pioneering papers by Bartoli and Litovitz [18,19]. Collecting the Raman spectra at polarized (VV) and depolarized (HV) scattering geometries, one obtains isotropic and anisotropic Raman line profiles,

$$I_{iso}(\nu) = I_{VV}(\nu) - \frac{4}{3} I_{VH}(\nu), \tag{1}$$

$$I_{aniso}(\nu) = I_{HV}(\nu). \tag{2}$$

where ν is the frequency measured in wavenumbers. Vibrational relaxation can be investigated from isotropic line profiles, and rotational relaxation from anisotropic ones. For instance, at certain limiting conditions, the characteristic times of vibrational relaxation τ_V and rotational relaxation τ_{2R} can be estimated as

$$\tau_V = (\pi c \Gamma_{iso})^{-1}, \qquad (3)$$

$$\tau_{2R} = [\pi c (\Gamma_{aniso} - \Gamma_{iso})]^{-1}, \qquad (4)$$

where Γ_i with respective indices are the full widths at half height of the isotropic and anisotropic spectral lines, and c is the speed of light.

To gain a deeper insight in the dynamics of liquids, one has to pass from the frequency domain to the time domain and to examine the TCFs of vibrational and rotational relaxation, $G_V(t)$ and $G_{\ell R}(t)$, obtainable by means of the Fourier transform of IR and Raman line profiles,

$$G_V(t) = \int_{-\infty}^{+\infty} I_{iso}(\nu)\exp(2\pi i c \nu t)d\nu, \qquad (5)$$

$$G_{1R}(t) = \int_{-\infty}^{+\infty} I_{IR}(\nu)\exp(2\pi i c \nu t)d\nu \Big/ G_V(t) = G_{IR}(t)/G_V(t), \qquad (6)$$

$$G_{2R}(t) = \int_{-\infty}^{+\infty} I_{aniso}(\nu)\exp(2\pi i c \nu t)d\nu \Big/ G_V(t) = G_{aniso}(t)/G_V(t), \qquad (7)$$

where the subscript at $G_{\ell R}(t)$ means that the rotation studied by different methods is described by the Legendre polynomials of the ℓ-th order, $\ell=1$ in IR and $\ell=2$ in Raman, and t is the time.

The study of how the TCF decays to equilibrium provides a useful information about the microscopic mechanisms of molecular interactions in the liquid in question. Fourier transforming line profiles one obtains information regarding the central part of the spectral line in the tail of the TCF, and the beginning of the TCF describes the far wings of the spectral line. In other words, small ν in the spectra correspond to great t in the TCF and *vice versa*; this fact is easily seen from the inverse proportionality of Γ_i and τ_i expressed by Eqs (3) and (4).

Eqs (5)-(7) reveal that the studies in the time domain (TCF analysis) and in the frequency domain (line profile analysis) are closely interrelated, and having knowledge on the TCF one can easily reproduce the corresponding line profile by means of Fourier transforms. There are numerous models enabling one to model vibrational and rotational relaxation in terms of TCFs. In turn, various mechanisms of vibrational relaxation exist, including vibrational dephasing, resonant transfer and energy relaxation (depopulation).

The treatment of vibrational dephasing is the most detailed. It is one of the main causes of the broadening of the isotropic Raman line contour, and arises due to time-dependent intermolecular interactions, which change the instantaneous vibrational frequency $\Delta\omega=f(t)$. Such a process of frequency modulation depends on how the phase memory decays. There are three main types of modulation events, *viz.*, Gaussian-Markovian, non-Markovian and purely discrete Markovian. They lead to vibrational profiles of significantly different form.

2.2 EXISTING MODELS OF VIBRATIONAL RELAXATION

2.2.1 Kubo-Rothschild theory of vibrational dephasing

If, because of intermolecular interactions, the instantaneous frequency changes to a new value by small steps, $\Delta\omega$ may be considered a Gaussian random process with zero mean. This takes place when the probe molecule is perturbed by interactions with few of its nearest neighbors, like in crystals. In such case the relaxation function can be written in the following way [6],

$$G_V(t) = \left\langle \exp\left(i \int_0^t dt' \Delta\omega(t') \right) \right\rangle = \exp\left(- \int_0^t dt' \int_0^{t'} G_\omega(t'')dt'' \right). \quad (8)$$

The first part of Eq. (8) is the relaxation function, the second part – its cumulant expansion, and $G_\omega(t)$ is the TCF of frequency modulation.

$G_\omega(t)$ may be taken as a simple exponential, $G_\omega(t)=\exp(-t/\tau_\omega)$, where τ_ω is the characteristic time of the frequency modulation process (perturbation correlation time). This means that the modulation process is considered Gaussian-Markovian, i.e. both Markovian (each value of the instantaneous frequency depends only on *one* previous frequency value, and the phase memory is short) and Gaussian at the same time. The use of a simple exponential in Eq. (8) yields the famous Kubo equation [21-23]

$$-\ln G_v(t)/M_2\tau_\omega^2 = \exp(-t/\tau_\omega) - 1 + t/\tau_\omega, \quad (9)$$

where $M_2 = \int \nu^2 I_{iso}(\nu)d\nu / \int I_{iso}(\nu)d\nu$ is the vibrational second moment (perturbation dispersion). Another parameter of the process, vibrational dephasing correlation time τ_V, is measured as the integral over $G_V(t)$.

The limiting conditions for $G_V(t)$ are easy to find. In one extreme, if $\tau_\omega \to \infty$ or $t \to 0$,

$$G_V(t) = \exp\left(-\frac{1}{2} M_2 t^2 \right) = \exp\left(-\frac{1}{4} \pi t^2 / \tau_V^2 \right), \quad (10)$$

where $\tau_V = (\pi/2M_2)^{1/2}$ and $G_V(t)$ is the Gaussian. In this "static" case, when modulation is slow and the environment of the probe particle could be considered frozen, the so-called inhomogeneous broadening occurs. The term "inhomogeneous" therefore means that the probe particle senses spatial inhomogeneities of the distribution of its neighbors. Eq. (10) can be Fourier transformed analytically giving the band shape of Gaussian form,

$$I(\nu) = \frac{\sqrt{\ln 2}}{\Gamma\sqrt{\pi}} \exp\left(-\frac{\nu^2}{\Gamma^2} \ln 2 \right). \quad (11)$$

In Eq. (11) Γ is the half width at the half height; for Gaussian lines it can be determined as $\Gamma = \sqrt{\pi \ln 2} / 2\pi c \tau_V \approx 0.678 / 2\pi c \tau_V$.

So, if $\tau_\omega \to \infty$, the Kubo TCF tends to be Gaussian and predicts the Gaussian form of the entire vibrational line. Recalling that, as mentioned in Sec. 2.1, small t in the TCF correspond to great ν in the spectrum, we can conclude that if τ_ω is finite, the limiting dependence of Eq. (9) at $t \to 0$ tells about the Gaussian form of far wings of the spectrum.

In another extreme, if $\tau_\omega \to 0$ or $t \to \infty$,

$$G_V(t) = \exp(-M_2\tau_\omega t) = \exp(-t/\tau_V), \quad (12)$$

where $\tau_V = (M_2\tau_\omega)^{-1}$ and $G_V(t)$ is exponential. In this case the so-called homogeneous broadening takes place, and – due to the fast modulation – the probe particle does not recognize inhomogeneities of the spatial distribution of its neighbors and senses it as being uniform. Fourier transforming Eq. (12) gives the Lorentzian band shape,

$$I(\nu)=\frac{1}{\pi\Gamma}\frac{1}{\Gamma^2+\nu^2}. \tag{13}$$

In Eq. (13) Γ is the half width at the half height of the Lorentzian profile; it is equal to $\Gamma=(2\pi c\tau_V)^{-1}$. Similarly, if $\tau_\omega\to0$, the Kubo TCF tends to be exponential and predicts the Lorentzian form of the entire vibrational line. If τ_ω is finite, the limiting dependence of Eq. (9) at $t\to\infty$ corresponds to the Lorentzian form of the centralmost part of the spectrum.

Considering now a line representing the Fourier transform of a general Kubo-type TCF with a finite value of τ_ω one can make the following conclusions. The line is characterized by a Lorentzian central part and Gaussian wings. Almost the whole line reflects the slow process, whereas the fastest process is buried in the far wings. The corresponding TCF has a Gaussian initial part and an exponential tail.

Interestingly, some authors (see, e.g., Refs [24,25]) employ Eq. (9) for the analysis of *rotational* relaxation. In that case, the time during which a particle moves as a free rotor τ_{FR} is used instead of τ_ω, τ_{2R} instead of τ_V, and hence the TCF is considered to reflect two definitely distinct mechanisms of molecular motion: a Gaussian initial part corresponds to free rotation at short times, when the particle experiences small perturbations, and an exponential tail reflects the Debye rotational diffusion, which occurs when the probe molecule is entrapped into a local structure and becomes influenced by its neighbors.

2.2.2 Gaussian, Lorentzian, and intermediate line profiles

Gaussian- and Lorentzian-type curves are quite often used for modeling real line profiles. The Kubo theory puts some restrictions to such kind of modeling. First of all, it demonstrates that Gaussian line profiles can be explained theoretically, whereas vibrational lines of a true Lorentzian profile, as well as TCFs of a true exponential form should be considered nothing more than a mere abstraction. The values of τ_ω can be assigned to various particular characteristic times of processes occurring in the system studied. Usually, τ_V are of the picosecond scale, $\tau_V\sim10^{-12}$ s. One can imagine that the $\tau_\omega\to\infty$ case may correspond to modulation caused by structural relaxation (molecular rearrangement) processes occurring in the nanosecond time scale, $\tau_{sr}\sim10^{-9}$ s; such a time can be safely considered infinite in the picosecond time domain. On the other hand, the $\tau_\omega\to0$ case can hardly be accessed. Modulation can never be faster than the fastest possible molecular process in the system. Even if modulation is caused by molecular collisions with characteristic times $\tau_{BC}\sim10^{-13}$ s, such times are too long for being regarded close to zero, if compared to τ_V.

Second, the Kubo theory clarifies the well-known problem [26] arising in the moment analysis of vibrational lines. Any vibrational line can be characterized by the central moments of the spectral distribution $M_n=\int\nu^n I(\nu)d\nu/\int I(\nu)d\nu$; the second vibrational moment M_2 has already been introduced in Eq. (9). In turn, any TCF can be represented in terms of the central moments of the spectrum [5],

$$G(t)=1-\frac{1}{2!}M_2t^2+\frac{1}{4!}M_4t^4\pm\ldots. \tag{14}$$

In Eq. (14) odd moments do not appear since they are of zero value for symmetric lines. Practice shows that the even moments of any real vibrational line are always finite. However, the even moments of Lorentzian lines are infinite. The Kubo theory resolves this dichotomy stating that any Lorentzian-like line must have Gaussian wings and hence its even moments are finite.

2.2.3 Rothshild-Perrot-Guillaume and Burshtein-Fedorenko-Pusep models

If a stretched exponential, $G_\omega(t)=\exp(-t/\tau_\omega)^\alpha$, where $0<\alpha\leq 1$, is used in Eq. (8), one gets the following TCF

$$-\ln G_V(t)/M_2\tau_\omega^2=\sum_{n=0}^{\infty}\frac{(-1)^n(t/\tau_\omega)^{2+n\alpha}}{n!(1+n\alpha)(2+n\alpha)}. \tag{15}$$

It has been introduced by Rothschild, Perrot and Guillaume [27-29] and corresponds to non-Markovian modulation. Markovian approximation is now invalid: relaxation times are distributed in some way reflecting many independent Poisson relaxation pathways, and the phase memory is long. Modulation events of this type are characteristic for strongly interacting liquids (liquid crystals with bulky molecules and some molten salts). It should be noticed that at $\alpha=1$ Eq. (15) reduces to Eq. (9).

In the theory elaborated by Burshtein, Fedorenko and Pusep [30-33], modulation events are purely discrete Markovian, the phase memory is completely absent, cumulant technique is no more applicable, and no TCF of the frequency modulation can be introduced directly. The instantaneous vibrational frequency subjects to changes suddenly, as in gases, and $\Delta\omega$ caused by each collision depends on the distance r to the probe particle and its neighbor as

$$\Delta\omega=f(r)=C_k r^{-k}. \tag{16}$$

Eq. (16) is crucial in this theory since it makes various types of potentials, which cause modulation, discernible. The k values from 3 to 12 have been probed either separately [31,32] or simultaneously [33]; it has been shown [33] that the exponential (Born-Meyer) repulsion gives the same result as Eq. (16) with $k=12$ (van der Waals repulsion).

In the so-called static case (no motion of particles takes place), the static TCF is defined in terms of Eq. (16),

$$G_V^0(t)=\exp\left(-4\pi N\int_{\rho_0}^{\infty}\{1-\cos[f(r)\cdot t]\}r^2dr\right), \tag{17}$$

where N is the number density of probe particles and ρ_0 is the closest approach distance. The transformation of the static TCF by molecular motion is expressed by an integral equation [31,32]

$$G_V(t)=G_V^0(t)\exp(-t/\tau_\omega)+\tau_\omega^{-1}\int_0^t G_V^0(t-t')\exp\left(-\frac{t-t'}{\tau_\omega}\right)G_V(t')dt', \tag{18}$$

and its approximate solution [34] reads as

$$-\ln G_V(t)/M_2\tau_\omega^2=\frac{3}{k}\sum_{n=0}^{\infty}\frac{(-1)^n(t/\tau_\omega)^{3/k+n+1}}{n!(3/k+n)(3/k+n+1)}. \tag{19}$$

Eq. (19) appears quite similar to Eq. (15) describing the Rothschild-Perrot-Guillaume TCF.

2.2.4 Time correlation functions and line profiles

Combining Eqs. (15) and (19), one gets a generalized scale-invariant expression [16,35] covering all three mechanisms of vibrational dephasing,

$$-\ln G_V(t)/M_2\tau_\omega^2 = \frac{3}{k}\sum_{n=0}^{\infty}\frac{(-1)^n (t/\tau_\omega)^{3/k+n\alpha+1}}{n!(3/k+n\alpha)(3/k+n\alpha+1)}. \quad (20)$$

For k=3 and α=1, Eq. (20) reduces to the Kubo equation, for k=3 and $\alpha \leq 1$ it describes non-Markovian processes, and for k>3 and α=1 purely discrete Markovian processes.

Using Eq. (20), it is easy to demonstrate the difference between three aforementioned types of TCFs in reduced coordinates, Fig. 1. The Kubo case is represented by a single curve. The fan of Rothschild-Perrot-Guillaume curves with their particular values of α intersects the Kubo curve; we show the one with α=0.3. The deviation of Rothschild-Perrot-Guillaume TCFs from the Kubo limiting law is as follows. In the short-time region, the vibrational decay appears slower than the Kubo equation predicts, and in the long-time region faster. The smaller the α value, the greater the deviations are. If the Burshtein-Fedorenko-Pusep formalism is valid, another fan of curves can be obtained, each with its particular k. These TCFs appear to decay faster in the short-time region and slower in the long-time region than the Kubo equation predicts, and the bigger the k value, the greater the deviations are.

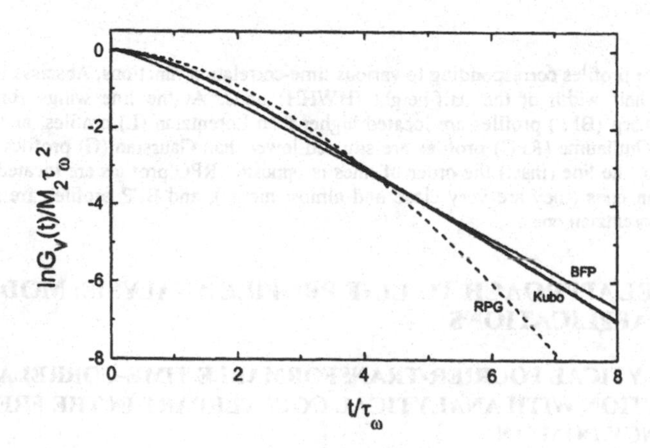

Figure 1. Time-correlation functions of vibrational dephasing in reduced coordinates. The Kubo function is shown by thick line; dashed line corresponds to the Rothschild-Perrot-Guillaume (RPG) curve with α=0.3; thin line represents the Burshtein-Fedorenko-Pusep (BFP) curve with k=6.

Differences in TCFs lead to very unlike line profiles (Fig. 2). It has been already noticed that, depending on the value of τ_{ω}, the Kubo TCF corresponds to vibrational lines whose profiles vary from Gaussian to Lorentzian. The Rothschild-Perrot-Guillaume TCF corresponds to vibrational lines of quite specific, over-Gaussian form. They are less sharp than Gaussian in their central part, and much faster fall to zero in the wings. The Burshtein, Fedorenko and Pusep TCF corresponds to over-Lorentzian line profiles. They are sharper than true Lorentzians in their central part, and broader in the wings.

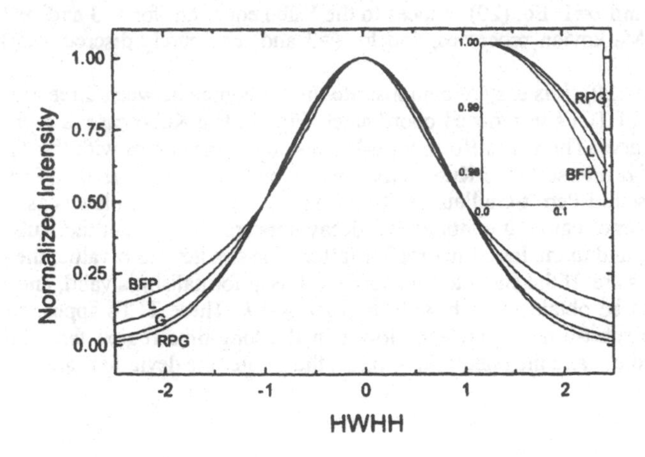

Figure 2. Line profiles corresponding to various time-correlation functions. Abscissa is measured in the half width at the half-height (HWHH) units. At the line wings Burshtein-Fedorenko-Pusep (BFP) profiles are located higher then Lorentzian (L) profiles, and Rothschild-Perrot-Guillaume (RPG) profiles are situated lower than Gaussian (G) profiles. At the central part of the line (inset) the order of lines is opposite: RPG profiles are located higher than Gaussian ones (they are very close and almost merge), and BFP profiles are situated lower than Lorentzian ones.

3. NOVEL APPROACH TO LINE PROFILE ANALYSIS: MODEL AND APPLICATIONS

3.1 ANALYTICAL FOURIER-TRANSFORMABLE TIME-CORRELATION FUNCTION WITH ANALYTICAL COUNTERPART IN THE FREQUENCY DOMAIN

As follows from the above-mentioned, the very appearance of TCFs describing possible causes of line broadening leaves no hope for an analytical interrelation between TCFs and spectral line profiles. Only Gaussian and Lorentzian profiles can be Fourier transformed analytically. However, real band shapes never represent these limiting cases and must be treated numerically.

Furthermore, from the point of view of mathematics, there are numerous well-known limitations to the Fourier transforms. First and foremost, numerical Fourier transformation of the spectrum registered in a finite wavenumber range is the so-called ill-posed problem [20]. Due to this fact, several precautions are to be made in order to obtain physically sound results [9,21]. For example, the length t of the TCF obtained by means of the Fourier transform is determined by the interval $\Delta\nu$ between the discrete data points in the spectrum as $t \leq (2c\Delta\nu)^{-1}$. Further, the time resolution Δt of the TCF depends on the finite wavenumber range ν of an experimental band contour (from the origin at ν_0 to $\nu_0 \pm \nu$) as $\Delta t \geq (2c\nu)^{-1}$. This means that the spectrum should be registered with "high density", and the far wings of the lines in question ought not to be missed.

As a result, in the overwhelming majority of papers dealing with TCFs, small molecules in pure liquids and concentrated solutions are the subject of discussion. This is easy to understand: in order to perform Fourier transforms in the most reliable way, the presence of a *single* and *strong* spectral line is needed. Results obtained for dilute solutions, where spectral lines are weak, and for complex molecules like polymers and biological systems, where spectral lines are often overlapped, are scarce and sometimes unreliable.

Similar problem exists in analyses of overlap spectral lines. Usually, in some commercial software, these lines can be empirically fitted by sums or products of Gaussians or Lorentzians. In reality, when a composite line is formed by lines of intermediate profiles, such approach is of no physical meaning, especially in the light of the moment analysis of the Lorentzian function to be outlined later.

Another problem to be mentioned here touches upon curve fits by sums or products of Gaussians or Lorentzians, or by Voigt-type curves before the Fourier transformation. An idea behind such procedures is to eliminate effects arising from discretization and the finiteness of the spectral interval (see examples in Ref. [16]). However, the advantage of its use seems to be feigned, no matter how flexible a fitting function may be. Should the fitting function be mistaken in reproducing the limiting behavior of the TCF, a missing part of the line contour would give an uncontrollable contribution to both Δt and t.

In Ref. [36], a novel method has been proposed enabling one to model real line profiles intermediate between Lorentzian and Gaussian by an analytical function, which has an analytical counterpart in the time domain. This means that one can find TCFs by fits in the frequency domain, without Fourier transforms, and fit overlapping lines, obtaining dynamic parameters at the same time.

The method is based on the model TCF written in the form

$$G_v(t) = \exp\{-[(t^2 + \tau_1^2)^{1/2} - \tau_1]/\tau_2\}, \tag{21}$$

whose Fourier transform may be performed analytically giving the vibrational line profile as

$$I(\nu) = 2nc\exp(\tau_1/\tau_2)(\tau_1^2/\tau_2)K_1(x)/x, \tag{22}$$

where $x = \tau_1[4\pi^2c^2(\nu - \nu_0)^2 + 1/\tau_2^2]^{1/2}$, ν_0 is the peak wavenumber, n=2 if ν_0=0 and n=1 if $\nu_0 \neq 0$, and $K_1(x)$ is the modified Bessel function of the second kind.

In Ref. [36] I noticed that this TCF was introduced by Egelstaff and Schofield in 1962 [37] and used in two works [38,39] for curve fittings in the *time* domain. However, preparing the manuscript of this paper and checking references I found that, to my surprise, Eqs (21) and (22) were first presented by Schofield at the Scottish Universities' Summer School 1961 [40], together with the Kubo theory of line shapes and relaxation [22], and published in the same volume. Schofield aimed at establishing an analytical inter-relation between the time-correlation function of molecular motion in a liquid and the line shape in the neutron scattering spectrum. Moreover, in his paper [40], an expression equivalent to the Kubo equation, Eq. (9), was also derived and rejected due to impossibility to perform its Fourier transform analytically! Regrettably, this profound contribution remained unnoticed up until now.

The limiting conditions for Eq. (21) are as follows: if $\tau_1 >> t$, it tends to

$$G_v(t) = \exp(-\frac{1}{2}\frac{t^2}{\tau_1\tau_2}); \tag{23}$$

and if $\tau_1 << t$, Eq. (21) tends to

$$G_v(t) = \exp(-t/\tau_2). \tag{24}$$

This means that Eq. (21) correctly reproduces limiting conditions existing for Eqs (9), (15), (19) and (20). Analysis shows [36] that τ_1 and τ_2 are close to τ_ω and τ_V, respectively (Table 1). Actually, treating the values of τ_1 and τ_2 as empirical parameters, one can completely reproduce any kind of TCF described by the Kubo, Rothschild-Perrot-Guillaume, or Burshtein theory.

TABLE 1. Parameters of Kubo functions and corresponding model functions.

Kubo function			Novel function	
τ_ω/ps	τ_V/ps	M_2/ps^{-2}	τ_1/ps	τ_2/ps
0.100	2.00	5.26	0.105	1.90
0.500	2.00	1.28	0.622	1.53
2.00	2.00	0.582	2.213	0.900

TABLE 2. Influence of cut-off on the parameters of the model TCF

Fitting	Spectral range *from/to* in cm^{-1}					
parameters	−10/+10	−20/+20	−30/+30	−40/+40	−50/+50	−8/+8
Quasi-Lorentzian spectrum, full width at half-hight ~5.2 cm^{-1}						
τ_1/ps	0.090	0.097	0.101	0.104	0.105	0.105
τ_2/ps	1.90	1.90	1.90	1.90	1.90	1.90
Quasi-Gaussian spectrum, full width at half-hight ~7.7 cm^{-1}						
τ_1/ps	2.277	2.215	2.214	2.213	2.213	2.213
τ_2/ps	0.085	0.900	0.900	0.900	0.900	0.900

Another advantage of the novel TCF is its ability to successfully fit the spectra after an artificial cut-off of their high-frequency tails. It has been already mentioned (see also Ref. [41]) that "short" spectra cannot be Fourier-transformed correctly, and their preferable length is to be more than 10 half-widths at half-height. As an example we demonstrate the fits of quasi-Gaussian and quasi-Lorentzian spectra after various cut-offs. As follows from Table 2, the cut-offs have practically no influence on the parameters of fits. This means that, unlike Fourier transform technique, direct fits in the frequency domain with Eq. (22) enable one to operate with "short" spectra.

For isotropic Raman spectra the computation procedure employing Eq. (22) is as follows. The composite spectrum under consideration is to be fitted to a set of Eq.

(22) in order to find τ_1 and τ_2 for all individual vibrational lines. After having found these values, each vibrational line is recovered, and its second moment $M_V(2)$ calculated. Then, corresponding time-correlation functions, $G_V(t)$, are to be computed according to Eq. (21). Further, τ_V values are to be found by integration of $G_V(t)$. Finally, Eqs (9), (15), (19) or (20) are to be adjusted to these time-correlation functions in order to obtain the value of τ_ω. Anisotropic Raman or IR spectra can be treated in a similar way.

3.2 OVERVIEW OF DATA FITS

3.2.1 Data fits in the low-frequency region

The first data fits using Eq. (22) were directed towards the studies of the low-frequency Raman features of amorphous solids and liquds, namely, the so-called quasi-elastic scattering and Boson peaks (a brief analysis of these features can be found in Ref. [42]), and the information on vibrational line profiles was not the primary concern of the authors. However, even at that stage some valuable results were obtained. For example, it was found that in glassy polystyrene [43,44], the low-frequency vibrational line corresponding to the damped librations of the phenyl ring can be easily recovered from the overlap line, which contains contributions from the QE scattering and Boson peak. Even more complicated overlap lines were decomposed in the case of molten and glassy mixtures of bismuth chloride with potassium chloride [45], and vibrational lines corresponding to $BiCl_4^-$ anions were retrieved.

Similarly, the profiles of vibrational lines showing up in the VV and HV low-frequency Raman spectra of liquid formamide have been recovered [46]. The parameters obtained for *depolarized* lines signify that their VV and HV profiles perfectly coincide (Table 3), thereby indicating that reorientation contribution to the HV component does not exist, and both VV and HV components are broadened by the same dephasing processes.

TABLE 3. Fitting results obtained for low-frequency vibrational lines of liquid formamide

Parameter	VV spectrum	HV spectrum
ν_0/cm^{-1}	86.7	87.0
τ_1/ps	0.0729	0.0735
τ_2/ps	0.154	0.155

3.2.2 Data fits and vibrational dynamics of isotopic species in liquid CCl_4

The Raman spectrum of liquid carbon tetrachloride in the region of the ν_1 (A_1) vibration demonstrates a remarkable isotopic structure. A successful fit of this structure has been presented in Ref. [36] as the first example of working ability of Eq. (22). In Ref. [47], the time-correlation functions of vibrational dephasing have been obtained for each isotopic species present in liquid carbon tetrachloride in their natural abundance, and vibrational dephasing and vibrational frequency modulation of the ν_1 mode of isotopic molecules have been analyzed in greater detail.

In Fig. 3 the results of data fits are shown. The isotopic lines centered at 462.8, 459.7, 456.7 and 453.1 cm^{-1} belong to the CCl_4^{35}, $CCl_3^{35}Cl^{37}$, $CCl_2^{35}Cl_2^{37}$, $CCl^{35}Cl_3^{37}$ species, respectively; three broad low-intensity side lines located at 449.9, 470.0 and 473.5 cm^{-1}, probably, represent binary combinations. Numerical parameters of data fits summarized in Table 4 reveal that for the CCl_4 isotopic molecules, the dephasing times are determined mainly by the frequency factor: the lower the line frequency, the slower is the dephasing time. This dependence entirely outweighs the opposite relationship between the dephasing times and masses. The frequency modulation times show the same trend, which contradicts to the simple collision models.

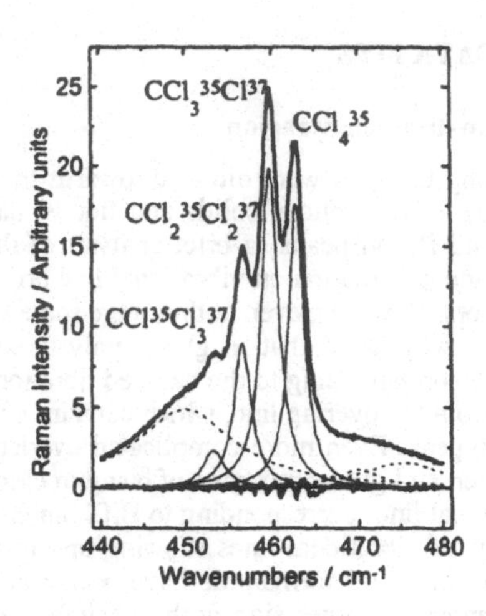

Figure 3. Isotropic Raman spectrum of CCl_4 and its fit. Thick lines: experimental data and residuals. Thin solid lines: isotopic components of the spectrum. Thin dashed lines: binary combinations.

TABLE 4. Vibrational dephasing parameters of CCl_4 isotopic molecules

Species	ν_1 / cm^{-1}	Γ / cm^{-1}	τ_1 / ps	τ_2 / ps	τ_ω / ps	τ_V / ps	M_2 / ps^{-2}
CCl_4^{35}	462.8	2.18	0.293	4.85	0.261	5.04	0.787
$CCl_3^{35}Cl^{37}$	459.7	2.35	0.264	4.50	0.235	4.69	0.940
$CCl_2^{35}Cl_2^{37}$	456.7	2.41	0.262	4.37	0.233	4.56	0.976
$CCl^{35}Cl_3^{37}$	453.1	2.97	0.236	3.56	0.209	3.76	1.34

3.2.3 Purely discrete Markovian modulation processes in ionic melts. Possible inadequacy of Kubo model

A detailed comparative analysis performed in Ref. [16] reveals that the major cause of vibrational dephasing in molecular liquids is Gaussian-Markovian frequency modulation (short phase memory), whereas in ionic liquids, all kinds of modulation processes (from non-Markovian to purely discrete Markovian, from long to completely absent phase memory) are equally possible. In Ref. [48] the novel approach described in previous sections has been applied to molten salts containing quasi-spherical complex MCl_4^{2-} anions (M=Mn^{+2} and Zn^{+2}) in order to make quantitative estimates of the spectroscopically active part of the interaction potential based on dephasing studies.

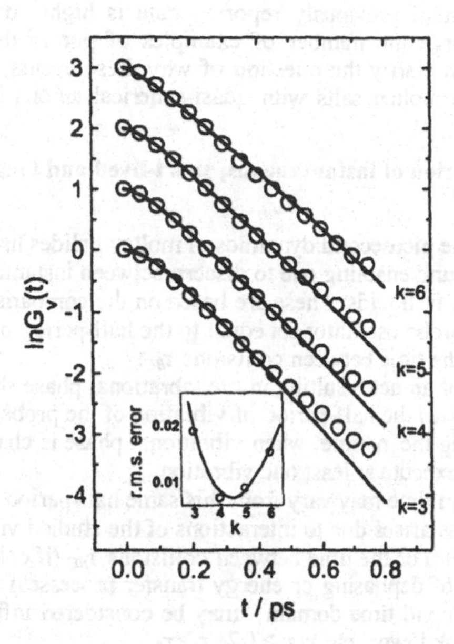

Figure 4. **Time-correlation functions of vibrational dephasing of the totally symmetric vibration of $MnCl_4^{2-}$ (points) and their fits to Eq. (6) (lines). Inset shows the dependence of the r.m.s. error on *k*.**

It follows from Fig. 4 that the best fit of the time-correlation functions of vibrational dephasing in ionic melts containing complex halides of manganese and zinc can be attained at k=4 in Eq. (16). This means that the frequency modulation in molten complex chlorides at high temperatures is the purely discrete Markovian process governed by the ion-induced dipole attraction:

$$V_{\text{I-ID}} = \left(\frac{\partial \alpha}{\partial Q_k}\right) Q_k \frac{ze^2}{r^4}. \tag{25}$$

On the other hand, according to Fig. 4, the difference between Gaussian-Markovian (Kubo-type) and purely discrete Markovian (Burshtein-type) modulation mechanisms for ionic substances is subtle. Time-correlation functions at different *k* look very similar, and their unlikeness becomes apparent only in comparative analyses, when the RMS errors for fits performed at different *k* values are inspected. This means that applying the same approach to molecular liquids, one can foresee much sharper distinction between modulation mechanisms, since for non-polar vibrations ($\partial\mu/\partial Q$=0) the $k\geq6$ value in Eqs (16) and (20) is expected.

Furthermore, in a long body of experimental facts collected to now, the Kubo approach was employed as the only way of data treatment. Again, as well seen from

Fig. 4, it is not so easy to distinguish between Gaussian-Markovian and purely discrete Markovian modulation schemes. This indicates that our knowledge regarding modulation mechanisms in liquids may be significantly model-dependent, and a careful re-examination of previously reported data is highly desirable. Such re-examination can enlarge the number of examples of purely discrete Markovian modulation events and clarify the question of why these events may happen in so different systems like molten salts with quasi-spherical anions [48,49] and liquid acetonitrile [50-52].

3.2.4 Dynamic criterion of instantaneous, short-lived and long-lived species in liquids

In Refs [53,54], the picosecond dynamics of molten halides has been studied and relations have been found enabling one to discern between instantaneous, short-lived and long-lived species in liquids. These are based on the comparison of the minimal damping time of the probe oscillator set equal to the half-period of vibration $T/2$, τ_V, τ_ω, and τ_d, as well as the time between collisions τ_{BC}:

(1) The duration of an act resulting in the vibrational phase shift must be longer than (or at least equal to) the half-period of vibration of the probe oscillator, $\tau_V \geq T/2$. This means that during the τ_V time, when vibrational phase is changing, the particle itself must be able to execute at least one vibration.

(2) The modulation time may vary from this same half-period of vibration $T/2$ (if the modulation process arises due to interactions of the studied vibrator with a similar neighboring particle) or the time between collisions τ_{BC} (if collision mechanisms are the motive force of dephasing or energy transfer processes) to very long times which - in the picosecond time domain - may be considered infinite (structural relaxation, network breakdowns, etc.): $\tau_\omega \geq T/2$, $\tau_\omega \geq \tau_{BC}$.

(3a) If vibrational dephasing and modulation processes elapse during the time equal to the half-period of vibration $T/2$, $\tau_\omega \approx T/2$, $\tau_V \approx T/2$, respective configurations should be considered *instantaneous*. This means that the bond energy in the configuration in question is very small, and just the first vibration disengages adjacent particles.

(3b) For *short-lived* complexes, the collision duration time τ_d must be sufficient for a probe particle to execute several vibrations. In this case, the longest of two characteristic times describing the phase decay cannot exceed possible duration of collision, which hence determines the lifetime of a complex: $\tau_\omega \leq \tau_d$, $\tau_V \leq \tau_d$.

(3c) To distinguish between long-lived and short-lived particles in liquids the ratio between the modulation and dephasing times and the period of vibration can be helpful. As follows from Table 5, for short-lived species the following relations hold: $0.5 \leq \tau_\omega/T \leq 2$ or $0.5 \leq \tau_V/T \leq 2$, whereas for long-lived species $\tau_\omega/T \geq 2$ or $\tau_V/T \geq 2$. The more stable the complex particle, the greater the τ_ω/T or τ_V/T ratio, and for the "infinitely" long-lived particle (CCl_4 [47]) it reaches the value of about 70. Therefore the ratios between the modulation and dephasing times and the period of vibration can serve as the dynamic criterion of *long-lived* particles in molecular and ionic liquids: $\tau_\omega/T \geq 2$ or $\tau_V/T \geq 2$.

TABLE 5. Dynamics of complex species in liquids (all times in ps).

Complex particle	T/K	$T/2$	τ_{BC}	τ_ω	τ_V	τ_d	τ_ω/T or τ_V/T
			Instantaneous species; $\tau_\omega \approx T/2$ or $\tau_V \approx T/2$ ($\tau_\omega/T \approx 0.5$ or $\tau_V/T \approx 0.5$)				
$CsCl_4^{3-}$	623–1123	0.10	0.03	0.03	0.10	0.25	0.5
$MgCl_4^{2-}$ [a]	1018	0.08	0.07	0.03	0.10	1.3	0.6
$CaCl_4^{2-}$ [a]	1073–1173	0.10	0.08	0.06	0.08	0.8–1.5	0.4
			Short-lived species; $\tau_\omega \leq \tau_d$ or $\tau_V \leq \tau_d$; $0.5 \leq \tau_\omega/T \leq 2$ or $0.5 \leq \tau_V/T \leq 2$				
$LiCl_4^{3-}$	623–1123	0.05	0.026	0.17	0.065	0.25	1.7
$MgBr_4^{2-}$ [a]	1023	0.12	0.14	0.46	0.28	1.7	1.9
$CaBr_4^{2-}$ [a]	1043	0.16	0.15	0.044	0.19	1.0	0.6
			Long-lived species; $\tau_\omega/T \geq 2$ or $\tau_V/T \geq 2$				
$MnCl_4^{2-}$	818–933	0.061	0.11	0.12	0.25	– [b]	2.0
$ZnCl_4^{2-}$	873	0.066	0.09	0.06	0.31	– [b]	2.3
$BiCl_4^-$	373–573	0.035	–	0.78	0.41	– [b]	11
CCl_4	302	0.036	0.39	0.26	5.04		70

[a] Calculated using experimental data by Bunten et al. [38,39].
[b] Lifetimes of the millisecond order of magnitude.

3.2.5 Studies of dynamics of glass-forming liquids, confined molecules, and reorientation

Of course, applications of the new time-correlation function for data fits in the time and frequency domain are in no way limited by those examples considered in this chapter. For example, vibrational dephasing and vibrational frequency modulation parameters can be used as an indicator of short-time interactions in glass-forming liquids. Overlap lines in glass-forming systems (As_2O_3 and $BiCl_3$-KCl) have been analyzed, and an unexpected discontinuity of the temperature dependence of the frequency modulation times at the glass transition temperature (τ_ω appears shorter in the glassy state than in the supercooled and liquid regime!) has been interpreted in terms of cooperative dynamics [55].

In Ref. [56] the new method is applied to the Raman spectra of molecular glass former salol in bulk, in a dilute CCl_4 solution and in restricted geometries after confining it in nanoporous silica glasses of various pore sizes. The important finding is that the vibrational dynamics of the confined molecules becomes faster with decreasing the pore size. An attempt is made to rationalize this effect by invoking the cooperativity issue related to the sluggish dynamics as the glass transition is approached. The removal of many-body effects by trapping the molecules in less-crowded environments seems to be the key factor.

The line profile analysis described in this chapter can be successfully employed in studies of reorientation. As an example, we have investigated a composite spectrum of liquid benzene, specifically, isotropic and anisotropic profiles of the ν_1 (A_{1g}, C–C vibrations) and ν_2 (A_{1g}, C–H vibrations) lines, which are badly overlapped with neighboring lines. The profiles of all individual lines have been recovered, and the time-correlation functions of vibrational and rotational relaxation have been built and analyzed [57]. A part of these data will be discussed in next Sections, in connection with the so-called non-coincidence effect.

4. FREQUENCY SHIFTS CAUSED BY REPULSION FORCES

4.1 INTRODUCTORY REMARKS

It has been already mentioned that interactions between particles in condensed media significantly affect the position of vibrational levels of molecules, leading to vibrational spectral shifts relative to those in the gas phase. Features of interest include also the solvent, temperature, and pressure (density) dependencies of the line shifts. Spectral shifts are the most striking signature of molecular interactions; they attract keen interest of theoreticians since 1935, when Buchheim [58] has published his first treatment of this phenomenon.

An approach employed by Buchheim can be briefly outlined as follows. For the molecule in a solution, in comparison with the gaseous state, the vibrational potential energy U is written as $U=U^{(0)}+V$, where $U^{(0)}$ is the potential energy of the unperturbed molecule, and V characterizes the perturbation caused by solute-solvent interactions, which is treated in terms of the sole attraction. Then U, $U^{(0)}$ and V are represented by Taylor series with respect to the vibrational normal coordinate, and frequency shifts are obtained in terms of $d^n\mu/dQ^n$ and $d^n\alpha/dQ^n$, the first and second derivatives of the dipole moment and polarizability of the particle studied with respect to the normal coordinate, which appear after proper differentiation of U. Since that time, the origin of red (blue) shifts is associated with the positive (negative) sign of $d^n\mu/dQ^n$ and $d^n\alpha/dQ^n$.

The extension of Buchheim's approach based on the use of the attractive part of interparticle potential has led to a plethora of very similar formulae for the calculation of harmonic frequencies. At the end of this chapter we enumerate some of them [59-68], however, this list is far from being complete; the book by Bakhshiev [1] may serve as a detailed survey of references up to 1969. Further, a commonly accepted conclusion has been drawn that red (blue) shifts are characteristic to weakened (strengthened) and elongated (shortened) bonds. However, such a conclusion is wrong [69]. For example, dissolution of HCl in CCl_4 leads to the red shift and the decrease in the bond length, but the bond dissociation energy increases.

This means that approaches elaborated by Buchheim and his followers fail to find out the right values of anharmonicity constants or to calculate changes in particle dissociation energy accompanying the transition from the gas phase to a condensed medium, and to predict the correct pressure dependencies of line frequencies. According to Bulanin et al. [69], the reason of these contradictions is due to neglecting repulsive forces. Really, purely attractive potentials do not satisfy the equilibrium conditions since the resultant forces on atoms (or molecules and ions as well) in a liquid are not zero. The use of the hard-sphere repulsion does not solve the problem because the respective potential function cannot be differentiated properly. Since 1976, when obvious shortcomings of simple models based on sole attractive forces have been put under focus [69], only one calculation scheme has been proposed [70] enabling one to explicitly take repulsion forces into account and calculate repulsion contributions to spectral shifts.

Another fascinating phenomenon occurring for liquids with strong molecular interactions is splitting of non-degenerate vibrational levels showing up in the non-coincidence of the vibrational frequency in isotropic (ν_{iso}) and anisotropic (ν_{aniso}) Raman and IR (ν_{IR}) spectra [71]. Such a non-coincidence effect closely reminds Davydov splitting in crystals [72] and is usually explained (see, *e.g.*, Refs [73-78]) in terms of resonance energy transfer, probably via transition dipole-transition dipole interactions. In the case of Davydov-like splitting, one can expect that non-coincidence effects must be the most pronounced for dipole-active (polar) vibrations having great values of $d\mu / dQ$ and therefore great intensities in IR spectra.

As a rule, $\nu_{iso}<\nu_{aniso}$, ν_{IR}; this observation is peculiar to the overwhelming majority of molecular liquids. If $\nu_{iso}>\nu_{aniso}$, ν_{IR}, the non-coincidence effect is considered *anomalous*, though such a behavior is not formally forbidden in terms of the most detailed theory [76]. Interestingly, since non-coincidences depend, first of all, on $(d\mu / dQ)^2$, even blue-shifted lines with $d\mu / dQ<0$ should demonstrate "normal" non-coincidences. This means that anomalous non-coincidence effects may arise due to repulsion forces. Anomalous non-coincidence effects are actively pursued since the time of their discovery [79,80]; the latest reviews of the subject can be found in Refs [10-17]. However, repulsion forces have never been quantitatively taken into account in non-coincidence effect studies. In what follows we try to fill this gap.

4.2 INITIAL BACKGROUND

It seems worth-while to start with a brief overview of the methods of calculation of spectral shifts; to do so we follow Ref. [70]. We consider an anharmonic oscillator of the potential energy

$$U^{(i)} = \frac{1}{2}K_i Q_i^2 + g_i Q_i^3 + j_i Q_i^4 . \tag{26}$$

Hereinafter we will discern between different kinds of vibrational normal coordinates. We denote the standard (mass-weighed) vibrational normal coordinate as Q_i, $[Q]=g^{1/2}cm$. Another kind of vibrational normal coordinates employed in this chapter is denoted as q_i, $[q]$=cm. In Eq. (26) the energy U is measured in erg, and the superscript i indexes the phase state of the system; the reader is reminded that usually $g_i<0$.

To calculate the wavenumber of transition between, *e.g.*, the zeroth and first vibrational levels, the following formulae are used:

$$\nu_{0\to 1}^{(i)} = \nu_0^{(i)}(1-2x^{(i)}) , \tag{27}$$

$$x^{(i)} = (hc\nu_0^{(i)})^{-1}(30\overline{Q}^{(i)6} g^{(i)2} / hc\nu_0^{(i)} - 6\overline{Q}^{(i)4} j^{(i)}) = hc\nu_0^{(i)} / 4D^{(i)} , \tag{28}$$

where h is the Planck constant, c is the velocity of light, $\nu_0^{(i)} = (2\pi c)^{-1} K_i^{1/2}$ is the harmonic wavenumber, $x^{(i)}$ is the anharmonicity constant, $\overline{Q}^{(i)} = (h / 8\pi^2 c \nu_0^{(i)})^{1/2}$ is the mean square vibrational amplitude, and $D^{(i)}$ is the bond dissociation energy.

Below we use the superscript (o) to index the spectroscopic parameters in the gaseous state, and consider the values without superscripts as being valid for the molecule in a medium. The perturbation V caused by solute-solvent interactions can be written as the function of the dipole moment μ and the polarizability α of the particle studied as

$$V = \mu L_\mu^{(1)} + \mu^2 L_\mu^{(2)} + \alpha L_\alpha + V_{repuls} , \tag{29}$$

where L^i are constants, and the last member is the repulsive term. Then Eq. (29) can be represented as a multipole series up to the second-order terms in Q; respective derivatives are denoted as V^k ($V^k=\partial^k V/\partial Q^k$). In turn, V^k are the functions of electro-optic parameters, $\mu^k=\partial^k \mu/\partial Q^k$ and $\alpha^k=\partial^k \alpha/\partial Q^k$:

$$V^{\mathrm{I}} = \frac{d\mu}{dQ_j} L_\mu^{(1)} + 2\mu^{(0)} \frac{d\mu}{dQ_j} L_\mu^{(2)} + \frac{d\alpha}{dQ_j} L_\mu + V^{\mathrm{I}}_{\mathrm{repuls.}} \tag{30a}$$

$$= \mu^{\mathrm{I}} L_\mu^{(1)} + 2\mu^{(0)} \mu^{\mathrm{I}} L_\mu^{(2)} + \alpha^{\mathrm{I}} L_\mu + V^{\mathrm{I}}_{\mathrm{repuls.}}, \tag{30b}$$

$$V^{\mathrm{II}} = \frac{d^2\mu}{dQ_j^2} L_\mu^{(1)} + \left[2\mu^{(0)} \frac{d^2\mu}{dQ_j^2} + 2\left(\frac{d\mu}{dQ_j} \right)^2 \right] L_\mu^{(2)} + \frac{d^2\alpha}{dQ_j^2} L_\alpha + V^{\mathrm{II}}_{\mathrm{repuls.}}$$

$$+ \frac{d^2\mu}{dQ_j dQ_k} L_\mu^{(1)} + 2\mu^{(0)} \frac{d^2\mu}{dQ_j dQ_k} L_\mu^{(2)} + \frac{d^2\alpha}{dQ_j dQ_k} L_\alpha \tag{31a}$$

$$= \mu^{\mathrm{II}} L_\mu^{(1)} + \left[2\mu^{(0)} \mu^{\mathrm{II}^2} + 2\mu^{\mathrm{I}^2} \right] L_\mu^{(2)} + \alpha^{\mathrm{II}} L_\alpha + V^{\mathrm{II}}_{\mathrm{repuls.}} + V^{\mathrm{II}}_{\mathrm{res.}}. \tag{31b}$$

In Eq. (31a) the second line contains mixed second derivatives giving rise to the resonance effects.

Solving the equation $U=U^{(0)}+V$ with Eqs (26), (30) and (31), after some algebra, yields

$$\begin{aligned} \nu_0 &= \nu_0^{(0)} + (8\pi^2 c^2 \nu_0^{(0)})^{-1} [V^{\mathrm{II}} - 6g^{(0)} \Delta Q + 12 j^{(0)} (\Delta Q)^2] \\ &= \nu_0^{(0)} + (\Delta\nu)_1 + (\Delta\nu)_2 + (\Delta\nu)_3, \end{aligned} \tag{32}$$

$$g = g^{(0)} + 4 j^{(0)} \Delta Q, \tag{33}$$

$$j = j^{(0)}, \tag{34}$$

$$\Delta Q = Q - Q^{(0)} = \frac{V^{\mathrm{I}}}{K_0 + V^{\mathrm{II}}} \cong \frac{V^{\mathrm{I}}}{K_0}. \tag{35}$$

ΔQ determined by Eq. (35) is the increment of the normal coordinate caused by interparticle interactions. Notice that in Eqs (32)-(35) resonance effects are neglected.

4.3 WHY ACCOUNT FOR REPULSION FORCES IS SO SIGNIFICANT?

It is obvious that, if repulsion forces are neglected, Eq. (32) predicts red shifts for the cases when $\partial\mu/\partial Q$, $\partial^2\mu/\partial Q^2$, $\partial\alpha/\partial Q$ and $\partial^2\alpha/\partial Q^2$ are all positive and blue shifts if these derivatives are all negative. Numerous useful correlations are based on Eq. (32), and those of Kirkwood-Bauer-Magat type predicting linear dependencies of line shifts on $\frac{\varepsilon - 1}{2\varepsilon + 1}$ are the most known.

What do these correlations tell? Forces between molecules are often classified into two main types: physical, bulk, long-range, or universal, and chemical, local, short-range, or specific. From this point of view, additive and non-directed orientational, inductive, dispersive, or, in a broader sense, electrostatic forces, as well as repulsion forces are considered universal. All pairwise interactions, like complex formation or hydrogen bond, having specific attributes of the chemical bond (saturability and preferred direction) are considered specific. Indeed, universal interactions operate in any system, both with and without specific interactions. All formulae derived in previous Section are based on the simple electrostatic approach and therefore aim at the study of universal interactions. However, they can be helpful for elucidation of specific interactions being a useful tool for discriminating contributions caused by universal interactions from contributions caused by specific ones. Moreover, some potentials containing an orientation factor can be equally useful for studies of both universal and specific pairwise interactions: for example, Sechkarev and his followers [63,68] utilized such potentials for dipolar liquids with strong molecular interactions and molten salts.

As far as specific forces are concerned, the principles of spectroscopy of interparticle interactions, though without mentioning the works of its founders, have been employed in recent treatments of blue shifting hydrogen bonds. For instance, complexes of ethylene oxide and fluoroform demonstrate the blue shifting hydrogen bonds [81], and the authors suggest that this is probably due to the fact that $d\mu/dq$ for CHF_3 is negative. The role of the negative sign of $d\mu/dq$ in this same phenomenon has also been emphasized in Ref. [82].

A commonly accepted conclusion of qualitative spectroscopy of interparticle interactions states that red shifts are typical to weakened and elongated bonds, whereas blue shifts are characteristic to strengthened and shortened bonds. However, Bulanin et al. [69] have shown that such a conclusion is wrong. This statement can be easily proven [70] in the case of HCl molecule, whose vibrational levels are well characterized both in the gas phase and in solutions (Table 6).

TABLE 6. Experimental and calculated parameters of the HCl molecule in CCl_4 solution with respect to the gas phase. Repulsion is neglected.

Parameter	Gas	Solution in CCl_4 Experimental data	Calculations using Eqs (32)-(35)	Solution of the "inverse" problem
ν_0	2290	2929	2915.7	
$x_i/10^{-2}$	1.741	1.673	1.786	
$D_i/10^{-12}$ erg	8.534	8.700	8.111	
$(\Delta\nu)_1$/cm^{-1}			-3.4	26.0
$(\Delta\nu)_2$/cm^{-1}			-72.3	-89.0
$(\Delta\nu)_3$/cm^{-1}			1.4	2.0
$U^I/10^7$ din g$^{-1/2}$			-3.742	-4.672
$U^{II}/10^{27}$ s^{-2}			-7.297	55.34

Experimental data clearly demonstrate that dissolution of HCl in CCl_4 leads to the red shift and the decrease in the bond length, but the bond dissociation energy increases. Calculations performed in terms of Eqs (32)-(35) with experimental values of electro-optic parameters and model interaction potentials fail to prove this fact. On the other hand, one can use these same equations, electro-optic parameters and interaction potentials for solving an "inverse" problem, *viz.*, finding the values of $(\Delta\nu)_i$ and U^i being based on the experimental values of ν_0^i, x_i and D_i. Proper calculations reveal that the source of mistakes lies in the wrong sign of U^{II}. According to Ref. [70], the reason of these contradictions is due to neglecting repulsive forces.

4.4 SPECTRAL SHIFTS CAUSED BY REPULSION FORCES

As follows from previous Sections, the solution if the line-shift problem requires the differentiation of the potential with respect to the normal coordinate. If such differentiation of attraction potentials is obvious, similar procedure for repulsion has long been known a stumbling-block for spectroscopy of interparticle interactions. A general approach to the solution of the repulsion problem has first been given by Valiev [83] in terms of $\partial S_i/\partial Q_j$, the derivatives of the i-th atom displacement with respect to the normal coordinate of j-th vibration.

Following Valiev, one can write $R_{i,\beta B}$, the distance between the i-th atom in the probe particle and the β-th atom in the B-th adjacent particle, as

$$R_{i,\beta B} = \left|\mathbf{R}_B + \mathbf{A}_{\beta B} - \mathbf{A}_i - \mathbf{S}_i\right| = R_B + \mathbf{n}_B\mathbf{A}_{\beta B} + \mathbf{n}_B\mathbf{A}_i + \mathbf{n}_B\mathbf{S}_i \,. \tag{36}$$

In this equation R_B is the vector connecting the centers of mass of the molecule of interest and molecule B, $A_{\beta B}$ and A_i are the vectors defining the positions of β or i atoms with respect to the center of mass of molecule B or the particle studied, and $\mathbf{S}_i$ is the vibrational displacement of the i-th atom. It is easy to understand that only the $\mathbf{S}_i$ value depends on the normal coordinate, and $dR_{i,\beta B}/dQ \equiv d\mathbf{S}_i/dQ$. This formalism has been widely used by Valiev and his followers in the description of line broadening in liquids [83-87].

The most detailed account for the role of repulsion forces in the line shift problem has been presented in Ref. [70]. Repulsion between atoms has been written as

$$V^{\text{rep}}_{i,\beta B}\left(R_{i,\beta B}\right) = V_{i,\beta B}\exp\left(-\Delta_{i,\beta B}R_{i,\beta B}\right). \tag{37}$$

Considering displacement as being linear one gets

$$V^{\text{I}}_{\text{repuls.}} = -\sum_{i,\beta B}\Delta_{i,\beta B}V_{i,\beta B}\exp(-R_{i,\beta B}\Delta_{i,\beta B})\left|d\mathbf{S}_i / dQ_k\right| \tag{38a}$$

$$= -\sum_{i,\beta B}\left|d\mathbf{S}_i / dQ_k\right| A_{i,\beta B}\,, \tag{38b}$$

$$V^{\text{II}}_{\text{repuls.}} = \sum_{i,\beta B}\Delta^2_{i,\beta B}V_{i,\beta B}\exp(-R_{i,\beta B}\Delta_{i,\beta B})\left|d\mathbf{S}_i / dQ_k\right|^2 \tag{39a}$$

$$= \sum_{i,\beta B}\Delta_{i,\beta B}\left|d\mathbf{S}_i / dQ_k\right|^2 A_{i,\beta B}\,, \tag{39b}$$

$$A_{i,\beta B} = (\pi/3)\Delta^3_{i,\beta B}V^{\text{rep.}}_{i,\beta B}(a)\frac{N_B}{V}(l_i^2 + l_{\beta B}^2)\int_0^\infty \exp(-\Delta_{i,\beta B}R_B)g(R_B)R_B^2 dR_B \tag{40a}$$

$$= (\pi/3)\Delta^2_{i,\beta B}V^{\text{rep.}}_{i,\beta B}(a)\rho_n(l_i^2 + l_{\beta B}^2)\left(a^2 + \frac{2a}{\Delta_{i,\beta B}} + \frac{2}{\Delta^2_{i,\beta B}} + 3a^2\pi\rho_n\right) \tag{40b}$$

Eqs (38a) and (39a) correspond to pairwise interactions between the probe particle and the B-th adjacent particle, whereas Eqs (38b) and (39b) are the result of averaging over orientations and distances. Eq. (40b) is the result of the averaging procedure performed using a model distribution function taken from Ref. [88], ρ_n is the number density of the probe particles (in neat liquids $\rho_n = nN_A/V_M$, N_A is the Avogadro number and V_M the molar volume, $n=1$ for molecular liquids, $n=a+m$ for the M_aA_m molten salt), and $l_i(l_{\beta B})$ is the length of the vectors defining the positions of i or β atoms with respect to the center of mass of the molecule studied or molecule B.

To solve the problem of spectral shifts caused by repulsion forces one has to use Eqs (38)-(40) in the following equation:

$$\Delta \nu_{0,\text{repuls.}} \cong (8\pi^2 c^2 \nu_0^{(0)})^{-1} [V^{\text{II}}_{\text{repuls.}} - 6g^{(0)} (4\pi^2 c^2 \nu_0^{(0)2})^{-1} V^{\text{I}}_{\text{repuls.}}]. \tag{41}$$

Contributions arising due to $V^{\text{I}}_{\text{repuls.}}$ and $V^{\text{II}}_{\text{repuls.}}$ have different signs, and it is a difficult task to find the correct sign of the overall spectral shift caused by repulsion forces. Table 6 reveals that in the case of HCl dissolved in CCl_4, $V^{\text{II}}_{\text{repuls.}}$ plays the major role. A rough estimate of the sign of the spectral shift caused by repulsion forces can be made by means of the expression, which follows from Eq. (41),

$$1 + 6g^{(0)} (4\pi^2 c^2 \nu_0^{(0)2})^{-1} (\Delta_{i,\beta B} d\mathbf{S}_i / dQ_k)^{-1}. \tag{42}$$

It is positive if $V^{\text{II}}_{\text{repuls.}}$ prevails, and negative if $V^{\text{II}}_{\text{repuls.}}$ is smaller than the second member in square brackets in Eq. (41). Taking typical values of $g^{(0)} = -1 \cdot 10^{47}$ g$^{-1/2}$cm^{-1}c^{-2}, $\nu_0^{(0)} = 1000$ cm^{-1}, $\Delta_{i,\beta B} = 3.6 \cdot 10^8$ cm^{-1}, $d\mathbf{S}_i / dQ_k = 5 \cdot 10^{-12}$ g$^{-1/2}$, one gets 0.2. This means that the model predicts $\Delta \nu_{0,\text{repuls.}}$ to be positive (blue), and the spectral shift caused by repulsion forces can be reasonably estimated by means of the approximated formula,

$$\Delta \nu_{0,\text{repuls.}} \cong (8\pi^2 c^2 \nu_0^{(0)})^{-1} V^{\text{II}}_{\text{repuls.}}. \tag{43}$$

A similar (qualitative) conclusion regarding the blue shifts caused by repulsion forces in the case of *specific* interactions (hydrogen bond) has been recently made in Ref. [82].

Eqs (41) and (43) show the way to the correction of any calculation scheme merely based on attractive forces. Examples presented in Ref. [70] prove this statement. The correct description of a vibrational level system was given and an exact value of dissociation energy of the HCl molecule in solution was obtained. Other applications of Eqs (41) and (43) include calculations of the repulsion contribution to frequency shifts in molten nitrates [16,70].

5. NON-COINCIDENCE EFFECTS CAUSED BY REPULSION FORCES

5.1 NON-COINCIDENCE EFFECTS CAUSED BY RESONANCE ENERGY TRANSFER. DAVYDOV-LIKE SPLITTING IN LIQUIDS

For liquids with strong molecular interactions, splitting of non-degenerate vibrational levels often occurs. This phenomenon manifests itself in the non-coincidence of the vibrational frequency in isotropic and anisotropic Raman and IR spectra; it is common to measure non-coincidences as $\Delta\nu_{NCE} = \nu_{aniso} - \nu_{iso}$ or $\Delta\nu_{NCE} = \nu_{IR} - \nu_{iso}$. First observations of the non-coincidence effect [71] have shown that $\nu_{iso} < \nu_{aniso}$, ν_{IR}, and hence $\Delta\nu_{NCE} > 0$; such observation is peculiar to the vast majority of molecular liquids. The so-called "anomalous" or negative non-coincidences have first been reported for molecular liquids in 1987 [79], in this case $\nu_{iso} > \nu_{aniso}$, ν_{IR}, and $\Delta\nu_{NCE} < 0$. It should be mentioned, however, that a similar effect for molten nitrates has been described by Brooker and Papatheodorou much earlier, in 1983 [89]. A list of existing anomalous non-coincidences in molten salts can be found in Ref. [16].

The non-coincidence effect has been explained (see, *e.g.*, Refs [73-78]) in terms of resonance energy transfer, probably via transition dipole–transition dipole interactions, by analogy with Davydov splitting in crystals. Davydov-like coupling of a pair of adjacent vibrators leads to in-phase (ω_+, low-frequency) and antiphase (ω_-, high-frequency) components, and ω_+ and ω_- differ from ω_0, the frequency of the unperturbed vibrator (Fig. 5):

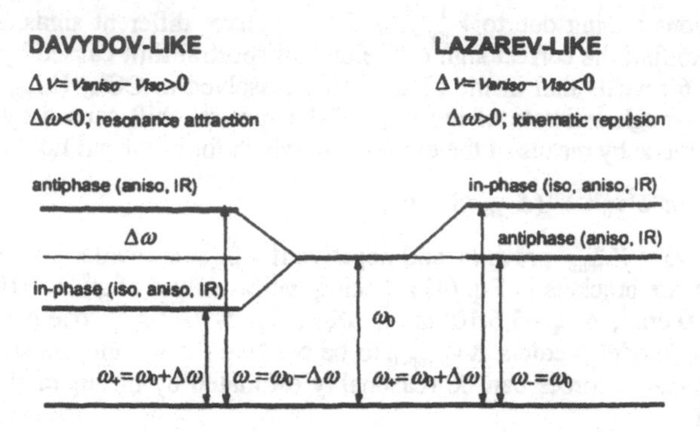

Figure 5. Schematic representation of energy levels for Davydov-like and Lazarev-like coupling in liquids.

$$\omega_+ = \omega_0 + \Delta\omega\,, \tag{44a}$$

$$\omega_- = \omega_0 - \Delta\omega\,. \tag{44b}$$

As follows from the theory proposed by Döge [73], both components of a split mode are present in both anisotropic Raman and IR spectra of a liquid, whereas the antiphase component ω_- is inactive in the isotropic spectrum. In this case $\nu_{iso} = \omega_+ = \omega_0 + \Delta\omega$, $\nu_{aniso} = (\omega_+ + \omega_-)/2 = \omega_0$, and $\Delta\nu_{NCE} = -\Delta\omega$.

In terms of transition dipole–transition dipole interactions, the sign of $\Delta\omega$ is determined by (i) mutual orientation of interacting dipoles and (ii) the sign of $d^2\mu/dq_jdq_k$, the second derivative of the dipole moment with respect to the vibrational coordinates of interacting vibrators; $d^2\mu/dq_jdq_k = (d\mu/dq_j)^2$ if adjacent vibrators are the same and hence $j=k$. Here we use q, the normal coordinate measured in cm; the values of $(d\mu/dq_j)^2$ are proportional to the IR intensities of the lines corresponding to the j-th vibrations and measured in D/Å. An exact treatment of the Davydov splitting in crystals leads to the expression [90,91]

$$\Delta\omega = \frac{\chi_\mu}{M\omega_0}\left(\frac{d\mu}{dq_j}\right)^2, \tag{45}$$

where χ_μ is the respective lattice sum (usually negative due to attraction) and M is the reduced mass of the vibrator. The χ_μ value is the same for vibrators of the same symmetry species. For negative values of χ_μ one gets $\Delta\omega<0$, and therefore the non-coincidence effect is normal, $\Delta\nu_{NCE}>0$.

An apparatus employed in numerous quantitative explanations of non-coincidences in liquids [73-78] is just as same as in the case of determining line shifts, Eqs. (32), (41) or (43), and only the V^{II} part of the interaction potential, which depends on the second derivatives with respect to the vibrational coordinate, is taken into account. It leads to formulae more or less similar to Eq. (45); naturally, pairwise interactions have been considered instead of lattice sums required for solids. Almost all of these formulae treat positive non-coincidences, $\Delta\nu_{NCE}>0$, though it should be noticed that in terms of the most detailed theories, the sign of $\Delta\omega$ is determined by the angle between the main axes of the scattering tensors of interacting vibrators [74], and negative (anomalous) non-coincidences with $\Delta\nu_{NCE}<0$ are not formally forbidden [76].

If vibrations are dipole-inactive (forbidden in IR) and therefore have $d\mu/dq_j\cong 0$, other resonant energy transfer mechanisms causing 'normal' non-coincidence effects should come into play. If dispersion forces operate between adjacent molecules, non-coincidences are proportional to the values of $d\alpha/dq_k$ (proportional to Raman intensities) and can be calculated as

$$\Delta\omega = \frac{\chi_\alpha}{M\nu_0}\left(\frac{d\alpha}{dq_j}\right)^2 . \tag{46}$$

Similarly to Eq. (45), χ_α is usually negative due to attraction; for negative values of χ_α one gets $\Delta\omega<0$, and therefore non-coincidences are normal, $\Delta\nu_{NCE}>0$.

It is easy to realize that, on the one hand, the description of non-coincidence effects leads to the formulae, which closely resemble those accounting for vibrational line shifts. In the other hand, methods of spectroscopy of molecular interactions give a cue of how non-coincidences can be gradually depleted. This can be done by means of dilution with a suitable inert solvent. The use of respective isotopomers as a solvent is especially preferable. In such case all ***non-resonant interactions*** in the solution ***remain the same*** as compared to the pure liquid, whereas ***resonant interactions are vanishing*** since the distance between molecules of the solute increases due to dilution. Isotopic dilution is therefore one of the most popular tools for studying non-coincidences [77,78].

5.2 ANOMALOUS NON-COINCIDENCES (LAZAREV SPLITTING) IN IONIC CRYSTALS AND MELTS

In the studies of crystals, a purely electrostatic resonance treatment of Davydov splittings of internal modes has been repeatedly stated to be insufficient. In the case of complex anions (like SO_4^{2-}, MoO_4^{2-}, WO_4^{2-}, etc.) in a lattice, the splitting of non-polar internal vibrations has been found to be sharply sensitive to the interactions between atoms of neighboring anions, and anomalous non-coincidences can be as large as about 10 % of the unperturbed frequency of high-frequency internal modes. In the earliest works in the field [92,93], close agreement of experimental and calculated splittings has been achieved if short-range forces (repulsion, in the first place) are taken into account.

A considerable number of examples in which the greatest splittings are peculiar just for non-polar or weakly polar ($d\mu_i/dq_j\cong 0$) symmetric-type motions has been found, analyzed and interpreted by Lazarev et al. [90,91] in terms of a new force-constant method, explicitly resolving the interactions originating from substantially short-range effects. Numerous normal-coordinate calculation performed by Lazarev et al. are in excellent agreement with experiment.

Lazarev has drawn a distinction between resonant and non-resonant splitting and shown that the latter depends on kinematic factors. Distances between the atoms of adjacent vibrators vary during in-phase motion and do not change when antiphase motion occurs (Fig. 6).

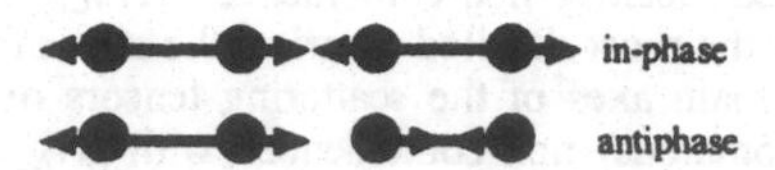

Figure 6. When vibrators execute in-phase motion, terminal atoms interfere each other experiencing strong repulsion; no changes in repulsion occur at antiphase motion.

Therefore one can write

$$\omega_+ = \omega_0 + \Delta\omega \tag{47a}$$

$$\omega_- = \omega_0 . \tag{47b}$$

This treatment remains valid both for atoms in direct contacts and interacting through intervening atoms. The splitting value depends on the second derivative of the interatomic potential energy with respect to the closest approach distance, first of all, on repulsion terms. In this case $\Delta\nu_{NCE} = -\Delta\omega/2$, since $\nu_{iso} = \omega_+ = \omega_0 + \Delta\omega$ and $\nu_{aniso} = (\omega_+ + \omega_-)/2 = \omega_0 + \Delta\omega/2$. As clearly seen from Fig. 5, Davydov and Lazarev effects are characterized by the different sign of $\Delta\omega$.

It has been already mentioned that anomalous non-coincidences are peculiar for non-polar vibrations both in ionic crystals and in melts [16,89-93]. Naturally, dilution experiments are difficult to perform for crystalline ionic solids. However, dilution data available for molten salts seem to demonstrate that resonance effects in ionic systems are negligible [16,94,95]. In view of this fact, the non-resonant nature of (anomalous) non-coincidences in ionic systems seems quite probable.

In the Lazarev approach, no general equations for non-coincidences have been proposed; direct computations have been preferred using model potentials of interparticle interactions. Based on the model [70] described above, Kirillov [96] proposed to calculate $\Delta\omega$ using Eq. (41):

$$\Delta\omega \cong (8\pi^2c^2\nu_0^{(0)})^{-1}[V^{\rm II}_{\rm repuls.} - 6g^{(0)}(4\pi^2c^2\nu_0^{(0)2})^{-1}V^{\rm I}_{\rm repuls.}] . \tag{48}$$

If the influence of $V^{I}_{\rm repuls.}$ is neglected,

$$\Delta\omega \cong (8\pi^2c^2\omega_0)^{-1}\sum_{i,\beta B}\Delta_{i,\beta B}\left|dS_i/dQ_j\right|^2 A_{i,\beta B} , \tag{49}$$

where $A_{i,\beta B}$ is now expressed as follows

$$A_{i,\beta B} = \Delta_{i,\beta B}V_{i,\beta B}(a)\left(1+\Delta_{i,\beta B}l_i\cos\theta_i\right)\left(1-\Delta_{i,\beta B}l_{\beta B}\cos\theta_{\beta B}\right), \tag{50}$$

a is the distance between the centers of masses of the adjacent particles, θ_k defines the angle between l_k and the vector connecting the centers of mass of the molecule studied and molecule B. It has been already noticed in Sec. 2.2 [see Eq. (39)] that $V^{\rm II}_{\rm repuls.}$ is positive; just due to this fact we have the positive value of $\Delta\omega$, as expected for the anomalous (negative) non-coincidence effect.

Unfortunately, the correct values of $V_{i,\beta B}$ and $\Delta_{i,\beta B}$ for liquids are difficult to find, and the use of any empirical parameters may strongly misrepresent the situation. Relative comparisons of non-coincidences are therefore preferable at this stage of our knowledge. There are two possible ways for doing so: comparison of non-coincidences of two vibrations of the same symmetry type of one and the same molecule, and comparison of non-coincidences of one vibration of one and the same molecule dissolved in various solvents. The first method has been applied to the ν_1 (Σ^+, C–N stretch) and ν_3 (Σ^+, C–S stretch) vibrations of the SCN^- anion in molten potassium thiocyanate [96], and the second to the ν_1 (A'_1) vibration of the NO_3^- anions in molten lithium and rubidium nitrates [16] showing that repulsion gives a valuable contribution to the overall value of the non-coincidence effect.

5.3 RESONANCE ENERGY TRANSFER (DAVYDOV-LIKE SPLITTING) CAUSED BY REPULSION FORCES IN LIQUIDS

Resonance energy transfer can arise not only in the attraction potential region. It can exist in the repulsive potential region causing small but distinct line broadening effects [84]. However, no attempts have been undertaken in order to estimate the role played by repulsion forces in resonance (Davydov-like) splitting in liquids. Such effect should be governed by the mixed derivatives of vibrational displacements of i-th atoms with respect to the vibrational coordinates of interacting vibrators, d^2S_i/dQ_jdQ_k; since adjacent vibrators are the same, $d^2S_i/dQ_jdQ_k=(dS_i/dQ_j)^2$. In this case, Eq. (45) can be rewritten as follows:

$$\Delta\omega = \sum_i \frac{\chi_S}{\nu_0}\left(\frac{dS_i}{dQ_j}\right)^2 \qquad (51)$$

In Eq. (51), the M value in the denominator does not appear since mass-weighed normal coordinates Q_j are used.

If pairwise interactions are taken into account explicitly, one can apply considerations similar to those employed in derivations of Eq. (49) and obtain the following value of $\Delta\omega$:

$$\Delta\omega = (8\pi^2c^2\omega_0)^{-1}\sum_{i,\beta B}\Delta_{i,\beta B}\left|dS_i/dQ_j\right|^2 A_{i,\beta B}\,. \qquad (52)$$

The $A_{i,\beta B}$ can be found using Eq. (50).

The difference between Eqs (49) and (52) is as follows. Eq. (49) gives an *approximate* value of $\Delta\omega$ due to the neglect of the second derivative of the interaction potential with respect to the vibrational coordinate, $V^{\mathrm{I}}_{\mathrm{repuls.}}$, whereas Eq. (52) operates solely with the second derivatives and therefore gives the *exact* value of $\Delta\omega$. From this point of view, since $V^{\mathrm{II}}_{\mathrm{repuls.}}$ is positive, the value of $\Delta\omega$ is positive *by definition* and leads to the *anomalous* (negative) non-coincidence effect, $\Delta\nu_{NCE}<0$.

The question can arise of how to discern between non-coincidence effects caused by non-resonance (Lazarev-like) and resonance (Davydov-like) effects. We believe that dilution studies can be of great help for this purpose. It has been already mentioned that in molten salts dilution does not cause frequency shifts, and this can be considered as a signature of the Lazarev-like splitting caused by kinematic effects. On the other hand, dilution studies in numerous molecular liquids demonstrate clear concentration dependencies of line frequencies. This can be interpreted as a signature of the Davydov-like (resonance transfer) effects. In the next Section, our aim will be to separate non-coincidences caused by attraction and repulsion in benzene. To achieve this goal, we will be based on direct measurements of non-coincidences, dilution studies, and formulae derived in previous Sections, specifically, Eqs (46) and (52).

6. SEPARATION OF NON-COINCIDENCES CAUSED BY ATTRACTION AND REPULSION IN MOLECULAR LIQUIDS

As noticed in previous Sections, straightforward calculations of repulsion contributions to spectral shifts are difficult since Born-Mayer potentials are very sensitive to the choice of $V_{i,\beta B}$ and $\Delta_{i,\beta B}$; just due to this limitation, relative comparison is preferable. Organic molecules offer much more opportunities for such comparison than inorganic ions, since organic molecules quite often have two totally symmetric vibrations: one of them corresponds to the motion of the carbon skeleton, and another to the motion of hydrogen atoms or their substituents. This is especially true for molecules containing aromatic and aliphatic rings. Then, hydrogen atoms in organic molecules can be easily substituted by deuterium, and non-coincidences observed for the same vibrational modes of light and heavy substances can be compared. Furthermore, electro-optic parameters of numerous organic molecules are carefully measured, enabling one to estimate the value of coupling arising due to attraction forces in the liquid studied and subtract it from the value of coupling arising due to repulsion forces.

Benzene can be a good example of such organic molecules. It has two totally symmetrical planar stretching modes of the A_{1g} type. One of them (ν_1) situated at 993 cm^{-1} corresponds mainly to the breathing motion of carbon atoms; another one (ν_2) located at 3063 cm^{-1} corresponds to the breathing motion of hydrogen atoms. In Fig. 7 the Raman spectra of benzene [57] are shown. These spectra are badly overlapped; in order to find the individual ν_1 and ν_2 lines the method [36] has been employed, as described in previous sections. The shape of the components of the composite lines is presented in Fig. 7 by thin lines; the peak frequencies of isotropic and anisotropic components of ν_1 and ν_2 lines are listed in Table 7 demonstrating weak non-coincidences of different signs. Different signs of the splitting of two A_{1g} lines in liquid benzene can be caused by a delicate interplay between repulsion and attraction.

It should be noticed that non-coincidence effects in liquids containing ring molecules, especially benzene, attract much attention of experimentalists. Isotopic dilution studies of the isotropic component of the ν_1 line of benzene-h$_6$ and benzene-d$_6$ performed by Meinander et al [97] reveal the (blue) shift by –0.27 and –0.59 cm^{-1}, respectively. According to the measurements by Döge et al. [98], $\Delta\nu_{NCE}$ for the ν_1 line of benzene-h$_6$ and benzene-d$_6$ equals to –0.28 and –0.53 cm^{-1}, respectively, whereas isotopic dilution data give –0.28 and –0.53 cm^{-1}. Kamogava and Kitagava [99] have found that isotopic dilution causes the red shift of the isotropic component of the ν_2 line of benzene-h$_6$ by 0.70 cm^{-1}. All these observations well agree with our data. Furthermore, as follows from papers by Döge et al. [98] and Morresi et al. [100], vibrations corresponding to the breathing motions of carbon atoms in ring molecules usually demonstrate small anomalous non-coincidence effects.

Being based on the apparatus described in previous Sections, one can estimate the relative role of attraction and repulsion in non-coincidence effects in liquid benzene in the following way [101].

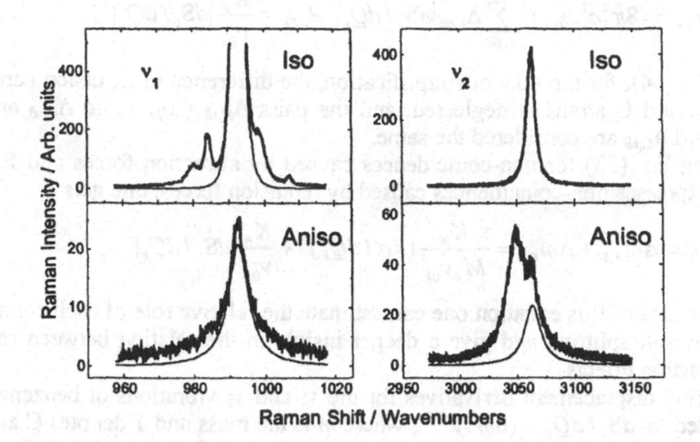

Figure 7. Raman spectra of benzene in the region of ν_1 and ν_2 vibrations. Data fits are shown by thin lines.

TABLE 7. Vibrational line frequencies in liquid benzene

Molecule and vibration	ν_{iso}/cm^{-1}	ν_{aniso}/cm^{-1}	$(\nu_{iso}-\nu_{aniso})$/cm^{-1}	Literature data
C_6H_6, ν_1 (A_{1g})	992.76±0.002	992.50±0.03	–0.26 (anomalous)	–0.24 to –0.28 [98], –0.27 [97]
C_6H_6, ν_2 (A_{1g})	3062.40±0.005	3063.0±0.1	+0.60 (normal)	+0.70 [99]
C_6D_6, ν_1 (A_{1g})				–0.50 to –0.53 [98], –0.59 [97]
C_6D_6, ν_2 (A_{1g})				+0.40 to +0.80 [99]

Dilution studies clearly show that non-coincidences of both the ν_1 and ν_2 lines in benzene are caused by resonance effects, and two distinct Davydov-like mechanisms operate. One of them arises due to resonance attraction, and another due to resonance repulsion.

Since the A_{1g} vibrations of benzene are dipole-inactive (forbidden in IR) and therefore have $d\mu/dQ_k=0$, other resonant energy transfer mechanisms, presumably due to dispersion forces between adjacent molecules, should be responsible for 'normal' non-coincidence effects. In this case non-coincidences are proportional to the values of $d\alpha/dq_k$, and can be calculated according to Eq. (46) written in the form

$$\Delta\omega_{A,k} = \frac{K_A}{M_k \nu_{0k}} (d\alpha / dQ_k)^2 . \tag{53}$$

Again, the reader is reminded that in crystals, the lattice sum is considered the same for vibrations of the same symmetry species. Most probably, an analogue of the lattice sum for a liquid K_A follows this trend.

To estimate the role of resonance repulsion, Eq. (52) can be used in the following form:

$$\Delta\omega_{R,k} = (8\pi^2 c^2 \omega_{0,k})^{-1} \sum_{i,\beta B} \Delta_{i,\beta B} \left| dS_i / dQ_k \right|^2 A_{i,\beta B} = \frac{K_R}{\nu_{0k}} \left| dS_i / dQ_k \right|^2 . \quad (54)$$

In Eq. (54), for the sake of simplification, the difference in repulsion parameters of H, D, and C atoms is neglected, and the pairs $\Delta_{H,iB}$ ($\Delta_{D,iB}$) and $\Delta_{C,iB}$ and $A_{H,iB}$ ($A_{D,iB}$) and $A_{C,iB}$ are considered the same.

Using Eq. (53) for non-coincidences caused by attraction forces and Eq. (54), which expresses non-coincidences caused by repulsion forces, one gets

$$\Delta\omega_k = \Delta\omega_{A,k} + \Delta\omega_{R,k} = \frac{K_A}{M_k \nu_{0k}} (d\alpha / dQ_k)^2 + \frac{K_R}{\nu_{0k}} \left| dS_i / dQ_k \right|^2 . \quad (55)$$

By means of this equation one can estimate the relative role of each contribution to the overall splitting and give a deeper insight in the relation between repulsion and attraction effects.

Atomic displacement derivatives for the ν_1 and ν_2 vibrations of benzene can be calculated as $dS_i / dQ_k = (6m_Y)^{-1/2}$, where m is the mass and Y denotes C atoms for ν_1 and H or D atoms for ν_2. In the literature, one can find the values of scattering coefficients which are equal to $d\alpha/dq_i$, the derivatives of polarisability with respect to the normal coordinate q_i measured in m. For benzene in *the gaseous state*, these values are tabulated in Ref. [102]. All parameters to input in Eq. (55) are listed in Table 8.

TABLE 8. Parameters of ν_1. and ν_2. vibrations in liquid benzene

Molecule and vibration	ν_0/cm^{-1}	M_i / 10^{-27} kg	$(d\alpha/dq_i)^2$/ 10^{-81} C^2 m^4 V^{-2}	$(dS_i/dQ_k)^2$/ 10^{24} kg$^{-1/2}$
C_6H_6, ν_1 (A_{1g})	993	3.32	36.1	8.35
C_6H_6, ν_2 (A_{1g})	3063	0.277	50.0	100.0
C_6D_6, ν_1 (A_{1g})	943	3.32	30.5	8.35
C_6D_6, ν_2 (A_{1g})	2303	1.81	24.3	50.20

First of all, let us perform calculations for the ν_1 and ν_2 vibrations of benzene-h_6 using the $\Delta\omega$ data obtained in our measurements of non-coincidences (recall that $\Delta\omega$ and $\Delta\nu_{NCE}$ are of different signs). The following system of equations can be obtained,

$$0.26 = 10.95 \cdot 10^{-3} K_A + 8.41 \cdot 10^{-3} K_R,$$

$$-0.60 = 58.93 \cdot 10^{-3} K_A + 32.65 \cdot 10^{-3} K_R,$$

which can be satisfied with K_A=–97.7 and K_R=158.5. Respective values of the attraction and repulsion contributions to the shift of each vibration studied are therefore as follows:

$$\Delta\omega_{A,1H} = -1.07 \text{ and } \Delta\omega_{R,1H} = +1.33,$$

$$\Delta\omega_{A,2H} = -5.77 \text{ and } \Delta\omega_{R,2H} = +5.17.$$

This means that for both totally symmetric vibrations of the benzene-h_6 molecule, non-coincidences are caused by a quite delicate interplay between repulsion and attraction forces.

Similar considerations can be applied to the ν_1 vibrations of benzene-h_6 and benzene-d_6. As $\Delta\omega$ for benzene-h_6 we use the non-coincidence value obtained in our experiments, and take similar $\Delta\omega$ for benzene-d_6 from Ref. [98]:

$$0.26 = 10.95 \cdot 10^{-3} K_A + 8.411 \cdot 10^{-3} K_R,$$
$$0.50 = 9.742 \cdot 10^{-3} K_A + 8.857 \cdot 10^{-3} K_R.$$

This system can be satisfied with K_A=–128.6 and K_R=198.5, which are quite close compared to those obtained for benzene-h_6. Respective values of the attraction and repulsion contributions to the shift of each vibration studied are:

$$\Delta\omega_{A,1H} = -1.41 \text{ and } \Delta\omega_{R,1H} = +1.67,$$
$$\Delta\omega_{A,1D} = -1.25 \text{ and } \Delta\omega_{R,1D} = +1.75.$$

This means that for the ν_2 vibration of the benzene molecule, resonance attraction is stronger in benzene-h_6, whereas resonance repulsion is more significant in benzene-d_6.

One can make a rough guess regarding the K_A and K_R values for the ν_2 vibration of benzene-h_6 and benzene-d_6. Such calculations are inexact since the value of $\Delta\omega$=–0.4 cm^{-1} for benzene-d_6 should be taken from the *non-linear* concentration dependence obtained in isotopic dilution studies [99]. The system of equations in this case,

$$-0.60 = 58.93 \cdot 10^{-3} K_A + 32.65 \cdot 10^{-3} K_R,$$
$$-0.40 = 5.829 \cdot 10^{-3} K_A + 2.180 \cdot 10^{-3} K_R,$$

can be satisfied with K_A=–188.2 and K_R=312.3, giving

$$\Delta\omega_{A,2H} = -11.09 \text{ and } \Delta\omega_{R,2H} = +10.49,$$
$$\Delta\omega_{A,2D} = -1.10 \text{ and } \Delta\omega_{R,2D} = +0.70.$$

In spite of the fact that the dispersion of the K_A and K_R values is too great, the mean K_A=–140 and K_R=226 perfectly reproduce relations existing between non-coincidences of two totally symmetric vibrational modes in benzene-h_6 and benzene-d_6 (Table 9). The calculated $\Delta\omega$ values and experimental ones are related as $\Delta\omega_{calc.} = (1.30 \pm 0.15)\Delta\omega_{expt.}$.

TABLE 9. Experimental and calculated non-coincidences in liquid benzene.

Vibration	$\Delta\omega$ / cm^{-1}		$\Delta\omega_A$ / cm^{-1}	$\Delta\omega_R$ / cm^{-1}
	expt.	calc.		
C_6H_6, ν_1 (A_{1g})	0.26	0.43	–1.53	1.90
C_6H_6, ν_2 (A_{1g})	–0.60	–0.88	–8.25	7.37
C_6D_6, ν_1 (A_{1g})	0.50	0.64	–1.36	2.00
C_6D_6, ν_2 (A_{1g})	–0.40	–0.33	–0.82	0.49

7. CONCLUSIONS

Studies of dynamics by means of line profile analysis have a lot to offer in in-depth understanding of the fundamental processes occurring in liquids and amorphous solids. One of the aims of this chapter is to demonstrate how an up-to-date treatment of a composite spectrum registered in a few minutes may help in an in-depth characterization of the structure and dynamics of the liquid studied. I trust that an approach presented here will give rise to a revival of the spectroscopic community's interest on the investigations of dynamics of complex liquids and solutions by means of vibrational line profile analyses, and will establish this method as one of the routinous methods of the spectroscopic experiment.

The spectroscopic investigation of molecular interactions in liquids using analysis of spectral shifts is an extremely complex problem. It is notoriously difficult to separate individual contributions to the spectral shift caused by different parts of the interparticle potential. In this chapter I tried to demonstrate that such problem could be solved, at least, in the case of the non-coincidence effects caused by attraction and repulsion. The resonance energy transfer in the repulsive potential region, by definition, results in the negative (anomalous) non-coincidences. On the other hand, the resonance energy transfer arising due to the transition dipole–dipole (and induced dipole–induced dipole) interactions causes mainly positive (normal) non-coincidences. Due to this fact, it appears possible to make a division of the contributions of different signs to the overall vibrational line splitting. I hope that a method proposed in this chapter will be helpful in spectroscopic studies of partially ordered liquids demonstrating non-coincidence effects of different signs.

It is worth mentioning as a concluding remark about recent startling advances in molecular dynamics simulations of time-correlation functions and vibrational spectra, especially, in the case of interaction-induced phenomena. It has been already noticed that an unusual line profile and time-correlation function of vibrational dephasing (Burshtein-Fedorenko-Pusep modulation scheme) of the ν_1 line in the vibrational spectrum of liquid acetonitrile [50,51] are reproducible in computational experiments [52]. At the present, the use of powerful computation facilities enables one to employ more realistic interaction potentials and perfectly model experimental time-correlation functions and vibrational line profiles (see, e.g., Refs [103,104]). Successful applications of molecular dynamics simulations to the study of the non-coincidence effects arising in ordinary [105] and interaction-induced [106] spectra seem very promising for deeper understanding the role of different types of interactions in these phenomena.

8. ACKNOWLEDGEMENTS

This work contains results achieved under the auspice of NATO [Science for Stability Program, *GR-Polyblend-SfS* Project, Outreach Dimension, 1996-1999, and Collaborative Linkage Grant *CLG.977358*, 2001-2003, in cooperation with Institute of Chemical Engineering and High-Temperature Chemical Processes (ICE/HT), Patras, Greece, as well as my NATO Senior Research Fellowship at ICE/HT, 2000]. Fruitful collaboration with Greek colleagues, especially, Prof. G. N. Papatheodorou, Dr. G. A. Voyiatzis, and Dr. S. N. Yannopoulos, is gratefully acknowledged. I am indebted to Prof. J. Samios for drawing my attention to the latest papers on interaction-induced phenomena and valuable comments.

REFERENCES

1. Bakhshiev, N.G. (1972) Spectroscopy of Intermolecular Interactions, Nauka, Leningrad.
2. Rothschild, W.G. (1984) Dynamics of Molecular Liquids, Wiley, New York.
3. Wang, C.H. (1985) Spectroscopy of Condensed Media. Dynamics of Molecular Interactions, Academic, Orlando.
4. Burshtein, A.I. and Temkin, S.I. (1994) Spectroscopy of Molecular Rotation in Gases and Liquids, Cambridge University Press, Cambridge.
5. Gordon, R.G. (1968) Correlation functions for molecular motion, Adv. Magn. Res. **3**, 1-42.
6. Oxtoby, D.W. (1979) Dephasing of molecular vibrations in liquids, Adv. Chem. Phys. **40**, 1-48.
7. Oxtoby, D.W. (1979) Vibrational relaxation in liquids, Annu. Rev. Chem. Phys. **32**, 77-101.
8. Ivanov, E.N. and Valiev, K.A. (1973) Rotational Brownian motion, Usp. Fiz. Nauk **109**, 31-64.
9. Keller, B. and Kneubühl, F. (1972) Experimental angular correlation functions of molecules in liquids and in crystals, Helv. Phys. Acta, **45**, 1127-1164.
10. Birnbaum, G., Guillot, B., and Bratos, S. (1982) Theory of collision-induced line shapes - Absorption and light scattering at low densities, Adv. Chem. Phys. **51**, 49-112.
11. Borysow, A. and Frommhold, L. (1989) Collision-induced light scattering: a bibliography, Adv. Chem. Phys. **75**, 439-505.
12. Collision- and Interaction-Induced Spectroscopy (1995) NATO ASI Series C: Mathematical and Physical Sciences, Vol 452, Tabisz, G.C. and Neuman, M.N. (eds.), Kluwer, Dodrecht.
13. Jarwood, J. (1979) Spectroscopic studies of intermolecular forces in dense phases, Annu. Rep. Progr. Chem., Sect. C **76**, 99-130; (1982) ibid. **79**, 157-197; (1987) ibid. **84**, 155-199; (1990) ibid. **87**, 75-118.
14. Faurskov Nielsen, O. (1993) Low-frequency spectroscopic studies of interactions in liquids, Annu. Rep. Progr. Chem., Sect. C **90**, 3-44.
15. Faurskov Nielsen, O. (1996) Low-frequency spectroscopic studies and intermolecular energy transfer in liquids Annu. Rep. Progr. Chem., Sect. C **93**, 57-99.
16. Kirillov, S.A. (1998) Interactions and picosecond dynamics in molten salts: a review with comparison to molecular liquids, J. Mol. Liq. **76**, 35-95.
17. Morresi, A., Paolantoni, M., and Sassi, P. (2000) The non-coincidence effect: a brief review of the structural and dynamical properties in liquid phase, Recent Res. Dev. Chem. Phys. **1**, 67-87.
18. Bartoli, F.J. and Litovitz, T.A. (1972) Analysis of orientational broadening of Raman lines, J. Chem. Phys. **56**, 404-412.
19. Bartoli, F.J. and Litovitz, T.A. (1972) Raman scattering: Orientational motion in liquids, J. Chem. Phys. **56**, 413-425.
20. Boldeskul, A.E., Esman, S.S., and Pogorelov, V.E. (1974) Vibrational and rotational relaxation of molecules in several liquids on the basis of Raman spectra, Opt. Spectr. **37**, 912-918.
21. Rothschild, W.G. (1976) Motional characteristics of large molecules from their Raman and infrared contours: vibrational dephasing, J. Chem. Phys. **65**, 455-462.
22. Kubo, R. (1962) A stochastic theory of line-shape and relaxation, In Fluctuations, Relaxation and Resonance in Magnetic Systems, Scottish Universities' Summer School 1961, ter Haar, G. (ed.), Oliver and Boyd, Edinburgh, p. 23-68.
23. Constant, M. and Fauquembergue, R. (1973) Raman scattering. I. Vibrational correlation in methyl iodide, J. Chem. Phys. **58**, 4030-4033.
24. Tanabe, K. and Jonas, J. (1977) Raman study of vibrational relaxation in liquid benzene-d_6 at high pressure, J. Chem. Phys. **67**, 4222-4228.

25. Ricci, M., Bartolini, P., Chelli, R., Gardini, G., Califano, S., and Righini, R. (2001) The fast dynamics of benzene in the liquid phase. Part 1. Optical Kerr effect experimental investigation, Phys. Chem. Chem. Phys. **3**, 2795-2802.
26. Seshadri K.S. and Jones, R.N (1963) The shapes and intensities of infrared absorption bands, Spectrochim. Acta **19**, 1013-1085.
27. Rothschild, W.G., Perrot, M., and Guillaume, F. (1986) Vibrational dephasing under fractional ("stretched") exponential modulation, Chem. Phys. Lett. **128**, 591-594.
28. Rothschild, W.G., Perrot, M., and Guillaume, F. (1987) On the vibrational T_2 processes in partially ordered systems, J. Chem. Phys. **87**, 7293-7299.
29. Kirillov, S.A. (1992) Fitting the stretched exponential model to experimental time correlation functions of vibrational dephasing, Chem. Phys. Lett. **200**, 205-208.
30. Burshtein, A.I. (1981) Motion-broadened and motionally narrowed spectra, Chem. Phys. Lett. **83**, 335-340.
31. Burshtein, A. I., Fedorenko, S. G., and Pusep, A. Yu. (1983) The lineshape of motion-averaged isotropic Raman spectra, Chem. Phys. Lett. **100**, 155-158.
32. Fedorenko, S.G., Pusep, A.Yu., and Burshtein, A.I. (1987) The transformation of inhomogeneously broadened spectra due to frequency migration, Spectrochim. Acta A **43**, 483-488.
33. Kirillov, S.A. and Kolomiyets, T.M. (1992) Repulsion forces in vibrational spectroscopy - II. Mutual effect of repulsion and attraction forces on the isotropic components of Raman line contours, with special reference to ionic systems, Spectrochim. Acta A **48**, 867-871.
34. Kirillov, S.A. (1993) Markovian frequency modulation in liquids. Analytical description and comparison with the stretched exponential approach, Chem. Phys. Lett. **202**, 459-463.
35. Kirillov, S.A. and Musiyenko, I.S. (1997) Estimation of the parameters of vibrational frequency modulation using generalized vibrational dephasing time correlation function, Khim. Fiz. (Russ.) **16**, 30-34 [Chem. Phys. Reports **16**, 1961-1966].
36. Kirillov, S.A. (1999) Time-correlation functions from band-shape fits without Fourier transform, Chem. Phys. Lett. **303**, 37-42.
37. Egelstaff, P.A. and Schofield, P. (1962) On the evaluation of the thermal neutron scattering law, Nucl. Sci. Eng. **12**, 260-270.
38. Bunten, R.A.J., McGreevy, R.L., Mitchell, E.W.J., Raptis, C., and Walker, P.J. (1984) Collective modes in molten alkaline-earth chlorides: I. Light Scattering, J. Phys. C: Solid State Phys. **17**, 4705-4724.
39. Bunten, R.A.J., McGreevy, R.L., Mitchell, E.W.J., and Raptis, C. (1986) Collective modes in molten alkaline-earth bromides. Light Scattering, J. Phys. C: Solid State Phys. **19**, 2925-2934.
40. Schofield, P. (1962) Neutron scattering and correlations in liquids, In Fluctuations, Relaxation and Resonance in Magnetic Systems, Scottish Universities' Summer School 1961, ter Haar, G. (ed.), Oliver and Boyd, Edinburgh, p. 207-217.
41. Tanabe, K. and Hiraishi J. (1980) Truncation effect on the second moment of vibrational band, Spectrochim. Acta A 36, 828-838.
42. Kirillov, S.A. (1999) Spatial disorder and low-frequency Raman pattern of amorphous solid, with special reference to quasi-elastic scattering and its relation to Boson peak, J. Mol. Struct. **479**, 279-284.
43. Kirillov, S.A., Perova, T.S., Faurskov Nielsen, O., Praestgaard, E., Rasmussen, U., Kolomiyets, T.M., Voyiatzis, G.A., and Anastasiadis, S.H. (1999) Fitting the low-frequency Raman spectra to Boson peak models: Glycerol, triacetin and polystyrene, J. Mol. Struct. **479**, 271-277.
44. Kirillov, S.A. and Kolomiyets, T.M. (2001) Disorder in polymer blends studied by low-frequency Raman spectroscopy, J. Phys. Chem. B **105**, 3168-3173.

45. Kirillov, S.A. and Yannopoulos, S.N. (2000) Charge-current contribution to low-frequency Raman scattering from glass-forming ionic liquids, Phys. Rev. B **61**, 11391-11399.
46. Kirillov, S.A. and Faurskov Nielsen, O. (2000) Boson peak in the low-frequency Raman spectra of ordinary liquids, J. Mol. Struct. **526**, 317-321.
47. Kirillov, S.A. (2002) Dephasing of the ν_1 vibration of isotopic molecules of carbon tetrachloride, J. Raman Spectr. **33**, 155-159.
48. Kirillov, S.A., Voyiatzis, G.A., Musiyenko, I.S., Photiadis, G.M., and Pavlatou, E.A. (2001) Ionic interactions in molten halides from vibrational dephasing, J. Chem. Phys. **114**, 3683-3691.
49. Kirillov, S.A. (1995) Purely discrete Markovian frequency modulation in molten alkali perchlorates, J. Mol. Struct. **349**, 21-25.
50. Schroeder, J., Schiemann, V.H., Sharko, P.T., and Jonas, J. (1977) Raman study of vibrational dephasing in liquid CH_3CN and CD_3CN, J. Chem. Phys. **66**, 3215-3226.
51. Nikiel, L., Hopkins, B., and Zerda, T.W. (1990) Rotational and vibrational relaxation of small molecules in porous silica gels, J. Phys. Chem. **94**, 7458-7464.
52. Westlund, P.-O. and Lynden-Bell, R.M. (1989) Separation of vibrational dephasing and reorientational contributions to the infrared and Raman lineshapes in a simulation of MeCN, Chem. Phys. Lett. **154**, 67-70.
53. Kirillov, S.A., Pavlatou, E.A., and Papatheodorou, G.N. (2002) Instantaneous collision complexes in molten alkali halides: Picosecond dynamics from low-frequency Raman data, J. Chem. Phys. **116**, 9341-9351.
54. Kirillov, S.A. (2003) Vibrational spectra of fused salts and dynamic criterion of complex formation in ionic liquids, J. Mol. Struct. accepted.
55. Kirillov, S.A. and Yannopoulos, S.N. (2002) Vibrational dynamics as an indicator of short-time interactions in glass-forming liquids and their possible relation to cooperativity, J. Chem. Phys. **117**, 1220-1230.
56. Kalampounias, A.G., Kirillov, S.A., Steffen, W., and Yannopoulos, S.N. (2003) Raman spectra and microscopic dynamics of bulk and confined salol, J. Mol. Struct. accepted.
57. Kirillov, S.A., Voyiatzis, G.A., Andrikopoulos, K.S., and Yannopoulos, S.N. (2002) Interactions and picosecond dynamics in liquid benzene from Raman line profile analysis, in preparation.
58. Buchheim, W. (1935) Beeinflussung des Ramaneffektes von Flüssigkeiten durch zwischenmoleculare Wirkungen, Physik. Z. **36**, 694-711.
59. Wolkenstein, M.V. (1937) Raman-effect and intermolecular interactions, Usp. Fiz. Nauk **18**, 153-202.
60. Frenkel, Ya.A., (1946) Kinetic Theory of Liquids, Oxford University Press, Oxford, Ch. VIII, § 1.
61. Buckingham, A.D. (1958) Solvent effect in IR spectroscopy, Proc. Roy. Soc. A **248**, 169-183.
62. Buckingham, A.D. (1960) Solvent effect in vibrational spectroscopy, Trans. Faraday Soc. **56**, 753-760.
63. Sechkarev, A.V. (1965) On the possible reason of line shift and broadening in vibrational spectra of polar organic substances without hydrogen bonding, Opt. Spektr. **19**, 721-730.
64. Bratos, S., Rios, J., and Guissani, Y. (1970) Infrared study of liquids. I. The theory of the IR spectra of diatomic molecules in inert solutions, J. Chem. Phys. **52**, 439-453.
65. Perrot, M., Turrell, G., and Huong, P.V. (1970) Vibrational anharmonicity of HCl in solution, J. Mol. Spectr. **34**, 47-52.
66. Rossi, I., Brodbeck, C., Bouanich, J.P., and Nguyen-van-Than (1975) Etude theorique du displacement de frequence des molecules diatomique en solution liquide.

Application a la frequence fondamentale des molecules HF, HCl, DCl, HBr, HJ, CO et NO en solution CCl_4, Spectrochim. Acta A **31**, 433-444.
67. Kolomiytsova, T.D., Melikova, S.M., and Shchepkin, D.N. (1975) Effect of intermolecular interactions on frequencies and anharmonicity constants, Opt. Spektr. **39**, 602-604.
68. Martin, R., Quinard, J., Pahin, J.-P., and de Gasquet, B. (1978) Diffusion Raman et Brillouin das nitrates monovalents (Li, Na, K, Rb, Cs, Tl, Ag), Rev. Chim. Minér. **15**, 79-92.
69. Bulanin, M.O., Kolomiytsova, T.D., and Shchepkin, D.N. (1976) On the effect of intermolecular interactions on vibrational spectra of molecules, Opt. Spektr. **41**, 201-213.
70. Kirillov, S.A. (1992) Repulsion forces in vibrational spectroscopy – I. Spectral shifts in vibrational spectra of condensed media caused by repulsion forces, Spectrochim. Acta A **48**, 861-866.
71. Fini, G., Mirone, P. and Fortunato, B. (1973) Evidence of short-range orientation effects in dipolar aprotic liquids from vibrational spectroscopy. Part I. - Ethylene and propylene carbonates, J. Chem. Soc. Faraday Trans. 2 **69**, 1243-1248.
72. Davydov, A.S. (1962) Theory of Molecular Excitons, Dower, London.
73. Döge, G. Moleculare Schwingungsexcitonen in Flüssigkeiten, Z. Naturforsch. A **28**, 919-932.
74. Korsunskii, V.I., Lavrik, N.L., and Naberukhin Yu.I. (1976) Different positions of isotropic and anisotropic components of Raman scattering in liquids as evidence of intermolecular vibrational coupling, Opt. Spektr. **41**, 794-798.
75. Wang, C.H. and McHale, J. (1980) Vibrational resonance coupling and the noncoincidence effect of the isotropic and anisotropic Raman spectral components in orientationally anisometric molecular liquids, J. Chem. Phys. **72**, 4039-4044.
76. McHale, J.L. (1981) The influence of angular dependent intermolecular forces on vibrational spectra of solution phase molecules, J. Chem. Phys. **75**, 30-35.
77. Logan, D.E., (1986) On the isotropic Raman spectra of isotopic binary mixtures, Mol. Phys. **58**, 97-129.
78. Logan, D.E., (1986) The non-coincidence effect in the Raman spectra of polar liquids, Chem. Phys. **103**, 215-225.
79. Zerda, T.W., Thomas, H.D., Bradley, M., and Jonas, J. (1987) High-pressure isotropic band widths and frequency shifts of the C-H and C-O modes of liquid methanol, J. Chem. Phys. **86**, 3219-3224.
80. Thomas, H.D. and Jonas, J. (1989) Hydrogen bonding and the Raman noncoincidence effect, J. Chem. Phys. **90**, 4632-4633.
81. Hobza, P. and Havlas, Z. (1999) The fluoroform…ethylene oxide complex exhibits a C–H…O anti-hydrogen bond, Chem. Phys. Lett. **303**, 447–452.
82. Hermansson, K. (2002) Blue-shifting hydrogen bonds, J. Phys. Chem. A **106**, 4695-4702.
83. Valiev, K.A. (1961) On the theory of dissipation of the energy of molecular vibrations energy in liquids, Zh. Eksp. Teor. Fiz. **40**, 1832-1837.
84. Tokuhiro, T. and Rothschild, W.G. (1975) Resonance vibrational energy transfer in the repulsive potential region, J. Chem. Phys. **62**, 2150-2154.
85. Oxtoby, D.W. (1979) Hydrodynamic theory of vibrational dephasing in liquids, J. Chem. Phys. **70**, 2605-2610.
86. Sarka, K. (1980) Contribution of interatomic repulsion forces to the broadening of vibrational bands of planar XY_3 molecules, Chem. Zvesti **34**, 721-725.
87. Sarka, K. and Kirillov, S.A. (1980) Line broadening in vibrational spectra of liquids caused by repulsion forces, Ukr. Fiz. Zhurn. **25**, 93-99.
88. Harmon, J.F. and Müller, B.H. (1969) Nuclear spin relaxation by translational diffusion in liquid ethane, Phys. Rev. **182**, 400-410.

89. Brooker, M.H. and Papateodorou, G.N. (1983) Vibrational spectroscopy of molten salts and related glasses and vapors, Adv. Molten Salt Chem. **5**, 26-184.
90. Lazarev, A.N., Mirgorodsky, A.P., and Ignatiev, I.S. (1985) Vibrational Spectra and Dynamics of Ionic-Covalent Crystals, Nauka, Leningrad.
91. Lazarev, A.N., Mirgorodsky, A.P., and Ignatiev, I.S. (1986) Vibrational spectra and dynamical properties of ionic-covalent crystals, Solid State Commun. **58**, 371-377.
92. Tsiashchenko, Yu.P, Krasnianskii, G.E., and Verlan, E.M. (1978) Intermolecular interactions and Davydov splitting in vibrational spectra of ionic-covalent crystals, Fiz. Tverd. Tela **20**, 864-870.
93. Tsiashchenko, Yu.P and Krasnianskii, G.E. (1979) Contribution of short-range interactions to the Davydov splitting of vibrational levels in Scheelite-type crystals, Opt. Spektr. **47**, 911-916.
94. Kirillov, S.A. and Voronin, B.M. (1974) Activation energies of orientational relaxation and conductivity in some molten nitrate-chloride mixtures, Teor. Eksp. Khim. **10**, 390-392.
95. Korniakova, I.D., Khokhlov, V.A., Khaimenov, A.P., and Kochedykov, V.A. (1993) Micro- and macrodynamics properties of carbonate-ion in molten LiCl-Li_2CO_3 system, Rasplavy, # 5, 35-41.
96. Kirillov, S.A. (1993) Repulsion forces in vibrational spectroscopy – III. Anomalous frequency non-coincidence effect in strongly interacting liquids, with special reference to molten salts, J. Raman Spectr. **24**, 167-172.
97. Meinander, N., Strube, M.M., Johnson, A.N., and Laane, J. (1987) Evidence for resonance intermolecular coupling in liquid benzene and pyridine from Raman difference spectroscopy of isotopic mixtures, J. Chem. Phys. **86**, 4762-4767.
98. Döge, G., Schneider, D., and Morresi, A. (1993) The negative non-coincidence effect of ring vibrations, Mol. Phys. **80**, 525-531.
99. Kamogawa, K. and Kitagawa, T. (1990) A new device for Raman difference spectroscopy and its application to observe frequency shifts due to isotope mixing, J. Phys. Chem. **94**, 3916-3921.
100. Morresi, A., Paolantoni, M., Sassi, P., Cataliotti, R.S., and Paliani, G. (2000) Non-coincidence effect of aromatic ring vibrations, J. Phys. Condens. Matter **12**, 3631-3637.
101. Kirillov, S.A. (2003) Separation of non-coincidences caused by attraction and repulsion in molecular liquids, in preparation.
102. Fernández-Sánchez, J.M. and Montero, S. (1989) Gas phase Raman scattering cross sections of benzene and perdeuterated benzene, J. Chem. Phys. **90**, 2909-2914.
103. Hatzis, G. and Samios, J. (2001) Estimation of the interaction-induced effects on the far-infrared and infrared correlation functions of HCl dissolved in CCl_4: a molecular dynamics study, J. Phys. Chem. A **105**, 9522-9527.
104. Medina, A., Roco, J.M.M., Calvo Hernández, A., and Velasco, S. (2003) Multipole-induced dipole contributions to the FIR spectra of diatomic in non-polar solvents, this volume.
105. Torii, H. (2003) Intermolecular vibrational interaction and its manifestation in vibrational spectra, this volume.
106. Ribeiro, M.C.C., Wilson, M., and Madden, P. (1999) Raman scattering in the network liquid $ZnCl_2$: relationship to the vibrational density of states, J. Chem. Phys. **110**, 4803-4811.

PICOSECOND DYNAMIC PROCESSES OF MOLECULAR LIQUIDS IN CONFINED SPACES – A REVIEW OF RESULTS IN POROUS GLASSES

JACK YARWOOD

Sheffield Hallam University, Materials Research Institute, City Campus, Howard Street, Sheffield S1 1WB, UK

Abstract: Techniques for the measurement of psec dynamic processes of liquids confined in silica-gel pores are reviewed in the context of 'non-wetting' (CS_2, $CHCl_3$) and wetting (CH_3CN,H_2O) liquids. The data presented show how 'physical' (geometry driven) and 'chemical' (interaction driven) confinement may be distinguished, i.e., how modifications to 'host' or guest chemistry may be used to elucidate the dominating thermodynamic drivers. It is shown how real time decay and band shape measurements, or light and neutron scattering techniques, can be used in a complementary fashion to provide fundamental information of importance for a variety of industrially important processes

1. INTRODUCTION AND BACKGROUND

Nanoconfined liquids are well recognised as important 'states' of a fluid in a number of important areas of technology. These include

(a) *heterogeneous catalysis (largely in the gas phase)[1-5] using a variety of porous materials including clays[6], zeolites[1,2,5] and vesicles[6-8],*
(b) *gas/liquid or gas/gas separation[9-11] using polymeric or composite membranes,*
(c) *Enhanced oil recovery[12,13] from clay-lined porous rock,*
(d) *enzyme behaviour via the protein/water nano-interface[14,15].*

It is thus[16,17] of considerable interest and importance to characterise such materials in the context of either host lattice structure or guest behaviour (or both). This paper presents a (very selective) review of the behaviour of nanoconfined liquids in one class of porous material; that of nanopore tailored silica-based sol-gel glass[18-20]. Such glasses have high pore surface area (often with fractal-like interconnectivity[21]) and a reasonably narrow (±10%) pore size distribution (0.5-100nm). They are usually chemically and mechanically stable and they are optically transparent (making them suitable for optical spectroscopic study). Such 'host' surfaces are produced[19] according to the following scheme which includes hydrolysis of an alkoxide (in the presence of an acid or base catalyst) and the subsequent con-

J. Samios and V.A. Durov (eds.), Novel Approaches to the Structure and Dynamics of Liquids: Experiments, Theories and Simulations, 229–247.

densation to produce a three dimensional network with the typical parameter set shown.

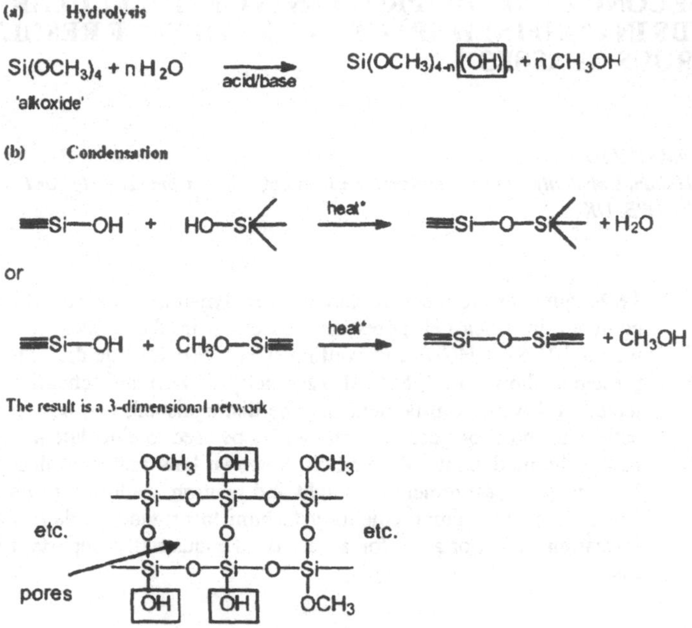

(a) Hydrolysis

$Si(OCH_3)_4 + n H_2O \xrightarrow{acid/base} Si(OCH_3)_{4-n}(OH)_n + n CH_3OH$

'alkoxide'

(b) Condensation

≡Si—OH + HO—Si< $\xrightarrow{heat^*}$ ≡Si—O—Si< + H_2O

or

≡Si—OH + CH_3O—Si≡ $\xrightarrow{heat^*}$ ≡Si—O—Si≡ + CH_3OH

The result is a 3-dimensional network

OCH_3 OH OCH_3

—Si—O—Si—O—Si—

etc. O O O etc.

—Si—O—Si—O—Si—

pores OH OH OCH_3

* (60 °C for 50 hr, dried at 180 °C; fired at 200-800 °C)

Figure 1. Schematic representation of silica glass pores (typical parameters are 40Å pore size, 5×10^{18} pores gm^{-1}, 1 OH group per nm^2, surface area ~450 m^2gm^{-1}).

The guest molecules may interact (by hydrogen bonding) or not! at the silanol, Si-OH groups shown. The nature of the pore surface may subsequently be modified by methylation (a) or chlorination (b) under the appropriate conditions[19]

(a) $\equiv Si — (OH) + CH_3OH \rightarrow \equiv Si - OCH_3 + H_2O$

(b) $\equiv Si — (OH) + CCl_4 \rightarrow \equiv Si - Cl + HCl + COCl_2$

It should be noted that there exist a wide variety of tailored glass-pore materials[22], some of which have better defined pore distribution than those covered here. The choice made for this review was based around the available literature, in particular on the spectroscopy of liquids confined in porous glasses.

2. STRATEGIES AND OVERVIEW

The principal fundamental question which has been asked by spectroscopists, is whether the molecules are confined by physical or chemical 'forces.' In the former case, properties are expected to be 'geometry' driven. In the latter case, the chemical interactions are expected to dominate the dynamic properties. Such a distinction between 'non-wetting' and 'wetting' liquids has been explored using a variety of techniques which measure fast dynamic processes; psec or femtosec laser spectroscopy[23-27], Infrared and Raman [28-31] nmr[32-34] Rayleigh scattering[34-40] and quasielastic neutron scattering[15,39,40,41-43]. The exploration methods used include,

1 Change 'interaction' strength of guest molecule of a similar size (chemical change).

2 Change pore size systematically (at given T) (physical change).

3 Change T systematically for same pore size.

4 Modify the 'host lattice' surface (-OCH_3 or OD or Cl).

All these have been tried over the last 10-15 years.

It is worthwhile reviewing, at this point, the general effect of confinement which have been found by these (and other) techniques. These are shown in table 1. As may be seen there are profound thermodynamic, dynamic and structural effects many of which have now been reasonably well established, especially for small molecules. In the case of polymeric liquids[45], surfactants, the situation is considerably less clear. Some specific results summarised in a broad sense are given in table 2, from where it is clear that there are key pointers to whether (or not) interactions at the surface silanol groups is of significant importance in controlling the dynamic properties.

TABLE 1. Effects of confinement

1 Slowing up of diffusional motion on psec scale
2 Increased micro-viscosity at interface ('super- cooling')
3 Shifting of phase transitions (T_m, T_f and T_g)
4 Stronger interactions at interface (weaker in bulk?)
5 Wider distribution of reorientation relaxation times
6 Changes in 'collective' intermolecular dynamics
7 Number density changes ($n_s < n_b$) (excluded volume effect)

TABLE 2. Behaviour of small molecules in silica based 'sol gel' glasses

Non Wetting Liquids (Physics)	***Wetting Liquids (Chemistry)***
(CS_2, CH_3I, SF_6, $CHCl_3$?)	(CH_3CN, $(CH_3)_2CO$, H_2O, glycols)
(i) No evidence of specific interactions at surface	(i) Evidence of molecular (HB) interaction (to Si-OH groups)
(ii) Clearly a dependence of τ^s_{OR} on pore size (physical trap)	(ii) No change of τ^s_{OR} as a function of 'pore filling' and size (chemical trap)
(iii) Two exponentials usually found (two τ^s_{OR} times) (psec scale) $\tau^s_{OR} > \tau^b_{OR}$	(iii) Observation of an intermediate relaxation time (due to translational motion at interface?)
(iv) No significant change in intermolecular dynamics (short time part of dynamics)	(iv) Significant change in intermolecular dynamics (and change of 'order' via g_2)
(v) Very little change in E_A for reorientation	(v) Small change in E_A reorientation/diffusion (~20%) (and E_A lower for intermediate time process)

3. RESULTS AND DISCUSSION

We present here an overview of the data obtained for selected non-wetting and wetting liquids (table 2). The specific aim is to illustrate the types of data on psec processes which are currently available and to assess the models available for interpretation.

A Carbon disulphide

Raman induced optical Kerr effect measurements (OKE) have been used [23,24,27] to assess the 'surface' as compared with 'bulk' (collective') reorientational relaxation times from measurements made in real time (figure 2). For the 'confined' liquid, the data decay on psec scale with two relaxation times, one of which corresponds to the reorientational dynamics of the bulk Liquid.

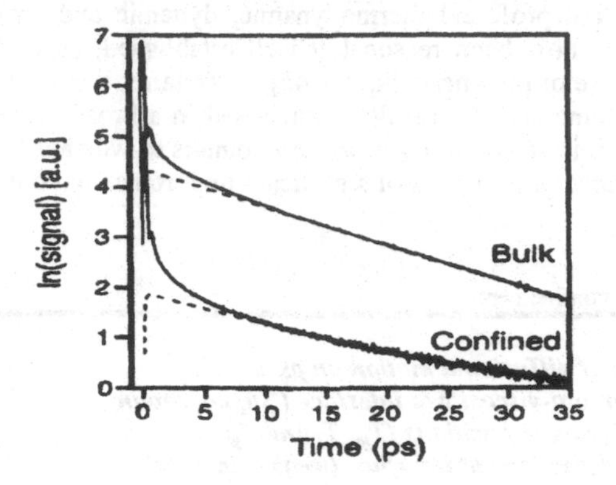

Figure 2 Representative CS_2 data obtained at 165 K in the bulk and confined in pores 24 Å in diameter. The hindered reorientational diffusion is apparent in the data from the confined liquid, especially at long times. Solid lines in the experimental data, and dashed lines are fits. *(Reproduced by permission from J.Phys.Chem. A., 101, p4007, (1997)).*

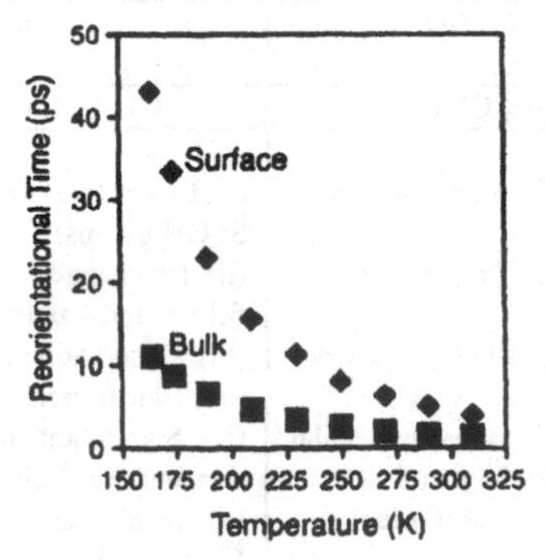

Figure 3 CS_2 reorientational correlation times as a function temperature. Squares are the correlation times in the bulk (and in bulk like liquid in the pores), and diamonds are the times for the surface layer in the pores. *(Reproduced by permission J.Phys.Chem. A., 101, p4007 (1999)).*

The data are summarised as a function of temperature in figure 3 from which the 'hindered' (or slowed) motion is discernable. Furthermore, this dynamic process is pore-sized dependent (figure 4) emphasising the point that CS_2 is geometrically confined. This was confirmed by an analysis of the temperature dependence of the microviscosity

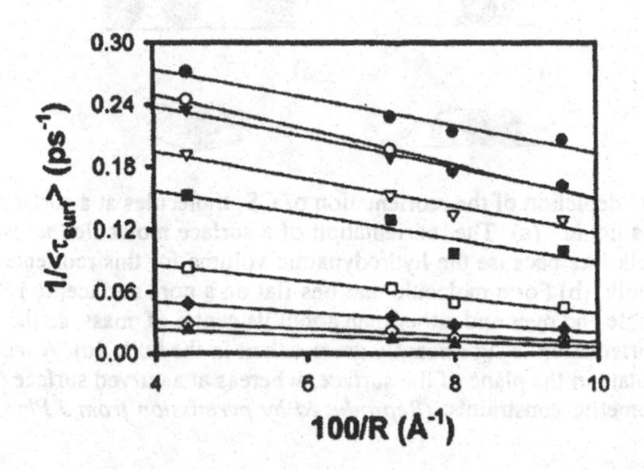

Figure 4 Surface reorientation rate of CS_2 versus pore curvature at, from bottom to top, 165, 174, 194, 271, 254, 272, 290, 293 and 310 K. *(Reproduced bypermission from J.Phys.Chem., B103, 6065, (1999)).*

For the 'surface' population, this was simply estimated from,

$$\eta_s = \frac{\tau^s}{\tau^b} + \eta b \tag{1}$$

assuming that the Stokes-Einstein-Debye (SED) theory is valid for both bulk and surface species. The two activation energies, obtained using the Arrhenius equation were found to be very similar. ($E_A{}^b$ = 5.32 kJ mol^{-1} and $E_A{}^S$ = 6.29 kJ mol^{-1}).

These data were further interpreted by attempting to assess the differences in hydrodynamic volume expected when CS_2 is confined at a flat or curved surface. The scheme used is shown in figure 5. An attempt was made to demonstrate that for a molecule lying on the surface (in particular a curved surface), the hydrodynamic volume should be increased, compared with that in the 'bulk' – with its long axis perpendicular to the surface (figure 5a). Since

$$\tau_{or} = f\eta V_h / kT \text{ by (SED)} \tag{2}$$

the product ηV_h should increase and $\tau_{or}{}^s$ thereby becomes slower (longer). This model is consistent with a confinement generated by geometric constraints with rather slow exchange of surface and bulk species by tumbling action.

Two effects of confinement on τ_s

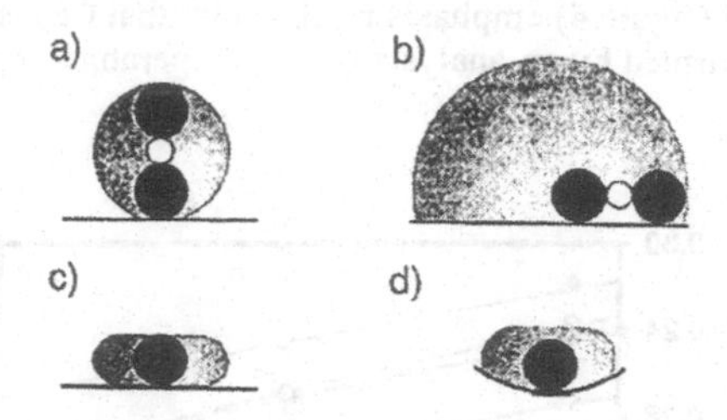

Figure 5 Schematic depiction of the reorientation of CS_2 molecules at a surface that is wetted only weakly by this liquid. (a) The reorientation of a surface molecule that is perpendicular to the surface is bulk-like because the hydrodynamic volume for this reorientation is equivalent to that in the bulk. (b) For a molecule that lies flat on a pore surface, to rotate off of the surface it must tumble end over end rather than about its centre of mass, so the hydrodynamic volume for this reorientation is significantly greater than in the bulk (c) A molecule at a flat surface is free to rotate in the plane of the surface, whereas at a curved surface (d) this motion is inhibited by geometric constraints. *(Reproduced by permission from J.Phys.Chem., B104, 5425 (2000)).*

B Chloroform, $CHCl_3$

A similar situation pertains for chloroform in similar pores (24-86Å)[26]. Two exponentials were again sufficient to describe the OKE data and the τ^s data were found to be weakly pore dependent (at a given temperature), but (within the experimental error) systematic variation is unclear. This may indicate some degree of hydrogen bonding to Si-OH (as for other proton acceptors with chloroform[46]). However, it is clear from the Bose-Einstein corrected Rayleigh wing (Figure 6) that there is little or no change in the intermolecular dynamics (as measured by the very short time part of the collective reorientational CF). This is quite different from the drastic alteration of the 'librational' and other external vibrational modes of the 'bulk' material for' wetting' liquids such as water[15]. To this extent, at least, chloroform is a 'non-interacting' guest molecule. Indeed, the corresponding activation energies for reorientation motion are $\Delta E_A{}^b = 7.3$ kJ mol^{-1} and $\Delta E_A{}^S = 8.1$ kJmol^{-1}; the difference being less than that for CS_2 (the molecule is more globular and less subject to confinement by pore curvature difference?).

It was established, nevertheless, some years ago using Raman spectroscopy[28] that the (single particle) reorientational and vibrational relaxation times both show distinct 'changes' on confinement (figures 7 and 8).

For an isolated Raman band, the correlation functions may be calculated using

$$G_{vib}(t) = \int (I_{VV} - {}^4/_3\, I_{VH}) \exp(-i\omega t) d\omega \,/ \int (I_{VV} - {}^4/_3\, I_{VH}) d\omega$$

$$= \int I_{iso} \exp(-i\omega t)\, d\omega \,/ \int I_{iso}\, d\omega \qquad (3)$$

and

$$G_{rot}(t) = \int I_{VH} \exp(-i\omega t)\, d\omega / G_{vib}(t) \tag{4}$$

where I_{VV} and I_{VH} are the VV and VH polarised components of the band and $I_{iso} = I_{VV} - {}^{4}/_{3} I_{VH}$ is the isotropic component. For Lorenztian band shapes, the corresponding times are given by equations 5 and 6.

$$\tau_{vib} = [2\pi c(\Delta\omega_{1/2})_{iso}]^{-1} \tag{5}$$

$$\tau_{rot}^{-1} = [2\pi c(\Delta\omega_{1/2})_{VH}] - \tau_{vib}^{-1} \tag{6}$$

where $\Delta\omega½$ is the band half width.

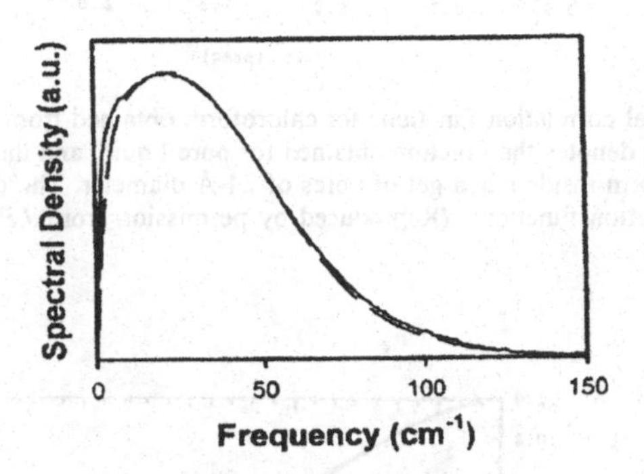

Figure 6 290-K Bose-Einstein-corrected Rayleigh-Wing spectra for chloroform in the bulk (-----) and confined in 24-Å pores (----). The reorientaitonal component has been removed from both spectra The resulting intermolecular spectra are identical to within the experimental accuracy. *(Reproduced by permission from Chem.Phys., 253, p327, Copyright 2000, by permission of Elsevier Science)*

Table 3 shows that, although the rate of reorientation decreases with decreasing pore size (compared with the bulk), the rate of vibrational relaxation is (marginally) faster in the pore. This would normally be regarded as an indicator of stronger or different interactions (presumably at the interface) – and a wider range of molecular environments[47,48] compared with the bulk. So it is not clear whether or not chloroform takes part in hydrogen bonding to Si-OH groups on glass. Such interactions are evidently of relatively minor significance for OKE measurement.

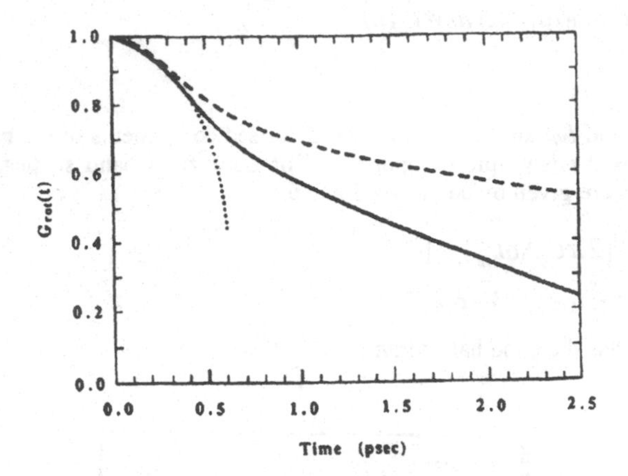

Figure 7 Rotational correlation functions for chloroform obtained from the C-H band analysis. The solid line denotes the function obtained for pure liquid, and the broken line that obtained for chloroform inside silica gel of pores of 24-Å diameter. The dotted line shows the free rotator correlation function. (Reproduced by permission from *J.Phys.Chem., 94, 7460 (1990)).*

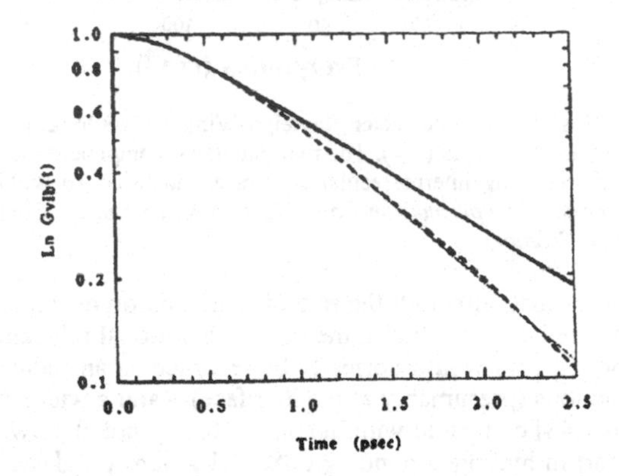

Figure 8. Vibration correlation functions for the C-H band of chloroform in the pure liquid (solid line) and inside pores 24 Å in diameter (broken line). The dotted lines represent the best fits of the experimental functions to the theoretical Kubo function. *(Reproduced by permission from J.Phys.Chem., 94, 7460, (1990)).*

TABLE 3 Rotational (τ_{rot}) and Vibrational, (τ_{vib},) Relaxation Times for Chloroform calculated from the ν_1 Bandwidth for Pure Liquid and inside Porous Glass[28].

Pore diameter of gel, Å	τrot, ps Non-treated gels	gels boiled in methanol	τvib, ps non-treated gels	gels boiled in methanol
20	4.21		0.75	
24	2.63	1.78	0.89	0.96
60	1.4 $\pm$ 0.4	1.90	1.05	0.95
Neat liquid	1.94		1.01	

C Acetonitrile, CH_3CN

Since acetronile is well-known[49] to form hydrogen bonds to proton donors (via the N lone pair), it was found[25] as expected to be a 'wetting' liquid with no dependence of τ^{OR} on pore size. What was not expected, however, was the observation of OKE-decay functions which can only be fitted with three(3) exponentials, as shown in figure 9. The data for the three different relaxation times are given in table 4.

TABLE 4. Decay times (τ) and amplitudes (A) from triexponential fits of OKE data for confined acetonitrile. All temperatures are in °K all times are in ps, and all amplitudes are relative to a bulk amplitude of unity. The uncertainties are approximately $\pm$2% for τ_{bulk}, $\pm$ 5% for τ_2, and $\pm$ 10% for τ_3 [25].

Temperature	τ_{bulk}	τ_2	A_2 (24 Å)	A_2 (44 Å)	τ_3	A_3 (24 Å)	A_3 (44 Å)
229	4.12	8.37	1.86	0.39	66.6	0.22	0.042
254	2.76	6.43	0.80	0.22	50.0	0.14	0.032
272	2.13	6.00	0.45	0.18	41.9	0.077	0.022
290	1.66	4.49	0.38	0.17	26.0	0.082	0.026
309	1.42	3.91	0.29	0.14	19.0	0.076	0.027
327	1.21	3.61	0.28	0.12	16.4	0.069	0.023
344	0.99	3.31	0.18	0.091	14.9	0.044	0.015

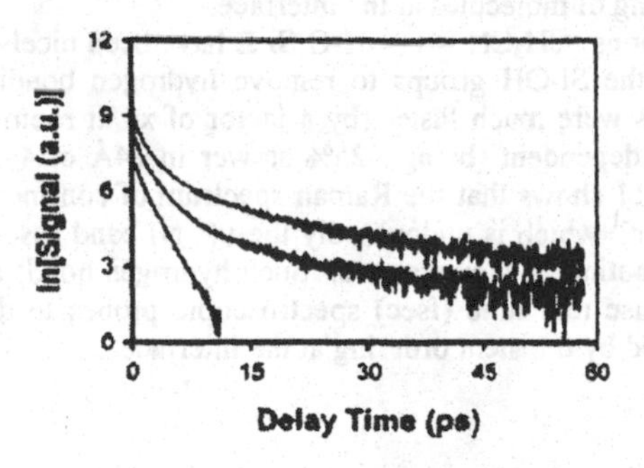

Figure 9. Representative OKE decays for acetonitrile-$d_3$at 290K obtained in the bulk liquid (bottom trace), confined in 44 Å pores (middle trace), and confined in 24 Å pores (top trace). *(Reproduced by permission from J.Chem. Phys., 111, 5117, (1999)).*

In figure 10 it is demonstrated that the activation energy for the intermediate τ_2 relaxation process (broken line) has a smaller activation energy than for the other two (bulk, τ_1) and surface (τ_3) processes.

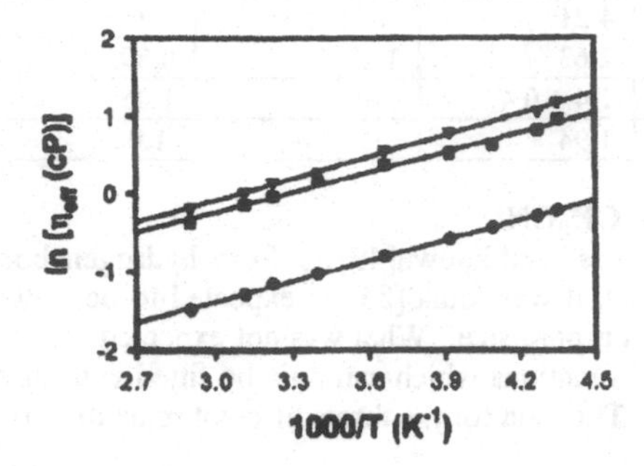

Figure 10 Arrhenius plots of the acetonitrile bulk viscosity (circles) and the effective surface viscosity in 44 Å pores (squares) and 24 Å pores (inverted triangles). *(Reproduced by permission from J.Chem.Phys., 111, 5117, (1999)).*

A model built around the supposed interchange of molecules between 'bulk' and 'surface' by translational diffusion was found to be consistent with the data.

The (collective) reorientational times obtained from OKE were also compared with (single particle) relaxation times obtained from NMR by Jonas et al[30,31]. Since the two (psec) times are related [48; p182] via the Kirkwood static g_2 (and dynamic, j_2) correlation factors, viz:

$$\tau^{OKE}_{Coll} = \left(\frac{g_2}{j_2}\right)\tau^{NMR}_{SP} \tag{7}$$

it is possible (assuming that $j_2 = 1$) to calculate the static orientational order. The values of g_2 1.6 and 2.0 respectively for 'bulk' and surface species demonstrates an increased ordering of molecules at the interface.

The interactions CH_3CN --------H-O Si Ξ have been nicely confirmed; firstly by methylation of the Si-OH groups to remove hydrogen bonding. In that case the OKE, τ^s_{or} values were much faster (by a factor of x4 at room temperature) and became pore-size dependent (being ~25% slower in 24Å cf 44Å pores). But, more directly, figure 11 shows that the Raman spectrum of confined CH_3CN has a 'new' band at 2270 cm^{-1} which is undoubtedly the ν(CN) band associated with molecules of CH_3CN interacting as shown above. Such hydrogen bonds are short lived but are sufficient to cause real time (fsec) spectroscopic probes to detect the polarisation difference caused by transient ordering at the interface.

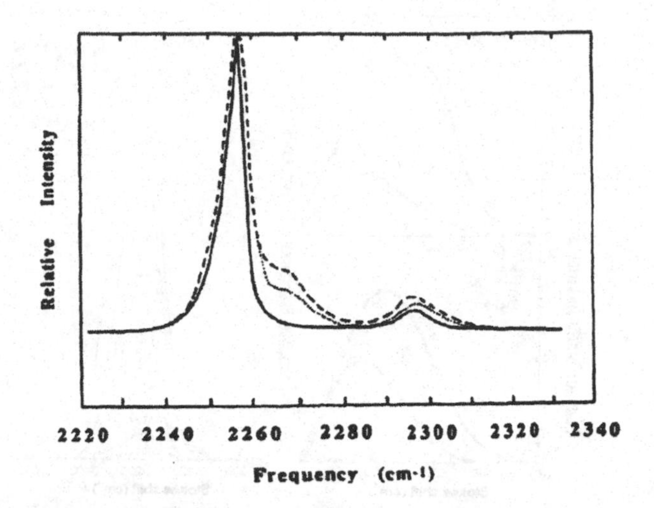

Figure 11 Isotropic component of the C Ξ N band of acetonitrile in pure liquid (solid line), inside the gel of pores 33 Å in diameter (dotted line, and inside pores 24 Å in diameter (broken line). The peak at about 2257 cm^{-1} is mainly due to the C Ξ N stretch while the peak at about 2297 cm^{-1} is mainly due to the ($\nu_3 + \nu_4$) combination. *(Reproduced by permission from J.Phys.Chem., 94, 7461 (1990)).*

D Water, H_2O

Water is of course the classical 'wetting' liquid and predictably the one studied most in confined spaces (although not, as far as I am aware, by OKE or other pulsed laser techniques). The two most important techniques are light scattering (Raman or Rayleigh) or quasielastic neutron scattering.

The ability to use polarised Raman measurements to measure the integrity of the water HB 'network' (considered extensive in the bulk at RT) under different circumstances was recognised by Green et al[50] and was developed by Hare and Sorensen[51] and by Kitano/Maeda[52] and their groups.

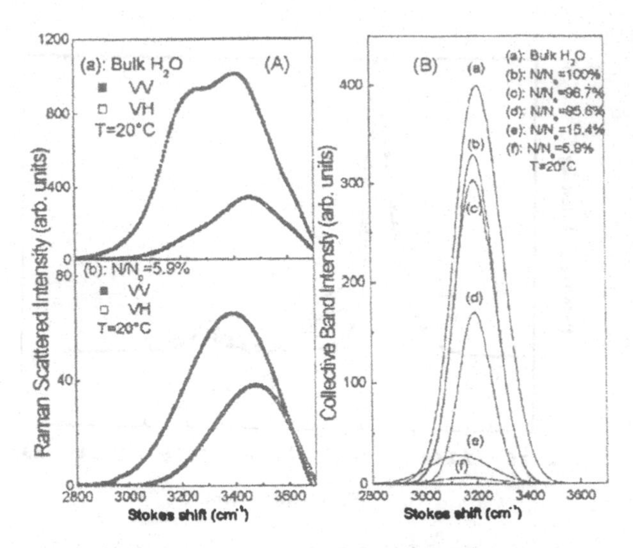

Figure 12.(A) Polarised (VV) and depolarised (VH), OH stretching Raman intensity for water in the bulk (a) and confined (b) state at N/N_o = 5.9%. (B) Collective band for water in the bulk and confined state at different relative water percentages. *(Reproduced by permission from Phys.Chem.Chem.Phys., 4, 2770 (2002)).*

Figure 12 shows a recent example[39] of how this technique has been used to demonstrate the destruction of the water network, measured by the intensity of the 'collective' band (I_c) of water (at ~ 3250 cm^{-1}), as the pore filling was reduced (giving an increasingly confined liquid). The process of spectral 'stripping' to recover the 'collective' spectral density shown in figure 12B enables[52] a calculation of the probability, P_d, that an OH oscillator is excluded from the tetrahedral network of HB water molecules because of an unfavourable interaction or orientation.

$$P_d = \frac{C_{w(T)} - C_{x(T)}}{C_{w(T)}} \tag{8}$$

where C_w and C_x are the collective band (relative) intensities in the bulk and confined liquid respectively (where $c = I_c/I_{vv}$). As may be seen, P_d approaches 1 for $x = N/N_o$ values at about 6%. This is a powerful technique for monitoring the psec dynamic correlations responsible for 'breaking' the (averaged) HB network in water. Kitano et al[52] give some very nice examples for water confined in polymeric networks.

Rayleigh scattering also provides a useful insight on the way in which the psec dynamics of liquid water are affected by confinement. Figure 13 demonstrates that the spectral density changes considerably for pore sizes in the 25Å regime and for filling fractions of ~6%. In figure 13B it is clear that the VDOS dramatically changes with N/N_o. In order to arrive at a quantitative interpretation of the data shown in figure 13A, it is necessary to use a suitable model (or models) for the relaxation processes involved in forming such spectra. The model most used for confined liquids is based around the well-established Kohlrausch, Williams, Watts expression[53] for a distribution of orientational correlation times. In the time domain

$$G^{or}(t) = \exp[-(^{t}/\tau_{kww})^{\beta}] \tag{9}$$

where β is a measure of the so called stretched exponential (β = 1 for Debye-like behaviour for a single correlation time τ).

In the frequency domain – shown in fig 13A – the corresponding function used was that of Havriliak-Negami[54] based on the concept of a 'relaxing cage' of water molecules with a lifetime in the psec regime,

$$HN(\omega) = -1/_{\omega}\ \mathrm{Im}[1 + (i\omega\tau_{HN})^{\alpha}]^{-\gamma} \tag{10}$$

where, again α and γ are shape parameters. The functions of equations (9) and (10) are not exact Fourier Transforms of each other; rather they are semi-dependent analytical forms attempting to express the same phenomenon (i.e. the water network breaking process on a psec time scale driving a profound alteration in spectral distribution). The two functions are, connectable (see equation 2 of ref 39) but it is more common to calculate a mean relaxation time <τ>, viz,

$$<\tau> = \left(\frac{\tau_{kww}}{\beta}\right)\Gamma\left(\frac{1}{\beta}\right) \tag{11}$$

(where Γ is the gamma function) by fitting depolarised Rayleigh spectra such as those in fig 13A to an expression of the form;

$$I_{VH}(\omega)=[\delta\omega+ HN(\omega) + V_g t_1(\omega) + V_g t_2(\omega)]*R(\omega) \tag{12}$$

Where

δ(ω) is the elastic contribution

HN(ω) is given by equation 10

R(ω) is the instrument function

and $V_g t_n$ (n = 1,2) are Voigt functions representing the two resonances (external vibrational modes of the water network) found at 70 cm^{-1} and 180 cm^{-1} in the low frequency collective reorientational spectra of liquid water (see figure 13A(a)). Values of the relevant parameters are shown in table 5 from which it may be seen that (compared with bulk water) the reorientational correlation time approximately doubles and the shape parameters all sharply drop from their bulk values, reflecting the much greater distribution of dynamic 'cages' in the pore-containing material. Equally important is the very profound reduction in the VDOS in the 50-120 cm^{-1} region (fig 13B) – caused by the virtual destruction of the bulk water network at the Si-OH interfaces.

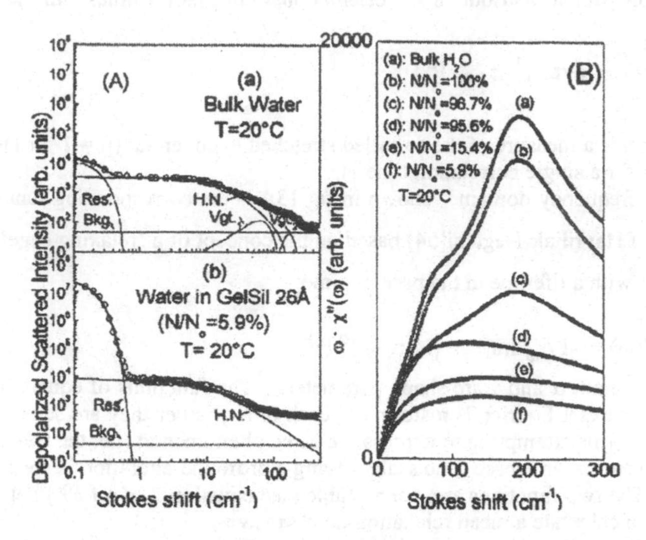

Figure 13 .(A) Rayleigh wing spectra of water in bulk ($\beta = 0.88 + 0.1$) and $<\tau> + 0.30 + 0.05$ ps (a) and confined ($\beta = 0.64 + 0.1$ and $<\tau>$ $+ 0.61 + 0.05$ ps) (b), state together with the total fit and the single components. (B) Raman effective VDOS for water in the bulk and confined states at different hydration levels.*(Reproduced, by permission, from Phys.Chem.Chem.Phys., 4, p2770 (2002))*

TABLE 5. Summary of parameters for bulk and confined water obtained from best fit to Rayleigh Wing[39]

Water	Porefill	τ_{HN}/ps	α	γ	β	$<\tau>$/ps
Bulk	100%	0.31	0.97	0.88	0.88	0.3
Water in Gelsil 25	100% 15.4% 6.0%	0.42 0.53 0.55	0.72 0.93 0.96	0.77 0.65 0.60	0.62 0.66 0.64	0.5 0.6 0.6

Quasi elastic neutron scattering[15, 35-45] provides yet another technique for the detailed study of fast dynamic processes in liquids. Since the glass pores discussed here are completely transparent to neutrons, the method provides an easily practicable approach; with variations, not only in energy (i.e. frequency) range and temperature, but also in momentum space ($^1/Q$) which allows the dynamic processes to be studied over different spatial ranges. Figure 14 shows typical scattering functions obtained for water in Gelsil 25Å pores at 20°C and Q = 0.946Å^{-1} (on intermediate spatial range). Such data have been interpreted via two different models.

(A) The relaxing case model of Havriliak and Negami[54] (equation 10). In this case , the scattering function was modelled by,

$$S_s(Q,\omega) = [HN(\omega)]^{*}\mathrm{Res}(\omega) + bkg(\omega) \tag{13}$$

The resulting fit is seen in figure 14(B) and the resulting parameter set is included in table 6.

(B) The confined diffusion model of Dianoux and Volino[55], equation 14. In this case the corresponding scattering function is, for bulk water,

$$S_s(Q,\omega) = e^{-Q^2\langle u^2\rangle/3}\left\{J_0^2(Qa)\frac{1}{\pi}\frac{\Gamma_T(Q)}{(\Gamma_T(Q))^2+\omega^2}\right.$$
$$\left.+\sum_{l=1}^{\infty}(2l+1)J_l^2(Qa)\frac{1}{\pi}\frac{\Gamma_T(Q)+l(l+1)D_r}{[\Gamma_T(Q)+l(l+1)D_r]^2+\omega^2}\right\}, \tag{14}$$

where J_ℓ are the Spherical Bessel functions, D_r the rotational diffusion coefficient and Γ_t the translational width.

and, for confined water,

$$S_s(Q,\omega) = e^{-Q^2\langle u^2\rangle/3}[A(Q)\delta(\omega) + B(Q)L_{tr}(Q,\omega) + C(Q)L_{rot}(Q,\omega)] \otimes R(Q,\omega) \tag{15}$$

where A(Q) is the Elastic Incoherent Scattering Factor (IESF); a form factor for the confining volume, given by,

$$A(Q) = \frac{3J_1(Qa)}{Qa} \approx \exp\left(-\frac{1}{3}Q^2a^2\right) \tag{16}$$

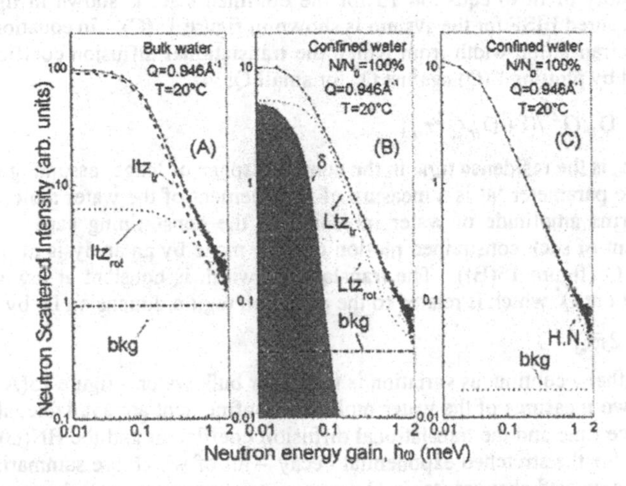

Figure 14. Experimental IQENS spectra of water in bulk (A) and confined (B) state fitted by equation (9) (CDM), together with the single components and the total best-fit. (C) Experimental IQENS spectra of water in confined state fitted by equation (12) (RCM), together with the single components and the total best fit. *(Reproduced by kind permission from Phys.Chem.Chem.Phys., 4, 2771(2002)).*

the single components and the total best fit. *(Reproduced by kind permission from Phys.Chem.Chem.Phys., 4, 2771(2002)).*

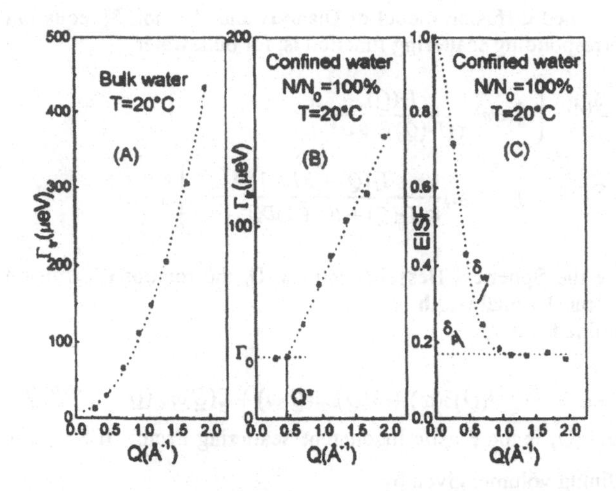

Figure 15. Translational Lorentzian contribution, Γ_{tr}, vs Q for water in bulk (A) and confined (B) states. (C) EISF term for confined water as obtained by CDM; δ_w and δ_p represent the contributions to the EISF coming from the confined water and the glass pores respectively. *(Reproduced by permission from Phys.Chem.Chem.Phys., 4, 2772 (2002)).*

The quality of fit to equation 15 for the confined state is shown in figure 14A and the measured EISF for the system is shown in figure 15(C). In equations 14-16, $\Gamma(Q)$ is the translation width from which the translational diffusion coefficient may be obtained by plotting $\Gamma(Q)$ against Q^2 for small Q; viz,

$$\Gamma_{tr}(Q) = D_{tr}Q^2/[1 + D_{tr}Q^2\tau_o] \tag{17}$$

where τ_o is the residence time in the confined space of 'cage' assuming jump diffusion. The parameter 'a' is a measure of confinement of the water molecules; formally the rms amplitude of water molecule in the constraining cage. A further measurement of such constrained motion may be made by an analysis of the plot of Γ_{tr} against Q (figure 15(B)). The translational width is constant at low Q until a 'break' point at Q^* which is related to the diffusion region dimension D^* by

$$D^* = 2\pi Q^{*-1} \tag{18}$$

(notice that a continuous variation is found for bulk water – figure 15(A)).

Thus, two measures of the water molecule confinement are available, along with the residence time and the translational diffusion coefficient and the HN(ω) function parameters for the stretched exponential decay – all of which are summarised in table 6. The expected changes (table 1) occur on confinement; i.e. a slowing up of the translational diffusion and reorientational processes and a much larger (x3) residence time in the 'cage'. The 'cage' dimension appears to be of the order of 12-15Å at 20°C in a pore of 25Å which means that a significant proportion of the water mole-

TABLE 6. Summary of the parameters for bulk and confined water obtained from QENS[39]

T	D_{tr}	τ	D'/Å	<τ>	β
Confined					
-12°C	0.7	6.2	6.6	0.5	0.62
20°C	1.9	3.3	13.9 (14.3)$^{+}$		
40°C	3.1	2.8	26.9		
Bulk				0.3	0.88
25°C	2.3	1.25	-		

$^{+}$the value of 'a' from EISF analysis (equation 16).

The QENS data may also be employed to examine, via the H/N model, the degree to which confined water behaves like 'super cooled' water with its attendant non-hydrodynamic behaviour, reminiscent of a distribution of relaxation times (β <1) around a kinetic glass transition. In the model, as in MCT[56], the mean relaxation time is

$$<\tau> \approx Q^{-\gamma'}$$

where γ' ≅ 2 for hydrodynamic behaviour, bulk water
and γ' > 2 for 'confined' water.

Figure 16 shows that γ approaches 2 and β approaches 1 only at low Q (i.e. over large spatial ranges) and at higher temperatures, when hydrodynamic behaviour is expected to be recovered.

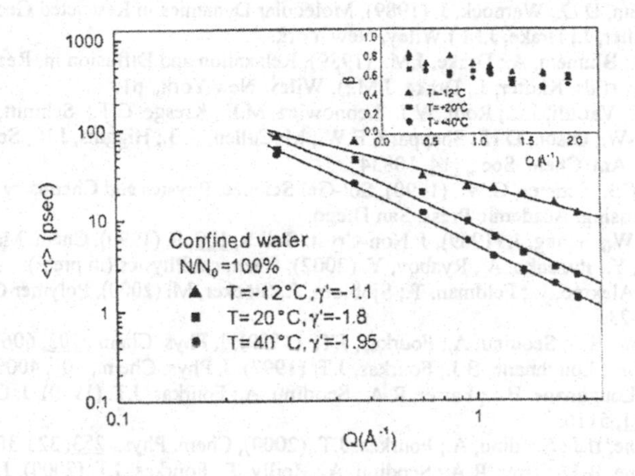

Figure 16. (τ) vs Q and, in the inset, β vs. Q for confined water at T = -12°C, 20°C, 40°C. (Reproduced from Physica A304, p64. Copyright 2002, with permission from Elsevier Science).

Finally, it is pertinent to draw attention to similar work[34,44,45] on glycols (which is outside the scope of this review but which is, nevertheless, important for attempts to separate the effects of physical confinement and Si-OH-guest (HB) interactions.

This is an essential requirement to achieve a better understanding of the behaviour of these interesting and important systems.

REFERENCES

1. Hansenne, C.; Jousse, F.; Leherte, L. Vercauteren (2001), J. Mol. Catalysis A; Chemical., 166, 147-165.
2. Tripathi, A.K.; Sahrasrabudhe, A.; Mitra, R.; Mukhopadhyay, R.; Gupta, M.; Kartha, V.B. (2001), Phys. Chem. Chem. Phys., 3, 4449-55.
3. Corma, A. (1995), Chem. Rev., 95, 559.
4. Corma, A.; Martinez-Soria, A.; Morton, J.B. (1995) J. Catal, 153, 25.
5. Guo, T.; Langley, K.H.; Karasz, F.E.(1994), Phys. Rev. B., 50, 3400.
6. Onori, G.; Santucci, A. (1993), J. Phys. Chem. 97, 5430 and references therein.
7. Giardano, R.; Migliardo, P.; Wanderlingh, U.; Bardez, E. (1995), J. Phys. B., 213, 585.
8. Bardez, E.; Giordano, R.; Jannelli, M.P.; Migliardo, P.; Wanderlingh, U. (1996), J. Mol. Struct., 383, 183.
9. Veith, W.R. Diffusion in and through polymers: principles and applications, (1991), Oxford University Press, New York.
10. Baker, R.W.; Cussler, E.L.; Eykamp, W.; Koros, W.J.; Riley, R.L.; Strathmann, H. (1991), Membrane separation systems: recent development and future directions, Noyes Data Corp., Park Ridge, NJ.
11. Gruger, A.; Regis, A.; Schmatko, T.; Colomban, P. (2001), Vib. Spectrosc., 26, 215-25.
12. Murad, M,A.; Cushman, J.H. (2000) Int. J. Eng. Sci., 38, 517-564 .
13. Smiles, D.E. (2000), Chem. Eng, Sci., 55, 773-781.
14. Petrescu, A-J.; Receveur, V.; Calmettes, P.; Durand, D.; Desmadril M.; Roux, B.; Smith, J.C. (1997), Biophysical J. 72, 335-42.
15. Bellissent-Funel, M.C.; Zanotti, J-M.; Chen, S.H. (1996), Faraday Diss. 103, 281-294; Bellissent-Funel, M.C. (2002), J. Mol. Liquids, 96-97, 287-304 (and references therein).
16. Awschalom, D.D.; Warnock, J. (1989), Molecular Dynamics in Restricted Geometries (Eds. Klafter, J.; Drake, J.M.) Wiley, New York.
17. Klafter, J.; Blumem, A.; Drake, J.M. (1989), Relaxation and Diffusion in Restricted Geometry, (Eds. Klafter, J.; Drake, J.M.;), Wiley, New York, p1.
18. Beck, J.S.; Vartuli, J.C.; Roth, W.J.; Leonowicz, M.E.; Kresge, C.T.; Schmitt, K.D.; Chu, C.T-W.; Olson, D.H.; Sheppard, E.W.; McCullen, S.B.; Higgins, J.B.; Schlenker, J. (1992), J. Am. Chem. Soc., 114, 10834.
19. Brinker, C.J.; Scherer, G.W.; (1990), Sol-Gel Science, Physics and Chemistry of Sol-Gel Processing, Academic Press, San Diego.
20. Zerda, T.W.; Hoang, J. (1989), J. Non-Cryst. Solids, 109, 9, (1990), Chem. Mat., 2, 372.
21. Feldman, Y.; Puzenko, A.; Ryabov, Y. (2002), Chemical Physics (in press).
22. Øye, G.; Alexrod, E.; Feldman, T.; Sjoblom, J., Stöcker, M. (2000), Polymer Coll. Sci., 278, 517-23.
23. Loughnane, B.J.; Scodinu, A.; Fourkas, J.T. (1999), J. Phys. Chem., 103, 6061.
24. Ferrar, R.A., Loughnane, B.J.; Fourkas, J.T. (1997), J. Phys. Chem., 101, 4005.
25. See also Loughnane, B.J.; Farrer, R.A.; Scodinu, A,; Fourkas, J.T. (1999), J. Chem. Phys., 111, 5116.
26. Loughnane, B.J.; Scodinu, A.; Fourkas, J.T. (2000), Chem. Phys., 253, 323-30.
27. Loughnane, B.J.; Farrer, R.A.; Scodinu, A.; Reilly, T.; Fourkas, J.T. (2000), J.Phys. Chem., 104, 5421-29.
28. Nikiel, L.; Hopkins, B.; Zerda, T.W. (1990), J. Phys. Chem., 94, 7458.
29. Mu, R.; Malhotra, V.M. (1991), Phys, Rev., B44, 4602.
30. Yi, J., Jonas, J. (1996), J. Phys. Chem., 100, 16789.
31. Hoang, G.C.; (2002), J. Korean Phys. Soc., 40, 224-31.
32. Liu, G.; Li, Y.; Jonas, J. (1991), J.Chem.Phys., 95, 6892; Zhang, J.; Jonas, J.; (1993), J. Phys. Chem., 97, 8812
33. Korb, J.P.; Malier, L.; Cross, F.; Xu, S.; Jonas, J. (1996), Phys. Rev. Lett., 77, 3212.

34. Asknes, D.W.; Gjerdaker, L.; Allen, S.G.; Booth, H.F.; Strange, J.H. (1998), Magn. Res. Imaging, 16, 579-81, Gjerdaker, L.; Sorland, G.H.; Aksnes, D.W. (1999), Micropor-ous/Mesoporous Mat., 32, 305-310.
35. Crupi, V.; Maisano, G.; Majolino, D.; Migliardo, P.; Venuti, V. (1998), J. Chem. Phys., 109, 7394.
36. Crupi, V.; Majolino, D.; Migliardo, P.; Venuti, V. (1998), Nuovo Cimento 20D, 2163.
37. Crupi, V.; Magazú, S.; Majolino, D.; Maisano, G.; Migliardo, P.; (1999), J. Mol. Liq., 80, 133.
38. Magazù, S.; Maisano, G.; Majolino, D.; Migliardo, P. (1995), Physical Chemistry of Aqueous Systems, White, H.J.; Sengers, J.; Neumann, D.; Bellows, J.; Eds.; Walling-ford: New York, p361.
39. Crupi, V.; Dianoux, A.J.; Majolino, D.; Migliardo, P.; Venuti, V. (2002), Phys. Chem. Chem. Phys., 4, 2768-2773.
40. Crupi, V.; Majolino, D.; Migliardo, P.; Venuti, V. (2002), Physica A, 304, 59-64, ibid 304, 249-52.
41. Mitra, S.; Mukhopadhyuy, R,; Tsukushi, I.; Ikeda, S.; (2001), J. Phys. Condensed Matter, 13, 8455-65.
42. Maisano, G.; Migliardo, P.; Fontana, M.P.; Bellissent-Funel, M. C.; Dianoux, A. J. (1985), J. Phys. C: 18, 1115 and references therein.
43. Zanotti, J.M.; Bellisent-Funel, M.C.; Chen, S. H. (1999), Phys. Rev. E., 59, 3084 and references therein.
44. Crupi, V.; Majolino, D.; Migliardo, P.; Venuti, V. (2000), J. Phys. Chem., 104, 11000-11012.
45. Crupi, V.; Venuti, V.; Majolino, D.; Migliardo, P. (1998), J. Mol. Structure, 482-483, 509-13.
46. Yarwood, J. (Editor), (1973), Spectroscopy and Structure of Molecular Complexes, Ple-num, London, p359.
47. Rothschild, W.G. (1984), Dynamics of Molecular Liquids, Wiley, New York.
48. Steel, D.; and Yarwood, J. (Eds) (1991), Spectroscopy and Relaxation of Molecular Liq-uids, Elsevier, Amsterdam.
49. Yarwood, J. (Editor), (1973), Spectroscopy and Structure of Molecular, Complexes, Plenum, London, p174.
50. Green, J.; Lacey, A.; Sceats, M. (1986), J. Phys. Chem., 90, 395,; ibid 1987, 87, 3603.
51. Hare, D.E.; Sorensen, C.M. (1990), J. Chem. Phys., 93, 25; ibid, (1990), 93, 6954.
52. Maeda, Y,; Ide, M.; Kitano, H. (1999), J. Mol. Liq., 80, 149-163 (and references therein).
53. Alvarez, F.; Alegria, A.; Colmenero, J. (1991) Phys, Rev. B44, 7306, ibid, 1994, 49, 14996.
54. Havriliak, S.; Negami, S. (1967), Polymer, 8, 101.
55. Volino, F.; Dainoux, A.J. (1980), Mol. Phys. 41, 271.
56. Gotze, W.; Sjogren, L. (1992), Rep. Prog. Phys., 55, 241.

PHOTOINDUCED REDOX PROCESSES IN PHTHALOCYANINE DERIVATIVES BY RESONANCE RAMAN SPECTROSCOPY AND TIME RESOLVED TECHNIQUES

H. ABRAMCZYK*, I. SZYMCZYK
Technical University, Department of Chemistry, Institute of Applied Radiation Chemistry, Wroblewskiego 15, 93-590 Lodz, Poland

1. INTRODUCTION

We will present the results on photoinduced redox processes in phthalocyanine derivatives by Resonance Raman Spectroscopy and femtosecond pump-probe absorption measurements.

Photochemical processes are only one of the subjects of our interest. The Laboratory of Laser Molecular Spectroscopy deals with the other topic including

a) vibrational relaxation in liquids, glasses and crystals
b) solvation dynamics of an excess electron
c) vibrational relaxation in H-bonded systems
d) correlation between vibrational dynamics and phase transitions
e) photochemistry

The paper will concentrate on the photochemical and photophysical properties of metal phthalocyanines.

Phthalocyanines: (Fig1) have a structure similar to porphyrins (Fig.2) They consists of four pyrrole units like porphyrins but one important difference between porphyrins and phthalocyanines is the replacements of the bridging metine groups (with carbon) by the azomethine groups (with nitrogen). Additional benzene rings are attached, which are absent in natural porphyrines (for example, in the heme protein - important compound of our blood).

Free phthalocyanine (without the metal inside) can be replaced by a metal cation. The metalophthalocyanine maintains the planarity of the molecule with the symmetry increase from D_{2h} for free phthalocyanine to the D_{4h} to metalophthalocyanine. The symmetry will drop to C_{4v} for metals that do not fit inside the ring. Peripheral substitution at the benzo group has a tremendous influence on chemical and physical properties. Substitution with strongly electron donating or accepting groups perturbs the excited state energies and can lead to the modified redox properties, new chemistry, and change of solubility. For example, most phthalocyanines are not soluble in water, but substitution with hydrophilic groups like the sulfono groups make them soluble in aqueous solutions. We will show the results for copper(II)phthalocyanine-3,4',4'',4'''-tetrasulfonic anion (Fig.1) with copper inside the ring, and the peripheral substitution with the sulfono groups in four positions. This

J. Samios and V.A. Durov (eds.), Novel Approaches to the Structure and Dynamics of Liquids: Experiments, Theories and Simulations, 249–264.

anion is obtained from a tetrasodium salt of the tetrasulfonic acid. After dissociation there are four negative charges at the substituent and four free positive sodium charges (cations). Further through the whole the paper we will use the abbreviation $Cu(tsPc)^{-4}$ for the structure presented in Fig.1.

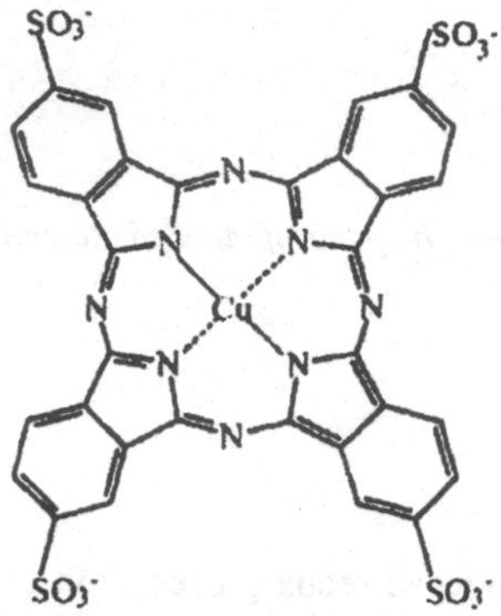

Fig.1 Structure of phthalocyanines, copper(II)phthalocyanine-3,4',4'',4'''-tetrasulfonic anion

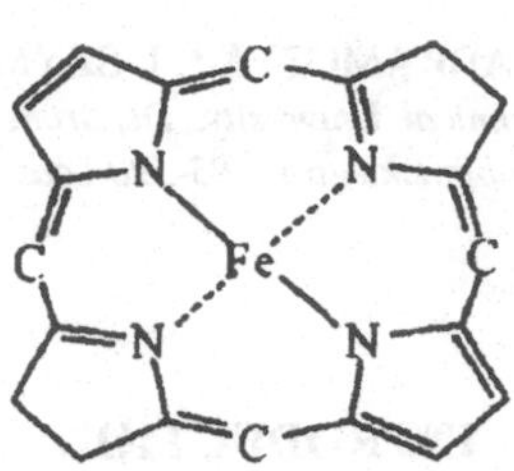

Fig.2 Structure of porphyrins

Why have we chosen phthalocyanines for our studies? Phthalocyanines have well established applications as green/blue pigments and dyes. However the page from an old book in Fig. 3 as well as "Les Parapluies" by Renoir in Fig.4 could not have been painted with phthalocyanine. The first one could have been painted with Egyptian blue (3rd millennium BC) while "Les Parapluies" with lazurite (lapis lazuri) (known since 1828). Phthalocyanine blue (known as Winsor blue) was synthesized in 1936.

Phthalocyanine derivatives act as catalysers in oxidaselike and catalaselike reactions. Phthalocyanines derivatives have a potential use (some of them have already been applied) as photodynamic reagents in cancer therapy and other medical applications. They may overcome the major drawbacks of the porphyrine-type materials: skin photosensitivity, low selectivity for tumour tissue, long period of clearance from the body. That is why we are interested in photochemistry of these substances.

Understanding photochemical mechanisms of phthalocyanines plays a crucial role in evaluation of their photodynamic activity and their potential application in photodynamic therapy (PDT). Answering the questions about the mechanism of photooxidation seems to be essential. Is it I type of photooxidation (with sentitiser reacting directly with another chemical entity (human tissue) by hydrogen or electron transfer to yield transient radicals which react further with oxygen) or II type of photooxidation (with sentitiser excited to tripled state, which interacts with oxygen, most commonly by energy transfer, to produce an electronically excited singlet state of oxygen which can react further with the chemical entity susceptible to oxidation)?

Fig.3

Fig.4

Futher potential uses of metal phthalocyanines are

a) sensing elements in chemical sensors
b) electronic display devices
c) applications to computer read/write discs
d) photovoltaic cell elements for energy generation
e) new red-sensitive photocopying applications
f) liquid crystal color display applications
g) molecular metals and conducting polymers.

In recent years there has been a growth in the number of laboratories exploring the fundamental academic aspects of phthalocyanines chemistry – their electronic structure, redox properties and their photocatalytic reactivities with different experimental methods.

2. EXPERIMENTAL METHODS

Here, we present the laser system that has been used for our measurements. It consists of the equipment for Raman scattering measurements, and for pump-probe femtosecond absorption measurements (Fig.5).

2.1 RAMAN SCATTERING

Raman scattering provides the information about vibrations like IR spectroscopy, but under certain conditions it also provides the information about the emissive electronic states. Raman scattering arises from the transient dipole moments induced in a system by the electric field of the incident electromagnetic radiation. The spontaneous Raman scattering is derived from inelastic light scattering which results from the transfer of radiative energy from an excitation beam at the frequency of ω_0 to the internal degrees of freedom (rotational or vibrational). The frequency of the exciting radiation ω_0 is usually chosen to be in the visible or ultraviolet region of the spectrum. The photons with the frequency at ω_0 excite the molecules from the ground vibrational state to the scattering state. When the excitation energy is near to the energy of the electronic transition from the ground to the excited state the system is in near-resonance conditions and this kind of scattering is called Resonance Raman scattering.

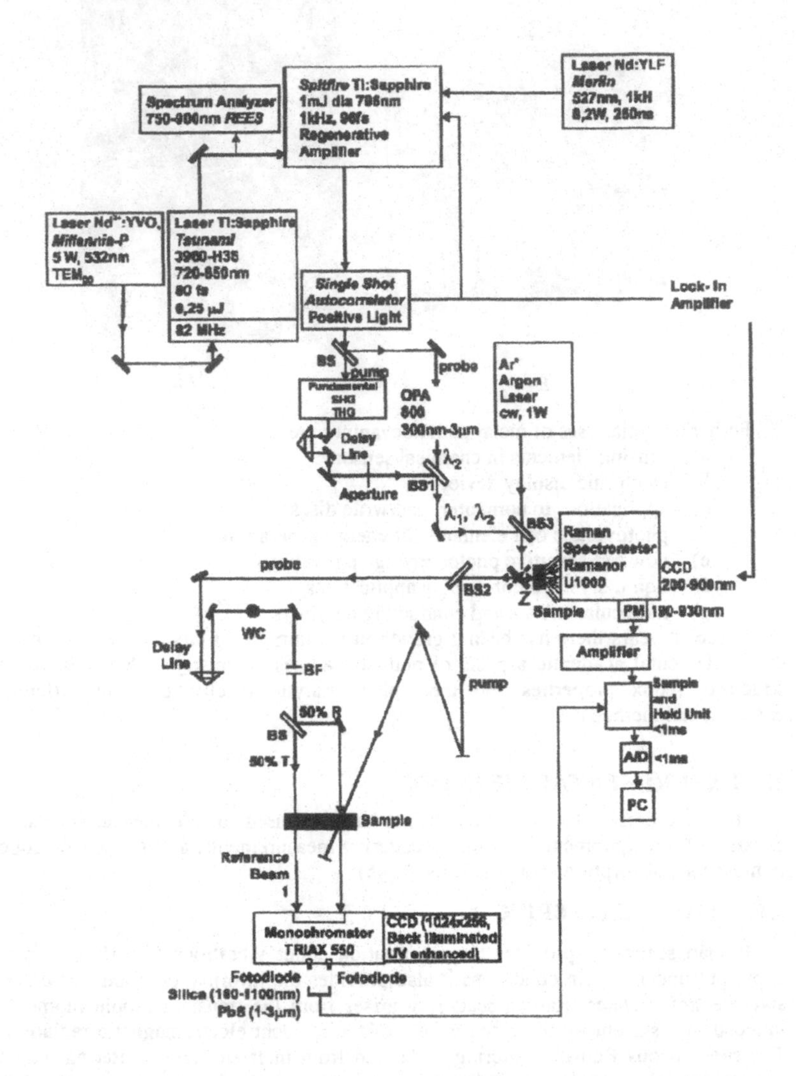

Fig. 5 Scheme of femtosecond laser system in the Laboratory of Laser Molecular Spectroscopy (LLMS)

The spontaneous Raman Resonance signal intensity is enhanced due to the coupling of the internal degrees of freedom to the molecular transient dipole moment. Additionally under resonance Raman conditions we can sometimes observe emission generated by the incident beam. Sometimes this emission creates the experimental problem, but very often it can be very valuable. After the excitation, the molecules return to a) the same ground vibrational state (Rayleigh scattering), b) to the first excited vibrational state (Raman scattering, Stokes component), c) the molecule excited from .the first excited state returns to the ground vibrational state (Raman scattering, anti-Stokes component) (Fig.6)

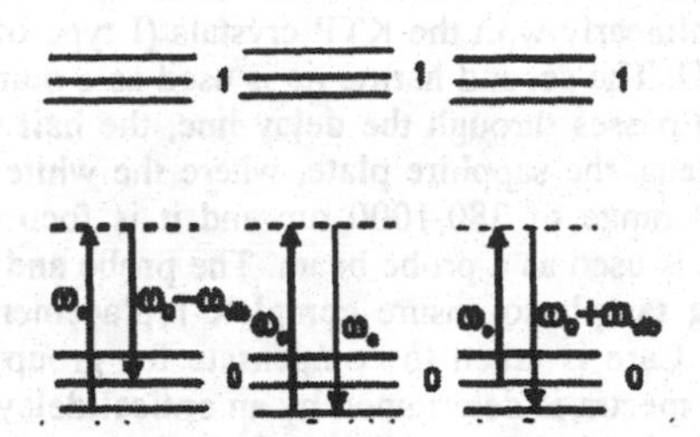

Fig. 6 Raman scattering (Stokes component), Rayleigh scattering, Raman scattering (anti-Stokes component)

Thus the resultant scattering appears over a range of energies (Fig.7) measured at lower energies ω_0-ω_{vib} (Stokes), at ω_0 for Rayleigh scattering and at higher energies ω_0+ω_{vib} (anti-Stokes), where ω_{vib} represents the vibrational frequency.

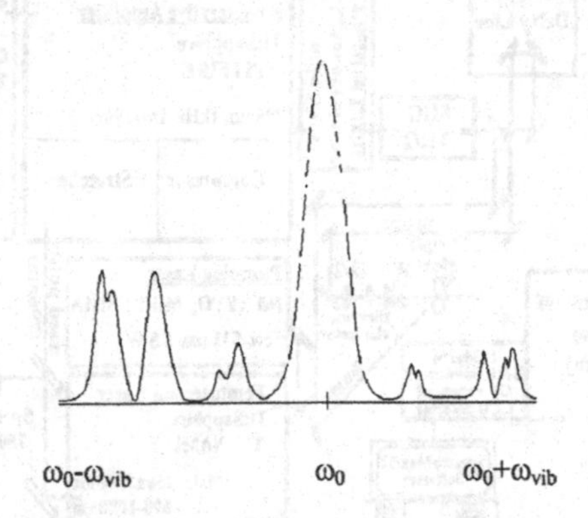

Fig.7 Raman and Rayleigh scattering spectra

2.2 PUMP-PROBE FEMTOSECOND ABSORPTION SPECTROSCOPY

Our femtosecond system is presented in Fig. 8. The femtosecond laser is pumped with the cw solid state laser (ytrium vanade crystal dopped with Neodymium). Its second harmonic (532 nm, power 5.5 W) is used as a pumping source for the Ti:Sapphire femtosecond laser working in the mode-locking regime at the repetition rate of 82 MHz . The femtosecond laser is tunable in a very broad range but we use only 796nm wavelength for further amplification. The seed pulse having the energy of 250 nJ and the duration of 80 fs is amplified in the regenerative amplifier. The small part of the seed pulse is sent by the beamsplitter to the spectrum analyser in the frequency domain to monitor the quality of the mode-locking. The regenerative amplifier is pumped by the solid state laser (Nd:YLF) working with the repetition of 1 kHz at the pulse duration of 250 ns emits at 527 nm (second harmonic from LBO nonlinear crystal inside the resonance cavity). The pulse mode is achieved by Q-switching with an optoacustic device. After the amplification the energy of the pulse is 1 mJ at the repetition rate 1 kHz, and the wavelength of 796nm. The pulse duration is measured with the single shot autocorrelator and was found to be 96 fs. The pulse interacts nonlinearly with the KTP crystals (I type of phase matching) to generate SHG and THG. The second harmonic is used as a pump beam. The fundamental beam (796 nm) passes through the delay line, the half wave plate, dichroic polarization analyser, lens, the sapphire plate, where the white continuum (WC) is generated in the broad range of 380-1000 nm and it is focused onto the flowing sample. The WC beam is used as a probe beam. The probe and pump beams are focused onto the flowing sample to ensure complete replacement of the sample between each laser shot. Care is taken to compensate for group velocity dispersion. The time delay for the spectra is determined by an optical delay line controlled by a motor-driven translational stage. The radiation from the sample enters the entrance slit of a 0.5 m single dispersion spectrometer (1800g/mm grating). The dispersed radiation is detected by a multichannel (256 x 1024) charge-coupled device detector, back illuminated, UV enhanced attached to the exit slit of the spectrometer. The data acquisition and the full computer control is provided by the software (Spectra Max/32).

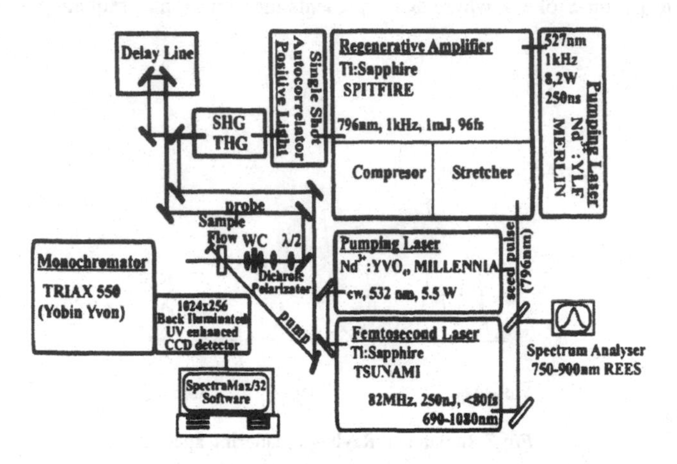

Fig.8 The laser system for femtosecond pump-probe absorption measurements.

We measure the difference absorption signal $\Delta A = A - A_0$, where A_0 is the absorption without pump pulse, A – the absorption when both pump and probe pulses excite the sample. ΔA is measured as a function of a time delay between the pump and the probe pulses for a given wavelength (Fig.9). The difference absorption signal can be negative or positive and may indicate the photo bleaching, stimulated emission or generation of the transient products that can be monitored with the femtosecond time resolution.

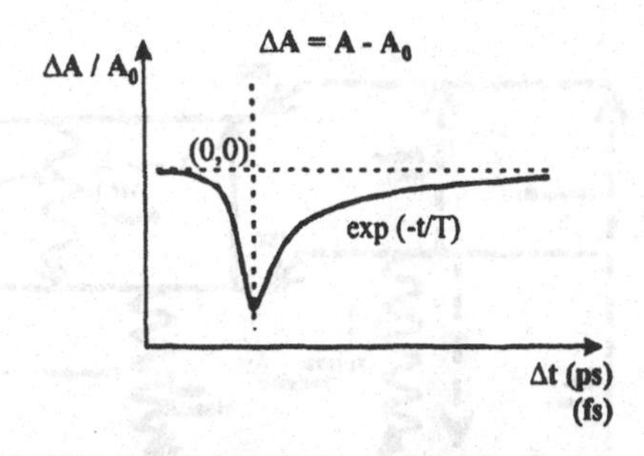

Fig. 9 The difference absorption signal as a function of the time delay (negative signal corresponds to photobleaching)

3. EXPERIMENTAL RESULTS AND DISCUSSION

If we want to work in Resonance Raman conditions we have to know something about the electronic states (Fig.10). Phthalocyanines have two major electronic absorption bands, namely the B (or Soret band) band at about 350 nm in ultraviolet, and the Q band at 600-700 nm (visible, red). As an extensively conjugated aromatic chromophore the electronic transitions have $\pi \rightarrow \pi^*$ character. In the same energy range there are electronic levels of the metal centre for many metal phthalocyanines and redox/oxidation processes may occur between ligand and metal centre that lead to ligand-metal, metal-ligand charge transfers (especially the d levels for the transition metals). For copper (II) phthalocyanine, which is a subject of this paper, the metal-ligand transitions are less important, especially in aqueous solutions where the metal-oxidized or metal-reduced species are less stable than the reduced radicals of the ligand.

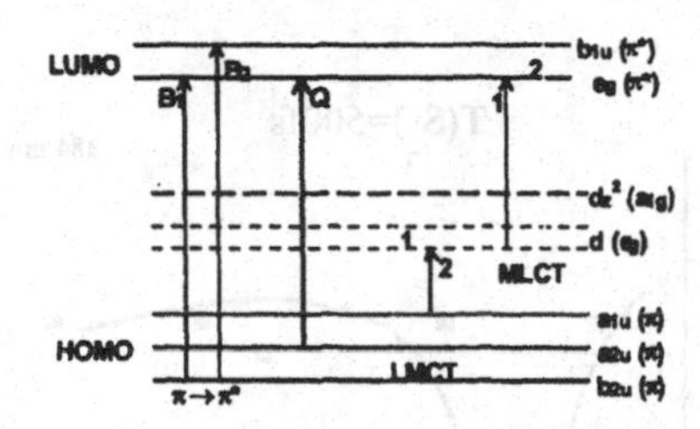

Fig.10 Schematic diagram of electronic states of phthalocyanines

The schematic energy diagram for the Q transition for tetrasulfonated copper phthalocyanine is given in the next Fig.11. One can see that with the laser system for Raman and pump-probe femtosecond measurements presented in the experimental section we can excite nearly all interesting transitions $S_0 \rightarrow S_1$, $S_0 \rightarrow S_n$, $S_1 \rightarrow S_n$, $T_1 \rightarrow T_n$.

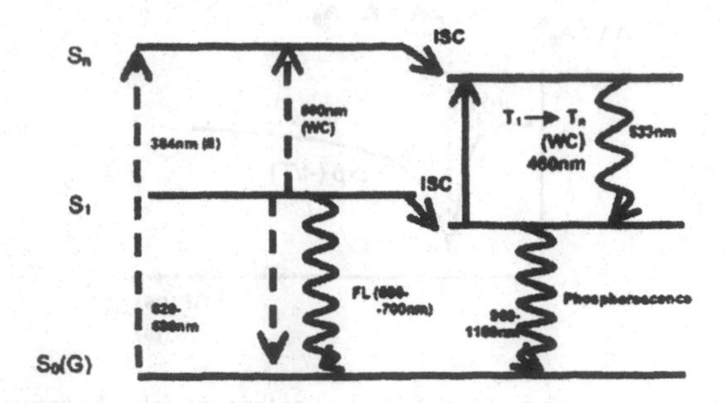

Fig.11. Schematic Energy Diagram for the Q transition for $Cu(tsPc)^{-4}$

In most photochemical reactions higher excited states are not involved in the reaction. They usually deactivate very rapidly to the lowest exciting state, which deactivates to the ground state exhibiting emission or via radiationless process (internal conversion) or undergoes reactions competing with the deactivation. This pattern of behaviour is known as a Kasha's rule). However, there are some groups of compounds contradicting Kasha's rule (azulene). It was suggested [1] that some porphyrines undergo electron transfer in the S_2 excite state. Taking into account the similar structure of porphyrines and phthalocyanines it was essential to determine the life time of the S_n state.

Femtosecond pump-probe measurements showed that the S_n life time is shorter than 500fs. We analysed the difference absorption signal $\Delta A = A - A_0$, as a function of a time delay between the pump and the probe pulses at different wavelengths (Fig.12) that was very useful to establish the mechanism of deactivation. Femtosecond pump-probe measurements showed that the S_n life time is shorter than 500fs.

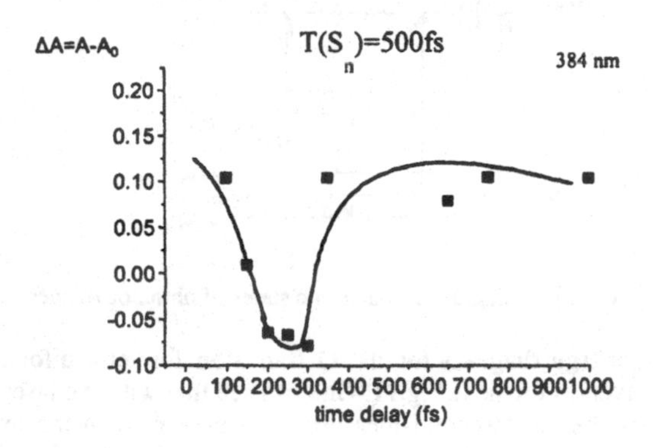

Fig.12 The difference absorption signal as a function of the time delay at 384 nm for $Cu(tsPc)^{-4}$ in aqueous solution for the concentration of 10^{-2} mol/dm^3

The Figure 13 shows the Raman spectrum of the $Cu(tsPc)^{-4}$ in aqueous solution for the concentration of 10^{-2} mol/dm^3. The similar pattern of behaviour has been recorded in the broad concentration range from 10^{-2} to 10^{-6} mol/dm^3. The narrow peaks in the region 200-2000cm^{-1} correspond to the internal vibrations of the $Cu(tsPc)^{-4}$, the peaks at around 3000cm^{-1} correspond to the stretching (symmetric and asymmetric) modes of water. In the region 4000-6000cm^{-1} we can see a very broad, structureless band with the maximum at 4800cm^{-1} corresponding to the emission of unknown origin. The intensity of the emission increases with temperature decreasing. The emission at 4800 cm^{-1} corresponds to the wavelength of 682nm. Many phthalocyanines were reported to have fluorescence $S_1 \rightarrow S_0$ for the Q transition in this spectral region [2-3]. Can the emission at 682 nm in Fig 13 be assigned to $S_1 \rightarrow S_0$ fluorescence also in this case? At this stage it is hard to decide but we will show later that this emission does not come from the fluorescence but rather from the emission of the transient radicals generated in the photoredox dissociation. They live shortly at room temperatures and cannot be detected but at lower temperatures when the molecular motions become slower the transient species are trapped in the rigid environment and they can be easily detected.

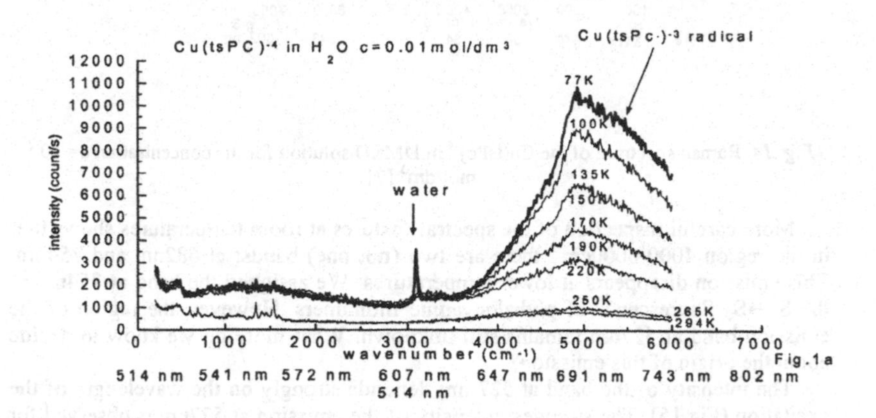

Fig. 13 Raman spectrum of the $Cu(tsPc)^{-4}$ in aqueous solution for the concentration c= 10^{-2} mol/dm^3.[6]

We have found that $Cu(tsPc)^{-4}$ dimerizes readily in aqueous solutions. We can see the band at 620nm for dimer and at 670nm for monomer. The dimerization equilibrium constant is shifted significantly towards the dimeric form and was found to be 8.75×10^4 in aqueous solution at 294K. In contrast to aqueous solutions, $Cu(tsPc)^{-4}$ in DMSO solution exist almost entirely as monomer.

So, it is interesting to study the photochemistry of Cu(tsPc)-4 in DMSO and compare with the behaviour in aquoeus solution to learn about the photochemical properties for the dimeric and monomeric forms. One can see from Fig.14 that the broad, structureless band at 682nm is also observed in DMSO like in water with intensity increasing when temperature decreases. However, in contrast to the aqueous solutions we observe an additional, very intensive emission with the maximum at around 527nm. The emission is recorded for liquid solutions. At lower temperatures when the solutions becomes a frozen matrix-this emission disappears whereas the emission at 682nm begins to increase. The isobestic point is found at around 620nm.

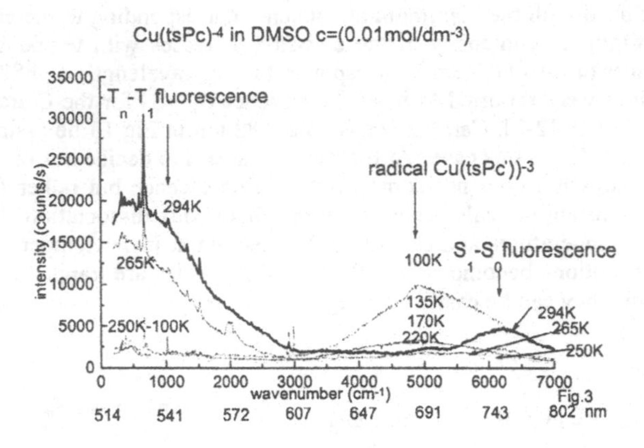

Fig. 14 Raman spectrum of the $Cu(tsPc)^{-4}$ in DMSO solution for the concentration c= 10^{-2} mol/dm^3 [7]

More careful inspection of the spectral features at room temperatures shows that in the region 4000-6000cm^{-1} there are two (not one) bands: at 682nm and 754nm. This emission disappears at lower temperatures. We assigned the band at 754nm to the $S_1 \rightarrow S_0$ fluorescence of phthalocyanine monomers. However, the region of the emission band at 527nm remains still unknown. What more do we know to decide about the origin of this emission?

The intensity of the band at 527 nm depends strongly on the wavelength of the excitation (Fig.15). The strongest intensity of the emission at 527nm is observed for the excitation with the wavelength of 465.8nm.

Some papers [4-5] based on the theoretical calculations suggest that the triplet $T_1 \rightarrow T_n$ transition is expected in this region. Following many experiments with electron and hole scavengers and various energy transfer experiments we have assigned the emission at 527nm to the $T_1 \rightarrow T_n$. This assignment has been supported by the sensitisation of oxygen triplets (Fig.16)

Indeed, the fluorescence has been suppressed significantly in nondeareated aqueous solutions when compared with degassed samples, confirming oxygen involvement in the process of deactivation of the triplet state. It indicates that II Type of photooxidation plays a role in $Cu(tsPc)^{-4}$ leading to generation of the singlet oxygen in the excited state that is very reactive and toxic leading to the necrosis of the human tissue in the photodynamic therapy.

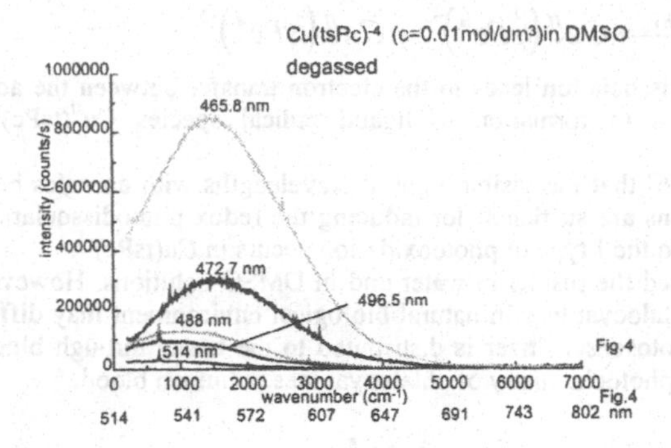

Fig.15 **Resonance Raman Spectrum as a function of wavelength excitation for $Cu(tsPc)^{-4}$ in DMSO solution for the concentration c= 10^{-2} mol/dm^3 [6].**

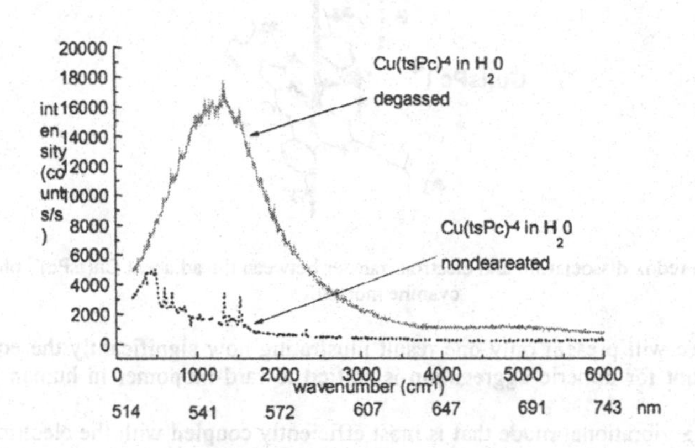

Fig.16. **Resonance Raman Spectrum as a function of wavelength excitation for $Cu(tsPc)^{-4}$ in aqueous solution for the concentration c= 10^{-2} mol/dm^3 in degassed and nondeareated samples [6].**

However, even in degassed samples the intensity of the fluorescence decreases with time. It indicates that there must exist another deactivation mechanism (or mechanisms) of the T_n triplet state (or the T_1 state) that is competitive to the $T_n \rightarrow T_1$ fluorescence. What mechanism? It is well known that mechanisms of quenching (deactivation) are very sensitive to the degree of aggregation. .It was reported [5] that the irradiation of phtalocyanines with the wavelengths corresponding to the resonance with the B transition or with shorter wavelengths induces the photoredox dissociation.

$$Cu(tsPc)^{-4} \xrightarrow{h\nu} Cu^{II}(tsPc^{\bullet})^{-3} + Cu^{II}(tsPc^{\bullet})^{-5} \quad (1)$$

The photoredox dissociation leads to the electron transfer between the adjacent molecules that results in formation of ligand radical species $Cu^{II}(tsPc)^{-3}$ and $Cu^{II}(tsPc)^{-5}$ (Fig.17).

We have shown [6] that the visible light at wavelengths with energies between the B and Q transitions are sufficient for inducing the redox photodissociation (1). This indicates that also the I type of photooxidation occurs in $Cu(tsPc)^{-4}$.

So far we presented the results in water and in DMSO solutions. However, the photochemistry of phtalocyanines in natural biological environment may differ significantly. As the photosensensitizer is distributed to the tissue through blood, we have studied also the photochemistry of phtalocyanines in human blood.

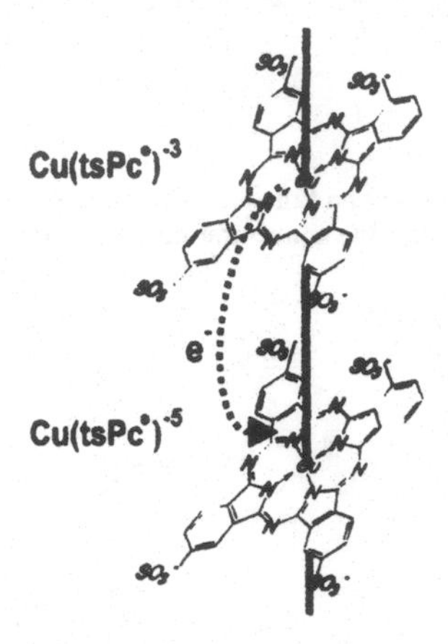

Fig.17 Photoredox dissociation and electron transfer between the adjacent $Cu(tsPc)^{-4}$ phthalocyanine molecules

Here, we will present only one result illustrating how significantly the equilibrium constant for dimeric aggregation is shifted toward monomer in human blood (Fig.18).

This is a vibrational mode that is most efficiently coupled with the electric transitions in resonance Raman scattering. The peak at $1530cm^{-1}$ corresponds to monomer, and at $1540cm^{-1}$- to dimer. We can see that in human blood the intensity of the band at 1540 cm^{-1} decreases significantly indicating that in the human blood there are much less dimers than in the aqueous solutions. This result is very promising for the photodynamic therapy since when the photosensitizer aggregate is excited by light no useful photochemistry (from the photobiological point of view) can be indicated, since the rate of deactivation by internal conversion to the ground state greatly exceeds that of the monomer.

So far we presented the results for the liquid phase and the crystal state. To provide further evidence about the photochemistry of phtalocyanines and the electron transfer between the adjacent molecules we want to show the results for glasses.

It is known that distinct structure of solids matrices are generated at rapid and slow cooling rate. Rapid cooling usually generates the glassy matrices, while the slow cooling rate leads to the generation of crystal matrices. The phases that are generated can be easily monitored by the low frequency Raman spectra where phonon peaks signalise the existence of translationally ordered crystal structure whereas the absence of any phonon peaks indicates that the glassy phase has been formed.

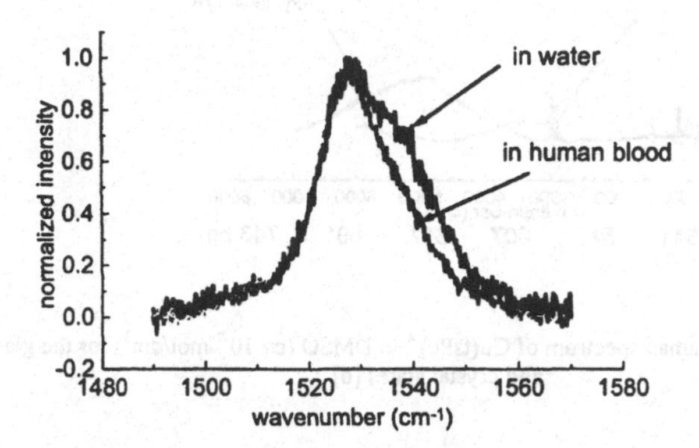

Fig.18 Resonance Raman spectrum of the internal vibration (ν_4) for $Cu(tsPc)^{-4}$ in aqueous solution for the concentration $c= 10^{-2}$ mol/dm^3 and in human blood [7]

Is the photochemistry for crystal and glasses the same? Fig.19 shows the Raman spectrum of $Cu(tsPc)^{-4}$ in DMSO for the rapid cooling when the glassy structure has been generated [6].

Striking behaviour is revealed when we compare the spectrum for the glassy phase with the spectrum for the crystal phase at the same temperature. First, there is no band at 682nm for the glassy phase, in contrast to the crystal phase at 77K. Second, the band at 527nm that exists in liquid solutions and disappears at lower temperatures for the crystal phase, still exists in the glassy phase at 77K. It looks like the photochemical properties of the liquid state have been preserved in the amorphous glassy phase. It clearly indicates that the structure has a profound influence on the photochemical behaviour of $Cu(tsPc)^{-4}$. The crystal structures of $Cu(tsPc)^{-4}$ form ring-stacked columns (Fig.20) close enough in space to overlap efficiently between the π-electronic clouds and as a consequence the photo induced electron transfer between the adjacent macromolecule rings occurs leading to the Pc radical generation.

In contrast, to the crystal phases, the liquid solutions and amorphous glassy phases are only partially organised (Fig.21).

As a consequence the distances between the adjacent phtalocyanine rings are larger and the overlapping between π-electronic clouds is much less efficient resulting in much less efficient electron transfer.

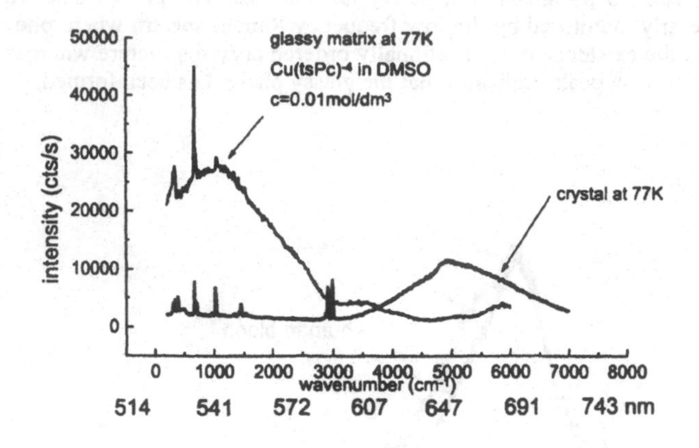

Fig.19 Resonance Raman spectrum of $Cu(tsPc)^{-4}$ in DMSO (c= 10^{-2} mol/dm^3) for the glassy and crystal states [6]

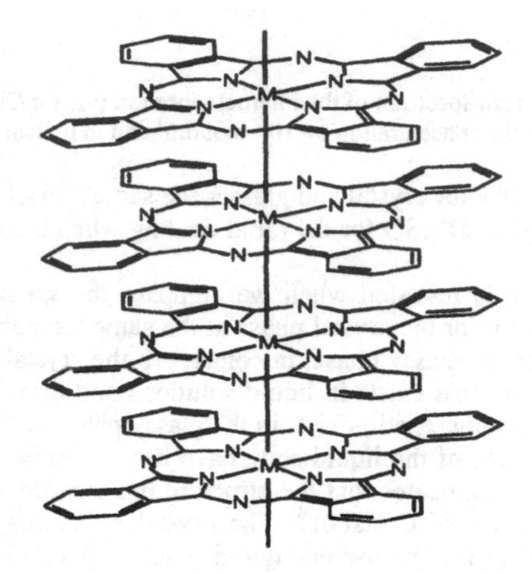

Fig.20. Schematic representation of ring stacked phthalocyanines

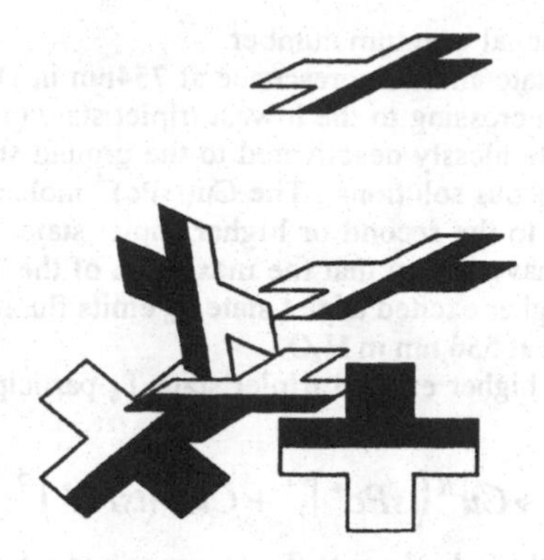

Fig.21 **Schematic representation of liquid and glassy phases in phthalocyanines**

4. CONCLUSIONS

From the results presented for crystal and glassy phases it is evident that the 682nm emission must be related somehow to the electron transfer photoredox dissociation and it is not related to the $S_1 \rightarrow S_0$ fluorescence. Finally, we have assigned the emission at 527nm to the $T_n \rightarrow T_1$ fluorescence, the emission at 682nm- to the $Cu^{II}(tsPc)^{-3}$ radical and the emission at 752nm to the $S_1 \rightarrow S_0$ fluorescence. We have proposed [6] the following mechanisms of photophysical and photochemical behaviour for the VIS light irradiation of the copper(II)phthalocyanine-3,4',4'',4'''-tetrasulfonic anion $Cu(tsPc)^{-4}$ with the wavelengths of energy between the B transition (350 nm) and the Q transition (600-700 nm) (Fig.22).

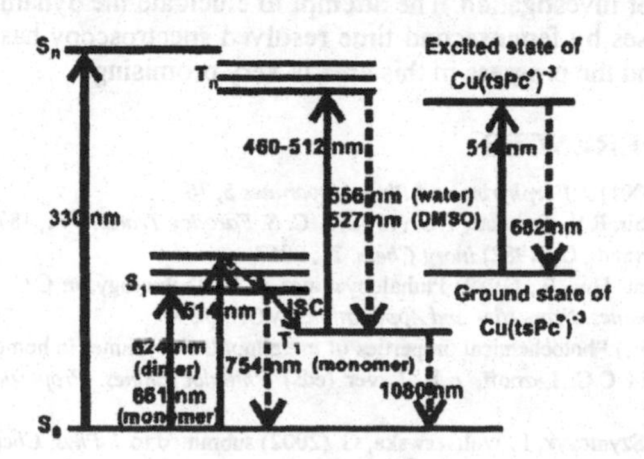

Fig.22 **Mechanisms of photophysical and photochemical behaviour of the copper(II)phthalocyanine-3,4',4'',4'''-tetrasulfonic anion $Cu(tsPc)^{-4}$ for the VIS light irradiation (514 nm)**

1) VIS light excites the ground state S_o of the molecule to the vibrationally ex cited lowest singlet excited state (S_1) for the Q transition

$$S_o \xrightarrow{h\nu} S_1(\upsilon > 0)$$

where υ is the vibrational quantum number.

2) The resulting S_1 state emits fluorescence at 754nm in DMSO liquid solution. or undergoes intersystem crossing to the lowest triplet state (T_1) that emits at 900 – 1100 nm [5], or is radiationlessly deactivated to the ground state. The $S_1 \rightarrow S_n$ fluorescence is absent in aqueous solutions . The $Cu(tsPc)^{-4}$ molecules that populate the T_1 state may be excited to the second or higher triplet state T_n ($T_1 \rightarrow T_n$ transition) with the VIS light. We have shown that the maximum of the $T_1 \rightarrow T_n$ absorption occurs at 463.8 nm. The higher excited triplet state T_n emits fluorescence at 533 nm for $Cu(tsPc)^{-4}$ in DMSO and at 556 nm in H_2O.

3) Competively, the higher excited triplet state T_n participates in photoinduced redox dissociation

$$Cu(tsPc)^{-4} \xrightarrow{h\nu} Cu^{II}\left(tsPc^{\bullet}\right)^{-3} + Cu^{II}\left(tsPc^{\bullet}\right)^{-5}$$

The excited state of the radical emits fluorescence and returns to the ground state of the radical. We have assigned the emission at 682 nm to the fluorescence of the excited state of the radical $Cu^{II}(tsPc^{\bullet})^{-3}$. The emission is recorded only at lower temperatures where the other competive processes in the excited state of the radicals become uneffective due to slow molecular motions.

We have shown that the photoinduced dissociation with the electron transfer between the molecules of $Cu(tsPc)^{-4}$ is determined by the distance between the adjacent rings and the structure. The electron transfer that lead to the generation of ligand-centered radicals may occur only for the ring-stacked structures with strong overlapping between the π-electronic cloud. These structures exist in highly concentrated aqueous solutions and in the crystal phases. The photoinduced electron transfer does not occur for the monomers of $Cu(tsPc)^{-4}$ in liquid DMSO solution and in the glassy phases.

The photochemistry of metallophtalocyanines is a very exiting and promising subject for further investigation. The attempt to elucidate the dynamics of the photochemical processes by femtosecond time resolved spectroscopy has been making in our laboratory and the progress in this area is very promising.

REFERENCES

1. Tokumaru, K. (2001) *J. Porphyrins and Phthalocyanines* 5, 76.
2. Mc Vie, J., Sinclair, R.S., Fruscott, T.G. (1978) *J. C. S. Faraday Trans.II*, 74, 1870
3. Prasad, Q.R., Ferraudi, G. (1982) *Inorg.Chem.* 21, 2967
4. Rosenthal, I., Ben- Hur, E., (1989) Phthalocyanines in Phothobiology, in C.C. Leznoff, A.B. Lever (eds) *Phthalocyanines. Properties and Applications*, VCH, p.393
5. Ferraudi, G. (1989) Photochemical properties of metallophthalocyanines in homogeneous solution, in Phothobiology, in C.C. Leznoff, A.B. Lever (eds) *Phthalocyanines. Properties and Applications*, VCH, p.291
6. Abramczyk, H., Szymczyk, I., Waliszewska, G. (2002) submitted to *J. Phys. Chem.*
7. Abramczyk, H., Szymczyk, I. (2002) submitted to *J. Mol. Liquids*

DIELECTRIC SPECTROSCOPY OF SOLUTIONS

RICHARD, BUCHNER
Institut für Physikalische und Theoretische Chemie,
Universität Regensburg, D-93040, Regensburg, Germany,
Email: Richard.Buchnerchemie.uni-regensburg.de

Abstract: This contribution gives a short introduction into the basic principles of dielectric spectroscopy and its application in solution chemistry. Exemplified by recent results it is shown that precise complex permittivity spectra obtained by the use of time domain and frequency domain methods in the mega- to gigahertz range yield specific information on solvent dynamics as well as on ion-solvation and ion-association phenomena that is not readily accessible with other techniques.

1. Introduction

Solvation and association of ions are the dominating features of aqueous and nonaqueous electrolyte solutions and determine their physicochemical properties. The importance of these phenomena in many fields, from geochemistry via biological processes to technical applications, leads to the development of solution chemistry as the interdisciplinary science between chemistry, physics, biology and chemical engineering devoted to the investigation of ion-solvent and ion-ion interactions [1–3]. Although important new insights were also coming from thermodynamics, transport studies, various spectroscopies and theory, it is probably fair to say that in the last decade the major impetus for our current understanding of electrolyte structure came from the rapid development of scattering methods and computer simulations [4,5].

However, many open questions remain. First of all, one has to note that x-ray and neutron scattering, but also (at least up to now) many simulation studies, give an essentially static picture of the solutions. The

J. Samios and V.A. Durov (eds.), Novel Approaches to the Structure and Dynamics of Liquids: Experiments, Theories and Simulations, 265–288.

coordination numbers (CN) determined by these techniques often differ considerably from the so-called primary solvation numbers (PSO [1]) determined from thermodynamic or transport properties. This is understandable because CN are essentially obtained by counting the number of solvent molecules within a certain distance from the ion, whereas PSO reflect the balance between ion-solvent and solvent-solvent interactions and thus monitor the 'fluid' character of the solutions. Of course, the difference between CN and PSO is not a mere flaw of one or the other approach, but significant information for the understanding of solutions and in the next decade a major task of solution chemistry will be to find a coherent description which accomodates CN and PSO and thus allows a prediction of the concentration and temperature dependence of thermodynamic and transport properties. The second focus will be on chemical speciation. The investigation of complex formation to e.g. ion pairs has always been a major field of solution chemistry, but due to experimental problems was usually restricted to association constants $K_A \gtrsim 20$. However, equilibria involving weak complexes are widespread, ranging from the importance of sodium oxalate ion pairs on kidney stones [6] to the Bayer process for the recovery of purified $Al(OH)_3$ from bauxitic ores [7,8], and there will be an increasing demand in the determination of the stability constants and the identification of the involved species.

In this review I would like to show what contributions can be expected from dielectric [relaxation] spectroscopy (DRS) on ion solvation and association. After a short introduction to the basic principles of the technique, the major features of the dielectric spectra of electrolyte solutions will be presented. Mainly exemplified with results from recent investigations of aqueous systems, the focus will be on equilibrium properties, that is effective solvation numbers and ion-pair concentrations. But also some of the accessible dynamic information will be highligted. For a comprehensive review of dielectric relaxation studies of solutions the reader is referred to Ref. [9]; an encompassing compilation of literature data in this field is available with Ref. [10].

2. Principles of Dielectric Spectroscopy

DRS monitors the response of a sample, its polarization $\vec{P}(t)$, towards an applied time-dependent electric field, $\vec{E}(t)$, as a function of time, t, or (alternatively) the dependence of $\vec{P}$ on the frequency, ν, of harmonic fields. The technique is able to probe dynamical processes on a time scale ranging from tens of femtoseconds to hours and is therefore a widely used tool in material science to characterize solids [11], polymers [12],

TABLE 1. Some applications of DRS and typical frequencies, ν, of associated relaxation processes.

field	$\log(\nu/\text{Hz})$	remark
solids		
plastic crystals	−3 ... 3	
ion conductors	0 ... 8	impedance spectroscopy
electrode processes	−3 ... 6	
liquid crystals	1 ... 8	switching time
polymers		
polymerisation kinetics	−4 ... 4	
cooperative dynamics	−3 ... 5	α-relaxation ⇔ mechanical properties
polar side chains	4 ... 8	β-relaxation
glass formation	−3 ... 9	
colloids, micelles, emulsions, biological systems		
counterion diffusion	−1 ... 7	particle size & shape
phase diagram	4 ... 9	
intermolecular interact.	4 ... 9	headgroup dynamics
bound water	6 ... 10	
"simple" liquids & solutions		
ionpair relaxation	6 ... 10	↷ speciation, kinetics
solvent dynamics	8 ... 12	↷ solvation

or meso-phases [13]. The technique also receives increasing interest in fields like biophysics [14] or pharmacy [15] and has a long tradition in the investigation of structure and dynamics of electrolyte solutions [2, 9, 16]. Table 1 summarizes some typical applications of DRS and the frequency range where characteristic relaxation processes are observed. For a general introduction to the method see Ref. [17].

For a nonmagnetic sample composed of individual components k with number densities $\rho_k = N_k/V$ at the temperature T the polarization, $\vec{P}(t)$, can be essentially devided into three additive contributions.

Always present is the induced polarization

$$\vec{P}_\alpha = \sum_k \rho_k \alpha_k (\vec{E}_{\text{int}})_k = \varepsilon_o(\varepsilon_\infty - 1)\vec{E} \tag{1}$$

from the intramolecular charge distortion induced by the local field $(\vec{E}_{\text{int}})_k$ in molecules of polarizability α_k. The relation between $(\vec{E}_{\text{int}})_k$ (resp. $(\vec{E}_{\text{dir}})_k$, see below) and $\vec{E}$ depends on the theoretical level adopted for the calculation of $\vec{P}$ [18]. Compared to molecular motions the induced polarization fluctuates very rapidly so that for most applications of DRS $\vec{P}_\alpha$ can be assumed to be in equilibrium with the external field and expressed by the material quantity ε_∞, typically around 2-3; ε_o is the vacuum permittivity.

Molecules with a permanent dipole moment, $\vec{\mu}_k$, are aligned by the local electric field, $(\vec{E}_{\text{dir}})_k$, against thermal motion. At equilibrium this orientational polarization is given by

$$\vec{P}_\mu^{\text{eq}} = \sum_k \rho_k \frac{\mu_k^2}{3k_{\text{B}}T} (\vec{E}_{\text{dir}})_k = \varepsilon_o(\varepsilon - \varepsilon_\infty)\vec{E} \tag{2}$$

defining thus the relative (dielectric) permittivity, ε of the specimen (obviously, $\varepsilon = \varepsilon_\infty$ for nonpolar samples); k_{B} is the Boltzmann constant. However, equilibrium between $\vec{P}$ and $\vec{E}$ is only reached for slowly varying fields. With increasing frequency the dipoles are more and more unable to follow $\vec{E}(t)$ without delay due to frictional forces arising from intermolecular interactions. Formally, this is expressed by the response function F_{P}^{or}

$$F_{\text{P}}^{\text{or}}(t) = \frac{\langle \vec{M}(0) \cdot \vec{M}(t) \rangle}{\langle \vec{M}(0) \cdot \vec{M}(0) \rangle} \tag{3}$$

of the total dipole moment

$$\vec{M} = \sum_k \sum_i^{N_k} \vec{\mu}_{ik}$$

i.e. the vector sum over all molecular moments. The brackets $\langle \ldots \rangle$ designate ensemble averages, $F_{\text{P}}^{\text{or}}(0) = 1$ and $\lim_{t\to\infty} F_{\text{P}}^{\text{or}} = 0$. Thus,

$$\vec{P}_\mu(t) = \vec{P}_\mu^{\text{eq}} F_{\text{P}}^{\text{or}}(t) \tag{4}$$

provides information on the dynamics of the intermolecular processes connected with fluctuations of $\vec{M}$, be they connected to the rotation of individual molecules or to the formation and destruction of supramolecular entities.

In the frequency domain, connected to $\vec{P}_\mu(t)$ via

$$\begin{aligned}\hat{\varepsilon}(\nu) - \varepsilon_\infty &= (\varepsilon - \varepsilon_\infty) \int_0^\infty \left(-\frac{\partial F_P^{or}(t)}{\partial t}\right) \exp[-i2\pi\nu t]\, dt \\ &= (\varepsilon - \varepsilon_\infty)\, \tilde{F}(\nu)\end{aligned} \tag{5}$$

this is manifested by the dispersion of permittivity from the static value ε down to ε_∞ at $\nu \to \infty$, yielding the *dispersion curve* $\varepsilon'(\nu)$, see Fig. 1. Simultaneously dissipation of energy over a broad range of frequencies is observed and this is characterized by the *dielectric loss spectrum* $\varepsilon''(\nu)$. ε' and ε'' together define the complex permittivity spectrum of the sample,

$$\hat{\varepsilon}(\nu) = \varepsilon'(\nu) - i\varepsilon''(\nu) \tag{6}$$

which contains the entire information on the interaction of electromagnetic radiation with a non-magnetic, non-conducting sample. The relaxation function $\tilde{F}(\nu)$ is the frequency domain analogue of the autocorrelation function $F_P^{or}(t)$ and contains equivalent information on the relaxation dynamics.

The third contribution to $\vec{P}$, migration of charge carriers, like ions, under the influence of the electric field, is specific to conducting samples. Formally, the contribution can be expressed in terms of the complex conductivity, $\hat{\kappa}(\nu)$, and from the structure of *Maxwell's* equations it follows that only the generalized permittivity

$$\hat{\eta}(\nu) = \hat{\varepsilon}(\nu) + \frac{\hat{\kappa}(\nu)}{i2\pi\nu\varepsilon_o} \tag{7}$$

can be determined by DRS [18]. However, experience shows that for electrolyte solutions the dispersion of $\hat{\kappa}$ is small so that

$$\eta'(\nu) = \varepsilon'(\nu) \tag{8}$$

$$\eta''(\nu) = \varepsilon''(\nu) + \frac{\kappa}{i2\pi\nu\varepsilon_o} \tag{9}$$

where $\kappa = \lim_{\nu\to 0} \kappa'(\nu)$ is the conductivity determined under quasi-static conditions [9, 16, 20].

Figure 1 shows a typical electrolyte spectrum. Generally, the spectrum is dominated by the contribution of the solvent in the gigahertz region (H_2O in Fig. 1), which is affected by ion-solvent interactions. Additionally, a contribution from the solute, generally due to the formation

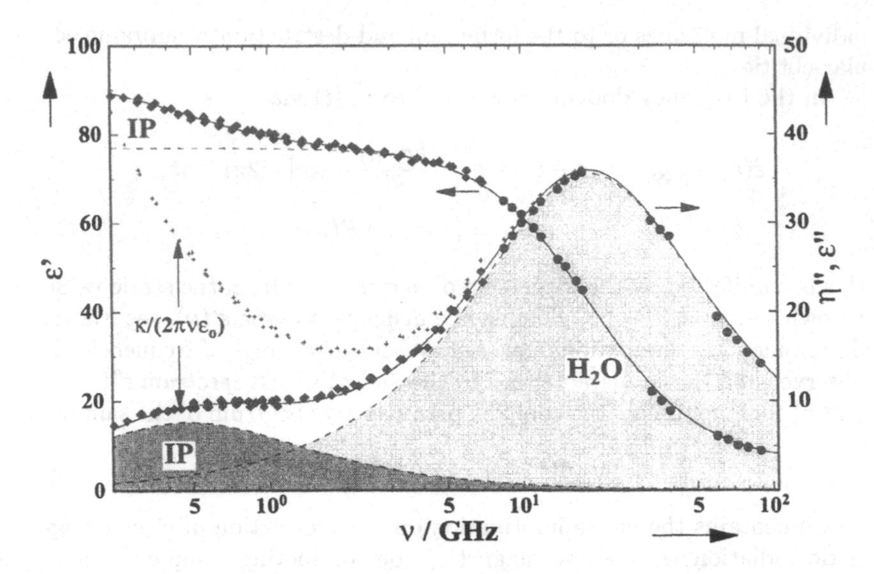

Figure 1. Dielectric permittivity, $\varepsilon'(\nu)$, and loss spectrum, $\varepsilon''(\nu)$, of 0.0345 mol dm^{-3} La[Fe(CN)$_6$] in water at 25 °C. Experimental data obtained with two different techniques (coaxial line TDR: ♦, waveguide: •) are fitted to a superposition of two Debye relaxation processes, solid lines, attributed to the ion-pair (IP) and to the solvent (H_2O), broken lines. Also indicated (+) is the total loss, $\eta''(\nu)$, of the solution [21].

of ion pairs (IP), may appear around a few hundered megahertz. Also indicated in Fig. 1 is the conductivity contribution to η''. It is obvious from eq. (9) that κ dominates at low frequencies and (since $\lim_{\nu\to 0} \varepsilon''(\nu) = 0$) thus determines the minimum frequency to which $\hat{\varepsilon}(\nu)$ can be reliably determined. For concentrated electrolytes this may seriously hamper the detection of small solute relaxation processes. Note, that often in the literature $\hat{\varepsilon}$ is used as the symbol for the total permittivity, related to a generalized conductivity via $\hat{\kappa} = 2\pi\nu\varepsilon_o\hat{\varepsilon}$, see e.g. Ref. [17].

3. Determination and Analysis of Solution Spectra

Figure 2 gives an overview of possible contributions to the dielectric spectrum of electrolyte solutions. At temperatures far above the glass transition temperature, T_g, molecular motions connected to fluctuations of $\vec{M}(t)$ from the terahertz down to the megahertz range may be observed. These processes range from the fast librations of a molecule in the ephemeral cage formed by its neighbours via molecular rotations to slow modes connected with the cooperative rearrangement of supramolecular

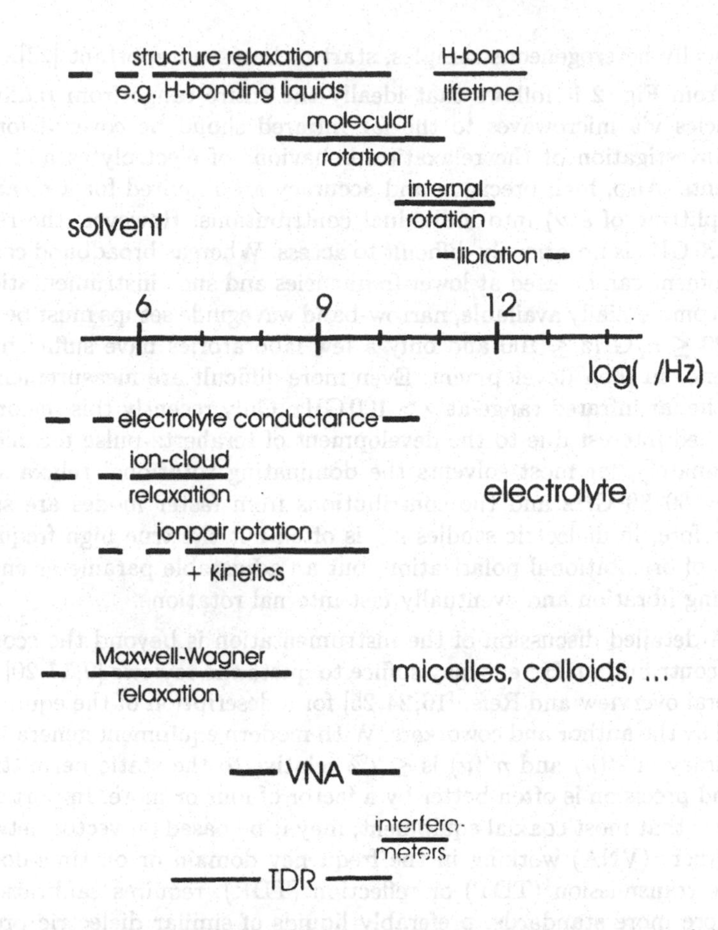

Figure 2. **Frequency scale of typical solvent and solute contributions to the dielectric spectrum, $\hat{\eta}(\nu)$, of electrolyte solutions at temperatures far from glass transition. Also indicated is the type of instruments used in recent works of the author and their frequency range.**

structures in hydrogen-bonding solvents like water or alcohols. The major electrolyte contribution (besides the conductivity term in η'') arises from the formation of ion pairs. Such species can be detected by DRS if their lifetime is at least comparable to the time required for dipole rotation. Ion-cloud relaxation — the Debye-Falkenhagen effect — is negligible for common electrolytes, but important for charged colloids [22]. Here also *Maxwell-Wagner*-type relaxation mechanisms, typical for macroscopically heterogeneous samples, start to become important [23].

From Fig. 2 it follows that ideally the entire range from radio frequencies via microwaves to the far infrared shoud be covered for the full investigation of the relaxation behaviour of electrolytes and their solvents. Also, high precision and accuracy are required for a meaningful splitting of $\hat{\varepsilon}(\nu)$ into individual contributions. However, the region $\nu > 20\,\mathrm{GHz}$ is notoriously difficult to access. Whereas broadband coaxial equipment can be used at lower frequencies and such instrumentation is now commercially available, narrow-band waveguide setups must be used for $20 \leq \nu/\mathrm{GHz} \leq 100$ and only a few laboratories have sufficient experience in their development. Even more difficult are measurements in the the far-infrared range at $\nu > 100\,\mathrm{GHz}$. Only recently this region has regained interest due to the development of terahertz-pulse techniques. Fortunately, for most solvents the dominating rotational relaxation is below 60–80 GHz and the contributions from faster modes are small. Therefore, in dielectric studies ε_∞ is often not the true high-frequency limit of orientational polarization, but an adjustable parameter encompassing libration and eventually fast internal rotations.

A detailed discussion of the instrumentation is beyond the scope of this contribution. Here it may suffice to quote the reviews [9,17,20] for a general overview and Refs. [16,24,25] for a description of the equipment used by the author and coworkers. With modern equipment generally the accuracy of $\varepsilon'(\nu)$ and $\eta''(\nu)$ is $\leq 2\%$ relative to the static permittivity, ε, and precision is often better by a factor of four or more. Important to note is that most coaxial equipment, may it be based on vector network-analyzers (VNA) working in the frequency domain or on time-domain pulse transmission (TDT) or reflection (TDR), requires calibration to one ore more standards, preferably liquids of similar dielectric properties. This is a serious problem if high accuracy is required, because the number of suitable reference fluids is limited [25] and systematic distortions arising from insufficient calibration are highly reproducible, which may mislead unexperienced users in the intrepretation of the spectra. Misalignment of cell and probe waveguide, higher-order modes or multiple reflections within the cell may lead to instrument-specific systematic errors of the four waveguide interferometers used in the author's laboratory to cover $8.5 \leq \nu/\mathrm{GHz} \leq 89$ [26]. However, this type of instruments does not require calibration with standard liquids and involves only the measurement of pathlength and signal intensity relatively to some arbitrary starting point. Therefore, these setups provide a reliable 'high-frequency anchor' for the VNA and TDR data.

For the interpretation of the complex permittivity spectrum a suitable relaxation model for the orientational polarization is fitted to $\hat{\varepsilon}(\nu)$, see Fig. 1. Its choice is far from trivial and partly dictated by the experimental accuracy and the frequency range covered by the data. The primary criterion in favour of a particular relaxation model is the variance of the fit. However, model selection should be guided by physical intuition of the possible relaxation processes, see Fig. 2, because only in this way a molecular interpretation of the fitting parameters and hence a check with results from other methods is feasible. Even the most elegant and best fitting relaxation model must be questioned if its parameters cannot be connected to the existing body of information accumulated with other techniques for the same or related systems.

For electrolyte solutions far from the glass-transition temperature it is generally possible to express $(\varepsilon - \varepsilon_\infty)\tilde{F}(\nu)$ as a sum of n individual relaxation processes j of amplitude S_j and characteristic relaxation time τ_j, eq.(12).

$$\hat{\varepsilon}(\nu) = \sum_{j=1}^{n} S_j \tilde{F}_j(\nu) + \varepsilon_\infty \tag{10}$$

where

$$\varepsilon = \sum_{j=1}^{n} S_j + \varepsilon_\infty \tag{11}$$

defines the relative static permittivity of the sample.

The relaxation functions $\tilde{F}_j(\nu)$ of the individual dispersion steps may generally be represented by modifications of the *Havriliak-Negami* equation

$$\tilde{F}_j(\nu) = \left[1 + (\mathrm{i}2\pi\nu\tau_\mathrm{j})^{1-\alpha_\mathrm{j}}\right]^{-\beta_j} \tag{12}$$

with relaxation time τ_j and relaxation time distribution parameters, $0 \leq \alpha_j < 1$ and $0 < \beta_j \leq 1$. Special cases of eq.(12) are the asymmetric *Cole-Davidson* relaxation time distribution, $\alpha_j = 0$, and the *Cole-Cole* equation, $\beta_j = 1$. A *Debye* relaxation process is determined by a single relaxation time, *i.e.* $\alpha_j = 0$ and $\beta_j = 1$ and represents an exponential decay of polarization. Although not mathematically equal, for certain combinations of α_j and β_j the *Havriliak-Negami* function is a good approximation to the *Kohlrausch-Williams-Watts* function which describes the stretched exponential decay of orientational polarization commonly found for supercooled or dissordered systems in the time domain [27]. For *Debye* and *Cole-Cole* equations with their symmetrical loss curves, relaxation time τ_j and frequency of maximum loss $\nu_j^{\max}$ are related by $\tau_j = 1/(2\pi\nu_j^{\max})$.

Provided the relaxation process j can be attributed to a chemical species, its amplitude S_j can be quantitatively interpretated with the help of the generalized *Cavell* equation [28]

$$c_j = \frac{3(\varepsilon + (1-\varepsilon)A_j)}{\varepsilon} \times \frac{k_B T \varepsilon_o}{N_A} \times \frac{(1-\alpha_j f_j)^2}{g_j \mu_j^2} \times S_j \tag{13}$$

or similar functions, that link S_j to the concentration c_j of the relaxing species. In eq. (13) μ_j is the dipole moment and α_j the polarizability of the species. The reaction-field factor, f_j, and the cavity-field factor, A_j, are defined by the size and shape of the dipole, the static (Kirkwood) dipole-dipole correlation factor, g_j, accounts for possible orientational correlations of neighbouring molecules of the same kind. It should be noted that generally c_j and g_j cannot be determined independently for solutions. For dilute solutions of j, like generally for ion pairs, $g_j \approx 1$ is a reasonable assumption. For the solvent c_j and g_j may be simultaneously affected by the solute, see below, so that it is convenient to normalize eq. (13) to the pure solvent [25].

The quantitative analysis of the relaxation time, τ_j, is less straightforward. It depends on the relaxation mechanism and generally data are less accurate. For rotational diffusion of individual dipoles, like e.g. for acetonitrile or N,N-dimethylformamide in their solutions or for long-lived ion pairs, τ_j can be converted to the rotational correlation time of the molecule, τ_j^{or} via the *Powles-Glarum* equation [29]

$$\tau_j = \frac{3\varepsilon}{2\varepsilon + \varepsilon_\infty} \times \frac{g_j}{\dot{g}_j} \times \tau_j^{\mathrm{or}} \tag{14}$$

where it is generally assumed that the dynamical correlation factor $\dot{g}_j \approx 1$.

τ_j^{or} can be compared with rotational correlation times from other methods (e.g. $\tau_j^{\mathrm{or}} \approx 3\tau_{\mathrm{NMR}} \approx 3\tau_{\mathrm{Raman}}$ for rotational diffusion) and is connected to molecular volume, V_{m}, and solution viscosity, η, via the *Stokes-Einstein-Debye equation*

$$\tau_j^{\mathrm{or}} = \frac{3V_{\mathrm{m}} f_\perp C}{k_B} \times \frac{\eta}{T} \tag{15}$$

where $f_\perp$ accounts for deviations from spherical shape, and C for the hydrodynamic coupling of molecular rotation to the environment [30] ($C = 1$ for *stick*, $C = 1 - f_\perp^{-2/3}$ for *slip* boundary conditions). For systems dominated by rotational diffusion τ_j generally exhibits Arrhenius

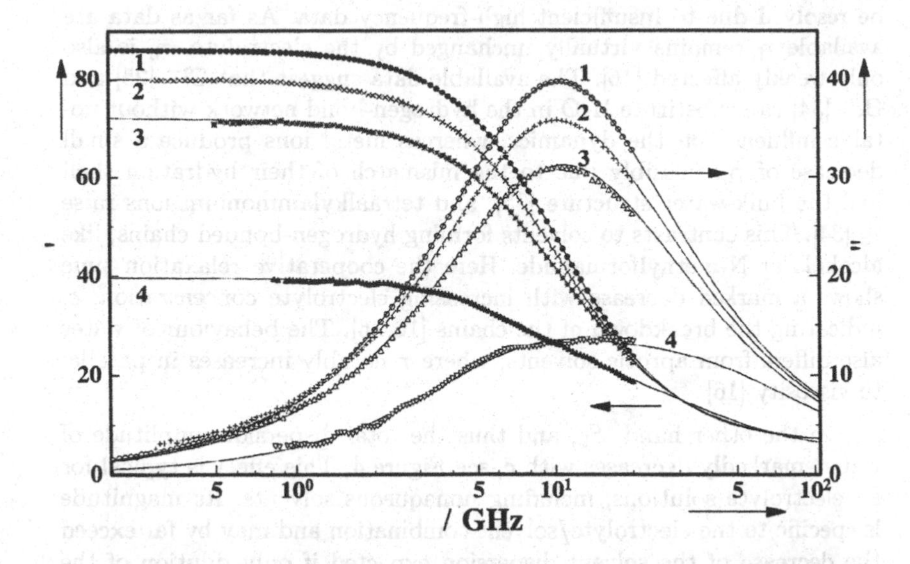

Figure 3. Dielectric dispersion, $\varepsilon'(\nu)$, and loss spectrum, $\varepsilon''(\nu)$, of NaCl solutions in water at 5 °C; 1 pure water, 2 $c = 0.400\,\mathrm{mol\,dm^{-3}}$, 3 $c = 0.990\,\mathrm{mol\,dm^{-3}}$; 4 $c = 4.643\,\mathrm{mol\,dm^{-3}}$ [25]. Experimental spectra 1...3 (symbols) are fitted to a single Cole-Cole equation (lines), spectrum 4 to a superposition of two Debye processes.

behaviour at $T \gg T_g$ and *Vogel-Fulcher-Tammann* behaviour at low temperatures [16, 17].

4. Solvent Relaxation

$\hat{\varepsilon}(\nu)$ of pure water exhibits a large dispersion step around 18 GHz ($\tau_b =$ 8.27 ps, 'b' is for *bulk*) plus a small high-frequency contribution around 400 GHz ($\tau_f \approx 0.4$ ps, $S_f \approx 0.025(\varepsilon - \varepsilon_\infty)$, 'f' for *fast*) [31, 32]. τ_b reflects the cooperative dynamics of the hydrogen-bond network and can be interpreted as the dwelling time a water molecule has to wait on average until all hydrogen bonds except (at most) one are broken, so that it can rapidly rotate with τ_f into a new H-bond configuration with similar energy but different $\vec{M}$. From the activation enthalpy of τ_b the average number of $\bar{n} = 2.5$ H-bonds per water molecule can be deduced, which is in good agreement with results of other methods [33].

Dissolution of inorganic electrolytes, but also of tetramethyl- or tetraethylammonium halides does not change this pattern. The τ_b and τ_f relaxations remain the only water contributions, though S_f cannot always be resolved due to insufficient high-frequency data. As far as data are available τ_f remains virtually unchanged by the electrolyte. τ_b is also only weakly affected [16]. The available data suggest that Cl^- [25] and Br^- [34] can substitute H_2O in the hydrogen-bond network without notable influence on the dynamics, whereas metal ions produce a small decrease of τ_b, possibly due to the mismatch of their hydration shell and the bulk-water structure [35], and tetraalkylammonium ions raise τ_b [34]. This contrasts to solvents forming hydrogen-bonded chains, like alcohols or N-methylformamide. Here the cooperative relaxation time shows a marked decrease with increasing electrolyte concentration, c, indicating the breakdown of the chains [16, 36]. The behaviour of water also differs from aprotic solvents, where τ notably increases in parallel to viscosity [16].

On the other hand, S_b, and thus the total dispersion amplitude of water, markedly decreases with c, see Figure 4. This effect is typical for all electrolyte solutions, including nonaqueous solvents. Its magnitude is specific to the electrolyte/solvent combination and may by far exceed the decrease of the solvent dispersion expected if only dilution of the solvent dipole density and kinetic depolarization were active [16]. The reason is (an at least partial) 'freezing' of solvent molecules due to strong ion-solvent interactions so that they cannot contribute to $\vec{P}_\mu$. With the help of eq. (13) the number of apparently frozen solvent molecules per equivalent of electrolyte, Z_{ib}, can be determined. For aqueous solutions the available data suggest that kinetic depolarization with *slip* boundary conditions applies. Additionally, it can be argued that $Z_{ib}(Cl^-) = 0$ and $Z_{ib}(Br^-) = 0$ because the hydrogen-bond strength between the halide ions and water is similar to the $O\cdots H{-}O$ interactions [34, 37], so that ionic Z_{ib} values can be derived. As an effective solvation number depending on the strength of ion-solvent interactions, Z_{ib} is generally different from coordination numbers determined with scattering techniques, but compares favourably with some of the primary solvation numbers obtained from thermodynamic or transport properties, see Table 2. For the investigated monovalent ions $Z_{ib} \leq$ CN, i.e. only the first hydration shell is affected by the field of the ion, whereas for the divalent ions also the second shell is significantly different from bulk water. It is interesting to note for Na^+ and OH^- that Z_{ib} decreases only weakly with c, whereas for SO_4^{2-} and CO_3^{2-} a pronounced breakdown is observed, see Fig. 4.

At low concentration, $c < 1\,\mathrm{mol\,L^{-1}}$, the decrease of the water dispersion amplitude induced by tetramethyl- or tetraethylammonium halides is entirely due to dilution and kinetic depolarization, i.e. $Z_{ib} = 0$ [41],

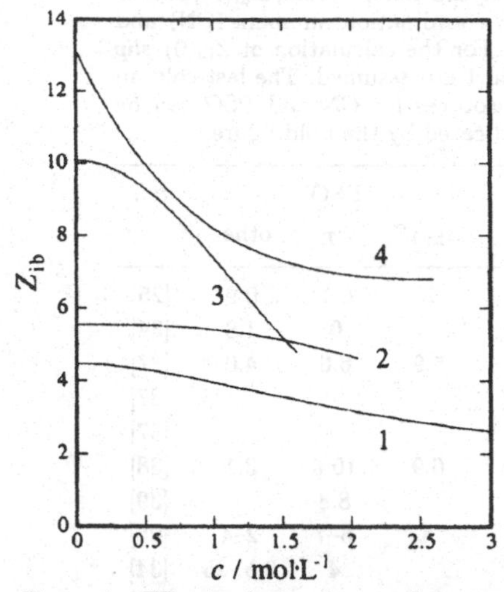

Figure 4. Effective hydration numbers, Z_{ib}, of Na^+ (1, [25]), OH^- (2, [37]), SO_4^{2-} (3, [38]), and CO_3^{2-} (4, [39]) as a function of electrolyte concentration, c, in water at 25 °C.

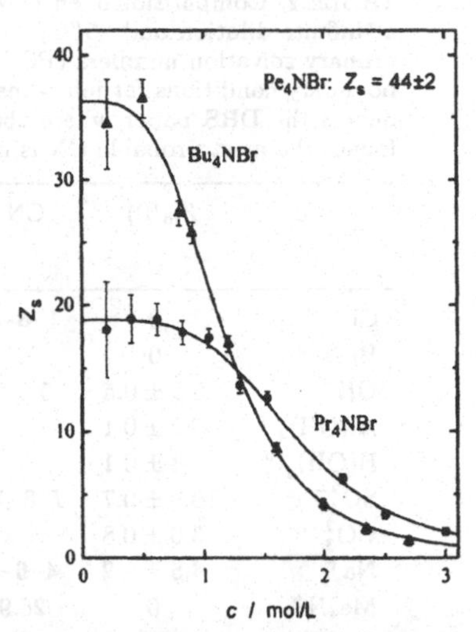

Figure 5. Effective 'hydrophobic' hydration numbers, Z_s, of tetraalkylmammonium bromides as a function of electrolyte concentration, c, in water at 25 °C [34].

whereas at high concentration the breakdown of the bulk-water structure is monitored by a strong decrease of the Kirkwood factor, g [34]. This is also true for the higher homologues. However, from tetrapropylammonium bromide onwards [34], as well as for long-chain alkyltrimethylammonium halides [22], an additional relaxation process appears around 6-8 GHz ($\tau_s \approx 20 - 25$ ps, 's' for *slow* water), that can be attributed to H_2O molecules close to the alkyl groups. Apparently, a sufficiently large hydrophobic surface area is required to observe 'slow' water. At $c \rightarrow 0$ the hydration number Z_s calculated from S_s with eq. (13), see Fig. 5, is smaller than the maximum hydration number possible from geometric criteria, but compares favourably with data derived from compressibility measurements. Z_s markedly decreases as soon as the hydration shells of the ions begin to overlap. This may be taken as an indirect evidence of cation-cation aggregation, as it was shown to occur for aqueous Pr_4NBr by Polydorou *et al.* with a combined neutron diffraction and reverse Monte Carlo study [42]. The observed decrease of Z_s indicates that in concentrated solutions the total hydrophobic surface is minimised.

The temperature dependence of τ_s shows *Eyring* behaviour [34], like

TABLE 2. Comparision of effective hydration numbers, $Z_{ib}(0)$, of ions at infinite dilution and 25 °C, with coordination numbers (CN) and primary solvation numbers (PSO). For the calculation of $Z_{ib}(0)$ *slip* boundary conditions for ion transport are assumed. The last column quotes the DRS paper where the sources for CN and PSO can be found; the most probable CN is indicated by the bold figure.

	$Z_{ib}(0)$	CN	PSO			ref.
			$\Delta_{hyd}S$	κ_T	other	
Cl^-	0	1–**6**–8	3	0.1	0.9	[25]
Br^-	0	6	2	0	1.8	[34]
OH^-	5.5 ± 0.5	5.8 [a]	5.9	6.6	4.0	[37]
$Al(OH)_4^-$ [b]	5.2 ± 0.1					[37]
$B(OH)_4^-$ [b]	5.3 ± 0.1					[37]
SO_4^{2-}	10.0 ± 0.7	7–**8**–12	6.9	10.6	3.1	[38]
CO_3^{2-}	13.0 ± 0.8			8.8		[39]
Na^+	4.5 ± 0.2	4–**6**–8	4	6–7	2–4	[25]
Me_4N^+	0	26.9		4	3.5, 25	[34]
Et_4N^+	0	34.5		8	4.5, 30	[34]

[a] Ref. [40]; [b] in 1.0 M NaOH

τ_b of pure water [33] and the investigated solutions do. However, the enthalpy, $\Delta H^{\neq}$, and entropy, $\Delta S^{\neq}$, of activation are significantly larger than for τ_b. Whereas for the bulk water of the solutions the same average number of hydrogen bonds, $\bar{n}_b = 2.5$, is found as for pure water, a significantly larger value, $\bar{n}_s = 3.5$, is found for the water close to the hydrophobic surface. Probably, the large $\bar{n}_s$ is a consequence of the screening of the hydrogen-bond connections within the 'hydrophobic hydration shell' against the disturbing 'fifth-neighbour' H_2O molecules essential for the dynamics of liquid water.

5. Solute Relaxation

There is considerable evidence that chemical speciation through association of ions to pairs or larger aggregates is a widespread phenomenon and the formation of contact ion pairs (CIP) is regarded as a prerequisite to chemical reactions between ions in solution [1,2]. According to Eigen and Tamm [43] ion-pair formation is generally a three-step process, see Fig. 6. Initial step is the essentially diffusion-controlled formation of an encounter complex from the free solvated cations and anions (state

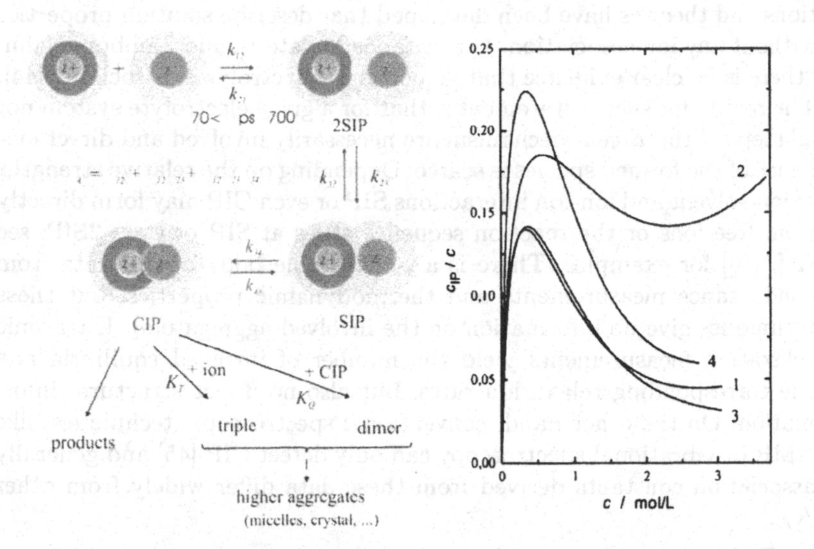

Figure 6. *Eigen* mechanism of ion association [43] and possible subsequent steps.

Figure 7. Relative ion-pair concentrations, c_{IP}/c, in aqueous solutions of (1) Me_4NBr, (2) Et_4NCl, (3) Pr_4NBr, and (4) Bu_4NBr at 25 °C [34].

1). In the formed solvent-separated ion pair (2SIP, 2) both ions keep their first solvation shell. In the subsequent steps solvent molecules are expelled from the region between the two ions to form solvent-shared (SIP, 3) and contact ion pairs (4). Each step is characterized by its stability constant β_{lm} and its forward, k_{lm}, and backward reaction rate, k_{ml} ($l = 1\ldots3$, $m = 2\ldots4$). The overall equilibrium constant is given by

$$K_A = \lim_{I\to0}\beta_A = \lim_{I\to0}(\beta_{12} + \beta_{12}\beta_{23} + \beta_{12}\beta_{23}\beta_{34}) \tag{16}$$

where I is the ionic strength. For symmetrical electrolytes β_A can be written as

$$\beta_A = \frac{c_{IP}}{(c - c_{IP})^2} \tag{17}$$

with $c_{IP} = c_{2SIP} + c_{SIP} + c_{CIP}$ as the total ion-pair concentration.

The mechanism sketched in Fig. 6 was established almost half a century ago to interpret the ultrasonic relaxation data of aqueous 2:2 electrolytes [43]. Nevertheless, there is still considerable discussion in how far ion-pair formation is relevant for the understanding of electrolyte solutions and theories have been developed that describe solution properties without any ion association. For instance Malatesta and Zamboni claim: "there is no clear evidence that [aqueous 2:2 electrolytes] associate" [44]. The reason for such a statement is that for a given electrolyte system not all steps of the *Eigen* mechanism are necessarily involved and direct evidence of the formed species is scarce. Depending on the relative strengths of ion-solvent and ion-ion interactions SIP or even CIP may form directly from free ions or the reaction sequence stops at SIP or even 2SIP, see Ref. [16] for examples. There is a considerable body of K_A data from conductance measurements and thermodynamic properties, but these techniques give no information on the involved aggregate(s). Ultrasonic relaxation measurements yield the number of involved equilibria and the corresponding relaxation rates, but also no direct structural information. On the other hand, conventional spectroscopic techniques, like NMR or vibrational spectroscopy can only detect CIP [45] and generally association constants derived from these data differ widely from other K_A.

Dielectric spectroscopy is sensitive to any species with $\mu_k > 0$ provided its lifetime is at least comparable to the rotational correlation time, τ_k^{or} . Since the dipole moment of an ion pair is essentially determined by the separation, d, of the charges z_- and z_+ of anion and cation and the corresponding dispersion amplitude is proportional to μ_k^2, see eq. (13), the sensitivity towards ion pairs increases in the sequence CIP < SIP < 2SIP. Thus, DRS is — in contrast to other techniques — able to detect SIP and 2SIP unambigously. One may even say that the contributions SIP and especially 2SIP to $\hat{\varepsilon}(\nu)$ are amplified with respect to the dispersion amplitudes of CIP and the solvent.

The ion-pair relaxation time is essentially governed by the *Stokes-Einstein-Debye equation* (15) and thus determined by the molecular volume, so that for electrolyte solutions of common solvents ion-pair relaxation processes may be expected in the 0.3 to 5 GHz range, see Fig. 1. Since the volumes of CIP, SIP and 2SIP are significantly different, their contributions to $\varepsilon''(\nu)$ peak at different frequencies and the experimental relaxation time can be used as a hint to the ion-pair species present. With the now accessible accuracy of the spectra it is even possible to extract the amplitudes, S_j, and relaxation times, τ_j, of simultaneously present ion-pair species from $\hat{\varepsilon}(\nu)$ [38, 39].

From S_j the corresponding ion-pair concentration is obtained with eq. (13). Since the involved species is usually not known *a priori* the required input parameters μ_j, α_j, f_j, and A_j are calculated from geo-

TABLE 3. Association constants, K_A, ion-pair species, rate constants of ion-pair formation, k_{1m}, and ion-pair lifetimes, $\tau_L = \ln 2/k_{m1}$, of aqueous electrolytes at 25°. The last column quotes the DRS paper where the references for K_A can be found. For the electrolytes where kinetic data are available $m = 4$, see text.

	K_A		species	$k_{1m} \times 10^{-9}$	τ_L	ref.
	DRS	literature		$L \cdot mol^{-1} s^{-1}$	$10^{-9}s$	
Me_4NBr	3.0 ± 0.4	1.24, 1.49	CIP	1.7	1.2	[34]
Et_4NCl	3.1 ± 0.6	2.3, 2.5	PIP	6	0.4	[34]
Et_4NBr	3.74^a	3.3, 3.4, 3.74, 7.5	PIP	4.0	0.65	[34]
Pr_4NBr	2.8 ± 0.4	3.1, 4.2, 4.7	PIP	1.12	1.74	[34]
Bu_4NBr	5.6 ± 0.8	2.9, 4.6, 4.8	PIP	1.2	3.2	[34]
Pe_4NBr	7.4^a	7.4	PIP	5	1	[34]
$LaFe(CN)_6$	5300^a	5300	CIP	3.9	96	[21]
Na_2SO_4	6.7	5.8, 6.6, 6.82	2SIP, SIP			[38]
Na_2CO_3	9.5	9.1, 18.6, 19.4	2SIP, SIP			[39]

[a] fixed to literature value

metric data for all conceivable ion pairs [28] and the resulting $\beta_A(I)$ are extrapolated to K_A with the help of a *Guggenheim*-type equation

$$\log \beta_A = \log K_A - \frac{2A_{DH}\,|z_+z_-|\,\sqrt{I}}{1+B\sqrt{I}} + b_\beta I + c_\beta I^{3/2} \qquad (18)$$

A_{DH} is the *Debye-Hückel* parameter of the solvent [38]. Depending on salt and solvent, usually only one or two of the empirical parameters B, b_β and c_β are required.

The obtained model-specific association constants are then compared with data from thermodynamic or conductance measurements to infer the ion-pair species. Due the high sensitivity of c_j to μ_j^2 and to the uncertainty of the hydrodynamic coupling of ion-pair rotation to viscosity — generally the experimental C is between *stick* and *slip* limits — this approach is more accurate than the analysis of the relaxation time. Table 3 summarizes the results of recently investigated aqueous electrolytes. Keeping in mind that for conventional methods values of $K_A < 20$ are difficult to access, the agreement with DRS data is very good.

For tetraalkylammonium halide solutions [34] as well as for $LaFe(CN)_6$ [21] a single ion-pair relaxation process is observed. Based on K_A and τ_{IP}^{or} the dispersion step of Me_4NBr and the $LaFe(CN)_6$ can be unequivocally attributed to CIP. For these electrolytes the intermediate steps 2SIP

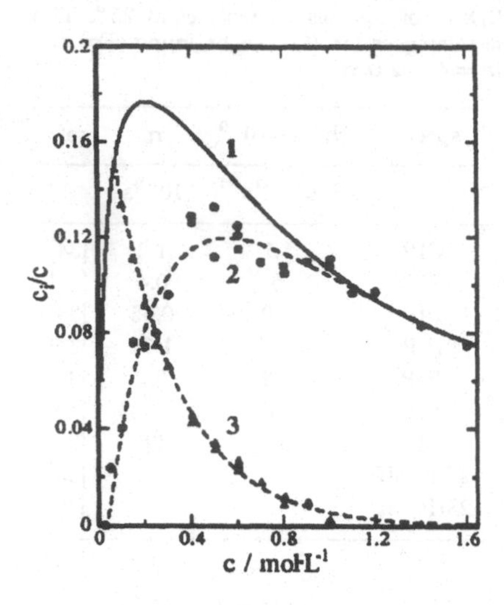

Figure 8. Relative concentrations, c_{IP}/c, of all ion-pairs (curve 1), of double solvent-separated (2SIP, curve 2) and of solvent-shared ionpairs (SIP, curve 3), in aqueous Na_2SO_4 solutions at 25 °C [38].

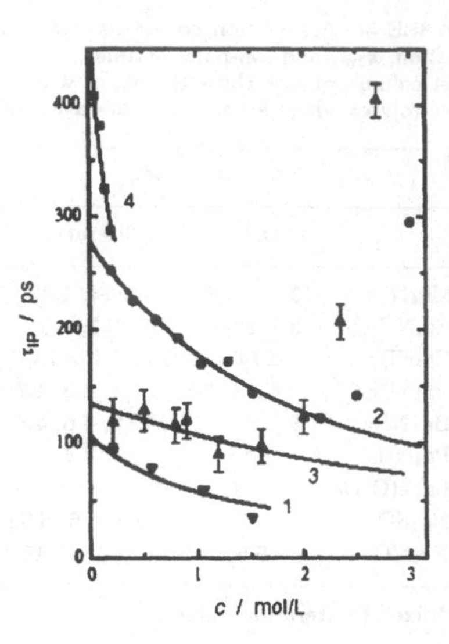

Figure 9. Ion-pair relaxation times, τ_{IP}, of (1) Et_4NBr, (2) Pr_4NBr, (3) Bu_4NBr, and (4) Pe_4NBr in aqueous solutions at 25 °C [34]. The lines are calculated with eq. 21.

and SIP of the Eigen association scheme, Fig. 6, are not stable species. In the case of the larger tetraalkylammonium ions DRS association constants and literature data can only be reconciled if it is assumed that the anion penetrates the cation, reaching the α-methylene group. Such penetration ion pairs (PIP), again directly formed from the free ions, are consistent with previous NMR results [46]. Figure 7 shows that despite the small association constants between 14 and 22 % of the ions are associated at $c \approx 0.3\,\mathrm{mol\,L^{-1}}$ and thus ion pairs cannot be neglected in the modeling of solution properties. At higher concentrations c_{IP}/c decreases significantly, as expected from the concentration dependence of activity coefficients. Unexpectedly, c_{IP}/c rises again for large concentrations of Et_4NCl and possibly Me_4NBr. An explanation for this behaviour is still lacking.

For sodium sulfate [38] and carbonate [39] two ion-pair relaxation processes can be detected and assigned to 2SIP and SIP. According to Figure 8 there is an initial rapid rise of the 2SIP concentration in Na_2SO_4

solutions. But after $\approx 0.08\,\mathrm{mol\,L^{-1}}$ this contribution decreases and 2SIP disappear when the solubility limit ($\approx 1.7\,\mathrm{mol\,L^{-1}}$) is approached. At intermediate and high concentrations SIP is the dominating species. It reaches the level of 2SIP around $0.3\,\mathrm{mol\,L^{-1}}$ and passes through a maximum at $0.7\,\mathrm{mol\,L^{-1}}$. For Na_2CO_3 a similar pattern is observed. From the data of Fig. 8 the stability constants $\beta_{12} = c_{2SIP}/(c_+c_-)$ and $\beta_{23} = c_{SIP}/c_{2SIP}$ are immediately availble. Up to the saturation limit no CIP are detected for these 2:1 electrolytes. Probably, this is not only due to strong Na^+–H_2O interactions, $Z_{ib}(Na^+, 0)$ is only 4.5, but also favoured by hydrogen-bonding of the 'squeezed' water molecule to sulfate. For $LaFe(CN)_6$ obviously ion-solvent interactions are not sufficient to outweigh the strong *Coulomb* attraction of the trivalent ions.

Provided that subsequent steps are absent or at least slow compared to the establishment of the first equilibrium [47], it can be shown [48] that the relaxation time τ_{IP1} of the ion pair formed from the free ions (usually $\tau_{IP1} = \tau_1$) contains a kinetic contribution, τ_{ch}, in addition to τ^{or}_{IP1} from the tumbling motion of the species

$$(\tau_{IP1})^{-1} = (\tau^{or}_{IP1})^{-1} + (\tau_{ch})^{-1} \tag{19}$$

with

$$(\tau_{ch})^{-1} = k_{m1} + 2k_{1m}(c_+ + c_-) \tag{20}$$

$m = 2$, 3 or 4 depending on the ion pair formed.

Neglecting the viscosity dependence of τ^{or}_{IP1} [49], the rate constant of formation, k_{1m}, and the effective volume of rotation, $V_e = V_m f_\perp C$ of the ion pair can be determined for a symmetrical electrolyte with the equation

$$\frac{1}{\tau_{IP1}} = \frac{3\eta(0)V_e}{k_B T} + k_{1m}\left[\frac{1}{K_A} + 2(c - c_{IP})\right] \tag{21}$$

$\eta(0)$ is the viscosity of the pure solvent. Figure 9 shows an example for such an analysis and Table 3 lists the data of k_{1m} and the ion-pair lifetimes, $\tau_L = \ln 2/k_{m1}$, of aqueous tetraalkylammonium halides and $LaFe(CN)_6$. For these salts $m = 4$ and $K_A = \lim_{I\to 0}(k_{1m}/k_{m1})$ since no 2SIP or SIP are detected. In all cases the experimental formation-rate constant is roughly by a factor of 2-10 smaller than the value expected for diffusion-controlled ion-pair formation, indicating that the rearrangement of the hydration shell has some influence on the kinetics of ion association in these systems. Although tetraalkylammonium halide ion pairs are rather labile their lifetime generally exceeds τ^{or}_{IP1} by a factor of 2 or more and is significantly larger than the value expected for diffusion-controlled decay. From these data it is obvious that even

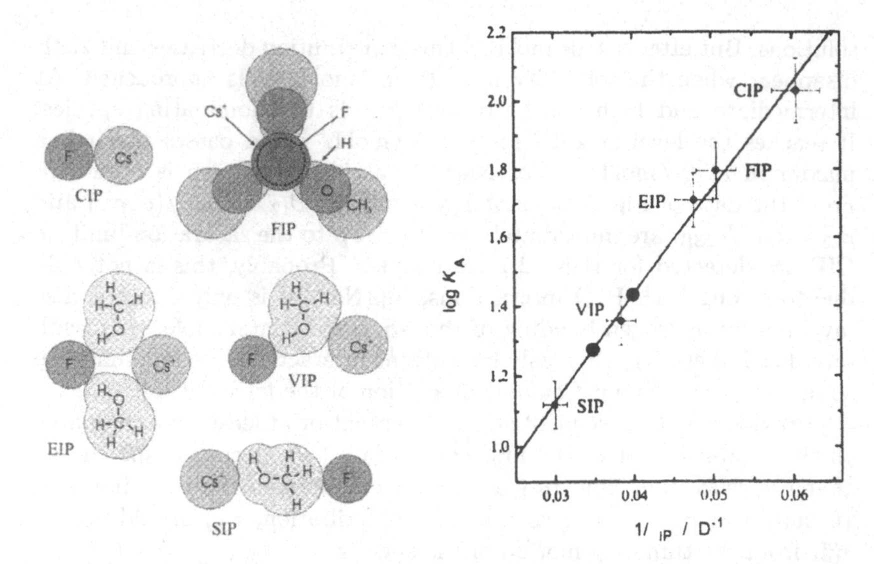

Figure 10. Representations of ion-pair models for CsF in methanol in order of increasing charge separation d: contact (CIP); face-centred (FIP, view along Cs^+–F^- axis); edge-centred (EIP); vertex-centred (VIP); and solvent-shared (SIP) ion pairs [36]. Only the methanol molecules relevant for d are shown around F^- which is assumed to have an octahedral solvation shell.

Figure 11. $\log K_A$ *versus* the inverse ion-pair dipole moment, μ_{IP}^{-1}, of various ion-pair models for CsF in methanol at 25 °C [36]. Also indicated are the values of $\log K_A$ determined by Hefter and Salomon [51] from conductivity measurements (•) with ($\log K_A = 1.443$) and without ($\log K_A = 1.274$) inclusion of the Chen effect.

in weakly associating electrolytes ion pairs are not ephemeral entities solely based on random encouters of cations and anions but more or less stable chemical species. Otherwise also NaCl or similar non-associating electrolytes should exhibit solute relaxation processes in DRS, which is not the case [25, 50].

As a final example for the potential of DRS in solution chemistry CsF in methanol [36] is presented. From conductance measurements it was known that the electrolyte is weakly associated with $K_A \approx 25$, but (as common for fluoride solutions) few further information was available.

The dielectric data revealed a marked decrease of the solvent dispersion amplitude and a small, statistically well defined ion-pair relaxation process. From the methanol relaxation it can be inferred that F^+ is strongly coordinated to 6 alcohol molecules, which has a pronounced effect on the 'bulk' methanol structure. The dipole correlation factor rapidly breaks down from $g \approx 3$ in the pure solvent to about 2 at the saturation limit ($\sim 1\,\mathrm{mol\,L^{-1}}$), indicating a decrease of the average length of the H-bonded methanol chains. Comparison of association constants from dielectric and conductance data reveals that the ion pair is neither contact nor 'conventionally' solvent-shared, see Figure 10. According to Figure 11 DRS values of K_A derived for SIP and CIP are either significantly too small or too large. Model calculations for other conceivable ion-pair structures show that in methanol CsF forms vertex-centered (VIP) ion pairs where anion and cation coordinate via the OH-group of the alcohol.

6. Concluding Remarks

The previous pages give a flavour of what kind of information can be obtained with dielectric spectroscopy in the field of solution chemistry. Obviously, the focus of the presentation as well as the selection of the examples where guided by the preferences of the author. But we may reiterate the following points to highlight the key issues:

– Time scale $0.1 \leq \tau/\mathrm{ps} \leq 10000$

  At least in principle (though the terahertz range is still difficult to access) the entire time scale relevant to solution chemistry can be probed.

– Solvent dynamics

  Information on molecular rotation and cooperative motions can be obtained. Especially, the cooperative dynamics of hydrogen-bonding liquids gives rise to distinct relaxation processes.

– Solvent / solution structure

  DRS yields the Kirkwood factor, describing the static dipole-dipole correlations, and/or the effective solvation number of the solvent as a function of electrolyte concentration. A careful comparison of Z_{ib} with coordination numbers from scattering experiments and with results from molecular dynamics simulations aimed at the solvation-shell dynamics should yield important information on the relative strengths of ion-solvent and solvent-solvent interactions. Water close to sufficiently large hydrophobic solutes exhibits a distinct relaxation process.

– Ion-pair formation

  In contrast to other techniques DRS can unambiguously detect not only contact ion pairs, but also solvent-separated and solvent-shared species. With the help of model calculations and literature data for K_A DRS allows the identification of the relaxing ion-pair species, yields the corresonding concentrations as a function of ionic strength and allows the calculation of stepwise and overall stability constants. In favourable cases the technique can provide the rate constants of the first association step.

– Micelles

Due to limitations of time and space micelles were not treated in the presentation, but an example for a DRS study is quoted [22]. This paper shows that nowadays information on the size of the micelles, the extent of counter-ion binding, micelle hydration and inter-micellar interactions can be deduced.

Although already a classical technique, the full potential of DRS in solution chemistry and related fields is just emerging with the recent advances in microwave technology, especially the development of accurate vector networkanalyzers. It is hoped that this contribution not only gives an overview suitable for solution chemists using other techniques, but also stimulates researchers to use and improve this method.

ACKNOWLEDGEMENTS

It is a pleasure to thank J. Barthel, G. Hefter, W. Kunz and the other coworkers involved in the investigations presented in this review for the fruitful and enjoyable collaboration. Financial support by the Deutsche Forschungsgemeinschaft and the Australian Research Council is gratefully acknowledged.

References

1. J.O'M. Bockris and A.K.N. Reddy, *Modern Electrochemistry 1: Ionics*, Plenum, New York, 2nd. Ed. (1998).
2. J. Barthel, H. Krienke, and W. Kunz, *Physical Chemistry of Electrolyte Solutions*, Steinkopff/Springer, Darmstadt/New York (1998).
3. A.R. Burkin, *Chemical Hydrometallurgy*, Imperial College Press, London (2001).
4. H. Ohtaki and T. Radnai, *Chem. Rev.* **93**, 1157 (1993).
5. H. Krienke and J. Barthel, *Ionic Fluids*, in: J.V. Sengers, R.F. Kayser, C.J. Peters, and H.J. White Jr. (Eds.), *Equations of State for Fluids and Fluid Mixtures*, Elsevier, Amsterdam (2000).
6. E. Königsberger and L.-C. Königsberger, *Pure Appl. Chem.* **73**, (2001) 785.
7. P. Sipos, P.M. May, G.T. Hefter, and I. Kron, *J. Chem. Soc., Chem. Commun.*, 2355 (1994).
8. R. Buchner, P. Sipos, G. Hefter, and P.M. May, *J. Phys. Chem. A*, **106**, 6527 (2002).
9. R. Buchner and J. Barthel, *Annu. Rep. Prog. Chem., Sect. C* **97**, 349 (2001).
10. a) J. Barthel, R. Buchner, and M. Münsterer, *Electrolyte Data Collection, Part 2: Dielectric Properties of Water and Aqueous Electrolyte Solutions*, in: G. Kreysa (Ed.), *Chemistry Data Series*, Vol. XII, DECHEMA, Frankfurt (1995).
    b) J. Barthel, R. Buchner, and M. Münsterer, *Electrolyte Data Collection, Part*

2a: Dielectric Properties of Nonaqueous Electrolyte Solutions, in: G. Kreysa (Ed.), *Chemistry Data Series*, Vol. XII, DECHEMA, Frankfurt (1996).
11. a) A.K. Jonscher, *Dielectric Relaxation in Solids*, Chelsea Dielectric Press, London (1983);
b) Special Issue on Dielectric Properties of Ferroelectrics (dedicated to A.R. von Hippel), Eds. B. Hilczer, T. Mitsui and V.H. Schmidt, *Ferroelectrics* **135** (1992).
12. C.C. Ku and R. Liepins, *Electrical Properties of Polymers*, Hanser, München (1987).
13. L.M. Blinov, *Electro-optical and Magneto-optical Properties of Liquid Crystals*, Wiley, Chichester (1983).
14. E.H. Grant, R.J. Sheppard and G.P. South, *Dielectric Behaviour of Biological Molecules in Solution*, Clarendon, Oxford (1978).
15. a) G. Smith, A.P. Duffy, J. Shen and C.J. Olliff, *J. Pharm. Sci.* **84**, 1029 (1995);
b) D.Q.M. Craig, *Dielectric Analysis of Pharmaceutical Systems*, Taylor & Francis, London (1995).
16. J. Barthel, R. Buchner, P.-N. Eberspächer, M. Münsterer, J. Stauber, and B. Wurm, *J. Mol. Liq.* **78**, 82 (1998).
17. F. Kremer and A. Schönhals (eds.), *Broadband Dielectric Spectroscopy*, Springer, Berlin (2002).
18. a) C.F.J. Böttcher, *Theory of Electric Polarization*, Vol. 1, 2nd ed., Elsevier, Amsterdam (1973);
b) C.F.J. Böttcher and P. Bordewijk, *Theory of Electric Polarization*, Vol. 2, 2nd ed., Elsevier, Amsterdam (1978).
19. B.K.P. Scaife, *Principles of Dielectrics*, Clarendon, Oxford (1989).
20. M.R. Moldover, K.N. Marsh, J. Barthel and R. Buchner, *Relative Permittivity and Refractive Index*, in: A.R.H. Goodwin, K.N. Marsh, and W.A. Wakeham (eds.), *Measurement of the Thermodynamic Properties of Single Phases*, Ch.9, *in press*.
21. R. Buchner, J. Barthel and J.B. Gill, *Phys. Chem. Chem. Phys.* **1**, 105 (1999).
22. C. Baar, R. Buchner, and W. Kunz, *J. Phys. Chem. B* **105**, 2906 and 2914 (2001).
23. Yu. Feldman, T. Skodvin, and J. Sjöblom, *Dielectric Spectroscopy on Colloidal Systems — A Review*, in: P. Becher (ed.), *Encyclopedia Handbook of Emulsion Technology*, Vol. 5, Marcel Dekker, New York (2001).
24. R. Buchner and J. Barthel, *Ber. Bunsenges. Phys. Chem.* **101**, 1509 (1997).
25. R. Buchner, G.T. Hefter, and P.M. May, *J. Phys. Chem. A* **103**, 1 (1999).
26. J. Barthel, K. Bachhuber, R. Buchner, H. Hetzenauer, and M. Kleebauer, *Ber. Bunsenges. Phys. Chem.* **95**, (1991) 853.
27. F. Alvarez, A. Alegria, and J. Colmenero, *Phys. Rev. B* **60**, 984 (1991).
28. J. Barthel, H. Hetzenauer, and R. Buchner, *Ber. Bunsenges. Phys. Chem.* **96**, 1424 (1992).
29. P. Madden and D. Kivelson, *Adv. Chem. Phys.* **56**, 467 (1984).
30. a) J.L. Dote, D. Kivelson, and R.N. Schwartz, *J. Phys. Chem.* **85**, 2169 (1981);
b) J.L. Dote and D. Kivelson *J. Phys. Chem.* **87**, 3889 (1983).
31. J. Barthel, K. Bachhuber, R. Buchner, and H. Hetzenauer, *Chem. Phys. Lett.* **165**, 369 (1990).
32. C. Rønne, L. Thrane, P.O. Åstrand, A. Wallqvist, K.V. Mikkelsen, and S.R. Keiding, *J. Chem. Phys.* **107**, 5319 (1997).
33. R. Buchner, J. Barthel, and J. Stauber, *Chem. Phys. Lett.* **306**, 57 (1999).
34. R. Buchner, C. Hölzl, J. Stauber, and J. Barthel, *Phys. Chem. Chem. Phys.* **4**, 2169 (2002).
35. T. Chen, G. Hefter, and R. Buchner, submitted to *J. Phys. Chem. A*.
36. R. Buchner and G. Hefter, *J. Solution Chem.* **31**, 517 (2002).
37. R. Buchner, G.T. Hefter, P.M. May, and P. Sipos, *J. Phys. Chem. B* **103**, 11186

(1999).
38. R. Buchner, S.G. Capewell, G.T. Hefter, and P.M. May, *J. Phys. Chem. B* **103**, 1185 (1999).
39. S.G. Capewell, R. Buchner, G.T. Hefter, and P.M. May, *Phys. Chem. Chem. Phys.* **1**, 1933 (1999).
40. M.E. Tuckerman, D. Marx, and M. Parinello, *Nature* **417**, 925 (2002).
41. Due to lacking low-frequency data the Z_{ib} discussed in [16] result from an incorrect separation of the ion-pair relaxation process, see [34].
42. N.G. Polydorou, J.D. Wicks, and J.Z. Turner, *J. Chem. Phys.*, 1997, **107**, 197.
43. M. Eigen and K. Tamm, *Z. Elektrochem.* **66**, 93 and 107 (1962).
44. F. Malatesta and R. Zamboni, *J. Solution Chem.* **26**, 791 (1997).
45. M.H. Brooker, *Raman Spectroscopic Measurements of Ion Hydration*, in R.R. Dogonadze, E. Kálmán, A.A. Kornyshev, and J. Ulstrup (Eds.), *The Chemical Physics of Ion Hydration*, Part B, Ch. 4, Elsevier, Amsterdam (1986).
46. M. Krell, M. Symons, and J. Barthel, *J. Chem. Soc., Faraday Trans. 1* **83**, 3419 (1987).
47. This is not the case for aqueous Na_2SO_4 and Na_2CO_3.
48. R. Buchner and J. Barthel, *J. Mol. Liq.* **63**, 55 (1995).
49. Compared with experimental accuracy this effect is small, but can be included [28].
50. K. Nörtemann, J. Hilland, and U. Kaatze, *J. Phys. Chem. A* **101**, 6864 (1997).
51. G.T. Hefter and M. Salomon, *J. Solution Chem.* **25**, 541 (1996).

DYNAMICS IN INTRA-MOLECULAR POLYMER MIXTURES

P. HOLMQVIST AND G. FYTAS
FORTH-Institute of Electronic Structure and Laser, P.O. Box 1527, 71110 Heraklion, Crete, Greece

Abstract: A symmetric and asymmetric high molecular mass diblock copolymer are utilized to experimentally address the main parameters controlling the dynamic structure factor, S(q,t), in this class of self-assembled materials. The spatio-temporal variation of S(q,t) can be theoretically described in the two regimes of the phase diagram.

1. INTRODUCTION

Covalently linking of two chemically distinct polymeric chains (A,B) produces intra-molecular mixtures A-B as opposed to the binary mixture of homopolymers i.e. a blend A/B. In the latter, the relevant composition fluctuations are long-range and become long-lived near the critical point for macro-phase separation [1]. Unfavorable enthalpic interactions between monomers, when summed over the chains, lead to strong slowing-down of the inter-diffusion of the chemically dissimilar chains (A,B) over long length-scales [2]. This phenomenon bears analogies to binary molecular liquid mixtures. Alternatively, intra-molecular mixtures, i.e. diblock copolymers A-B, result in a micro-phase separation when the enthalpic interactions dominate[3,4] and there are no analogues in molecular liquids.

The static structure factor, S(q), has been thoroughly utilized to investigate disordered diblock copolymers in the bulk and in solutions [5]. The evolution of the maximum of the static structure factor, S(q*), and the characteristic spacing, $2\pi/q^*$, with either temperature (melt) or concentration (solutions) approaching the ordered to disordered transition (ODT) is well established. The phenomenon of internal diffusion in diblock copolymer has only recently received attention, especially for wave-vectors, q, near q* [6-10]. The understanding of the dynamic response that influences the rheological properties of AB matrices requires sensitive measurements of the intermediate scattering function S(q,t) over a broad time range in the relevant q-region.

Sufficiently high molecular mass AB block copolymers allows for an employment of dynamic light scattering at $q \cong O(q^*)$. The determination of the dynamic response from photon correlation spectroscopy (PCS) relies on the identification of the dominant mechanisms to relax the order parameter (composition) fluctuations, $\phi_q(t)$, manifested in the intermediate scattering function (dynamic structure factor) S(q,t) of the system. Two main modes can be identified as the overall chain relaxation with characteristic rate $\Gamma_2 \sim 1/\tau_0$ (τ_0 is the longest chain relaxation time) and chain self-

J. Samios and V.A. Durov (eds.), Novel Approaches to the Structure and Dynamics of Liquids: Experiments, Theories and Simulations, 289–294.

diffusion with $\Gamma_1=D_sq^2$ (D_s is the self diffusion coefficient). The development of these modes with the wave vector, q, and concentration, ϕ, and their dependence of polydispersity, composition of block A, f_A, and block type will be briefly presented.

2. EXPERIMENTAL S(Q,T)

Two anionic polymerised ultra high molecular weight block copolymers were chosen as reference systems in this report [8]. The two samples are characterised as a symmetric (50 % styrene and M_w=950 kg/mol) polystyrene-polyisoprene diblock copolymer, SI50, and an asymmetric (85 % styrene and M_w=1.6 Mg/mol) polystyrene-poly(isoprene -random-ethylene-alt-propylene) diblock copolymer, SEP80.

The characteristic diblock copolymer peak in the $S(q)\propto I(q)/\phi$ falls within the light scattering q's as shown in Figure 1 for solutions of SEP80 in toluene at $\phi<\phi_{ODT}$ (where ϕ_{ODT} is the concentration at which the ODT occurs at 20°C). The reduced intensity, $I(q^*)/\phi$, at q^* increases strongly with ϕ beyond about 4 wt% that signifies the effect of composition fluctuations approaching the ODT. From the static I(q), both the q^* and the intensity at q^*, $I(q^*)/\phi$, can be determined at each concentration and their ϕ-dependence are shown in the two insets to Figure 1. Like the situation in the melt [5], the characteristic spacing, $d=2\pi/q^*$, increases towards the ODT due to chain stretching. The variation of $\phi/I(q^*)$ with concentration also mimics the dependence of $1/I(q^*)$ vs. $1/T$ in undiluted diblock copolymers. The inflection point in the $\phi/I(q^*)$ vs. ϕ plot announces the onset of composition fluctuations and the validity limit of the mean-field region. The ϕ-dependence of these two experimental quantities is at odds with the theoretical predictions [11].

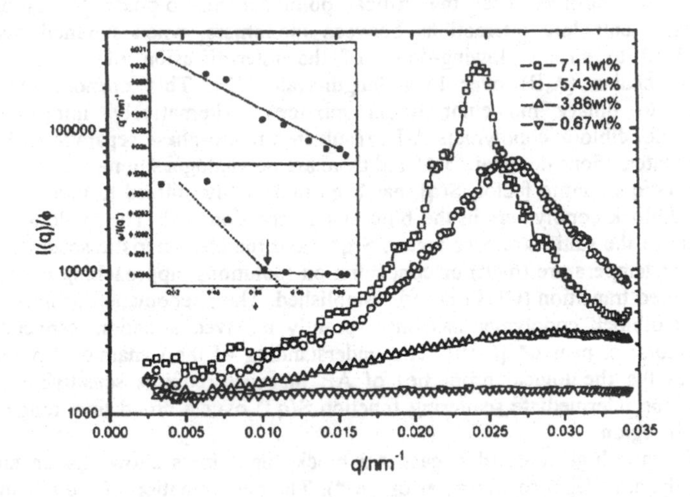

Figure 1. Light scattering intensity distribution $I(q)/\phi$ for disordered solutions of SEP80 in toluene at four different concentrations ϕ at 20°C. Insets: q^* vs. ϕ and $\phi/I(q^*)$ vs. ϕ. The lines are guide for the eye and the arrow indicates the onset of non-mean-field behaviour.

From the intermediate scattering function C(q,t) ($\propto S(q,t)$) the rate, Γ_k, and intensity, I_k, of the different relaxation processes (k=1,2) was extracted via inverse,

Laplace transformation [8]. Three relaxation processes were found [6]. The fastest process relates to the cooperative diffusion [12] responsible for the relaxation of the total diblock concentration fluctuations and is not diblock copolymer specific; hence, it is not addressed in this report. The other two processes relate to the composition fluctuations $\phi_q(t)$ and are identified with the chain relaxation (k=2) and chain self diffusion (k=1) in the mean field regime.

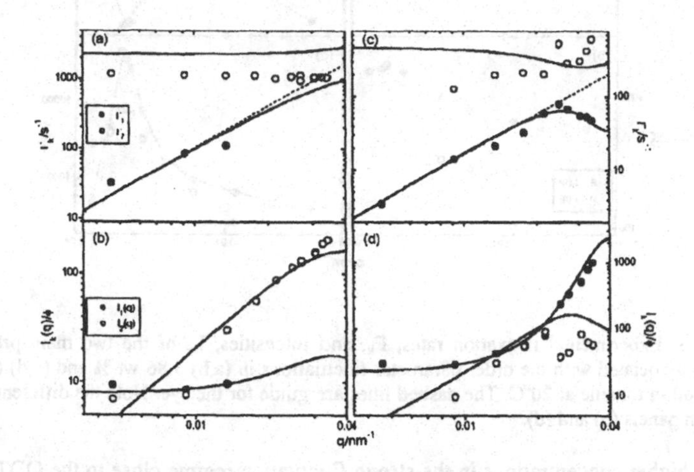

Figure 2. Experimental (symbols) and theoretical (lines) relaxation rates, Γ_k, and intensities, I_k, of the two main processes (k=1,2) associated with the order parameter fluctuations in (a,b) 4.85 wt % and (c,d) 8.76 wt % SI50 in toluene at 20°C. The dashed lines indicate a q^2-dependence for Γ_1. Note the different scales between panels (b) and (d).

To address the effect of diblock copolymer concentration (ϕ), composition polydispersity ($\kappa_0 = \langle f_A^2 \rangle - \langle f_A \rangle^2$), composition of block A (f_A), block type on the q-dependence of the relaxation rates, $\Gamma_k(q)$, and the reduced intensities, $I_k(q)/\phi$ (k=1,2) we utilized SI50 (low polydispersity Figure 2) and SEP80 (high polydispersity Figure 3). We present two concentrations for each system one falling into the mean-field regime (panels a,b) and the other into the non-mean-field regime (panels c,d). For both systems at the low concentration (a in Figures 2 and 3), Γ_2 is a relaxational and Γ_1 a diffusive rate characteristic for an overall chain conformational relaxation and a chain self-diffusion mechanism respectively. The corresponding reduced intensities, $I_k(q)/\phi$, of the two processes are shown in panel (b) of Figures 2 and 3. In the mean-field regime, the same two main processes can be found independently of block type, polydispersity or f_A. However, the extent of composition polydispersity, κ_0, affects the relative contribution of the two processes to relax the thermal collective composition fluctuations. As is evident from Figures 2b and 3b between the two systems the contribution of the slow diffusive process to S(q,t) increases with increasing polydispersity, κ_0. For SI50, the second process (2) is responsible for the peak in I(q*) in contrast to the situation in the more polydisperse (larger κ_0) SEP80. At even higher polydispersity, the slow diffusive process becomes dominant and obscures the relaxational mode (k=2) leading to mono-modal diffusive S(q,t). Theoretically, the slow diffusive process should be light scattering inactive in an ideally "monodisperse" sample.

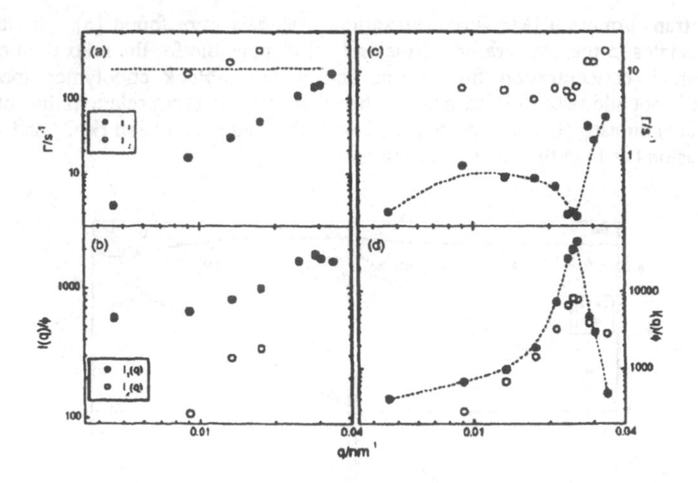

Figure 3. Experimental relaxation rates, Γ_k, and intensities, I_k, of the two main processes (k=1,2) associated with the order parameter fluctuations in (a,b) 3.86 wt % and (c,d) 6.10 wt % SEP80 in toluene at 20°C. The dashed lines are guide for the eye. Note the different scales between panels (b) and (d).

At higher concentrations in the strong fluctuation regime close to the ODT (panels c,d in Figures 2 and 3) S(q,t) is again bimodal (cooperative diffusion is not considered) with pertinent features. The q-dependence of the two rates shown in panels (c) is modified. While far from the ODT, the two processes merge in the vicinity of q* (panels a) thus making their resolution in the time scale increases close to the ODT (panels c) making the resolution feasible near q*. A second pertinent feature is the slowing down of the diffusive process, Γ_1, at q* fully displayed in the SEP80 sample (Figure 3). This behaviour is general for all investigated block copolymers and independent of the polydispersity, f_A, and block type [8,9]. The intensities $I_1(q)$, $I_2(q)$ (panel d in Figures 2 and 3) have exchanged their role close to q*. In the polydisperse sample, SEP80, the diffusive mode still dominates at all q's and is responsible for the strong peak of S(q) at q*. In the less polydisperse SI50, the diffusive mode has also gained intensity by several orders of magnitude at q* (Figure 3b,d) becoming again the dominant process of S(q,t) in the vicinity of q*. On the contrary, the picture at low q's is the same as in the mean-field regime. Note that it is the diffusive mode that is responsible for the increase of S(q*) near the ODT independently of polydispersity, block composition, f_A, and block type. From a dynamic point of view, the low q dynamics are virtually insensitive to the proximity to the ODT.

3. THEORETICAL S(Q,T)

In the framework of the random phase approximation Semenov et al [9,10] have theoretically obtained the S(q,t) for polydisperse diblock copolymer solutions in neutral solvent. For systems following the classic statistics the dynamic structure factor, S(q,t), is related to the generalized linear susceptibility, $\kappa_{ij}(q,t)$. Using the Flory-Huggins model for an incompressible diblock copolymer AB:

$$\kappa_{ij}(q,p)=\kappa(q,p)(2\delta_{ij}-1) \tag{1}$$

$$\frac{1}{\kappa(q,p)}=\frac{1}{\kappa^{(0)}(q,p)}-2\chi \tag{2}$$

$$\kappa^{(0)}(q,p)=\frac{\kappa_{AA}^{(0)}\kappa_{BB}^{(0)}-(\kappa_{AB}^{(0)})^2}{\kappa_{AA}^{(0)}+2\kappa_{AB}^{(0)}+\kappa_{BB}^{(0)}} \tag{3}$$

where δ_{ij} is the Kronecker delta, p the Laplace variable and χ the Flory-Huggins interaction parameter. For compositionally polydisperse copolymers:

$$\kappa_{ij}^{(0)}(q,p)=\int_{-1}^{1}\kappa_{ij}^{0,f}(q,p)\rho(f)df \tag{4}$$

where δ(f) is the volume fraction of kind f and $f=(N_A-N_B)/(N_A+N_B)$ with N_A and N_B being the number of segment of block A and B. The dynamic structure factor of a homogeneous system can be expressed in terms of a superposition of exponentials

$$S(q,t)=\sum_k I_k(q)\exp[-\Gamma_k(q)t] \tag{5}$$

In this case the linear susceptibility is a meromorphic function (all singularities are poles) and all poles are simple and located in the real half-plane, Re(p):

$$\kappa(q,p)=\sum_k\frac{I_k(q)\Gamma_k(q)}{p+\Gamma_k(q)} \tag{6}$$

And the intensities, $I_k(q)$, and rates, $\Gamma_k(q)$, can be calculated from:

$$1/\kappa(q,-\Gamma_k)=0\,,I_k=\frac{1}{\Gamma_k}\mathrm{Res}_{p=-\Gamma_k}\left[\kappa(q,p)\right] \tag{7}$$

The computation of I_k and Γ_k requires the knowledge of κ_0, the size of the chain (R_g), the proximity parameter $\varepsilon=(\phi_{ODT}-\phi)/\phi_{ODT}$, the longest chain relaxation time and the optical contrast. Using a polydispersity of $M_w/M_n=1.03$, the experimentally determined radius of gyration ($R_g=50$ nm) and the proximity parameter ε for SI50, a good agreement between experimental and theory is found as shown by the solid lines in Figure 2. The theoretical predictions capture well the evolution of S(q,t) when it changes from a dominant relaxational to a dominant diffusive character approaching ODT in vicinity to q* (a to c and b to d in Figure 2). The thermodynamic slowing down at q* approaching the ODT is also well described theoretically (c, in Figure 2). Note that the theory overestimates the rate Γ_2 for the overall chain motion.

4. CONCLUDING REMARKS

A good understanding of the dynamic response of block copolymer solutions in neutral solvent is now established. Among the various parameters, it is the polydispersity (κ_0) and proximity (ε) to the ODT that affects the bimodal nature of the dynamic structure factor. Both κ_0 and ε bias the slow diffusive mode (1) and hence determine which of the two processes is responsible for the characteristic peak in the static structure $S(q^*)$.

For sufficiently low κ_0, the increase of the contribution of the diffusive mode (k=1) leads to the intriguing dynamic crossover from a relaxational to the diffusive behaviour of $S(q,t)$. This dynamic crossover occurs in the concentration region where the static $S(q)$ deviates from the mean-field behaviour.

A major advantage of employing ultra high molecular weight block copolymers is the use of PCS at wave vector close to q* albeit in solutions. To extend these studies in the melt, the development of the x-ray-photon spectroscopy has to be awaited for.

REFERENCES

1. Klein, J. *Science* **1990**, *250*(4981), 640.
2. Fytas, G. *Macromolecules* **1987**, *20*, 1430.
3. Bates, F.S. *Science* **1991**, *251*(4996), 898.
4. Leibler, L. *Macromolecules* **1980**, *13*, 1602.
5. Rosedale, J.H.; Bates, F.S.; Almdal, K.; Mortensen, K.; Wignall, G.D. *Macromolecules* **1995**, *28*, 1429.
6. Sigel, R.; Pispas, S.; Hadjichristidis, N.; Vlassopoulos, D.; Fytas, G. *Macromolecules* **1999**, *32*, 8447.
7. Holmqvist, P.; Pispas, S.; Hadjichristidis, N.; Fytas, G.; Sigel, R. *Macromolecules* **2002**, *35*, 3157.
8. Holmqvist, P.; Pispas, S.; Hadijichristidis, N.; Fytas, G.; Sigel, R. *Macromolecules* **2003**, *In press.*
9. Chrissopoulou, K.; Pryamitsyn, V.A.; Anastasiadis, S.H.; Fytas, G.; Semenov, A.N.; Xenidou, M.; Hadjichristidis, N. *Macromolecules* **2001**, *34*, 2156.
10. Semenov, A.N.; Anastasiadis, S.H.; Boudenne, N.; Fytas, G.; Xenidou, M.; Hadjichristidis, N. *Macromolecules* **1997**, *30*, 6280.
11. Fredrickson, G.H.; Leibler, L. *Macromolecules* **1989**, *22*, 1238.
12. De Gennes, P.G., *Scaling concept in Polymer Physics*: Cornell University press, Ithaca NY, 1979.

VISCO-ELASTIC BEHAVIOR AND SMALL ANGLE SCATTERING OF COMPLEX FLUIDS

H. VERSMOLD*, S. MUSA and H. KUBETZKI
Institut für Physikalische Chemie der RWTH, 52062 Aachen, Germany

Author to whom correspondence should be addressed.

Abstract: In this contribution we investigate whether the structure of concentrated shear-ordered dispersions as determined by small angle synchrotron x-ray or neutron scattering can be rationalized in terms of viscoelastic flow behavior as in rheological investigations. Although so far scattering experiments have contributed little to the understanding of rheological systems we are convinced that rheological investigations will profit considerably from a better knowledge of the micro-structure of the dispersions. Sheared dispersions are usually ordered in layers. There are two kinds of ordering of interest: (a) The structure in a layer (b) The structure between the layers. One interesting point of mesogenic systems is their viscoelastic nature. There is strong evidence that the structure in a layer (a) can be treated as elastic or solid-like, whereas the structure (b) between the layers seems to be fluid-like. With scattering experiments one is in the excellent position that two experiments exist with which the two effects can be investigated separately: The solid-like microstructure (a) can be determined at perpendicular incidence. The fluid-like microstructure (b) is given by the scattering intensity along certain Bragg rods (3n±1-rods). In particular this second micro-structure may change in time by aging.

1. INTRODUCTION

In this paper the structure of shear ordered dispersions is considered and how it can be analyzed by scattering techniques [1,2]. In order to understand charge stabilized shear ordered dispersions two states must be considered: (A) an intermediate layered mesogenic and (B) the final crystalline state. This paper is mainly concerned with the structure of the layered state. There are several scattering experiments which show the existence of the two phases. If shear is applied, usually a random stacking, layered system exists first, the stacking structure of which changes slowly to the final crystalline state. A light scattering (LS) observation of this behavior was reported years ago by Clark et al. [3] for a dilute dispersion. The stacking structure of a similar system was investigated years later by Dux et al. [4]. When studying the LS intensity along the c*axis of a $3n\pm 1$-rod, they observed for a somewhat more concentrated dispersion a slow transition from random stacking layers to fcc. Reus et al. [5] studied the structure of dilute dispersions by LS and by small angle x-ray

J. Samios and V.A. Durov (eds.), Novel Approaches to the Structure and Dynamics of Liquids: Experiments, Theories and Simulations, 295–304.

scattering. In her LS investigation she identified the layered structure by missing diffraction lines and was able to relate it to the final bcc crystal structure.

At present the interest has shifted to more concentrated systems, for which LS is no longer the adequate method of analysis because usually such dispersions are non-transparent. Fortunately, meanwhile small angle neutron scattering and small angle synchrotron x-ray diffraction are well developed for structural investigations. These methods also possess the necessary penetration strength to analyze turbid materials.

It seems to be of considerable advantage if the dispersions can be manipulated before the performance of scattering experiments. For example shear flow can be used to orient and order a sample which in turn allows more specific evaluations of the scattering data. Electro- and magneto-rheology have a very interesting potential but no examples are known at present. This paper is organized as follows: In the next chapter the reciprocal space and the Ewald sphere which in the case of small angle scattering becomes Ewald planes is introduced. In order to reconstruct the reciprocal space but also for ordering the sample we used two types of shear cells both for neutron and for x-ray scattering: A disk shear cell (resembling a plate-plate rheometer) and a Couette cell. Two kinds of rotation about two mutual perpendicular axes (called α- and β-axes) can be performed which is described in the next section. In the following chapter experimental results and their discussion are presented. Conclusions which can be drawn from our experiments are given in the final chapter.

2. THEORETICAL CONSIDERATIONS: SMALL ANGLE SCATTERING, RECIPROCAL SPACE, EWALD PLANES, AND α- OR β-ROTATIONS

It is known from rheological as well as scattering investigations that mesogenic systems are visco-elastic, i.e. in certain situations they are viscous like a fluid, in others they are elastic like a crystal. In order to understand the Bragg scattering from such a dispersion we begin with its reciprocal space, a concept well known to solid state physicists.

There is no doubt that energetically the lowest state is crystalline. However, every mechanical manipulation like flow, brings a dispersion into a layered state. For mesogenic substances these layered states can be very long-lived. To be able to describe a dispersion from the very beginning of a shear experiment we assume that the system is layered and close packed in each layer. For charge stabilized dispersions both facts can be shown experimentally, Fig. 1. According to Kittel [1] the reciprocal lattice of such a two-dimensional hexagonal layer is a system of hexagonally arranged Bragg rods which is rotated, however, by 90° with respect to the original particle layer. A cut through the Bragg rods at the height l=0 is shown in Fig. 2a.

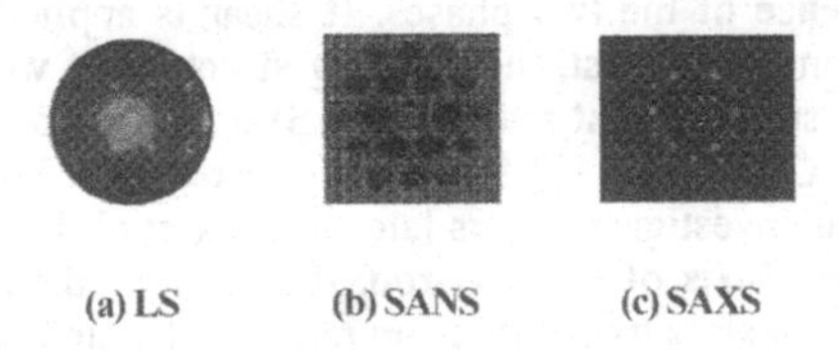

Figure 1: (a) LS, (b) SANS, and (c) Synchrotron SAXS of a layered dispersion.

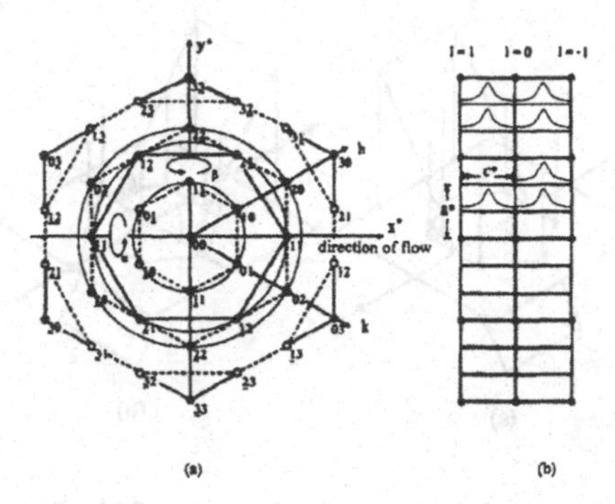

Figure 2 : (a) Top view on the Bagg rods of a layered dispersion ordered by shear with Miller indices *h,k*. (b) View from the side on the Bragg rods after 90° rotation about the axis (1,1)-(0,0)-(1,1).

The rods are hexagonally arranged and two coordinates *h, k* can be used for their enumeration. The coordinates *h, k* are identical with the Miller indexes and can posses the values *h, k=0, ±1, ±2, ...* There are two typesof Bragg rods [2]: With *n* a natural number, rods for which *(h-k)*=3*n* are drawn as filled (black) circles in Fig. 2a. On the other hand rods for which *(h-k)*=3*n*±1 are drawn as open (white) circles in Fig. 2a. If Fig. 2a is rotated by 90° about the axis (-1,1)-(0,0)-(1,-1) one obtains a view on the rods from the side. The scattering intensity along the two types of rods is shown in Fig. 2b. Although it will be discussed later, it is important to note here that it is different for the two types of rods: For rods with *(h-k)*=3*n* (black rods) there are true Bragg reflections. On the other hand rods with *(h-k)*=3*n*±1 (white rods) have a much broader intensity distribution with a maximum shifted in the simplest case by 0.5 *l*. For both types of rods there should be the same period c*.

Next, the occurrence of Bragg reflections will be considered. In terms of the reciprocal lattice a Bragg reflection will occur as soon as two or more points of the reciprocal space are simultaneously situated on the Ewald sphere. This is a sphere in reciprocal space with radius $r^*=2\pi/\lambda_i$, where λ_i is the wave length of the incident radiation. Now, in a small angle experiment the radius $r^*=2\pi/\lambda_i$, is much larger than any reciprocal lattice vector g_{hk}. This means that the curvature of the Ewald sphere is negligible and it can be replaced by a plane.

Thus, for small angle scattering one has the simplified situation that a Bragg reflection occurs as soon as two or more points of the reciprocal lattice are positioned on the Ewald plane. A lattice rotation is equivalent to a rotation of the Ewald plane. Since we want to consider the ordering by flow, the direction of flow is taken as one axis and rotations about it are called α-rotations. Similarly rotations in plane but orthogonal to the previous one are called β-rotations. In Fig. 3 we show how the inner ring of Bragg rods is intersected by α- and by β-rotations in a slightly different manner. For completeness we consider one special case of β-rotation. If β=90° Fig. 4b applies. Now the Ewald plane slices several Bragg rods along the y* axis such that their scattering intensity along the *l* coordinate is visible.

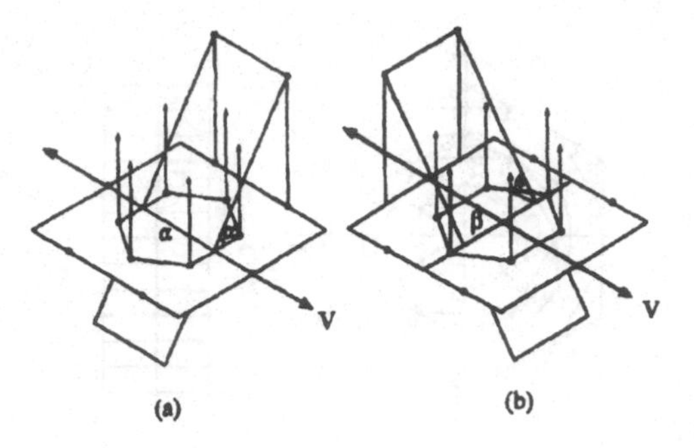

Figure 3 : Inner ring of Bragg rods intersected (a) by α- and (b) by β-rotations

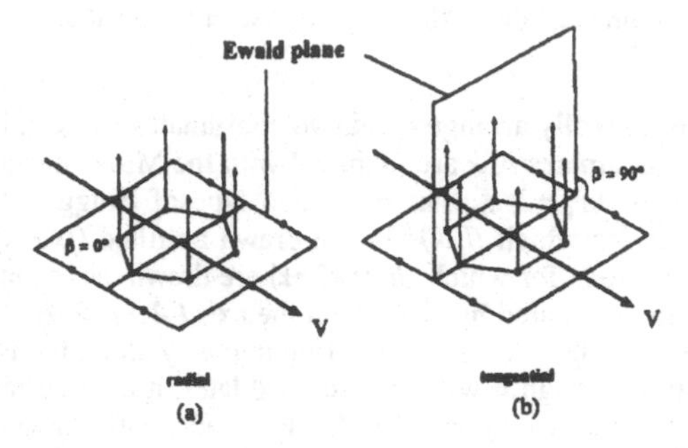

Figure 4: (a) β=0°, Perpendicular incidence, (b) β=90°: Now the Ewald plane slices the Bragg rods along the y* axis. See also Fig. 13

3. EXPERIMENTAL

3.1 Disk shear cell and Couette cell

Schematically, in Fig. 5 our disk shear cell is shown. It consists of a shear disk powered by a little motor. By selecting the rotational speed the desired shear rate can be chosen. For neutron scattering the shear disk and the front- and the backside windows are made from quartz glass free of boron. In a second cell for synchrotron x-ray scattering, almost perfect transmission is obtained with poly-carbonate windows and disk. The shear disk rotates in an Aluminum case coated with Teflon on the inner side which is in contact with the dispersion. At perpendicular incidence, shown in Fig. 7 the Bragg peaks of the solid-like ordered layers become visible. In Fig. 5 the two axes for α- and β-rotations are indicated. Reorientation of our disk shear cell about these axes is possible and necessary for a determination of the intensity along all the Bragg rods. It should be noted that with α-rotations but not with β-rotations the intensity along the rods (-1,1) and (1,-1) is accessible in the usual step by step way. However, as we have shown recently [6], for the rods (-1,1) and (1,-1) or in general for (n,-n)-rods the intensity can be obtained by tangential scattering from a Couette cell.

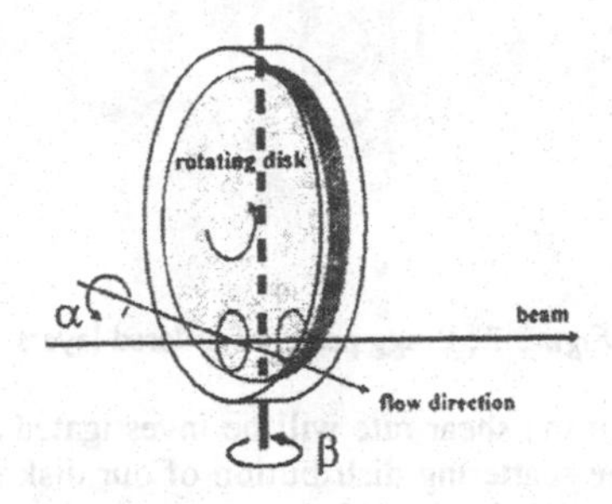

Figure 5 : Our disk shear cell

In order to get familiar with the Couette cell we consider Fig. 6.

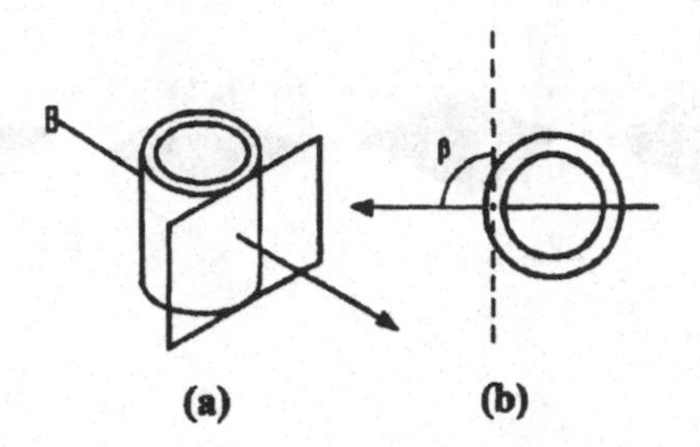

Figure 6 : Couette cell

Shear is usually generated by the outer rotating cup. For neutron scattering the cup and the central static cylinder are usually made from quartz glass, free of boron. For synchrotron x-ray scattering a similar construction with poly-carbonate cup and cylinder is used. The colloidal sample is kept in the gap between the cylinder and the cup. As the outer cup starts rotating a linear shear gradient is set up in the gap filled with the sample. Usually, a translation perpendicular to the beam is carried out with this cell which is equivalent to a β-rotation. If the beam passes through the middle of the cell the scattering is called *radial*, if on the other hand the beam just touches the gap from the side the scattering is called *tangential*.

3.2 Experimental Results and Discussion

As shown in Fig. 7 at perpendicular incidence with synchrotron x-ray scattering many Bragg reflections can be seen. Since all these reflections can also be observed at other sample orientations we conclude that the scattering is due to Bragg rods, i.e. the sample must be layered. The example shown was obtained at the ESRF synchrotron beam line ID02 in Grenoble, France. Our disk shear cell and a dispersion with particles of diameter σ = 94nm and a particle concentration Φ = 34% by volume fraction were used.

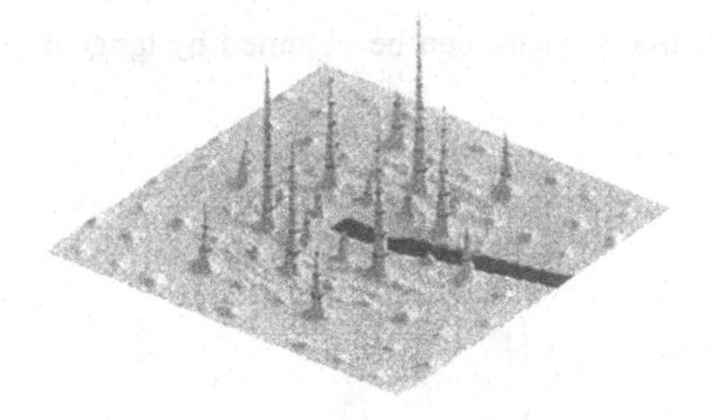

Figure 7 : **Bragg peaks of ordered layers**

Next, the influence of the shear rate will be investigated at normal incidence. In this case α=β=0° and the scattering distribution of our disk shear cell and the Couette cell should be similar. We begin with neutron scattering data as shown in Fig. 8.

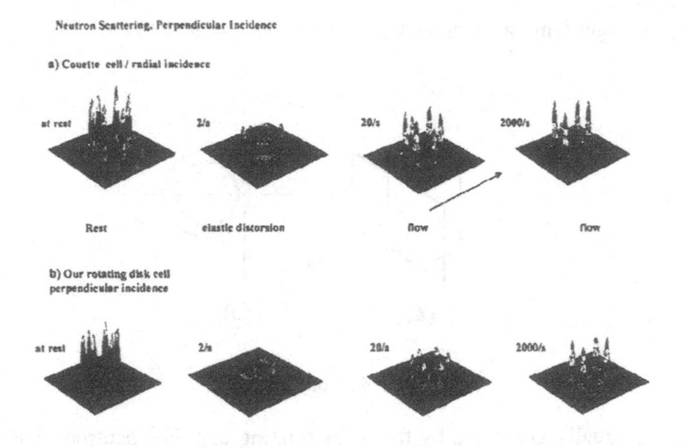

Figure 8 : **(a) Neutron scattering ILL, D11: Couette cell. (b) Neutron scattering ILL, D11: Disk shear cell**

These were determined with the same sample as the one described above but with the D11 small angle neutron spectrometer of the ILL again at Grenoble. Fig. 8a shows data which were obtained with the Couette cell of that research center. Fig. 8b shows the corresponding scattering data, obtained however, with our disk shear cell. With the exception of the amplitudes the two scattering distributions are considered as identical. An interpretation, however, will be given after the presentation of the corresponding synchrotron x-ray results which are of better resolution.

In Fig. 9 a compilation of the synchrotron data is given. At rest we see Bragg peaks, with the smallest shear rate rings and double rings are obvious, with further increase of the shear rate single Bragg reflections are visible again. Such a scenario was first described by Chen et al. [7] in neutron scattering. The rings were assumed to result from randomly oriented crystallites. A different interpretation of the scattering pattern was given by Versmold et al. [8] They assumed that a twist of the layered

system is the reason for the occurrence of the rings. For the layered system one finds a linear relation between Q^2 and *(h2+k2+hk)* [9]. A plot of Q^2 versus *(h2+k2+hk)* should result in a straight line if the assumption of a twisted 2D-system is correct. Fig. 10 shows such a plot and convincingly demonstrates the linear relation between Q^2 and *(h2+k2+hk)*.

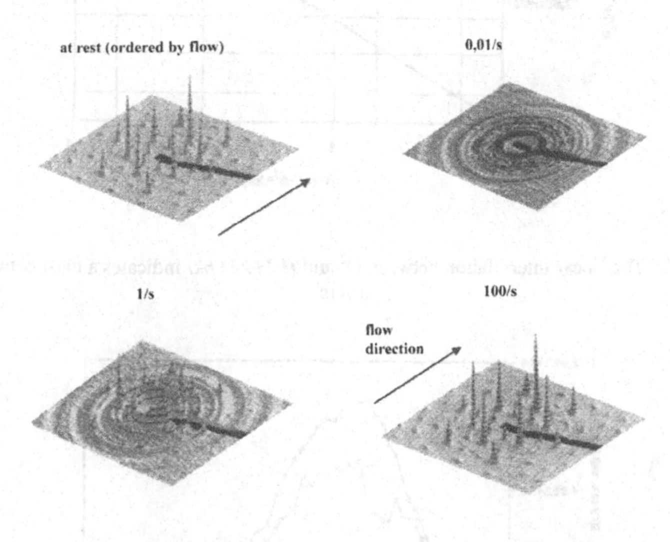

Figure 9 : Synchrotron x-ray scattering

So far due to the spectacular Bragg reflections of a solid-like ordering in the layers this has been investigated [10-13] mainly. Much less spectacular are the liquid-like intensity distributions along the *l* direction of certain rods [14]. Although they are important for mesogenic systems almost no experimental investigations exist. They will be treated next.

First we discuss how the intensity distribution along a Bragg rod can be measured. According to the theoretical section given above, the rod must be intersected by the Ewald plane and the intensity must be determined. There are several ways how this can be done. We consider Fig. 3a again which shows the intersection of the white Bragg rods of the inner ring by the α-plane. By varying, step by step, the angle α the rods are intersected at different heights and the scattering intensity along the rod can be determined. Fig. 3b shows the intersection of the innermost ring by the β-plane. If we vary the angle β the rods will be intersected at different height and the step by step method described above can be applied. As an experimental example we consider Fig. 11 where the scattering power along the Bragg rod (02) for three shear rates is given.

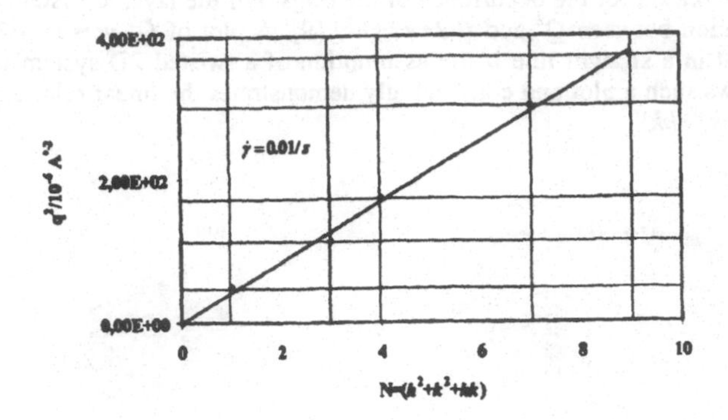

Figure 10 : The linear interrelation between Q^2 and *(h2+k2+hk)* indicates a twist between the layers.

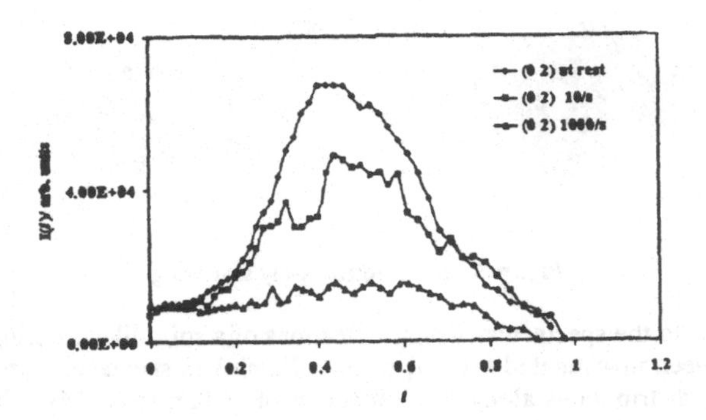

Figure 11 : Step by step measured intensity along the rod (02)

A complication occurs for the rods on the β-rotation axis. They are intersected at l=0 irrespective of the value of β. Here, tangential scattering comes into play. Fig. 4 shows again two kinds of β-intersection. Fig. 4a refers to perpendicular incidence. If on the other hand, β=90° is chosen as in Fig. 4b, the rods on the rotation axis will be sliced and their scattering power all along the rod can be determined in a single shot [6]. Fig. 12 gives a schematic view concerning the scattering power of the rods on the y* axis.

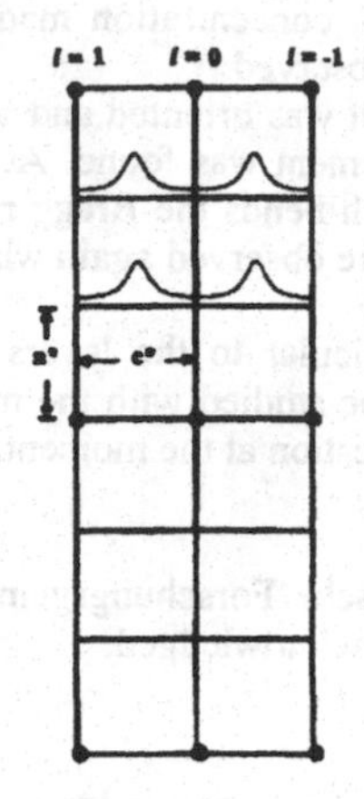

Figure 12 : Schematic view on the rods of the y* axis

The same sequence of the rods, ……black-white-white-black..... is found in tangential scattering from a Couette cell which is shown in Fig. 13.

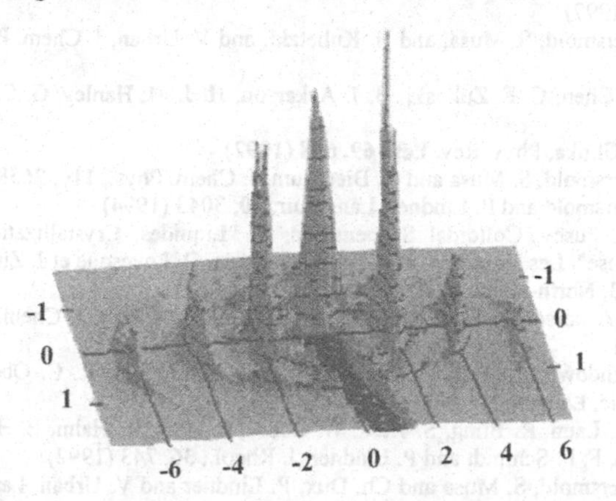

Figure 13 : Experimental tangential scattering from a Couette cell

4. CONCLUSIONS

In this contribution the determination of two kinds of ordering in mesogenic systems have been treated. These were:

(a) The structure in a layer and

(b) The structure between the layers.

By small angle scattering and sample rotation one is able to characterize both structures. How this can be done is described in the paper.

For a dispersion of 34 Vol. % concentration made from polymer-particles of 94nm diameter the following was observed:

(a) After shearing the sample it was oriented and layered at rest. In the layers a hexagonal particle Arrangement was found. At low shear rate an elastic deformation took place, which bends the Bragg reflections to rings. At higher shear rates Bragg spots were observed again which were attributed to sliding layers.

(b) Random stacking perpendicular to the layers was found. The kinetics of crystallization which can be studied with the methods presented remains an interesting though open question at the moment.

Acknowledgments
Financial support of the Deutsche Forschungsgemeinschaft and the Fonds der Chemischen Industrie is gratefully acknowledged.

REFERENCES

1. C. Kittel, Introduction to Solid State Physics, Wiley, New York, 1967
2. A. Guiniers, X-Ray Diffraction, Freeman, London, 1963
3. N. H. Clark, A. J. Hurd and B. J. Ackerson, Nature, **281**, 57 (1979)
4. Ch. Dux and H. Versmold, Physica A, **235**, 75 (1997), Phys. Rev. Lett., **78**, 1811 (1997)
5. V. Reus, L. Belloni, T. Zemb, N. Lutterbach and H. Versmold, J. Phys. II, France **7**, 603 (1997)
6. H. Versmold, S. Musa, and H. Kubetzki, and V. Urban, J. Chem. Phys., in preparation
7. L. B. Chen, C. F. Zukoski, B. J. Ackerson, H. J. M. Hanley, G. C. Straty, J. Baker and
8. C. J. Glinka, Phys. Rev. Lett. **69**, 688 (1992)
9. H. Versmold, S. Musa and A. Bierbaum, J. Chem. Phys., **116**, 2658 (2002)
10. H. Versmold and P. Lindner, Langmuir, **10**, 3043 (1994)
11. P. N. Pusey, Colloidal Suspensions, in "Liquides, Crystallization et Transition Vitreuse", Les Houches 1989, Ed. J.P Hansen, D. Levesque et J. Zinn-Justin, Vol. **II** p. 763, North-Holland, 1991
12. B. J. Ackerson, J. B. Hayter, N. A. Clark and L. Cotter, J. Chem. Phys., **84**, 2344 (1984)
13. S. Ashdown, I. Markovic, R. H. Ottewill, P. Lindner, R. C. Oberthür and A. R. Rennie, Langmuir, **6**, 303 (1990)
14. H. M. Laun, R. Bung, S. Hess, W. Loose, 0. Hess, K. Hahn, E. Hädicke, R. Hingmann, F. F. Schmidt and P. Lindner, J. Rheol., **36**, 743 (1992)
15. H. Versmold, S. Musa and Ch. Dux, P. Lindner and V. Urban, Langmuir, **17**, 6812 (2001)

COMPUTER SIMULATION STUDIES OF SOLVATION DYNAMICS IN MIXTURES

BRANKA M. LADANYI
Department of Chemistry, Colorado State University, Fort Collins, CO 80523, U.S.A.

Abstract: Solvation dynamics (SD) in liquid mixtures often occurs on a slower time scale than predicted on the basis of the mixture dielectric properties. In one-component polar liquids, the SD mechanism is dominated by reorientation of solvent molecules in response to the change in solute charge distribution. If the solute polarity changes upon electronic excitation, the mixed solvent response includes a change in local composition. SD in mixtures will usually depend on the time scale of this process. The relative importance of the solvent redistribution step varies with solvent component polarity, composition, the extent of change in preferential solvation and other factors. Considerable progress towards understanding SD in mixtures has been made in recent years, through experiments, computer simulation and theory. I will review some of the recent results, focusing primarily on computer simulation studies.

1. INTRODUCTION

Solvation dynamics (SD) refers to the time-evolution of the solvatochromic shift in fluorescence spectra of solutes [1]. Large shifts are observed for large changes in solute-solvent interactions arising from solute electronic excitation. For typical SD chromophores such as Coumarin 153 (C153), the dipole moment changes by about 8 D when the molecule undergoes the $S_0 \rightarrow S_1$ electronic transition.[2]. A large Stokes shift is therefore observed in highly polar solvents, although sizable shifts occur even for solvents lacking permanent dipoles, but with reasonably large quadrupole moments [3]. A great deal of work has been done in recent years to characterize the solvation response, especially in one-component polar solvents [1, 4-10]. In these media, it has been found that the response to a change in solute dipole is due primarily to collective solvent reorientation and that it can be predicted reasonably well using information on pure solvent dipolar reorientation, for example, from dielectric permittivity measurements, as input [1, 8-10]. It has been found more recently that a similar solvation mechanism applies to SD in nondipolar solvents, provided that the change in solute-solvent interactions is primarily electrostatic in nature [3, 11-13].

However, evidence is accumulating that this physical picture is inadequate in explaining SD in mixtures. It is often found that changes in preferential solvation play a role, leading to a response that is not simply related to the dielectric properties of the individual components or of their mixture [14-24]. Preferential solvation can

J. Samios and V.A. Durov (eds.), Novel Approaches to the Structure and Dynamics of Liquids: Experiments, Theories and Simulations, 305–321.

also contribute to the steady-state solvatochromic shifts in solute electronic spectra, leading to a lack of correlation between the solvent dielectric properties and the shift magnitude [15, 16, 23-28].

While it has long been recognized that preferential solvation can play an important role in chemical reactivity in mixed solvents [29], its contributions to SD, and consequently, to reaction dynamics, are less well characterized. Several studies of SD in mixtures have recently been carried out and insights into the contributions of local concentration fluctuations to the solvation mechanism and time scale are starting to emerge [14-24, 30-35].

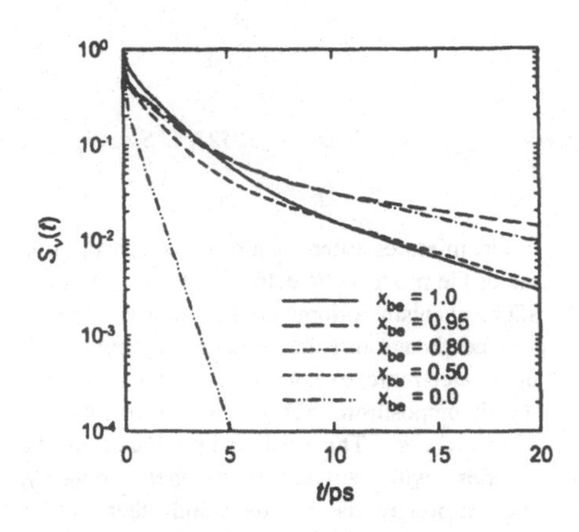

Figure 1. Multiexponential fits to experimental SD data for C153 in room-temperature benzene-acetonitrile mixtures (Data from Luther, B.M., Kimmel, J.R., and Levinger, N.E. (2002) *J. Chem. Phys.* **116**, 3370-3377 and (for pure acetonitrile) Gardecki, J.A., and Maroncelli, M. (1999) *Chem. Phys. Lett.* **301**, 571-578).

As noted above, typical SD chromophores are significantly less polar in their ground than in their electronically excited states. In mixed solvents, the ground state is then likely to be solvated predominantly by the less polar of the two solvent components and the excited state by the more polar one. Because a change in solvation shell composition needs to occur, equilibrium solvation of the excited state cannot be reached solely through reorientation of solvent molecules, the main mechanism of SD in one-component solvents. Solvent molecule translation, in addition to rotation, has to contribute to the SD mechanism, leading to the overall time scale that is not predictable solely on the basis of solvent component SD response nor from the mixture dielectric permittivity.

The SD response is reported usually in terms of the frequency $\nu(t)$ at the peak of the fluorescence band [1]. The experimental solvation response function is given by

$$S_\nu(t) = \frac{\nu(t) - \nu(\infty)}{\nu(0) - \nu(\infty)}, \tag{1}$$

where t = 0 corresponds to the time of electronic excitation (which occurs essentially instantaneously on the time scale of nuclear motions) and $\nu(\infty)$ corresponds to the peak frequency of the steady-state fluorescence.

An example of $S_\nu(t)$ measured in a simple solvent mixture, that of acetonitrile and benzene, is shown in Figure 1. As can be seen from the figure, the short-time dynamics of solvation follow a trend that is predictable in terms of the component responses, while the long-time SD in benzene-rich mixtures (x_{be} = 0.80 and 0.95) is considerably slower than in pure benzene, the more slowly relaxing solvent component. The slow component is also present at x_{be} = 0.50, but its overall contribution to $S_\nu(t)$ is quite small, indicating that the slowly relaxing component gains in importance as the mole fraction of the solvent component with a stronger electrostatic attraction to the excited state chromophore decreases. This strongly suggests that that a new SD mechanism, associated with the build-up of an increase in local concentration of acetonitrile, is responsible for the slow-down in the solvation response. Further support for this interpretation is provided by the composition dependence of the steady-state solvatochromic shift, $\Delta\nu = \nu(0) - \nu(\infty)$, which in the mixtures is larger than the value predicted by assuming ideal behavior [23]:

$$\Delta\nu > \Delta\nu_{id} = x_{be}\Delta\nu_{be,p} + (1 - x_{be})\Delta\nu_{ac,p} \, , \tag{2}$$

where the subscript 'p' denotes pure solvent components. In the case of the C153 chromophore, $\Delta\nu_{ac,p} \cong 2.5\Delta\nu_{be,p}$ [3]. Therefore $\Delta\nu > \Delta\nu_{id}$ signals the presence of an excess local concentration of acetonitrile surrounding the excited state chromophore.

Nonideal behavior of solvatochromic shifts has been observed in a number of mixed solvents [15, 23, 25, 27, 28]. It is usually attributed to preferential solvation if the deviation from ideality is greater than the nonideality of the mixture dielectric properties [25, 28]. For example, in continuum dielectric theory, the solvation free energy of a dipolar solute is proportional to the 'Onsager function' [15, 25, 29]

$$f(\varepsilon) = \frac{2(\varepsilon - 1)}{2\varepsilon + 1} \, , \tag{3}$$

where ε is the mixture dielectric constant.

In a binary mixture of A and B, an ideal Onsager function would be

$$f_{id} = x_A f(\varepsilon_A) + (1 - x_A) f(\varepsilon_B) \, , \tag{4}$$

where ε_A and ε_B are the dielectric constants of the pure A and B components. The deviation of $f(\varepsilon)$ from f_{id} as a measure of nonideality in the dielectric properties relevant to solvation. In cases where a larger nonideality in $\Delta\nu$ than in $f(\varepsilon)$ vs x_A is observed, preferential solvation is suspected.

While this procedure is quite reasonable, it has several drawbacks. One of them is the fact that continuum dielectric theory does not always accurately predict electrostatic solvation free energies. It is especially poor for solvents that are nondipolar, but possess appreciable quadrupole moments. In such systems, the dielectric constant is not well-correlated with the strength of dipole-quadrupole interactions that largely determine the solvation energetics [3, 11, 13].

Unraveling the contributions of different SD mechanisms by experimental means is also a major challenge, especially in mixtures in which both solvent components actively contribute to the solvatochromic shift. In a mixture of a polar and an apolar component, for example such as alcohol-alkane, the task is a little easier since one can focus largely on the motion of one of the solvent components and the long-time decay of $S_\nu(t)$ can then be approximately correlated with its diffusion rate in the solvent mixture.[15, 22, 25, 26, 34] However even there the extent of the change in local concentration and its relative contribution to $S_\nu(t)$ are difficult to predict on the basis of mixture properties.

Computer simulation studies make it possible to monitor separately the behavior of the two solvent components, thus providing a way to unravel their contributions to the different mechanisms and time scales of SD. They have therefore been quite helpful in elucidating the key mechanistic steps in SD in mixed solvents.

I will start with a review of the basics of MD simulations of SD in mixtures and then discuss some of the findings that have emerged from these studies.

2. SOLVATION DYNAMICS - THEORETICAL BACKGROUND

Connection between the peak frequency of the fluorescence spectrum and solvation can be made by noting that $\nu(t)$ contains a contribution from the isolated-molecule transition energy (E_{el}) and a time-dependent contribution, $\Delta E(t)$, due to the presence of the solvent [8, 36-38]

$$h\nu(t) = E_{el} + \overline{\Delta E(t)}, \tag{5}$$

where the overbar indicates an average over all the solute molecules contributing to the observed signal. The solvation response can therefore be expressed as

$$S_\nu(t) = S(t) = \frac{\overline{\Delta E(t)} - \overline{\Delta E(\infty)}}{\overline{\Delta E(0)} - \overline{\Delta E(\infty)}} \tag{6}$$

When S(t) is calculated from computer simulation on a system containing a single solute molecule, the overbar is interpreted as an average over statistically independent nonequilibrium trajectories. The total Stokes shift can also be obtained via equilibrium statistical mechanical theory or simulation,

$$\Delta\tilde{\nu} = \frac{\overline{\Delta E(0)} - \overline{\Delta E(\infty)}}{hc} = \frac{\langle \Delta E \rangle_0 - \langle \Delta E \rangle_1}{hc} \tag{7}$$

where $\langle \ldots \rangle_n$ denotes an equilibrium ensemble average for the solvent in the presence of the ground state (n = 0) or electronically excited (n = 1) solute.

For relatively rigid chromophores typically used to measure SD, the main effect of the change in the solute electronic state on the solvent environment comes from the change in the solute charge distribution [39]. Other changes, involving solute geometry, polarizability, solute-solvent dispersion and short-range repulsion occur as well but their effects on the time-evolution of the solvatochromic shift and its steady-state value are usually less pronounced. In theoretical and simulation of SD in mixtures containing a dipolar component, the focus has so far been the electrostatic perturbation in solute-solvent interactions. $\Delta E(t)$ is then represented as a change in solute-solvent Coulomb interactions due to changes in the solute partial charges. Thus for a system of one solute molecule (molecule 0) and N solvent molecules

$$\Delta E = \sum_{j=1}^{N} \sum_{\alpha\in 0} \sum_{\beta} \frac{\Delta q_{0\alpha} q_{j\beta}}{4\pi\varepsilon_0 r_{0\alpha,j\beta}}, \tag{8}$$

where $\Delta q_{0\alpha}$ is the change in the partial charge of the solute site α, $q_{j\beta}$ is the partial charge on the site β of the jth solvent molecule and $r_{0\alpha,j\beta}$ is the scalar distance between these two sites. Since the above form of ΔE is pairwise-additive, the contributions to it from different solvent components can be readily identified. Thus in a binary mixture of components A and B.

$$\Delta E = \Delta E_A + \Delta E_B. \tag{9}$$

The solvation responses of each solvent component m (= A, B) can be defined as

$$S_m(t) = \frac{\overline{\Delta E_m(t)} - \overline{\Delta E_m(\infty)}}{\overline{\Delta E_m(0)} - \overline{\Delta E_m(\infty)}} \tag{10}$$

and the total solvation response expressed in terms of contributions of the two components:

$$S(t) = \chi_A S_A(t) + \chi_B S_B(t), \tag{11}$$

where χ_A and $\chi_B = 1 - \chi_A$ are 'solvation energy difference fractions',

$$\chi_m = \frac{\overline{\Delta E_m(0)} - \overline{\Delta E_m(\infty)}}{\overline{\Delta E(0)} - \overline{\Delta E(\infty)}} \tag{12}$$

If ΔE can be considered to be a small perturbation in system properties, the solvation response can be estimated using the linear response approximation (LRA), which relates S(t) to the time correlation function (TCF) $C_0(t)$ of fluctuations $\delta\Delta E = \Delta E - \langle \Delta E \rangle$ of ΔE in the unperturbed system, [38]

$$C_0(t) = \langle \delta\Delta E(0)\,\delta\Delta E(t) \rangle_0 \Big/ \langle [\delta\Delta E]^2 \rangle_0 . \tag{13}$$

Previous studies indicate that LRA is not always applicable to solvation dynamics [20, 30, 37, 40-44]. For SD in mixtures, the LRA is expected to break down if the change in the solute charge distribution triggers a substantial change in local concentrations of the solvent components. A comparison of solvation responses calculated using Eqs. (6) and (13) would reveal how much this source of nonlinearity influences SD.

When the SD response is not linear and the final solute state differs from the initial one (as for perturbations that lead to dipole enhancement as opposed to dipole reversal), it has been found that $C_0(t)$ provides a good approximation to S(t) at short times, but that the longer time decay is approximated more accurately by [2, 42, 45]

$$C_1(t) = \langle \delta\Delta E(0)\,\delta\Delta E(t) \rangle_1 \Big/ \langle [\delta\Delta E]^2 \rangle_1 , \tag{14}$$

the TCF of $\delta\Delta E$ for the solvent in the presence of the excited-state (S_1) solute. This reflects the fact that at the longer, diffusive, time scales more information on the change in solute-solvent interaction has been transmitted to the surrounding solvent. In mixtures, we expect that the changes in preferential solvation, which affect primarily the longer-time portion of S(t),[14, 18, 20, 24] will be reflected more accurately in $C_1(t)$ than in $C_0(t)$.

3. MD SIMULATION RESULTS ON SD IN MIXTURES

The simulations of SD in mixtures can be classified into several categories. One of them concerns the type of perturbation in solute charge distribution. The first simulation of SD in a realistic model of a solvent mixture involved solute dipole reversal in mixtures of water and methanol at room temperature [30, 31, 46]. In this case, the equilibrium solvation energies for the initial and final solute states are the same and only the reorganization of the nearby solvent molecules plays an important role in the SD mechanism. Nevertheless, preferential solvation does play a role, resulting in a solute-dependent solvation response. This is illustrated in Figure 2, which depicts the total $S(t)$ and its contributions from the two solvent components, methanol (M) and water (W), at $x_M = 0.5$. It can be seen from the figure that in the case of the 'large' solute, which turns out to be preferentially solvated by methanol, the solvation response is dominated by $S_M(t)$, while the responses of the two solvent components are of approximately equal importance in the case of the 'small' solute, consistent with a lack of preferential solvation in this case.

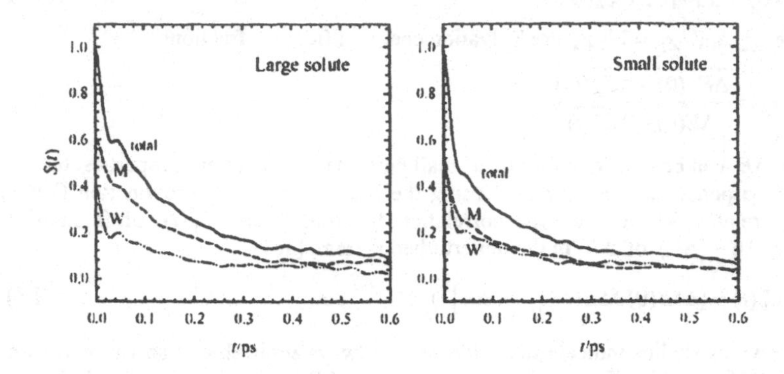

Figure 2. SD in response to a dipole reversal in a diatomic solute. Depicted are the results for the equimolar methanol (M) – water (W) mixture. In addition to the total response, $S(t)$, $\chi_M S_M(t)$ and $\chi_W S_W(t)$ are shown. The left panel is for a large solute (Lennard-Jones site diameter, σ_{LJ} =4.20 Å) and the right panel for a small solute (σ_{LJ} =3.08 Å) (Data from Skaf, M.S., and Ladanyi, B.M. (1996) J. Phys. Chem. **100**, 18258-18268.)

For most SD chromophores, the strength of solute-solvent interactions differs in the ground and excited electronic states of the solute. Subsequent simulation studies of SD in mixtures have focused on models which correspond to this situation [14, 16, 18, 20, 21, 24]. Although SD experiments correspond more closely to perturbations in which the solute dipole increases in magnitude, charge creation, which represents the simplest electrostatic perturbation, has often been considered in computational and theoretical SD studies [1].

Day and Patey have simulated this type of perturbation first in a series of mixtures of Stockmayer molecules [14] and then in methanol-water and dimethyl sulfoxide (DMSO) – water mixtures [18]. Charge creation in DMSO-water mixtures was also simulated by Laria and Skaf [20].

The Stockmayer fluid simulations involved mixtures of solvent molecules which differed only in the sizes of their dipole moments, with the Lennard-Jones (LJ) potential parameters σ_{LJ} and ε_{LJ}, masses and moments of inertia the same for both solvent species. The mixtures were simulated at constant volume and temperature. A representative set of results for $S(t)$ values at different compositions is depicted in Figure 3.

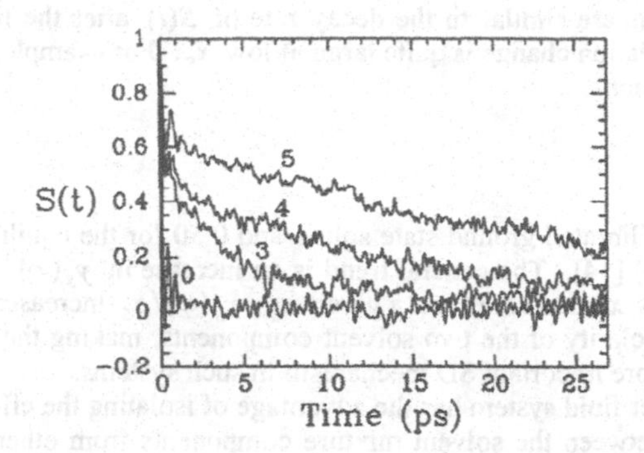

Figure 3. SD in response to charge creation in Stockmayer fluids at the reduced temperature $T^* = k_B T / \varepsilon_{LJ} = 1.35$ and reduced density $\rho^* = \rho\sigma_{LJ}^3 = 0.8$. The four sets of curves represent results for different compositions of mixtures of solvent molecules S and W with dipole moments $\mu_s^* = \mu_s / (\varepsilon_{LJ}\sigma_{LJ}^3)^{1/2} = 2.0$ and $\mu_w^* = 0.5$. The systems 0, 3, 4 and 5 correspond to x_s = 1.0, 0.25, 0.10 and 0.05, respectively. (Reprinted with permission from Day, T.J.F., and Patey, G.N. (1997) *J. Chem. Phys.* **106**, 2782-2791. Copyright, American Institute of Physics.)

These results show a dramatic slowing down of the solvation response as the mole fraction x_S of the more polar solvent component decreases. In all cases, the fast (around 0.1 ps) component occurs on a similar time scale, while the slowly relaxing portion grows in amplitude and decreases in relaxation rate as x_S decreases. The slowing down of SD reflects the increasing time scale associated with the build-up of enhanced local concentration of S in the vicinity of the excited state solute as the bulk concentration of this species decreases.

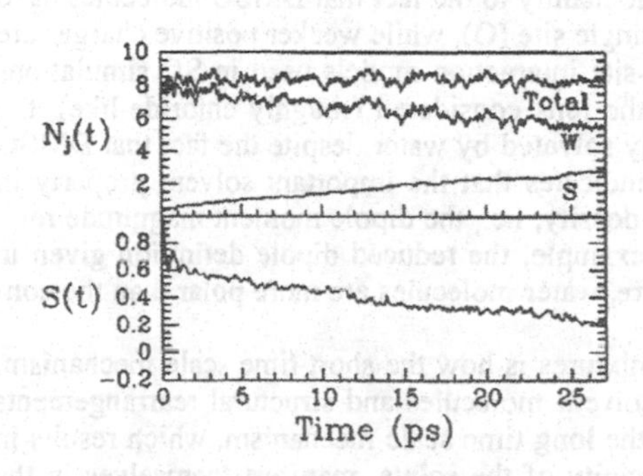

Figure 4. The top panel depicts the time-evolution of the coordination numbers N_s and N_w and of their sum (total); the bottom panel depicts the solvation response $S(t)$, all for system 5 (see Figure 3 for its description). (Reprinted with permission from Day, T.J.F., and Patey, G.N. (1997) *J. Chem. Phys.* **106**, 2782-2791, Copyright, American Institute of Physics.)

Since both solvent components are polar, they both actively participate in the local concentration change. This is illustrated in Figure 4, which shows the time-evolution of the coordination numbers $N_S(t)$ and $N_W(t)$ for system 5 (x_S = 0.05), obtained by counting the number of solvent molecules within a radius of 1.3 σ_{LJ} around the solute. Note that the total number of solvent molecules in the first coordination shell $N_S(t) + N_W(t)$ increases very slightly as time progresses, reflecting the fact that the two solvent species have the same LJ diameters. The slow time scales associated with the first shell population changes of the two components are similar to the decay rate of $S(t)$ after the first 3 or 4 ps. The extent of population change is quite large at low x_S. For example, for system 5, the local mole fraction

$$y_S = \frac{N_S}{N_S + N_W} \tag{15}$$

is 0.05 for the equilibrated ground state solute and 0.50 for the equilibrated excited state solute ($t = \infty$) [14]. The general trend is an increase in $y_S(\infty)/x_S$ for the excited-state solute as x_S decreases. At a given x_S, $y_S(\infty)/x_S$ increases with increasing difference in polarity of the two solvent components, making the local concentration change a more important SD mechanism in such systems.

The Stockmayer fluid system has the advantage of isolating the effects of the polarity difference between the solvent mixture components from other properties of polar molecules such as their shapes and charge distributions. Thus one might expect that mixtures of real polar molecules would not follow all the trends predicted by simulations for this model system. However, the observation that local concentration change is an important solvation mechanism in mixtures remains true for model solvents designed to resemble closely real polar molecules such as water, methanol, and DMSO [18, 20]. The observation holds even for a polar-apolar mixture designed to resemble methanol-hexane [21] and a dipolar-quadrupolar mixture representing acetonitrile-benzene [24].

In fact, despite the more complicated intermolecular interactions that include hydrogen (H) bonding, SD accompanying charge creation in water-methanol mixtures [18] was found to resemble quite closely the behavior observed in Stockmayer fluid mixtures. Because of the asymmetry in the charge distrubution of the solvent molecules, SD in water-methanol mixtures differs for positive and negative ion creation, but the differences affect mainly the relatively short solvation time scale. A much larger asymmetry in response to anion and cation creation is found in water-DMSO mixtures [18, 20]. This is due mainly to the fact that DMSO molecules have a large negative partial charge on a single site (O), while weaker positive charges are distributed over three sites in the 4-site interaction models used in SD simulations [18, 20]. Further, for the sizes of the ions considered (roughly chloride-like), the excited state solute was preferentially solvated by water despite the fact that DMSO has a larger dipole moment. This indicates that the important solvent property in determining its polarity is its dipole density, i.e., the dipole moment magnitude relative to the molecular size (see, for example, the reduced dipole definition given in the Figure 3 caption). By this measure, water molecules are more polar than the considerably larger DMSO molecules.

An interesting aspect of SD in mixtures is how the short time scale mechanism, which consists of reorientation of solvent molecules and structural rearrangements within the first solvation shell, and the long time scale mechanism, which results in the concentration changes in the vicinity of the solute, manifest themselves in the component solvation responses.

This behavior is illustrated in Figures 5 and 6, for two very different mixtures and perturbations in solute charge distribution. Figure 5 depicts the responses of the components of water (W) – DMSO mixtures of different composition to negative charge creation in a solute with chlorine-like LJ parameters.

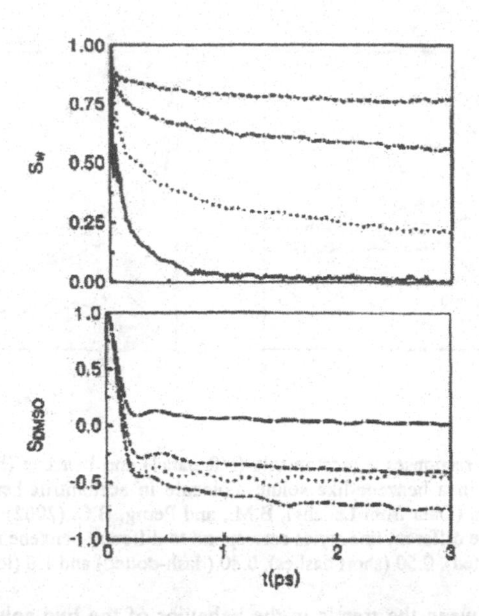

Figure 5. Solvation responses of water (top panel) and DMSO (bottom panel) to Cl → Cl⁻ charge creation in water-DMSO mixtures of different composition. The different line styles correspond to the DMSO mole fractions x_{DMSO} =0.0 (full), 0.25 (dotted), 0.50 (dash-dotted), 0.75 (short dashes) and 1.0 (long dashes). (Reprinted with permission from Laria, D., and Skaf, M.S. (1999) *J. Chem. Phys.* 111, 300-309. Copyright, American Institute of Physics.).

Figure 6 depicts component responses to dipole creation in a benzene-like solute in mixtures of acetonitrile (ac) and benzene (be).

In both cases, the excited state solute is solvated preferentially by the more polar solvent component, water in the mixures shown in Figure 5 and acetonitrile in mixtures depicted in Figure 6. The response of this component, $S_W(t)$ (Fig. 5) or $S_{ac}(t)$ (Fig. 6), exhibits a more pronounced slowly-decaying component as its mole fraction decreases. For both $S_W(t)$ and $S_{ac}(t)$, the fast and the slow portions of the decay add constructively to contribute to progress towards excited-state equilibrium solvation, i.e., the responses approach zero from above. A different behavior is seen for the less polar solvent components, $S_{DMSO}(t)$ (Fig. 5) and $S_{be}(t)$ (Fig. 6). In both cases, the short-time decay in the mixtures results in negative values of these response functions. On a much longer time scale, these functions then decay to zero from below. This means that the short-time solvation mechanisms, which correspond to structural rearrangements within the first solvation shell lead to 'oversolvation' by the less polar component, which then has to be partially destroyed to reach equilibrium. The similarities in the composition-dependence of the behavior of $S_{DMSO}(t)$ and $S_{be}(t)$ are quite striking. They both include a larger dip into negative range of values for the equimolar mixture than for mixtures that are both rich and poor in the less polar component.

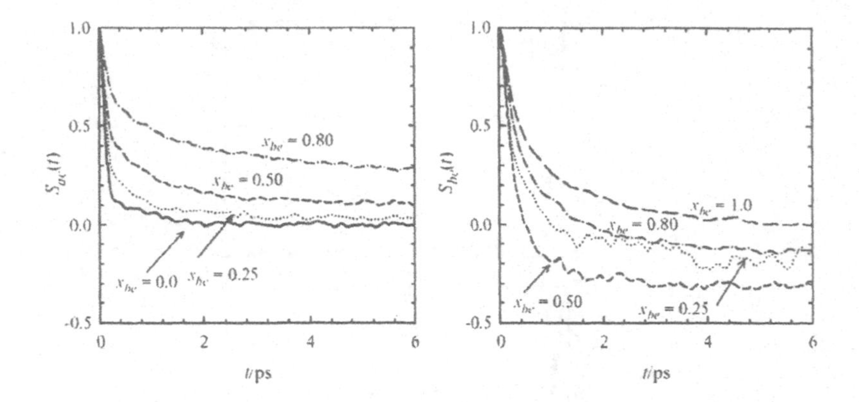

Figure 6. Solvation responses of acetonitrile (left panel) and benzene (bottom panel) to a creation of a dipole in a benzene-like solute molecule in acetonitrile-benzene mixtures of different composition. (Data from Ladanyi, B.M., and Perng, B.C. (2002). J. Phys. Chem. A **106**, 6922-6934.) The different line styles correspond to different benxene mole fractions: x_{be} = 0.0 (full), 0.25 (dotted), 0.50 (short dashes), 0.80 (dash-dotted) and 1.0 (long dashes).

Similarities between the trends in the behavior of the two solvent components seen in Figs. 5 and 6 are surprising in view of the fact that benzene is not even dipolar, but despite this behaves in a way that essentially mimicks the composition dependence of $S_{DMSO}(t)$. This demonstrates that electrostatic solvation by dipolar and quadrupolar solvent molecules is quite similar, even in dipolar-quadrupolar mixtures.

The dip below zero of the less polar component response was also seem for $S_M(t)$ in the case of charge creation in methanol- water mixtures, so it appears to be common for SD in mixtures in which both solvent components are active participants and in which solute excitation leads to a change in the solvent component that preferentially solvates the chromophore.

The 'oversolvation' by the less polar component in mixtures in which this component initially has a high local concentration, manifests itselt also in the time-evolution of the local population of this component. This is illustrated in Figure 7 for SD accompanying anion creation in an equimolar methanol-water mixture. In addition to the total and component solvation responses, the first shell coordination numbers, based on the center-center distance $\leq$ 4.2 Å and the first shell population response are displayed.

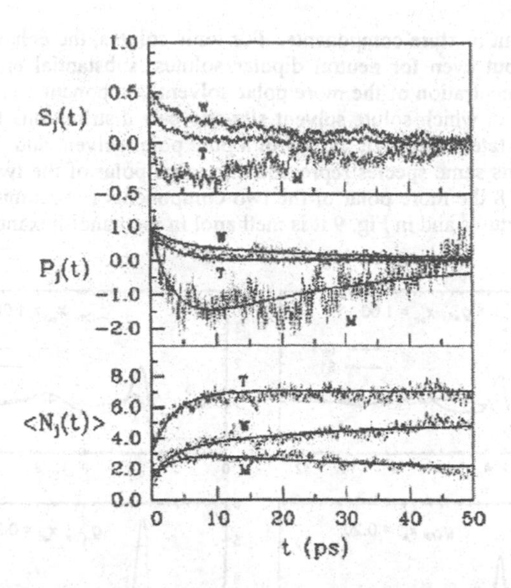

Figure 7. SD in response to $Cl^{0} \rightarrow Cl^{-}$ in a room-temperature methanol-water mixture at $x_M = 0.5$. The top panel depicts the solvation response ($S_j(t)$), the middle the first shell population response ($P_j(t)$), and the bottom the coordination numbers for methanol (M), water (W), and total (T). Smooth lines are fits of $P_j(t)$ MD data to biexponential decay. (Reprinted with permission from Day, T.J.F., and Patey, G.N. (1999). *J. Chem. Phys.* **110**, 10937-10944. Copyright, American Institute of Physics.)

Denoting the instantaneous coordination number for solvent species m as $N_m(t)$, the population response $P_m(t)$ is defined as

$$P_m(t) = \frac{\overline{N_m(t)} - \overline{N_m(\infty)}}{\overline{N_m(0)} - \overline{N_m(\infty)}} \tag{16}$$

which amounts to replacing ΔE_m by $N_m(t)$ in Eq. (10). (Note that the authors of Ref. [18] use $\langle \ldots \rangle$ instead of an overbar for nonequilbrium averages.)

In this case there is a striking similarity in the behavior of $S_m(t)$ and $P_m(t)$ for each of the two solvent components for most of the time interval that is displayed. Oversolvation by methanol manifests itself by the presence of minima in $S_M(t)$ and $P_M(t)$ and a maximum in $N_M(t)$. A rise and subsequent decay in the local population of the less polar solvent component has also been observed in MD studies of SD in water-DMSO [18, 20] and acetonitrile-benzene [24] mixtures. Day and Patey [18] have shown that a quasi-chemical association-dissociation model for the solute with both solvent components fits the of $P_m(t)$ behavior reasonably well. The Patey group has also developed a model based on density functional theory which can account for the differences in the long-time decay $P_m(t)$ and $S_m(t)$ [17, 32].

Solute-solvent pair correlations illustrate perhaps more dramatically than do the coordination numbers the extent of preferential solvation and also provide information on the distance range over which the local concentration is enhanced. The extent of local concentration enhancement depends on solvent composition and the difference in the strength and range of electrostatic attraction between the solute and each of the solvent mixture components. For ionic solutes, the enhancement can be quite dramatic, but even for neutral dipolar solutes, substantial enhancement can occur at low concentration of the more polar solvent component. This is illustrated in Figs. 8 and 9 in which solute-solvent site-site pair distributions for ground (S_0) and excited (S_1) state solutes are shown for a pure polar solvent and for solvent mixtures in which this same species represents the more polar of the two solvent components. In Fig. 8 the more polar of the two components in acetonitrile in acetonitrile-benzene miztures and in Fig. 9 it is methanol in methanol-hexane mixtures.

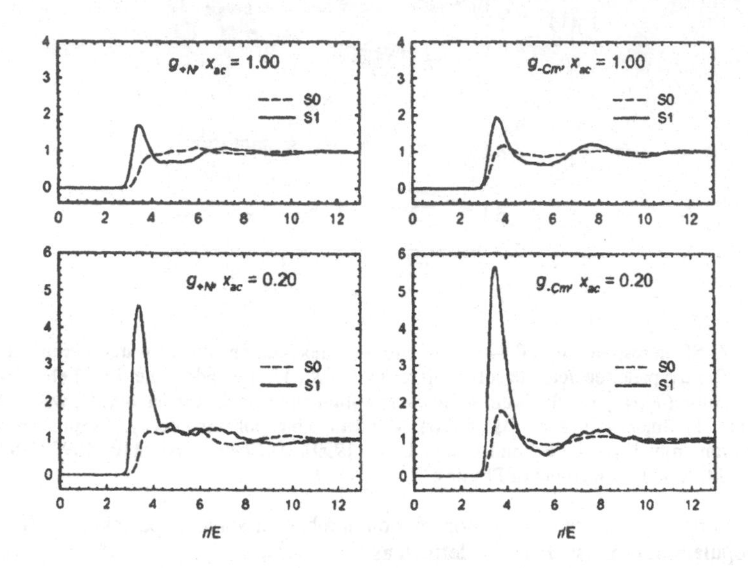

Figure 8. Acetonitrile-solute pair distribution functions in pure acetonitrile (top) and in acetonitrile-benzene mixture at x_{ac} = 0.20 (bottom) (Data from Ladanyi, B.M., and Perng, B.C. (2002). J. Phys. Chem. A **106**, 6922-6934). Depicted are g_{+N} (left) and g_{-Cm} (right), where + and – denote benzene-like solute sites for which partial charges change by +e/2 and –e/2, respectively, and N and Cm are the nitrogen and methyl carbon sites of acetonitrile.

In Figure 8, as in Figure 6, the solute is benzene-like in its ground state. The partial charges of two of its C sites at relative para positions change by +e/2 and –e/2 in the S_1 state. The figure depicts the pair correlations $g_{+N}(r)$ and $g_{-Cm}(r)$, where + and – denote these two solute sites and N and Cm are nitrogen and the carbon of the CH_3 group in acetonitrile. These are depicted at mole fractions $x_{ac} = 1.0$ and 0.20. No preferential solvation by acetonitrile exists in the S_0 solute state, so the mole fraction dependence of $g_{+N}(r)$ and $g_{-Cm}(r)$ is quite modest. On the other hand, a large increase in the height of the first peaks of both functions with decreasing x_{ac} occurs for the S_1 state, with some additional increase extending into the second solvation shell in the case of $g_{+N}(r)$, providing dramatic evidence of preferential solvation.

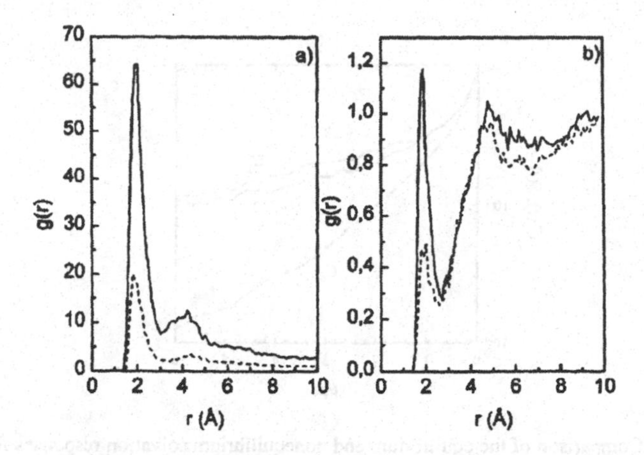

Figure 9. The pair correlation between the carbonyl oxygen site of C153 and hydroxyl H of methanol at $x_M = 0.044$ in a methanol-hexane mixture (panel a) and in pure methanol (panel b). Dashed lines represent S_0 state and full line S_1 state results. (Reprinted with permission from Cichos, F., Brown, R., Rempel, U., and Von Borczyskowski, C. (1999) J. Phys. Chem. A **103**, 2506-2512. Copyright, American Chemical Society,.)

Even more dramatic evidence of preferential solvation is displayed in Figure 9 in which the C153 chromophore-methanol pair correlations in a very dilute methanol-hexane mixture, $x_M = 0.044$, are compared with pure methanol results [21]. The figure depicts pair correlations between O of the carbonyl group in C153 and hydroxyl H of methanol. A solute-solvent H-bond is formed in both S_0 and S_1 states of C153. The H-bond peak height increases in the S_1 state, signaling an increase in the strength of this solute-solvent H-bond on electronic excitation. The increase is modest for pure methanol, but very large in the case of the $x_M = 0.044$ mixture. Comparison of the results in the two panels further indicates that preferential solvation by methanol exists for both solute electronic states, but that it increases substantially in size and range when the solute is electronically excited.

Given that large changes in the local composition of the ground and excited state solutes, the linear response approximation, which is often fairly accurate for one-component polar solvents [2], is less accurate in the case of mixtures [20, 24]. A typical result, taken from the MD data of Ref. [24] for SD in benzene-rich (x_{be} = 0.80) acetonitrile-benzene mixtures, is shown in Figure. 10. The results for $S(t)$, obtained from nonequilibrium MD trajectories, are compared to the time correlations, $C_0(t)$ and $C_1(t)$, Eqs.(13) and (14), obtained from equilibrium MD data containing the solvent in the presence of the ground and excited state solutes, respectively.

It is evident from the figure that $C_0(t)$ is a very poor approximation to $S(t)$ at long times. Interestingly, $C_1(t)$ represents a much better approximation to the long-time portion of $S(t)$, despite the fact that translational diffusion that leads to the build-up of the enhanced concentration of acetonitrile presumably plays a less important role in the dynamics contributing to $C_1(t)$ than it does for $S(t)$.

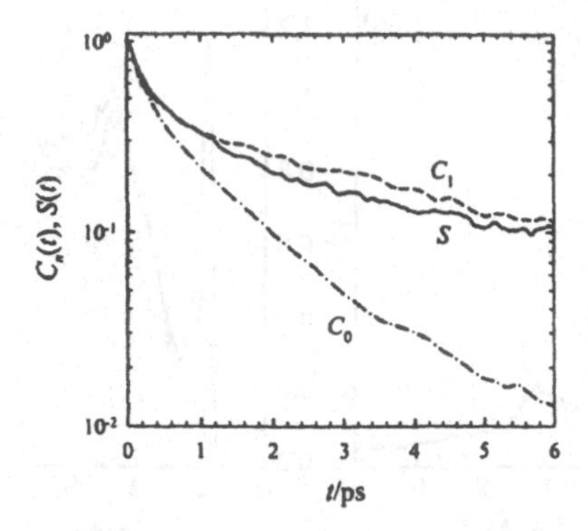

Figure 10. Comparison of the equilibrium and nonequilibrium solvation responses for dipole creation in a benzene-acetonitrile mixture at x_a = 0.20, based on the data from Ladanyi, B.M., and Perng, B.C. (2002). *J. Phys. Chem. A* **106**, 6922-6934. Shown are the nonequilibrium response $S(t)$ and the time correlations $C_0(t)$ and $C_1(t)$.

Comparison between MD and experimental results for SD in acetonitrile-benzene [23, 24] and methanol-hexane [15, 21] mixtures resulted in good agreement between the two sets of results and confirmed that the slow solvent redistribution is an important mechanistic SD step in these mixtures, especially at low concentrations of the more polar solvent component. A recent simulation of SD involving C153 in water-DMSO mixtures [16] also produced results in good agreement with experiment [47]. However, these results were quite different from the ion creation data [18, 20], some of which are displayed in Figure. 5. C153, despite a large increase in its dipole moment, remains hydrophobic in its S_1 state, so the change in local solvent composition surrounding it turns out to be relatively modest, resulting is a less important role of the solvent redistribution mechanism to SD in this case. This points to the fact that interactions of the two solvent components with the ground and excited state solute are an extra source of complexity in determining the mechanisms contributing to SD in mixed solvents.

4. SUMMARY AND CONCLUSION

I have reviewed here the results of recent MD simulations of SD in mixtures in which at least one solvent component is polar. Since interactions of the solute with different mixture components are different, preferential solvation usually plays a role, giving rise to a solvation response that is more complex than for one-component polar solvents. When the strength of solute-solvent electrostatic interactions changes substantially upon electrostatic excitation, dynamical processes associated with the change in local solvent oncentration surrounding the solute contribute significantly, in addtion to the SD mechanisms primarily associated with orientational relaxation of solvent molecules, the main SD mechanism in one-component polar liquids. Preferential solvation change introduces into SD a new, usually slower, time scale, dependent on the solute-solvent mutual diffusion rate. This mechanism was theoretically predicted by Suppan [25] and first characterized via MD simulation by the Patey group [14, 17, 32]. Although originally proposed for polar-apolar binary mixtures [25] and first simulated for a two-component Stockmayer fluid [14], it has since been shown to be relevant to a variety of mixtures containing both protic and nonprotic solvents and quadrupolar as well as dipolar molecules [16, 18, 20, 21, 24]. It has been observed in a number of experimental studies of SD [15, 19, 22, 23, 47]. It is thus quite likely that it also plays an important role in the dynamics of chemical reactions, especially those involving charge transfer, in mixed solvents. One might expect that it will lead to considerably slower reaction rates than those n the pure components or based on the rates of molecular relaxation processes in the mixtures in the absence of the reacting solutes.

Acknowledgments

I am grateful to Profs. Ross Brown, Daniel Laria, Gren Patey and Munir Skaf for granting me permission to reproduce figures from their papers and to Prof. Nancy Levinger for helpful discussions. I would also like to acknowledge the contributions of my coworkers and scientific collaborators, especially Baw-Ching Perng, Munir Skaf and Ivana Borin, to the work discussed here. This research was supported by in part by grants from the U.S. National Science Foundation.

References

1. Maroncelli, M. (1993). The dynamics of solvation in polar liquids. J. Mol. Liq. **57**, 1-37.
2. Kumar, P.V., and Maroncelli, M. (1995). Polar solvation dynamics of polyatomic solutes: simulation studies in acetonitrile and methanol. J. Chem. Phys. **103**, 3038-3060.
3. Reynolds, L., Gardecki, J.A., Frankland, S.J.V., Horng, M.L., and Maroncelli, M. (1996). Dipole Solvation in Nondipolar Solvents: Experimental Studies of Reorganization Energies and Solvation Dynamics. J. Phys. Chem. **100**,10337-10354.
4. Simon, J.D. (1988). Time-resolved studies of solvation in polar media. Acc. Chem. Res. **21**, 128-134.
5. Bagchi, B. (1989). Dynamics of solvation and charge transfer reactions in dipolar liquids. Annu. Rev. Phys. Chem. **40**, 115-141.
6. Maroncelli, M., MacInnis, J., and Fleming, G.R. (1989). Polar solvent dynamics and electron-transfer reactions. Science **243**, 1674-1681.
7. Rossky, P.J., and Simon, J.D. (1994). Dynamics of chemical processes in polar solvents. Nature (London) **370**, 263-269.
8. Stratt, R.M., and Maroncelli, M. (1996). Nonreactive Dynamics in Solution: The Emerging Molecular View of Solvation Dynamics and Vibrational Relaxation. J. Phys. Chem. **100**, 12981-12996.
9. Fleming, G.R., and Cho, M. (1996). Chromophore-solvent dynamics. Annu. Rev. Phys. Chem. **47**, 109-134.
10. Ladanyi, B.M. (2000). Mechanistic Studies of Solvation Dynamics in Liquids. In Theoretical Methods in Condensed Phase Chemistry, S.D. Schwartz, ed. (Dordrecht, the Netherlands: Kluwer), pp. 207-233.
11. Ladanyi, B.M., and Stratt, R.M. (1996). Short-Time Dynamics of Solvation: Relationship between Polar and Nonpolar Solvation. J. Phys. Chem. **100**, 1266-1282.
12. Ladanyi, B.M. (1997). Molecular mechanisms of solvation dynamics in polar and nonpolar liquids. In Electron Ion Transfer Condens. Media, A.A. Kornyshev, M. Tosi and J. Ulstrup, eds. (Singapore: World Scientific), pp. 110-129.
13. Ladanyi, B.M., and Maroncelli, M. (1998). Mechanisms of solvation dynamics of polyatomic solutes in polar and nondipolar solvents: A simulation study. J. Chem. Phys. **109**, 3204-3221.
14. Day, T.J.F., and Patey, G.N. (1997). Ion solvation dynamics in binary mixtures. J. Chem. Phys. **106**, 2782-2791.
15. Cichos, F., Willert, A., Rempel, U., and von Borczyskowski, C. (1997). Solvation Dynamics in Mixtures of Polar and Nonpolar Solvents. J. Phys. Chem. A **101**, 8179-8185.
16. Martins, L.R., Tamashiro, A., Laria, D., and Skaf, M.S. (2003). Solvation dynamics of coumarin 153 in dimethylsulfoxide-water mixtures: Molecular dynamics simulations. J. Chem. Phys. **118**, 5955-5963.

17. Yoshimori, A., Day, T.J.F., and Patey, G.N. (1998). An investigation of dynamical density functional theory for solvation in simple mixtures. J. Chem. Phys. **108**, 6378-6386.
18. Day, T.J.F., and Patey, G.N. (1999). Ion solvation dynamics in water-methanol and water- dimethylsulfoxide mixtures. J. Chem. Phys. **110**, 10937-10944.
19. Nishiyama, K., and Okada, T. (1998). Relaxation dynamics of inhomogeneous spectral width in binary solvents studied by transient hole-burning spectroscopy. J. Phys. Chem. A **102**, 9729-9733.
20. Laria, D., and Skaf, M.S. (1999). Solvation response of polar liquid mixtures: Water-dimethylsulfoxide. J. Chem. Phys. **111**, 300-309.
21. Cichos, F., Brown, R., Rempel, U., and Von Borczyskowski, C. (1999). Molecular dynamics simulations of the solvation of coumarin 153 in a mixture of an alkane and an alcohol. J. Phys. Chem. A **103**, 2506-2512.
22. Petrov, N.K., Wiessner, A., and Staerk, H. (1998). Transient dynamics of solvatochromic shift in binary solvents. J. Chem. Phys. **108**, 2326-2330.
23. Luther, B.M., Kimmel, J.R., and Levinger, N.E. (2002). Dynamics of polar solvation in acetonitrile-benzene binary mixtures: Role of dipolar and quadrupolar contributions to solvation. J. Chem. Phys. **116**, 3370-3377.
24. Ladanyi, B.M., and Perng, B.C. (2002). Solvation dynamics in dipolar-quadrupolar mixtures: A computer simulation study of dipole creation in mixtures of acetonitrile and benzene. J. Phys. Chem. A **106**, 6922-6934.
25. Suppan, P. (1987). Local Polarity of Solvent Mixtures in the Field of Electronically Excited Molecules and Exciplexes. J. Chem. Soc. Faraday Trans. I **83**, 495-509.
26. Suppan, P. (1988). Time-Resolved Luminescence Spectra of Dipolar Excited Molecules in Liquid and Solid Mixtures - Dynamics of Dielectric Enrichment and Microscopic Motions. Faraday Discussions, 173-184.
27. Khajehpour, M., and Kauffman, J.F. (2000). Dielectric enrichment of 1-(9-anthryl)-3-(4-N,N-dimethylaniline) propane in hexane-ethanol mixtures. J. Phys. Chem. A **104**, 7151-7159.
28. Khajehpour, M., Welch, C.M., Kleiner, K.A., and Kauffman, J.F. (2001). Separation of dielectric nonideality from preferential solvation in binary solvent systems: An experimental examination of the relationship between solvatochromism and local solvent composition around a dipolar solute. J. Phys. Chem. A **105**, 5372-5379.
29. Reichardt, C. (1988). Solvents and Solvent Effects in Organic Chemistry, 2nd Edition (New York: VCH).
30. Skaf, M.S., and Ladanyi, B.M. (1996). Molecular Dynamics Simulation of Solvation Dynamics in Methanol-Water Mixtures. J. Phys. Chem. **100**, 18258-18268.
31. Skaf, M.S., Borin, I.A., and Ladanyi, B.M. (1997). Simulation of solvation dynamics in H-bonding solvents: dynamics of solute-solvent H-bonds in methanol-water mixtures. Mol. Eng. **7**, 457-472.
32. Yoshimori, A., Day, T.J.F., and Patey, G.N. (1998). Theory of ion solvation dynamics in mixed dipolar solvents. J. Chem. Phys. **109**, 3222-3231.
33. Gardecki, J.A., and Maroncelli, M. (1999). Solvation and rotational dynamics in acetonitrile propylene carbonate mixtures: a binary system for use in dynamical solvent effect studies. Chem. Phys. Lett. **301**, 571-578.
34. Petrov, N.K., Wiessner, A., and Staerk, H. (2001). A simple kinetic model of preferential solvation in binary mixtures. Chem. Phys. Lett. **349**, 517-520.
35. Agmon, N. (2002). The dynamics of preferential solvation. Journal of Physical Chemistry A **106**, 7256-7260.
36. Bader, J.S., and Chandler, D. (1989). Computer simulation of photochemically induced electron transfer. Chem. Phys. Lett. **157**, 501-504.
37. Fonseca, T., and Ladanyi, B.M. (1991). Breakdown of linear response for solvation dynamics in methanol. J. Phys. Chem. **95**, 2116-2119.
38. Carter, E.A., and Hynes, J.T. (1991). Solvation dynamics of an ion pair in a polar solvent: Time-dependent fluorescence and photochemical charge transfer. J. Chem. Phys. **94**, 5961-5979.
39. Horng, M.L., Gardecki, J.A., Papazyan, A., and Maroncelli, M. (1995). Subpicosecond Measurements of Polar Solvation Dynamics: Coumarin 153 Revisited. J. Phys. Chem. **99**, 17311-17337.

40. Ando, K., and Kato, S. (1991). Dielectric relaxation dynamics of water and methanol solutions associated with the ionization of N,N-dimethylaniline: theoretical analyses. J. Chem. Phys. **95**, 5966-5982.
41. Phelps, D.K., Weaver, M.J., and Ladanyi, B.M. (1993). Solvent dynamic effects in electron transfer: molecular dynamics simulations of reactions in methanol. Chem. Phys. **176**, 575-588.
42. Fonseca, T., and Ladanyi, B.M. (1994). Solvation dynamics in methanol: solute and perturbation dependence. J. Mol. Liq. **60**, 1-24.
43. Re, M., and Laria, D. (1997). Dynamics of Solvation in Supercritical Water. J. Phys. Chem. B **101**, 10494-10505.
44. Aherne, D., Tran, V., and Schwartz, B.J. (2000). Nonlinear, Nonpolar Solvation Dynamics in Water: The Roles of Electrostriction and Solvent Translation in the Breakdown of Linear Response. J. Phys. Chem. B **104**, 5382-5394.
45. Maroncelli, M., and Fleming, G.R. (1988). Computer simulation of the dynamics of aqueous solvation. J. Chem. Phys. **89**, 5044-5069.
46. Skaf, M.S., and Ladanyi, B.M. (1995). Computer simulation of solvation dynamics in hydrogen-bonding liquids. THEOCHEM **335**, 181-188.
47. Luther, B.M. (2000). Ph. D. thesis, Colorado State University, Fort Collins, CO, U.S.A.

USING SIMULATIONS TO STUDY VIBRATIONAL RELAXATION OF MOLECULES IN LIQUIDS

R.M. LYNDEN-BELL AND F.S. ZHANG
Atomistic Simulation Group, Queen's University Belfast
Belfast BT7 1NN, UK

Abstract: The aim of this contribution to the summer school is to show how atomistic computer simulations can be used to study and interpret vibrational relaxation in solutions. In the first part of the article the three distinct relaxation rates (population relaxation T_1^{-1}, decoherence rate T_2^{-1} and the pure dephasing rate $(T_2^*)^{-1}$) are introduced and theoretical expressions for the rates involving solvent-solute forces and solute-solvent energy derivatives are developed from perturbation theory. In the second part the way in which relaxation rates can be determined from simulations of flexible molecules is illustrated using the example of the stretching modes of the triiodide ion. The origin and explanation of the variations in rate are then discussed combining data from simulations of rigid solute molecule and the expressions from perturbation theory.

1. Introduction

The relaxation rates for vibrational transitions of molecules in the liquid phase provide a probe for the local interactions of the molecules with the solvent bath. They may be determined experimentally by methods ranging from the simple measurement of spectral line widths to sophisticated femtosecond laser methods. There are two distinct relaxation rates which give different information about the magnitude and dynamics of the solvent-solute interaction processes. As these processes are local and molecular in origin, atomistic computer simulation may usefully be used to complement experimental studies by providing insight into the mechanisms at the molecular level. Atomistic simulation has been used to study liquids and solutions for many years. The methods are described in Allen and Tildesley [1] and there a many programmes available for

J. Samios and V.A. Durov (eds.), Novel Approaches to the Structure and Dynamics of Liquids: Experiments, Theories and Simulations, 323–341.

use in this field. The first calculations of vibrational dephasing were performed about twenty five years ago [2, 3, 4] for liquid nitrogen and hydrogen chloride. This was followed by a number of studies of neat liquids [5, 6, 7, 8], while more recently it became possible to study solutions. All the early studies used rigid molecules in the simulation combined with perturbation theory to estimate the relaxation rates. However if one has a flexible model for the molecule which describes the harmonic and anharmonic parts of the vibrational Hamiltonian, it is possible to study the vibrational relaxation directly, albeit for a classical rather than a quantum model. This has been done by a number of authors[9, 10, 11, 12]. Here we show how both types of calculation can be used and how the relevant quantities are extracted. The information from the two types of calculation (flexible and rigid) is complementary; the first giving the relaxation rates and the second allowing one to identify different contributions to the relaxation. As an illustrative example we take I_3^- in a Lennard-Jones solvent.The solvent was parametrised to model xenon and the calculations are carried out at a state point (280 K and 411 bar) where the solvent is liquid. The flexible model of the ion is a semi-empirical valence bond model which we have developed [13] and used in earlier work on solvent-induced symmetry breaking[14]. Recently we have applied this model to the study of vibrational relaxation [15, 16, 17]. The details of the model are not important for understanding the methods of obtaining and analysing vibrational relaxation data which is the aspect of the work that we emphasise in this article.

2. Vibrational Relaxation

The energies of molecules in solution can be described by the Hamiltonian

$$H = H_0 + H_{Mb} \tag{1}$$

where H_0 is the Hamiltonian of the isolated molecule and H_{Mb} is the interaction between a particular molecule M and the bath of all other molecules in the solution. The molecule-bath interaction is a time dependent quantity which fluctuates as the molecules move relative to each other. The vibrations of the molecule in the gas phase can be described in terms of normal modes ζ_i found from diagonalising the vibrational Hamiltonian H_{v0} which can be derived from the full molecular Hamiltonian H_0. In most cases the perturbation due to the solvent is small and the vibrations in solution can be described by the same normal modes as in the gas phase. The main effect of the solvent is to change the relaxation rates, although the solvent also causes small shifts in the vibrational frequencies.

There are three different relaxation times associated with each mode, T_1, T_2 and T_2^*. These terms have been taken over from the NMR literature and have the same meaning. In many contexts it is more useful to talk about rates rather than times. T_1^{-1} is the rate at which energy is dissipated from a particular vibration to the surroundings. It is associated with changes in the population of the vibrational energy levels and is therefore known as the population relaxation rate or the energy relaxation rate. T_2^{-1} is the rate at which the vibrations in the ensemble lose coherence or become decorrelated, and is known as the decoherence rate or sometimes the total dephasing rate. This has two contributions, one from the population relaxation rate and a new term, the pure dephasing rate, $(T_2^*)^{-1}$, which depends on fluctuations in the instantaneous frequency. The relationaship is

$$T_2^{-1} = \frac{1}{2}T_1^{-1} + (T_2^*)^{-1}. \tag{2}$$

T_2^{-1} is related to the half width of the vibrational spectral line by $\Delta_{1/2} = 1/(2\pi T_2)$. In many situations the decoherence of intramolecular vibrational normal modes is dominated by the pure dephasing term, but this is not necessarily so. In the NMR situation, for example, the contributions from the pure dephasing and the population relaxation are equal, so that $T_1 = T_2$.

This discussion has assumed conditions of motional narrowing, which implies that the dephasing rate is large compared to the spread of instantaneous frequencies. If this is not true then the line is said to be inhomogeneously broadened and is composed of a superposition of lines from different environments. In a liquid environment one normally expects the conditions of motional narrowing to hold, but in solids or glasses this is not necessarily so. Pulsed laser experiments such as hole-burning can be used to probe the dynamics of inhomogeneously broadened lines.

Useful reviews of the theory of vibrational relaxation are found in [18, 19].

3. Perturbation Theory and Vibrational Relaxation

Vibrational relaxation is caused by fluctuations in the interaction between the molecule and the solvent. The effect of these can be treated by perturbation theory using the equation of motion of the density matrix σ for the whole system

$$i\hbar\dot{\sigma} = [(H_{0v} + H_{bv} + H_b), \sigma], \tag{3}$$

where H_{0v} is the Hamiltonian for the vibrations of the molecule, H_b is the Hamiltonian of the bath (the solvent) and H_{bv} describes the interaction between the solvent and solute vibrations.

If we concentrate on the vibrational variables and their density matrix ρ it is possible to treat the interactions with the bath as a perturbation and expand in powers of H_{bv} to obtain [20]

$$\dot{\rho}^\dagger = -\hbar^{-2}\int_0^t [(H_{bv}^\dagger(\tau), [H_{bv}^\dagger(t-\tau), \rho^\dagger]]d\tau, \tag{4}$$

where

$$H^{\dagger}_{bv}(t) = \exp(iH_{0v}t)H_{bv}(t)\exp(-iH_{0v}t) \tag{5}$$

and

$$\rho^{\dagger}(t) = \exp(iH_{0v}t)\rho\exp(-iH_{0v}t). \tag{6}$$

To illustrate the properties of a harmonic system we consider a single harmonic oscillator with eigenstates $|j\rangle$ and eigenvalues $E_j = (j+1/2)\hbar\nu$. We expand the bath-vibration interaction in powers of the normal coordinate ζ and retain the first term

$$H_{bv} = -F(t)\zeta + \cdots, \tag{7}$$

where $F = -\partial H_{bv}/\partial\zeta$ is the instantaneous force of the bath on the normal mode. As ζ only connects states whose quantum number differs by one unit, within this level of approximation the fluctuations induce transitions between adjacent levels. Standard time dependent perturbation theory yields an expression for the rate of transition between level n and $n-1$

$$k_{n-1\leftarrow n} + k_{n\leftarrow n-1} = Q'\frac{2n}{m\omega}\int_0^{\infty}\cos(\omega t)\langle F(0)F(t)\rangle dt, \tag{8}$$

where Q' is a quantum correction factor which depends on the temperature and the nature of the bath. In this expression m is the effective mass of the normal mode and ω is the frequency in units of radians per unit time.

A remarkable property of the harmonic oscillator is that, no matter what the distribution of energy, the decay is described by a single exponential at this level of approximation (second order perturbation theory; linear coupling to the bath). We may write the rate of decay of the vibrational energy E as

$$\dot{E} = T_1^{-1}(E - E^{eq}). \tag{9}$$

This simple equation may be derived as follows. The change of energy is made up of the sum of terms in which single quanta are lost to or absorbed from the solvent. Thus

$$\dot{E} = \hbar\omega\sum_n(-k_{n-1\leftarrow n} + k_{n+1\leftarrow n})\rho_{nn}, \tag{10}$$

where the first term describes the loss of a quantum and the second term is the gain of the quantum of vibrational energy. The diagonal element of the density matrix ρ_{nn} is just the population of state n. Using the expression for $k_{n-1\leftarrow n}$ above together with the fact that, to maintain thermal equilibrium, the detailed balance condition, $k_{n\leftarrow n-1} = \exp(-\hbar\omega/k_BT)k_{n-1\leftarrow n}$, must apply we obtain

$$T_1^{-1} = Q'\left(\frac{\tanh(\hbar\omega/2k_BT)}{\hbar\omega/2k_BT}\right)(1/mk_BT)\int_0^{\infty}\cos(\omega t)\langle F(0)F(t)\rangle dt. \tag{11}$$

In this expression the correlation function in the integral is a classical correlation function which can be determined from a classical atomistic simulation. The term in brackets,

$$Q_0 = \left(\frac{\tanh(\hbar\omega/2k_BT)}{\hbar\omega/2k_BT} \right), \tag{12}$$

accounts for detailed balance and the finite temperature of the bath. Until recently this term was thought to include all the necessary corrections needed to account for the differences between the classical force correlation function used in the integral and the true quantum force correlation function so that Q' was taken to be equal to unity. However recent work by Berne and coworkers, Skinner and co-workers and others [21, 22, 23, 24, 25] suggests that in some cases Q' may be significantly different from unity.

The rate of dephasing of the response of the harmonic oscillator to an oscillating field is given by the equation of motion of a linear combination of off-diagonal elements of the density matrix, ρ. The signal is proportional to S where

$$S = \sum_n \sqrt{n}\, Real(\rho_{n,n-1}). \tag{13}$$

As in the case of energy relaxation, for a harmonic oscillator this function decays as a single exponential with a rate equal to the decoherence rate. Equation 2 shows that this is the sum of half the energy relaxation rate plus the pure dephasing rate. The perturbation expression for the pure dephasing rate is [26]

$$(T_2^*)^{-1} = \int_0^\infty \langle \delta\omega(0)\delta\omega(t) \rangle dt, \tag{14}$$

where $\delta\omega(t)$ is the value of the difference between the *instantaneous* frequency and the *average* frequency of the oscillator in the bath. If the oscillator were truly harmonic, solvent frequency shifts would only occur as a result of quadratic and higher terms in the solvent-solute interaction. Hence the expansion of the molecule-bath interaction must be taken to second order

$$V_{bv} = -\sum_i F_i\zeta_i + \sum_{ij} \frac{1}{2} K_{ij}\zeta_i\zeta_j + \ldots, \tag{15}$$

where $K_{ij}(t)$ is the curvature term (the second derivative of the molecule-bath interaction with respect to the normal coordinates). However in most real situations this is not the most important term and it is necessary to include anharmonic terms in the molecular potential. The vibrational potential energy of an isolated molecule expressed as a function of the mass-weighted normal coordinates ζ_i is

$$V_{0v} = \sum_i \frac{1}{2} k_{ii}\zeta_i^2 + \sum_{ijk} \frac{1}{6} f_{ijk}\zeta_i\zeta_j\zeta_k + \ldots \tag{16}$$

where f_{ijk} are coefficients of cubic anharmonicity. Combining equations (15) and (16) gives an instantaneous potential energy which is a cubic function of the normal coordinates and which includes linear terms. The nearest minimum is found by equating the derivative of the potential energy to zero, which gives to lowest order

$$\zeta_j^{min} = F_j/k_{jj}. \tag{17}$$

The instantaneous frequency $\omega_j(t)$ is related to the second derivative of the total potential energy at the minimum, and is given by

$$m\omega_j^2 = \left(\frac{\partial^2(V_{0v}+V_{vb})}{\partial\zeta_j^2}\right)_{\zeta_j^{min}} = k_{jj} + K_{jj}(t) + \sum_k f_{jjk}\zeta_k^{min}(t). \tag{18}$$

When the value for ζ_j^{min} is inserted, we can rewrite this as

$$m\omega_j^2 = k_{jj}[1 + K_{jj}(t)/k_{jj} + \sum_k f_{jjk}F_k/(k_{kk}k_{jj}) + \ldots]. \tag{19}$$

Using the fact that the instantaneous solvent frequency shift is small compared to the gas phase frequency, and applying the expansion $(1+x)^{1/2} = 1 + x/2 + \ldots$ we obtain

$$\omega_j(t) = \omega_{j0}[1 + K_{jj}(t)/2k_{jj} + \sum_k f_{jjk}F_k/2(k_{kk}k_{jj}) + \ldots]. \tag{20}$$

Hence the average solvent shift is (to lowest order)

$$\langle\omega_j(t)\rangle - \omega_{j0} = \frac{\langle K_{jj}(t)\rangle}{2\sqrt{mk_{jj}}} + \frac{\sum_k f_{jjk}\langle F_k\rangle}{2(k_{kk}\sqrt{mk_{jj}})}, \tag{21}$$

and the instantaneous fluctuation in frequency due to the solvent fluctuations is (to lowest order)

$$\delta\omega_j(t) = \frac{\sum_k f_{kjj}(F_k - \langle F_k\rangle)}{2k_{kk}\sqrt{mk_{jj}}} + \frac{(K_{jj} - \langle K_{jj}\rangle)}{2\sqrt{mk_{jj}}}. \tag{22}$$

There are two types of terms in this last expression; the first term is the sum of the cross terms between cubic anharmonicities and solvent forces on the various modes while the second term is the solvent contribution to the curvature of the potential. Thus we may write

$$\delta\omega_j(t) = \Delta_F + \Delta_K \tag{23}$$

where Δ_F is the shift due to the solvent force and Δ_K is the shift due to the curvature term. It should be noted that the forces on both the same mode and other modes may contribute to the instantaneous solvent force shift because of cross terms f_{jjk} in the cubic anharmonicity.

Using this division of the instantaneous frequency shift into solvent force and solvent curvature terms we obtain

$$\frac{1}{T_2^*} = \int_0^\infty \langle \Delta_F(0)\Delta_F(t)\rangle dt + \int_0^\infty \langle \Delta_K(0)\Delta_K(t)\rangle dt + 2\int_0^\infty \langle \Delta_F(0)\Delta_K(t)\rangle dt, \tag{24}$$

which can be rewritten in terms of a sum of terms from solvent forces, solvent curvature and their cross interaction as

$$\frac{1}{T_2^*} = \langle \Delta_F^2 \rangle \tau_F + \langle \Delta_K^2 \rangle \tau_K + 2\langle \Delta_F \Delta_K \rangle \tau_{FK}. \tag{25}$$

The correlation times in this equation are defined by, for example,

$$\tau_F = \int_0^\infty \langle \Delta_F(0)\Delta_F(t)\rangle dt / \langle \Delta_F^2 \rangle. \tag{26}$$

These equations can now be used in the analysis of simulation results.

4. Vibrational relaxation from simulations of flexible molecules

The examples that we use to illustrate this article are the two stretching modes of the the triiodide ion I_3^- which is linear and centro-symmetric in the gas phase. It has three distinct normal modes which are determined by symmetry to be the symmetric stretch, the antisymmetric stretch and the bend; the last of these is doubly degenerate. Figure 1 shows the stretching normal modes.

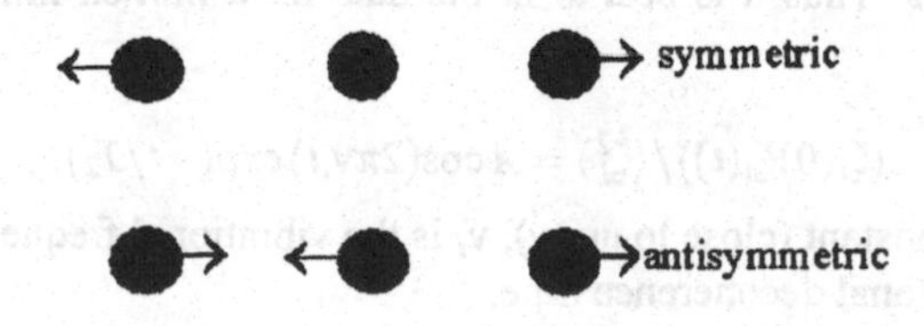

Figure 1. Normal modes of linear and symmetric I_3^- in the gas phase.

There are two ways of estimating vibrational relaxation rates from simulations. One way is to measure the energy relaxation rate and the decoherence rate from a simulation in which the molecules are flexible and vibrate. The accuracy of this method depends on the accuracy of the intramolecular potential and the intermolecular potential. The other way is to carry out a simulation with rigid molecules and evaluate the correlation functions needed for the perturbation theory expressions. These methods are complementary as flexible simulations include all anharmonic effects, while perturbation theory allows one to investigate the relative importance of different contributions. We first consider the direct determination of the relaxation rates from simulations with flexible molecules.

4.1. DETERMINING DECOHERENCE RATES

The most convenient way to determine T_2^{-1} is to examine the time correlation function of the derivative of the normal modes, as this derivative is just a linear combination of atomic velocities:

$$\dot{\zeta}_i = \sum_{p,\alpha} c_{ip\alpha} v_{p\alpha}, \tag{27}$$

where $v_{p\alpha}$ is the α component of the velocity of atom p in the *molecule-fixed frame* and $c_{ip\alpha}$ are the coefficients relating the atomic displacements of atom p in direction α to the normal modes i. If the time correlation function of the velocity of the ith normal mode has the form of an exponentially decaying oscillation

$$\langle \dot{\zeta}_i(0)\dot{\zeta}_i(t)\rangle = \langle \dot{\zeta}_i^2\rangle \cos(\omega_i t)\exp(-t/T_2), \tag{28}$$

then the value of T_2 can be determined.

Time correlation functions should be constructed using averages over many time origins for a long simulation [1]. The accuracy of the results depends on the quality of the fit as well as the accuracy of the correlation function and can be estimated by the method of block averages. In practice the decay of the normal mode velocity correlation function may not be exponential at very short times or at long times. Thus it is best to fit the data for a limited time range to the function

$$\langle \dot{\zeta}_i(0)\dot{\zeta}_i(t)\rangle / \langle \dot{\zeta}_i^2\rangle = A\cos(2\pi\nu_i t)\exp(-t/T_2), \tag{29}$$

where A is a constant (close to unity), ν_i is the vibrational frequency and T_2 the required vibrational decoherence time.

Figure 2 shows an example of the velocity correlation functions for the two modes at a particular state point and the function obtained by fitting to equation (29) while Table 1 shows the values of the parameters determined from this data.

The fit to the data in figure 2 is excellent for the time range shown; it is only just possible to distinguish the dashed curve of the fit from the solid line of

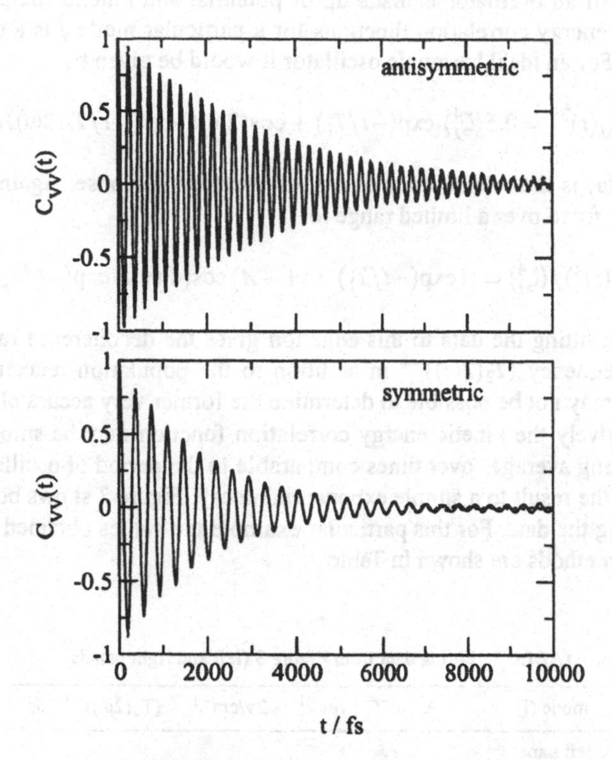

Figure 2. **Normalised velocity time correlation functions (solid line) and their fit (dashed line) for I_3^- in Xe at 280 K. Above: anti-symmetric mode; below: symmetric mode.**

TABLE 1. Fitting data in Figure 2 by equation (29).

mode (i)	A	ν/cm^{-1}	T_2^{-1} / ps^{-1}
symmetric (s)	0.97	90.6	0.52
antisymmetric (a)	1.02	139.4	0.32

the observed curve at the longest times in the figures. At even longer times one sees a revival of the oscillations. This is a well known problem in simulations of phonons and is probably due to the presence of periodic boundaries.

4.2. DETERMINING ENERGY RELAXATION RATES

The energy of an oscillator is made up of potential and kinetic energy terms. The kinetic energy correlation functions for a particular mode j is a decaying oscillation. For an ideal harmonic oscillator it would be given by

$$\langle \dot{\zeta}_j(0)^2 \dot{\zeta}_j(t)^2 \rangle = 0.5 \langle \dot{\zeta}_j^4 \rangle [\exp(-t/T_1) + \cos(2\omega_j t) \exp(-t/T_2(2\omega))], \quad (30)$$

where $T_2(2\omega)$ is the relaxation time for the overtone response. Again the real data may be fitted over a limited range to

$$\langle \dot{\zeta}_j(0)^2 \dot{\zeta}_j(t)^2 \rangle / \langle \dot{\zeta}_j^4 \rangle = A \exp(-t/T_1) + (1-A) \cos(4\pi \nu_j t) \exp(-t/T_2(2\omega)). \quad (31)$$

In principle, fitting the data to this equation gives the decoherence rate of the overtone frequency $(T_2(2\omega))^{-1}$ in addition to the population relaxation rate, although it may not be possible to determine the former very accurately.

Alternatively the kinetic energy correlation function may be smoothed by taking running averages over times comparable to the period of oscillation and then fitting the result to a simple exponential decay. Figure 3 shows both methods of fitting the data. For this particular example the values obtained from the two fitting methods are shown in Table 2.

TABLE 2. Fitting data from Figure 3 (left and right panels).

mode (i)	A	T_1^{-1} /ps^{-1}	2ν/cm^{-1}	$(T_2(2\omega))^{-1}$ / ps^{-1}
left panel				
symmetric (a)	0.47	0.46	178.8	1.5
antisymmetric (s)	0.50	0.10	278.0	0.9
right panel				
symmetric (s)	0.46	0.45		
antisymmetric (a)	0.50	0.10		

Once the decoherence and energy relaxation rates have been determined the pure dephasing rate $(T_2^*)^{-1}$ can be found from equation (2), which can be rewritten

$$(T_2^*)^{-1} = T_2^{-1} - 0.5 T_1^{-1}. \quad (32)$$

The corresponding equation for the overtone relaxation is

$$(T_2^*(2\omega))^{-1} = T_2^{-1}(2\omega) - T_1^{-1}. \quad (33)$$

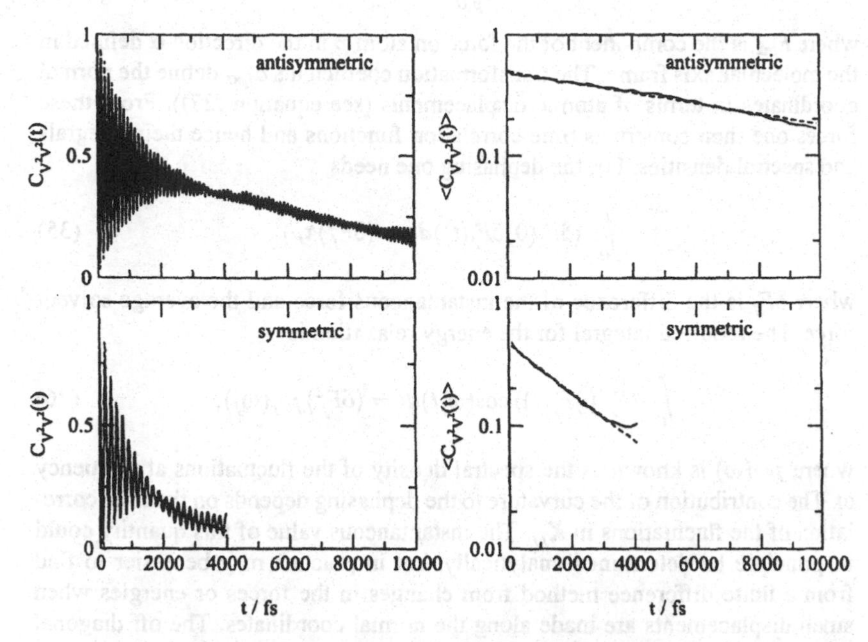

Figure 3. Kinetic energy time correlation functions (solid line) and their fit (dashed line) for I_3^- in Xe at 280 K for the two modes. The graphs on the left show the original data, while those on the right show the data averaged over a vibrational oscillation and plotted logarithmically. The straight lines in the latter show that the energy relaxation is exponential.

From the data in Table 1 and Table 2, the pure dephasing rates for symmetric and antisymmetric modes are found to be 0.29 ± 0.04 and $0.27 \pm 0.04\ \mathrm{ps}^{-1}$ respectively. The pure dephasing rates for symmetric and antisymmetric overtone frequencies are $1.1 \pm 0.1\ \mathrm{ps}^{-1}$ and $0.8 \pm 0.2\ \mathrm{ps}^{-1}$ respectively. The ratios of the rates of relaxation of the overtone to that of the fundamental are 3.6 ± 1.1 and 3.1 ± 1.1 for the symmetric and antisymmetric modes respectively.

4.3. USING PERTURBATION THEORY

The advantage of using perturbation theory is that one can analyse different contributions to the relaxation. The perturbation theory expressions for both energy relaxation and pure dephasing have been given above in equations (11) and (25). In order to use these equations one carries out a simulation with no vibrations, that is with a rigid molecule, and monitors the force on each normal mode j

$$F_j = \sum_{p,\alpha} c_{jp\alpha} F_{p\alpha}, \tag{34}$$

where $F_{p\alpha}$ is the component of the force on atom p in the direction α defined in the molecular axis frame. The transformation coefficients $c_{jp\alpha}$ define the normal coordinates in terms of atomic displacements (see equation (27)). From these forces one then constructs time correlation functions and hence their integrals and spectral densities. For the dephasing one needs

$$\int_0^\infty \langle \delta F_j(0)\delta F_j(t)\rangle dt = \langle \delta F_j^2\rangle \tau_{Fj}, \tag{35}$$

where δF_j is the difference of the instantaneous force and the average solvent force. The required integral for the energy relaxation is

$$\int_0^\infty \langle F_j(0)F_j(t)\rangle \cos(\omega_j t)dt = \langle \delta F_j^2\rangle j_{Fj}(\omega_j), \tag{36}$$

where $j_{Fj}(\omega)$ is known as the spectral density of the fluctuations at frequency ω. The contribution of the curvature to the dephasing depends on the time correlation of the fluctuations in K_{jj}. The instantaneous value of this quantity could in principle be determined analytically, but in practice may be easier to find from a finite difference method from changes in the forces or energies when small displacements are made along the normal coordinates. The off diagonal elements such as K_{ij} can be neglected as their effect is small. Finally the cross terms between the force and curvature terms must be taken into account.

4.3.1. *Energy relaxation*

Figure 4 shows the fluctuating force distributions for antisymmetric and symmetric mode respectively. The first thing to note is that, although the distribution of fluctuating forces on the antisymmetric mode is symmetric about zero as is required by symmetry, there is a mean force acting on the symmetric mode which tends to stretch the molecule. The mean square fluctuating forces $\langle \delta F^2\rangle = \langle (F^2 - \bar{F}^2)\rangle$ can be determined from this data; values are given in Table 3.

It is more difficult to determine the spectral densities accurately. Figure 5 shows the integral

$$j(\omega,t) = \int_0^t \langle \delta F_j(0)\delta F_j(t)\rangle \cos(\omega_j t)dt / \langle \delta F_j^2\rangle, \tag{37}$$

whose asymptotic value at long times gives the required spectral density.

One can see that although there is quite a bit of noise in the integral, it is possible to estimate the asymptotic values for each mode. The frequencies

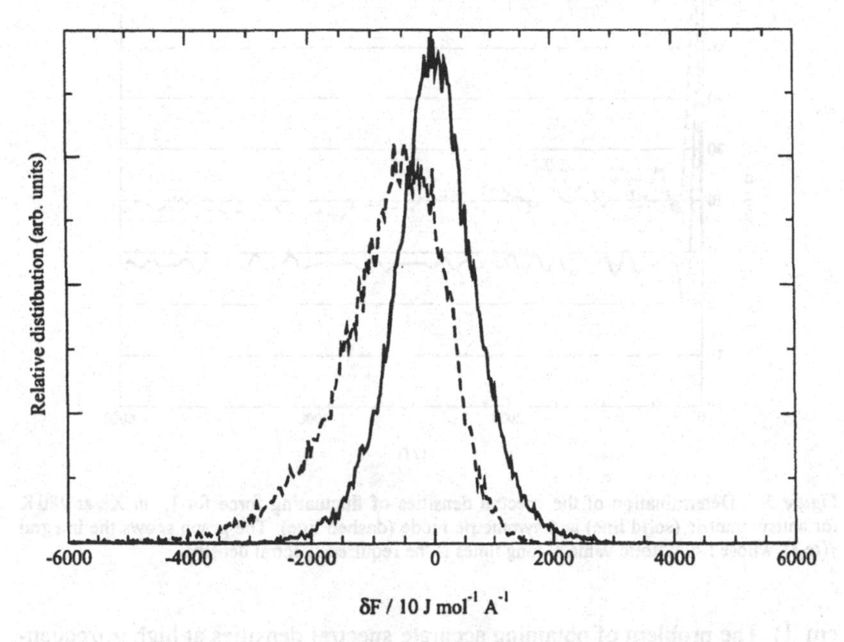

Figure 4. **Fluctuating force distributions for I_3^- in Xe at 280 K for antisymmetric (solid line) and symmetric mode (dashed line). Note the shift for the symmetric mode.**

used here are the vibrational frequencies obtained from the earlier fitting of flexible molecule simulations. The energy relaxation rates for symmetric and antisymmetric modes are shown in Table 3. As the ratio $\hbar\omega/k_B T$ is less than one

TABLE 3. Components of equation (11) from Figure 4 and Figure 5.

mode (i)	Q_0	$\langle \delta F^2 \rangle$ /10^3 J^{2}mol^{-2}Å^{-2}	$j(\omega)$ / ps	T_1^{-1} /ps^{-1}
symmetric (s)	0.98	9.0	0.019	0.57
antisymmetric (a)	0.96	5.3	0.008	0.14

the quantum correction for detailed balance $Q_0 \approx 1$, and the additional quantum correction Q' is also expected to be unimportant.

It should be remembered that the vibrational frequencies of the triiodide ion are much lower than most molecular vibrations (100 cm^{-1} rather than 1000

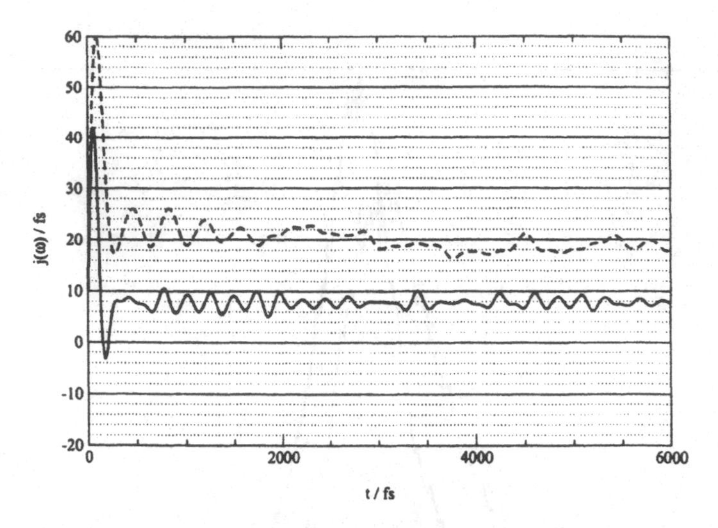

Figure 5. Determination of the spectral densities of fluctuating force for I_3^- in Xe at 280 K for antisymmetric (solid line) and symmetric mode (dashed line). The graph shows the integral $j(\omega, t)$, whose asymptotic value at long times is the required spectral density.

cm^{-1}). The problem of obtaining accurate spectral densities at higher frequencies is considerable and methods have been suggested for extrapolation of the spectral density from low frequencies [27].

4.3.2. *Pure dephasing*

Figure 6 shows the distribution of the instantaneous solvent frequency shifts together with the contributions from the fluctuating force and the fluctuating curvature. All the contributions give a net high frequency solvent shift (blue shift). Although the total frequency distributions are very similar for the two cases, there is a marked contrast between the two modes. Although curvature, force and their cross terms contribute to the mean shifts of both modes, the fluctuations in the antisymmetric mode frequency are due almost entirely to the force contribution. The values of the mean square frequency fluctuations which are required for evaluating the pure dephasing rates from equation (25) are given in Table 4. The correlation times can be found from the integrals of the normalised frequency correlation functions. Figure 7 shows these correlation functions and their integrals for each contribution and for the total. It can be seen that it is difficult to determine the correlation times accurately although in every case the integrals reach a plateau. The data from Figure 6 and Figure 7

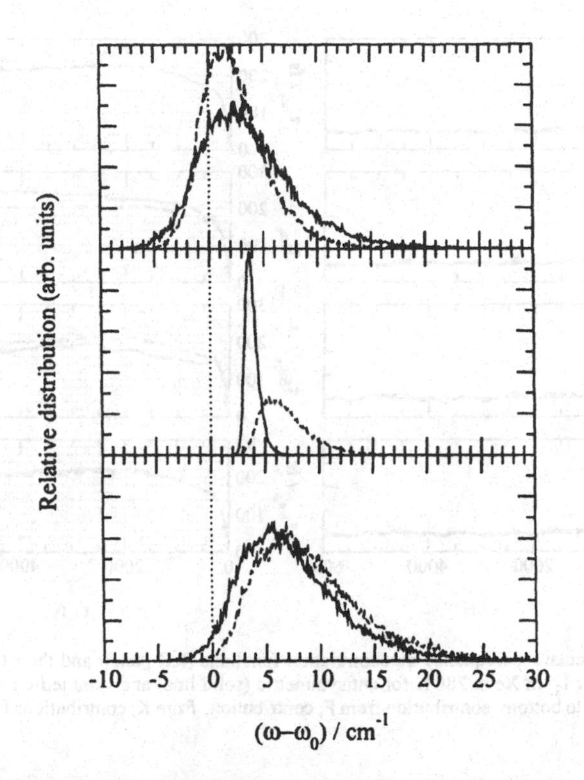

Figure 6. **Instantaneous frequency distributions for I_3^- in Xe at 280 K for antisymmetric (solid line) and symmetric mode (dashed line). Above: contribution from F; middle: contributions from K; below: totals. The frequencies are measured relative to the gas phase values.**

used in the determination of the pure dephasing rate are collected together in Table 4.

One finds that for both antisymmetric and symmetric modes, the time correlation functions for the different contributions, that is the fluctuations from the force F, the curvature K, and their cross term FK, have almost the same form and very similar correlation times. For the antisymmetric mode, the force term contributes about 90% to the total pure dephasing rate, the curvature term 1%, and the cross term about 9% to the total pure dephasing rate. However, for symmetric modes, the force term contributes only about 36%, while the curvature term contributes about 21%, and the cross term about 43 %. Most molecular

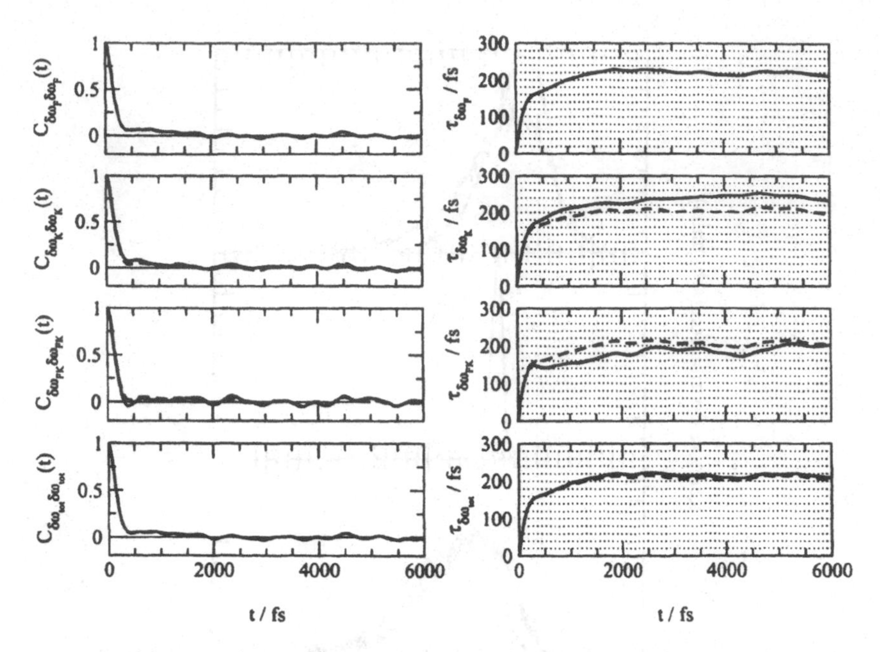

Figure 7. Fluctuating frequency autocorrelation functions (left panel) and their time integrals (right panel) for I_3^- in Xe at 280 K for antisymmetric (solid line) and symmetric modes (dashed line). From top to bottom: contribution from F; contributions from K; contributions from the cross FK; totals.

vibrations that have been studied have a very small contribution from the curvature term which is often neglected.

The analysis of the pure dephasing presented in this article is based on the premise that there is motional narrowing of the vibrational spectral lines so that they are homogeneously broadened. We can now check the self consistency of this assumption. If it is correct then $T_2^{-1} < \tau^{-1}$ or equivalently the dimensionless quantity $\alpha = \tau\sqrt{\Delta^2} < 1$. Using the data in table 4 and converting to consistent units one obtains $\alpha_a = 0.20$ and $\alpha_s = 0.21$ for antisymmetric and symmetric modes, confirming that the criterion for motional narrowing is met. Hence the spectral line shapes are predicted to be Lorentzian. Another criterion for the self consistency of the analysis is that the relaxation rates from flexible molecule simulations and from perturbation theory are in agreement. If this is not so it is likely that the effects of anharmonicity are larger than expected and the perturbation theory should be taken to higher order. Table 5 compares the relaxation

TABLE 4. Components of equation (25) from Figure 6 and Figure 7.

mode (i)	symmetric (s)	antisymmetric (a)
τ_F / ps	0.22	0.22
$\langle\Delta_F^2\rangle$ / cm^{-2}	9.7	21.2
$(T_2^*)_F^{-1}$ / ps^{-1}	0.08	0.17
τ_K / ps	0.21	0.24
$\langle\Delta_K^2\rangle$ / cm^{-2}	6.0	0.4
$(T_2^*)_K^{-1}$ / ps^{-1}	0.045	0.003
τ_{FK} / ps	0.20	0.18
$\langle\Delta_F\Delta_K\rangle$ cm^{-2}	6.4	1.7
$(T_2^*)_{FK}^{-1}$/ps^{-1}	0.09	0.02
τ_{tot} / ps	0.21	0.22
$\langle\Delta_{tot}^2\rangle$ / cm^{-2}	28.6	25.0
$(T_2^*)^{-1}$ / ps^{-1}	0.21	0.19

TABLE 5. Comparison of the rates from flexible and rigid (perturbation theory) simulations.

mode (i)	T_1^{-1} /ps^{-1} rigid	T_1^{-1} /ps^{-1} flexible	$(T_2^*)^{-1}$ /ps^{-1} rigid	$(T_2^*)^{-1}$ /ps^{-1} flexible
symmetric (s)	0.57 ±0.09	0.45 ± 0.06	0.21±0.02	0.29 ± 0.04
antisymmetric (a)	0.14±0.04	0.10 ±0.05	0.19±0.04	0.27 ± 0.04

rates from the two types of simulation, which are in agreement within the estimated uncertaintics. The latter are estimated from comparing the values from independent blocks of each simulation.

5. Conclusions

In this article we have shown, using the vibrations of a flexible model of the triiodide ion as an example, how information about the different types of vibrational relaxation rates can be determined from simulations and how perturbation theory can be used to interpret the results.

Acknowledgements

We are grateful to EPSRC for financial support (grant GR/N38459/01).

References

1. Allen, M. P. and Tildesley, D. J. (1994) Computer Simulation of Liquids, Clarendon Press, Oxford.
2. Oxtoby, D. W., Levesque, D. and Weis, J. J. (1978) A molecular dynamics simulation of dephasing in liquid nitrogen, J. Chem. Phys. 68, 5528-5533.
3. Levesque, D. and Weis, J. J. (1980) A molecular dynamics simulation of dephasing in liquid nitrogen. II. Effect of the pair potential on dephasing, J. Chem. Phys. 72, 2744-2749.
4. Levesque, D., Weis, J. J. and Oxtoby, D. W. (1983) A molecular dynamics simulation of rotational and vibrational relaxation in liquid HC, J. Chem. Phys. 79, 917-925.
5. Chesnoy, J. and Weis, J. J. (1986) Density dependence of the dephasing and energy relaxation times by computer simulation, J. Chem. Phys. 84, 5378-5388.
6. Westlund, P. O. and Lynden-Bell, R. M. (1987) A study of vibrational dephasing of the A_1 modes of CH_3CN in computer simulation of liquid phase, Mol. Phys. 60, 1189-1209.
7. Lynden-Bell, R. M. and Westlund, P. O. (1987) The effects of pressure and temperature on vibrational dephasing in a simulation of liquid CH_3CN, Mol. Phys. 61, 1541-1547.
8. Postma, J. P. M., Berendsen, H. J. C. and Straatsma, T. P. (1984) Intramolecular vibrations from molecular dynamics simulations of liquid water, J. Phys. C4, 31-40.
9. Chorny, I., Vieceli, J. and Benjamin, I. (2002) Molecular dynamics study of the vibrational relaxation of OClO in bulk liquids, J. Chem. Phys. 116, 8904-8911.
10. Poulsen, J., Nymand, T. M. and Keiding, S. R. (2001) Asymmetric stretch vibrational energy relaxation of OClO in liquid water, Chem. Phys. Lett. 343, 581-587.
11. Morita, A. and Kato, S. (1998) Vibrational relaxation of azide ion in water: The role of intramolecular charge fluctuation and solvent-induced vibrational coupling, J. Chem. Phys. 109, 5511-5523.
12. Diraison, D., Guissani, Y., Leicknam, J. C. and Bratos, S. (1996) Femtosecond solvation dynamics of water: solvent response to vibrational excitation of the solute, Chem. Phys. Lett. 258, 348-351.
13. Margulis, C. J., Coker, D. F. and Lynden-Bell, R. M. (2001) A Monte Carlo study of symmetry breaking of I3- in aqueous solution using a multistate diabatic Hamiltonian, J. Chem. Phys. 114, 367-376.
14. Margulis, C. J., Coker, D. F. and Lynden-Bell, R. M. (2001) Symmetry breaking of the triiodide ion in acetonitrile solution, Chem. Phys. Lett. 341, 557-560.
15. Zhang, F. S. and Lynden-Bell, R. M. (2002) A simulation study of vibrational relaxation of I_3^- in liquids, submitted for publication.
16. Zhang, F. S. and Lynden-Bell, R. M. (2002) Pure vibrational dephasing of triiodide in liquids and glasses, Mod. Phys. Lett. in press.
17. Lynden-Bell, R. M. and Zhang, F. S. (2002) in preparation.
18. Oxtoby, D. W. (1979) Dephasing of molecular vibrations in liquids, Adv. Chem. Phys. 40, 1-48.
19. Okazaki, S. (2001) Dynamical approach to vibrational relaxation, Adv. Chem. Phys. 118, 191-270.
20. Rothschild, W. G. (1984) Dynamics of Molecular Liquids Wiley-Interscience, New York.
21. Bader, J. S. and Berne, B. J. (1994) Quantum and classical relaxation rates from classical simulations, J. Chem. Phys.100, 8359-8366.
22. Egorov, S. A. and Skinner, J. L. (1996) A theory of vibrational energy relaxation in liquids, J. Chem. Phys. 105, 7047-7058.
23. Egorov, S. A., and Berne, B. J. (1997) Vibrational energy relaxation in the condensed phases: Quantum vs classical bath for multiphonon processes, J. Chem. Phys. 107, 6050-

6061.
24. Cherayil, B. J. and Fayer, M. D. (1997) Vibrational relaxation in supercritical fluids near the critical point, J. Chem. Phys.107, 7642-7650.
25. Rostkier-Edelstein, D., Graf, P. and Nitzan, A. (1997) Computing vibrational energy relaxation for high-frequency modes in condensed environment, J. Chem. Phys. 107, 10470-10479.
26. Kubo R. (1963) Stochastic processes in chemical physics, Adv. Chem. Phys. 13, 101-127.
27. Whitnell, R. M., Wilson, K. R. and Hynes J. T. (1990) Fast vibrational relaxation for a dipolar molecule in a polar solvent, J. Phys. Chem. 94, 8625-8628.

COMPUTATIONAL METHODS FOR ANALYZING THE INTERMOLECULAR RESONANT VIBRATIONAL INTERACTIONS IN LIQUIDS AND THE NONCOINCIDENCE EFFECT OF VIBRATIONAL SPECTRA

HAJIME TORII
Department of Chemistry, School of Education, Shizuoka University,836 Ohya, Shizuoka 422-8529, Japan

1. INTRODUCTION

In the liquid phase, vibrational dynamics and spectra of molecules are affected more or less by intermolecular interactions. There are two types of such effects. One of them is responsible for the modulation of the vibrational frequencies of each molecule, and is called "diagonal". This effect is operating even for a solute molecule in a dilute solution, and gives rise to a solvation-induced vibrational frequency shift. This effect also induces vibrational dephasing, since the magnitude of the solvation-induced frequency shift is modulated as time evolves according to the liquid dynamics. The other type of the effects of intermolecular interactions on vibrational dynamics and spectra arises from the direct coupling of vibrational modes of different molecules in the system. It is called "off-diagonal", because it is represented by the off-diagonal terms of the force constant matrix. This effect manifests itself most clearly in the resonant case, where the intrinsic frequencies of the interacting vibrational modes are sufficiently close to each other as compared with the magnitude of the coupling.

In the frequency-domain picture, intermolecular resonant vibrational interaction (off-diagonal according to the above classification) gives rise to delocalization of vibrational modes. When the vibrations of two molecules are resonantly coupled as shown in Figure 1 (a), we obtain two delocalized normal modes. One of them is the symmetric (in-phase) linear combination of the vibrations of the two molecules, and the other is the antisymmetric (out-of-phase) counterpart. The symmetric one is located on the lower-frequency side when the vibrational coupling is negative, and on the higher-frequency side otherwise. When the vibrations of many molecules are coupled with varying magnitudes in the liquid, the vibrational pattern of each delocalized normal mode is not so simple. However, even in this case, the frequency positions of delocalized normal modes are affected in a similar way by the vibrational patterns and the vibrational coupling constants.

J. Samios and V.A. Durov (eds.), Novel Approaches to the Structure and Dynamics of Liquids: Experiments, Theories and Simulations, 343–360.

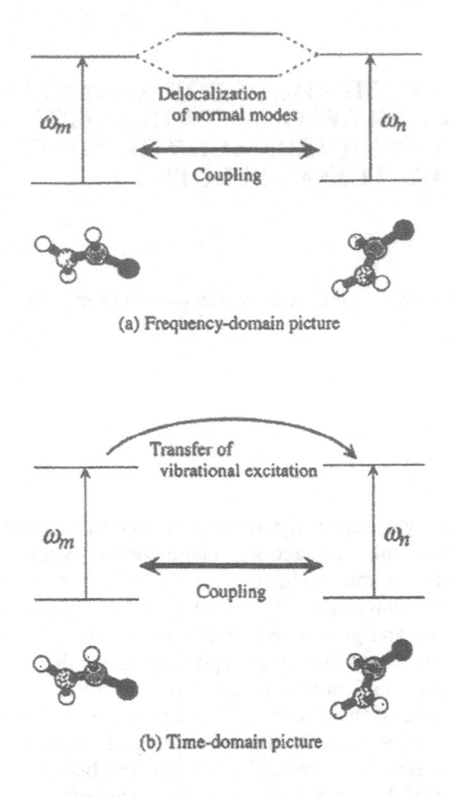

Figure 1. Intermolecular resonant vibrational interaction in (a) the frequency-domain picture and (b) the time-domain picture.

The noncoincidence effect (NCE) of vibrational spectra [1–4] is a phenomenon that is relevant to the frequency-domain picture of intermolecular resonant vibrational interaction. This effect refers to the spectral feature that the vibrational frequency positions of the infrared (IR), isotropic Raman, and anisotropic Raman components of a vibrational band do not coincide. In particular, the difference in the frequency positions between the isotropic and anisotropic Raman components is examined most often, and is called Raman noncoincidence. The NCE is observed in the cases where the vibrational patterns generating strong isotropic Raman intensities are different from those generating strong anisotropic Raman intensities or strong IR intensities, and the modes with these particular vibrational patterns are located at different frequency positions. Generally, the modes with large contribution from the in-phase linear combination of the vibrations of the molecules in the system have strong isotropic Raman intensities. However, the vibrational patterns generating strong anisotropic Raman intensities or strong IR intensities depend on the relative orientation of molecules [5]. In many cases, the intermolecular resonant vibrational interaction that gives rise to the NCE is determined by the transition dipole coupling (TDC) mechanism [3,4,6]. As described below in section 2, a TDC constant depends on the distance and relative orientation of the molecules involved in the coupling. Therefore, analysis of the sign and the magnitude of the NCE elucidates the distances and relative orientations of the molecules in the liquid system, i.e., the liquid structures.

In the time-domain picture of intermolecular resonant vibrational interaction, a vibrational excitation localized on one or a few molecules cannot be regarded as an eigenstate of the liquid system, and is rapidly transferred to other molecules, as shown schematically in Figure 1 (b). When the molecules involved in the coupling are oriented randomly in various directions, the polarization of the vibrational excitations initially made will decay rapidly as time evolves. Experimentally, such a phenomenon has been observed recently for the OH stretching mode of liquid water [7]. In this experiment, the transient absorption of the OH stretching band is measured after it is excited by an ultrashort IR pump pulse (~200 fs), and it is found that the anisotropy of the transient absorption decays in a very short time (almost instantaneously). It has been shown [8] that the decay rate of the transient absorption anisotropy in the time domain has some relation with the magnitude of the NCE in the frequency domain.

In the present paper, we review the computational methods for analyzing the intermolecular resonant vibrational interactions in liquids and the NCE of vibrational bands. Although there exists a useful analytical method to treat the NCE [3,9,10], it is applicable only to dipolar (pure and mixed) liquids and dipolar–nonpolar liquid mixtures. To examine the cases of hydrogen-bonding liquids, numerical methods are essential. Numerical methods are also useful to treat the time-dependent phenomena arising from intermolecular vibrational interactions, and to examine the competing roles of the diagonal and off-diagonal effects. The present paper shows the ways of using numerical methods to solve such problems with some examples.

2. TRANSITION DIPOLE COUPLING: THE ROLE OF DIPOLE DERIVATIVES

Transition dipole coupling (TDC) is an electrostatic interaction between molecular vibrations. It is harmonic in the sense that it is represented by the second derivative (not by a higher derivative) of the interaction potential energy with respect to vibrational coordinates. The explicit formula for the TDC may be derived as follows. The potential energy arising from the dipole–dipole interaction between molecules m and n is expressed as

$$V_{mn} = \frac{\mu_m \cdot \mu_n - 3(\mu_m \cdot \mathbf{n}_{mn})(\mu_n \cdot \mathbf{n}_{mn})}{R_{mn}^3} \tag{1}$$

where μ_m and μ_n are the dipole moments of molecules m and n, $\mathbf{n}_{mn}$ is the unit vector along the line connecting the two dipoles, and R_{mn} is the distance between the two dipoles. The formula for the TDC is obtained as the second derivative of this equation with respect to the vibrational coordinates of molecules m and n, which are denoted as q_m and q_n. It is given as

$$F_{mn} = \frac{\partial^2 V_{mn}}{\partial q_m \partial q_n} = \frac{\left(\frac{\partial \mu_m}{\partial q_m}\right)\cdot\left(\frac{\partial \mu_n}{\partial q_n}\right) - 3\left[\left(\frac{\partial \mu_m}{\partial q_m}\right)\cdot \mathbf{n}_{mn}\right]\left[\left(\frac{\partial \mu_n}{\partial q_n}\right)\cdot \mathbf{n}_{mn}\right]}{R_{mn}^3} \tag{2}$$

[For high-frequency intramolecular modes, we can safely neglect the contribution of the cross terms ($\partial\mu_m/\partial q_n$ and $\partial\mu_n/\partial q_m$) and the derivatives of R_{mn} and $\mathbf{n}_{mn}$ with respect to q_m and/or q_n.] This mechanism is called *transition dipole* coupling, because the dipole derivatives $\partial\mu/\partial q$ appearing in this formula are related within the harmonic approximation to the transition dipole for the vibrational transition $v = 0 \rightarrow 1$ (or $v = 1 \rightarrow 0$) as

$$<0|\mu|1> = \left(\frac{\partial\mu}{\partial q}\right)<0|q|1> \quad (3)$$

On the basis of this relation, the dipole derivative is obtained from the IR intensity, which is proportional to the square of the transition dipole. A large TDC constant is expected for a mode with a strong IR intensity.

More generally, dipole derivatives represent the response of molecular vibrations to external electric field [11]. In the case of IR absorption or emission, the electric field of radiation interacts with molecular vibrations. The formula for the TDC may be regarded as representing the response of a molecular vibration (of molecule *m*) to the dipole field of another molecule (*n*). Since the dipole field is modulated by molecular vibration (of molecule *n*), we obtain a direct coupling between the vibrations of different molecules (*m* and *n*). If, on the other hand, the dipole field is modulated more slowly, e.g., by molecular translations and rotations, we obtain a displacement of molecular structure along vibrational coordinate. This phenomenon, called vibrational polarization, is an origin of the modulation of vibrational frequencies [12–14] and other molecular properties [15–19] in condensed phases.

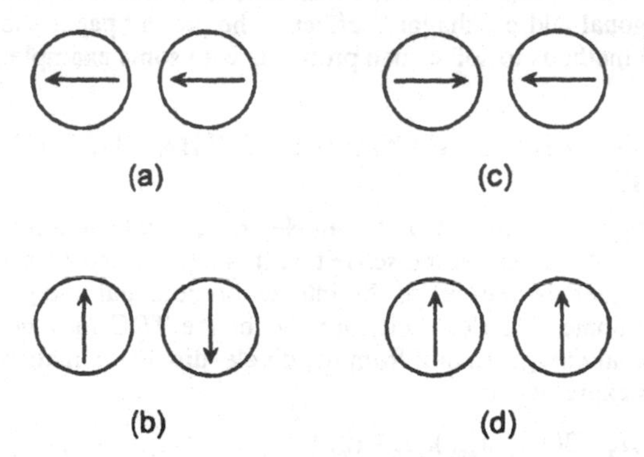

Figure 2. Scheme of typical cases where negative [(a) and (b)] and positive [(c) and (d)] intermolecular resonant vibrational interactions are operating according to the TDC mechanism. The transition dipoles of molecules are denoted by arrows. Reprinted with permission from H. Torii, *J. Phys. Chem. A* **106**, 3281–3286. Copyright (2002) American Chemical Society.

According to Eq. (2), the numerator depends on the relative orientation of the molecules involved in the coupling, and the denominator depends on the intermolecular distance. As a result, a TDC constant is sensitive to the liquid structure. Typical cases where negative and positive TDC constants are obtained are shown in Figure 2, where transition dipoles are represented by arrows. We obtain $F_{mn} < 0$ in the case of Figure 2 (a), since we have $0 < (\partial\mu_m/\partial q_m) \cdot (\partial\mu_n/\partial q_n) < 3\,[(\partial\mu_m/\partial q_m) \cdot \mathbf{n}_{mn}]\,[(\partial\mu_n/\partial q_n) \cdot \mathbf{n}_{mn}]$. In the case of Figure 2 (b) also, we obtain $F_{mn} < 0$, because we have $(\partial\mu_m/\partial q_m) \cdot (\partial\mu_n/\partial q_n) < 0$ and $(\partial\mu_m/\partial q_m) \cdot \mathbf{n}_{mn} = (\partial\mu_n/\partial q_n) \cdot \mathbf{n}_{mn} = 0$. By contrast, $F_{mn} > 0$ is obtained in the cases of Figure 2 (c) and (d).

When the transition dipole of the mode in question is parallel to the molecular permanent dipole, as in the case of the C=O stretching mode of acetone, the conformations shown in Figure 2 (a) and (b) are more stable than those shown in Figure 2 (c) and (d) because of the dipole–dipole interaction of permanent dipoles [Eq. (1)], and hence negative TDC is dominant when thermal average is taken. Then, the normal modes with large contribution from the in-phase linear combination of the vibrations of the molecules in the system, which tend to have strong isotropic Raman intensities, are likely to be located at lower-frequency positions than those with other vibrational patterns. A positive value of ν_{NCE} (= $\nu_{aniso} - \nu_{iso}$) is obtained as a result. (Here, ν_{iso} and ν_{aniso} are the vibrational frequencies of the isotropic and anisotropic Raman components.) However, even when the intermolecular vibrational interaction represented by the second derivative of the potential energy is dominated by the dipole–dipole interaction [TDC shown in Eq. (2)], it is possible that the potential energy itself is dominated by another type of interaction, such as hydrogen-bonding interaction. In such a case, a large number of closely located pairs of molecules may be oriented in a way that their vibrational transition dipoles are positively coupled to each other, as shown in Figure 2 (c) and (d). Therefore, various behavior of NCE may be considered on the basis of the TDC mechanism.

3. COMPUTATIONAL METHOD OF ANALYSIS 1: MD/TDC METHOD

3.1 BASIC CONCEPT AND COMPUTATIONAL SCHEME

For liquids consisting of polyatomic molecules, calculations of vibrational spectra may be computationally demanding if all the vibrational degrees of freedom are taken into account. However, when the vibrational mode of interest in each molecule is sufficiently well separated in frequency from the other modes, it is considered reasonable to construct the vibrational Hamiltonian only on the basis of the vibrational mode of interest, neglecting any possible intramolecular mode mixing induced by intermolecular interactions. In this case, each molecule is treated as an oscillator playing the role of a vibrational exciton, and the force constant matrix (F matrix) of the liquid system is constructed by properly taking into account the intermolecular resonant coupling between those vibrational excitons. Since the vibrational bands exhibiting large NCE generally have strong IR intensities (and hence large dipole derivatives), it is reasonable to expect that the TDC mechanism plays a major role in the intermolecular vibrational coupling.

The computational method developed on the basis of this concept, called MD/TDC method [4,20], is shown schematically in Figure 3. In this method, molecular dynamics (MD) simulations are carried out first, and hundreds of configurations of liquid structure are sampled with reasonable time intervals. (Monte Carlo simulations may also be used for this purpose. The method is called MC/TDC in this case [4].) Then, each molecule is treated as an oscillator, representing the vibrational mode of interest. Each oscillator has its own intrinsic vibrational frequency, dipole derivative (transition dipole), and polarizability derivative (Raman tensor). These properties may depend on the environment of each molecule, e.g., the hydrogen-bonding condition. The F matrix of the coupled-oscillator system is then constructed for each configuration of liquid structure. Each diagonal term is calculated as the square of the intrinsic vibrational frequency (after appropriate conversion of unit). The off-diagonal terms are determined by the TDC mechanism. The F matrix thus constructed is diagonalized to obtain the normal modes of the liquid system. Each eigenvector represents the vibrational pattern of the normal mode $\partial q_m/\partial Q_k$, i.e., the relative amplitude (including the sign) of the molecular vibrations (q_m) in the normal mode (Q_k). The IR and Raman intensities of the normal modes are obtained from the product of the eigenvectors with the tensors of dipole derivatives ($\partial \mu_m/\partial q_m$) and polarizability derivatives ($\partial \alpha_m/\partial q_m$). From these intensities and the vibrational frequencies calculated as the square root of the eigenvalues, we obtain the IR and Raman spectra.

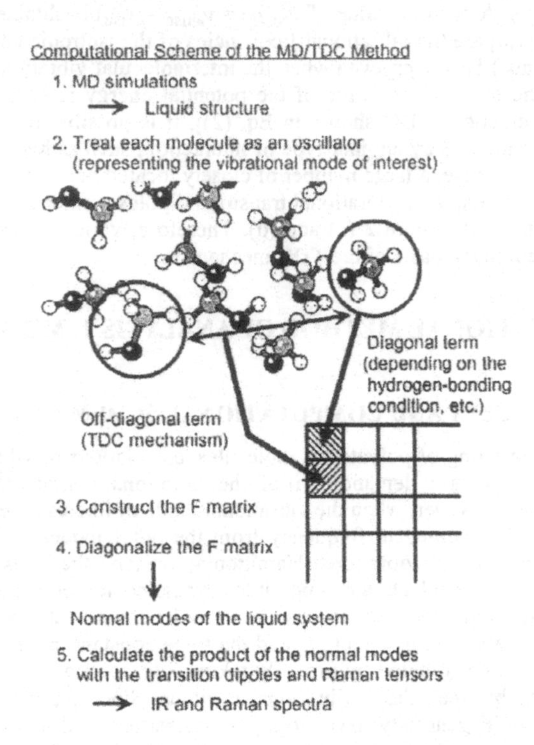
Computational Scheme of the MD/TDC Method
1. MD simulations
→ Liquid structure
2. Treat each molecule as an oscillator
(representing the vibrational mode of interest)
Diagonal term
(depending on the
hydrogen-bonding
condition, etc.)
Off-diagonal term
(TDC mechanism)
3. Construct the F matrix
4. Diagonalize the F matrix
Normal modes of the liquid system
5. Calculate the product of the normal modes
with the transition dipoles and Raman tensors
→ IR and Raman spectra

Figure 3. Computational scheme of the MD/TDC method.

The spectra obtained in this way are 'stick spectra' even after taking the statistical average of the liquid structure, because the effects of vibrational and rotational relaxation are not taken into account. An effective way to implicitly include these effects is to broaden each stick with appropriate band shape functions (in Gaussian, Lorentzian, or other forms).

Since liquid structures are sampled by MD, this computational method is applicable to any type (nonpolar, dipolar, or hydrogen-bonding; pure or mixed) of liquids if the intermolecular interactions are represented in a way that can be treated by MD. Hydrogen-bonding liquids like liquid methanol and its mixture [4,20,21] are good examples showing the usefulness of this method. However, since the mechanism of intermolecular resonant vibrational interaction is limited to TDC, it is not applicable to a case such as liquid 1,2,5-thiadiazole [22], where the NCE is observed for a mode (ν_6) with weak IR intensity.

3.2 INCLUSION OF THE EFFECT OF LIQUID DYNAMICS

For vibrational bands of the solute in a dilute solution, it is well known that the IR and anisotropic Raman band profiles are broadened to some extent by rotational relaxation that arises from liquid dynamics. This effect may be treated explicitly in a simple way by using the time correlation functions of the molecular orientation. For pure liquids and concentrated solutions also, the IR and anisotropic Raman bands tend to be broader than the isotropic Raman band. In the presence of intermolecular resonant vibrational interaction, however, theoretical treatment of the effect of rotational relaxation is not so simple. For example, suppose that two molecules are located as shown in Figure 2 (a). When the molecule on the left-hand side rotates by 180°, we obtain the situation shown in Figure 2 (c), with a positive instead of negative intermolecular vibrational interaction. It is therefore understandable that molecular rotations in liquids give rise to changes in the intermolecular vibrational interactions and, hence, in the vibrational patterns of the delocalized normal modes of the liquid. If the changes in the vibrational patterns are significant, effects of rotational relaxation on vibrational spectra cannot be simply treated by time correlation functions of the orientation of individual molecules.

To include explicitly the effects of liquid dynamics in the presence of intermolecular resonant vibrational interaction, the time evolution of vibrational excitations should be calculated [23]. For modes with $\hbar\omega >> kT$, the IR spectrum is expressed as

$$I^{(\mathrm{IR})}(\omega) = Re \int_0^\infty dt \exp(i\omega t) \left\langle \sum_{r=1}^{3} <0|\, \mu_r U(t) \mu_r \,|0> \right\rangle \tag{4}$$

where $|0>$ is the wave function of the ground state, μ_r is the dipole operator (with r = 1, 2, and 3 correspond to the x, y, and z directions), the large bracket denotes statistical average, and $U(t)$ is the time evolution operator [24] expressed as

$$U(t) = \exp_+ \left[-\frac{i}{\hbar} \int_0^t d\tau\, H(\tau) \right] \tag{5}$$

where $H(\tau)$ is the time-dependent vibrational Hamiltonian, and $\exp_+$ denotes the time-ordered exponential. The quantum mechanical bracket appearing on the right-hand side of Eq. (4) may be evaluated as

$$<0|\, \mu_r U(t) \mu_r \,|0> = \sum_{\xi,\zeta=1}^{N} <0|\, \mu_r \,|\zeta_t><\zeta_t \,|\, U(t) \,|\, \xi_0><\xi_0 \,|\, \mu_r \,|0> \tag{6}$$

where $|\zeta_t>$ and $|\xi_t>$ are eigenstates of the liquid system at time t. The wave function defined as

$$|\psi_r^{(\mathrm{IR})}(0)> = \sum_{\xi=1}^{N} |\xi_0><\xi_0 \,|\, \mu_r \,|0> \tag{7}$$

is the one formed at time $t = 0$ by the interaction with the radiation polarized in the rth direction. The time evolution of this wave function is controled by the operator $U(t)$ as

$$|\psi_r^{(IR)}(t)\rangle = U(t)\,|\psi_r^{(IR)}(0)\rangle \tag{8}$$

In numerical calculations, $U(t)$ is expressed as a product of short-time evolutions, with $\Delta\tau$ taken as the time step of the MD simulation. We have

$$|\psi_r^{(IR)}(\tau+\Delta\tau)\rangle = \exp\left[-\frac{i}{\hbar}\Delta\tau H(\tau)\right]|\psi_r^{(IR)}(\tau)\rangle$$

$$= \sum_{\xi=1}^{N} |\xi_\tau'\rangle \exp\left[-i\omega_\xi(\tau)\,\Delta\tau\right]\langle\xi_\tau|\psi_r^{(IR)}(\tau)\rangle \tag{9}$$

where $\omega_\xi(\tau)$ is the vibrational frequency for the eigenstate $|\xi_\tau\rangle$, and $|\xi_\tau'\rangle$ is the wave function with the same amplitudes of molecular vibrations as $|\xi_\tau\rangle$ but with the molecular orientations evaluated at time $\tau+\Delta\tau$. In this method, the vibrational Hamiltonian $H(\tau)$ should be constructed and diagonalized every time step to obtain $\omega_\xi(\tau)$ and $|\xi_\tau\rangle$. To obtain a frequency resolution of about 0.5 cm^{-1}, with the time step of $\Delta\tau = 2$ fs, the time correlation function should be calculated for 32768 time steps for one sample in the statistical average. A sufficiently large number of samples should be taken to get good statistics.

The formula for the Raman spectrum includes the polarizability operator instead of dipole operator in Eq. (4). The time evolution of the Raman excitation is evaluated in a similar way.

By using the method described above (called MD/TDC/WFP method [23], where WFP stands for 'wave function propagation'), it is possible to treat explicitly the effects of liquid dynamics on the diagonal and off-diagonal terms of the vibrational Hamiltonian and the resultant changes in the IR and Raman band profiles. An example of the application of this method is shown in section 5.2.

4. COMPUTATIONAL METHOD OF ANALYSIS 2: *AB INITIO* MOLECULAR ORBITAL CALCULATIONS FOR MOLECULAR CLUSTERS

The MD/TDC and MD/TDC/WFP methods described above and the analytical method developed by Logan [3,9,10] may be regarded as "direct" in the sense that they deal with liquid systems themselves. However, they assume TDC as the mechanism of intermolecular resonant vibrational interaction, so that the magnitude, direction in the molecular frame, and intramolecular location of the transition dipoles are needed as parameters. In addition, the parameters of potential energy functions used in MD simulations (for the MD/TDC and MD/TDC/WFP methods) or the permanent dipoles and hard sphere diameters of molecules (for the analytical method) are also needed as the factors that determine the liquid structures. In most cases, these parameters are determined directly or indirectly by referring to the results of some experiments, such as those obtained from thermodynamic and diffraction measurements. Nevertheless, it may be said that these methods inevitably contain some "empirical" character, in the same way as many other theoretical methods of this type.

In contrast, it may be conceivable to limit the computed systems to clusters of a few molecules but to calculate the vibrational spectra from first principles by using the *ab initio* molecular orbital (MO) method. The merit of this method is that parameters are no longer required for the calculations of vibrational frequencies and intensities because the calculations are carried out from first principles. This method may also be used to examine the validity of the parameters used in the methods described in section 3, or to improve the quality of those parameters. However, it seems practically impossible at present to do high-quality *ab initio* MO calculations for large clusters containing many molecules, so that the calculated systems are inevitably limited to small clusters. As a result, by using this method, it is hard to deal with properties characteristic of liquid systems, such as randomness and fluctuation, so that it tends to provide some hints but not the solution for the problems concerning the interpretation of the vibrational spectra of liquid systems. The *ab initio* MD method developed by Car and Parrinello [25] is promising in this respect, but its application to molecular liquids is limited at present to systems consisting of small molecules such as water [26,27].

An example of the application of the *ab initio* MO method to the problem of the NCE in solutions is shown in section 5.3.

5. EXAMPLES

5.1 PRESSURE-INDUCED CHANGE IN THE LIQUID STRUCTURE AND THE NONCOINCIDENCE EFFECT OF LIQUID METHANOL

Liquid methanol is one of the simplest hydrogen-bonded systems. An interesting subject in relation to the structural formation of hydrogen-bonded liquids is the competing role of the OH groups, which participate in hydrogen bonds, and the other parts in the molecules, which interact with each other much more weakly. Studies on the variation of the structure of liquid methanol as a function of thermodynamic variables will be helpful to our understanding on this point. Since methanol has two vibrational modes (OH and CO stretching modes) that show the NCE [28–31], studies on the vibrational spectra provide us with much information on the liquid structure.

In contrast to many other cases where the NCE is observed, the CO stretching band of liquid methanol shows a negative NCE [29], defined as $\nu_{NCE} = \nu_{aniso} - \nu_{iso}$. An important point concerning this NCE is that its magnitude (absolute value) decreases with increasing pressure, although the NCE arises from intermolecular resonant vibrational interactions. To elucidate the structural origin of this phenomenon, the MD/TDC (or MC/TDC) method is considered to be most useful [4,32].

The vibrational frequencies (first moments) of the isotropic and anisotropic Raman components (ν_{iso} and ν_{aniso}) and the values of ν_{NCE} calculated for the CO and OH stretching bands of liquid methanol of the density $d = 0.782$ and 0.942 g cm^{-3} at 303 K are shown in Table I. It is seen that the magnitude of ν_{NCE} of the CO stretching band decreases by 1.3 cm^{-1} with increasing density, in agreement with the experimental result (1.5 cm^{-1} [29]). For the OH stretching band, both the isotropic and anisotropic components shift to the low-frequency side, and the value of ν_{NCE} increases, as the density increases. This calculated low-frequency shift is also in agreement with the experimental result [33].

TABLE I. Vibrational frequencies (first moments) of the isotropic and anisotropic Raman components (ν_{iso} and ν_{aniso}) and the values of Raman noncoincidence (ν_{NCE}) calculated for the CO and OH stretching bands of liquid methanol of the density d = 0.782 and 0.942 g cm^{-3} at 303 K.

	d = 0.782 g cm^{-3}		d = 0.942 g cm^{-3}	
	CO stretch	OH stretch	CO stretch	OH stretch
ν_{iso} / cm^{-1}	1036.7	3332.3	1035.3	3314.4
ν_{aniso} / cm^{-1}	1032.7	3370.7	1032.6	3357.4
ν_{NCE} / cm^{-1}	−4.0	38.4	−2.7	43.0

The relation between the behavior of ν_{NCE} and the liquid structure is more clearly seen by evaluating the pair distribution functions [20,21,34,35] $g(\boldsymbol{R}_{mn}; \boldsymbol{\Omega}_m, \boldsymbol{\Omega}_n)$ of the CO and OH bonds, expanded to the second order as

$$g(\mathbf{R}_{mn}; \mathbf{\Omega}_m, \mathbf{\Omega}_n) = g_0(R_{mn}) + h_\Delta(R_{mn})\mathbf{\Omega}_m\mathbf{\Omega}_n$$

$$+h_D(R_{mn}) [3(\mathbf{R}_{mn}\mathbf{\Omega}_m)(\mathbf{R}_{mn}\mathbf{\Omega}_n)/R_{mn}^2 - \mathbf{\Omega}_m\mathbf{\Omega}_n] \qquad (10)$$

where $\boldsymbol{R}_{mn}$ is the vector connecting the relevant (CO or OH) bonds of molecules m and n, R_{mn} is the length of this vector, and $\boldsymbol{\Omega}_m$ and $\boldsymbol{\Omega}_n$ are the unit vectors in the direction of the relevant bonds of the two molecules. When this second-order expansion is a good approximation, ν_{NCE} is proportional to $\rho H_D(\infty)$, where ρ is the number density and the function $H_D(r)$ is defined as

$$H_D(r) = \int_0^r dR\, h_D(R) / R \qquad (11)$$

In addition to the total profiles, the contributions of the following three classes of molecule pairs to $g(\boldsymbol{R}_{mn}; \boldsymbol{\Omega}_m, \boldsymbol{\Omega}_n)$ are calculated: (1) molecule pairs which are directly hydrogen bonded to each other, (2) molecule pairs which are *not* directly hydrogen bonded to each other but hydrogen bonded to a common molecule, and (3) all the other pairs of molecules, such as the pairs of molecules belonging to different hydrogen-bonded chains.

The functions $g_0(r)$ and $H_D(r)$ calculated for the CO bond of liquid methanol at d = 0.782 and 0.942 g cm^{-3} are shown in Figure 4 (a–d). The negative contribution of the class 1 pairs to $H_D(r)$ (and hence to ν_{NCE} of the CO stretching band) is partially canceled by the positive contribution of the class 3 pairs. As the density increases, the number of the class 3 pairs [shown as $g_0(r)$] increases in the r = 4–5 Å region, resulting in significantly larger contribution of these pairs to $H_D(r)$ (and ν_{NCE}). Although the contribution of the class 1 pairs to $\rho H_D(\infty)$ increases slightly, the increase of the contribution of the class 3 pairs is more significant so that the cancellation between these classes becomes more severe. For this reason, the value of ν_{NCE} of the CO stretching band is smaller at the higher density.

The pair distribution functions calculated for the OH bond of liquid methanol at d = 0.782 and 0.942 g cm^{-3} are shown in Figure 4 (e–h). In this case, the contribution to $H_D(r)$ (and hence to ν_{NCE}) is dominated by that of the class 1 pairs. The value of $\rho H_D(\infty)$ increases slightly as the density increases, in accord with the behavior of ν_{NCE} shown in Table I. This is considered to be due to the slight shortening of the hydrogen-bond lengths ($r_{O...H}$) occurring upon increasing density. The low-frequency shift of the OH stretching band shown in Table I is also considered to originate from this slight shortening of $r_{O...H}$.

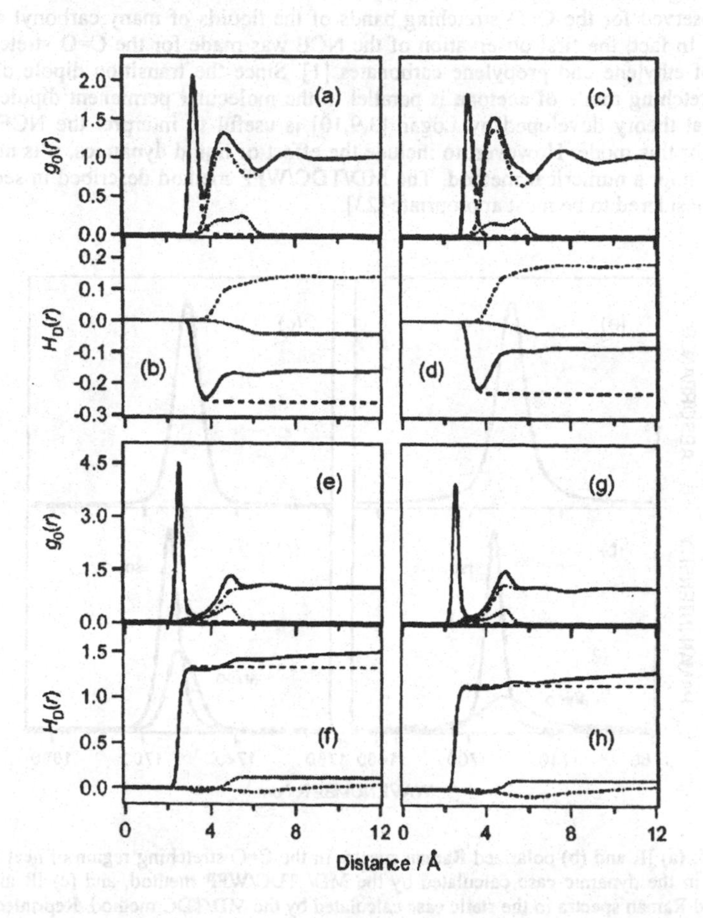

Figure 4. The functions $g_O(r)$ and $H_D(r)$ calculated for the CO bond (a to d) and the OH bond (e to h) of liquid methanol of the density $d = 0.782$ g cm^{-3} (a, b, e, f) and 0.942 g cm^{-3} (c, d, g, h) at 303 K. Solid line: total profiles; dashed line: contribution from molecule pairs that are directly hydrogen bonded to each other (class 1); dotted line: contribution from molecules that are not directly hydrogen bonded to each other but hydrogen bonded to a common molecule (class 2); dot-dashed line: contribution from "all the other" pairs of molecules (class 3).

5.2 EFFECT OF ROTATIONAL RELAXATION ON THE RAMAN BAND WIDTH IN THE PRESENCE OF INTERMOLECULAR RESONANT VIBRATIONAL INTERACTIONS: THE CASE OF LIQUID ACETONE

The C=O stretching band of liquid acetone, as the simplest ketone compound, may be regarded as a typical case where the NCE is observed, since the NCE has been observed for the C=O stretching bands of the liquids of many carbonyl compounds. In fact, the first observation of the NCE was made for the C=O stretching bands of ethylene and propylene carbonates [1]. Since the transition dipole of the C=O stretching mode of acetone is parallel to the molecular permanent dipole, the analytical theory developed by Logan [3,9,10] is useful to interpret the NCE observed for this mode. However, to include the effect of liquid dynamics, it is necessary to adopt a numerical method. The MD/TDC/WFP method described in section 3.2 is considered to be most appropriate [23].

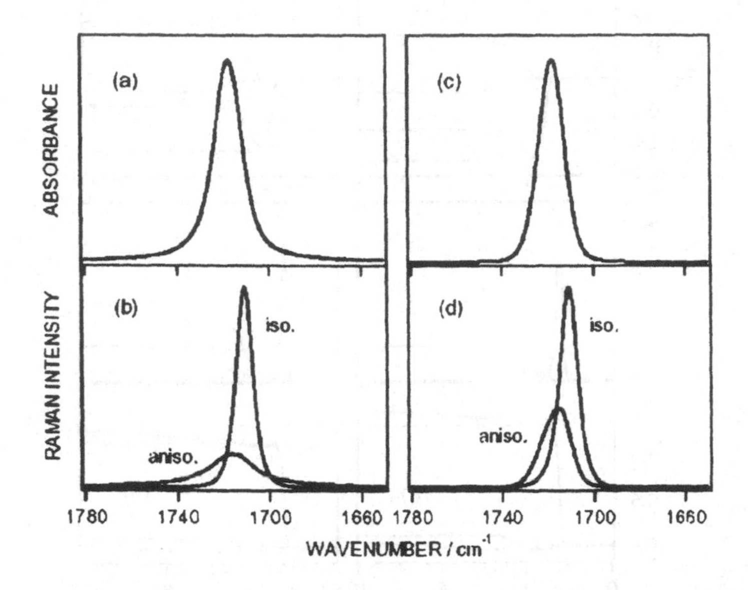

Figure 5. (a) IR and (b) polarized Raman spectra in the C=O stretching region of neat liquid acetone in the dynamic case calculated by the MD/ TDC/WFP method, and (c) IR and (d) polarized Raman spectra in the static case calculated by the MD/TDC method. Reprinted with permission from H. Torii, *J. Phys. Chem. A* **106**, 3281–3286. Copyright (2002) American Chemical Society.

The IR and Raman spectra of liquid acetone in the C=O stretching region calculated with the MD/TDC/WFP method are shown in Figure 5 (a) and (b). In this calculation, the effect of liquid dynamics is taken into account only for the off-diagonal terms of the F matrix, leaving the diagonal terms constant and common to all the molecules. At present, the effect of vibrational dephasing (modulation of the diagonal terms) is effectively included with a band shape function of FWHM = 7.8 cm^{-1}, considering that this effect is common to the IR, isotropic Raman, and anisotropic Raman bands, but it should be explicitly considered in the future after the mechanism of the modulation of diagonal terms is clarified. In Figure 5 (c) and (d), the spectra calculated for the static case with the MD/TDC method are shown for comparison.

TABLE II: Calculated and observed widths of the C=O stretching band of neat liquid acetone.

	method of calculation	width (FWHM, cm^{-1})		
		IR	iso.	aniso.
calculated	dynamic	14.6	8.4	23.9
	static	13.4	8.8	15.0
observed		~14 [a]	8.5 [b]	20.3 [b]

[a] Reference [30]. The spectrum of α_m".
[b] References [36] and [37].

It is seen in these spectra that significant magnitude of NCE is calculated in both cases. The calculated ν_{NCE} is 5.6 cm^{-1} in the dynamic case and 5.7 cm^{-1} in the static case, in good agreement with the observed value (5.2 cm^{-1} [36]). This means that the NCE remains almost the same when the liquid dynamics is taken into account in the calculations. By contrast, the bandwidths are evidently different between the two cases. The calculated bandwidths (FWHM) are summarized on the first and second rows of Table II. The broadening of the IR and anisotropic Raman bands obtained in the dynamic case as compared to the static case is considered to originate mainly from rotational relaxation. The observed bandwidths taken from the literatures [30,36,37] are listed on the third row. It is clearly recognized from the comparison of the observed and calculated bandwidths that a better agreement is obtained when the liquid dynamics is taken into account in the calculations. A little too large width of the anisotropic Raman band calculated in the dynamic case suggests that the liquid dynamics in the MD simulation is slightly too fast. A better result may be obtained by improving the potential energy function used in the MD simulation.

To help understand how rapidly the delocalized vibrational eigenstates change their forms, the time evolution of the instantaneous vibrational frequencies of those eigenstates are shown in Figure 6. It is seen that the instantaneous vibrational frequencies are modulated on the time scale of 100 fs or less, with many real and avoided crossings with each other, due to the changes in the off-diagonal terms of the F matrix arising from liquid dynamics. The changes in the vibrational patterns of the delocalized eigenstates occur on the same time scale. In such cases, effects of rotational relaxation on vibrational spectra cannot be simply treated by time correlation functions of the orientation of individual molecules, and this is the reason for the necessity of the MD/TDC/WFP method.

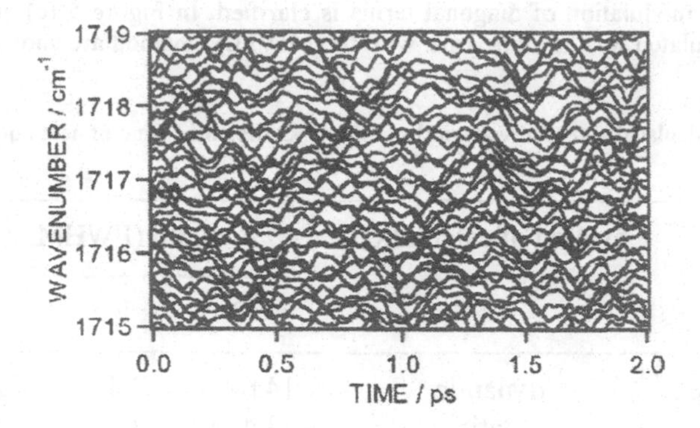

Figure 6. Time evolution of the instantaneous vibrational frequencies of delocalized eigenstates in the C=O stretching band of neat liquid acetone. Only the 1719–1715 cm^{-1} region is shown. A time interval of 2 ps is sampled. Reprinted with permission from H. Torii, *J. Phys. Chem. A* **106**, 3281–3286. Copyright (2002) American Chemical Society.

5.3 LIQUID STRUCTURE AROUND AN ANION AND THE NONCOINCIDENCE EFFECT OF AN ELECTROLYTE SOLUTION OF METHANOL

The liquid structures formed around ions in an electrolyte solution are most likely to be different significantly from those found in neat liquid. Since the NCE is sensitive to the relative orientation of closely located molecules, it is expected that information on the difference in the liquid structures is obtained by analyzing the NCE. An efficient way in such an analysis of locally formed liquid structure is to carry out *ab initio* MO calculations for small clusters consisting of an ion and a few solvent molecules.

As an example, we take the case of the methanol solution of LiCl [20]. An optimized structure of the $(methanol)_4Cl^-$ cluster is shown in Figure 7. It is seen that the four methanol molecules are located at tetrahedral positions around the chloride ion, and the OH groups of these methanol molecules are oriented toward the chloride ion because of their hydrogen-bonding ability. The relative orientation of the OH groups of the neighboring methanol molecules is somewhere between Figure 2 (c) and (d), resulting in positive intermolecular vibrational coupling according to the TDC mechanism. As a result, a negative value of ν_{NCE} is expected for the OH stretching band of this cluster. Indeed, we obtain $\nu_{NCE} = -25.8$ cm^{-1} at the MP2/6-31(+)G** level [20]. A different tetrahedral structure is formed around a lithium ion because of the strong electrostatic interactions between the ion and the oxygen atoms of methanol molecules (not shown). However, a very small value of ν_{NCE} is obtained in this case.

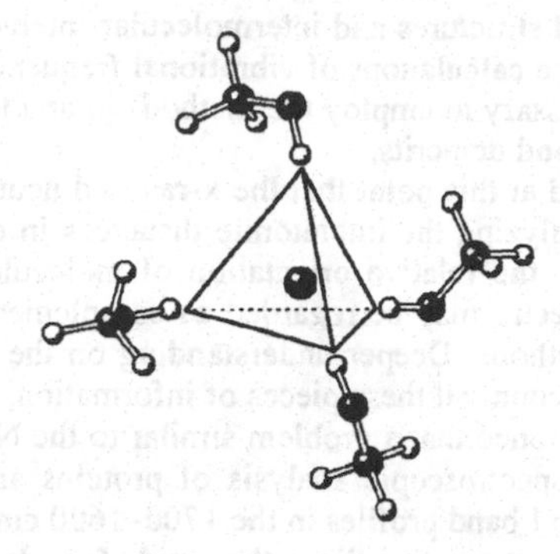

Figure 7. An optimized structure of the (methanol)$_4$Cl$^-$ cluster. A tetrahedron is drawn as the guide to the eye.

Experimentally, a large positive ν_{NCE} (of about 40 cm^{-1} when the value is obtained from the first moments) is observed for the OH stretching band of neat liquid methanol, but when LiCl is dissolved in it, the magnitude is reduced as the concentration of LiCl increases and becomes negative above a certain concentration [38]. At the LiCl/ methanol mole ratio of 0.18, ν_{NCE} is –44 cm^{-1}. A similar phenomenon is observed for the methanol solution of LiBr [39].

Concerning this experimental result, two different mechanisms were proposed. In one mechanism [39], it was suggested that this reversed Raman noncoincidence originates from a tetrahedral arrangement of methanol molecules around the halide ion, the vibrational interactions among which being determined by the TDC mechanism. In the other medchanism [38,40], it was suggested that indirect vibrational interactions between methanol molecules mediated by the polarization of the metal ion and/or the halide ion [41], rather than direct vibrational interactions, give rise to the negative Raman noncoincidence for the methanol–LiCl solution and some other electrolyte solutions. The optimized structures and the vibrational frequencies obtained in the *ab initio* MO calculation described above and the result of the vibrational analysis based on the TDC mechanism indicate that the former mechanism is more probable.

6. CONCLUDING REMARKS

In the present paper, we have discussed the computational methods for analyzing the intermolecular resonant vibrational interactions in liquids and the NCE of vibrational bands. The NCE is a spectral feature that originates from delocalization of vibrational modes. In many cases, the intermolecular resonant vibrational interactions that give rise to this effect are determined by the TDC mechanism, so that they are strongly dependent on liquid structures, especially on the relative orientation of closely located molecules. The relation between the vibrational patterns in delocalized vibrational modes and the spectral intensities is also dependent on the relative orientation of molecules. As such, analysis of the NCE will provide useful information on liquid structures. As explained above, the computational methods described in this paper and Logan's analytical method have their own merits and demerits concerning their applicability and the quality of the results, because of the differences in the ways of treating liquid structures and intermolecular interactions, and in the extent of approximation in the calculations of vibrational frequencies and intensities. It may be said that it is necessary to employ the methods in an efficient way by taking into account those merits and demerits.

It should be mentioned at this point that the x-ray and neutron diffraction methods are often used for analyzing the interatomic distances in condensed-phase systems. The information on the relative orientation of molecules obtained from the analysis of vibrational spectra may be regarded as complementary to that obtained from those diffraction methods. Deeper understanding on the liquid structures will be given by taking into account all these pieces of information.

It should also be mentioned that a problem similar to the NCE of liquid systems exists in the vibrational spectroscopic analysis of proteins and polypeptides. It is well known that the amide I band profiles in the 1700–1600 cm^{-1} region are strongly correlated to the secondary structures. Since the amide I mode has a large transition dipole, the amide I vibrations of peptide groups are coupled to each other by the TDC mechanism [42–45]. Since the peptide groups are connected covalently to each other along polypeptide chains, the vibrational interactions are not solely determined by the TDC mechanism [46–48]. However, a method similar to the MD/TDC method described in section 3.1 is useful to get insight into the structure–spectrum correlation, and quantitative agreement between experiment and calculation will be obtained when the parameters in the diagonal and off-diagonal terms of the F matrix are properly tuned.

Although the frequency- and time-domain pictures shown schematically in Figure 1 are related by Fourier transform, most of the experimental results concerning the delocalization of vibrational modes in liquids are obtained in the frequency domain. Experiments and theoretical analyses of delocalized vibrations in the time domain, or those with the time-domain picture taken into account (such as the MD/TDC/WFP method described in section 3.2), may give us new insight into the vibrational dynamics, especially concerning the vibrational excitation transfer, in condensed-phase systems.

REFERENCES

1. Fini, G., Mirone, P., and Fortunato, B. (1973) Evidence for short-range orientation effects in dipolar aprotic liquids from vibrational spectroscopy Part 1. – Ethylene and propylene carbonates, *J. Chem. Soc. Faraday Trans. 2* **69**, 1243–1248.
2. Wang, C. H. and McHale, J. (1980) Vibrational resonance coupling and the noncoincidence effect of the isotropic and anisotropic Raman spectral components in orientationally anisometric molecular liquids, *J. Chem. Phys.* **72**, 4039–4044.
3. Logan, D. E. (1986) The non-coincidence effect in the Raman spectra of polar liquids, *Chem. Phys.* **103**, 215–225.
4. Torii, H. and Tasumi, M. (1993) Local order and transition dipole coupling in liquid methanol and acetone as the origin of the Raman noncoincidence effect, *J. Chem. Phys.* **99**, 8459–8465.
5. Torii, H. and Tasumi, M. (1998) Liquid structure, infrared and isotropic/anisotropic Raman noncoincidence of the amide I band, and low-wavenumber vibrational spectra of liquid formamide: Molecular dynamics and *ab initio* molecular orbital studies, *J. Phys. Chem. B* **102**, 315–321.
6. McHale, J. L. (1981) The influence of angular dependent intermolecular forces on vibrational spectra of solution phase molecules, *J. Chem. Phys.* **75**, 30–35.
7. Woutersen, S. and Bakker, H. J. (1999) Resonant intermolecular transfer of vibrational energy in liquid water, *Nature* **402**, 507–509.

8. Torii, H. (2000) Ultrafast anisotropy decay of coherent excitations and the noncoincidence effect for delocalized vibrational modes in liquids, *Chem. Phys. Lett.* **323**, 382–388.
9. Logan, D. E. (1986) On the isotropic Raman spectra of isotopic binary mixtures, *Mol. Phys.* **58**, 97–129.
10. Logan, D. E. (1989) The Raman noncoincidence effect in dipolar binary mixtures, *Chem. Phys.* **131**, 199–207.
11. Torii, H. (2002) The role of electrical property derivatives in intermolecular vibrational interactions and their effects on vibrational spectra, *Vib. Spectrosc.* **29**, 205–209.
12. Park, E. S. and Boxer, S. G. (2002) Origins of the sensitivity of molecular vibrations to electric fields: carbonyl and nitrosyl stretches in model compounds and proteins, *J. Phys. Chem. B* **106**, 5800-5806.
13. Torii, H. (2002) Field-modulating modes of solvents for describing electrostatic intermolecular vibrational interactions in solution, *J. Phys. Chem. A* **106**, 1167–1172.
14. Torii, H. (2002) Locally strong polarity in the solvent effect of nonpolar solvent carbon tetrachloride: The role of atomic quadrupoles, *Chem. Phys. Lett.* **365**, 27–33.
15. Bishop, D. M. (1998) Molecular vibration and nonlinear optics, *Adv. Chem. Phys.* **104**, 1–40.
16. Torii, H., Furuya, K., and Tasumi, M. (1998) Raman intensities induced by electrostatic intermolecular interaction and related nonlinear optical properties of a conjugated π-electron system: A theoretical study, *J. Phys. Chem. A* **102**, 8422–8425.
17. Kirtman, B., Champagne, B., and Luis, J. M. (2000) Efficient treatment of the effect of vibrations on electrical, magnetic, and spectroscopic properties, *J. Comput. Chem.* **21**, 1572–1588.
18. Torii, H. (2002) Intensity-carrying modes important for vibrational polarizabilities and hyperpolarizabilities of molecules: Derivation from the algebraic properties of formulas and applications, *J. Comput. Chem.* **23**, 997–1006.
19. Torii, H. (2002) Vibrational polarization and opsin shift of retinal Schiff bases: Theoretical study, *J. Am. Chem. Soc.* **124**, 9272–9277.
20. Torii, H. (1999) Liquid structures and the infrared and isotropic/anisotropic Raman noncoincidence in liquid methanol, a methanol–LiCl solution, and a solvated electron in methanol: Molecular dynamics and *ab initio* molecular orbital studies, *J. Phys. Chem. A* **103**, 2843–2850.
21. Musso, M., Torii, H., Ottaviani, P., Asenbaum, A., and Giorgini, M. G. (2002) Noncoincidence effect of vibrational bands of methanol/CCl_4 mixtures and its relation with concentration dependent liquid structures, *J. Phys. Chem. A* **106**, 10152–10161.
22. Jones, D. R., Wang, C. H., Christensen, D. H., and Nielson, O. F. (1976) Raman and depolarized Rayleigh scattering studies of molecular motions of liquid 1,2,5-thiadiazole, *J. Chem. Phys.* **64**, 4475– 4483.
23. Torii, H. (2002) Influence of liquid dynamics on the band broadening and time evolution of vibrational excitations for delocalized vibrational modes in liquids, *J. Phys. Chem. A* **106**, 3281–3286.
24. Mukamel, S. (1995) *Principles of Nonlinear Optical Spectroscopy,* Oxford University Press, New York.
25. Car, R. and Parrinello, M. (1985) Unified approach for molecular dynamics and density-functional theory, *Phys. Rev. Lett.* **55**, 2471–2474.
26. Silvestrelli, P. L. and Parrinello, M. (1999) Structural, electronic, and bonding properties of liquid water from first principles, *J. Chem. Phys.* **111**, 3572–3580.
27. Izvekov, S. and Voth, G. A. (2002) Car–Parrinello molecular dynamics simulation of liquid water: New results, *J. Chem. Phys.* **116**, 10372–10376.
28. Perchard, C. and Perchard, J. P. (1975) Liaison hydrogene en phase liquide et spectrometrie Raman. I: Alcools liquides purs, *J. Raman Spectrosc.* **3**, 277–302.
29. Zerda, T. W., Thomas, H. D., Bradley, M., and Jonas, J. (1987) High pressure isotropic bandwidths and frequency shifts of the C–H and C–O modes of liquid methanol, *J. Chem. Phys.* **86**, 3219–3224.

30. Bertie, J. E. and Michaelian, K. H. (1998) Comparison of infrared and Raman wave numbers of neat molecular liquids: Which is the correct infrared wave number to use?, *J. Chem. Phys.* **109**, 6764–6771.
31. Kecki, Z., Sokolowska, A., and Yarwood, J. (1999) The influence of molecular local order on the non-coincidence Raman spectra of methanol in liquid mixtures, *J. Mol. Liq.* **81**, 213–223.
32. Torii, H., Pressure Dependence of the Liquid Structure and the Raman Noncoincidence Effect of Liquid Methanol Revisited, *Pure Appl. Chem.*, in press.
33. Arencibia, A., Taravillo, M., Pérez, F. J., Núñez, J., and Baonza, V. G. (2002) Effect of pressure on hydrogen bonding in liquid methanol, *Phys. Rev. Lett.* **89**, #195504.
34. Blum, L. (1972) Invariant expansion. II. Ornstein-Zernike equation for nonspherical molecules and an extended solution to the mean spherical model, *J. Chem. Phys.* **57**, 1862–1869.
35. Torii, H. (1994) Approximate theories of the Raman noncoincidence effect: A critical evaluation in the case of liquid acetone, *J. Mol. Struct. (Theochem)* **311**, 199–203.
36. Musso, M., Giorgini, M. G., Döge, G., and Asenbaum, A. (1997) The non-coincidence effect in highly diluted acetone–CCl_4 binary mixtures. I. Experimental results and theoretical predictions, *Mol. Phys.* **92**, 97–104.
37. Musso, M., Torii, H., Giorgini, M. G., and Döge G. (1999) Concentration dependence of the band profile parameters for the ν_3 ($^{12}C{=}O$) Raman band of acetone in acetone–CCl_4 binary mixtures. Experimental and Monte Carlo simulation results and their interpretation, *J. Chem. Phys.* **110**, 10076–10085.
38. Sokolowska, A. and Kecki, Z. (1993) Crossing of anisotropic and isotropic Raman components in the intermolecular resonance coupling of vibrations, *J. Raman Spectrosc.* **24**, 331–333.
39. Perchard, J. P. (1976) Characteristics of the OH (OD) bands of methanol and ethanol in lithium salt-alcohol mixtures: Saturated solutions and crystals, *Chem. Phys. Lett.* **44**, 169–172.
40. Kecki, Z. and Sokolowska, A. (1994) Crossing of anisotropic and isotropic Raman components in the intermolecular resonance coupling of vibrations. II – $LiClO_4$ and LiI solutions in acetone, *J. Raman Spectrosc.* **25**, 723–726.
41. Craig, D. P. and Thirunamachandran, T. (1989) Third-body mediation of resonance coupling between identical molecules, *Chem. Phys.* **135**, 37–48.
42. Krimm, S. and Abe, Y. (1972) Intermolecular interaction effects in the amide I vibrations of β polypeptides, *Proc. Natl. Acad. Sci. USA* **69**, 2788–2792.
43. Moore, W. H. and Krimm, S. (1975) Transition dipole coupling in amide I modes of β polypeptides, *Proc. Natl. Acad. Sci. USA* **72**, 4933–4935.
44. Torii, H. and Tasumi, M. (1992) Model calculations on the amide-I infrared bands of globular proteins, *J. Chem. Phys.* **96**, 3379–3387.
45. Torii, H. and Tasumi, M. (1996) Theoretical analyses of the amide I infrared bands of globular proteins, in H. H. Mantsch and D. Chapman (eds.), *Infrared Spectroscopy of Biomolecules*, Wiley-Liss, New York, pp. 1–18.
46. Torii, H. and Tasumi, M. (1998) *Ab initio* molecular orbital study of the amide I vibrational interactions between the peptide groups in di- and tripeptides and considerations on the conformation of the extended helix, *J. Raman Spectrosc.* **29**, 81–86.
47. Hamm, P., Lim, M., Degrado, W. F., and Hochstrasser, R. M. (1999) The two-dimensional IR nonlinear spectroscopy of a cyclic penta-peptide in relation to its three-dimensional structure, *Proc. Natl. Acad. Sci. USA* **96**, 2036–2041.
48. Choi, J.-H., Ham, S., and Cho, M. (2002) Inter-peptide interaction and delocalization of amide I vibrational excitons in myoglobin and flavodoxin, *J. Chem. Phys.* **117**, 6821–6832.

MULTIPOLE - INDUCED DIPOLE CONTRIBUTIONS TO THE FAR-INFRARED SPECTRA OF DIATOMIC MOLECULES IN NON-POLAR SOLVENTS

A. MEDINA
ETSII de Béjar, Universidad de Salamanca,
37700 Béjar, Salamanca, Spain
AND
J.M.M. ROCO, A. CALVO HERNÁNDEZ AND S. VELASCO
Departamento de Física Aplicada, Facultad de Ciencias
Universidad de Salamanca, 37008 Salamanca, Spain

Abstract: In this paper we overview our work on far-infrared spectroscopy of both hetero- and homonuclear diatomic molecules in nonpolar fluids. Special attention is paid to electric multipolar induced contributions. From comparison between theoretical and experimental spectra estimation of the leading multipoles of CO and N_2 molecules has been obtained. Also a temperature and density dependence analysis of many-body cancellation effects in the different pure induced components of the far-infrared integrated absorption coefficient of CO in liquid Ar is reported.

1. Introduction

Dilute solutions of diatomic molecules in dense nonpolar, particularly monatomic, solvents form one of the simplest classes of high-density molecular systems which attracted great deal both of theoretical and experimental spectroscopic work during past decades [1]. Some features that make these systems specially challenging are: 1) interactions among solute molecules can be disregarded and their mutual spatial correlations neglected; 2) far-infrared (FIR) and near-FIR experimental spectra show usually well separated pure rotational and vibration-rotational bands free of many complications, such as the Coriolis coupling, Fermi and Darlin-Dennison resonances, anharmonic splittings,

J. Samios and V.A. Durov (eds.), Novel Approaches to the Structure and Dynamics of Liquids: Experiments, Theories and Simulations, 361–385.

etc., arising in the spectra of polyatomics; 3) nonpolar fluids cause the smallest perturbations to the intramolecular dynamics of solute substances and present a very low absorption over a broad frequency range; 4) absorption spectra give information on molecular constants, as well as on the mechanisms to which absorption can be ascribed; 5) the changes in such spectra upon variation of the bulk parameters (density, temperature, composition) of the system provide information on intermolecular interactions and on their influence on the dynamics of molecular motions.

Far-infrared (FIR) absorption spectra of heteronuclear diatomic molecules dissolved in dense nonpolar solvents have two main contributions. The first one is due to the permanent dipole moment (permanent contribution) of the diatomic itself. The second one comes from the induced dipole moment (induced contribution) of the solution which appears from the interaction between unlike molecules. The relative importance of these contributions depends mainly on the value of the permanent dipole moment of the diatomic and on the density of the fluid. If the permanent dipole moment of the diatomic is high enough, the permanent contribution dominates the FIR absorption spectrum even at liquid densities [2]. However, for low enough permanent dipole moments this term will not be overwhelming and induction absorption mechanisms must be considered, especially for high densities [3, 4]. Moreover, in the case of homonuclear diatomics dissolved in rare-gas fluids there is no permanent contribution so the whole FIR absorption spectrum arises from induced contributions [5, 6].

One of our aims in the last years was to theoretically analyze this kind of interaction-induced effects. This analysis has three fundamental problems [1]:

i) The coupling between optically active intramolecular and intermolecular degrees of freedom. In these spectra the relevant degrees of freedom contributing to absorption are the rotation of the diatomic and the translation of the solvent atoms and the diatomic itself. These degrees of freedom are coupled via the anisotropic part of the solute-solvent interaction, but if it is assumed that they are uncorrelated (a usual hypothesis in most spectral theories) the spectra can be expressed as a convolution of a translational part and a rotational one.

ii) The difficulty to determine which induction mechanisms are relevant in each particular system. Interactions among molecules cause shifts in their charge distributions that provoke induced dipole moments depending on the intermolecular separation, and vary as the molecules rotate and vibrate. These interaction-induced dipoles are responsible for several kinds of spectra (in particular FIR absorption) of symmetric non-polar species and contribute to allowed spectra of dipolar molecules [7]. In fact, the interaction-induced component is present in the spectrum of any dense molecular system, although it can be vanishingly small compared to the permanent component. The leading long-range mechanism for the dipole moment induction in molecular systems has its origin in the linear polarization of the solvent molecules by the field due to electric multipoles of the diatomic probe. Besides this direct induction, other contributions exist due to back induction (i.e., the dipole moment induced on the solvent polarizes the solute), due to the field-gradient effects, and due to hyperpolarizabilities of the solvent molecules. Ordinarily these are second-order effects at best. Possible though apparently unexplored exceptions could be expected for the systems with resonantly enhanced polarizabilities including usually neglected vibrational polarizability [8].

iii) The usual difficulties arising in all theoretical studies involving many-body effects. A significant intensity decrease with increasing solvent density have been observed in the integrated absorption coefficient (IAC) of both hetero- and homonuclear diatomic molecules in spherical solvents. This decreasing has been attributed to the existence of static cancellation effects between two- and three-body components of the pure induced dipole spherical components. The extent of these cancellation effects represents a measure of the isotropy of the environment around a molecule in the fluid. Furthermore, these cancellation effects decrease when going to high multipolar induction mechanisms, and even for hexadecapole-induced dipole cancellation disappears and transforms into enhancement effects at least for some particular thermodynamical conditions. The nature of these effects remains still not well understood. A significant part of our work in the last years has been devoted to analyze how the translational and rotational degrees of freedom of the system affect the existence of cancellation or enhancement effects [9, 10].

This work briefly compiles our main contributions to the study of electric multipole induced contributions to the FIR spectra of diatomic molecules in nonpolar solvents.

2. Dipole absorption coefficient

We deal with a diluted solution of diatomic molecules in a nonpolar fluid. The concentration of diatomic molecules is assumed to be low enough so that a diatomic molecule (absorber) interacts with the solvent molecules (perturbers) independently of other diatomics. Under this condition the linear response theory gives the FIR absorption coefficient per unit path length, $\alpha(\omega)$, of a sample at temperature T, as [11]

$$\alpha(\omega) = \frac{4\pi\rho_A\omega}{3\hbar c} f(n)\left(1-e^{-\beta\hbar\omega}\right) Re\left\{\int_0^\infty dt\, e^{i\omega t}\,\langle\vec{M}(t)\cdot\vec{M}(0)\rangle\right\} \quad (1)$$

where ρ_A is the number density of absorbers, $f(n) = (1/n)(n^2+2)^2/9$ is the Polo-Wilson correcting factor for internal field effects (n is the optical refractive index) [6], $\beta = 1/(k_BT)$, $\vec{M}(t)$ is the dipole moment operator in the Heisenberg picture of the system associated to the presence of each diatomic molecule in the solution, and the angular brackets denote a canonical equilibrium average. In Eq. (1), the refractive index, n, can be obtained from the Lorenz-Lorenz equation [12] $(n^2-1)/(n^2+2) = (4\pi/3)\rho_B\alpha_B$, where ρ_B is the number density of the solvent molecules and α_B its spherical polarizability.

The dipole moment is the sum of the permanent dipole moment of the diatomic molecule and the induced dipole moment due to the diatomic-solvent interaction. Assuming that the induced dipole moment arises from a sum of binary diatom-atom interactions, one can write,

$$\vec{M}(t) = \vec{\mu}_p[\vec{r}(t)] + \sum_{k\in \text{solvent}} \vec{\mu}[\vec{r}(t), \vec{R}_k(t)] \tag{2}$$

where $\vec{\mu}_p(\vec{r})$ is the permanent dipole moment of the diatomic, $\vec{\mu}(\vec{r},\vec{R}_k)$ represents the induced dipole moment in a diatom-atom collision, $\vec{r} = (r,\Omega)$ is the vector describing the internuclear separation and the orientation of the diatomic molecule, and $\vec{R}_k = (R_k,\Omega_k)$ is the vector joining the centers of mass of the k-th solvent molecule and the diatomic one. This assumption implies that eventual irreducible triplet or higher order dipole moments are neglected. In the FIR, the vibrational degrees of freedom are not optically active and we can suppose that the diatomic molecule remains on its ground vibrational state. Then, the dependence on r in Eq. (2) disappears. Besides, the induced dipole moment can be expanded in terms of spherical harmonics in the following way [13]:

$$\mu_M(\Omega,\vec{R}_k) = \frac{4\pi}{\sqrt{3}}\sum_{\lambda,L} B_{\lambda,L}(R_k) \sum_{m_\lambda,m_L} C(\lambda,L,1;m_\lambda,m_L,M)\, Y_L^{m_L}(\Omega_k)\, Y_\lambda^{m_\lambda}(\Omega) \tag{3}$$

where the subscript $M=-1,0,1$ denotes spherical tensor components of the dipole moment, $C(\lambda,L,1;m_\lambda,m_L,M)$ are Clebsch-Gordan coefficients, Y denotes spherical harmonics, and $B_{\lambda,L}(R_k)$ are radial expansion coefficients specifying various dipole induction mechanisms.

The analytical calculation of the absorption coefficient from Eqs. (1)-(3) is unattainable without the consideration of some simplifying hypotheses. We would like to mention the three main assumptions considered in our works (besides the pairwise additivity of the intermolecular interactions):

1) The induction mechanisms responsible for the absorption in the system under study are essentially electrostatic in nature. With this assumption only radial coefficients, $B_{\lambda,\lambda+1}(R)$, will occur in the expansion (3) of the induced dipole moment, and one has [14]-[16]

$$B_{\lambda,L}(R_k) = \sqrt{\lambda+1}\,\frac{\alpha_B Q_\lambda}{R_k^{\lambda+2}}\,\delta_{\lambda+1,L} \quad (\text{for}\,\lambda \geq 1) \tag{4}$$

where Q_λ is the λ-order multipole moment of the diatomic molecule (i.e., $Q_1 \equiv \mu_p$, $Q_2 \equiv \Theta$, $Q_3 \equiv \Omega$, and $Q_4 \equiv \Phi$ are, respectively, the permanent dipole, quadrupole, octupole and hexadecapole moments). Note that, in general, $B_{\lambda,\lambda+1}(R)$ has both an overlap (short range) and an electric multipole (long range) contribution. The asymptotic approximation (4) seems to be adequate for low-density gases and highly polarizable solvents, for which the induction mechanisms are fundamentally long range, and can be used as a reasonable starting point for the evaluation of the multipole induced effects in high-density gases and liquids.

2) We consider that the dynamics of the total system is governed by the Hamiltonian $H = H_R + H_B + H'$, where H_R is the free Hamiltonian for the rotational motion of the diatomic molecule (system R), H_B is the Hamiltonian for the translational motions of the solute and solvent molecules (system B), and H' is the Hamiltonian describing the interaction $R - B$. Then, we assume that the translational degrees of freedom are not affected by the rotational motion of the diatomic molecule, i.e., by the anisotropic part of the interaction Hamiltonian H', but the opposite does not hold and the translational part behaves as a thermal bath for the rotational relaxation. We shall refer to this assumption as the ***translational decoupling approximation***. It allows to perform separate averages in Eq. (1), but it breaks down if the density fluctuations in the fluid give rise to large, rapid fluctuations of the electric field at the site of the diatomic molecule [17].

3) As carefully studied by Borysow *et al.* [18], spectral moments have in general two kinds of contributions: the pure terms coming from each spherical component of the induced dipole moment and the mixing terms coming from the superposition of induced dipole spherical components of different symmetry. Strictly, last terms vanish only when interactions are considered isotropic. We have restricted our study to the spectral contributions arising from pure electric induced terms.

The above assumptions allows us to express the FIR absorption coefficient (1) as the sum of a permanent (P) and an electrostatic induced (I) contribution,

$$\alpha(\omega) = \alpha^{(P)}(\omega) + \alpha^{(I)}(\omega) \tag{5}$$

with

$$\alpha^{(P)}(\omega) = \frac{4\pi\rho_A\omega}{3\hbar c} f(n) \left(1 - e^{-\beta\hbar\omega}\right) \mu_p^2 \widehat{C}_{rot}^{(1)}(\omega), \tag{6}$$

$$\alpha^{(I)}(\omega) = \sum_{\lambda=1}^{\infty} \alpha^{(Q_\lambda)}(\omega) \tag{7}$$

where $\alpha^{(Q_\lambda)}(\omega)$ is given by the convolution of a λ-translational spectrum and a λ-rotational spectrum,

$$\alpha^{(Q_\lambda)}(\omega) = \frac{4\rho_A \omega}{3\hbar c} f(n)\left(1 - e^{-\beta\hbar\omega}\right) \int_{-\infty}^{\infty} \widehat{C}_{tr}^{(\lambda,\lambda+1)}(\omega)\, \widehat{C}_{rot}^{(\lambda)}(\omega - \omega')d\omega', \quad (8)$$

where

$$\widehat{C}_{tr}^{(\lambda,\lambda+1)}(\omega) \equiv Re\left\{\int_0^{\infty} dt\, e^{i\omega t}\, C_{tr}^{(\lambda,\lambda+1)}(t)\right\}, \quad (9)$$

$$\widehat{C}_{rot}^{(\lambda)}(\omega) \equiv Re\left\{\int_0^{\infty} dt\, e^{i\omega t}\, C_{rot}^{(\lambda)}(t)\right\}, \quad (10)$$

with

$$C_{tr}^{(\lambda,\lambda+1)}(t) = \langle \sum_{k,k'} B_{\lambda,\lambda+1}(R_k(t))\, B_{\lambda,\lambda+1}(R_{k'}(0))\, P_{\lambda+1}(\cos\theta_{k,k'}(t)) \rangle, \quad (11)$$

$$C_{rot}^{(\lambda)}(t) = \frac{4\pi}{2\lambda+1} \sum_{m_\lambda} (-1)^{m_\lambda} \langle\langle Y_\lambda^{m_\lambda}(\Omega(t))\, Y_\lambda^{-m_\lambda}(\Omega(0)) \rangle\rangle, \quad (12)$$

where P_L are the Legendre Polynomials, $\theta_{k,k'}$ is the angle between $\vec{R}_k$ and $\vec{R}_{k'}$, $\langle\cdots\rangle$ denote a canonical equilibrium average over the translational variables, and $\langle\langle\cdots\rangle\rangle$ denote a canonical equilibrium average over both translational and rotational degrees of freedom.

In the case of homonuclear diatomic molecules there is no permanent contribution and only induced contributions with λ even must be taken into account in Eq. (7).

2.1. TRANSLATIONAL SPECTRAL CONTRIBUTION

For low- and moderate-density gas mixtures the translational spectral contribution $\widehat{C}_{tr}^{(\lambda,\lambda+1)}(\omega)$ can be successfully calculated using the Zwanzig-Mori (or memory function) approach [19]-[21] based on a continued-fraction representation for the Laplace transform of the translational time-correlation function (11), truncated at the first or second stage, and assuming a Gaussian form for the corresponding memory function. This enables one to obtain the first three even spectral moments of $\widehat{C}_{tr}^{(\lambda,\lambda+1)}(\omega)$. In particular, the zeroth order moment (the integrated intensity) is given by

$$\begin{aligned} C_{\lambda,\lambda+1} \equiv\ & C_{tr}^{(\lambda,\lambda+1)}(0) = \rho_B \int g_{AB}(R_k) \left[B_{\lambda,\lambda+1}(R_k)\right]^2 d^3\vec{R}_k \\ &+ \rho_B^2 \int\int g_{ABB}(R_k, R_{k'}, R_{k,k'}) \times \\ &\times B_{\lambda,\lambda+1}(R_k)\, B_{\lambda,\lambda+1}(R_{k'})\, P_{\lambda+1}(\cos\theta_{k,k'})\, d^3\vec{R}_k\, d^3\vec{R}_{k'} \end{aligned} \quad (13)$$

where $g_{AB}(R_k)$ and $g_{ABB}(R_k, R_{k'}, R_{k,k'})$ are, respectively, the two- and three-body distribution functions, and $R_{k,k'} = |\vec{R}_k - \vec{R}_{k'}|$.

For low-gas density systems, the pair distribution functions $g_{AB}(R_k)$ and $g_{BB}(R_{k,k'})$ can be evaluated from the low density limit expressions:

$$g_{AB}(R_k) = \exp\left[-\frac{V_0^{(AB)}(R_k)}{kT}\right]; g_{BB}(R_{k,k'}) = \exp\left[-\frac{V_0^{(BB)}(R_{k,k'})}{kT}\right], \quad (14)$$

where $V_0^{(AB)}(R_k)$ and $V_0^{(BB)}(R_{k,k'})$ are the respective solute-solvent and solvent-solvent isotropic intermolecular potentials, while the three-body distribution function can be evaluated in the Kirkwood superposition approximation [22]:

$$g_{ABB}(R_k, R_{k'}, R_{k,k'}) = g_{AB}(R_k) g_{AB}(R_{k'}) g_{BB}(R_{k,k'}), \quad (15)$$

with $g_{AB}(R_k)$ and $g_{BB}(R_{k,k'})$ given by Eqs. (14).

A similar formalism can also be applied to calculate the translational spectral contributions for high-density gas and liquid phases, but one then needs the knowledge of appropriate two- and three-body distribution functions. Guillot *et al.* [21, 23] have used this method for the study of the FIR spectrum of dense N_2 and N_2-rare gas compressed mixtures. These authors performed the calculation of the spectral moments only for the quadrupole-induced dipole contribution. In their work the two-body distribution function was evaluated from the analytical density dependent approximation provided by Chesnoy [24], whereas the three-body contributions were evaluated making use of a lattice-gas model. Within this model the three-body distribution functions were implemented by assuming that a multiple occupation of a given site of the lattice is forbidden, all the remaining locations being equally probable. Values of the spectral moments obtained by these authors allow for satisfactorily reproducing experimental FIR spectra for systems with density less than about 650 amagat. However, for systems at higher densities the predicted moments disagree with the experimental ones.

Furthermore, we also intend to evaluate the effect of other (higher than quadrupole) multipole-induced dipole contributions. Since each of these contributions has a characteristic range of interaction, a suitable evaluation of the two- and three-body distribution functions is needed in each case. Concretely, if the lattice-gas model is used to evaluate the three-body contributions, the weight of each configuration must be different in order to account for the particular range of the considered multipole-induced contribution. The latter complicates the abovementioned method for the calculation of the moments for the different multipole-induced dipole contributions. Therefore, in our work, the translational functions $\widehat{C}_{tr}^{(\lambda,\lambda+1)}(\omega)$ for the high-density gas and liquid solutions have been obtained from molecular dynamics (MD) calculations.

2.2. ROTATIONAL SPECTRAL CONTRIBUTION

The λ-rotational contribution $\widehat{C}_{rot}^{(\lambda)}(\omega)$ was calculated in the framework of a quantum theory describing the rotational dynamics of the diatomic molecule. This theory is based upon the following assumptions:

(i) The unperturbed rotational states of the diatomic (system R) are taken as the eigenstates of a free quantum rigid rotor, $\{|i>\equiv|j_i,m_i>\}$, with eigenvalues $E_{j_i}=Bhcj_i(j_i+1)$, where B is the rotational constant of the diatomic molecule on its ground vibrational state, and other symbols have their usual meaning. A non-rigid rotor including centrifugal distortion of the solute molecules has been also considered [25]. Deviations from the rigid rotor model become noticeable only for the transitions from the eigenstates with large enough values of j.

(ii) All translational degrees of freedom, both of absorber and perturber species, behave as a classical thermal bath (system B) at the temperature T of the solution.

(iii) The anisotropic part of the absorber-perturber intermolecular potential V is given by a truncated series in Legendre polynomials, in the form

$$V=\sum_{k\in\text{solvent}}\sum_{J=1}^{J_{max}}V_J(R_k)P_J(\cos\Omega_k), \qquad (16)$$

where Ω_k is the angle between $\vec{R}_k$ and the diatomic internuclear axis. Then, the interaction hamiltonian H' is given by $H'=V-\langle V\rangle$.

(iv) The time evolution of the radial part $V_J(R_k)$ of the intermolecular potential (16) is described by means of a classical stochastic process with a time autocorrelation function (TCF) given by

$$\Phi_J(t)=\langle V_J(R_k(t))\,V_J(R_k(0))\rangle\rangle-\langle V_J(R_k(t))\rangle^2=\kappa_J^2\exp[-|t|/t_J], \qquad (17)$$

where κ_J and t_J are characteristic parameters of the system measuring the mean strength of the anisotropic J contribution to the potential (16) and its correlation time, respectively.

Using these assumptions, it has been shown elsewhere that the λ-rotational contribution $\widehat{C}_{rot}^{(\lambda)}(\omega)$ may be written as a sum of two contributions [26]

$$\widehat{C}_{rot}^{(\lambda)}(\omega)=\widehat{C}_{0,rot}^{(\lambda)}(\o mega)+\widehat{C}_{1,rot}^{(\lambda)}(\omega) \qquad (18)$$

The first one, called secular profile, has the form

$$\widehat{C}_{0,rot}^{(\lambda)}(\omega)=\sum_{j_i}\sigma_{j_i}^o\sum_{j_f}(2j_f+1)A_{j_f,j_i}(\lambda)Re\left\{\Lambda_{j_f,j_i}^{(\lambda)}(\omega)\right\} \qquad (19)$$

where $\sigma^o_{j_i}$ is the usual rotational population factor of state $| j_i >$, the coefficients $A_{j_f,j_i}(\lambda)$ are given by

$$A_{j,j'}(\lambda) = (2j'+1)\begin{pmatrix} j & \lambda & j' \\ 0 & 0 & 0 \end{pmatrix}^2, \tag{20}$$

and

$$\Lambda^{(\lambda)}_{j_f,j_i}(\omega) = \frac{i}{\omega - \omega_{j_f,j_i} + i\overline{W}^{(\lambda)}_{j_f,j_i}(\omega)} \tag{21}$$

with $\overline{W}^{(\lambda)}_{j_f,j_i}(\omega) = \sum_{J=1}^{J_{max}} \overline{W}^{(\lambda,J)}_{j_f,j_i}(\omega)$. Explicit expressions for the functions $\overline{W}^{(\lambda,J)}_{j_f,j_i}(\omega)$ are given in Ref. [26] in terms of the Fourier-Laplace transform of the TCF (17).

The second contribution $\widehat{C}^{(\lambda)}_{1,rot}(\omega)$ in (18), called interference profile, has the form:

$$\widehat{C}^{(\lambda)}_{1,rot}(\omega) = \sum_{J=1}^{J_{max}} \widehat{C}^{(\lambda,J)}_{1,rot}(\omega). \tag{22}$$

Explicit expressions for the anisotropic contributions $\widehat{C}^{(\lambda,J)}_{1,rot}(\omega)$ are reported in Ref. [26] in terms of the rotational lines (21).

The secular approximation becomes invalid [27, 28] when the spectrum is significantly affected by the overlap among the basic secular resonances ($\Lambda^{(\lambda)}_{j_f,j_i}(\omega)$ and $\Lambda^{(\lambda)}_{j_{f'},j_{i'}}(\omega)$) causing an intensity redistribution inside the absorption band. This line-mixing (nonadditivity) effect is taken into account by means of the interference contribution, $\widehat{C}^{(\lambda)}_{1,rot}(\omega)$ [Eq.(22)]. The number of basic resonances involved in the interference depends on the order of Legendre polynomials in the interaction Hamiltonian.

Besides interference contribution, the spectral theory incorporates non-markovian (or memory) effects that mathematically stem from the ω-dependence of the basic resonances. When the markovian approach is valid, each secular resonance $\Lambda^{(\lambda)}_{j_f j_i}(\omega)$ becomes a Lorentzian line with halfwidth given by $Re[\overline{W}^{(\lambda)}_{j_f j_i}(\omega_{j_f j_i})]$. ¿From the point of view of the involved solute-solvent autocorrelation functions, this means that the autocorrelation time t_J of $\Phi_J(t)$ is much smaller that the inverse of the mean-square interaction strength, $\kappa_J t_J/h << 1$ [27, 28].

3. Far-infrared spectra of CO in Ar

Comparison between theoretical and experimental FIR spectra of HCl-rare gas solutions at different thermodynamic states show that the permanent dipole contribution is sufficient to explain the whole spectrum of this system both at low and high densities [27]. Roco et *al.* [3] have also applied the outlined spectral theory to investigate the FIR absorption spectra of CO in Ar gas at various densities. Comparison with the experimental spectra measured by Buontempo et *al.* [4] showed that, as in the case of HCl-rare gas mixtures, the permanent dipolar contribution is overwhelming at low-gas densities. However, at difference of HCl-rare gas cases, as the Ar density increases the permanent dipolar contribution does not account for the observed spectrum, specially for the existence of a long tail extending towards the high frequency range. This fact clearly shows the existence of significant induced dipole contributions to the FIR absorption spectra of CO-Ar solutions as density increases.

The spectral theory has been applied to calculate the FIR spectra of CO dissolved in Ar at three different thermodynamic states: 1) a low-density gas solution at $T = 129$ K and 13×10^{20} cm^{-3} (48.3 amagat); 2) a high-density gas at $T = 152$ K, and $\rho_{Ar} = 118 \times 10^{20}$ cm^{-3} (438.7 amagat); and 3) a liquid solution at $T = 90$ K and $\rho_{Ar} = 200 \times 10^{20}$ cm^{-3} (743.6 amagat). Evaluation of the FIR absorption spectra of CO-Ar at the considered temperatures and densities requires the knowledge of the multipole moments of CO on its ground vibrational state. These parameters have been subject of many experimental and theoretical investigations, and the results cover a moderately wide range of values. In this context, a delicate point is the fact that, although the absolute value of the electric dipole moment is independent of the location of the coordinate origin, all other multipole moments are not reference-frame invariant. Consequently, a shift of the coordinate origin from, say, the center of mass to the center of charge changes the values of all $|Q_\lambda|$ with $\lambda \geq 2$. The values available in the literature for the molecular multipole moments of CO lie in the following ranges

$$|\mu_p| = [0.1096, 0.112]\,\mathrm{D} \qquad |\Theta| = [0.92, 2.81]\,\mathrm{D\AA}$$
$$|\Omega| = [2.34, 3.56]\,\mathrm{D\AA}^2 \qquad |\Phi| = [2.92, 4.74]\,\mathrm{D\AA}^3. \tag{23}$$

In the present work, we consider these ranges for a first estimation of the translational strength parameters $C_{\lambda,\lambda+1}$. However, for comparison purposes, we take $|\mu_p| = 0.1097D$ while we consider the other multipole moments as adjustable parameters. An estimation of $|\Theta|$, $|\Omega|$ and $|\Phi|$ can be then made by resolving the FIR absorption spectra of CO in Ar into a permanent and a multipole-induced dipole absorption band and by fitting to the experimental profiles.

The theoretical spectra are obtained from the expressions reported in the previous Section with $B_{CO} = 1.92$ cm^{-1}[4], $\alpha_{Ar} = 1.62\,\mathrm{\AA}^3$ [32], and $J_{max} = 2$. Then, taking into account Eq. (17), four parameters (κ_1, κ_2, t_1, t_2) are required

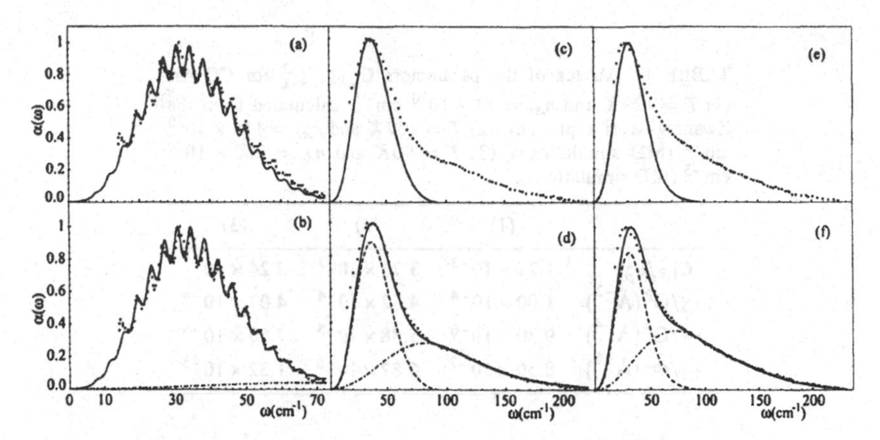

Figure 1. FIR spectra of CO in Ar at: $T = 129$ K and $n_{Ar} = 13 \times 10^{20}$ cm^{-3} [(a) and (b)], $T = 152$ K, and $n_{Ar} = 118 \times 10^{20}$ cm^{-3} [(c) and (d)], and $T = 90$ K, and $n_{Ar} = 200 \times 10^{20}$ cm^{-3} [(e) and (f)]. (a), (c), and (e): Theoretical profiles obtained considering only the permanent dipole contribution (solid line). (b), (d), and (f): Theoretical profiles (solid line) obtained as a sum of the permanent (dashed line) and the induced contributions (dash-dotted line). Experimental profiles (dotted lines).

in order to evaluate the theoretical FIR absorption coefficient. These parameters are considered as a set of adjustable parameters whose values are chosen by comparison with experimental spectra.

Experimental FIR ($15 - 75$ cm^{-1}) spectrum of CO in Ar gas at low enough densities present a well resolved rotational structure. Calculated and experimental absorption coefficients are represented in Fig. 1. Since the experimental solute number density ρ_{CO} is unknown, each spectrum is normalized to the respective maximum value. The theoretical absorption coefficient obtained just considering the permanent contribution is represented in Fig. 1.a. This permanent contribution, $\alpha^{(P)}(\omega)$, has been calculated from Eq. (6). From Fig. 1.a it is clear that the low-density gas FIR spectrum of this system is essentially due to permanent dipole absorption. However, on the high-frequency region it is observed that the theoretical permanent FIR spectrum underestimates experimental absorption. It is reasonable to assume that this slight discrepancy can be evaluated in terms of the induced absorption band.

In order to evaluate the significant induced contributions to the absorption profile we have calculated the translational strength parameters $C_{\lambda,\lambda+1}/Q_{\lambda}^{2}$ from Eq. (13) by using the long-range induction functions (4) and the two- and three-body radial distribution funtions (14) and (15) with isotropic intermolecular potentials $V_0^{(AB)}(R_k)$ and $V_0^{(BB)}(R_{k,k'})$ taken as $6-12$ Lennard-Jones surfaces with parameters [6]: $\sigma_{CO-Ar} = 3.483$ Å, $\varepsilon_{CO-Ar} = 117.06$ K,

TABLE 1. Values of the parameters $C_{\lambda,\lambda+1}/Q_\lambda^2$ for CO-Ar: (1) $T = 129K$ and $n_{Ar} = 13 \times 10^{20}$ cm^{-3} (calculated from the Zwanzig-Mori approach); (2) $T = 152K$ and $n_{Ar} = 118 \times 10^{20}$ cm^{-3} (MD simulations); (3) $T = 90K$ and $n_{Ar} = 200 \times 10^{20}$ cm^{-3} (MD simulations).

	(1)	(2)	(3)
$C_{1,2}/\mu_p^2$	1.21×10^{-3}	3.73×10^{-3}	2.24×10^{-3}
$C_{2,3}/\Theta^2\,(\text{Å}^{-2})$	1.00×10^{-4}	4.52×10^{-4}	4.07×10^{-4}
$C_{3,4}/\Omega^2\,(\text{Å}^{-4})$	9.30×10^{-6}	5.88×10^{-5}	7.50×10^{-5}
$C_{4,5}/\Phi^2\,(\text{Å}^{-6})$	8.50×10^{-7}	6.87×10^{-6}	1.32×10^{-5}

and $\sigma_{Ar-Ar} = 3.405$ Å, $\varepsilon_{Ar-Ar} = 119.8$ K. The calculated values are given by the first column of Table 1. We note the following points: a) $C_{1,2}/\mu_p^2 \ll 1$, which indicates that the dipole-induced dipole contribution can be neglected; b) since the permanent contribution presents a maximum absorption at the rotational transition line $j = 7 \rightarrow 8$, the maximum absorption of the quadrupole-, octupole- and hexadecapole-induced dipole contributions respectively occurs at $\omega \approx 65$ cm^{-1} [$\omega_{j+2,j} = 2B(2j+3)$ with $j = 7$], $\omega \approx 104$ cm^{-1} [$\omega_{j+3,j} = 6B(j+2)$ with $j = 7$], and $\omega \approx 146$ cm^{-1} [$\omega_{j+4,j} = 4B(2j+5)$ with $j = 7$], then, one can assume that in the observed frequency range ($15 - 75$ cm^{-1}), the octupole ($\lambda = 3$) and hexadecapole ($\lambda = 4$) contributions can be also neglected. Thus, the FIR spectrum of CO in Ar gas at low density has been obtained from the sum of permanent and quadrupole-induced dipole contributions,

$$\alpha(\omega) = \alpha^{(P)}(\omega) + \alpha^{(\Theta)}(\omega). \tag{24}$$

where $\alpha^{(P)}(\omega)$ has been evaluated from Eq. (6) and $\alpha^{(\Theta)}(\omega)$ from Eq. (8) for $\lambda = 2$. The translational spectral function $\widehat{C}_{tr}^{(2,3)}(\omega)$ was calculated using the Zwanzig-Mori approach while the rotational spectral functions $\widehat{C}_{rot}^{(1)}(\omega)$ and $\widehat{C}_{rot}^{(2)}(\omega)$ were calculated from Eq. (18). The best agreement with the experimental lineshapes was obtained for $\kappa_1 = 4.5 \times 10^{-22}$ J, $\kappa_2 = 8.0 \times 10^{-22}$ J, $t_1 = t_2 = 1.1 \times 10^{-14}$ s, and $C_{2,3} = 3.80 \times 10^{-4}$ D^2. Taking into account the value of $C_{2,3}/\Theta^2$ reported in the first column of Table 1, this fit leads to

$$|\Theta| = 1.95\,\text{D Å} \tag{25}$$

which favorably compares with the result 1.94(4) DÅ measured by Meerts *et al.* [29] by using a molecular beam electric resonance method. In Fig. 1.b we have plotted the theoretical spectrum obtained from Eq. (24), with Θ given by (25). As can be observed in this figure, the above discrepancies between permanent and experimental profiles disappear when the quadrupole-induced contribution is considered.

The FIR ($15 - 220$ cm^{-1}) spectra of CO in Ar gas at $T = 152$ K and $\rho_{Ar} = 118 \times 10^{20}$ cm^{-3} and in Ar liquid at $T = 90$ and $\rho_{Ar} = 200 \times 10^{20}$ cm^{-3} have been also measured by Buontempo *et al.* [4] and they are devoid of any rotational structure. The corresponding calculated and experimental absorption coefficients are also represented in Fig. 1. As before, the spectra are normalized to the respective maximum value of the total spectrum considered in each case. The theoretical absorption coefficients obtained by considering only the permanent contribution, $\alpha^{(P)}(\omega)$, are represented in Figs. 1.c and 1.e. These figures clearly show that, in both cases, permanent dipole absorption can not account for the whole FIR profile. This is particularly valid for the long tail appearing at intermediate and high frequencies, and becomes more evident as solvent density increases.

We have calculated the FIR spectra at these conditions as the sum of the permanent contribution and an electrostatic induced contribution including its dipole, quadrupole, octupole and hexadecapole components,

$$\begin{aligned} \alpha(\omega) &= \alpha^{(P)}(\omega) + \alpha^{(I)}(\omega) \\ \alpha^{(I)}(\omega) &= \alpha^{(\mu)}(\omega) + \alpha^{(\Theta)}(\omega) + \alpha^{(\Omega)}(\omega) + \alpha^{(\Phi)}(\omega) \end{aligned} \qquad (26)$$

where the induced contributions, $\alpha^{(\mu)}(\omega)$, $\alpha^{(\Theta)}(\omega)$, $\alpha^{(\Omega)}(\omega)$ and $\alpha^{(\Phi)}(\omega)$ have been obtained from Eq. (8) with $\lambda = 1, 2, 3$, and 4.

Because of the difficulties in applying the Zwanzig-Mori approach to high density systems, the translational functions $\widehat{C}_{tr}^{(\lambda,\lambda+1)}(\omega)$ for the high-density gas and liquid cases have been obtained from MD calculations. In all computer simulations we consider a sample of 250 solvent atoms (Ar) and one diatomic molecule (CO) in a cubic box with the usual boundary conditions. The CO molecule has been considered as isotropic and the interactions (CO-Ar and Ar-Ar) were considered to be of the Lennard-Jones type with parameters previously reported. The equations of motion were integrated by means of a leap-frog Verlet algorithm with coupling to a thermal bath to keep the system temperature at the desired values. The time step was $\Delta t = 0.5 \times 10^{-14}$ *s*. After a long equilibration period, the dynamics of the particles was followed during 8×10^5 time steps in all cases, and the averages of the time correlation functions relevant for the problem were performed. The values of the four leading ($\lambda = 1, 2, 3$, and 4) strength parameters $C_{\lambda,\lambda+1}/Q_\lambda^2$ obtained from MD calculations are listed Table 1.

The rotational spectral functions $\widehat{C}_{rot}^{(\mu)}(\omega)$, $\widehat{C}_{rot}^{(\Theta)}(\omega)$, $\widehat{C}_{rot}^{(\Omega)}(\omega)$, and $\widehat{C}_{rot}^{(\Phi)}(\omega)$ were calculated from Eq. (18) and with values of the statistical parameters $\kappa_1 = 10.1 \times 10^{-22}$ J, $\kappa_2 = 16.2 \times 10^{-22}$ J, and $t_1 = t_2 = 1.1 \times 10^{-14}$ s for the high-density gas case, and $\kappa_1 = 13.2 \times 10^{-22}$ J, $\kappa_2 = 20.9 \times 10^{-22}$ J, and $t_1 = t_2 = 1.1 \times 10^{-14}$ s for the liquid.

In Fig. 1.d we have plotted the overall absorption coefficient for the high-density gas case with $C_{1,2} = 4.49 \times 10^{-5}$ D^2, $C_{2,3} = 2.03 \times 10^{-3}$ D^2, $C_{3,4} = 5.15 \times 10^{-4}$ D^2 and $C_{4,5} = 6.47 \times 10^{-5}$ D^2, which give the best agreement with the experimental profile. Taking into account the values of $C_{2,3}/\Theta^2$, $C_{3,4}/\Omega^2$ and $C_{4,5}/\Phi^2$ given in the second column of Table 1, this fit leads to:

$$|\Theta| = 2.12\,\mathrm{D\AA}, \quad |\Omega| = 2.96\,\mathrm{D\AA}^2, \quad |\Phi| = 3.07\,\mathrm{D\AA}^3. \tag{27}$$

The value of $|\Theta|$ is within the experimental range (23), but it is a 7.7% larger than the value (25) obtained from the low-density gas case. The value of $|\Omega|$ is in excellent agreement with the theoretical values reported by Diercksen and Sadlej [30] and by Bounds and Wilson [31], while the value of $|\Phi|$ is in the low region of the interval (23).

Fig. 1.f contains the theoretical shape of $\alpha(\omega)$ for the liquid case with $C_{1,2} = 2.61 \times 10^{-5}$ D^2, $C_{2,3} = 2.08 \times 10^{-3}$ D^2, $C_{3,4} = 7.54 \times 10^{-4}$ D^2 and $C_{4,5} = 5.75 \times 10^{-5}$ D^2, which give the best agreement with the experimental profile. Taking into account the values of $C_{2,3}/\Theta^2$, $C_{3,4}/\Omega^2$ and $C_{4,5}/\Phi^2$ given in the third column of Table 1, this fit leads to:

$$|\Theta| = 2.26\,\mathrm{D\AA}, \quad |\Omega| = 3.17\,\mathrm{D\AA}^2, \quad |\Phi| = 6.60\,\mathrm{D\AA}^3. \tag{28}$$

The value of $|\Theta|$ is still into the range (23), but it is a 16% greater than the value (26) obtained from the fit of the low-density gas spectrum. The value of $|\Omega|$ is also into the range (23), and it is a 7% greater than the value (27) obtained from the fit of the high-density gas spectrum. Conversely, the value of $|\Phi|$ is now higher than the reported by Bounds and Wilson [31] and it is 2.15 times the value (27) obtained from the fit of the high-density gas spectrum.

As we have commented before, the consideration of only long-ranged induction mechanisms could be a reasonable assumption for low gas densities. However, when the density increases, other contributions could be significantly involved in the FIR absorption spectrum. For example, the influence of short ranged (overlap) induction mechanisms for the liquid nitrogen has been pointed out by Guillot *et al.* [6, 23]. Their consideration in our work would require to carry out first principles calculations in order to provide expressions for the involved functions $B_{\lambda,\lambda\pm1}$. Furthermore, since the hexadecapole-induced dipole contribution is specially significant in the high-frequency region, the calculated increasing of Φ at the liquid density can be also attributed to higher order multipole mechanisms or to spectral manifestations of triplet induced dipoles in this particular frequency region. Therefore, in our spectral theory, the multipole moments $|Q_\lambda|$ must be considered as effective parameters, providing all the induced mechanisms which have not been considered.

4. Far-infrared spectra of N_2 in Xe

The theory outlined in Sec. II was also applied to describe the FIR-spectrum ($50-250$ cm^{-1}) of a N_2–Xe gaseous mixture at $T = 295$ K, $\rho_{N_2} = 3$ amagat (0.81×10^{20} cm^{-3}) and $\rho_{Xe} = 350$ amagat (94.15×10^{20} cm^{-3}) which has been measured by Guillot et *al.* [6]. The theoretical FIR spectrum was obtained only from an electrostatic induced contribution decomposed in its quadrupole and hexadecapole components,

$$\alpha(\omega) = \alpha^{(I)}(\omega) = \alpha^{(\Theta)}(\omega) + \alpha^{(\Phi)}(\omega), \tag{29}$$

where the induced contributions $\alpha^{(\Theta)}(\omega)$ and $\alpha^{(\Phi)}(\omega)$ have been obtained from Eq. (8) with $\lambda = 2$ and 4. We have taken the values $B_{N_2} = 1.999$ cm^{-1} [33], $\alpha_{Xe} = 4.11$ Å^3 [6], and $n = 1.257$. Only contributions arising from the anisotropic term $J = 2$ of the intermolecular potential (16) are considered. Furthermore, most recent experimental and theoretical values available in the literature for the quadrupole and hexadecapole moments of N_2 in its ground vibrational state lie in the ranges:

$$|\Theta| = [1.2, 1.65]\,\text{DÅ}, \qquad |\Phi| = [2.1, 4.1]\,\text{DÅ}^3. \tag{30}$$

We have considered these ranges for a first estimation of the translational strength parameters $C_{2,3}$ and $C_{4,5}$.

The calculation of the translational functions, $\widehat{C}_{tr}^{(\lambda,\lambda+1)}(\omega)$, with $\lambda = 2$ and 4, was carried out from MD simulations. The N_2 molecule has been considered as isotropic and the interactions N_2–Xe and Xe–Xe were considered to be of the 12–6 Lennard-Jones type with parameters [6]: $\sigma_{N_2-Xe} = 3.74$ Å, $\varepsilon_{N_2-Xe} = 166.3$ K, and $\sigma_{Xe-Xe} = 3.89$ Å, $\varepsilon_{Xe-Xe} = 282.4$ K.

The λ-rotational functions $\widehat{C}_{rot}^{(\lambda)}(\omega)$, with $\lambda = 2$ and 4, were calculated from Eqs. (18). Therefore, the evaluation of the quadrupolar, $\alpha^{(\Theta)}(\omega)$, and the hexadecapolar, $\alpha^{(\Phi)}(\omega)$, contributions to $\alpha(\omega)$ can be performed, for given values of the quadrupole $|\Theta|$ and hexadecapole $|\Phi|$ moments of N_2. We have considered the statistical parameters as a set of adjustable parameters with values chosen by comparison with the experimental spectrum. In this fitting procedure we have considered that the main contribution to absorption comes from the quadrupolar inductionmechanism, and we have added the hexadecapolar contribution in order to reproduce the high frequency behavior of the experimental spectrum without exceeding its maximum intensity. We have found a satisfactory agreement between the theoretical and experimental spectra for $\kappa_2 \geq 30 \times 10^{-22}$ J, $t_2 \leq 2.6 \times 10^{-14}$ s, and values of the quadrupole and hexadecapole moments lying in the ranges:

$$|\Theta| = [1.47, 1.60]\,\text{DÅ}, \qquad |\Phi| = [0, 5.20]\,\text{DÅ}^3. \tag{31}$$

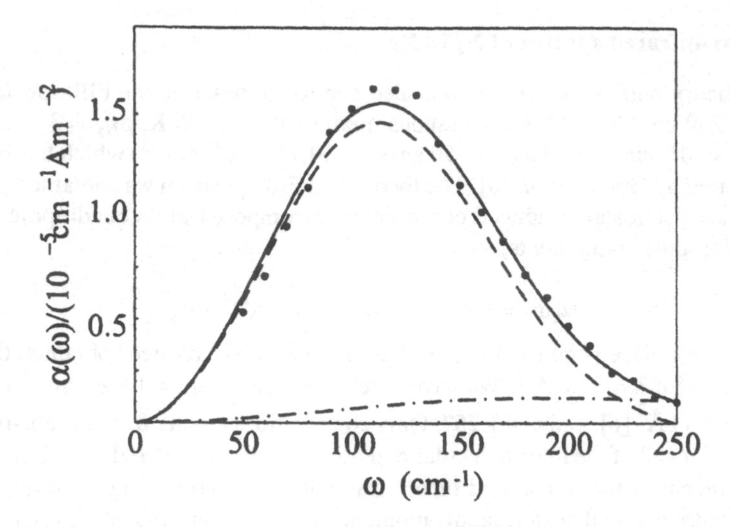

Figure 2. FIR spectrum of a N_2-Xe gaseous mixture at $T = 295$ K, $\rho_{N_2} = 3$ amagat and $\rho_{Xe} = 350$ amagat. Theoretical absorption coefficient obtained by considering the quantum λ-rotational spectra with $|\Theta| = 1.52$ DÅ, $|\Phi| = 3.22$ DÅ^3, $\kappa_2 = 58.8 \times 10^{-22}$ J, and $t_2 = 7.2 \times 10^{-15}$ s: total line shape (solid line), quadrupolar contribution $\alpha^{(\Theta)}(\omega)$ (dashed line), and hexadecapolar contribution $\alpha^{(\Phi)}(\omega)$ (dash-dotted line). Experimental spectrum (dots).

Regarding the quadrupole moment we must underline that the obtained range is contained in the corresponding interval given in (30). However, the interval of values obtained for the hexadecapole is wider than the expressed in (30). In order to attain a more precise estimation for the value of the hexadecapole, we have considered the quadrupole moment as a fixed parameter, and performed the fitting procedure only for the remaining three parameters of the theory, i.e., $|\Phi|$, κ_2 and t_2. Assuming the value $|\Theta| = 1.52$ DÅ reported by Dagg et *al.* [34], we have found the best agreement between the theoretical and the experimental spectra for $\kappa_2 = (60 \pm 20) \times 10^{-22}$ J, $t_2 = (7.2 \pm 0.8) \times 10^{-15}$ s, and $|\Phi| = 3.2 \pm 0.3$ DÅ^3.

The calculated spectrum represented in Fig. 2 corresponds to the last case. The good agreement between the experimental and the calculated spectra together with the drastic reduction of the interval of values corresponding to the hexadecapole moment allows to consider this interval as a fairly good estimation of the value of this quantity. The obtained value for the hexadecapole moment is very close to the value $|\Phi| = 3 \pm 1$ DÅ^3 reported by Birnbaum and Cohen [33], and greater than the one reported by Maroulis and Thakkar [35] obtained from many-body perturbation theory calculations.

The quadrupolar, $\alpha^{(\Theta)}(\omega)$, and the hexadecapolar, $\alpha^{(\Phi)}(\omega)$, contributions to the calculated spectrum $\alpha(\omega)$ are also sketched in Fig. 2. A glance to this figure allows to corroborate that $\alpha^{(\Theta)}(\omega)$ is the most significant contribution to the absorption intensity, while $\alpha^{(\Phi)}(\omega)$ is revealed as important in order to describe the behavior of the experimental spectrum at high frequencies.

5. Translational/rotational contributions and static cancellation effects

The FIR spectra of CO-Ar at high-gas and liquid densities show that the consideration of electric multipolar induction is essential in order to properly reproduce the available experimental results for these systems. The relevant degrees of freedom contributing to induced absorption are the rotation of the diatomic and the translation of the solvent atoms and the diatomic itself. These degrees of freedom are coupled via the anisotropic part of the solute-solvent interaction, but assuming the translational decoupling approximation, the spectra can be expressed as a convolution of a translational part and a rotational one. On the other hand, the translational part can be expressed as a sum of two- and three-body components, which allows one to analyze the possible existence of many-body static cancellation effects in the total integrated absorption coefficient (IAC) and their dependence with the range of the induction mechanism. The study of how the different degrees of freedom of the system affect the existence of cancellation or enhancement effects, and the influence of thermodynamic conditions on these effects have been also object of our interest in the last years.

With these goals in mind we have considered the general expression for the IAC associated to each λL term in expansion (4), $\alpha_{1,\lambda L}$. The mentioned expression was firstly derived by Poll and van Kranendonk [36], and reads:

$$\alpha_{1,\lambda L} \equiv \alpha_{1,\lambda L}^{(r)} + \alpha_{1,\lambda L}^{(t)}, \tag{32}$$

where the indices (r) and (t) stands for rotation and translation respectively. Moreover, it is possible to split each contribution in two-body $(2b)$ and three-body $(3b)$ terms, $\alpha_{1,\lambda L}^{(r)} = \alpha_{1,\lambda L}^{(r,2b)} + \alpha_{1,\lambda L}^{(r,3b)}$ and $\alpha_{1,\lambda L}^{(t)} = \alpha_{1,\lambda L}^{(t,2b)} + \alpha_{1,\lambda L}^{(t,3b)}$ with:

$$\alpha_{1,\lambda L}^{(r,2b)} = \frac{2\pi^2 \rho_A}{3c} \frac{\lambda(\lambda+1)}{I_A} \sum_k < B_{\lambda L}^2(R_k) >, \tag{33}$$

$$\alpha_{1,\lambda L}^{(r,3b)} = \frac{2\pi^2 \rho_A}{3c} \frac{\lambda(\lambda+1)}{I_A} \sum_{k \neq k'} < B_{\lambda L}(R_k) B_{\lambda L}(R_{k'}) P_L(\cos\theta_{k,k'}) >, \tag{34}$$

$$\alpha_{1,\lambda L}^{(t,2b)} = \frac{2\pi^2 \rho_A}{3c} \sum_k \frac{1}{m_{AB}} \left[< \left(\frac{dB_{\lambda L}(R_k)}{dR_k} \right)^2 > + L(L+1) < \frac{B_{\lambda L}^2(R_k)}{R_k^2} > \right], \tag{35}$$

$$\alpha_{1,\lambda L}^{(t,3b)} = \frac{2\pi^2\rho_A}{3c}\frac{1}{m_A}\left[\frac{L+1}{2L+1}\sum_{k\neq k'} < C_{\lambda L}(R_k)C_{\lambda L}(R_{k'})P_{L+1}(\cos\theta_{k,k'}) > + \right.$$

$$\left. + \frac{L}{2L+1}\sum_{k\neq k'} < D_{\lambda L}(R_k)D_{\lambda L}(R_{k'})P_{L-1}(\cos\theta_{k,k'}) > \right]. \qquad (36)$$

In these equations m_{AB} is the solute-solvent reduced mass, I_A is the moment of inertia of the diatomic molecule, m_A its mass, and the remaining symbols have their usual meaning. Besides, we have defined the functions $C_{\lambda L}(R_k)$ and $D_{\lambda L}(R_k)$ as:

$$C_{\lambda L}(R_k) = \frac{dB_{\lambda L}(R_k)}{dR_k} - L\frac{B_{\lambda L}(R_k)}{R_k}, \qquad (37)$$

$$D_{\lambda L}(R_k) = \frac{dB_{\lambda L}(R_k)}{dR_k} + (L+1)\frac{B_{\lambda L}(R_k)}{R_k}. \qquad (38)$$

The details of the calculations and the necessary hypotheses to obtain Eqs. (32)-(36) can be found elsewhere in the literature [27],[36]-[38] and specially in Ref. [9]. We only remark here that, in general, in the FIR spectra of these systems, there exist a non-negligible contribution coming from the interference between pure induced and pure allowed spectra. Only when an isotropic diatom-atom potential is assumed, interference terms vanish. One of the assumptions underneath the above equations is precisely the isotropy of the interactions, and thus these expressions allow to calculate only the integrated absorption coefficients of the pure terms contributing to absorption. We restrict our discussion only to those pure induced spectral moments with radial functions $B_{\lambda L}(R_k)$ given by Eq. (4).

The thermodynamic conditions we have elected to perform a systematic study of $\alpha_{1,\lambda L}$ for CO in liquid Ar are the following:

(1) Evolution with temperature for a fixed solvent density, $\rho = 1.3$ g/cm^3, and temperatures between 90 and 150 K. This interval in the temperature approximately covers all the range between the temperature of the triple and critical points for Ar.

(2) Evolution with density for a fixed temperature, $T = 130$ K and densities between 1.1 and 1.5 g/cm^3. This density interval roughly covers all the region between the solid-liquid and liquid-gas coexistence curves.

For all these thermodynamic states we performed molecular dynamics simulations with the same technical details that in Sec. 3. The results have been normalized in such a way that they are independent of the solute density and the values of the electric multipole moments of CO. This normalization reads:

$$\bar{\alpha}_{1,\lambda L} \equiv \frac{1}{4\pi\rho_B}\frac{\alpha_{1,\lambda L}}{\alpha_{1,\mu_p}}\frac{\mu_p^2}{Q_\lambda^2}, \qquad (39)$$

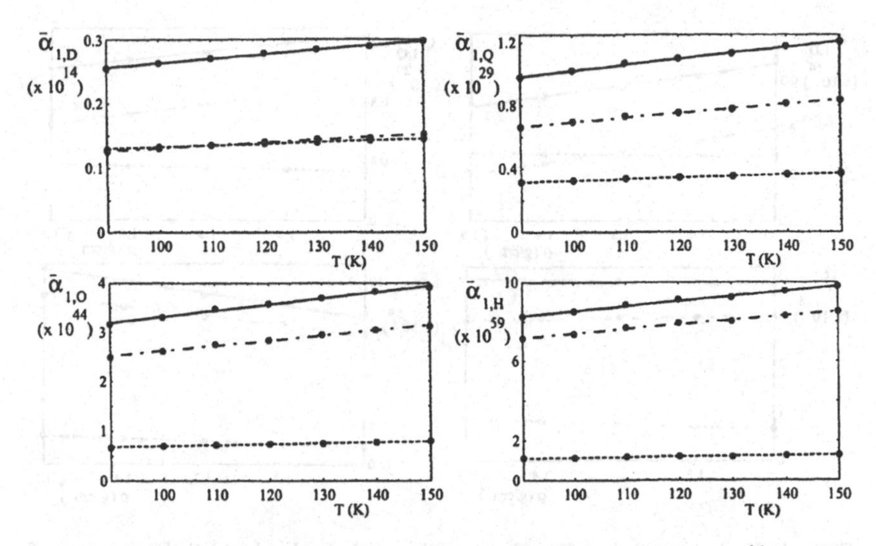

Figure 3. Evolution with temperature of the translational, $\bar{\alpha}^{(t)}_{1,\lambda L}$, and rotational, $\bar{\alpha}^{(r)}_{1,\lambda L}$, components of $\bar{\alpha}_{1,\lambda L}$. Lines represent linear fittings for each term: solid line, $\bar{\alpha}_{1,\lambda L}$; dashed line, $\bar{\alpha}^{(t)}_{1,\lambda L}$; and lines-points, $\bar{\alpha}^{(r)}_{1,\lambda L}$. Units are $cm^{3-2\lambda}/g$ for each multipolar order.

where $\alpha_{1,\mu_p} = \rho_A(2\pi\mu_p^2)/(3cI_A)$ is the integrated permanent dipole absorption.

5.1. EVOLUTION WITH TEMPERATURE AND DENSITY

The evolution with temperature of the translational, $\bar{\alpha}^{(t)}_{1,\lambda L}$, and, rotational, $\bar{\alpha}^{(r)}_{1,\lambda L}$, contributions is displayed in Fig. 3. For dipole induced-dipole ($\lambda = 1$), both contributions are very similar in magnitude, but when we go to higher multipolar terms the rotational contribution is always clearly greater than the translational one for any temperature. This can be understood in terms of the range of the multipolar induced mechanisms: higher multipolar terms correspond to shorter ranges and at short ranges translation is more hindered than rotation. Respect to the relative evolution with T, both contributions increase almost linearly with temperature, and $\bar{\alpha}^{(r)}_{1,\lambda L}$ always increase with T clearly faster than $\bar{\alpha}^{(t)}_{1,\lambda L}$.

The evolution with density of the translational, $\bar{\alpha}^{(t)}_{1,\lambda L}$, and, rotational, $\bar{\alpha}^{(r)}_{1,\lambda L}$, contributions is displayed in Fig. 4. The translational component of $\bar{\alpha}_{1,\lambda L}$ always increase linearly with density. This increase is greater for higher multipolar orders. Nevertheless, the density behavior of $\bar{\alpha}^{(r)}_{1,\lambda L}$ is more complicated.

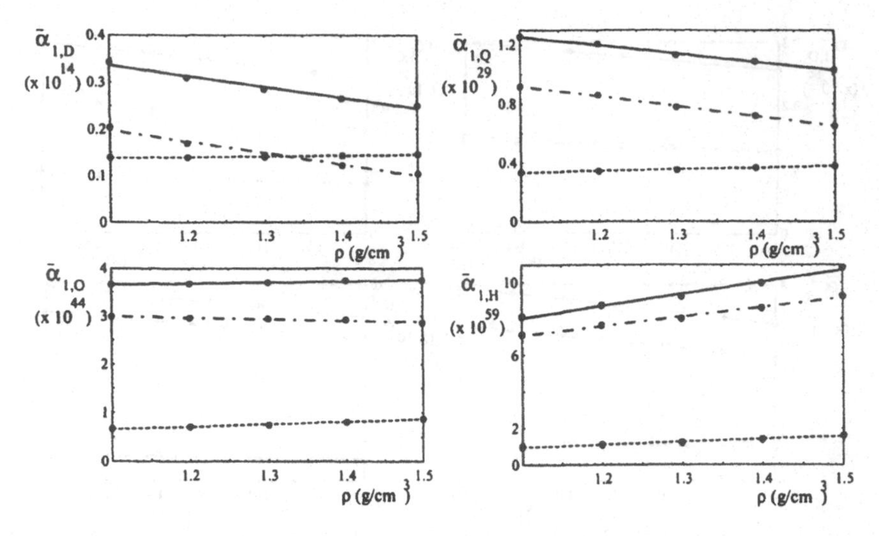

Figure 4. Evolution with the solvent density of the translational and rotational components of $\bar{\alpha}_{1,\lambda L}$. Notation and units are as in Fig. 3.

For D, Q and O, it decreases with density (decreasing that diminishes with the multipolar order), but for H it increases with density. In any case the density behavior of $\bar{\alpha}^{(t)}_{1,\lambda L}$ is smoother than that of $\bar{\alpha}^{(r)}_{1,\lambda L}$ and this causes that the density evolution of the total integrated coefficient $\bar{\alpha}_{1,\lambda L}$ is mainly associated to $\bar{\alpha}^{(r)}_{1,\lambda L}$. With respect to magnitudes, $\bar{\alpha}^{(r)}_{1,12}$ and $\bar{\alpha}^{(t)}_{1,12}$ are similar, with $\bar{\alpha}^{(r)}_{1,12} > \bar{\alpha}^{(r)}_{1,12}$ for low densities and the opposite for high densities. For all other multipolar terms, the rotational component is always greater than the translational one.

Then, as a first conclusion, we obtain that for the CO-Ar system, the evolution of the pure induced components of the IAC with the thermodynamic conditions of the solvent is essentially due to the evolution of the rotation of the diatomic, specially for high-order electric multipolar terms.

5.2. CANCELLATION EFFECTS AND DEGREES OF FREEDOM

We have also evaluated from MD simulations the two- and three-body components of $\bar{\alpha}_{1,\lambda L}$. The obtained results are reported in Table 2. These results show that the sign of the three-body components (i.e., the existence of destructive or constructive interferences) does not change in all the liquid phase, and so, the existence of cancellation or enhancement effects is independent of temperature or density. It seems to be only affected by the multipolar order, that is, the range of the functions to be averaged to compute $\bar{\alpha}_{1,\lambda L}$. Cancellation effects for long-range electric multipolar induction (*D*, *Q*, and *O*) progressively decrease and even get into enhancement when the range is low enough (for *H*). Furthermore, our results also show that cancellation effects for *D*, *Q* and *O* are more complete (the absolute values of two- and three-body terms are more similar) as the isotropy of the fluid increase, that is, cancellation increases as density increases and temperature decreases.

TABLE 2. Magnitude of cancellation effects for the translational and rotational components of $\bar{\alpha}_{1,\lambda L}$ for the system $CO-Ar$ at different temperatures and densities. The upper part of the Table corresponds to a systematic study with temperature for a fixed density ($1.3 g/cm^3$) and the lower part to a systematic study with density for a fixed temperature ($130 K$). Under the symbol of each contribution its radial decay is displayed. Note that for any component cancellation is absent when the decay is equal or faster than R_k^{-12}.

	$\bar{\alpha}^{(t,3b)}_{1,\lambda L}/\bar{\alpha}^{(t,2b)}_{1,\lambda L}$				$\bar{\alpha}^{(r,3b)}_{1,\lambda L}/\bar{\alpha}^{(r,2b)}_{1,\lambda L}$			
T (K)	*D* (R_k^{-8})	*Q* (R_k^{-10})	*O* (R_k^{-12})	*H* (R_k^{-14})	*D* (R_k^{-6})	*Q* (R_k^{-8})	*O* (R_k^{-10})	*H* (R_k^{-12})
90	-0.44	-0.27	0.02	0.25	-0.87	-0.74	-0.46	0.04
120	-0.42	-0.25	0.03	0.21	-0.85	-0.72	-0.43	0.06
150	-0.41	-0.23	0.04	0.19	-0.84	-0.70	-0.40	0.06
ρ (g/cm^3)	*D*	*Q*	*O*	*H*	*D*	*Q*	*O*	*H*
1.1	-0.37	-0.19	0.04	0.15	-0.78	-0.63	-0.33	0.06
1.3	-0.41	-0.24	0.03	0.20	-0.85	-0.71	-0.41	0.06
1.5	-0.46	-0.30	0.01	0.29	-0.90	-0.78	-0.51	0.02

From the results of Table 2 one can also conclude that the translational components present cancellation effects for dipole induced-dipole absorption and quadrupole induced-dipole absorption. For octupole and hexadecapole this cancellation moves to enhancement (notice that in the total integrated coefficient there exist enhancement only for hexadecapole). Moreover, the existence of cancellation or enhancement for the translational component seems to be independent of the thermodynamic state in the range we are considering. The thermodynamic conditions seems to be important only with respect to the completeness of cancellation or enhancement. For the rotational component the situation is slightly different, because there exist constructive interference between many-body terms only for hexadecapole. In any other case and for any temperature or density, cancellation is clear. We also note that cancellation (when it exists) is always greater in rotational components than in translational ones.

6. Concluding remarks

In the last years we have devoted special attention to the study of FIR absorption spectra of both polar diatomic molecules with a low enough permanent dipole moment and nonpolar diatomic molecules in nonpolar fluids. In this context, our studies were focused on two main aspects.

First, FIR spectra give valuable information about the nature of the induction processes arising from the diatom-atom interactions. In the case of a very diluted solution at low and moderate gas densities and a highly polarizable solvent, the main induced contributions to the absorption intensity come from the multipole-induced dipole mechanisms. Hence, this type of spectra can be useful to obtain values of the leading multipole moments of the diatomic molecule. This is the case of CO-Ar and N_2-Xe solutions, for which experimental FIR spectra at different densities are available. We have analyzed theoretically these spectra in order to estimate the leading multipoles of CO and N_2. The theoretical line shapes were obtained from the convolution of translational and rotational components. The former have been derived both from classical approachs and MD simulations, while the latter have been evaluated by assuming that the translational degrees of freedom behave as a thermal bath and the rotational dynamics of the diatomic is described by means of a quantum rigid rotor stochastically interacting with the bath.

Second, we have shown the existence, at dense-gas and liquid densities, of static cancellation effects between many-body components of the pure induced terms of the FIR integrated absorption coefficient. These cancellation effects decrease when going to high multipolar induction mechanisms, and even for hexadecapole induced-dipole cancellation disappears and transforms into enhancement. We have been interested in the influence of solvent temperature and density in the completeness of these effects and we have also tried to elucidate how the active degrees of freedom of the system affect the existence of cancellation or enhancement effects. In particular, for CO-Ar at liquid densities, we concluded that when it exists, cancellation is more complete in these conditions where the liquid is more isotropic: low temperatures and high densities. Furthermore, the magnitude of cancellation is always more complete in rotational components. Moreover in any case rotational components are more important than translational ones, except for the dipole induced-dipole term, in which the importance of rotation and translation is similar.

Also, we would like to stress the following points:

(i) A crucial assumption in our spectral theory is the translational decoupling approximation. It is a bit difficult to assess to what extent our estimates of the multipole moments of CO and N_2 depend on this approximation. In the low-gas density case, the excellent agreement between the value (25) and the measured by Meerts *et al.* [29] for the quadrupole moment of CO seems to justify *a posteriori* such approximation. In the liquid case, the small density fluctuations make more consistent the decoupling approximation and thus one can hope that it does not practically influence on the corresponding estimates. Dense gas cases are subtler. Only comparisons with MD calculations seem to be the way of checking this approximation. In any case, we would like to remark that the translational decoupling picture works properly in order to reproduce all the considered spectra.

(ii) Our spectral theory only considers the linear isotropic polarizability of the solvent, α_B. Our estimates for Q_λ have been made for a fixed value of α_B. Any change in its numerical value implies a modification of Q_λ values, so that $\alpha_B \cdot Q_\lambda = cte$ in each considered case. In this context, it would be of interest to notice that interatomic interactions can induce an anisotropic polarizability. We have checked [37] that for the CO-Ar mixture at low-gas density anisotropic polarizability correction is negligible. As density increases, one can also expect that the anisotropic polarizability of the solvent takes a very small value compared to the corresponding isotropic polarizability due to many-body cancellation effects.

(iii) Contribution due to the higher order electric multipoles was found to appreciably increase with density. This interesting effect, tentatively attributed to possible influence of the irreducible three-body dipoles or of the short-ranged (overlap) induction, certainly deserves further study.

(iv) We have assumed that only electrostatic long-ranged induction mechanisms are responsible for the FIR absorption in the systems under study. Other contributions due to back induction, field-gradient effects, and nonzero second order (fourth rank tensor) hyperpolarizability of the solvent molecules are not considered. Ordinarily these effects can be neglected. Recently Samios *et al.* [39] calculated back-induction contributions in the infrared spectra of HCl in CCl_4.

(v) Interference contributions arising from permanent-induced and induced-induced cross-correlations have not been considered. Classical MD simulations [40] for CO-Ar show that at high-gas and liquid densities these interferences can reach a 3% of the total FIR integrated absorption. Unfortunately, a quantum spectral theory taken into account these interference effects is yet not available.

Finally, we feel important to mention that in the last years interesting spectral theories [41] capable to take into account interferences among absorption lines and among spectral branches at any order have been proposed. These theories improve the secular-interference scheme by a new one based upon diagonal and non-diagonal terms, in which interferences can be calculated in an exact way. Moreover, they are formulated independently of the spectral region considered and were succesfully applied to far- and near-infrared, and Raman absorption [42]. It would be a challenging work in the future to extend this formalism to study the importance of interaction-induced effects in some open problems, as the appearance of Q-branch absorption on the central range of the near-infrared bands of hydrogen halides in simple liquid solvents [43].

7. Acknowledgments

We would like to thank Prof. M.O. Bulanin for many helpful and elucidating discussions. We thank financial support from the Ministerio de Ciencia y Tecnología of Spain under Grant BFM 2002-01225 FEDER, and by Junta de Castilla y León under Grant SA097/01.

References

1. M. O. Bulanin, S. Velasco, and A. Calvo Hernández, *J. Mol. Liq.* **70**, 107 (1996).
2. A. Calvo Hernández, S. Velasco, and F. Mauricio, *Phys. Rev. A* **31**, 3419 (1985).
3. J. M. M. Roco, A. Medina, A. Calvo Hernández, and S. Velasco, *Chem. Phys. Lett.* **216**, 593 (1993).
4. U. Buontempo, S. Cunsolo, and G. Jacucci, *J. Chem. Phys.* **59**, 3750 (1973).
5. U. Buontempo, S. Cunsolo, and G. Jacucci, *Mol. Phys.* **21**, 381 (1971).
6. B. Guillot, Ph. Marteau, and J. Obriot, *Mol. Phys.* **65**, 765 (1988).
7. G. Birnbaum and B. Guillot in *'Spectral Line Shapes'*, edited by M. Zoppi and L. Ulivi (Firenze, Italy, 1996).
8. M. O. Bulanin, *Opt. Spectrosc.* **72**, 45 (1992).
9. J. M. M. Roco, A. Medina, A. Calvo Hernández and S. Velasco, *J. Chem. Phys.* **107**, 4844 (1997).
10. A. Medina, J. M. M. Roco, A. Calvo Hernández and S. Velasco, *J. Chem. Phys.* **108**, 9480 (1998).
11. R. G. Gordon, *Adv. Magn. Reson.* **3**, 1 (1968).
12. V. V. Bertsev, in *Molecular Cryospectroscopy*, edited by R. J. H. Clark and R. H. Hester (John Wiley & Sons, 1995), pp 1-19.
13. J. D. Poll and J. L. Hunt, *Can. J. Phys.* **54**, 461 (1976); G. Birnbaum, G. Bachet, and L. Frommhold, *Phys. Rev.* A **36**, 3729 (1987).
14. J. D. Poll and J. Van Kranendonk, *Can. J. Phys.* **39**, 189 (1961).
15. J. D. Poll and J. L. Hunt, *Can. J. Phys.* **54**, 461 (1976).
16. G. Birnbaum, G. Bachet, and L. Frommhold, *Phys. Rev. A* **36**, 3729 (1987).
17. D. Frenkel, PhD. Thesis. University of Amsterdam (1977).
18. A. Borysow and M. Moraldi, *J. Chem. Phys.*, **99**, 8424 (1993).
19. H. Mori, *Prog. Theor. Phys.* **33**, 423 (1965); **34**, 399 (1965).
20. U. Balucani, V. Tognetti, and R. Vallauri, *Phys. Rev. A* **19**, 177 (1979).
21. B. Guillot and G. Birnbaum, *J. Chem. Phys.* **79**, 686 (1983).
22. J. G. Kirkwood, *J. Chem. Phys.* **3**, 300 (1935).
23. Ph. Marteau, J. Obriot, F. Fonder and B. Guillot, *Mol. Phys.* **59**, 1305 (1986).
24. J. Chesnoy, *Chem. Phys.* **83**, 283 (1984).
25. W. A. Herrebout, B. J. Van der Veken, A. Medina, A. Calvo Hernández, and M. O. Bulanin, *Mol. Phys.* **96**, 1115-1124 (1999).
26. J. M. M. Roco, A. Medina, A. Calvo Hernández and S. Velasco, *J. Chem. Phys.* **103**, 9161-9174 (1995).
27. A. Medina, S. Velasco, and A. Calvo Hernández, *Phys. Rev.*, **A44**, 3023 (1991).
28. A. Medina, S. Velasco, and A. Calvo Hernández, *Phys. Rev.*, **A45**, 5289 (1992).

29. W. L. Meerts, F. H. Leeuw and A. Dynamus, *Chem. Phys.* **22**, 319 (1977).
30. G. H. F. Diercksen and A. J. Sadlej, *Chem. Phys.* **96**, 17 (1985).
31. D. G. Bounds and S. Wilson, *Mol. Phys.* **54**, 445 (1985).
32. R. M. van Aalst and J. van der Elsken, *Chem. Phys. Lett.* **23**, 198 (1973).
33. G. Birnbaum and E. R. Cohen, *Mol. Phys.* **32**, 161 (1976).
34. I. R. Dagg, A. Anderson, S. Yang, W. Smith, and L. A. A. Read, *Can. J. Phys.* **63**, 625 (1985).
35. G. Maroulis and A. J. Thakkar, *J. Chem. Phys.* **88**, 7623 (1988).
36. J. D. Poll and J. van Kranendonk, *Can. J. Phys.* **39**, 189 (1961).
37. J. M. M. Roco, A. Medina, A. Calvo Hernández and S. Velasco, *J. Chem. Phys.* **103**, 9175 (1995).
38. L. Frommhold, *'Collision-Induced Absorption in Gases'* (Cambridge University Press, Cambridge, 1993) Chap. 6.
39. G. Chatzis, and J. Samios, *J. Phys. Chem.*, **A105**, 9522 (2001).
40. A. Medina, A. Calvo Hernández, J. M. M. Roco, and S. Velasco, *J. Mol. Liq.*, **70**, 169 (1996).
41. A. Padilla, J. Pérez, and A. Calvo Hernández, *J. Chem. Phys.* **111**, 11015 (1999).
42. A. Padilla, J. Pérez, and A. Calvo Hernández, *J. Chem. Phys.* **111**, 11026 (1999); **113**, 4290 (2000).
43. A. Medina, J. M. M. Roco, A. Calvo Hernández, S. Velasco, M. O. Bulanin, W. A. Herrebout and B. J. van der Veken, *J. Chem. Phys.* **116**, 5058 (2002).

RECENT ADVANCES IN THE UNDERSTANDING OF HYDROPHOBIC AND HYDROPHILIC EFFECTS: A THEORETICAL AND COMPUTER SIMULATION PERSPECTIVE

RICARDO L. MANCERA[1], MICHALIS CHALARIS[2] and JANNIS SAMIOS[2]
[1]De Novo Pharmaceuticals Ltd., Compass House, Vision Park, Chivers Way, Histon, Cambridge CB4 9ZR, United Kingdom: [2]Laboratory of Physical Chemistry, Department of Chemistry, University of Athens, Panepistimiopolis 157-71, Athens, Greece

Abstract: The study of solvation phenomena is fundamental to the local microscopic structure and dynamics of liquids and aqueous solutions and the chemical reactions that take place in them. Here we review some of the recent theoretical and computer simulation advances in the study of hydrophobic effects in small relatively simple non-polar molecules in aqueous solution. We also summarize recent efforts in understanding aqueous DMSO solutions as a system where both hydrophobic and hydrophilic effects seem to be present.

Key words: Hydrophobic, hydrophilic, DMSO, water, aqueous solution, computer simulation

1. INTRODUCTION

In recent years there has been considerable academic and technological interest in the microscopic structure and dynamics of liquids and aqueous solutions. The hydrophobic and hydrophilic solvation properties of water are key to our understanding of the formation, structure and stability of micelles, bio-membranes and proteins. Over the last 50 years or so a vast amount of research has been carried out to try to rationalize a number of aqueous phenomena ranging from the low solubility of inert gases in water to the assembly of micelles, the folding of proteins and ligand-protein and protein-protein interactions [1-10].

Liquid water is a unique molecular system whose solvation properties are probably central not only to explaining the molecular-scale forces responsible for the structure, stability, interactions and function of biomolecules, but also to explaining the origin of life itself. Water has a number of peculiar properties, such as having a density maximum above the melting point and a very low compressibility [11,12]. It is widely thought that the hydrogen-bonding interactions between water molecules play a crucial role in determining these physico-chemical properties and as well as water's solvation properties.

Much of the more fundamental work in this area has been carried out hand in hand with the development of models and theories of the unique thermodynamic,

J. Samios and V.A. Durov (eds.), Novel Approaches to the Structure and Dynamics of Liquids: Experiments, Theories and Simulations, 387–396.

structural and dynamical properties of neat water itself. We give here a brief account of some of the theoretical and computer simulation efforts that have recently appeared, in an attempt to provide a concise picture of the current standing of the theory and computer simulation of hydrophobic and hydrophilic effects.

2. THE HYDROPHOBIC EFFECT

The low solubility of non-polar gases in water is the simplest example of the hydrophobic effect. Other, more complex, and certainly much more important, phenomena in which the hydrophobic effect plays an important role are the assembly of micelles and bilayer membranes, the folding, stability and aggregation of proteins and drug delivery [1-10].

The hydrophobic effect exhibits characteristic thermodynamic fingerprints that distinguish it from other solvation effects [5,6,9]. In contrast to hydrophilic hydration, which is enthalpically driven, hydrophobic hydration is entropically driven at room temperature and characterized by a large positive heat capacity change. Crucially, the hydrophobic effect manifests itself by an unusual temperature dependency. For example, the solubility of non-polar solutes in water often decreases with increasing temperature until a minimum is reached, and some proteins undergo what is called cold denaturation upon cooling [10]. These properties have no known counterparts in organic solvents and are assumed to arise from the unique structure of liquid water. Several properties of liquid water, such as its low compressibility, the packing of water molecules and the existence of hydrogen bond networks have been proposed to play a part in explaining the hydrophobic effect, but their precise role remains unclear [9,13].

The hydrophobic effect has been studied from a wide range of experimental, theoretical and computational perspectives. Experimental determinations have provided the fundamental thermodynamic phenomenological description of the hydrophobic effect [6,10], allowing for the development of empirical models for the prediction of aqueous solubilities or hydrophobic contributions to protein folding [9,10,14]. Analytical theories have been put forward to try to explain the hydrophobic effect at a molecular level [15-19]. Computer simulations have been used to reproduce measured characteristic properties of simple hydrophobic phenomena, to provide a microscopic description of the solvent around hydrophobic species and to determine which factors are responsible for the hydrophobic effect in its various forms [20-25].

3. AN INFORMATION THEORY OF HYDROPHOBIC EFFECTS

A number of analytical and semi-analytical approaches to the theory of hydrophobic effects have appeared over the years. Most of them give hydrogen bonding a central role [1-3,18,26]. A number of recent theoretical developments have shown that the microscopic nature of the hydrophobic effect is dependent on the length scales involved [27,28]. We concentrate here on the weak, short-ranged attractions experienced by small hydrophobic species in aqueous solution.

The analysis of the dependence of the excess chemical potential of a solute on the probability distribution of cavity sizes in water and other molecular liquids has provided a deep insight into the hydrophobic effect [15,16]. The comparison of distributions of cavity sizes in water and non-polar liquids (such as hexane) arising from density fluctuations has revealed that one is less likely to find a cavity of atomic or small molecular size in water than in a hydrocarbon liquid under the same thermodynamic conditions, even though water has more free volume [20,21]. As a consequence, a non-polar solute is less soluble in water than in a non-polar liquid. The success of such scaled-particle theories revealed that a specific knowledge about the arrangement of hydrogen bonds between water molecules around non-polar solutes is not needed to capture the main features of the hydrophobic effect. Instead, the role of hydrogen bonds arises by influencing the bulk properties of water, such as the density and the radial distribution functions [17].

In recent years an information theory has been developed on the basis that knowledge of the density and the radial distribution functions is essential for the correct description of the hydrophobic effect. This theory allows for the calculation of the excess chemical potential of cavities of molecular sizes and arbitrary shapes. As in the case of small cavities, the model requires only results of simulations of the neat liquid. Computer simulations have been used to analyze the statistics of transient cavities in water, showing that the above probability distribution is accurately predicted by a maximum entropy model using the two moments that are obtained from the experimental liquid density and the experimental radial distribution of oxygen atoms [19]. This two-moment model can predict the solubility of non-polar molecules, potentials of mean force, and hydrophobic effects on conformational equilibria [29]. This theory has also provided an interpretation of the hydration effects influencing protein stability, as is the case of the pressure denaturation of proteins [30] and the convergence of the entropy of transfer [31]. However, this model still fails to describe the effect of dewetting large hydrophobic surfaces [13,28].

4. COMPUTER SIMULATION OF HYDROPHOBIC EFFECTS

One of the main goals in the computer simulation of hydrophobic effects has been the study of the aggregation of simple non-polar solutes in aqueous solution and the associated potentials of mean force. These simulations are of great importance since the hydrophobic effects responsible for the interaction between non-polar solutes are likely to manifest themselves with the opposite thermodynamic fingerprints as those observed in hydrophobic hydration [9]. Furthermore, the clustering and aggregation of non-polar chemical groups is likely to be the mechanism underlying many biomolecular processes such as membrane formation, protein folding and ligand binding [5,6].

Early simulations of a number of non-polar particles in solution exhibited no clear tendency for aggregation [32-35]; however, these studies were carried out at low temperature. During the last ten years a number of molecular dynamics simulations have attempted to measure directly the tendency for aggregation of small non-polar solutes in solution and, in particular, its temperature dependence [36-39]. These simulations found the existence of a temperature maximum after which aggregation decreased, as expected from the temperature dependence of the solubility of these solutes.

The pairwise interaction between non-polar solutes in solution is best described by the potential of mean force (PMF) acting between them. Early studies showed the existence of well-defined contact and solvent-separated configurations. The contact configuration was seen having a significantly lower relative free energy [40,41] and becoming deeper as temperature was increased [42,43]. The entropic contribution to the free energy was found to produce an attraction between the solutes at short distances, which increases as temperature rises [42,43]. Particle-insertion methods confirmed that the contact configuration was more stable than the solvent-separated one [44]. However, the use of polarisable water models produced some ambiguous results about the relative stability of the contact and solvent-separated configurations [45,46]. A thorough examination of the effect of temperature revealed that the contact configuration becomes more populated in the range 300-350 K as temperature increases, reaching a maximum after which the tendency for aggregation decreases [47,48]. It was confirmed that the hydrophobic aggregation is entropy driven at low temperature and enthalpy driven at higher temperatures [47,48].

The effect of density on the above PMF calculations has also been a matter of some controversy. A study at two different temperatures with corresponding different densities seemed to contradict earlier studies by showing little temperature dependence of the tendency for aggregation [49]. Some of the above studies also compared two PMFs at the same temperature but with different water densities but found no significant pressure effect [47,48]. Further PMF calculations have been shown to be pressure dependent [50,51]. However, a calculation using a polarisable model of water not only confirmed the increased tendency for aggregation with increasing temperature and its entropic nature, but also concluded that the PMF calculated under constant pressure conditions should give similar results if calculated under constant volume conditions if the correct volume is used [52]. Further PMF calculations showed yet again that a higher water density raises the desolvation barrier between the contact and solvent-separated configurations and stabilises the former [53], with the pressure effects being amplified with larger solutes [54]. It was then found that the entropy and enthalpy of the contact configuration are rather pressure insensitive, while solvent-separated configuration is increasingly stabilised at higher pressures by enthalpic contributions that offset the slightly unfavourable entropic contributions to the free energy [55].

Calculations of the heat capacity change upon aggregation of two non-polar particles in solution gave a small and *positive* value, contrary to what would be expected if hydrophobic aggregation were the simple partial thermodynamic reversal of hydrophobic hydration [56]. However, other calculations have yielded large negative changes in the heat capacity upon two-particle association [57].

The issue of cooperativity in hydrophobic interactions has recently become the subject of controversy. The comparison of three-methane and multiple-methane PMFs with two-methane PMF showed that the effective interactions when the methanes are in contact is anti-cooperative [58], later confirmed through a modified PMF expansion of the free energy [59]. However, another investigation concluded that the three-methane interaction had small cooperative effects at the contact configuration [60], and this cooperativity in the hydrophobic interaction increased with larger solutes [61]. A subsequent thorough study of the spatial dependence of three-body effects then showed that a majority of three-methane configurations exhibited anti-cooperativity [62], and a discussion about the actual details of the PMF calculations ensued [63,64]. A recent study confirmed the anti-cooperativity of the hydrophobic interaction for both the contact and solvent-separated configurations, while cooperativity was observed at the desolvation barrier [65]. Increasing the solute size or adding salt increased this anti-cooperativity, while increasing the pressure had the opposite effect [65].

5. HYDROPHOBIC *VERSUS* HYDROPHILIC EFFECTS: THE CASE OF AQUEOUS DMSO SOLUTIONS

Mixtures of water and dimethyl sulfoxide (DMSO) provide an interesting example of how a rather aprotic substance can affect the structure of water. DMSO is a bifunctional molecule, having both polar and non-polar groups and thus exhibiting hydrophilic and hydrophobic effects simultaneously.

Aqueous DMSO solutions have unique physical chemical and biological properties [66,67]. DMSO and water can mix in all proportions, and the excess thermodynamic properties of their solutions exhibit strong deviations from ideality, such as in the density [68,69], viscosity [69,70], adiabatic and isothermal compressibility [71], relative dielectric permittivity [72,73], surface tension [73,74] and heats of mixing [69,75-77]. Solutions of DMSO in water have very low freezing points compared to the neat liquids [78]. These solutions can act as cryoprotectant for membranes and proteins [79,80]. The molecular mechanisms responsible for these physical and biological properties are poorly understood.

Aqueous DMSO solutions have been investigated using X-ray and neutron diffraction [81,82], as well as optical [76,83-88], acoustic [71,89-91], NMR [92-98] and dielectric [99-102] spectroscopies. The commonly accepted picture that has emerged from these studies is that at low concentration DMSO behaves as a strong structure maker by rigidifying the water structure, possibly through the hydrophobic hydration of its methyl groups. However, some recent combined quantum chemical and computer simulation studies have challenged this view [103,104]

Early simulations showed a sharpening in the water-water pair correlation functions, the existence of 1DMSO:2H_2O hydrogen-bonded aggregates and a hydrophobic interaction between DMSO molecules, in DMSO concentrations of up to 0.2 mole fraction [105]. Other simulations at concentrations of up to 0.35 mole fraction have revealed the structuring of water with increasing DMSO concentration with a simultaneous decrease in the average number of water-water hydrogen bonds [106], in agreement with neutron diffraction data [107,108]. However, no evidence of hydrophobic interactions between the methyl groups of DMSO was found, despite the characteristic hydrophobic hydration orientational correlations of water around these non-polar groups [105,109]. The strong DMSO-water correlations observed in these mixtures were explained as arising from the stronger DMSO-water hydrogen bonds.

Molecular dynamics simulations across the entire DMSO composition range in water revealed the existence of a 1DMSO:2H2O aggregate in water-rich mixtures and a 2DMSO:1H_2O aggregate with a central water molecule making hydrogen bonds to two DMSO molecules in DMSO-rich mixtures [110]. Other kinds of aggregates have also been observed [111,112].

We have recently performed a series of simulations at different concentrations [113] and temperatures [114]. The increase in DMSO concentration from 0.055 to 0.19 mole fraction saw the enhancement not only of the water-water structure but also of the hydrophobic hydration shell around the methyl groups of DMSO, while at the same time the strength of the hydrogen bonds between water and DMSO was increased [112]. Figure 1 shows the Me-OW and Me-H pair correlations between the methyl groups of DMSO and the water sites. The sharpening of the peaks in these pair correlations indicate that an increase in DMSO concentration enhances the structuring of water around the non-polar Me groups. The positions of the first peaks reveal that water molecules adopt a nearly tangential arrangement in the vicinity of the Me groups, which is characteristic of hydrophobic hydration. Figure 2 shows the OS-OW and OS-HW pair correlations between the oxygen in the sulphonyl group of DMSO and the water sites. The increase in DMSO concentration produces a significant sharpening of the peaks of these pair correlations, with water molecules establishing a linear hydrogen bond to the OS atom in DMSO. At a DMSO concentration of 0.19 mole fraction there are practically no bulk water molecules left. As a consequence, there is an apparent loss of hydrogen bonds for water molecules in the first hydration shell of DMSO, explaining their increased structuring to compensate for this loss [112].

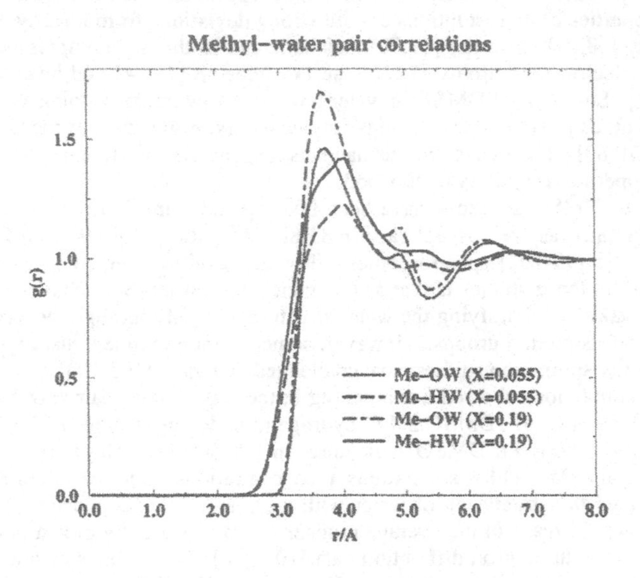

Figure 1. Me-water radial distribution functions at different DMSO concentrations at 298 K.

Similar hydration patterns were also observed as the temperature was gradually increased at the lower DMSO concentration of 0.055 mole fraction [113]. However, no temperature-dependent hydrophobic interactions were detected. Figure 3 shows the Me-Me pair correlation between the methyl groups of DMSO, which portrays the average tendency for aggregation of these non-polar groups at different temperatures. There does not seem to be any clear temperature trend, which is the main fingerprint of hydrophobic interactions [36-39], since the tendency for aggregation seems to decrease from 298 to 318 K before increasing significantly at 338 K [113]. This evidence seems to reveal that the hydrophilic hydration properties of DMSO dominate over any hydrophobic effects that may arise as temperature is increased.

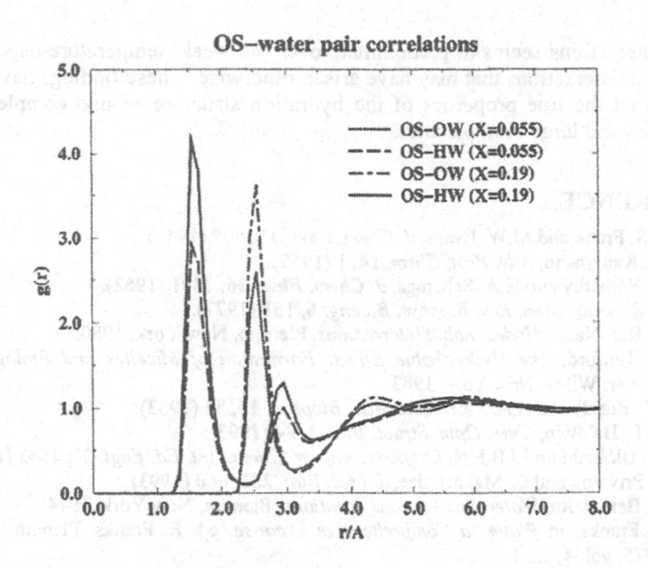

Figure 2. OS-water radial distribution functions at different DMSO concentrations at 298 K.

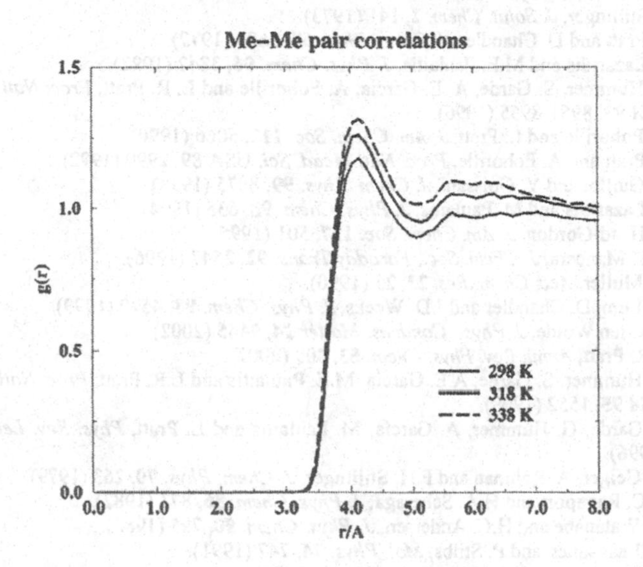

Figure 3. Me-Me radial distribution functions for different temperatures at a DMSO concentration of 0.055 mole fraction.

Aqueous solutions of DMSO provide a good example where both hydrophobic and hydrophilic hydration is observed. However, the stronger nature of the DMSO-water interactions seems to predominate over the weak, temperature-dependent hydrophobic interactions that may have arisen otherwise. These findings have implications about the true properties of the hydration structure around complex organic molecules and large biomolecules.

REFERENCES

1. H.S. Frank and M.W. Evans, *J. Chem. Phys.* **13**, 507 (1945)
2. W. Kauzmann, *Adv. Prot. Chem.* **14**, 1 (1959).
3. G. Némethy and H.A. Scheraga, *J. Chem. Phys.* **36**, 3401 (1962).
4. F. Richards, *Ann. Rev. Biophys. Bioeng.* **6**, 151 (1977).
5. A. Ben-Naim, *Hydrophobic Interactions*, Plenum, New York, 1980.
6. C. Tanford, *The Hydrophobic Effect: Formation of Micelles and Biological Membranes*, Wiley, New York, 1982.
7. J.T. Edsall and H.A. McKenzie, *Adv. Biophys.* **16**, 53 (1983).
8. R. L. Baldwin, *Curr. Opin. Struct. Biol.* **3**, 84 (1993).
9. W. Blokzihl and J.B.F.N. Engberts, *Angew. Chem., Int. Ed. Engl.* **32**, 1545 (1993).
10. P. Privalov and G. Makhatadze, *J. Mol. Biol.* **232**, 660 (1993).
11. A. Ben-Naim, *Water and Aqueous Solutions*, Plenum, New York, 1974.
12. F. Franks, in *Water, a Comprehensive Treatise*, ed. F. Franks, Plenum, New York, 1975, vol. 4, ch. 1.
13. N.T. Southall, K.A. Dill and A.D.J. Haymet, *J. Phys. Chem. B* **106**, 521 (2002).
14. M Costas, B. Kronberg and R. Silveston, *J. Chem. Soc., Faraday Trans.* **90**,1513 (1994).
15. R. Pierotti, *J. Phys. Chem.* **67**, 1840 (1963).
16. F. Stillinger, *J. Solut. Chem.* **2**, 141 (1973).
17. L. Pratt and D. Chandler, *J. Chem. Phys.* **67**, 3683 (1977).
18. T. Lazaridis and M.E. Paulaitis, *J. Phys. Chem.* **96**, 3847 (1992).
19. G. Hummer, S. Garde, A. E. Garcia, A. Pohorille and L. R. Pratt, *Proc. Natl. Acad. Sci. USA* **93**, 8951-8955 (1996).
20. A. Pohorille and L. Pratt, *J. Am. Chem. Soc.* **112**, 5066 (1990).
21. L. Pratt and A. Pohorille, *Proc. Natl. Acad. Sci.* USA **89**, 2999 (1992).
22. B. Guillot and Y. Guissani, *J. Chem. Phys.* **99**, 8075 (1993).
23. T. Lazaridis and M. Paulaitis, *J. Phys. Chem.* **98**, 635 (1994).
24. T. Head-Gordon, *J. Am. Chem. Soc.* **117**, 501 (1995).
25. R.L. Mancera, *J. Chem. Soc., Faraday Trans.* **92**, 2547 (1996).
26. N. Muller, *Acc. Chem. Res.* **23**, 23 (1990).
27. K. Lum, D. Chandler and J.D. Weeks, *J. Phys. Chem.* **98**, 4570 (1999).
28. P.R. ten Wolde, *J. Phys.: Condens. Matter* **14**, 9445 (2002).
29. L.R. Pratt, *Annu. Rev. Phys. Chem.* **53**, 409 (2002).
30. G. Hummer, S. Garde, A.E. García, M.E. Paulaitis and L.R. Pratt, *Proc. Natl. Acad. Sci. USA* **95**, 1552 (1998).
31. S. Garde, G. Hummer, A. García, M. Paulaitis and L. Pratt, *Phys. Rev. Lett.* **77**, 4966 (1996).
32. A. Geiger, A. Rahman and F.H. Stillinger, *J. Chem. Phys.* **70**, 263 (1979).
33. D.C. Rapaport and H.A. Scheraga, *J. Phys. Chem.* **86**, 873 (1982).
34. K. Watanabe and H.C. Andersen, *J. Phys. Chem.* **90**, 795 (1986).
35. A. Laaksonen and P. Stilbs, *Mol. Phys.* **74**, 747 (1991).
36. N.T. Skipper, *Chem. Phys. Lett.* **207**, 424 (1993).
37. R.L. Mancera and A.D. Buckingham, *Chem. Phys. Lett.* **234**, 296 (1995).
38. N.T. Skipper, C.H. Bridgeman, A.D. Buckingham and R.L. Mancera, *Faraday Discuss.* **103**, 141 (1996)
39. R.L. Mancera, A.D. Buckingham and N.T. Skipper, *J. Chem. Soc., Faraday Trans.* **93**, 2263 (1997).
40. C. Pangali, M. Rao and B.J. Berne, *J. Chem. Phys.* **71**, 2982 (1979).

41. W.L. Jorgensen, J. Chandrasekhar, J.D. Madura, R.W. Impey and M.L. Klein, *J. Chem. Phys.* **89**, 3742 (1988).
42. D.E. Smith, L. Zhang and A.D.J. Haymet, *J. Am. Chem. Soc.* **114**, 5875 (1992).
43. D.E. Smith and A.D.J. Haymet, *J. Chem. Phys.* **98**, 6445 (1993).
44. J. Forsman and B. Jönsson, *J. Chem. Phys.* **101**, 5116 (1994).
45. D. van Belle and S.J. Wodak, *J. Am. Chem. Soc.* **115**, 647 (1993).
46. M.H. New and B.J. Berne, *J. Am. Chem. Soc.* **117**, 7172 (1995).
47. S. Lüdemann, H. Schreiber, R. Abseher and O. Steinhauser, *J. Chem. Phys.* **104**, 286 (1996).
48. S. Lüdemann, R. Abseher, H. Schreiber and O. Steinhauser, *J. Am. Chem. Soc.* **119**, 4206 (1997).
49. L.X. Dang, *J. Chem. Phys.* **100**, 932 (1994).
50. V.A. Payne, N. Matubayasi, L.R. Murphy and R.M. Levy, *J. Phys. Chem. B* **101**, 2054 (1997).
51. G. Hummer, S. Garde, A.E. Garcia, M.E. Paulaitis and L.R. Pratt, *Proc. Natl. Acad. Sci. USA*, **95**, 1552 (1998).
52. S.W. Rick and B.J. Berne, *J. Phys. Chem. B* **101**, 10488 (1997).
53. S. Shimizu and H.S. Chan, *J. Chem. Phys.* **113**, 4683 (2000).
54. T. Ghosh, A.E. García and S. Garde, *J. Am. Chem. Soc.* **123**, 10997 (2001).
55. T. Ghosh, A.E. García and S. Garde, *J. Chem. Phys.* **116**, 2480 (2002).
56. S. Shimizu and H.S. Chan, *J. Am. Chem. Soc.* **123**, 2083 (2001).
57. S.W. Rick, *J. Phys. Chem. B* **104**, 6884 (2000).
58. J.A. Rank and D. Baker, *Protein Sci.* **6**, 347 (1997).
59. G. Hummer, *J. Am. Chem. Soc.* **121**, 6299 (1999).
60. C. Czaplewski, S. Rodziewics-Motowidlo, A. Liwo, D.R. Ripoll, R.J. Wawak and H.A. Scheraga, *Protein Sci.* **9**, 1235 (2000).
61. C. Czaplewski, D.R. Ripoli, A. Liwo, S. Rodziewciz-Motowidlo, R.J. Wawak and H.A. Scheraga, *Int. J. Quantum Chem.* **88**, 41 (2002).
62. S. Shimizu and H.S. Chan, *J. Chem. Phys.* **115**, 1414 (2001).
63. C. Czaplewski, S. Rodziewics-Motowidlo, A. Liwo, D.R. Ripoll, R.J. Wawak and H.A. Scheraga, *J. Chem. Phys.* **116**, 2665 (2002).
64. S. Shimizu and H.S. Chan, *J. Chem. Phys.* **116**, 2668 (2002).
65. T. Ghosh, A.E. García and S. Garde, *J. Phys. Chem. B* **107**, 612 (2003).
66. S.W. Jacobs, E.E. Rosenbaum and D.C. Wood (eds.), Dimethyl Sulfoxide, Marcel Dekker, New York, 1971.
67. D. Martin and H.G. Hauthal, Dimethyl Sulfoxide, Wiley, New York, 1975.
68. G.V. Roshkovskii, R.A. Ovchinnikova and N.V. Penkina, Zh. Prikl. Khim. 55 (1982) 1858.
69. J.M.G. Cowie and P.M. Toporowski, Can. J. Chem. 39 (1964) 2240.
70. J. Mazurkiewicz and P. Tomasik, J. Phys. Org. Chem. 3 (1990) 493.
71. U. Kaatze, M. Brai, F.-D. Sholle and R. Pottel, J. Mol. Liq. 44 (1990) 197.
72. M.Y. Doucet, F. Calmes-Perault and M.T. Durand, C.R. Acad. Sci. 260 (1965) 1878.
73. E. Tommila and A. Pajunen, Suom. Kemistil. B41 (1969) 172.
74. A. Luzar, J. Chem. Phys. 91 (1989) 3603.
75. H.L. Clever and S.P. Pigott, J. Chem. Thermodyn. 3 (1971) 221.
76. M.F. Fox and K.P. Whittingham, J. Chem. Soc. Faraday Trans. 75 (1974) 1407.
77. J. Kenttammaa and J.J. Lindberg, Suom. Kemistil. B33 (1960) 32.
78. D.H. Rasmussen and A.P. Mackenzie, Nature 220 (1968) 1315.
79. J.E. Lovelock and M.W.H. Bishop, Nature 183 (1959) 1394.
80. T.J. Anchordoguy, C.A. Ceccini, J.N. Crowe and L.M. Crowe, Cryobology 28 (1991) 467.
81. H. Bertagnolli, E. Schultz and P. Chieux, Ber. Bunsenges. Phys. Chem. 93 (1989) 88.
82. G.J. Safford, P.C. Schaffer, P.S. Leung, G.F. Doebbler, G.W. Brady and E.F.X. Lyden, J. Chem. Phys. 50 (1969) 2140.
83. G. Brink and M. Falk, J. Mol. Struct. 5 (1970) 27.

84. A. Burneau, J. Mol. Liq. 46 (1990) 99.
85. Y. Higashigaki, D.H. Christensen and C.H. Wang, J. Phys. Chem. 85 (1981) 2531.
86. H. Kelm, J. Klowoski and E. Steger, J. Mol. Struct. 28 (1975) 1.
87. A. Bertulozza, S. Bonora, M.A. Battaglia and P. Monti, J. Raman Spectrosc. 8 (1979) 231.
88. A. Allerhand and P. von R. Schleyer, J. Am. Chem. Soc. 85 (1963) 175.
89. D.E. Bowen, M.A. Priesand and M.P. Eastman, J. Phys. Chem. 78 (1974) 2611.
90. W.M. Madigsky and R.W. Warfield, J. Chem. Phys. 78 (1983) 1912.
91. T. Kondo, L.L. Kirschenbaum, J. Kim and P. Riesz, J. Phys. Chem. 97 (1993) 522.
92. T. Tokuhiro, L. Menafra and H.H. Szmant, J. Chem. Phys. 61 (1974) 2275.
93. J.A. Glasel, J. Am. Chem. Soc. 92 (1970) 372.
94. K.J. Packer and D.J. Tomlinson, Trans. Faraday Soc. 67 (1971) 1302.
95. T. Freech and H.G. Hertz, Z. Phys. Chem. (Munich) 142 (1984) 43.
96. B.C. Gordalla and M.D. Zeidler, Mol. Phys. 59 (1986) 817.
97. B.C. Gordalla and M.D. Zeidler, Mol. Phys. 74 (1991) 975.
98. E.S. Barker and J. Jonas, J. Phys. Chem. 89 (1985) 1730.
99. W. Feder, H. Dreizler, H.D. Rudolph and V. Trypke, Z. Naturforsch. A 24A (1969) 266.
100. E.S. Verstakov, P.S. Yastremskii, Y.M. Kessler, V. Goncharov and V.V. Kokovin, Zh. Strukt. Khim. 21 (1980) 91.
101. U. Kaatze, R. Pottel and M. Schafer, J. Phys. Chem. 93 (1989) 5623.
102. T.A. Novskova, V.I. Gaiduk, V.A. Kudryashova and Y.I. Khurgin, Sov. J. Chem. Phys. 8 (1991) 1636.
103. B. Kirchner, D.J. Searles, A.J. Dyson, P.S. Vogt and H. Huber, J. Am. Chem. Soc. 122 (2002) 5379.
104. H. Huber, B. Kirchner and D.J. Searles, J. Mol. Liq. 97-98 (2002) 71.
105. I.I. Vaisman and M.L.Berkowitz, J. Am. Chem. Soc. 114 (1992) 7889.
106. A. Luzar and D. Chandler, J. Chem. Phys. 98 (1993) 8160.
107. A.K. Soper and A. Luzar, J. Chem. Phys. 97 (1992) 1320.
108. A.Luzar, A.K.Soper and D.Chandler, J.Chem.Phys. 99 (1993) 6836.
109. A.K. Soper and A. Luzar, J. Phys. Chem. 100 (1996) 1357.
110. I.A Borin and M.S. Skaf, J. Chem. Phys. 110 (1999) 6412.
111. A. Vishnyakov, A.P. Lyubartsev and A. Laaksonen, J. Phys. Chem. 105 (2001) 1702.
112. B. Kirchner and J. Hutter, Chem. Phys. Lett. 364 (2002) 497.
113. R.L. Mancera, M. Chalaris and J. Samios, J. Mol. Liquids, 2003, in the press.
114. R.L. Mancera, M. Chalaris, K. Refson and J. Samios, submitted for publication.

MOLECULAR SIMULATIONS OF NAFION MEMBRANES IN THE PRESENCE OF POLAR SOLVENTS

D.A. MOLOGIN
Department of Physical Chemistry, Tver State University, Sadovy per. 35, 170002 Tver, Russia

P.G. KHALATUR
Department of Polymer Science, University of Ulm, Albert-Einstein-Allee 11, Ulm, D-89069, Germany

A.R. KHOKHLOV
Physics Department, Moscow State University, 117234 Moscow, Russia

Abstract: **We describe molecular dynamics computer simulations coupled with quantum chemical calculations of relevant geometries and interaction constants for studying the detailed behavior of solvent-containing Nafion® membranes. Our attention here is focused on the effect of different solvent additives on the equilibrium structure of micellar aggregates. Taking into account the practical importance of methanol membrane fuel cells, methanol-containing systems are the subject of our primary interest. Also, we study mixed aggregates containing alcohols $H(CH_2)_nOH$ with longer hydrocarbon chain, up to $n = 7$. In the case of the bicomponent solvent-containing aggregates we have suggested how events occurring at the molecular level produce polymorphic transitions in these mixed aggregates from spherical structures toward cylindrical and rather exotic toroidal micelle structures, stabilized by a more uniform compact packing of the hydrophobic groups in the micelle exterior. These transitions predicted in the present study illustrate how intermolecular forces such as solvation as well as solvent chain length can affect overall aggregate structure and result in polymorphism.**

1. INTRODUCTION

Nafion® represents a novel family of ionic copolymers (ionomers) which consists of a poly(tetrafluoroethylene) backbone with short perfluoro- polyether pendant side chains terminated by ionizable sulfonate groups. When swollen in polar solvents (water, alcohols), the structure of Nafion® resembles that of a three-dimensional array of inverse micelles in which the ionic end groups of side chains are clustered together in solvent-containing aggregates within the apolar perfluoro-

J. Samios and V.A. Durov (eds.), Novel Approaches to the Structure and Dynamics of Liquids: Experiments, Theories and Simulations, 397–425.

carbon matrix. It is assumed that the mixed aggregates are interconnected by short channels acting as transient crosslinks in the polymer. This organized structure exhibits excellent ionic conductivity and electrochemical and chemical resistance [1,2]. Due to these features, the solvent-containing Nafion® membranes are widely used in a lot of electrochemical devices such as water electrolyzers for hydrogen energetics, batteries, and methanol membrane fuel cells, which are seen as the most promising energy suppliers for vehicles [1–5].

The equilibrium structure of mixed aggregates formed in ionomer systems was shown to be governed by both the interaction between associating groups and the nature of solvent [6–8]. In principle, one can distinguish the following three aggregation regimes: (i) When the characteristic energy ε of attraction between associating groups is comparable to k_BT, the aggregates have broad, thermally fluctuating, interfaces between their inner and outer regions. This regime is known as the weak segregation regime [9–11]. (ii) If the attractive forces become more intensive (or the temperature is lowered), the thermal fluctuations of interfaces decrease leading to the aggregates with well-separated inner and outer regions. Such a structure is analogous to the microphase separated structure observed in two-letter AB block copolymers in a selective solvent that is an extremely poor solvent for A blocks and good for B blocks. In the context of phase separation phenomena, this regime corresponds to the strong segregation regime. (iii) With a further increase of the effective energy ε, one can observe the crossover transition from the strong segregation regime to the so-called superstrong segregation regime [12] in which specific rearrangements of the aggregate structure take place. The superstrong segregation is difficult to achieve for usual block copolymers with long A and B chain sections. However, this regime is easily realized for ordinary ionomers in the medium of low polarity due to the very strong dipolar attraction between ionic groups (in this case, the characteristic energy is $|\varepsilon| \approx 10 \div 25\ k_BT$ [13]). The perfluorinated ionomers, including Nafion®, are corresponding to this category.

One of the most interesting features of the ionic aggregates (multiplets) emerging in the superstrong segregation regime is their non-spherical geometry. Non-spherical aggregates have been observed in solutions and melts of ionomers and associative polymers [14–21]. In particular, disklike [17,18] and lamellar [19] microstructures have been reported. Small-angle X-ray and neutron scattering (SAXS and SANS) studies of the perfluorinated ionomers have revealed the presence of non-spherical aggregates in these systems [14,15]. Depending mainly on the nature of the solvent, Aldebert et al. [14,15] and Loppinet et al. [21] observed disklike and rodlike shapes for the mixed multiplets whose radius was found to vary from ≈20 to ≈25 Å. Recently, the existence of rodlike shape aggregates formed in short pendant chain prefluorinated ionomers in the presence of polar solvents was confirmed using SAXS and SANS techniques [22]. The aggregate radius was found to be ≈17 Å for water-containing systems.

In our previous work [8,23,24], using Monte Carlo computer simulation, we have performed a study of the structure of mixed solvent-containing aggregates formed by one-end-functionalized associating chains. In the present work, we will pursue this line further, using a more sophisticated model and molecular dynamics simulation technique. Our attention here will be focused on the effect of different solvent additives on the equilibrium structure of micellar aggregates. Taking into account the practical importance of methanol membrane fuel cells, methanol-containing systems will be the topic of our primary interest.

2. COMPUTATIONAL TECHNIQUES AND MODEL

2.1 QUANTUM MECHANICAL CALCULATIONS

The side chains in Nafion® are perfluorinated double ether chains with the following structure

$$-CFOCF_2CF(CF_3)OCF_2CF_2SO_3H$$

To study interactions between the hydrophilic sites in Nafion® and solvent molecules, in the previous papers [23,24] we performed semi-empirical and *ab initio* self-consistent-field (SCF) molecular orbital calculations for some model systems, including the complexes with the trifluoromethane sulfonic acid (CF_3SO_3H) and single "probe" water and methanol molecules. Also, the same calculations were carried out for the bimolecular complexes CF_3SO_3H/CF_3SO_3H, H_2O/H_2O, and CH_3OH/CH_3OH.

In the present study, the computational scheme was implemented as a two-step process. In the first of these steps, fully optimized geometries were computed using the standard semi-empirical method AM1, which predicts reasonable results for fluorine-containing compounds [23], and the Polak-Ribiere conjugate gradient method. The second step then consisted of the calculation of the potential energy of ground state using the Hartree-Fock MO–LCAO approximation with the conventional *ab initio* basis sets such as STO-3G, D95**, 6-31G**, 6-311G**, etc.. Of course, choosing a proper basis set in *ab initio* calculations is critical to the reliability and accuracy of the calculated results. Electronic potential derived atom centered partial charges were found on the basis of commonly-used schemes (see below and Ref. [25]). The characteristic potential energies of attraction, ε, were obtained as a difference between the potential energies of the geometry optimized molecular complexes and the same two noninteracting molecules, at infinite separation. As an example, the values of ε found using the methods mentioned above are presented in Table 1.

In principle, the electrostatic potential, $\psi(r)$, at a given distance r from a point charge is given by Coulomb's law. However, for multiple charges in three-dimensional space (both point nuclear charges and electrons in molecular orbitals), the equation becomes too complex. We defined the 3D function $\psi(r) = \psi(x,y,z)$ as the potential energy of a unit positive charge interacting with the bimolecular systems under consideration. In calculating the interaction with electrons in molecular orbitals, we found the $\psi(r)$ function using a full expression (i.e., without neglecting components due to diatomic differential overlap) when an *ab initio* method was chosen.

The minimum energy configurations of the systems CH_3OH/CH_3OH and CH_3OH/CF_3SO_3H together with the corresponding 3D electrostatic potentials are shown in Fig. 1. The analogous data for water-containing molecular systems have been presented in our previous study [23,24]. As seen from Fig. 1, the optimized molecular complex CH_3OH/CF_3SO_3H demonstrates the presence of a typical noncovalent hydrogen bonding interaction between the sulfonic acid group and the methanol molecule which is located near the OH group of the trifluoromethane sulfonic acid. It should be noted that the distances separating the oxygen atoms in this complex are significantly smaller as compared to these found for the minimum energy structures corresponding to the perfluorinated ether, CF_3OCF_3 with a single probe molecule CH_3OH and the complex $CH_3OH/H(CF_2)_5H$ which can be considered as a model of the solvated backbone chain in Nafion®.

TABLE 1. The interaction energies, ε, found for bimolecular complexes using different quantum mechanical methods

Quantum Mechanical Method	Characteristic Energy ε, Kcal/mol
H_2O/H_2O	
AM1	-5.46[a)]
STO-3G	-5.87[a)]
STO-6G*	-5.14
6-21G**	-8.70
5-31G**	-5.62
6-31G**	-5.53[a)]
D95**	-4.94
6-311G**	-5.54
CH_3OH/CH_3OH	
AM1	-3.74[a)]
STO-3G	-3.00[a)]
3-21G	-4.62[a)]
STO-6G*	-2.02
6-21G**	-6.56
5-31G**	-4.76
6-31G**	-4.96[a)]
D95**	-3.78
6-311G**	-5.70
CF_3SO_3H/H_2O	
AM1	-12.9[a)]
STO-3G	-6.50
CF_3SO_3H/CH_3OH	
AM1	-9.77[a)]
STO-3G	-7.35
CF_3SO_3H/CF_3SO_3H	
AM1	-12.4[a)]
STO-3G	-14.0

a) Fully optimized geometries

This indicates that methanol absorbed by the Nafion® membrane should be preferentially localized near the polar terminal parts of the pendant chains. The same is true for water molecules which demonstrate even slightly more extensive hydrogen bonding [23]. On the contrary, both the ether groups of the pendant chains and perfluorinated main chain should be considered as hydrophobic. However, compared to water, methanol shows more extensive attraction to the hydrophobic polytetrafluoroethylene skeleton of Nafion®. As a result, it can be expected to be a better solvent for Nafion® compared to water.

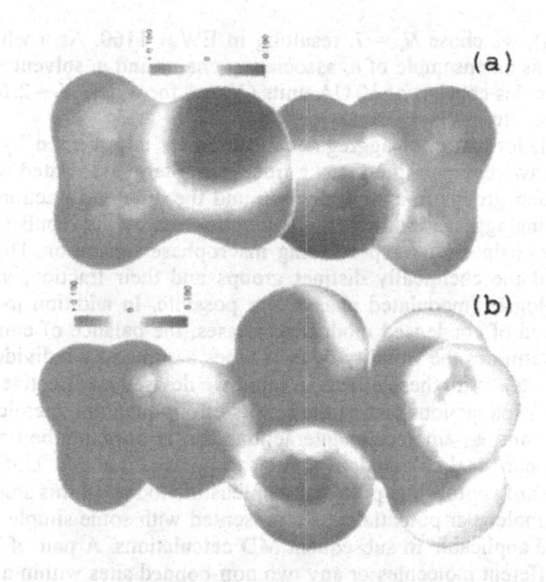

Figure 1. The minimum energy configurations of the interacting bimolecular systems (a) CH_3OH/CH_3OH and (b) CH_3OH/CF_3SO_3H. The corresponding 3D electrostatic potentials are shown.

2.2 MOLECULAR DYNAMICS SIMULATIONS

It is clear that detailed atomistic description of complex solvent-containing aggregates with explicitly represented chemical groups and all structural features is practically impossible in any simulation techniques, and *ab initio* calculations are currently out of the question for such large systems. Taking into account these circumstances, we will employ in our molecular dynamics (MD) calculations a coarse-grained model of the system.

Elliot et al. [26] have used MD method to simulate water-containing ionomer membranes. In this study, the system was modeled in a simplified way – as a mixture of single sulfonate fragments rather than a fully connected comblike Nafion® copolymer. Such a representation was taken in order to produce higher molecular mobilities, so that the system under study could undergo more significant structural reorganization over the course of a simulation. In our MD study we decided to use the same representation. Moreover, we further simplified the model by introducing the united atom (UA) approximation, that is, all the hydrogen-containing groups were assumed to be single spherical units with "collapsed" hydrogen atoms. In this case, water molecule is viewed as an one-center spherical particle, and methanol molecule is considered as a two-center (dumb-bell) particle.

The model system as a whole was similar in spirit to the one used in our previous works [8,23,24,27]. We consider a three-component system containing chain molecules with strongly attracting sulfonate end-groups, apolar external medium, and polar low-molecular-weight solvent (water or alcohols). Each chain molecule consists of N_c UA units in the main chain. Since most of the experimental studies of solvated Nafion® were performed for Nafion® 117 having equivalent weight EW = 1200 g/equiv (EW is the mass of solvent-free polymer containing one mole of sulfonate groups), we chose $N_c = 7$, resulting in EW ≈ 1160. As a whole, the system was modeled as an ensemble of n_c associating chains and n_s solvent molecules, each of these molecules consists of N_s UA units ($N_s = 1$ for water, $N_s = 2$ for methanol, $N_s = 3$ for ethanol, etc.).

In general, formation of aggregates in ionomers is governed by a delicate balance between two competing effects: a free energy term associated with the dissolution of the polar groups in apolar medium and the dipolar attraction between these groups, favoring aggregation, which are counteracted by the repulsion between connected apolar chain sections, preventing macrophase separation. Depending on the distribution of the chemically distinct groups and their fraction, various types of final morphologies (modulated phases) are possible. In addition to controlling the type and period of condensed modulated phases, the balance of competing interactions also determines the stability of the shapes assumed by individual ionic aggregates (multiplets). With these effects in mind we devised an effective potential based on site-site UA interactions and on the results of our quantum-chemical calculation.

The inter- and intramolecular interactions can in principle be obtained from direct quantum-chemical calculations similar to presented above. Unfortunately, such a straightforward approach is practically unfeasible today. In this study, the resulting inter-(intra-) molecular potential was represented with some simple functional form in order to be applicable in subsequent MD calculations. A pair of sites i and j belonging to different molecules or any two non-bonded sites within a molecule interact via a shifted (repulsive) Lennard-Jones (LJ) function, $U_{LJ}(r)$, plus a screened attractive Coulomb term, $U_c(r)$

$$U_{LJ}(r) = \begin{cases} 4\varepsilon_{LJ}\left[\left(\dfrac{\sigma_{LJ}}{r}\right)^{12} - \left(\dfrac{\sigma_{LJ}}{r}\right)^{6} + \dfrac{1}{4}\right], \text{ if } r < r_{LJ} \\ 0, \text{ if } r > r_{LJ} \end{cases} \quad (1)$$

$$U_c(r) = \begin{cases} \dfrac{\varepsilon\sigma}{r}\left[1 - \left(\dfrac{r}{r_c}\right)^{2}\right]^{2}, \text{ if } r < r_c \\ 0, \text{ if } r > r_c \end{cases} \quad (2)$$

where $r = |\mathbf{r}_i - \mathbf{r}_j|$ is the separation between sites i and j, ε_{LJ} and σ_{LJ} are usual LJ parameters, $r_{LJ} = 2^{1/6}\sigma_{LJ}$ denotes the distance at which the potential (1) is cut, ε and σ are the corresponding energy and distance parameters of the attractive potential (2), and r_c is the screening length, i.e., the cutoff distance. The internal bond vibrations were treated explicitly using the following potential [28]

$$U_b(r) = \begin{cases} \varepsilon_b\left[\left(\dfrac{b}{r}\right)^{12} - 2\left(\dfrac{b}{r}\right)^{6} + 1\right], \text{ if } r < b \\ \varepsilon_b\left[\left(\dfrac{b}{2b-r}\right)^{12} - 2\left(\dfrac{b}{2b-r}\right)^{6} + 1\right], \text{ if } r > b \end{cases} \quad (3)$$

which links the respective neighboring sites i and j ($|i-j| = 1$) into a chain with given bond lengths b. This potential has the form of a simple shifted LJ function for $r < b$ and has the symmetric form for $r > b$, being positive for all r. Note that at $\varepsilon_b =$

ε_{LJ} the time scale corresponding to the intramolecular vibrational motion, $\tau_b = b(m/\varepsilon_b)^{1/2}$, is not shorter than that of the translational motions of the atomic groups.

It is clear that the main energy parameters determining the structure of solvent-containing aggregates and their stability are the values $\varepsilon_{\alpha\beta}$ which characterize the attractive interaction between the "cluster-forming" polar terminal groups α and β belonging to sulfonate fragments or solvent molecules. Below, these hydrophilic groups are denoted by the subscripts "t" and "s", respectively. The semi-empirical and *ab initio* quantum mechanical calculations have shown that the absolute values of these energy parameters are much greater than those characterizing the interactions with the participation of hydrophobic (h) groups (CH_2, CF_2, etc.). Thus, it was decided to set $\varepsilon_{hh} = \varepsilon_{ht} = \varepsilon_{hs} = 0$, because the physical background for these interactions is quite similar. As a result, in the model under consideration the hydrophobic/hydrophobic and hydrophobic/hydrophilic interactions are represented as pure excluded-volume effects determining by the LJ potential (1). This condition corresponds to such external medium which acts as a good (athermal) "solvent" for hydrophobic polymer sections and hydrophobic solvent groups.

Below, all the parameters and the results of calculations are expressed in standard reduced units, i.e., lengths and energies are measured in LJ units of σ_{LJ} and ε_{LJ}, temperature T in units of ε_{LJ}/k_B, and time t in units of $\tau = \sigma_{LJ}(m/\varepsilon_{LJ})^{1/2}$, where m is the mass of the corresponding unit and k_B denotes the Boltzmann constant. In accordance with our previous study [23,24] and the quantum mechanical calculations presented here, we put $\varepsilon_{tt} = \varepsilon_{ts} = -10$ and $\varepsilon_{ss} = -5$. For convenience, features associated with position along the chain molecule – pseudoatoms and bonds – are numbered in increasing order from the terminal polar group to the opposite hydrophobic end, i.e., the associative hydrophilic group is UA #1. There are no additional angular constraints imposed on adjacent bonds other than those which are a consequence of the potential (1). We also ignore dihedral-angle potentials.

The evolution of the system was followed by numerical solving the Newtonian equations of motion using the Verlet leapfrog algorithm and Gaussian isokinetic (GI) scheme to achieve constant kinetic energy [29,30]. It should be noted that the implementation of GI constant temperature scheme is not easy within the framework of a simple Verlet algorithm. The computational procedure used in this study is described in Appendix A.

We present the following type of MD simulation. The molecular aggregates (i.e., single micelles) were prescribed *a priori* and then the micelle organization was studied as a function of temperature, micelle size, and the amount of solvent particles incorporated into the micellar core. The same computational scheme has been used in one of our previous papers [8].

The simulations were started from a structure with the sulfonate chain fragments in all-*trans*-conformation and extending radially from the coordinate origin surrounded by a hole. This corresponds to a fairly loose micellar assembly. The solvent molecules were placed in the hole inside the aggregate. Initial configurations were annealed and then averages were calculated. Associating hydrophilic groups (SO_3H, H_2O, OH), belonging both to the sulfonate fragments and the solvent molecules, were considered as a single micellar core if they form a connected cluster. A cluster was defined as such an aggregate in which each polar group was not further away than a radius $R_{cl} = 2^{1/6}$ from at least one other polar group in the aggregate. This means that the cluster could assume any possible geometry. We call a cluster "stable" if the set of contributing particles fulfills the criterion given above over a full time period of simulation. We have experimented with values of R_{cl} ranging from

$2^{1/6}$ to 2.8 and found that the properties of the clusters are relatively insensitive to R_{cl} which lies well within the region of strong attraction of the potential (2) involving energy parameters $|\varepsilon_c/k_BT| \geq 5$. A total aggregation number, p, of a given system is the number of groups forming the micellar core and is a constant coinciding with the number of chains in the system, n_c, plus n_s solvent molecules.

3. RESULTS AND DISCUSSION

3.1 SOLVENT EFFECT ON AGGREGATE STABILITY

As has been mentioned above, the effective potential (2) is, of course, only a prototype of a realistic potential for interactions involving hydrophilic groups. Simplifications are introduced partly for computational efficiency but also because of our limited knowledge concerning treatment of the corresponding molecular forces in the framework of classical approach. Our preliminary simulations have shown that the most essential properties of aggregates and their stability do not practically depend on the detailed functional form of the interaction potential, but depend quite strongly on the radius r_c of the interaction between aggregating groups. At a given temperature, this parameter plays a main role in determining the stability of aggregates in the strong and superstrong segregation regimes. Thus, first of all we will examine the behavior of mixed solvent-containing aggregates at different values of r_c and solvent content, n_s. To this end, the following calculation scheme was used (see also Ref. [8]).

A given number n_s of solvent particles was incorporated into the core of an aggregate of a given size, n_c. The system was equilibrated at $T = 1$ and $r_c = 3$, which corresponds to strong interaction between particles and provides micelle stability during a long period (about 10^6 time steps and more). Then, the parameter r_c was decreased gradually. This is accompanied by a decrease in the aggregation number from the starting value p_0, which coincides with the starting number of chains n_c and solvent molecules n_s in the micelle, to a certain value p. This effect is due to the detachment of some of the molecules from the core. If after the detachment of this portion of molecules the value p remained constant for 3×10^5 time steps, then we considered this value p as a new total aggregation number corresponding to the stable state of a micelle at given r_c. For each value of r_c we repeated calculations for 3÷5 times and took averages over the results of these independent calculations. As a result, we obtained the dependence of p on r_c for different values of n_s. Most of the results were obtained for the micelle with $n_c = 64$ as the initial state, containing water-like and methanol-like molecules (below, they will be denoted by letters "W" and "M", respectively).

Figure 2 shows the results of these simulations in the form of the $r_c - n_s$ state diagram, where the region above the curve corresponds to stable (micellar) state.

As seen, the general tendency is that the critical value of r_c increases with increasing the solvent content, n_s, and approaches some limiting value for both solvents at large n_s. In the case of methanol-containing aggregates, the limiting value of r_c is larger as compared to bicomponent aggregates with incorporated water-like particles. For the latter system, r_c demonstrates nonmonotonous behavior in the region of relatively small n_s. Taking into account these observations, in all the subsequent calculations we put $r_c = 2.8$.

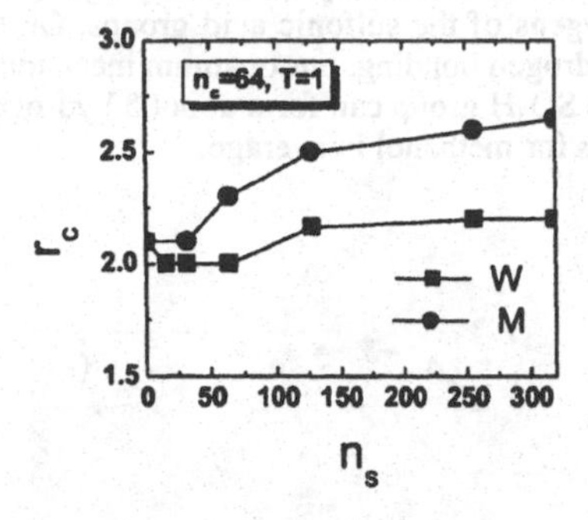

Figure 2. The r_c–n_s state diagrams for the water-containing and methanol-containing systems, at $R_{cl} = r_c$, $n_c = 64$, and $T = 1$. The regions above the curve correspond to stable (micellar) state.

Let us now consider the temperature stability of the mixed aggregates. The general strategy in this series of simulations was analogous to that described above. At the equilibrated stage of the simulation, aggregates containing a given number of solvent molecules were typically aged for $t = 4\times10^5$ time steps at $T = 1$. Temperature was then gradually decreased starting from $T = 4$. If at a given T a decrease in the total aggregation number was observed, temperature was reduced by a small increment $\Delta T = 0.1$ until the initial aggregation number remained constant for 2×10^5 time steps. Evidently, this computational scheme does not provide strict thermodynamic information on aggregate stability. This information is given by the corresponding phase diagram, which can be determined from the behavior of the free energy $\mathcal{F}$ as a function of T and n_s. To a large extent our criterion for aggregate stability has a kinetic meaning [8]. Nevertheless, we can treat our data as a suitable starting point.

We present the results on temperature stability of mixed aggregates as the T–n_s state diagram in Fig. 3.

In this figure, the curves separate stable and unstable states from each other; the region which lies below these curves corresponds to bonded (micellar) states. One can conclude that, in the case of methanol-containing aggregates, the critical temperature always increases with increasing n_s, that is, incorporation of these solvent particles inside the micellar core decreases the stability of aggregates as compared to the system without solvent. On the other hand, for water-containing system we observe a more complex behavior: at approximately equimolar solvent/polymer composition, the bicomponent aggregates become more stable compared to the corresponding dry (solvent-free) aggregates and do not disintegrate for a very long period. This observation is in line with our previous data obtained for simple coarse-grained (lattice) models of reverse bicomponent micelles [8]. As seen from Fig. 3, however, when solvent content becomes sufficiently large, the critical temperature begins to decrease with n_s. One of the most important observations following from the simulation is that the water-containing aggregates are always more stable at the same number of incorporated solvent molecules than their methanol-containing counterparts. We can explain this feature by steric factors, because methanol molecules are larger than the molecules of water and contain hydrophobic CH_3 groups. In addition, stability of the mixed clusters can be characterized by the average number of hydrogen bonds which are formed by the corresponding solvent molecules with negatively charged oxygens of the sulfonic acid groups. On the basis of a standard geometry criteria of hydrogen bonding, our quantum mechanical calculations predict that the oxygens of one SO_3H group can form about 5 hydrogen bonds for water and about 3 hydrogen bonds for methanol in average.

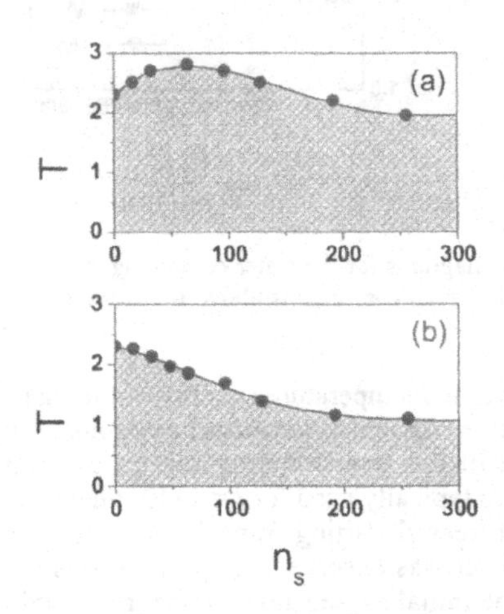

Figure 3. The T–n_s state diagrams for the water-containing (a) and methanol-containing (b) systems, at $R_{cl} = r_c = 2.8$ and $n_c = 64$. The regions above the curves correspond to stable (micellar) state.

3.2 POTENTIAL ENERGY

Figures 4 and 5 show the time-averaged potential energy per particle U/N (the total number of particles is given by $N = N_c n_c + N_s n_s$) as a function of the total aggregation number, $p = n_c + n_s$, and temperature for clusters containing water and methanol molecules, at $n_c = 64$.

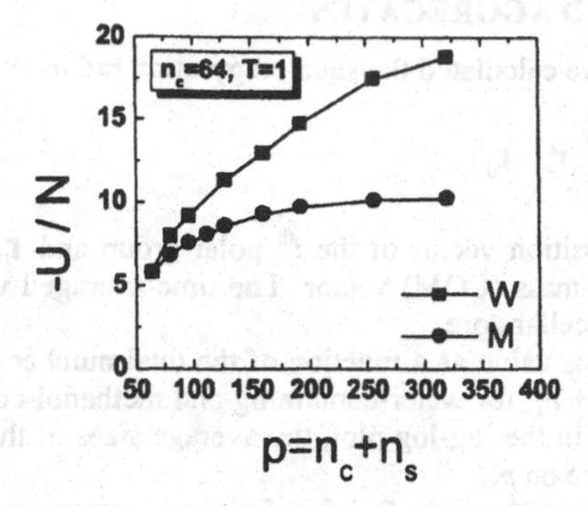

Figure 4. The time-averaged potential energy per particle *U/N* (the total number of particles is given by $N = N_c n_c + N_s n_s$) as a function of the total aggregation number, $p = n_c + n_s$, at $n_c = 64$ and $T = 1$.

One can see that, at fixed temperature, the value of *U/N* decreases in both cases as solvent content increases. Evidently, this is due to the reduction of the role of surface effects in the polar aggregate core. For the same solvent content, the potential energy of water-containing aggregates is lower than in the case of the aggregates with incorporated methanol. This difference is quite visible even for the relatively small number of incorporated solvent molecules. Therefore, we can speculate that the methanol-containing core is more friable as compared to its water-containing counterpart or has some different geometry (indeed, this is the case, as we will see below). Since the aggregate stability decreases with temperature increasing, one can expect that this will result in an increase in the time-averaged potential energy. Such a behavior is clearly seen in Fig. 5 for both systems under discussion. We see that, at T = *constant*, the water-containing aggregates have a lower energy, as compared to the system containing methanol.

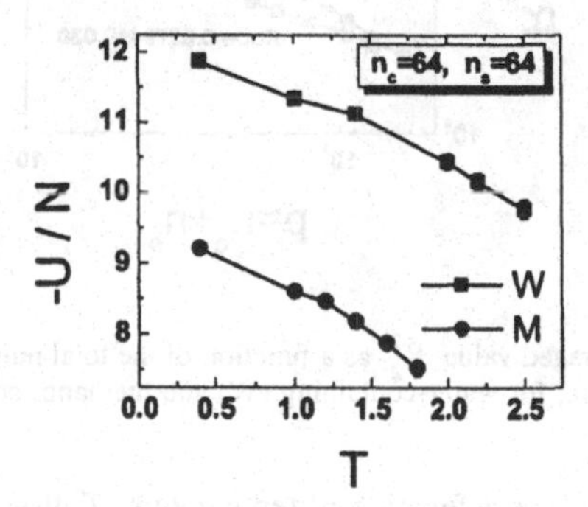

Figure 5. The time-averaged potential energy per particle *U/N* as a function of temperature for the water-containing (W) and methanol-containing (M) systems, at $n_c = 64$ and $n_s = 64$.

3.3 SIZE OF MIXED AGGREGATES

During the runs, we calculated the squared gyration radius R_g^2 of the polar core

$$R_g^2 = \frac{1}{n_c + n_s} \sum_{i=1}^{n_c+n_s} (\mathbf{r}_i - \mathbf{r}_0)^2 \tag{4}$$

where $\mathbf{r}_i$ is the position vector of the ith polar group and $\mathbf{r}_0 = (n_c + n_s)^{-1} \sum_i \mathbf{r}_i$ denotes the center-of-mass (COM) vector. The time-averaged value R_g^2 characterizes the size of the micellar core.

Figure 6 shows this value as a function of the total number of polar groups in a micellar core, $p = n_c + n_s$, for water-containing and methanol-containing aggregates at $T = 1$ and $n_c = 64$. In the log-log plot, the average sizes of the micellar cores exhibit linear dependence on p.

One can expect that the size R_g of a finite aggregate considered as a liquid spherical p-particle droplet without internal cavities must vary as $R_g \propto p^{1/3}$. The data presented in Fig. 6 for the water-containing system show that this expected dependence is followed rather approximately. Deviations from the expected power law $R_g \propto p^{1/3}$ (or $R_g^2 \propto p^{2/3}$) probably indicate a change in the equilibrium structure of the polar core as the water content increases. In the case of the methanol-containing system, we observe very strong deviations from the power law $R_g^2 \propto p^{2/3}$ when the value of n_s becomes $\gtrsim$ 80, i.e., when the solvation level $n_s/n_c \gtrsim 1$. An apparent exponent found from the log-log plot of R_g vs. p is close to 0.65. Such a strong dependence of R_g on p indicates that an increase in the cluster size is accompanied by a dramatic rearrangement in the cluster shape. This feature will be discussed below.

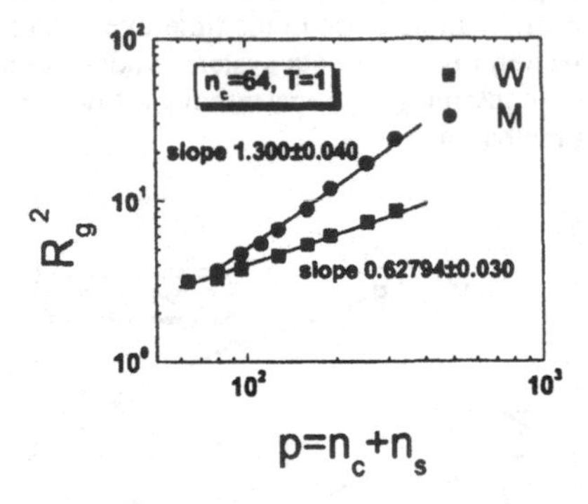

Figure 6. The time-averaged value R_g^2 as a function of the total number of polar groups in a micellar core, $p = n_c + n_s$, for water-containing (W) and methanol-containing (M) aggregates at $T = 1$ and $n_c = 64$.

Figure 7 shows R_g^2 as a function of temperature, T. It is seen that the solvated aggregates swell with T increasing. This behavior is more pronounced in the case of the methanol-containing system when a rapid growth of R_g^2 begins at $T \gtrsim 1$.

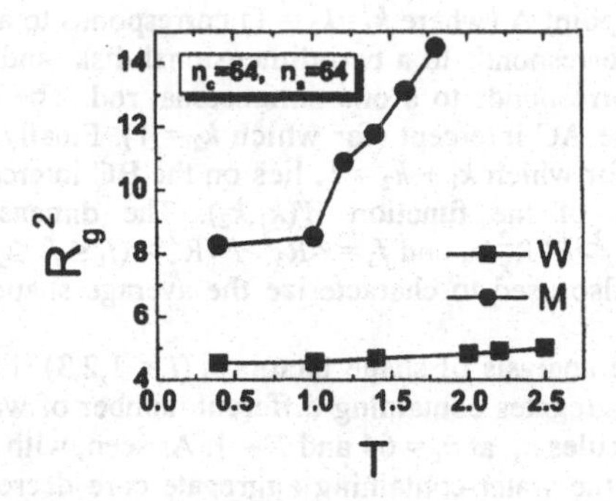

Figure 7. R_g^2 as a function of temperature, T, at $n_c = 64$ and $n_s = 64$.

3.4 MICELLAR-CORE SHAPE

To study the shape of the micellar core, we computed the principal moments of the inertia tensor and the squared gyration radius. For each configuration we found the symmetrical tensor of inertia T with the components

$$T_{ij} = (\delta_{ij}\delta_{kl} - \delta_{ik}\delta_{jl})\sum_{m=1}^{n_c+n_s}(x_{mi} - x_{0i})(x_{mj} - x_{0j}) \quad (i,j,k,l = 1,2,3; i \neq l, j \neq k) \tag{5}$$

where δ is the Kronecker delta, x_{mi} (i = 1,2,3) are the components of the position vector $\mathbf{r}_m$ of the mth chain sulfonate group or the mth solvent polar group, and x_{0i} (i = 1,2,3) are the components of the center-of-mass vector $\mathbf{r}_0$. For each micelle configuration generated, T was diagonalized to yield the moments t_1, t_2, and t_3 along principal axes 1, 2, and 3 of the core. Then these moments were ordered according to their magnitude, so that $t_1 \geq t_2 \geq t_3$. Using t_i, we can obtain the corresponding principal moments R_i^2 of the squared radius of gyration

$$R_i^2 = \frac{1}{2(n_c + n_s)}(t_j + t_k - t_i) \quad (i,j,k = 1,2,3; i \neq j \neq k) \tag{6}$$

Note that $R_g^2 = R_1^2 + R_2^2 + R_3^2 = \mathrm{Tr(T)}/[2(n_c+n_s)]$. In the coordinate system with the coordinate axes coinciding with principal axes 1, 2, and 3, we can treat the distribution function $W(t_1,t_2,t_3)$ as a scalar value lying between three planes $t_3 - t_2 = 0$ (the 0AB plane), $t_2 - t_1 = 0$ (the 0AC plane), and $t_2 + t_3 - t_1 = 0$ (the 0BC plane). Intersections of these planes with each other give the vectors A, B, and C. The points belonging to the vector A (where $t_1 = t_2 = t_3$) correspond to the bodies with spherical symmetry. The points lying on the vector B ($t_2 = t_3 = t_1/2$) correspond to two-dimensional disklike bodies. The points of the vector C ($t_1 = t_2$, $t_3 = 0$) correspond to one-dimensional rodlike bodies oriented along principal axis 3. The points belonging to the OBC plane characterize two-dimensional objects with the shape different from that of a perfect disk. It is more convenient to use a pair distribution function $W(k_1,k_2)$ of the ratios $k_1 = t_3/t_1$ and $k_2 = t_2/t_1$ ($k_1 \leq k_2$). This function is defined in a trihedral prism of unit height. The base of the prism is formed by the ABC plane; k_1 varies in the AC intercept, and the k_2 axis is perpendicular to the unit vector e oriented along AC. The point A (where $k_1 = k_2 = 1$) corresponds to a sphere; the point B (where $k_1 = k_2 = 1/2$) corresponds to a two-dimensional disk; and the point C (where $k_1 = 0$ and $k_2 = 1$) corresponds to a one-dimensional rod. The region of elongated ellipsoids lies near the AC intercept (for which $k_2 = 1$). Finally, the region of two-dimensional bodies, for which $k_1 + k_2 = 1$, lies on the BC intercept. Further we will discuss the behavior of the function $W(k_1,k_2)$. The dimensionless ratios $f_1 = \langle R_1^2\rangle / \langle R_g^2\rangle$, $f_2 = \langle R_2^2\rangle / \langle R_g^2\rangle$, and $f_3 = \langle R_3^2\rangle / \langle R_g^2\rangle$ ($f_1 \leq f_2 \leq f_3$), which are called shape factors, were also used to characterize the average shape of the hydrophilic micellar core.

We start from the analysis of shape factors f_i ($i = 1,2,3$). Figure 8 presents the values of f_i for the aggregates containing different number of water-like and methanol-like solvent molecules n_s, at $n_c = 64$ and $T = 1$. As seen, with an increase in n_s the shape asymmetry of the water-containing aggregate core decreases (we recall that for a spherical object one has $f_1 = f_2 = f_3 = 1/3$). For $n_c = constant$, with increasing n_s the average shape of the core must approach that of a liquid droplet formed by water molecules. It turns out, however, that the f_i values found for a droplet and a micellar core at the same total number of particles in these two objects are slightly different when n_s is not too large; the water-containing micellar core has a slightly larger asymmetry of the shape, as compared to the water droplet with the same number of particles. It is clear that this is due to the difference in interaction energies as well as to the influence of hydrophobic chain sections surrounding the hydrophilic micellar core. Mutual repulsion of the hydrophobic sections in the outer part of aggregate and partial penetration of the hydrophobic units into the core results in some increase of the core size and affects the core shape. It should be noted that the average shape asymmetry of the aggregate core becomes less noticeable as the interaction radius r_c for the core-forming potential U_c increases. At large values of r_c, the dependence of f_i on the total number of hydrophilic groups p in the core becomes very weak. The same conclusion has been drawn in Ref. [8].

The shape factors of the methanol-containing aggregates demonstrate quite different behavior, as compared to the water-containing counterparts. It is seen from Fig. 8 that in this case the values of f_i change as a function of n_s nonmonotonically and show very large fluctuations. Moreover, contrary to the water-containing system, for this system we observe an opposite global tendency with respect to the shape asymmetry: the mixed hydrophilic core becomes more asymmetric as the volume fraction of methanol molecules in the core increases. We find that the smallest component of the squared gyration radius, f_1, decreases rapidly with n_s increasing, while the middle component f_2 remains almost constant ($f_2 \approx 1/3$). This means that at the values of p studied the increase in the micellar core size is accompanied by relative contraction of the minimum (transverse) size of the core; in other words, it becomes more flat. On the other hand, when n_s is not too large, the largest component f_3 increases with n_s, indicating the increase in the maximum (longitudinal) micellar core size. By contrast, at very small n_s, an opposite tendency takes place. All these features can be clearly traced by viewing the corresponding pair distribution function $W(k_1,k_2)$ of the rations of the principal moments of the inertia tensor.

In Fig. 9 we present the set of the $W(k_1,k_2)$ functions calculated for the methanol-containing aggregates at different n_s for $n_c = 64$ and $T = 1$.

As apparent from the diagram, the most probable shape of the solvent-free core corresponds to a slightly elongated ellipsoid of revolution. Indeed, the domain of definition of the $W(k_1,k_2)$ function is localized near the AC intercept, where $k_2=1$, and shifted to the point A corresponding to spherical bodies.

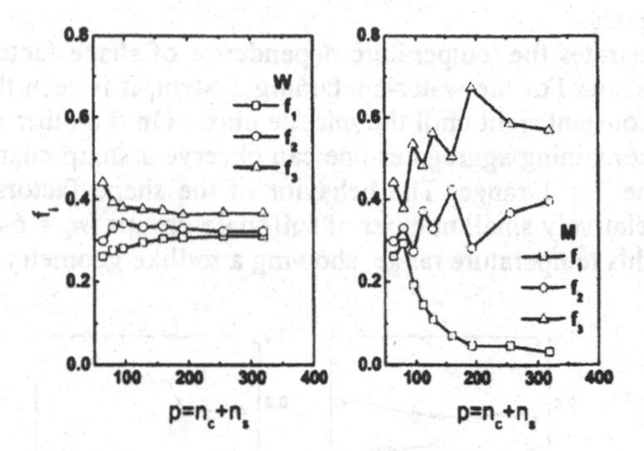

Figure 8. Shape factors f_i (i = 1,2,3) for the aggregates containing the different number of water-like (W) and methanol-like (M) solvent molecules n_s, at n_c = 64 and T = 1.

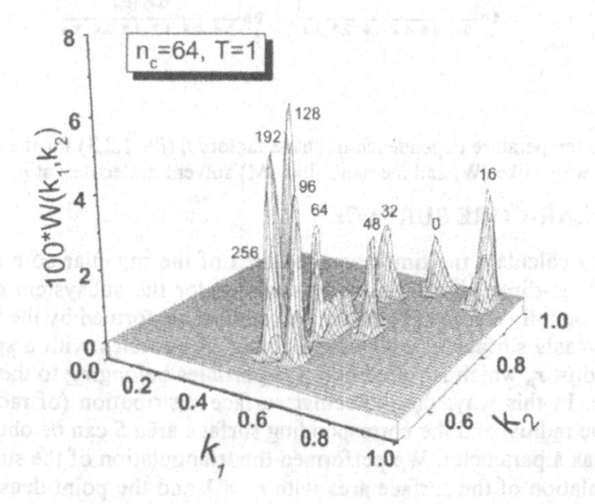

Figure 9. The $W(k_1,k_2)$ functions calculated for the methanol-containing aggregates at different n_s (shown near the corresponding peaks) for n_c = 64 and T = 1.

With increasing n_s the $W(k_1,k_2)$ peak shifts along the AC intercept to the point A. This means that the micellar-core shape approaches a spherical one as a result of contraction of the longest axis of the equivalent ellipsoid. Nevertheless, objects with axis ratios k_1 = 1 and k_2 = 1 occur extremely rarely, as do bodies with perfectly symmetrical shapes, i.e., spheres. As the value of n_s further increases, one first observes an elongation of the core along longitudinal axis (at n_s = 32) and then the core begins to flatten out (at the solvation level $n_s/n_c \gtrsim 1$), demonstrating large shape fluctuations resulting in a larger domain of definition of $W(k_1,k_2)$. When the solvent content becomes sufficiently high ($n_s/n_c \gtrsim 3$), the core shape becomes disklike. All these shape evolutions are in sharp contrast to those observed for the water-containing aggregates.

Figure 10 illustrates the temperature dependence of shape factors for both systems under discussion. For the water-containing system, it is seen that the values of f_i remain almost constant right until the micelle crash. On the other hand, in the case of the methanol-containing aggregates one can observe a sharp change in the aggregate's shape in the $T \gtrsim 1$ range. The behavior of the shape factors shows that the core formed by relatively small number of sulfonate groups ($n_c = 64$) becomes more asymmetrical in this temperature range, showing a rodlike geometry.

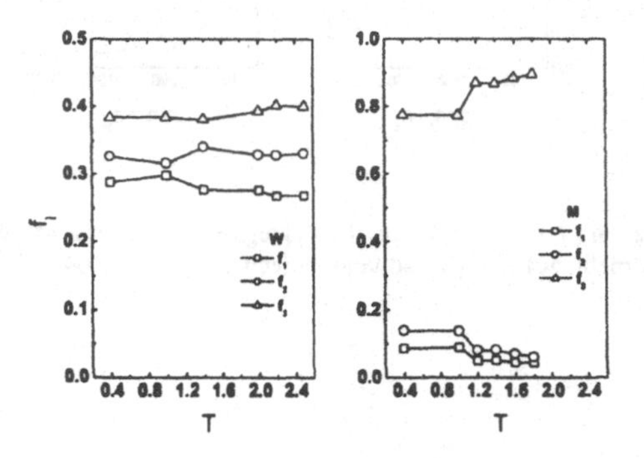

Figure 10. The temperature dependence of shape factors f_i (i = 1,2,3) for the aggregates containing $n_s = 64$ water-like (W) and methanol-like (M) solvent molecules, at $n_c = 64$.

3.5 MICELLAR-CORE SURFACE

In order to calculate the time-averaged area of the micellar-core surface, S, we constructed three-dimensional Connolly surfaces for the subsystem of hydrophilic particles. A Connolly surface [31] is the locus of points formed by the intersection of the van der Waals surface of a given ensemble of particles with a spherical probe particle of radius r_p which is rolled over the particles belonging to the system under consideration. In this way, the molecular surface distribution (of radius r_s = atom radius + probe radius) and the corresponding surface area S can be obtained with the probe radius as a parameter. We performed the triangulation of the surface and subsequent calculation of the surface area with $r_p = 1$ and the point density which was equal to 20. The value of S is measured in units of σ^2.

Figure 11 shows the log-log plot of S vs. $p = n_c + n_s$ for the systems containing water and methanol, at $n_c = 64$ and $T = 1$.

As can be seen, these dependencies are quite different for both systems. In the first case, the value of S as a function of p gives a straight line for all p's considered. On the other hand, for the methanol-containing aggregates we observe a change in the dependence in the $n_s/n_c \approx 1$ region. It is clear that in the case of a compact hydrophilic cluster the cluster volume V must be proportional to p (it should be stressed that, for the superstrong segregation regime discussed in the present study, the penetration of hydrophobic (CH_3, CF_2, etc.) groups into the hydrophilic core can be ignored). For a solid spherical object, therefore, the surface area must be proportional to $p^{2/3}$. Approximately such a behavior is observed for the water-containing aggregates, showing a nearly spherical geometry of their shape and high compactness. In the case of the methanol-containing clusters, however, similar behavior is observed only for low solvent content; at sufficiently high solvation level, the best fit to the calculated data points yields $S \propto p$. This can indicate that the cluster surface is very crumble. On the other hand, such a dependence is characteristic for the objects with cylinder-like and toroidal-shaped geometries. We postpone a more detailed discussion of these results. Here, it suffices to say that the dependence $S \propto p$ begins to be visible after strong structural rearrangements take place.

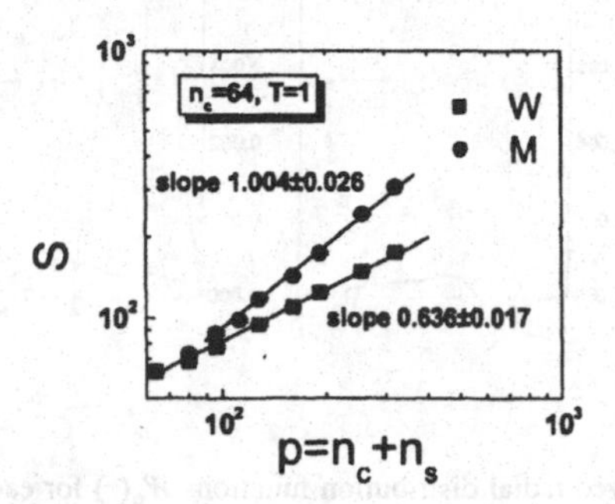

Figure 11. The log-log plot of S vs. $p = n_c + n_s$ for the systems containing water (W) and methanol (M), at $n_c = 64$ and $T = 1$.

Figure 12 shows the value of S as a function of temperature for the aggregates having $n_c = 64$ sulfonate chain fragments. We see that the temperature variation of S observed for the methanol-containing aggregate with $n_s = 64$ is more pronounced as compared to the water-containing system at the same solvent content.

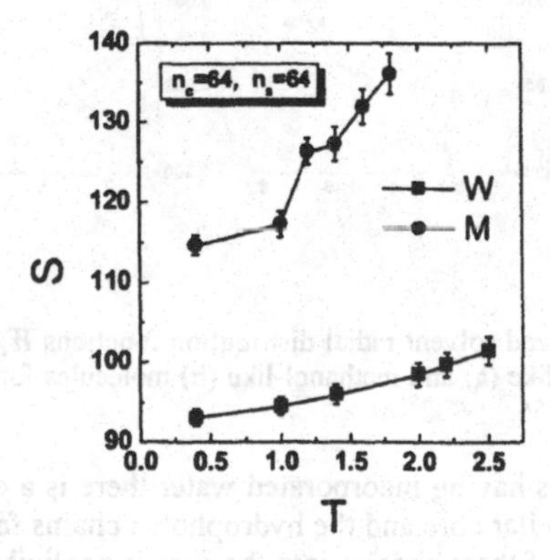

Figure 12. The value of S as a function of temperature for the aggregates having $n_c = 64$ sulfonate chain fragments and $n_s = 64$ water (W) or methanol (M) solvent molecules.

3.6 DISTRIBUTION OF PARTICLES INSIDE SOLVATED AGGREGATES

Figures 13 and 14 present the normalized radial distribution functions $W_\alpha(r)$ for each component of the solvated aggregates (the index α corresponds to hydrophobic units (c), terminal hydrophilic chain groups (t), or solvent hydrophilic groups (s)). The center of mass of the micellar core is taken as the origin.

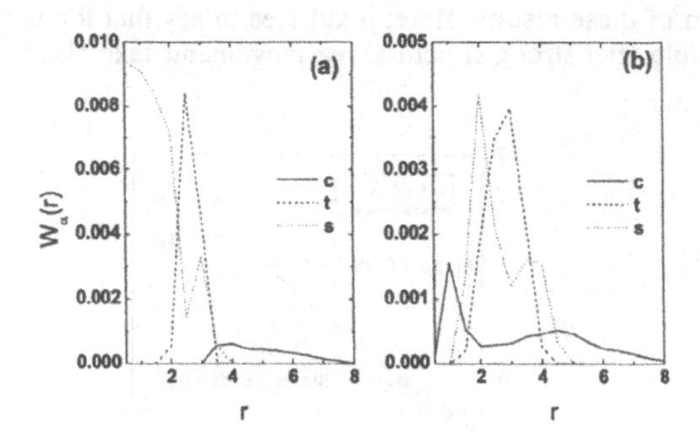

Figure 13. The normalized radial distribution functions $W_\alpha(r)$ for each component of the solvated aggregates containing water-like (a) and methanol-like (b) molecules, at $n_c = 64$, $n_s = 96$, and $T = 1$. The index α corresponds to hydrophobic units (c), terminal hydrophilic chain groups (t), or solvent hydrophilic groups (s).

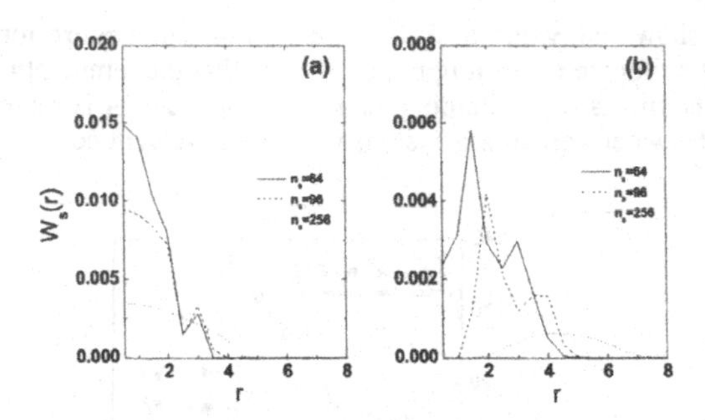

Figure 14. The normalized solvent radial distribution functions $W_s(r)$ for the solvated aggregates containing water-like (a) and methanol-like (b) molecules for $n_c = 64$ and $T = 1$, at different $n_s = 64$, 96, and 256.

For the aggregates having incorporated water there is a quite well-defined interface between the micellar core and the hydrophobic chains forming the outer corona. The interpenetration of these species into the core is negligible. This is a characteristic feature of the strong and superstrong segregation regimes [27,32]. Next, we can see an inner (central) region filled with absorbed water molecules. This region is surrounded by an outer layer which contains mainly polar sulfonate groups. Thus, for a water-containing aggregate there are quite distinct interfaces between the components. This result is consistent with the recent theoretical predictions for the superstrong segregation regime [12,32,33]. As the size of the inner part of the bicomponent core increases, the relative thickness of the outer layer decreases. In the case of large excess of water molecules over hydrophilic chain groups, both components are mixed in the outer layer of the core.

For the methanol-containing system we observe a quite different behavior. In this case, the internal structure of the sufficiently large bicomponent aggregates looks more homogeneous. In particular, as seen from Fig. 13, the distribution of hydrophobic groups inside the aggregate with $n_s = 96$ is almost homogeneous. However, near the center of the aggregates, one observes a well-pronounced dip in the hydrophilic-group density distribution (see Figs. 13 and 14). This dip arises at the solvation level $n_s/n_c \gtrsim 1$ and is absent for lower solvent content (see Fig. 14). Thus, one can conclude that the sufficiently large methanol-containing multiplets have an inner cavity which is filled with hydrophobic groups. At the same time, the hydrophilic groups are pushed out toward the outer region of the aggregates. Such a density distribution may correspond to curved – toroidal or vesicular – geometries.

3.7 SCATTERING FUNCTIONS

The SAXS and SANS spectra of solvated perfluorinated ionomer membranes [34–40] demonstrate the existence of a small-angle scattering maximum, called "ionomer peak". This peak can be described as an oscillation of the form-factor which has interparticle origin and is associated with the internal structure of the ionic clusters considered as noninteracting particles with a specific spatial distribution.

Scattering experiments allow measurements of the scattered intensity from molecular systems over a wide range of wave numbers q. Experimentally

$$I(q)/I(0) = Nw(q)S(q) \tag{7}$$

where $I(q)$ and $I(0)$ denote the scattered intensity at q and $q = 0$, respectively, and N is the total number of scattering centers in a system. The function $w(q)$ is the particle scattering factor ("form-factor") which characterizes scattering from individual particles and $S(q)$ is a collective static structure factor associated with the presence of intermolecular correlations. In this subsection, the discussion is confined to the intramicellar form-factor $w(q)$. Note that this function is very sensitive to the equilibrium structure of complex systems and is related by Fourier transform to appropriate pair distribution function of distances between scattering centers, $w(r)$. For a system of N point scatterers, one has

$$w(q) = \frac{1}{N}\sum_{\alpha,\beta}\left\langle \exp(\mathbf{iqr}_{\alpha\beta})\right\rangle \tag{8}$$

where $\mathrm{i} = \sqrt{-1}$ and $\mathbf{r}_{\alpha\beta}$ is the vector connecting scattering centers α and β belonging to the system considered. On averaging over all space orientations Eq. (8) reduces to

$$w(q) = \frac{1}{N}\sum_{\alpha,\beta}\left\langle w_{\alpha\beta}(q)\right\rangle \tag{9}$$

where $w_{\alpha\beta}(q) = \sin(qr_{\alpha\beta})/(qr_{\alpha\beta})$ is the corresponding site-site contribution to the total scattering function and $\langle\bullet\rangle$ means an ensemble average. At low q, in the so-called Guinier regime ($qR_g << 1$), Eq. (9) further simplifies to $Nw(q) = 1-q^2R_g^2/3$. It is well-known, however, that this last expression does not adequately represent the form of $w(q)$ at high and intermediate q (when $qR_g > 1$).

In general, calculations of $w(q)$ require the numerical evaluation of $w(q)$ through Eq. (9) for the large number of system configurations. It is a very time-consuming stage of simulation, because the direct calculation of $w(q)$ needs $\sim N^2$ intersite distances $r_{\alpha\beta}$ for each configuration, the use of sufficiently fine grid for q, and averaging over the large number of configurations of the system. Taking into account these circumstances, to find $w(q)$ we used the very efficient procedure based on the fast Fourier transformation (FFT) technique (see Appendix B). In this subsection we assume that scattering centers are located on all hydrophilic groups.

Figure 15 shows some example of the calculations. Calculated values of $w(q)$ can conveniently be presented with the so-called Kratky plot of $q^2w(q)$ vs. q. Note that we are interested only in distances larger than the UA size, or $q \lesssim \pi/\sigma$.

Shown in Fig. 15 are the $w(q)$ functions obtained for the following three systems: (1) the water-containing aggregate with $n_c = 64$ and $n_s = 96$, (ii) the methanol-containing aggregate with the same values of n_c and n_s, and (iii) a liquid droplet consisting of 64 sulfonate groups and 96 water-like particles. First of all, we observe that the internal structures of the core of the water-containing cluster and the corresponding equivalent droplet are practically identical for any wave numbers. This fact indicates that the effect of hydrophobic chain sections on the internal arrangement of hydrated ionic aggregates is very weak in the superstrong segregation regime. This feature has been mentioned above. On the other hand, there are noticeable differences between these functions and those calculated for the micellar core with incorporated methanol molecules. For the latter system, one can observe the presence of two characteristic peaks in the small-q region.

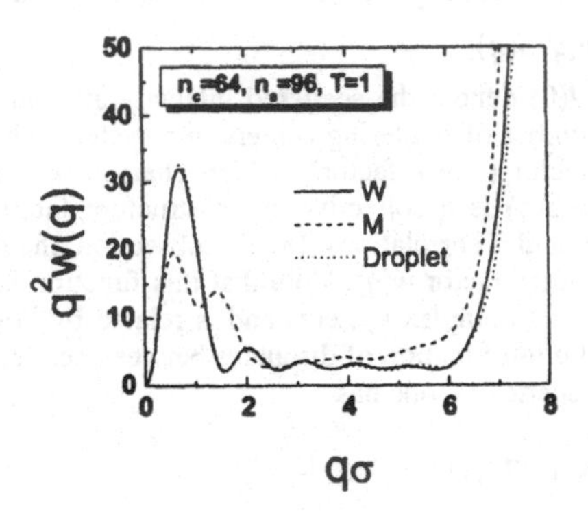

Figure 15. The $w(q)$ functions obtained for the water-containing aggregate (W) with $n_c = 64$ and $n_s = 96$, the methanol- containing aggregate (M) with the same values of n_c and n_s, and a liquid droplet consisting of 64 sulfonate groups and 96 water-like particles at $T = 1$.

3.8 VISUAL ANALYSIS

In this subsection, we will discuss the general morphological features of the solvated aggregates. Computer simulations provide a significant advantage in these studies, including geometrical information about the three-dimensional organization of the self-assembling ionic copolymers, as well as the morphology of the system. Furthermore, the graphical output from the simulation allows a direct visualization of these complex structures. In this way, we performed a visual analysis of molecular images obtained in the simulation.

Below, the following atomic color scheme is used: sulfonate groups are depicted in light tone, solvent hydroxyl group in dark tone, and hydrophobic groups (including carbon or fluorine) in light-gray tone. For hydrophilic micellar cores, we constructed the Connolly surfaces which provide a more direct way of visualizing of the system organization; they are shown in medium gray tone or depicted using a color scheme that is related to the proximity of the surface to the corresponding atomic group.

Shown in Figs. 16 and 17 are examples of the snapshots obtained for the systems with the different number of water and methanol molecules, n_s, at $n_c = 64$ and $T = 1$. As seen from Fig. 16, the water-containing multiplets swell with n_s increasing and maintain nearly spherical shape for any n_s. At the same time, in the case of the methanol-containing aggregates, we distinctly see the formation of toroidal-shaped structures at the solvation level $n_s/n_c \gtrsim 1$ (Fig. 17).

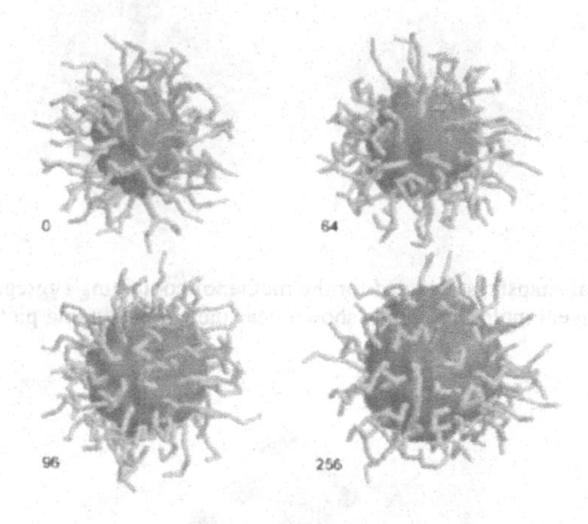

Figure 16. Typical snapshots obtained for the water-containing aggregates with the different number of solvent molecules, n_s (as shown near the corresponding pictures), at $n_c = 64$ and $T=1$.

In Fig. 18 we present the Connolly surfaces obtained for the micellar cores with $n_c = 64$ and $n_s = 96$ using a small particle probe radius of 0.4; the remaining hydrophobic groups are hidden for clarity. Figure 19 shows the Connolly surface of the large methanol-containing cluster with $n_c = 64$ and $n_s = 256$; here, the methanol CH_3 groups are presented, but the hydrophobic groups of sulfonate fragments are hidden.

As seen, the cluster surface of the water-containing core is rather smooth while the toroidal-shaped methanol-containing clusters are characterized by a high degree of surface irregularities and look as very crumble.

From the direct visualization and also from the analysis of the density radial distribution functions (see Figs. 13 and 14) we can conclude that the hole in the methanol-containing toroidal-shaped aggregates is partly filled with the hydrophobic solvent CH_3 groups while the hydrophobic chains belonging to the sulfonate fragments are directed to the micelle exterior. Such a spatial organization provides a more uniform compact packing of the hydrophobic groups. It should be noted that the locally-cylindrical toroidal structure is more favorable as compared to alternative bilayer-shaped (disklike) microstructures where strong interchain repulsion takes place, resulting in short-ranged orientational correlations and chain stretching (see, e.g., Ref. [27]).

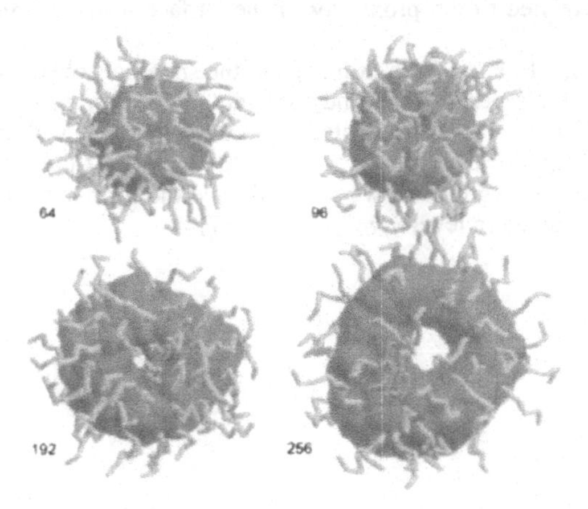

Figure 17. Typical snapshots obtained for the methanol-containing aggregates with the different number of solvent molecules, n_s (as shown near the corresponding pictures), at n_c =64 and T = 1.

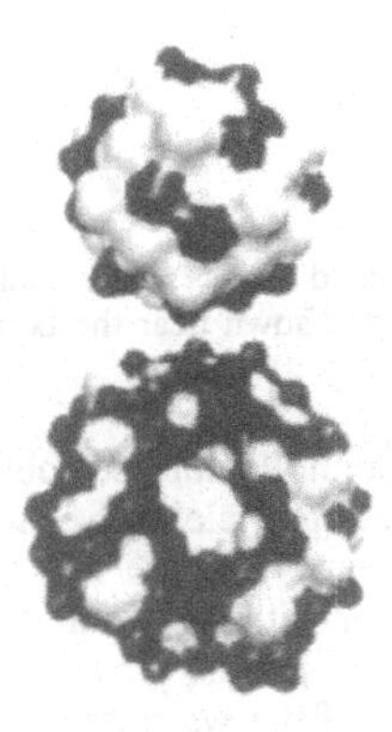

Figure 18. The Connolly surfaces obtained for the water-containing (upper panel) and methanol-containing (lower panel) micellar cores with n_c = 64 and n_s = 96 using a small particle probe radius of 0.4; the remaining hydrophobic groups are hidden for clarity.

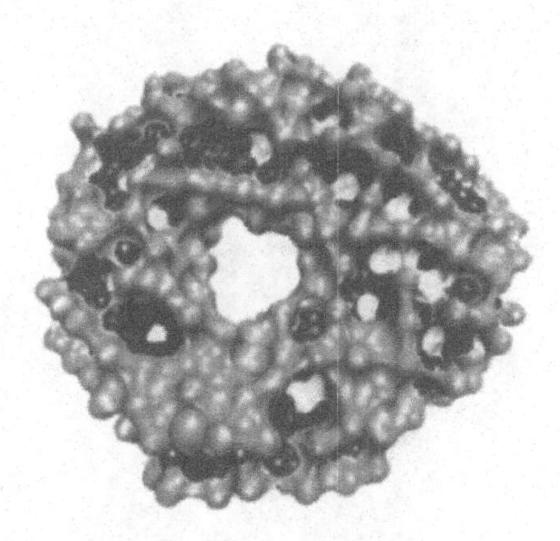

Figure 19. The Connolly surface of the methanol-containing cluster with $n_c = 64$ and $n_s = 256$. In this picture, methanol CH_3 groups are presented, but the hydrophobic groups of sulfonate fragments are hidden.

3.9 EVOLUTION OF TOROIDAL-SHAPED STRUCTURES

In this subsection, we present some additional data concerning the behavior of the methanol-containing aggregates forming toroidal structures. In particular, we are interesting in the surface area to volume ratio for such systems.

The volume, V, and surface area, S, of the torus are given by

$$V = 2\pi^2 R r^2 \tag{10}$$

$$S = 4\pi^2 R r \tag{11}$$

where the values of R and r are defined in Fig. 20.

Hence, the radii R and r as functions of V and S are given by the following relationship

$$R = S^2/8\pi^2 V \tag{12}$$

$$r = V/S \tag{13}$$

Let us suppose that, for the model system under consideration, the volume of a micellar core is proportional to the total number of particles, $p = n_c + n_s$, forming the core. Next, let us define the following two functions of p

$$\phi_1(p) = S^2/p \tag{14}$$

$$\phi_2(p) = p/S \tag{15}$$

As seen from Eqs. (12) and (13), $R \propto \phi_1(p)$ and $r \propto \phi_2(p)$. These functions calculated for the methanol-containing aggregates are shown in Fig. 21 (the values of S were calculated via the Connolly surfaces with $r_p = 1$ as it has been described above). We find that $\phi_2(p) \approx$ *constant* and $\phi_1(p) \propto p$ at $p \gtrsim 10^2$. From these results one can conclude that an increase in the micellar-core volume is accompanied by an increase in the large torus radius R, whereas the small torus radius r remains almost constant.

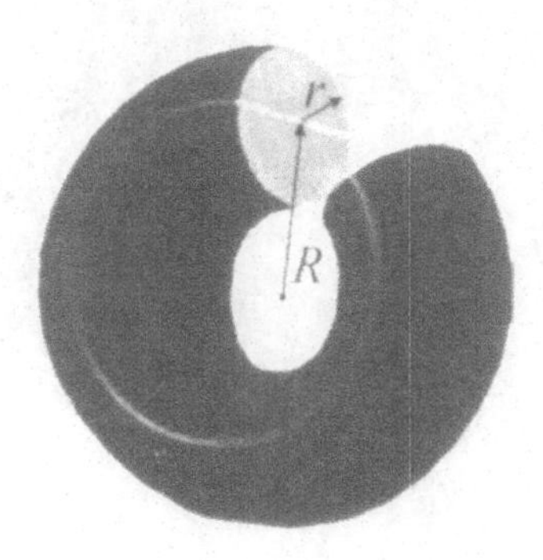

Figure 20. Parameters of torus.

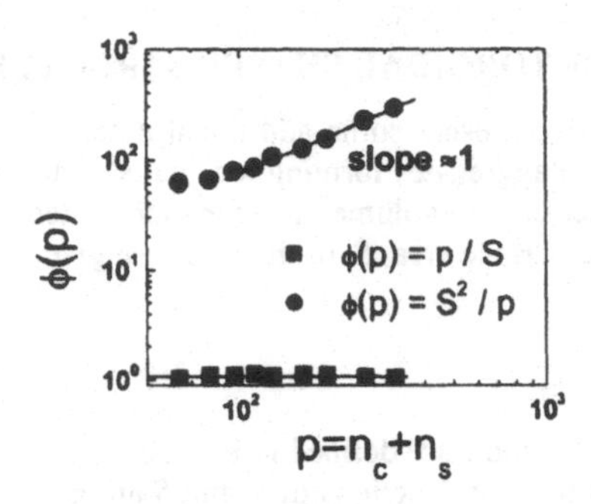

Figure 21. The $\phi_2(p)$ and $\phi_1(p)$ functions for the methanol-containing aggregates, at $n_c = 64$ and $T = 1$.

3.10 DEPENDENCE ON THE ALCOHOL-CHAIN LENGTH

In addition to the methanol-containing system discussed above, we investigated mixed aggregates containing alcohols $H(CH_2)_nOH$ with longer hydrocarbon chain, up to $n = 7$.

The results of the calculation illustrating the temperature stability of these bi-component aggregates are shown in Fig. 22 in the form of the T–N_s state diagram where $N_s = n_{CH_2} + 1_{OH}$ and $n_c = n_s = 64$. It is seen that the micelle stability decreases as the solvent chain length increases. This is due to steric factors which are a result of repulsion between hydrocarbon chains of the alcohols.

Figure 23(a) shows the time-averaged squared gyration radius R_g^2 of the solvent-containing core as a function of N_s, at $T = 1$ and $n_c = n_s = 64$. One can observe an increase in R_g^2 with N_s increasing. It turns out that this is accompanied by the elongation of the aggregates; for sufficiently large N_s, we observe a polymorphic transition in these aggregates from near-spherical toward cylindrical microstructures.

Such a behavior is clearly seen in Fig. 23(b), where the corresponding shape factors are presented.

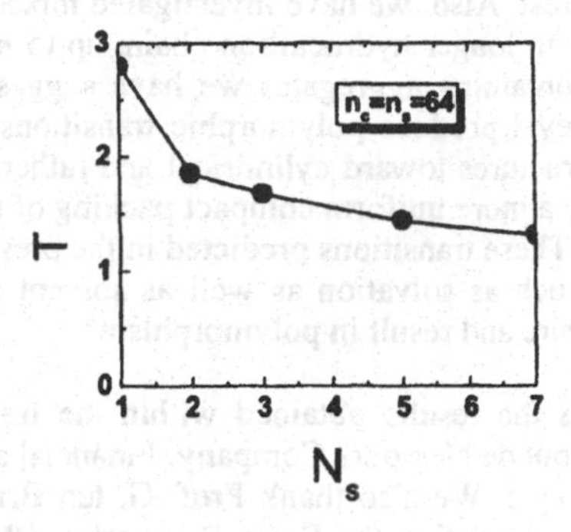

Figure 22. The T–N_s state diagram at $n_c = n_s = 64$.

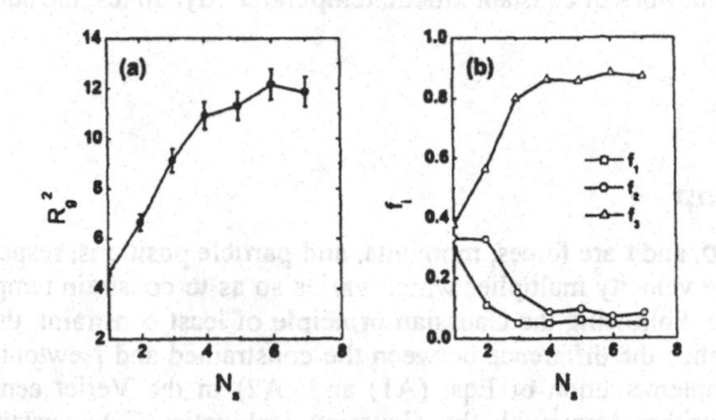

Figure 23. The time-averaged squared gyration radius R_ξ^2 (a) and shape factors f_i (b) of the solvent-containing micellar core as functions of N_s, at $T = 1$ and $n_c = n_s = 64$.

4. CONCLUDING REMARKS

Polymorphic changes in solvent-containing ionomer systems, including water-swollen and alcohol-swollen Nafion® membranes, are supposed to cause the transient formation of solvated complexes, pores, or channels involved in many aspects of spatial organization and function (e.g., transport properties) of these systems. The structural origins of these transitions are poorly understood at the molecular level. This lack of detail primarily reflects the complex (dynamic) character of solvent-containing aggregates, which limit X-ray or neutron scattering experiments to the determination of time-averaged and spatial-averaged molecular features. Here we have described molecular dynamics computer simulations that provide an alternative approach for studying the detailed behavior of self-assembling ionomers. Our attention here has been focused on the effect of different solvent additives on the equilibrium structure of micellar aggregates. Taking into account the practical importance of methanol membrane fuel cells, methanol-containing systems have been the subject of our primary interest. Also, we have investigated mixed aggregates containing alcohols $H(CH_2)_nOH$ with longer hydrocarbon chain, up to $n = 7$. In the case of the bicomponent solvent-containing aggregates we have suggested how events occurring at the molecular level produce polymorphic transitions in these mixed aggregates from spherical structures toward cylindrical and rather exotic toroidal micelle structures, stabilized by a more uniform compact packing of the hydrophobic groups in the micelle exterior. These transitions predicted in the present study illustrate how intermolecular forces such as solvation as well as solvent chain length can affect overall aggregate structure and result in polymorphism.

Acknowledgment

This paper presents the results obtained within the framework of the project sponsored by E.I. Du Pont de Nemours Company. Financial support of this company is gratefully acknowledged. We also thank Prof. G. ten Brinke for helpful discussions and the Russian Foundation for Basic Researches (Projects ## 01-03-32154 and 01-03-32152) for financial support.

APPENDIX A. CONSTANT KINETIC TEMPERATURE DYNAMICS

In the framework of constant kinetic temperature dynamics, the equations of motion are

$$\dot{\mathbf{r}} = \mathbf{p}/m \tag{A1}$$

$$\dot{\mathbf{p}} = \mathbf{f} - \alpha\mathbf{p} \tag{A2}$$

where $\mathbf{f}$, $\mathbf{p}$, and $\mathbf{r}$ are forces, momenta, and particle positions, respectively, and α is a Lagrange velocity multiplier which varies so as to constrain temperature to the desired value. Following the Gaussian principle of least constraint, the multiplier α should minimize the difference between the constrained and Newtonian trajectories [29]. The implementation of Eqs. (A1) and (A2) in the Verlet central-difference leapfrog algorithm form with the Gaussian isokinetic (GI) constant temperature scheme can be realized in the following way

$$\dot{\mathbf{r}}'(t) = \dot{\mathbf{r}}(t - \Delta t/2) + \mathbf{f}\Delta t/2m \tag{A3}$$

$$\dot{\mathbf{r}}(t) = \dot{\mathbf{r}}'(t)/(1 + \alpha\Delta t/2) \tag{A4}$$

where α is given by

$$\alpha = \frac{\sum_{i=1}^{N} \dot{\mathbf{r}}_i(t) \cdot \mathbf{p}_i(t) - \beta}{\sum_{i=1}^{N} m_i \dot{\mathbf{r}}(t) \cdot \dot{\mathbf{r}}(t)} \tag{A5}$$

and β is defined by the time derivative of the kinetic energy K; β can be written as

$$\dot{K} = \sum_{i=1}^{N} \dot{\mathbf{p}}\mathbf{p}/m_i = \beta \qquad \text{(A6)}$$

During production runs β = 0, but prior to this, it is adjusted several times to provide the desired temperature [30]. To this end, Eqs. (A3)–(A6) are iterated several times to establish a self-consistent value for α(t). Having established α(t) then one can find the velocities

$$\dot{\mathbf{r}}(t+\Delta t/2) = c_1\dot{\mathbf{r}}(t-\Delta t/2) + c_2\mathbf{f}(t)\Delta t/m \qquad \text{(A7)}$$

where

$$c_1 = 1 - \frac{\alpha\Delta t}{1+\alpha\Delta t/2};\ c_2 = 1 - \frac{\alpha\Delta t/2}{1+\alpha\Delta t/2} \qquad \text{(A8)}$$

Having $\dot{\mathbf{r}}(t+\Delta t/2)$ we obtain the particle positions

$$\mathbf{r}(t+\Delta t) = \mathbf{r}(t) + \dot{\mathbf{r}}(t+\Delta t/2)\Delta t \qquad \text{(A9)}$$

APPENDIX B. CALCULATION OF FORM-FACTOR USING FFT

Let us group the $r_{\alpha\beta}$ values corresponding to any given system configuration into histogram h with elements h_i (i = 1,2,...,L) equal to the number of distances within the range [$i\Delta r$, $(i+1)\Delta r$], where Δr is the step size. Then, we have $w(r_i) = 2h_i/[4\pi N\Delta r(i\Delta r)^2]$, where $w(r_i)$ is the value of the step-wise density distribution function $w(r)$ at the distance $r_i = i\Delta r$. Thus, we have

$$w(q_j) = 1 + \frac{2}{N}\sum_{i=2}^{L-1} h_i \frac{\sin[\pi ij/L]}{(\pi ij/L)},\ (j=1,2,\ldots,L) \qquad \text{(B1)}$$

where $w(q_j)$ is the corresponding value of the discretized function of $w(q)$ at $q_j = j\Delta q$ ($\Delta q = \pi/\Delta rL$). To perform the FFT, $w(r_i) \xrightarrow{FFT} w(q_j)$, one should: (i) multiply all values $w(r_i)$ by $i\Delta r$; (ii) set $w(r_1)$ = 0; (iii) transform the obtained array $w'(r_i)$: $w'(r_i) \xrightarrow{FFT} w'(q_j)$; (iv) convert $w'(q_j)$ to $w(q_j)$: $w(q_j) = 1 + w'(q_j)/(j\Delta q)$.

REFERENCES

1. Yeager, H.L. and Eisenberg, A. (1982) Introduction, in A. Eisenberg and H.L. Yeager (eds.), *Perfluorinated Ionomer Membranes*., ACS Symp. Series 180; Amer. Chem. Soc., Washington, DC.
2. Tant, M.R., Mauritz, K.A., and Wilkes, G.L. (1997) *Ionomers: Synthesis, Structure, Properties and Applications*, Blackie Academic and Professional, London, 514 pp.
3. Polymer Electrolyte Fuel Cells (1995) *Electrochim. Acta* 40, 283.
4. *Ionomers: Characterization, Theory and Applications* (1996) S. Schlick (ed.), CRC Press, Boca Raton, FL.
5. Norby T. (1999) Solid-state protonic conductors: Principles, properties, progress and prospects, *Solid State Ionics* 125, 1-11.
6. Gebel, G. and Moore, R.B. (2000) Small angle scattering study of short pendant chain perfluorosulfonated ionomer membranes, *Macromolecules* 33, 4850-4859.

7. Semenov, A.N., Nyrkova, I.A., and Khokhlov, A.R., (1996) Statistics and dynamics of ionomer systems, in S. Schlick (ed.), *Ionomers: Characterization, Theory and Applications*, CRC Press, Boca Raton, FL, Chapter 11, pp. 251-272.
8. Khalatur, P.G., Khokhlov, A.R., Mologin, D.A., and Zheligovskaya, E.A. (1998) Computer simulation studies of aggregates of associating polymers: Influence of low-molecular-weight additives solubilizing the aggregates, *Macromol.Theory Simul.* 7, 299-323.
9. Helfand, E. and Wasserman, Z.R. (1982) Microdomain structure and the interface in block copolymers, in I. Goodman (ed.), *Developments in Block Copolymers*, Applied Science, London, vol. 1.
10. Fredrickson, G.H. and Bates, F.S. (1997) Design of bicontinuous polymeric microemulsions, *J. Polymer Sci. B*, 35, 2775-2781.
11. Fredrickson, G.H. and Bates, F.S. (1996) Dynamics of block copolymers: Theory and experiment, *Annu. Rev. Mater. Sci.* 26, 501-550.
12. Nyrkova, I.A., Khokhlov, A.R., and Doi, M. (1993) Microdomains in block-copolymers and multiplets in ionomers: Parallels in behavior, *Macromolecules* 26, 3601-3611.
13. Khokhlov, A.R. and Philippova, O.E. (1996) Self-assemblies in ion-containing polymers, in S. Webber (ed.), *Solvents and Polymer Self-Organization*, Kluwer, New York, pp. 5-45.
14. Aldebert, P., Dreyfus, B., and Pineri, M. (1986) Small-angle neutron scattering of perfluorosulfonated ionomers in solution, *Macromolecules* 19, 2651-2653.
15. Aldebert, P., Dreyfus, B., Gebel, G., Nakamura, N., Pineri, M., and Volino, F. (1988) Rod-like micellar structures in perfluorinated ionomer solutions, *J. Phys. (Paris)* 49, 2101-2109.
16. Rebrov, A.V., Ozerin, A.N., Svergun, D.I., Bobrova, D.L., and Bakeyev, N.F. (1990) Study of aggregation of macromolecules of perfluorosulfinated ionomer in solution by the small-angle X-ray-scattering method, *Polym. Sci. U.S.S.R.* 32, 1593-1599.
17. Hilger, C., Drager, M., and Stadler, R. (1992) Molecular origin of supramolecular self-assembling in statistical copolymers - Cooperative structure formation by combination of covalent and association chain polymers, *Macromolecules* 25, 2498-2505.
18. Hilger, C. and Stadler, R. (1992) Cooperative structure formation by directed non-covalent interactions in an unpolar polymer matrix: Differential scanning calorimetry and small-angle X-ray scattering, *Macromolecules* 25, 6670-6677.
19. Lu, X., Steckle, W.P., and Weiss, R.A. (1993) Morphological studies of a triblock copolymer ionomer by small angle X-ray scattering, *Macromolecules* 26, 6525-6530.
20. Kim, J.-S., Jackman, J., and Eisenberg, A. (1994) Filler and percolation behaviour of ionic aggregates in styrene-sodium methacrylate ionomers, *Macromolecules* 27, 2789-2803.
21. Loppinet, B., Gebel, G., and Williams, C.E. (1997) Small-angle scattering study of perfluorosulfonated ionomer solutions, *J. Phys. Chem.* B 101, 1884-1892.
22. Loppinet, B. and Gebel, G. (1998) Rodlike colloidal structure of short pendant chain perfluorinated ionomer solutions, *Langmuir* 14, 1977-1983.
23. Khalatur, P.G., Talitskikh, S.K., and Khokhlov, A.R. (2002) Structural organization of water-containing Nafion: The integral equation theory, *Macromol. Theory Simul.* 11, 566–586.
24. Mologin, D.A., Khalatur, P.G., and Khokhlov, A.R. (2002) Structural organization of water-containing Nafion: A cellular-automaton-based simulation, *Macromol. Theory Simul.* 11, 587–607.
25. (a) Breneman, C.M. and Wiberg, K.B. (1990) Determining atom-centered monopoles from molecular electrostatic potentials - The need for high sampling density in formamide conformational-analysis, *J. Comput. Chem.* 11, 361-373; (b) Wiberg,

K.B., Hadad, C.M., Breneman, C.M., Laidig, K.E., Murcko, M.A., and LePage, T.J. (1991) How do electrons respond to structural changes? *Science* 252, 1266-1272.

26. Elliot, J.A., Hanna, S., Elliot, A.M.S., and Cooley, G.E. (1999) Atomistic simulation and molecular dynamics of model systems for perfluorinated ionomer membranes, *Phys. Chem. Chem. Phys.* 1, 4855-4863.
27. Khalatur, P.G., Khokhlov, A.R., Nyrkova, I.A., and Semenov, A.N. (1996) Aggregation processes in self-associating polymer systems: Computer simulation study of micelles in the superstrong segregation regime, *Macromol. Theory Simul.* 5, 713-748.
28. Khalatur, P.G., Papulov, Yu.G., and Pavlov, A.S. (1986) The influence of solvent on the static properties of polymer-chains in solution - A molecular-dynamics simulation, *Molec. Phys.* 58, 887-895.
29. Allen, M.P. and Tildesley, D.J. (1987) *Computer Simulation of Liquids*, Claredon Press, Oxford.
30. Evans, D.J and Morriss, G.P. (1990) *Statistical Mechanics of Nonequilibrium Liquids*, Academic Press, London.
31. Connolly, M.L. (1983) Solvent-accessible surfaces of proteins and nucleic acids, *Science* 221, 709-713.
32. Semenov, A.N., Nyrkova, I.A., and Khokhlov, A.R. (1995) Polymers with strongly interacting groups: Theory for non-spherical multiplets, *Macromolecules* 28, 7491-7499.
33. Khalatur, P.G., Khokhlov, A.R., Nyrkova, I.A., and Semenov, A.N. (1996) Aggregation processes in self-associating polymer systems: A comparative analysis of theoretical and computer simulation data for micelles in the superstrong segregation regime, *Macromol. Theory Simul.* 5, 749-756.
34. Marx, C.L., Caulfield, D.F., and Cooper, S.L. (1973) Morphology of ionomers, *Macromolecules* 6, 344-353.
35. Yarusso, D.J. and Cooper, S.L. (1983) The microstructure of ionomers: Interpretation of small-angle X-ray scattering data, *Macromolecules* 16, 1871-1879.
36. Ding, Y.S., Hubbard, S.R., Hodgson K.O., Register, R.A., and Cooper, S.L. (1988) Anomalous small-angle X-ray scattering from a sulfonated polystyrene ionomer, *Macromolecules* 21, 1698-1703.
37. Ishioka, T. and Kobayashi, M. (1990) Small-angle X-ray-scattering study for structural changes of the ion cluster in a zinc salt of an ethylene methacrylic-acid ionomer on water-absorption macromolecules, *Macromolecules* 23, 3183-3186.
38. Kutsumizu, S., Nagao, N., Tadano, K., Tachino, H., Hirasawa, E., and Yano, S. (1992) Effects of water sorption on the structure and properties of ethylene ionomers, *Macromolecules* 25, 6829-6835.
39. Chu, B., Wang, J., Li, Y., and Peiffer, D.G. (1992) Ultra-small-angle X-ray-scattering of a zinc sulfonated polystyrene, *Macromolecules* 25, 4229-4231.
40. Gebel, G. and Lambard, J. (1997) Small-angle scattering study of water-swollen perfluorinated ionomer membranes, *Macromolecules* 30, 7914-7920.

COMPUTER SIMULATION OF MESOGENS WITH AB INITIO INTERACTION POTENTIALS

An application to oligophenyls

I. CACELLI, G. CINACCHI, G. PRAMPOLINI AND A. TANI
Dipartimento di Chimica e Chimica Industriale,
Università di Pisa via Risorgimento 35, I-56126 Pisa, Italy

1. Introduction

Computer simulation methods such as Monte Carlo (MC) and molecular dynamics (MD) have proven to be a powerful tool to study liquid crystals, despite the computational problems due to the wide range of length and time scales that characterizes their dynamics. The latter feature, combined with the complex nature of typical liquid crystal forming molecules, has suggested to adopt rather simplified descriptions of the intermolecular interactions. After the Lebwohl-Lasher lattice model [1], where even translational freedom was missing, anisotropic interaction models have been considered, either with hard [2] or continuous potential functions, the most widely employed being the Gay-Berne model [3, 4, 5]. In all these cases, molecules are considered single-site interaction centers and no molecular flexibility is taken into account. Despite their simplicity, these models have proven valuable to study both the general structure-property relationships and the basic features responsible of the liquid crystal behavior. However, their simplicity becomes a drawback when the interest focuses on a specific liquid crystal, with a well defined molecular composition.

Actually, along with the impressive increase of computational capabilities, there has been a growing interest, in the last ten years or so [6]-[11], in computer simulations of liquid crystals with models of higher, physical, realism. Thus, the problem becomes the construction of potential functions able to accurately describe the interactions between large molecules. This is a particularly hard task as the complex phase behavior of mesogenic materials is the result of a delicate balance between subtle energetic (including electrostatic, dispersion, induction and excluded volume interactions) and entropic effects of positional, orientational and conformational nature [11]. In fact, seemingly

J. Samios and V.A. Durov (eds.), Novel Approaches to the Structure and Dynamics of Liquids: Experiments, Theories and Simulations, 427–454.

modest changes of chemical structure can lead to significantly different positional and orientational organization in a given phase and hence to largely different macroscopic properties. This means that it is not possible to separate weak attractive interactions from excluded volume repulsive forces, as effectively as in the perturbation theory approach to high density simple liquids, if the model is to keep chemical specificity.

Traditionally, there have been two main routes towards 'realistic' potentials for computer simulations. The first, the empirical route, relies on a variety of experimental data to fix the parameters of a chosen potential function. The second route samples the potential energy surface (PES) of the dimers with *ab initio* calculations to collect a sufficiently large number of configurations with their interaction energy values. The latter is the database to be matched adjusting the parameters of the model. A fairly straightforward application of this approach is currently feasible for most small-to-medium size molecules at a rather high level of quantum chemical sophistication, but so far, out of question for large, complex systems as potentially mesogenic molecules. A sort of intermediate way also exists to build potentials, successfully pursued by various groups, leading to a number of widely used force fields, (see [11] for a discussion and references). In this case, an initial set of parameters derived from *ab initio* calculations is modified until the simulation results reproduce their experimental counterpart for a few properties, e.g. density and internal energy. It turns out that the best results from the latter approach are obtained if the potentials are tailored on a specific class of molecules, showing that transferability of parameters for atom-atom interactions is far from perfect. In addition, the above mentioned force fields do not provide data for all classes of materials of interest. As a consequence, it seems that it may be necessary to develop accurate potentials designed for the specific application at hand. Quantum chemical approaches look as the most sensible choice, especially when experimental data are not available and/or one wishes to set up a procedure with predictive capabilities. In principle, the Car-Parrinello method [12] has the potentiality of solving the problem. However, its actual implementation based on local functionals appears problematic for systems where dispersion forces are dominant. Moreover, its application to simulation of long-time dynamics of large and complex molecular systems remains a future objective. The same reasons rule out a straightforward application of *ab initio* calculations to mesogenic molecules (route two to potentials).

To circumvent the problem, we have recently proposed an approach [13] whereby interaction potentials for mesogenic molecules can be constructed from *ab initio* calculations. The basic idea behind this approach is that a very large number of potentially mesogenic molecules can be thought as composed of a rather small number of moieties, e.g. phenyl rings and hydrocarbon chains.

The smaller size (and possible higher symmetry) of the latter fragments makes them amenable to accurate *ab initio* calculations at a reasonable computational cost. From the PES between these 'building blocks' that of the whole dimer can be obtained through a suitable recombination procedure. Among the pro's of this method is that the fragments we consider are still fairly large which might reduce problems of transferability. In addition, the PES we obtain can be used to best fit models of different complexity and 'realism'. The single PES underlying the models makes the comparison of the results they produce in simulations a consistent test of their capabilities. This is of particular interest as intermediate level or 'hybrid' models are still a sensible choice to overcome the computational problems due to the long times required for equilibrating mesophases with 'large' systems of particles (see [14] for a thorough discussion of these issues). Moreover, the comparison of results from different levels of modeling can also help to shed light on the physical origin of the observed behavior.

As a test case for our approach we have chosen the series of *p*-phenyls, with a number of rings from two to five. These systems offer unique advantages from our point of view, as their structure is very simple, with just one type of fragment (the phenyl ring) and basically a single internal degree of freedom, the torsional inter ring angle. Yet the higher members of the series are able to form mesophases, such as a nematic phase for the *p*-quinquephenyl and even a smectic phase for the six ring member. Our main goal is to study the phase diagram of these systems, and in particular to assess the level of modeling required to observe the nematic phase of *p*-quinquephenyl.

In this Chapter we describe how to decompose a molecule into fragments and how the PES of the dimer can be recombined from that of the fragment pairs (Section 2). The model potentials adopted are also discussed in Section 2, together with some computational details of the MC simulations carried out. The results we obtain are presented in Section 3, while Section 4 collects the main conclusions and suggestions for future developments.

2. Theory

2.1. THE FRAGMENTATION APPROACH

Our approach relies on the assumption that the interaction energy of a dimer can be approximated to a good accuracy as a sum of energy contributions between each pair of fragments into which the molecule can be decomposed. The basic criterion behind the fragmentation scheme, is that the ground state electronic density around the atoms of each fragment has to be as close as possible to that around of the same atoms in the whole molecule.

Therefore, if we consider a molecule AB as composed of two fragments A and B linked by a single bond, we can symbolically write

$$A{-}B = AH_a + H_bB - H_a - H_b \qquad (1)$$

where the "operators" +/– are used to underline the disappearance of the H "intruder" atoms in going from the fragments to the whole molecule. One can write the interaction between the molecule AB and a generic molecule X as

$$E(A{-}B\cdots X) = E([AH_a + H_bB - H_a - H_b]\cdots X) =$$
$$= E(AH_a\cdots X) + E(H_bB\cdots X) - E(H_a\cdots X) - E(H_b\cdots X) \qquad (2)$$

If a fragmentation scheme is also applied to the X molecule, the interaction energy between the AB and X molecules can be reconstructed in terms of interaction contributions between all possible pairs of fragments in which each molecule has been decomposed.

Once the decomposition has been applied, the interaction energy between each couple of fragments can be calculated by quantum mechanical methods. Since the electronic correlation is an essential feature of the interaction between Van der Waals complexes, beyond Hartee-Fock methods have to be considered. In this work we have used the Möller-Plesset second order perturbation theory (MP2) in the supermolecule approach with polarized basis sets. Although the validity of this method in the calculation of the interaction energy of aromatic molecules is still matter of discussion [15, 16], results for the benzene-benzene interaction [13] are in good agreement with more computationally expensive methods [16]. Thus, in view of the large number of configurations to be considered, our MP2 calculations, with a carefully calibrated 6-31G* basis-set seems to date a good compromise between computational cost and accuracy.

All the calculations were carried out with the GAUSSIAN98 package [17]. The counterpoise correction scheme was applied in all calculations to take care of the basis set superposition error. More details of the fragmentation-reconstruction method can be found in Ref. [13].

2.2. THE MODEL POTENTIAL FUNCTIONS

Once the PES of the dimer under study has been reconstructed as a sum of *ab initio* fragment-fragment interactions, it can be fitted with a model potential function, suitable for computer simulations. This is done through a non linear least squares fitting procedure, *i.e.* looking for the set of parameters **P** that minimizes the functional,

$$I = \sum_{i}^{N_{geom}} w_i \Big(E(X_i, Y_i, Z_i, \alpha_i, \beta_i, \gamma_i) - U(\mathbf{P}, X_i, Y_i, Z_i, \alpha_i, \beta_i, \gamma_i) \Big)^2 \qquad (3)$$

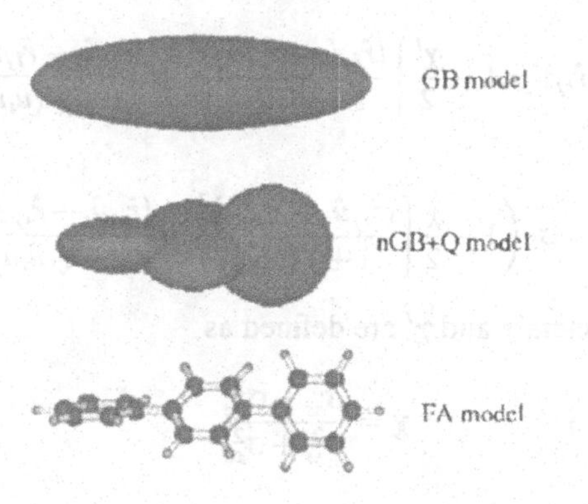

Figure 1. Models adopted for the *p*-oligophenyls: schematic representation of the models for *p*-terphenyl.

where N_{geom} is the number of dimer geometries considered and w_i is the applied weighting function. E and U are, respectively, the reconstructed interaction energy and the fitting function values for the dimer geometries. Each geometry is defined by the center of mass coordinates X_i,Y_i and Z_i and the three Eulerian angles $\alpha_i,\beta_i,\gamma_i$.

Studying *p*-oligophenyls as a test case, we used three different levels of modeling for the series: a single site Gay-Berne (GB) model, an 'hybrid' model (nGB+Q) in which each phenyl ring is represented by a GB oblate ellipsoid plus a quadrupole moment (Q), and a full atomic (FA) representation. All these models are sketched in Fig. 1 for *p*-terphenyl.

2.2.1. *GB model*

The first model employed is a modified form of the anisotropic single-site Gay-Berne potential [3], *i.e.*

$$U_{ij}^{GB}(X,Y,Z,\alpha,\beta) = 4\varepsilon_0\varepsilon_1(\hat{u}_i,\hat{u}_j)^{\nu}\varepsilon_2(\hat{u}_i,\hat{u}_j,\hat{r}_{ij})^{\mu} \tag{4}$$

$$\times\left[\left(\frac{\sigma_0\xi}{r_{ij}-\sigma(\hat{u}_i,\hat{u}_j,\hat{r}_{ij})+\sigma_0\xi}\right)^{12}-\left(\frac{\sigma_0\xi}{r_{ij}-\sigma(\hat{u}_i,\hat{u}_j,\hat{r}_{ij})+\sigma_0\xi}\right)^{6}\right]$$

where $\hat{u}_i$ and $\hat{u}_j$ are the unit vectors defining the orientation of sites i and j respectively, and $\hat{r}_{ij}$ is the unit vector of the distance between the two interacting GB sites. The anisotropy of the interactions is contained in the expressions for $\varepsilon_1(\hat{u}_i,\hat{u}_j)$, $\varepsilon_2(\hat{u}_i,\hat{u}_j,\hat{r}_{ij})$ and $\sigma(\hat{u}_i,\hat{u}_j,\hat{r}_{ij})$ which are:

$$\varepsilon_1(\hat{u}_i,\hat{u}_j) = [1-\chi^2(\hat{u}_i\hat{u}_j)^2]^{-\frac{1}{2}} \tag{5}$$

$$\varepsilon_2(\hat{u}_i,\hat{u}_j,\hat{r}_{ij}) = 1 - \frac{\chi'}{2}\left[\frac{(\hat{r}_{ij}\hat{u}_i+\hat{r}_{ij}\hat{u}_j)^2}{1+\chi'(\hat{u}_i\hat{u}_j)} + \frac{(\hat{r}_{ij}\hat{u}_i-\hat{r}_{ij}\hat{u}_j)^2}{1-\chi'(\hat{u}_i\hat{u}_j)}\right] \tag{6}$$

$$\sigma(\hat{u}_i,\hat{u}_j,\hat{r}_{ij}) = \sigma_0\left(1 - \frac{\chi}{2}\left[\frac{(\hat{r}_{ij}\hat{u}_i+\hat{r}_{ij}\hat{u}_j)^2}{1+\chi(\hat{u}_i\hat{u}_j)} + \frac{(\hat{r}_{ij}\hat{u}_i-\hat{r}_{ij}\hat{u}_j)^2}{1-\chi(\hat{u}_i\hat{u}_j)}\right]\right)^{-\frac{1}{2}} \tag{7}$$

where the parameters χ and χ' are defined as

$$\chi = \frac{\sigma_{ss}^2 - \sigma_{ee}^2}{\sigma_{ss}^2 + \sigma_{ee}^2} \tag{8}$$

and

$$\chi' = \frac{\varepsilon_{ss}^{\frac{1}{\mu}} - \varepsilon_{ee}^{\frac{1}{\mu}}}{\varepsilon_{ss}^{\frac{1}{\mu}} + \varepsilon_{ee}^{\frac{1}{\mu}}} \tag{9}$$

In these equations ε_{ee} and σ_{ee} are the well depth and the zero-energy distance for a configuration where the GB axes and the interparticle vector r_{ij} are all parallel; the *ss* under-scripts refer to the configurations in which r_{ij} is orthogonal to the axis of both the GB sites.

It is also customary to define the quantities:

$$\kappa = \frac{\sigma_{ee}}{\sigma_{ss}} \tag{10}$$

and

$$\kappa' = \frac{\varepsilon_{ee}}{\varepsilon_{ss}} \tag{11}$$

to describe the shape and well depth anisotropy, respectively. Finally, added to the standard GB potential form, the parameter ξ provides the possibility of adjusting the width of the potential well. In summary the GB model potential is completely defined by the set of parameters **P**,

$$\mathbf{P} = [\sigma_0, \sigma_{ee}, \sigma_{ss}, \varepsilon_0, \varepsilon_{ee}, \varepsilon_{ss}, \mu, \nu, \xi] \tag{12}$$

which have to be optimized by the fitting procedure.

Attention must be paid to the fact that, due to its rotational symmetry about its directional axis, the GB model function does not depend on the Eulerian angle γ (Eq. 4). The γ-dependence of the PES must be averaged out in some way passing from the biaxial function $E(X,Y,Z,\alpha,\beta,\gamma)$ to the uniaxial $U^{GB}(X,Y,Z,\alpha,\beta)$. Details on the averaging procedure can be found in a previous publication [18].

2.2.2. *nGB+Q model*

Increasing the level of description, we employed as fitting function the nGB+Q model potential, where n stands for the number of aromatic rings. In this model, each phenyl ring is represented by a Gay-Berne disc (Fig. 1) plus a linear quadrupole along the orientational axis of the site. It is worth mentioning that the parametrization conforms to the molecular symmetry i.e. equivalent rings are represented by identical GB+Q potential terms. Since no cross term is taken into account, the nGB+Q potential results in a sum of two terms, namely a generalized Gay-Berne interaction [19] U^{gGB} between non equivalent GB sites and a quadrupole-quadrupole contribution [20] U^{QQ} whose asymptotic expression is

$$U^{QQ} = \frac{3}{4}Q_iQ_j[1+2(\hat{u}_i\hat{u}_j)^2 - 5(\hat{u}_i\hat{r}_{ij})^2 - 5(\hat{u}_j\hat{r}_{ij})^2 - \\ -20(\hat{u}_i\hat{u}_j)(\hat{u}_i\hat{r}_{ij})(\hat{u}_j\hat{r}_{ij}) + 35(\hat{u}_i\hat{r}_{ij})^2(\hat{u}_j\hat{r}_{ij})^2]/r_{ij}^5 \quad (13)$$

where the unit vectors were already defined for the GB interaction energy and Q's are the quadrupole values of the interacting sites.

For U^{gGB}, we adopt the form proposed by Cleaver *et al.* [19]. For two interacting molecules A and B, defined by the parameter arrays

$$\mathbf{P}_A = [\sigma_A^0, \sigma_A^{ee}, \sigma_A^{ss}, \varepsilon_A^0, \varepsilon_A^{ee}, \varepsilon_A^{ss}, \mu_A, \nu_A, \xi_A] \quad (14)$$

and

$$\mathbf{P}_B = [\sigma_B^0, \sigma_B^{ee}, \sigma_B^{ss}, \varepsilon_B^0, \varepsilon_B^{ee}, \varepsilon_A^{ss}, \mu_B, \nu_B, \xi_B] \quad (15)$$

respectively, the generalized GB interaction is written as

$$U_{AB}^{gGB} = 4\varepsilon_{AB}^0(\varepsilon_{AB}^{(1)}(\hat{u}_A,\hat{u}_B))^{\nu_{AB}}(\varepsilon_{AB}^{(2)}(\hat{u}_A,\hat{u}_B,\hat{r}_{AB}))^{\mu_{AB}} \times \\ \times\left[\left(\frac{\sigma_{AB}^0\xi_{AB}}{r_{AB}-\sigma_{AB}(\hat{u}_A,\hat{u}_B,\hat{r}_{AB})+\sigma_{AB}^0\xi_{AB}}\right)^{12} - \\ -\left(\frac{\sigma_{AB}^0\xi_{AB}}{r_{AB}-\sigma_{AB}(\hat{u}_A,\hat{u}_B,\hat{r}_{AB})+\sigma_{AB}^0\xi_{AB}}\right)^{6}\right] \quad (16)$$

where

$$\sigma_{AB}(\hat{u}_A,\hat{u}_B,\hat{r}_{AB}) = \sigma_{AB}^0 \times \quad (17)$$

$$\times\left[1-\chi_{AB}\frac{(\alpha_{AB}^2(\hat{r}_{AB}\hat{u}_A)^2+\alpha_{AB}^{-2}(\hat{r}_{AB}\hat{u}_B)^2-2\chi_{AB}(\hat{r}_{AB}\hat{u}_A)(\hat{r}_{AB}\hat{u}_B)}{1-\chi_{AB}^2(\hat{u}_A\hat{u}_B)^2}\right]^{-\frac{1}{2}};$$

$$\varepsilon^{(1)}_{AB}(\hat{u}_A,\hat{u}_B) = [1-\chi^2_{AB}(\hat{u}_A\hat{u}_B)^2]^{-\frac{1}{2}} \tag{18}$$

and

$$\varepsilon^{(2)}_{AB}(\hat{u}_A,\hat{u}_B,\hat{r}_{AB}) = \tag{19}$$

$$1-\chi'_{AB}\left[\frac{(\alpha'^2_{AB}(\hat{r}_{AB}\hat{u}_A)^2+\alpha'^{-2}_{AB}(\hat{r}_{AB}\hat{u}_B)^2-2\chi'_{AB}(\hat{r}_{AB}\hat{u}_A)(\hat{r}_{AB}\hat{u}_B))}{1-\chi'^2_{AB}(\hat{u}_A\hat{u}_B)^2}\right]$$

In their work Cleaver *et al.* [19] suggest a combination rule to calculate χ_{AB} and α_{AB} (namely as they appear in the above equations) from the set of parameters $\mathbf{P}_A$, $\mathbf{P}_B$ characterizing each GB site:

$$\chi_{AB}\alpha^2_{AB} = \frac{(\sigma^{ee}_A)^2-(\sigma^{ss}_A)^2}{(\sigma^{ee}_A)^2+(\sigma^{ss}_B)^2} \tag{20}$$

$$\chi_{AB}\alpha^{-2}_{AB} = \frac{(\sigma^{ee}_B)^2-(\sigma^{ss}_B)^2}{(\sigma^{ee}_B)^2+(\sigma^{ss}_A)^2} \tag{21}$$

$$\chi^2_{AB} = \frac{[(\sigma^{ee}_A)^2-(\sigma^{ss}_A)^2][(\sigma^{ee}_B)^2-(\sigma^{ss}_B)^2]}{[(\sigma^{ee}_B)^2+(\sigma^{ss}_A)^2][(\sigma^{ee}_A)^2+(\sigma^{ss}_B)^2]} \tag{22}$$

No other mixing rule was found in the literature to calculate the interaction parameters from the arrays $\mathbf{P}_A$ and $\mathbf{P}_B$. Following the Lorentz-Berthelot approach for the standard Lennard-Jones mixing rules we have used

$$\varepsilon^0_{AB} = (\varepsilon^0_A\varepsilon^0_B)^{\frac{1}{2}} \;;\; \sigma^0_{AB} = \frac{1}{2}(\sigma^0_A+\sigma^0_B) \;; \tag{23}$$

The μ and ν exponents and the parameter ξ were combined as

$$\mu_{AB} = \frac{1}{2}(\mu_A+\mu_B) \;;\; \nu_{AB} = \frac{1}{2}(\nu_A+\nu_B) \;;\; \xi_{AB} = (\xi_A\xi_B)^{\frac{1}{2}} \tag{24}$$

Mixing rules for the remaining parameters were obtained by comparing the standard GB expressions for χ' with the expression found in literature [19] for $\chi\alpha^2$, $\chi\alpha^{-2}$ and χ^2, respectively.

$$\chi'_{AB}\alpha'^2_{AB} = \frac{(\varepsilon^{ss}_A)^{\frac{1}{\mu_A}}-(\varepsilon^{ee}_A)^{\frac{1}{\mu_A}}}{(\varepsilon^{ss}_A)^{\frac{1}{\mu_A}}+(\varepsilon^{ee}_B)^{\frac{1}{\mu_B}}} \;; \tag{25}$$

$$\chi'_{AB}\alpha'^{-2}_{AB} = \frac{(\varepsilon_B^{ss})^{\frac{1}{\mu_B}} - (\varepsilon_B^{ee})^{\frac{1}{\mu_B}}}{(\varepsilon_B^{ss})^{\frac{1}{\mu_B}} + (\varepsilon_A^{ee})^{\frac{1}{\mu_A}}}; \tag{26}$$

$$\chi'^2_{AB} = \frac{[(\varepsilon_A^{ss})^{\frac{1}{\mu_A}} - (\varepsilon_A^{ee})^{\frac{1}{\mu_A}}][(\varepsilon_B^{ss})^{\frac{1}{\mu_B}} - (\varepsilon_B^{ee})^{\frac{1}{\mu_B}}]}{[(\varepsilon_B^{ss})^{\frac{1}{\mu_B}} + (\varepsilon_A^{ee})^{\frac{1}{\mu_A}}][(\varepsilon_A^{ss})^{\frac{1}{\mu_A}} + (\varepsilon_B^{ee})^{\frac{1}{\mu_B}}]} \tag{27}$$

It's worth noting that the above expressions correctly reproduce the standard GB interaction in the limiting case A=B. Moreover, they also can be used to model the interaction between an anisotropic and a spherical site, *i.e.* a GB-LJ pair. To our knowledege, these are the first complete mixing rules for the interaction of two different GB sites, *e.g.* such as required to simulate mixtures of rod- and disk-like molecules.

With respect to the GB model, two new features are taken into account in the nGB+Q model: quadrupolar effects and internal flexibility. For the latter, the rotation of each disk around the main axis is driven by a torsional potential $V(\theta)$ which takes the form proposed by Tsuzuki *et al.* [21] for biphenyl

$$V(\theta)(\mathrm{kJ/mol}) = 5.19 + 3.10\cos(2\theta) + 5.19\cos(4\theta) + 0.88\cos(6\theta) \tag{28}$$

where θ is the angle between two adjacent rings.

2.2.3. *FA model*

The last model that we have investigated is a fully atomistic representation, in which all interactions between atoms of molecules A and B are modeled by a modified Lennard-Jones potential and a point charge term, as follows

$$U_{AB} = \sum_i^A \sum_j^B 4\varepsilon_{ij}\left[\left(\frac{\sigma_{ij}\xi_{ij}}{r_{ij} - \sigma_{ij}(1-\xi_{ij})}\right)^{12} - \left(\frac{\sigma_{ij}\xi_{ij}}{r_{ij} - \sigma_{ij}(1-\xi_{ij})}\right)^{6}\right] + \frac{q_i q_j}{r_{ij}}$$

where ε_{ij} and σ_{ij} can be obtained by the standard mixing rules, and $\xi_{ij} = (\xi_i\xi_j)^{\frac{1}{2}}$. Due to its computational cost, this last model is being used only on the biphenyl molecule, in a MD simulation still in progress.

2.3. COMPUTER SIMULATIONS

In most simulations performed so far, we have used the Monte Carlo method [22]. In order to study the phase diagram, we have chosen to carry out our MC simulations in the isothermal isobaric ensemble, MC NPT [23] which, although a bit computationally more expensive, possesses several advantages with respect to the canonical ensemble Monte Carlo, MC NVT [24]. Allowing the shape and the volume of the simulation box to fluctuate favors, in fact, the achievement of the natural structure of the system, in keeping with the periodic boundary conditions actually used. In addition, the variation of the density in constant pressure simulations reduces the risk that the system remains trapped in a metastable state whose density belongs to a coexistence region of the phase diagram. This, in turn, means that the smearing of first order transitions in finite systems, common in fixed density simulations, is reduced when density is allowed to change. Finally, we note that real experiments are usually performed at constant pressure, so that the isothermal isobaric appears to be the statistical ensemble closest to experimental conditions.

The numerical experiments have been performed according to the usual rules of the MC NPT scheme, using systems of 600 particles, at a pressure of 1 atm and at several temperatures, varied in a stepwise manner. Both in the GB and nGB+Q models, a particle has been selected at random and trial displacements of the center of mass and reorientations of the long axis have been executed. In the nGB+Q model an additional attempt to rotate a randomly selected disc of a molecule perpendicularly to its axis has been carried out, in order to sample the conformational space. The shape and the volume of the computational box have been changed during the simulations by attempting to vary a randomly selected edge of the box. The different moves have been selected randomly and not sequentially to preserve the detailed balance condition. The value of the maximum allowed changes has been chosen in order to reach a 30%-40% of acceptance ratio for every kind of move, but never modified during both equilibration and production runs, again to preserve detailed balance [25].

The equilibration has been assessed by monitoring the evolution of a number of observables such as enthalpy, density and orientational order parameter, η. The latter has been calculated following Veillard-Baron [26], that is diagonalizing the **Q** tensor:

$$Q_{\alpha\beta} = \frac{1}{N}\sum_{i=1}^{N}\left(\frac{3}{2}u_i^{\alpha}u_i^{\beta} - \frac{\delta_{\alpha\beta}}{2}\right) \tag{29}$$

where N is the number of molecules, α,β = x,y,z and u_i^{α} is the component along α of the unit vector describing the orientation of particle i, and picking the largest eigenvalue. Every equilibration has been followed by a production run in which averages of the thermodynamic properties have been collected and whose error bars have been calculated by standard block method. In addition, a family of spatial correlation functions have been computed in order to quantitatively characterize the translational and orientational structure of the phases involved. Particular attention has been paid to $g(r)$, the well known pair correlation function of the centers of mass, and $G_2(r)$

$$G_2(r_{ij}) = \langle P_2(\hat{u}_i \hat{u}_j) \rangle (r_{ij}) \qquad (30)$$

where P_2 denotes the second Legendre polynomial and $\hat{u}_i$ and $\hat{u}_j$ are the unit vectors defining the orientation of the long axis of the molecules i and j, separated by the intermolecular distance r_{ij}. This function describes the orientational correlation of two molecules as a function of the distance between their center of mass and it reaches the asymptotic value of η^2 at large r_{ij}.

3. Results and Discussion

3.1. POTENTIAL ENERGY SURFACES

Considering each p-n-phenyl as composed of n benzene fragments, only two fragment-fragment contributions suffice to reconstruct the PES of the whole dimers, namely the benzene-benzene and the benzene-hydrogen ones. We use the interaction energy already calculated and reported in Ref. [13]. Here we show (Table 1) only a few values for three local minima together with recent literature data.

TABLE 1. Comparison of calculated interaction energies for some local minima of the benzene dimer in the parallel sandwich (SS), T-shaped (TS) and parallel displaced (PD) configurations. One benzene lies in the XZ plane and x, y, z, α, β, γ coordinates fix the position of the other. (a) ref. [13], (b) ref. [15] and (c) ref. [16] .

Geom	X (Å)	Y (Å)	Z (Å)	α (deg)	β (deg)	γ (deg)	Energy (kcal/mol)
SS$^{(a)}$	0.0	3.9	0.0	0	0	0	-1.70
SS$^{(b)}$	0.0	3.9	0.0	0	0	0	-1.72
SS$^{(c)}$	0.0	3.9	0.0	0	0	0	-1.48
TS$^{(a)}$	0.0	0.0	5.2	0	90	90	-2.20
TS$^{(b)}$	0.0	0.0	5.2	0	90	90	-2.48
TS$^{(c)}$	0.0	0.0	5.2	0	90	90	-2.46
PD$^{(a)}$	0.0	3.3	1.7	0	0	0	-2.30
PD$^{(b)}$	0.0	3.3	1.7	0	0	0	-2.79
PD$^{(c)}$	0.0	3.3	1.7	0	0	0	-2.48

For each pair of fragments we have calculated interaction energies for almost 200 geometries, which where used to compute a fragment-fragment interaction energy expressed as a sum of atom-atom contributions. This form has been then used to recover the PES for the pair of molecules under study.

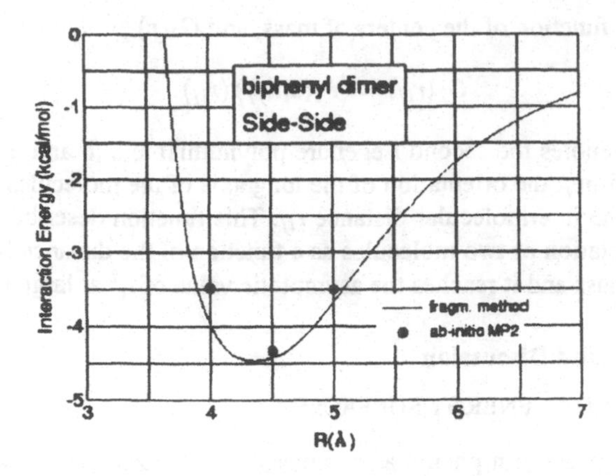

Figure 2. Cross section of the biphenyl dimer PES for the sandwich configuration. The angle between the two rings is kept fixed at 42° . Reconstructed curve (solid line) and direct *ab initio* data (dots) are reported.

The biphenyl molecule constitutes an appropriate case to test the validity of the reconstruction procedure: its dimensions, indeed, allow direct *ab initio* calculations at the same level of accuracy as used for the fragment-fragment interaction. For this purpose, the reconstructed PES of the biphenyl dimer and a few energy values, calculated directly at the MP2 level on the whole dimer, are compared in Fig. 2, for the sandwich configuration. It is apparent that the reconstruction approach does not introduce unwanted changes in the characteristic features of the dimer PES and a good accuracy in reproducing the interaction is achieved.

The PES of the other members of the series has been calculated in the same way. Cross sections are reported in Fig. 3 for some selected configurations. In all cases, the reported interaction energies have been obtained for the internal equilibrium geometry of the molecules. The interaction energy increases significantly along the series for the parallel configurations, while it is almost constant for others configurations, *e.g.* T-shaped, cross and end-to-end. This trend suggests that the increasing shape anisotropy, *i.e.* the elongation of the rod-like molecule, results in an augmented tendency of the dimer to prefer aligned conformations, thus leading to orientationally more ordered bulk phases. In particular, as already noticed for the benzene dimers, a displacement along the long axis direction allows the molecules to reach stable configurations, favoring the parallel displaced geometries to the detriment of the parallel sandwich ones. It is also worth noting that the T-shaped configuration, in contrast with the benzene results, are not favored with respect to the parallel ones. This is

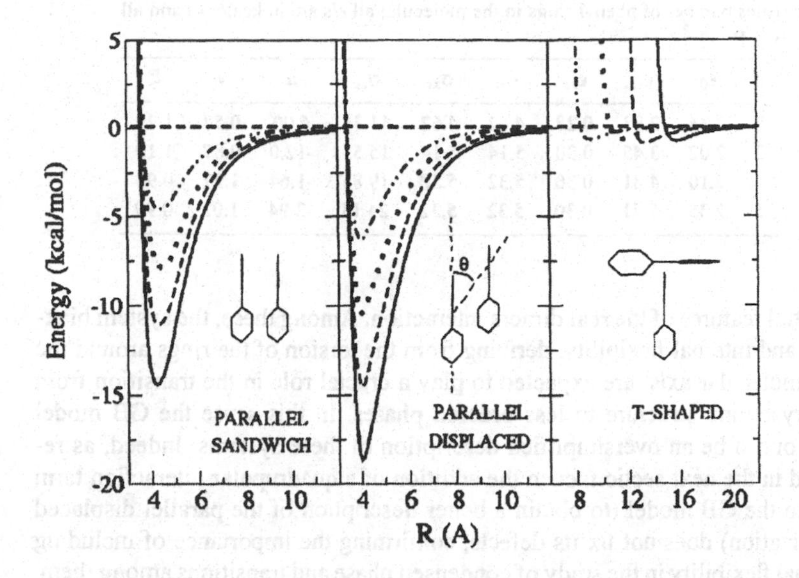

Figure 3. Cross section of oligophenyls dimer PES for some relevant geometries. Curves for biphenyl (dotted-dashed line), *p*-terphenyl (dotted line), *p*-quaterphenyl (dashed line) and *p*-quinquephenyl (solid line) are reported. R is the distance between the centers of mass. The PD configuration is defined by the polar angle θ=110° (drawn in the figure) and the azimuthal angle set to 70°

not surprising, considering that the stabilization of the parallel alignment of the oligophenyls series arises from an increasing number of sandwich (SS) and parallel displaced (PD) contributions of the phenyl-phenyl pairs (see Table 1), whereas in the T-shaped conformation, only a pair of fragments contribute significantly , regardless of the molecular size.

3.2. FITTING RESULTS

3.2.1. *GB model*

For the biphenyl molecule, all three models described in section 2.2 were applied both in the fitting and in the simulations.

As said in the previous section, details of the fitting procedure using the GB model have been given elsewhere [18]. In Table 2 we report the set of optimized parameters for all the series. Although the GB model accounts for the anisotropic nature of the intermolecular interactions, it misses some other

TABLE 2. Parameters of the GB model for the oligophenyls series. n indicates number of phenyl rings in the molecule; all ε's are in kcal/mol and all σ's are in Å.

n	ε_0	ε_{ss}	ε_{ee}	σ_0	σ_{ss}	σ_{ee}	μ	ν	ξ
2	2.34	2.93	0.28	4.62	4.62	11.31	6.00	0.57	1.29
3	2.02	3.45	0.30	5.14	5.14	15.53	-12.0	0.87	1.10
4	2.10	4.81	0.30	5.32	5.32	19.83	-1.64	1.27	0.93
5	2.35	5.71	0.30	5.32	5.32	24.14	-2.94	1.02	0.99

essential features of the real dimers interaction. Among these, the system biaxiality and internal flexibility, deriving from the torsion of the rings around the long molecular axis, are expected to play a crucial role in the transition from the crystalline structure to less ordered phases. In this sense the GB model turns out to be an oversimplified description of these systems. Indeed, as reported in the next section, even the addition of a quadrupolar interaction term [20] to the GB model (to obtain a better description of the parallel displaced stabilization) does not fix its defects, confirming the importance of including internal flexibility in the study of condensed phase and transitions among them.

3.2.2. *nGB+Q model*

The nGB+Q model can describe to a good level of accuracy the potential feature of the reconstructed *ab initio* PES and can take into account the changes in energy arising from the internal flexibility. In Fig. 4 we compare the cross sections of the calculated PES and the fitted potential for some conformations of *p*-quinquephenyl. Similar results have been obtained for the other members of the series.

From the parallel displaced cross section in Fig. 4, it can be noted the effect of the quadrupolar interaction which shifts the minimum toward the slipped configuration, in agreement with the quantum mechanical results.

The set of optimized GB parameters are reported in Table 3, while the optimized quadrupole (which were imposed to be the same for each phenyl ring) and the standard deviations are reported in Table 4. The weighting function w_i was

$$w_i = e^{-\alpha E_i} \tag{31}$$

where E_i is the energy at the i-th geometry and we have used $\alpha = 0.05$ (kcal/mol)$^{-1}$ as coefficient. Moreover, in view of the poor significance of very repulsive values in a freezed-geometry, all the energy values $E_i > 20$ kcal/mol were discarded from the fitting.

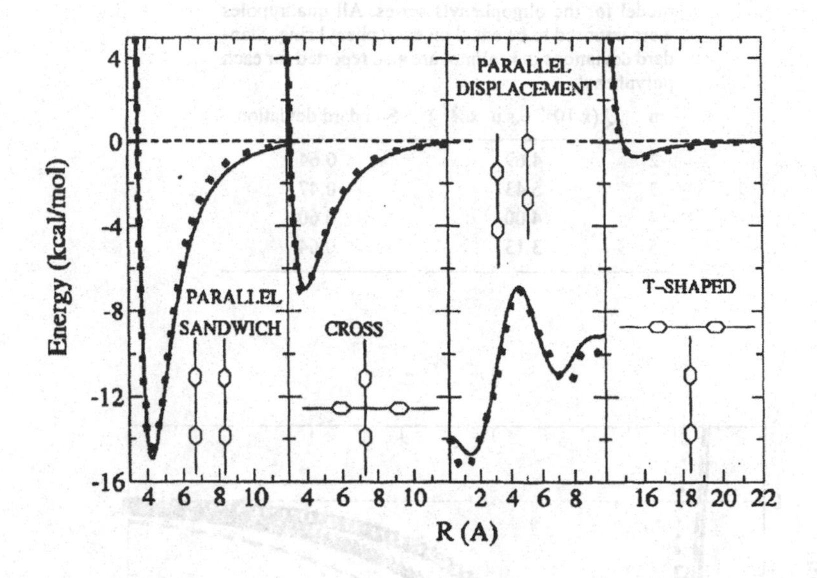

Figure 4. Comparison of reconstructed dimer PES (dots) and fitting potential of *p*-quinquephenyl (curves) with the nGB+Q model for some configurations.

TABLE 3. Parameters of the nGB+Q model for the oligophenyls series. n indicates number of phenyl rings in the molecule, and m indicates the position of the ring inside the polyphenyl ; ε's are in kcal/mol and σ's in Å.

n	m	ε_0	ε_{ss}	ε_{ee}	σ_0	σ_{ss}	σ_{ee}	μ	ν	ξ
2	1,2	0.30	1.00	6.78	6.83	6.83	3.18	1.0	1.20	0.86
3	1,3	0.38	1.20	3.25	6.69	6.69	3.41	1.0	2.07	0.89
3	2	0.17	0.43	11.48	6.86	6.86	2.83	1.0	-0.87	0.77
4	1,4	0.32	1.39	11.27	6.90	6.90	3.20	1.0	-0.86	0.83
4	2,3	0.40	0.56	5.29	6.56	6.56	3.16	1.0	-0.61	0.84
5	1,5	0.32	0.69	8.58	6.85	6.85	3.21	1.0	-1.24	0.79
5	2,4	0.54	0.56	5.99	6.51	6.51	3.19	1.0	-3.42	0.83
5	3	0.29	1.40	8.13	6.43	6.43	3.31	1.0	1.46	0.81

TABLE 4. Optimized quadrupoles of the nGB+Q model for the oligophenyls series. All quadrupoles were imposed to be equal on each phenyl ring. Standard deviations in kcal/mol are also reported for each polyphenyl.

n	Q (x 10^{26} e.s.u. x $Å^2$)	Standard deviation
2	4.69	0.64
3	5.43	0.47
4	4.00	0.60
5	3.13	0.64

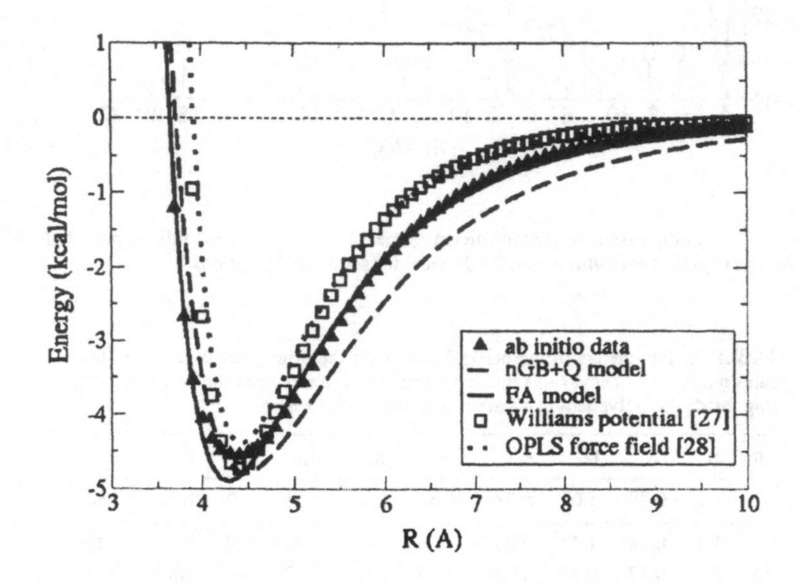

Figure 5. Interaction energy of a biphenyl dimer in a sandwich configuration, from this work and literature potentials.(a) Williams potential Ref. [27] and (b) OPLS force-field [28].

3.2.3. *FA model*

Finally, to further improve the matching to the *ab initio* reconstructed PES, we have used the FA model for the fitting of biphenyl and compared the results with the nGB+Q model and some literature force-fields (Fig. 5). Standard deviation, obtained with the same fitting procedure as for the nGB+Q model, was

0.2 kcal/mol, indicating an improved agreement with the *ab initio* data.

3.3. SIMULATION RESULTS

The oligophenyls series constitutes a natural and perhaps unique choice to test our procedure for calamitic molecules, for several reasons: the biphenyl and *p*-terphenyl form the core of many mesogenic molecules; the *p*-quinquephenyl is actually nematic at atmospheric pressure between $\simeq$655 and $\simeq$685 K [29, 30]; the n-phenyls have no flexible tails and a shape that makes them amenable to a single site modeling.

3.3.1. *GB model*

As mentioned in the Introduction, we wanted to compare results obtained from different models, all parameterized on the same set of *ab initio* interaction energy values. Let's start the discussion of our results from the single site Gay-Berne model which represents the simplest level of our hierarchy of descriptions of the PES of oligophenyls.

We recall that most computer simulation studies using GB particles are usually performed at reduced pressure, $P^* = P\sigma_0^3/\varepsilon_0$, and reduced temperature, $T^* = k_B T/\varepsilon_0$, of the order of unity. After inserting typical values for ε_0 and σ_0, one obtains real temperature T of the correct order of magnitude, but corresponding values of pressure of the order of 10^3 bar, *i.e.* remarkably higher than 1 atm [31, 32]. With the single site GB, in particular, our aim was to test its capability under conditions of pressure and temperature close to that of most experimental and synthetic works.

In principle this issue should be assessed through a systematic variation of the adjustable parameters occurring in the GB potential. As this would require a huge computational effort, we resort to the procedure proposed in Ref. [33], where a realistic pair potential, is built and mapped onto the GB potential. In Ref. [33] a set of Lennard-Jones sites was used to model *p*-terphenyl while here we use *ab initio* calculated potentials for the whole series of *p*-oligophenyls up to 5 phenyl rings. The parameters we find for the single site GB potential are given in Table 2.

Systems of 600 particles have been simulated at atmospheric pressure and several temperatures by Monte Carlo method in the isothermal-isobaric ensemble. Because the GB particles have an ellipsoidal character, the equilibration runs have started from an expanded fcc crystal structure, elongated in the [111] direction, chosen parallel to the z axis. During the production runs, thermodynamic and structural indicators have been monitored. The results have been presented in Ref. [18] so we give here only a brief summary.

When the simple GB model is applied to biphenyl, it proves to be able of yielding crystal, liquid and gas phases. For the successive members of the series, however, the condensed fluid phase completely disappears and the crystalline phase evolves directly toward the gas phase. These results are in accord with previous studies performed on Gay-Berne particles changing one of the original GB parameters [3] at a time. For instance, the effect of varying κ' and κ has been studied in Ref. [34] and [35] respectively. Increasing κ', a translationally and orientationally ordered phase is stabilized at low temperatures; the liquid region in the P-T plane is reduced till the critical point disappears, if κ also is augmented. In [34], [35] and elsewhere the ordered phase is characterized as a Smectic B phase. However, a recent study on the global phase diagram of the GB model with the original parameterization [36] has shown, by free energy computations, that such phase is actually a crystal, as in our investigations. It is likely that this conclusion extends to all members of the GB class of potentials.

Coming back to the question whether the GB model is able to describe the behavior of mesogens at atmospheric pressure, it seems that the answer, although not definite, should be negative, at least for calamitic liquid crystals. It is fair to notice, in fact, that typical values of ε_0 and σ_0 for discotic molecules lead to a pressure of 40 atm, if a reduced pressure of 1 is again considered [37]. Discotic GB particles seem therefore more suitable for a single site GB modeling under atmospheric pressure conditions. This might be traced back to the fact that the core of discotic liquid crystals is more rigid than that of calamitic molecules. It seems that some degree of internal flexibility is a key feature for a reasonably realistic model of rodlike mesogens. It is for such reason that we have turned our attention to the nGB+Q model, whose results are presented below.

3.3.2. *nGB+Q model*

The hybrid, multi-site nGB+Q model constitutes the next level of our hierarchy of models, lying between the single site GB and a fully atomistic pair potential, as far as realism and computational complexity are concerned. Here, each phenyl ring is described as an oblate GB ellipsoid with an axial quadrupole at its center. The quadrupole-quadrupole interaction is known to play an important role to determine the properties of systems of benzene molecules [16]. The internal rotations between discs along the molecular axis have been assumed regulated by a torsional potential coming from *ab initio* calculations [21]. Different torsional angles are assumed to be uncorrelated.

As for the GB model, we have simulated systems of 600 molecules at atmospheric pressure and several temperatures, making use of the MC-NPT technique described in Section 3. For consistency with the work with the single site model, we have started the equilibration runs from the same crystal structure, replacing each GB particle with a nGB+Q molecule, with dihedral

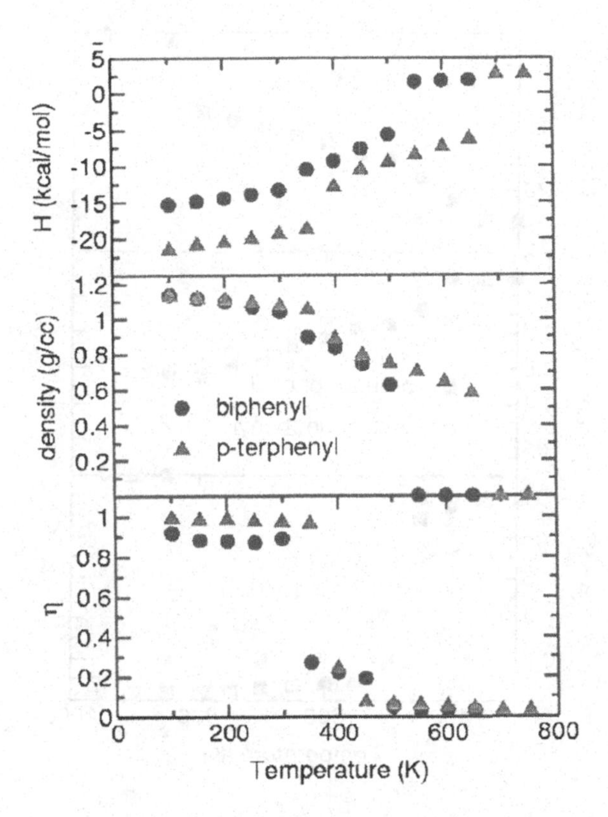

Figure 6. **nGB+Q model: thermodynamic properties as a function of temperature for biphenyl (black circles) and *p*-terphenyl (triangles).**

angles between two contiguous discs set to the value of ~45° , which corresponds to the minimum of the torsional potential as calculated in [21]. In Fig. 6 and 7 the enthalpy, density and orientational order parameter are shown as a function of temperature for the four systems. Some of the results reported in Fig. 6 and 7 are compared with experimental data in Tab. 5. The melting points (T_m) were determined in simulation by considering the average value between the highest temperature in the solid and the lowest in the liquid phase for each oligophenyl; clearing (T_{NI}) and boiling (T_b) points where determined analogously. Enthalpies of melting (ΔH_m), clearing (ΔH_{NI}) and vaporization (ΔH^0_{vap}) where determined taking the differences H^{liq} - H^{sol}, H^{liq} - H^N and H^{gas} - H^{liq}, respectively.

Broadly speaking, the range of stability of the phases is reproduced by

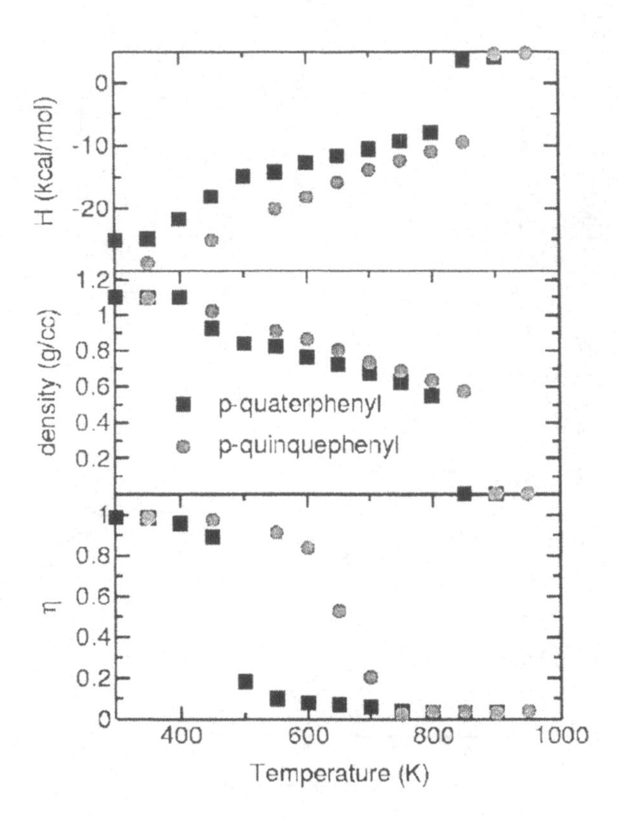

Figure 7. **nGB+Q model: thermodynamic properties as a function of temperature for *p*-quaterphenyl (black squares) and *p*-quinquephenyl (circles).**

the simulation, but melting temperatures are significantly underestimated for *p*-terphenyl and *p*-quaterphenyl. All transition enthalpies are also underestimated, with relative deviations increasing along the series.

Nevertheless, an improvement stands out clearly if these results are compared with the corresponding data for the single site model, which fails in giving a stable liquid phase. Besides a translationally and orientationally ordered phase, at low temperatures, the four systems show two isotropic fluid phases both with an almost vanishing orientational order parameter η, but with definitely different enthalpies and densities.

More importantly, we have obtained a value of orientational order parameter of ~ 0.5 typical of a nematic phase, for *p*-quinquephenyl at T = 650 K, not far the temperature range found experimentally for this mesophase, *i.e.*

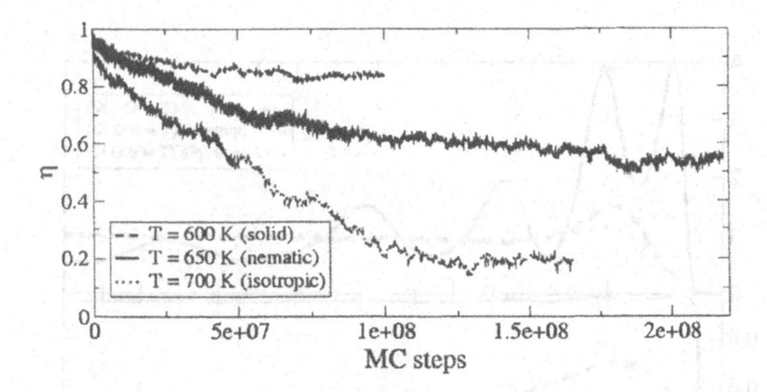

Figure 8. nGB+Q model: equilibration of the order parameter η for the *p*-quinquephenyl at 600 K (dashed line), 650 K (solid line) and 700 K (dotted line).

between 653 and 681 K [30] or 660 and 688 K [38].

TABLE 5. Thermodynamic results of the nGB+Q model. *n* indicates the number of phenyl rings in the *p*-n-phenyls. All temperatures are in K, all ΔH's are in kcal/mol. (a) Ref. [38] (b) Ref. [29] and references therein (c) Ref. [30].

n	T_m	T_{NI}	T_b	ΔH_m	ΔH_{NI}	ΔH°_{vap}
2	325	-	525	2.84	-	7.34
	344 (a)	-	529 (b)	4.49 (a)	-	12.91 (b)
3	375	-	675	5.85	-	9.00
	493 (a)	-	658 (b)	8.48 (a)	-	18.88 (b)
4	475	-	825	3.20	-	10.00
	587 (a)	-	773 (b)	9.03	-	27.96 (b)
5	625	675	875	2.4	2.0	14.10
	660 (a)	688 (a)	823 (b)	10.11 (a)	0.22 (a)	36.09 (b)
	653 (c)	681 (c)				

In view of the known tendency of these systems to remain trapped in metastable states, we have extended the equilibration run to over 220 million of steps, until the order parameter apparently reached a reasonably constant

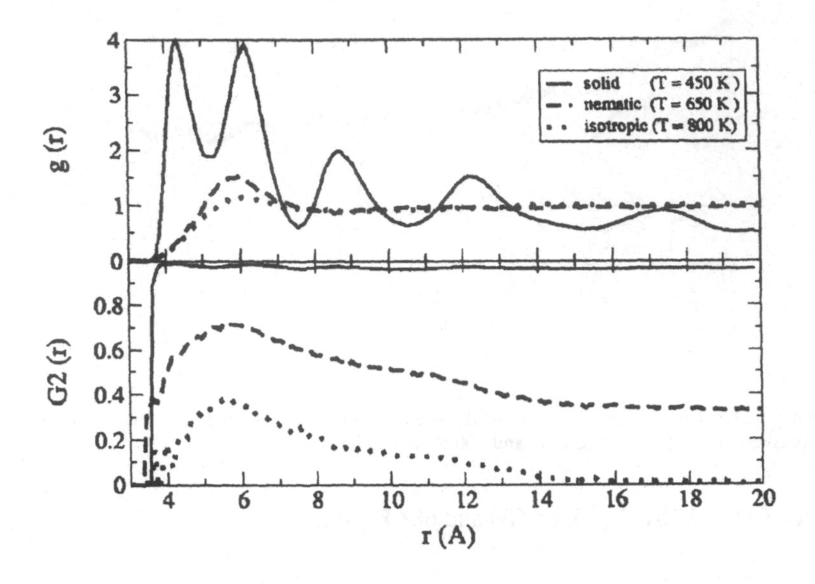

Figure 9. nGB+Q model: correlation functions for *p*-quinquephenyl in the crystal (solid line), in the nematic (dashed line) and liquid phase (dotted line) at T = 450 K, 650 K and 800 K respectively.

value (see Fig.8) for about 50 million configurations: the average value of η = 0.53, obtained during the production run, lead us to consider the equilibrium phase of *p*-quinquephenyl at T = 650 K as nematic. This conclusion is confirmed by the fluid-like character of both the pair correlation functions $g(r)$ and $G_2(r)$, whose asymptotic value is equal to η^2 (see Fig. 9). Nevertheless, further investigation is certainly warranted before the actual nature of the phase of *p*-quinquephenyl at 650 K can be reliably assessed. However, we consider this a promising result and an indication that the level of modeling entailed by the nGB+Q model might be sufficient for at least a semi-quantitative reproduction of the phase diagram of mesogens. Of particular importance when studying the oligophenyls series is the dependence of the dihedral angles between two contiguous discs on the phase thermodynamically stable at a given temperature. It is known in fact (see references in [39]) that for biphenyl the two phenyl rings are almost coplanar in the crystalline phase, have an angle of twist of $\sim 30°$ in the liquid phase and $\sim 45°$ in the gas phase. This trend is well reproduced by our computer simulation data, as one can see in Fig.10

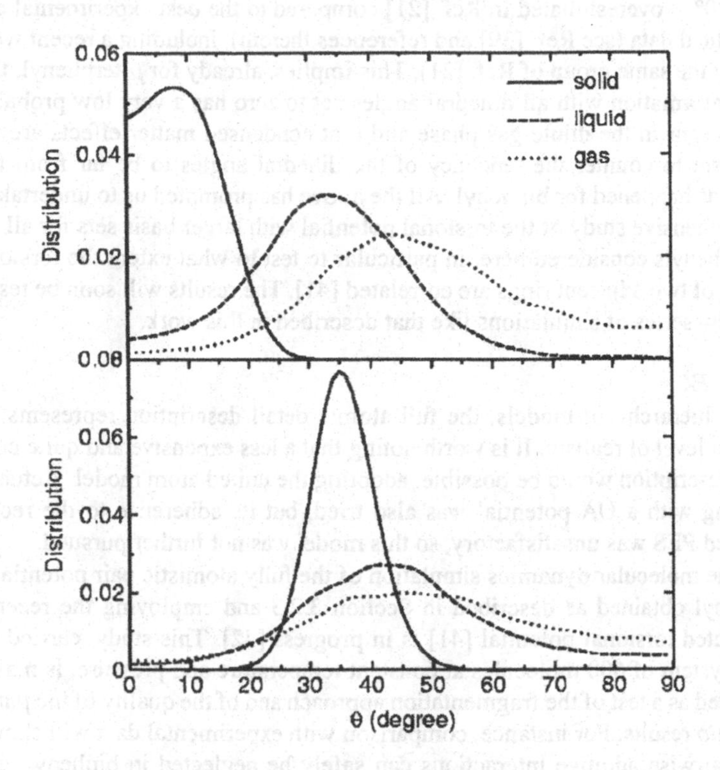

Figure 10. **nGB+Q model: phase dependence of distributions of the inter-ring torsional angle θ for biphenyl (upper panel) and *p*-terphenyl (lower panel).**

where the distribution function of the dihedral angle is reported at different temperatures in the three phases. The maximum of the distribution function is found at $\theta \sim 10°$ at T= 100 K and it moves toward $\theta \sim 35°$ in the liquid at T= 400 K and $\theta \sim 45°$ in the gas phase, at T= 550 K. A similar behavior could be also encountered in the other members of the series, However, as shown in Fig.10 for *p*-terphenyl, the maximum of the distribution function of the dihedral angle does not show a similar sensitivity to temperature as for biphenyl. It stays essentially fixed at $\theta = 45°$, which corresponds to the minimum of the torsional potential, whereas the average value found in the ordered phase (T= 100 K) is no lower than 35° . A similar behavior was observed for the higher homologues. A possible reason of this finding is that the computed torsional potential, though reasonably accurate for biphenyl, might not be transferable to the larger oligophenyls. Actually, it turns out that the height of the barrier at $\theta = 0°$ is overestimated in Ref. [21] compared to the best experimental and theoretical data (see Ref. [39] and references therein), including a recent work [40] by the same group of Ref. [21]. This implies, already for *p*-terphenyl, that the conformation with all dihedral angles set to zero has a very low probability to occur in the dilute gas phase and that condensed matter effects are not sufficient to counter the tendency of the dihedral angles to be far from this value, as happened for biphenyl. All the above has prompted us to undertake a comprehensive study of the torsional potential with larger basis sets for all the oligophenyls considered here, in particular to test to what extent the torsional angles of two adjacent rings are correlated [41]. The results will soon be tested in a new series of simulations like that described in this work.

3.3.3. *FA*

In our hierarchy of models, the full atomic detail description represents the highest level of realism. It is worth noting that a less expensive and quite common description would be possible, adopting the united atom model. Actually, a fitting with a UA potential was also tried, but its adherence to the reconstructed PES was unsatisfactory, so this model was not further pursued.

The molecular dynamics simulation of the fully atomistic pair potential of biphenyl obtained as described in Section 3.2.3 and employing the recently computed torsional potential [41] is in progress [42]. This study, carried out on a system of 600 molecules at constant temperature and pressure, is mainly intended as a test of the fragmentation approach and of the quality of the parent *ab initio* results. For instance, comparison with experimental data will show if non pairwise additive interactions can safely be neglected in biphenyl, as it has been proven true for benzene [43], at least in homogeneous, bulk environments. The results obtained so far, however, are still too preliminary to allow a reliable comparison with those relevant to the nGB+Q model, as well as with experimental data.

4. Conclusions

In this Chapter we have presented the first results of a recently proposed approach to the study of liquid crystal forming molecules, based on interaction potentials derived from *ab initio* calculations [13]. The basic idea is that of decomposing the molecule of interest into a number of fragments, whose interaction potential can be calculated with accurate *ab initio* quantum mechanical methods. These data are then used to obtain the PES of the whole parent molecule through a proper recombination scheme.

There are two main advantages of this approach. The first is that, since no experimental data is necessary to build the two-body potential, it can in principle be used with predictive purposes. The second is that it allows a consistent comparison of different models coming from the same PES, which is a source of useful information, both to assess the 'least- detail' level required for a satisfactory reproduction of experimental data and to trace back the observed results to features of the model.

To validate this approach, we have chosen the series of p-oligophenyls as a test case. These systems, despite a simple structure with no flexible tails, are able to form stable mesophases and can be considered a sort of 'fruit fly' among mesogenic molecules.

The PES obtained with the recombination scheme has been fitted with three different models of increasing realism. The first is a simple single-site GB model while the second (nGB+Q) describes each phenyl ring with an oblate GB ellipsoid bearing a linear quadrupole at its center and includes a potential term governing the interring torsional motions. Finally, the third is expected to be the most realistic representation, involving a LJ and a charge term on each atomic site (FA model). The latter is obviously a computationally very demanding model, so its use in simulation is currently restricted to biphenyl. On the other hand, this is also the most flexible description and the one that more closely reproduces the recombined PES. So its use is mostly meant to discriminate the capabilities of the *ab initio* plus recombination route to potentials.

The process of validation of any force field for simulations should be extended to the study of the largest possible set of structural, energetic and dynamical properties [44]. In this preliminary report, we mainly focused on the phase diagram also in view of its importance for potentially mesogenic materials.

All simulations have been carried out at constant one atmosphere pressure, irrespective of the model employed, also for an easier comparison with experimental results. Under these conditions, the single site GB model produced satisfactory results for melting and boiling temperatures (less so for the corresponding enthalpy changes) only in the case of biphenyl. For all longer p-phenyls considered, no stable liquid phase was obtained. Possible size effects have been tested running a simulation with a larger number of particles (2160) for p-quinquephenyl. However the results substantially reproduced that of the smaller system, again with no liquid phase.

From this point of view, the more realistic description provided by the nGB+Q model yields remarkable improvements, although the disagreement with experimental data is still significant. Stable liquid phases are obtained for all *p*-phenyls with better results also for transition enthalpies. More important, an order parameter consistent with a nematic phase has been found for *p*-quinquephenyl at 650 K, close to the nematic range determined experimentally for this system ($\sim$655$\div$ $\sim$685 K).

Unlike the single site GB, the nGB+Q model is a biaxial model in which both quadrupole moment and internal flexibility of the molecule are taken into account. As the difference between the two descriptions is not restricted to a single feature, it is less straightforward to point out the role of each change. However, we carried out some testing with a single site GB model including a quadrupole moment along the long molecular axis, but no significant progress was obtained with respect to the stability of the liquid phase. Hence, in our opinion, most of the credit for the improvement provided by the nGB+Q over the GB model is due to the explicit account of the torsional degrees of freedom. Actually, some kind of internal flexibility is present in all real mesogenic molecules, but is missing in the rigid single site GB model. So, the ability of the latter model to form mesophases seems to require rather high pressure in addition to proper values of shape and interaction well anisotropy.

Flexibility apparently introduces elements of disorder that, combined with the ordering forces related to shape anisotropy, can produce the balance of energetic and entropic effects necessary to stabilize mesophases. Due to the importance of this internal degree of freedom, a comprehensive study of the potential which drives the torsional motions in the *p*-phenyls and of the correlation among torsions in p-phenyls has been carried out [41]. The results will be used in a new series of simulations with the nGB+Q model as well as with the FA description, at least for biphenyl.

The results obtained for the latter molecule in the FA scheme are so far too preliminary to allow significant comparison with the nGB+Q model. It seems however that the nGB+Q model, of much reduced computational cost with respect to a full atomic detail description, contains all the basic features which are responsible for the mesomorphic behavior of real oligophenyls.

Hence, since biphenyl and *p*-terphenyl are very common building blocks of liquid crystal forming molecules, the results summarized above open the way to the development, from *ab initio* computed pair interaction data, of hybrid models of known or, more interestingly, novel materials, whose phase behavior and properties can be investigated by computer simulation techniques.

References

1. P.A. Lebwohl and G. Lasher (1972) Nematic-Liquid-Crystal Order - A Monte Carlo Calculation, *Phys. Rev. A*, **Vol. 6, No. 1**, pp. 426–429
2. M.P. Allen, G.T. Evans, D. Frenkel and B.M. Mulder (1993) Hard Convex Body Fluids, *Adv. Chem. Phys.*, **Vol. 86**, pp. 1-166
3. J.G. Gay and B.J. Berne (1981) Modification of the overlap potential to mimic a linear site-site potential, *J. Chem. Phys.*, **Vol. 74, No. 6**, pp. 3316–3319
4. M.A. Bates and G.R. Luckhurst (1999) Computer simulation of Liquid Crystal phases formed by Gay-Berne mesogens, *Struct. Bonding (Berlin)*, **Vol. 94**, pp. 65-138
5. C. Zannoni (2001) Molecular design and computer simulations of novel mesophases, *J. Mater. Chem.*, **Vol. 11**, pp. 2637-2646

6. S.J. Picken, W.F. Van Gusteren, P.T. Van Dujen and W.H. De Jeu (1989) A molecular dynamics study of nematic phase of 4-n-pentyl-4'-cyanobiphenyl, *Liq. Cryst.*, **Vol. 6**, pp. 357–371
7. M.R. Wilson and M.P. Allen (1992) Structure of *trans*-4-(*trans*-4-n-pentylcyclohexyl)-cyclohexylcarbonitrile (CCH5) in the isotropic and nematic phases: a computer simulation study, *Liq. Cryst.*, **Vol. 12**, pp.157–176
8. A.V. Komolkin, A. Laaksonen and A. Maliniak (1994) Molecular dynamics simulation of a nematic liquid crystal, *J. Chem. Phys.*,**Vol. 101, No. 5**, pp. 4103–4116
9. S. Hauptmann, T. Mosell, S. Reiling and J. Brickmann (1996) Molecular dynamics simulations of the bulk phases of 4-cyano-4'-n-pentyloxybiphenyl, *Chem. Phys.*,**Vol. 208**, pp. 57–71
10. C. McBride, M.R. Wilson and J.A.K. Howard (1998) Molecular dynamics simulations of liquid crystal phases using atomistic potentials, *Mol. Phys.*, **Vol. 93, No. 6**, pp. 955-964
11. Glaser, M.A. (2000) Atomistic simulation and modeling of smectic liquid crystals in *Advances in the computer simulations of liquid crystals*,P. Pasini and C. Zannoni (eds) Kluwer, Dordrecht, pp. 263-331.
12. R. Car and M. Parrinello (1985) Unified approach for Molecular Dynamics and Density-Functional theory, *Phys. Rev. Lett.*, **Vol. 55, No. 22**, pp. 2471-2474
13. C. Amovilli, I. Cacelli, S. Campanile and G. Prampolini (2002) Calculation of the intermolecular energy of large molecules by a fragmentation scheme: Application to the 4-n-pentyl-4'-cyanobiphenyl (5CB) dimer, *J. Chem. Phys.*,**Vol. 117, No. 7**, pp. 3003–3012
14. S. Y. Yakovenko, A. A. Muravski, F. Eikelschulte and A. Geiger (1998) Temperature dependence of the properties of simulated PCH5,*Liq. Cryst.*, Vol. 24, pp. 657-671.
15. P. Hobza, H.L. Sezle and E.W. Shlag (1996) Potential energy surface for the benzene dimer. Results of *ab initio* CCSD(T) calculations show two nearly isoenergetic structures: T-shaped and Parallel-Displaced, *J. Phys. Chem.*,**Vol. 100**, pp. 18790–18794
16. S. Tsuzuki, K. Honda, T. Uchimaru, M. Mikami and K. Tanabe (2002) Origin of attraction and directionality of the π/π interaction: model chemistry calculations of benzene dimer interaction, *J. Am. Chem. Soc.*,**Vol. 124, No. 1**, pp. 104–112
17. *Gaussian 98 (Revision A.1)* , M.J. Frisch *et al.* (1998), Gaussian, Inc., Pittsburgh, PA
18. I. Cacelli, G. Cinacchi, C. Geloni, G. Prampolini and A. Tani (2002) Computer simulations of *p*-phenyls with interaction potentials from *ab initio* calculations, *Mol. Cryst. Liq. Cryst.*, in press
19. D.J. Cleaver, C.M. Care, M.P. Allen and M.P. Neal (1996) Extension and generalization of the Gay-Berne potential, *Phys. Rev. E*,**Vol. 54, No. 1**,pp. 559–567
20. A.D. Buckingam (1978) Permanent and induced molecular moments and long-range interaction forces, *Adv. Chem. Phys.*, **Vol. 12, pp.** 107–142
21. S. Tsuzuki and K. Tanabe (1991) *ab initio* molecular orbital calculations of the internal rotational potential of biphenyl using polarized basis sets with electron correlation correction, *J. Phys. Chem.*,**Vol. 95**, pp. 139–144
22. N. Metropolis, A. W. Rosenbluth, M. H. Rosenbluth, A. H. Teller and E. Teller (1953) Equation of state calculations by fast computing machines, *J. Chem. Phys.*, **Vol. 21, No. 6**, pp. 1087–1093
23. W. W. Wood (1968) *Physics of Simple Liquids*, edited by H. N. V. Temperley, J. S. Rowlinson and G. S. Rushbrooke, North-Holland
24. H. F. King (1972) Isobaric versus Canonical Ensemble formulation for Monte Carlo studies of liquids, *J. Chem. Phys.*, **Vol. 57, No. 5**, pp. 1837-1841
25. M. A. Miller, L. M. Amon, W. P. Reinhardt (2000) Should one adjust the maximum step size in a Metropolis Monte Carlo simulation?, *Chem. Phys. Lett.*, **Vol. 331**, pp. 278-284
26. J. Veillard-Baron (1974) The equation of state of a system of hard spherocylinders, *Mol. Phys.* **Vol. 28, No. 3**, pp. 809-818
27. A. Baranyai and T.R. Welberry (1991) Molecular dynamics of solid biphenyl, *Mol. Phys.*, **Vol. 73, No. 6**, pp. 1317–1334
28. W.L. Jorgensen and D.L. Severance (1990) Aromatic-aromatic interactions: free energy

profiles for the benzene dimer in water,chloroform and liquid benzene, *J. Am. Chem. Soc.*,**Vol. 112**, pp. 4768–4774

29. P. A. Irvine, C. Wu, P.J. Flory (1984) Liquid-crystalline Transitions in Homologous *p*-Phenylenes and their Mixtures, *J. Chem. Soc., Faraday Trans. 1* **Vol. 80**, pp. 1795-1806
30. T.J. Dingemans, N.S. Murthy, E.T. Samulski (2001) Javelin, hockey stick and boomerang-shaped liquid crystals. Structural variations on *p*-quinquephenyl, *J. Phys. Chem B* **Vol. 105**, pp. 8845-8860
31. R. Hashim, G. R. Luckhurst, S. Romano (1995) Computer simulation studies of anisotropic systems XXIV: Constant pressure investigations of the Smectic B phase of the Gay-Berne mesogen, *J. Chem. Soc. Faraday Trans.* **Vol. 91 No. 14**, pp. 2141-2148
32. M. A. Bates, G. R. Luckhurst (1999) Computer simulation studies of anisotropic systems XXX: The phase behavior and structure of a Gay-Berne mesogen, *J. Chem. Phys.* **Vol. 110 No. 14**, pp. 7087-7108
33. G. R. Luckhurst, P. S. J. Simmonds (1993) Computer simulation studies of anisotropic systems XXI: Parameterization of the Gay-Berne potential for model mesogens, *Mol. Phys.* **Vol. 80 No. 2**, pp. 230-252
34. E. De Miguel, E. Martin Del Rio, J. T. Brown, M. P. Allen (1996) Effect of the attractive interactions on the phase behavior of the Gay-Berne liquid crystal model, *J. Chem. Phys.* **Vol. 105 No. 10**, pp. 4234-4249
35. J. T. Brown, M. P. Allen, E. Martin Del Rio, E. De Miguel (1998) Effects of elongation on the phase behavior of the Gay-Berne fluid, *Phys. Rev. E* **Vol. 57 No. 6**, pp. 6685- 6699
36. E. De Miguel, C. Vega (2002) The global phase diagram of the Gay-Berne model, *J. Chem. Phys.* **Vol. 117 No. 13**, pp. 6313-6322
37. M. A. Bates, G. R. Luckhurst (1996) Computer simulation studies of anisotropic systems XXVI: Monte Carlo investigations of a Gay-Berne discotic at constant pressure, *J. Chem. Phys.* **Vol. 104 No. 15**, pp. 6696-6710
38. G.W. Smith, (1979) Phase behavior of some linear polyphenyls, *Mol. Cryst. Liq. Cryst.* **Vol. 49**, pp. 207-209
39. A. Göller, U.W. Grummt, (2000) Torsional barriers in biphenyl, 2,2'-bipyridine and 2-phenylpyridine, *Chem. Phys. Lett.*, **Vol. 321**, pp. 399-405
40. S. Tsuzuki, T. Uchimaru, K. Matsumura, M. Mikami, K. Tanabe (1999) Torsional potential of biphenyl:*Ab initio* calculations with the Dunning correlation consisted basis sets, *J. Chem. Phys.* **Vol. 110 No. 6**, pp. 2858-2861
41. I. Cacelli, G. Prampolini, *work in progress*
42. I. Cacelli, G. Cinacchi, G. Prampolini and A. Tani, *work in progress*
43. L. X. Dang (2000) Molecular dynamics study of benzene-benzene and benzene-potassium ion interactions using polarizable potential models, *J. Chem. Phys* **Vol. 113 No. 1**, pp. 266-273
44. W.F. Van Gunsteren, A.E. Mark, (1998) Validation of molecular dynamics simulation, *J. Chem. Phys.*, **Vol. 108**, pp. 6109-6116

PHASE TRANSFORMATIONS AND ORIENTATIONAL ORDERING IN CHEMICALLY DISORDERED POLYMERS - A MODERN PRIMER

LORIN GUTMAN

Department of Chemistry and Chemical Biology, Harvard University, Cambridge, MA 02138, Department of Chemistry, Massachussets Insititute of Technology, Cambridge, MA, 02139-4307

AND

EUGENE SHAKHNOVICH

Department of Chemistry and Chemical Biology, Harvard University, Cambridge, MA 02138

Abstract: We review recent theoretical and experimental developments in the field of complex polymers that carry a chemical and topological disorder. The polymer classes discussed include chemically disordered crosslinked heteropolymers and liquid crystalline heteropolymers. Field theory and spin glass averaging methods are very useful to study complex disordered polymers. The field theory allows explicit account of chain conformation contributions to thermodynamical quantities and a faithful representation of experimentally observable order parameters. Spin glass averaging methods presented here are instrumental for averaging over chemical and topological disorders. The review centers on theory development, predictions for conformational and orientational ordering, phase diagram analysis and also comparison with experimental results when possible.

J. Samios and V.A. Durov (eds.), Novel Approaches to the Structure and Dynamics of Liquids: Experiments, Theories and Simulations, 455–484.

Random heteropolymers (RHPs) with physical crosslinks are shown to exhibit three globular phases: frozen-globular with micro-domain structure, random-globular and frozen-random-globular. For RHPs with chemical crosslinks our theory predicts three frozen-globular phases, and one random-globular phase; the intra-frozen transitions are conformational transitions which do not require any re-entrant passages via the random-globular phase. The phase diagram of crosslinked RHPs is systematically explored in parameter and thermodynamic variable space, and physical explanations for the conformational organization and the order of the phase transitions is provided.
The second class of disordered polymers we review are many-chain mesogen/flexible disordered copolymers (DLCP). A field theory and creation-annihilation summation rules are proposed to carry out coupled orientational and conformational averages of polymer chain conformations in these complex systems. Predictions for the effect of flexibility, stiffness and inter-segment alignment on orientational ordering, the nematic/isotropic density threshold and the segmental orientational ordering at the nematic/isotropic transition is discussed in close proximity to experimental studies.

1. Introduction

Gels and polymer networks play a central role in the fabrication of impermeable tubes, tires, sieving matrixes [1], hydrolic fracturing gels (petroleum industries) [2], and self-regulating insulin delivery systems [3]. In these applications they are sealers of electrical components, elastomers, crack promoters of rocks, designing oil cleaning products, electrophoresis and chromatography supports, adsorbers, and thermo-gel responsive polymer matrices.

In a broad sense, gels can be categorized by the size of crosslink constituents in two sub-classes: particle gels, formed by colloid, particulate or globular heteropolymer aggregation [4], and gels formed by reversible / irreversible crosslinking of polymer chains. From the view point of the gel crosslink nature polymer gels may be divided into two sub-classes. One category is comprised of systems wherein the crosslink junctions freely reorganize as dictated by existing experimental conditions; these are physically crosslinked gels. The second category is comprised of systems in which the identity of segments involved in the crosslinks become fixed upon the removal of the crosslinking agent; systems in this class are referred to as chemical or irreversible gels.

Gel phase features and macroscopic performance are determined by the chain conformations adopted by the polymer networks [5], and the response of these conformations to external disturbances as stresses, electric fields, and thermodynamic conditions set by solvent quality, pH and temperature. The gel responsiveness is set to a large extent by gel preparation conditions (pre-gel state) [6], which ultimately determine the network topology, the gel nature (reversible or irreversible), and the gel chemical constitution. Irreversible gels constructed via disparate processing pathways render networks with qualitatively different thermodynamic and mechanical properties. Consider for example the synthesis of soft homopolymer vulcanizates; properties such as tensile and tear strength, stiffness and hardness display a shallow maximum with increase of the cure time. Other properties as elongation and permanent set drop continuously, while cure at higher temperatures shift the stress - strain curves to shorter times [5].

Thermo-mechanical response of physical and chemical homopolymer gels, was studied via spectroscopic and mechanical methods [7] (and refs. herein), and theory/computation [8]–[9] (and refs. herein), and many aspects of their properties and phase behavior are now fairly understood. Recently, the advent of the hetero-dyne/non-ergodic methods made progress possible for the study of dynamical and kinetic features of sol/gel systems. Experiments can now probe non self-averaging, gel meso-speckles, and essentially distinguish crosslink effects in relaxation spectrum from scattering of uncrosslinked monomers; these studies focused on a broad range of experimental conditions for homopolymer sol/gels and the emergence of dynamical regimes across the sol-gel transition, for physical [10] and chemical gels of polysacharides [11].

The theoretical study of random heteropolymer (RHP) gels offer unique challenges which do not need to be confronted in homopolymer gels. The distinctly new challenges arise due to multiple sources of frustration present in RHP gels among microscopic constraints and microscopic interactions; the constraint types are the fixed distribution of sequences and chain connectivity in the pre-gel RHP ensemble, and the crosslink distribution determined by the pre-gel state. These constraints and the inter-segment, segment-solvent, solvent-solvent interactions compete and effect strongly the conformational gel organization; they are a source of multiple free energy minima, a complex phase diagram and thermo-mechanical gel response observed in experiments. The strong need for theory in this field to determine and quantify the effects of these competing forces was recognized in a seminal experimental paper published in Nature by A. Annaka and T. Tanaka [12].

The effects of crosslinking on phase ordering were studied in few classes of simpler multi-component gels. In homopolymer blends e.g. [13], [14], random chemical crosslinks formed in the homogeneous phase preclude macro-phase separation and stabilizes formation of microphases. Other systems studied include inter-penetrating polymer networks (IPNs) [15] and protein gels [16]. Experimental studies of thermo-mechanical response in RHP gels revealed unprecedented phase features that include the occurrence of multiple volume transitions which depend strongly on molecular interaction features and ph [12].

Equilibrium properties of linear RHPs, i.e. the uncrosslinked pre-gel state, were analyzed by lattice theories [17] and field theoretic approaches [18]–[19]. The Flory-Huggins approach provided insights in the phase behavior of RHPs immersed in critical solvents (e.g. [20]), yet this approach cannot predict meso-ordering observed in experiments [21] and computer simulations [22] such as the formation of microphases, and the reduction of the total number of chain conformations to few dominant folds.

A new, modern theoretical view is needed to disclose, rationalize and predict physics of complex disordered polymeric materials. Competition among interactions and constraints in these systems originate new forms of ordering and clearly a microscopic explicit theoretical representation of the forces and costraints is needed. Field theory is a robust framework that can express these competing forces and also their manifestations at a longer scale in the form of order parameters. This approach is primarily used in particle physics but it has been adapted also to the study of some simple polymeric systems. In this review we report progress made in our group in developing field theoretic methods for the study of technologically relevant disordered polymeric materials.

In our group, we developed theoretical methods that allow to average a field theory over two quenched disorders. In the first part of the review we describe application of our approach to the study of two broad classes of technologically relevant materials, i.e. physically and chemically crosslinked RHPs in compact state at synthesis conditions [10]–[11]. Results for phase predictions are reviewed here in detail. Recently we also developed new diagramatic methods that allowed us to study the thermo-mechanical response of crosslinked non-compact RHPs at ambient gel conditions, [25], [26].

In the second part of the review we discuss application of a new method we developed to compute coupled conformational and orientational averages of chain conformations in disordered semiflexible heteropolymer solutions. Phase diagram results from application of these methods to study disordered liquid crystalline polymers (DLCPs), a class of paramount technological importance, will be reviewed. DLCPs have been recently the focus of an increasing experimental attention due to opportunities in controlling material features via sequence statistics and interactions; material properties of interest include giant optical non-linearities (GON) in photo-refractive response for non-linear wave mixing processes, thermal reorientation, light induced anchoring, and photo-alignment in Langmuir-Blodgett films [27]. The control of mentioned phenomena is relevant to information processing and storage, and also to fiber optics technology.

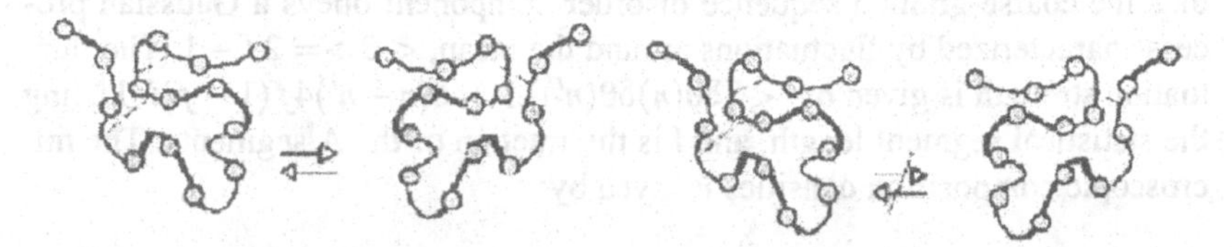

Fig. 1a Qualitative description of formation of composition specific physical crosslinks in random heteropolymers.

Fig 1b Qualitative description of topological contraints on conformations due to composition specific chemical crosslinks in random heteropolymers.

In sec. 2 we define Hamiltonians for physically and chemically crosslinked RHPS and also the detailed experimental aspects of the crosslinking scenarios treated. A brief description of the mathematical steps involved in the derivation will be given; the derivations were provided in great detail elsewhere [29], [30]. In sec. 3 we present the free energies equations of RHPs with physical and chemical crosslinks; in sec. 4 the phase diagrams are presented and the emerging physical picture is discussed while comparison is made among the manifestations of chemical and physical crosslinks in the phase behavior. In sec. 5 we introduced the theoretical modeling of DLCPs with emphasis on theoretical and experimental developments in the field. In sec. 6 we describe the physical motivation for the DLCP Hamiltonian and a description of the mathematical steps involved in the free energy for this system. Sec. 7 centers on broad theoretical predictions for nematic/isotropic ordering in disordered DLCPs in the context of recent optical microscopy measurements for these materials.

2. Theory development for physically/chemically crosslinked RHPs

We consider a solution of crosslinked RHP chains in globular states. Lets introduce first the Hamiltonian of a linear RHP:

$$H'_{RHP} = \frac{3}{2l}\int \dot{\mathbf{r}}(n)^2 dn + \frac{1}{2}\sum_{i,j=A,B}\int d\mathbf{r}\int d\mathbf{r}'\hat{\rho}_i^T(\mathbf{r})\mathbf{V}_{ij}(\mathbf{r}-\mathbf{r}')\hat{\rho}_j(\mathbf{r}) \quad (1)$$

r(n) designates the spatial location of the n'th segment, while $\theta(n)$ identifies the type of unit located at n. $\theta(n) = 1$ if at n there is an A segment; $\theta(n) = -1$ if at n there is a B segment. Here we consider ideal random heteropolymers, thus the coarse-grained sequence disorder component obeys a Gaussian process characterized by fluctuations around the mean, $< \theta >= 2f - 1$. The fluctuation strength is given by: $< (\delta\theta(n)\delta\theta(n')) >= \delta(n-n')4f(1-f)l$, l being the statistical segment length, and f is the fraction of the A segments. The microscopic composition densities is given by:

$$\hat{\rho}_A(\mathbf{r}) = \int dn\frac{1}{2}(1+\theta(n))\delta(\mathbf{r}-\mathbf{r}(n))$$

$$\hat{\rho}_B(\mathbf{r}) = \int dn\frac{1}{2}(1-\theta(n))\delta(\mathbf{r}-\mathbf{r}(n)) \quad (2)$$

and $\overline{\mathbf{V}}(ij)$, the binary inter-segment interactions matrix:

$$\overline{\mathbf{V}}_1 = \begin{pmatrix} V_{AA}(\mathbf{r}-\mathbf{r}') & V_{AB}(\mathbf{r}-\mathbf{r}') \\ V_{BA}(\mathbf{r}-\mathbf{r}') & V_{BB}(\mathbf{r}-\mathbf{r}') \end{pmatrix} \quad (3)$$

V_{AA}(r-r'), V_{AB}(r-r') and V_{BB}(r-r') are specific A-A, A-B, and B-B segment-segment interactions, respectively. For convenience, the Hamiltonian in eq. 1 is written in units of thermal energy, kT.

Imagine now a linear RHP chain described by the Hamiltonian in eq. 1 immersed in soluted composition specific crosslinking agents. First scenario, the cross-linking agent chemically equilibrated with the cross-linked moieties. For example proteins carrying cysteins and immersed in GSSG (glutathione), a linear crosslinking agent, become mixed di-sulfides by partial oxidation of the sulphydril groups on the cysteins; the crosslinking formation process is effectively a first order kinetic process since the second step of intra crosslink formation is very fast, 10^{-6}sec [28]. The two kinetic constants for formation and annihilation of the mixed disulfide defines an equilibrium constant for crosslink formation. This crosslinking scenario is representative of synthesis of physical crosslinks considered herein. The formation of physical crosslinks in linear RHPs is represented in fig. 1a. Clearly, two identical linear RHPs sequences can be continuously modified between distinct crosslink realizations without the requirement of chain connectivity modification to accommodate the crosslink shuffling [29]. Physical crosslinks can be accounted for as spatial constraints on the partition function of linear RHPs in a grand canonical ensemble formalism where the crosslink formation is controlled by generalized activities for homogeneous crosslinking a_{aa}, a_{bb} and heterogeneous crosslinking a_{ab} (viz. [29]).

Another type of crosslinks is possible as a continuation of the above described crosslinking scenario; upon the reach of chemical equilibrium, the crosslinking agent is either removed from solution or the crosslinks are frozen in space by acidification; under these chemical circumstances the crosslinks become fixed, and no further crosslink organization along the originally linear RHP is possible. The constraint setting determined by this crosslinking scenario is represented in fig. 1b. Clearly two identical RHP sequences having different fixed crosslink realizations are now topologically distinct, and no continuous deformation will transform one crosslinked RHP chain into the other in fig. 1b. a situation that is very different compared with the physical crosslinks depicted in fig. 1a. The fluctuations between different crosslink realizations for RHPs with chemical crosslinks are allowed by a generalized Poisson process: The probability of occurrence of M crosslinks of type A-A, J crosslinks of type A-B and K crosslinks of type B-B is linearly parameterized by μ_{AA}, μ_{BB} and μ_{AB}, and given by:

$$P[\mu's] = \frac{[\mu_{AA}]^M}{M!}\frac{[\mu_{AB}]^J}{J!}\frac{[\mu_{BB}]^K}{K!} exp[-(\mu_{AA}+\mu_{BB}+\mu_{AB})] \tag{4}$$

This description is consistent with the quenched crosslinking constraint description in homopolymer systems [31], [32]. In the present problem the average number of crosslinks are linearly parameterized by μ_{AA}, μ_{BB}, μ_{AB} (viz. [30]).

3. Outline of Solution

Regardless of the crosslink nature the free energy of the system has the following implicit form:

$$F = \sum_{[\theta],[r_i]} P_1([\theta])P_2([\theta],[r_i])log(Z([\theta],[r_i])) \tag{5}$$

$[\theta]$ represents one sequence realization while $[r_i]$ are the coordinates of one crosslink realization. $P_1([\theta])$ is the probability distribution of the quenched sequence, while $P_2([\theta],[r_i])$ is the conditional probability for one realization of composition specific cross-links at a given fixed sequence. Note that our definition of crosslink formation closely follows the crosslink preparation process;

for each given sequence realization, crosslinks can form by fixing composition specific inter-segment contacts from spontaneously occurring chain conformations adopted by the linear RHPs. The algebra for the chemical and physical gels made of disordered RHPs was carried out by us in detail in [10], [11].

The resulting implicit form of the free energy for crosslinked RHPs has the form of F=E-TS, explicitly:

$$E_{physical} = V(a_{i,j},f)\sum_k \int d\mathbf{r}_k \rho_k(\mathbf{r}_k)^2 - \frac{1}{2}Trlog(\frac{1}{2\chi_f(\mu' s)}\delta_{k,k'} - \frac{\sigma^2}{4}Q_{k,k'}) \quad (6)$$

where k is the replica index.

$$V(a_{i,j},f) = 0.5\overline{V_0} + 0.25(\overline{V_{AA}} - \overline{V_{BB}}\overline{\theta}) - 0.25\overline{\chi}\overline{\theta}^2 \quad (7)$$

where

$$\begin{aligned} \overline{V}_{AA} &= V_{AA} - a_{aa} \\ \overline{V}_{BB} &= V_{BB} - a_{bb} \\ \overline{V}_{AB} &= V_{AB} - a_{ab} \end{aligned} \quad (8)$$

for a description of the physical meaning of the overlaps $Q_{k,k'}$ viz. [29]. The composed interactions are:

$$\begin{aligned} \overline{V}_0 &= \frac{1}{4}(\overline{V}_{AA} + \overline{V}_{BB} + 2\overline{V}_{AB}) \\ \overline{\chi}_F &= \frac{1}{2}(\overline{V}_{AB} - \overline{V}_{AA} + \overline{V}_{BB}) \end{aligned} \quad (9)$$

The entropy of RHPs with physical crosslinks may be expressed in the following manner:

$$S_{physical} = log(<\prod_{k,k'} \delta(Q_{k,k'} - \hat{Q}_{k,k'})>) \quad (10)$$

<> is the chain conformation average.

For chemical crosslinks:

$$\begin{aligned} E_{chemical} = {} & V(\mu' s, f)\sum_{k,\alpha} \int d\mathbf{r}_k^\alpha \rho_k^\alpha(\mathbf{r}_k^\alpha)^2 - \\ & V(\mu' s)\sum_{k,\alpha,\beta} \int d\mathbf{r}_k^\alpha d\mathbf{r}_k^\beta Q_{k,k}^{\alpha,\beta}(\mathbf{r}_k^\alpha, \mathbf{r}_k^\beta)^2 + \\ & \frac{1}{2}Trlog(\frac{1}{2\chi_f(\mu' s)}\delta_{k,k'}\delta_{\alpha,\beta} - \frac{\sigma^2}{4}Q_{k,k'}^{\alpha,\beta}) \end{aligned} \quad (11)$$

with the interaction parameters defined as:

$$\chi_F = 0.5(2V_{AB} - V_{AA} - V_{BB})$$
$$V(\mu's) = 0.5(\mu_{AA} + \mu_{BB} + 2\mu_{AB})$$
$$V(f) = 0.5V_0 + 0.25(V_{AA} - V_{BB})\overline{\theta} - \chi_F\overline{\theta}^2)$$
$$\chi_{Fcl}(\mu's) = 0.5(2\mu_{AB} - \mu_{AA} - \mu_{BB})$$
$$V(f,\mu's) = 0.5\,V(\mu's) - 0.25(\mu_{AA} + \mu_{BB})\overline{\theta} -$$
$$0.25\chi_{Fcl}(\mu's)\overline{\theta}^2; \overline{\chi(\mu's)} = \chi_F - \chi_{Fcl}(\mu's) \tag{12}$$

for discussion on the physical meaning of overlap fields $Q_{k,k'}^{\alpha,\beta})$ for chemical crosslinks viz. [30]. The entropy of RHPs with chemical crosslinks may be expressed in the following manner:

$$S_{chemical} = log\left(\left\langle \prod_{k,k',\alpha,\beta} \delta\left(Q_{k,k'}^{\alpha,\beta} - \hat{Q}_{k,k'}^{\alpha,\beta}\right) \prod_{k,k',\alpha,\beta,\gamma} \delta\left(Q_{k,k'}^{\alpha,\beta,\gamma} - \hat{Q}_{k,k'}^{\alpha,\beta,\gamma}\right) \right.\right. \tag{13}$$
$$\left.\left. \prod_{k,k'',\alpha,\beta,\gamma,\delta} \delta\left(Q_{k,k'}^{\alpha,\beta,\gamma,\delta} - \hat{Q}_{k,k'}^{\alpha,\beta,\gamma,\delta}\right) \right\rangle\right)$$

$\langle\rangle$ is the chain conformation average.

A simple scaling argument can be made [30] following the lines of [33] (viz. references herein) which shows that a one step RSB calculation is proper for RHPs with physical crosslinks in compact states. This argument has been recently generalized for RHPs with chemical crosslinks, and it has been shown that one step calculation is also suitable.

The free energies calculated with Parisi Anstaz [34] are:

$$F_{physical} = -\frac{1}{x_0}(\gamma + log(1 - cx_0)) \tag{14}$$

with

$$c = \frac{\sigma^2}{2}\rho\overline{\chi_F} \;\; ; \;\; \overline{\chi_F} = \chi_F - \frac{1}{2}(2a_{ab} - a_{aa} - a_{bb}) \tag{15}$$

and

$$F_{chemical} = \frac{\gamma}{x_0 x_0'}(1 - 2(x_0 + x_0') +$$
$$(2x_0' - 1)(x_0 - 1)(x_0 - 2)^2)$$
$$+ \frac{log(1 - \mu_a x_0 x_0')}{x_0 x_0'} - V(\mu's, f)(x_0 - 1) \tag{16}$$

with:

$$\mu_a = \frac{\sigma^2}{2}\rho\chi\overline{(\mu's)}^2 \tag{17}$$

The order parameter which measures the total reduction of chain conformations to a few dominant folds, ξ_0, due to energetic and entropic constraints, is given by:

$$\xi_0 = 1 - \sum p_i^2 \tag{18}$$

p_i is the probability of the i'th fold. For many chain conformations, each conformation acquires a low probability, and ξ_0 practically equals one. When the chain is collapsed into one conformation, i, p_i=1 while other chain conformations with i$\neq$ j, p_j=0 implying that ξ_0=0. The proper one step parameterization of ξ for RHP with chemical crosslinks was shown [30] to be $\xi = x_0 x'_0$ For chemical crosslinks, when x'_0=1, the reduction to a few dominant folds occurs due to the quenched sequence only; under such circumstances $\xi_0 \to x_0$, and the free energy of RHP with chemical crosslinks given in eq. 17 becomes almost identical with that of RHPs with physical crosslinks. This is not surprising since the grand canonical description of physical crosslinks resembles to a large extent the Poisson distribution for the crosslinks with the assumption of the crosslink chemical equilibration with the crosslinking species. On the other hand, if $x_0 \to 1$ then $x'_0 = x_0$ and few dominant folds may occur solely due to the presence of chemical crosslinks: The free energy of such a scenario is:

$$F_{cl} = \frac{1}{x'_0}(\gamma(1 - 2x'_0 + (x'_0 - 1)(x'_0 - 2)^2) + log(1 - \mu_a x_0) - V(\mu's, f)x_0^2 \tag{19}$$

The values of x_0 and x'_0 in the pertinent free energy sectors, and the stability of F with respect to x_0 and x'_0 are computed numerically using equation 16, x_0, and x'_0 are computed first, and then from them ξ_0 is computed. The order of calculation is crucial; for example the situation of $x_0 > 1$, $x'_0 < 1$ requires to first set $x_0 = 1$, and only then ξ_0 is computed from $x'_0 x_0 = \xi_0$. While the value of ξ_0 obtained is used to determine the occurrence of freezing, the value and stability of the free energy sectors (sequence only, crosslink only or sequence+crosslinks) provides essential information on the sensitivity of the micro-domain structure of the frozen phase to the multi-form disorder components.

Below, we study numerically the phase diagram dependence on interaction parameters of the RHP, temperature, disorder fluctuations, and number of composition specific crosslinks. These phase diagrams provide important information on conformational organization of globular RHP gels.

4. Phase diagram analysis

Lets analyze the phase diagram of RHPs with physical crosslinks first; in fig. 2 we display the temperature dependence of RHPs with physical crosslinks in the presence of heterogeneous crosslinkers (the generalized activity for heterogeneous crosslinking is positive) and a positive, and relatively large χ_{Flory}. At low temperatures, and relatively small generalized activities of the hetero-

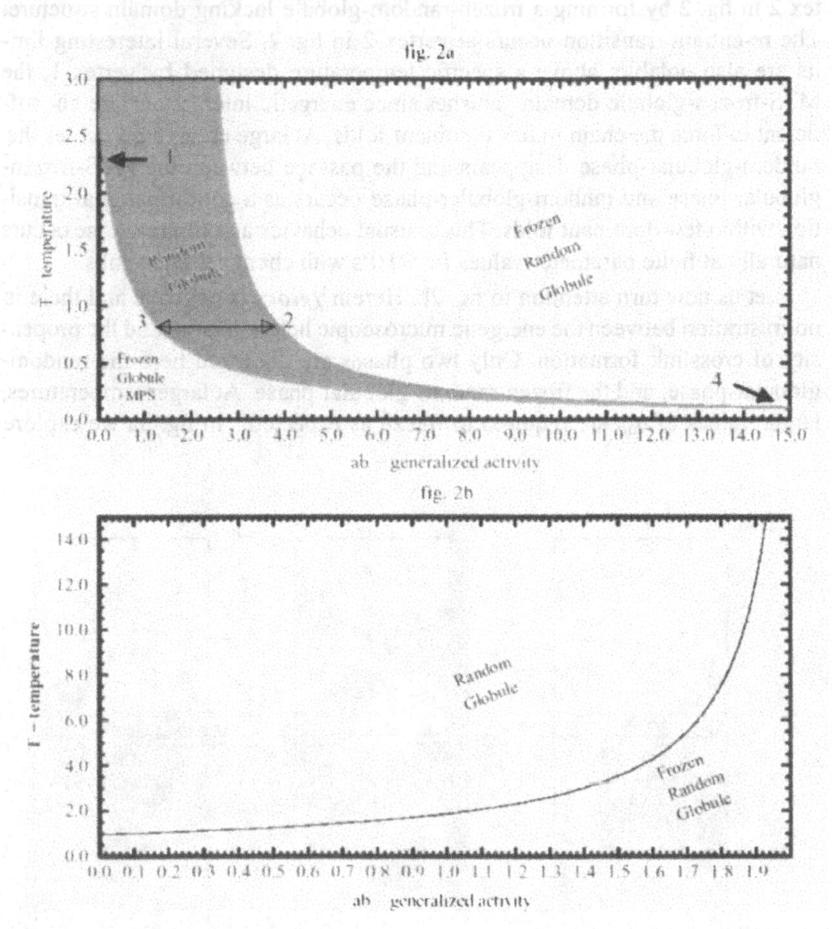

Fig. 2 Phase diagram analysis of RHPs with composition specific and annealed crosslinks; 2a - l=1, a_{aa}=0, a_{bb}=0, χ_F=3., ρ=1., v_0=0.8, f=0.5 ; 2b - l=1, a_{aa}=0, a_{bb}=0, χ_F=-4., ρ=1., v_0=0.8,f=0.5

geneous crosslinking agent, the RHP energetic interactions are dominant, and we expect formation of few dominant folds with segregated domain structures (viz. frozen globule + MPS phase). An isothermal increase in the generalized activity for heterogeneous crosslinking enhances frustration between the tendency of microphase formation and the domain mixing induced by the heterogeneous crosslink formation. When a_{ab} is large enough the chain takes advantage of many chain conformations and exits the frozen domain at vertex 3 in fig. 2. At larger a_{ab} values the chain reenters the frozen phase at vertex 2 in fig. 2 by forming a frozen-random-globule lacking domain structure. The re-entrant transition occurs at vertex 2 in fig. 2. Several interesting limits are also notable; above a specific temperature designed by vertex 1, the MPS-frozen-globule domain vanishes since energetic interactions are not sufficient to force the chain in few dominant folds. At large enough a_{ab} values the random-globular-phase disappears and the passage between the MPS-frozen-globular phase and random-globular-phase occurs as a conformational transition within few dominant folds. This unusual behavior and limiting case occurs naturally at finite parameter values for RHPs with chemical crosslinks.

Let us now turn attention to fig. 2b. Herein χ_{Flory} is negative and there is no frustration between the energetic microscopic heterogeneity and the propensity of crosslink formation. Only two phases are observed here the random-globular-phase, and the frozen-random-globular phase. At larger temperatures, larger values of a_{ab} are required to freeze as expected. In fig. 3a we explore

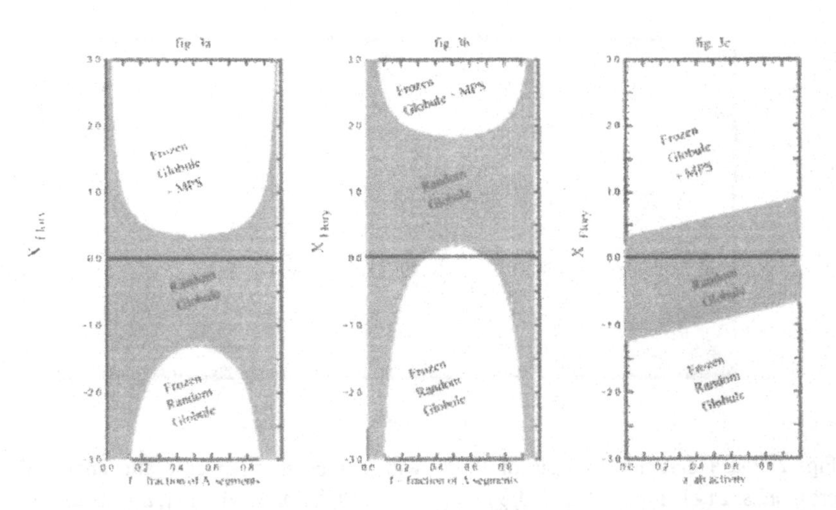

Fig. 3 Frozen phase analysis of RHPs with composition specific and annealed crosslinks. 3a - l=1, a_{aa}=0, a_{bb}=0, a_{ab}=0, ρ=1., v_0=0.8, T=0.3; 3b - l=1, a_{aa}=0, a_{bb}=0, a_{ab}=2.5, ρ=1., v_0=0.8, T=0.3; 3c - l=1, a_{aa}=0, a_{bb}=0, ρ=1., v_0=0.8, T=0.3, f=0.5

the explicit dependence of χ_F on the fraction of A segments. At large sequence fluctuations spatial segregation of micro-domains readily occurs. Thus freezing occurs at large enough χ_F with domain structure, and at very negative χ_F values with no domain structure. The χ_F region between these two extremes (the dark background region), is a random-globular-phase wherein many chain conformations are available. Fig. 3b has a higher value for the propensity for heterogeneous crosslink formation measured by a_{ab}. The phase diagram presented in fig. 3b is shifted upwards relative to fig. 3b; larger χ_F are required to freeze and counteract the mixing propensity due to heterogeneous crosslink formation. Note also the interesting inverted region which occurs at the bottom of the graph, and not predicted heretofore for any RHP system.

Lets look along the zero χ_{Flory} line. Under such conditions no freezing can occur due to energetic interactions. However at large enough sequence fluctuations the segments can spatially mix and crosslink leading to a few dominant folds. The form of this phase diagram is indeed inverted and the transition is between a random-globular phase to frozen-random-globular phase and again to random-globular-phase. Fig. 3c is a phase diagram analysis in the χ_F generalized activity and a_{ab} space at fixed fraction of A segments. Under these circumstances both vertical (fixed χ_F) and horizontal (fixed a_{ab}) re-entrant transitions occur. In fig. 4a the χ_F dependence on temperature at zero

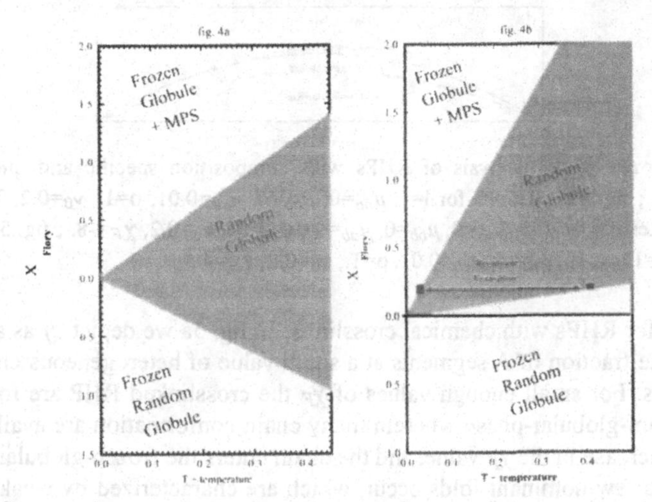

Fig. 4 Frozen phase analysis of RHPs with composition specific and annealed crosslinks. 4a - l=1, a_{aa}=0, a_{bb}=0, a_{ab}=1, ρ=1., v_0=0.8, f=0.5; 4b - l=1, a_{aa}=0, a_{bb}=0, a_{ab}=2.3, ρ=1., v_0=0.8, T=0.3

crosslinks is presented. No re-entrant transition along the horizontal axis is observed here as there insufficient heterogeneous crosslinks to induce freezing at a vanishing χ_F; thus it is not possible to reduce the total number of folds to few dominant ones that lack domain structure at high T's and positive χ_F. However a re-entrant transition along the vertical axis (fixed χ_f) is expected as for sufficiently large absolute values of χ_F segregation or significant mixing of the composition profiles will occur. In fig. 4b the propensity of heterogeneous crosslinks is large, and the re-entrant transition along the horizontal axis occurs here (the transition points are marked by the two rectangles bounding the horizontal double headed arrow). Let us now turn attention to the phase diagram

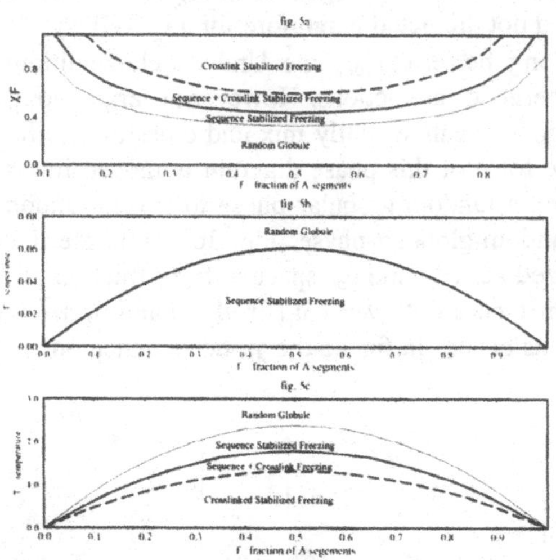

Fig. 5 Frozen phase analysis of RHPs with composition specific and quenched crosslinks ; fig. 5a - Results for l=1, μ_{aa}=0, μ_{bb}=0, μ_{ab}=0.01, ρ=1., v_0=0.2, T=0.2 ; fig. 5b - Results for l=1, μ_{aa}=0, μ_{bb}=0, μ_{ab}=0.01, ρ=1., v_0=0.2, χ_F=-8. ; fig. 5c - Results for l=1, μ_{aa}=0, μ_{bb}=0, μ_{ab}=0.01, ρ=1., v_0=0.2, χ_F=4.

analysis for RHPs with chemical crosslinks. In fig. 5a we depict χ_f as a function of the fraction of A segments at a small value of heterogeneous chemical crosslinks. For small enough values of χ_F the crosslinked RHP are found in the random-globular-phase wherein many chain conformation are available; a further increase in the χ_F value, and the chain enters the frozen-globular-phase wherein a few dominant folds occur, which are characterized by weak segregated domain formation (viz. thin solid line in fig. 5a). At even larger χ_F values the micro-domain regime becomes sharper and a significant domain reorganization wherein the heterogeneous crosslinks located themselves in the vicinity of the micro-domain interfaces; the interfacial region is not sharp enough to orient the heterogeneous crosslinks toward the favorable domain. This transition is continuous, as the order parameter changes continuously from one phase to another (viz. [30]). The transition line is depicted here by the thick line in fig. 5a. At a even larger χ_F the micro-domain interfacial region becomes sharper expectedly of the order of the crosslink size. Under these circumstances we expect that positioning the heterogeneous crosslinks at the interface as an entropic constraint significantly lowers the interfacial free energy and provide nucleation centers for inter-domain interfaces. This sudden drop in the interfacial free energy allows formation of a large quantity of interface and leads to the formation of a large number of smaller micro-domains bounded by crosslinks. Our numerical calculation shows that the occurrence of this conformational organization within the frozen phase occurs by a first order phase transition. This is clearly the case since the order parameter ξ presents a discontinuity at the transition threshold. These conformational transitions occur for chemical crosslinks within few dominant folds.

Fig. 5b depicts the scenario of χ_F is negative, and also a_{ab} differs from zero. There is no energetic-crosslinks induced frustration here, and one transition between random-globular phase to sequence-frozen globular phase is observed. Fig. 5c depicts the temperature dependence on the fraction of A segments f at positive inter-segment interaction dissimilarities and a_{ab}=0.01. Two conformational transitions occur here within few RHP globular folds but the order of occurrence is inverse compared with fig. 7a. By temperature increase we pass the crosslink-frozen-globular phase, mixed-frozen-globular phase, and the sequence-frozen-globular phase all the way toward the random-globule wherein many chain conformations are being sampled.

5. Liquid crystalline disordered heteropolymers

Below we review of recent theoretical, computational and experimental progress attained in understanding orientational ordering in homopolymer, alternating mesogen/flexible, and disordered LCPs. Liquid crystalline homopolymers were studied by lattice models, analytical variational methods and also by field theoretic methods. Flory and co. [35] used lattice models with some success to study phase separation in LCPs. A variational approach based on Onsager [36] and Mayer and Saupe [37] developments in liquid crystals (LC's), was also employed to study LCP's for several mechanisms of chain flexibility and anisotropic interaction potentials [38]; Onsager approach was also employed in conjunction with lattice models to correct for artifacts [39] in phase predictions by lattice models. The effect of an external alignment field on single chain properties was also investigated [40].

Few studies attempted to bridge predictions of analytical theories, matrix based computational methods [41], and computer simulations [42], [43] of LCP's. Using a united atom approach that employs the Amber code potential (LJ, rotational potential, Coulombic and harmonic bonds) Jung et al. [41] computed the effective chain persistence length and compared with predictions from the rotational isomeric state model. Based on Onsager theory, Zheng et al. [44] studied nematic ordering in many-chain nematic polymers with anisotropic interactions and compared with Khokhlov and Semenov (KS) approach [45].

More recently, a field theoretic formalism of many chain LCP's based on a microscopic Hamiltonian was proposed by Gupta and Edwards (GE) [57]. In GE calculation a microscopic formulation of the interactions is employed, and a non-perturbative solution of the free energy was obtained in limit of long chains, and the phase diagram elucidated. The GE approach was also employed to study orientational ordering in semi-flexible homopolymers embedded in flexible surfaces [46]; the field theoretic formalism is perhaps the most promising for quantitative comparison with experiments.

Sequence disordered LCPs received notable experimental attention and novel properties not present in liquid crystalline homopolymers, were predicted, few summarized below. Baharadwaj et al. [47] studied Vectra, i.e. a disordered liquid crystalline polymer made of aromatic co-polyesters of p-hydroxybenzoic acid (HBA) and 2-hydroxy 6-naphtoic acid HNA, 70/30 HBA/-HNA, by NPT ensemble simulation of 12 chains, 10 monomeric units each. In nematic LC state the chains were extended (RMS measured in orthorombic cell). At 298° ester groups were shown to line up in register despite the presence of chemical disorder. Butzbach et al. studied [48] the structure formation and registry in long, sequence disordered LCP of HBA(stiff)-HNA(flexible) with the N/I transition at 300 C. Unit lengths of HBA were 6.35 A and of HNA, 8.37 A. In melt state an increase in registry was also inferred by width of meridional reflections. A detailed calculations of scaling lengths, Kuhn segment length, C_∞ and persistence length of two letter LCPs with A segments stiff and B segments flexible was performed by Rutledge et al. [49], while Percek et al. [50] studied the even-odd effects in MBPE - methylene (17, 18, 19, 20) alternating ordered polymers made of stiff and flexible units. The even-odd effect is associated with large zigzag jumps in T_{IN} with change in size of the flexible spacers. In the present case T_{IN} maximum zig-zag gap found was to be $\sim 120^o$. This effect was shown to decrease with increasing length of flexible units. Thermotropic alternating polyesters, were studied by Blumstein et al. [51]; in their studies a first order N/I transition and a decrease in the even-odd effect with molecular weight was observed.

Finally Stupp et al. [52, 53] focused on ordering in alternating and sequence - disordered LCP. For the sequence disordered LCPs, optical mesophases were observed in a large temperature window $\sim 120^o$, and these finding were attributed to sequence-disorder effects and finite chain effects. While experimental knowledge on conformational orientational organization has accumulated, an understanding and study of these phenomena by theory is incipient; one theoretical study predicted that sequence disorder effects on ordering diminish with increase in chain length; this results was shown to be correct in sequence disordered LCP's with irrelevant dissimilarity in stiffness of disparate segments [54]. In order to set up a more general theoretical framework for providing with an understanding of the rich physics displayed in experiments for sequence disordered LCP's, we construct a field theory of LCPs made of stiff mesogens and flexible spacers randomly distributed on the chain. Our model is an adequate description of scenarios in which one segment is stiff and the other is flexible which is the case in many disordered LCP's synthesized [55].

6. Model Development for DLCPs

The microscopic Hamiltonian for a solution of many-chain sequence disordered LCP's made of mesogenic segments (A's) and flexible segments (B's) is given by:

$$H = \sum_i \int \left(\left(\frac{1-\theta(n_i)}{2} \right) \frac{3}{2l} \mathbf{u}^2(n_i) + \left(\frac{1+\theta(n_i)}{2} \right) \frac{\beta\varepsilon}{2kT} \dot{\mathbf{u}}^2(n_i) \right) dn_i$$
$$+ \frac{w}{8} \sum_{i,j} \int dn_i \int dn'_j (1+\theta(n_i))(1+\theta(n'_j)) \delta(\mathbf{r}(n_i) - \mathbf{r}(n'_j)) (\dot{\mathbf{r}}(n_i) \times \dot{\mathbf{r}}(n'_j))^2$$
$$+ \frac{u}{2} \sum_{i,j} \int dn_i \int dn'_j \delta(\mathbf{r}(n_i) - \mathbf{r}(n'_j)) \quad (20)$$

$\mathbf{r}(n_i)$ is the spatial location of the n'th segment of the i'th chain, $\mathbf{u}(n_i)$ is the chain tangent at n_i for the i'th chain and $\theta(n_i)$ is the chemical composition variable of the n_i'th segment on the i'th chain. $\theta(n_i) = 1$ for an A segment and $\theta(n_i) = -1$ for a B segment. The sequence heterogeneity, represented by fluctuations in the segment composition along the chain contour obeys a Gaussian process with mean, $<\theta>= 2f-1$, $\delta\theta = \theta(n_i) - <\theta>$, and sequence fluctuations $\overline{\delta\theta^2} =< (\delta\theta(n_i)\delta\theta(n'_j)) >= \delta(n_i - n'_j) 4f(1-f)l$ [56]; f is the fraction of A segments, and l is the statistical segment length. The first term in eq. 20 represents the nearest-neighbors harmonic interaction potential of flexible segments of type B, while the second term in eq. 20 is the bending potential of the stiff units, A's. The third term in eq. 20 represents the pair anisotropic potential (viz. also [57]) responsible for alignment of stiff A-A pairs; this energetic penalty is zero for 100 % aligned tangents of A segments adjacent in space, and w for normal tangents. The A-A anisotropic interaction potential, w, contains athermal interactions due to anisotropic hard core potential and also thermotropic contributions due to soft-core dispersion forces [57]. A schematic description of a many-chain sequence disordered LCP snapshot with interactions among mesogen segments emphasized by line contour is displayed in fig. 6. The fourth term in eq. 20, is the excluded volume inter-segment interactions.

Fig. 6 A solution of many-chain sequence disordered LCP

The Hamiltonian of eq. 20 is a rigorous description of the physical system, and accounts for microscopic connectivity, flexibility, mesogenity mechanisms and mesogen alignment in a disordered ensemble of chains; this Hamiltonian provide a description of mesoscopic and macroscopic orientational ordering which goes far beyond simple thermodynamical arguments.

The partition function of an ensemble of sequenced disordered LCPs is given by:

$$Z[\theta(n_i),\mathbf{u}(n_i)] = \underline{\int\!\!\int} \prod_{n_i} D\mathbf{u}_i(n_i) exp[H(\mathbf{u}(n_i),\theta(n_i))] \tag{21}$$

In order to make progress toward an analytical solution we re-express eq. 20 in a more compact form with tensor fields:

$$H = \sum_i \int \left(\left(\frac{1-\theta(n_i)}{2} \right) \frac{3}{2l} \mathbf{u}^2(n_i) + \left(\frac{1+\theta(n_i)}{2} \right) \frac{\beta\varepsilon}{2kT} \dot{\mathbf{u}}^2(n_i) \right) dn_i$$
$$+ \frac{w}{2} \int d\mathbf{r} [\hat{\sigma}^{ii}(\mathbf{r})\hat{\sigma}^{jj}(\mathbf{r}) - \hat{\sigma}^{ij}(\mathbf{r})\hat{\sigma}^{ji}(\mathbf{r})] + \frac{u}{2} \int d\mathbf{r} \hat{\rho}(\mathbf{r})^2 \tag{22}$$

with

$$\rho(\hat{\mathbf{r}}) = \sum_k \int dn_k \delta(\mathbf{r} - \mathbf{r}(n_k)) ;$$

$$\hat{\sigma}^{ij}(\mathbf{r}) = \sum_k \int dn_k \left(\frac{1+\theta(n_k)}{2} \right) \delta(\mathbf{r} - \mathbf{r}(n_k)) \mathbf{u}_k^j(n_k) \mathbf{u}_k^i(n_k) \tag{23}$$

where Einstein summation notation is employed. The partition function is expressed with continuous orientational tensors by imposing appropriate delta function constraints, while the fluctuations in the magnitude of the chain director of the stiff segments are constrained globally (viz. [57], [58]). The partition function at fixed sequence is given by:

$$Z[\theta(n_i),\mathbf{u}(n_i)] =$$

$$\int\!\!\int [\prod_{n_i,\mathbf{r}} D\mathbf{r}_i(n_i) D\sigma(\mathbf{r})] exp(H[\hat{\sigma}(\mathbf{r}),\theta(n_i)]) \tag{24}$$

$$\prod_{n_i,\mathbf{r}} \delta[\sigma(\mathbf{r})-\hat{\sigma}(\mathbf{r})]\delta[\mathbf{u}(n_i)^2\frac{(1+\theta(n_i))}{2}-1]$$

The delta function constraints in eq. 25 are now expressed with auxiliary fields.

$$\int\!\!\int [\prod_{\mathbf{r},jk} D\psi^{jk}(\mathbf{r})] exp\left(-i\int d\mathbf{r}\sum_{jk}\psi^{jk}(\mathbf{r})[\hat{\sigma}^{jk}(\mathbf{r})-\sigma^{jk}(\mathbf{r})]\right) \tag{25}$$

$\hat{\sigma}(\mathbf{r})$ is given in eq. 1; $\psi(\mathbf{r})$ are fields auxiliary to $\sigma(\mathbf{r})$.

The orientational global constraint on the magnitude of the segment tangents of the stiff segments sets the average magnitude of the tangent of the stiff segments to 1 for each chain i, via the auxiliary fields λ_i (viz. [58] for further details).

$$\int \prod_i d\lambda_i exp\left(-i\sum_i \int dn_i [\mathbf{u}(n_i)^2\left(\frac{1+\theta(n_i)}{2}-1\right)]\right) \tag{26}$$

The statistical mechanical problem expressed with the Hamiltonian in eq. 20 can also be viewed as a quantum problem expressed with the imaginary time action functional, i.e. $\frac{it}{\hbar}=n$, and $H(\frac{it}{\hbar}) \Leftrightarrow L_{im}(n)$. Quenched average over the sequence dependent free energy must be performed over all sequence realizations of the LCP. In order to carry the averaging procedure efficiently, the action functional is expressed by sequence fluctuations $\delta\theta(n)$ dependent terms and average $<\theta>$ terms:

$$\begin{aligned} L_{im} = \quad & \textstyle\sum_i \int dn_i \left(-A\dot{\mathbf{u}}^2(n_i) - \mathbf{u}(n_i)\cdot\mathbf{h}(\mathbf{r}(n_i))\cdot\mathbf{u}(n_i)\right) \\ & -\textstyle\sum_i \int dn_i \frac{\delta\theta(n_i)}{2}\left(C\dot{\mathbf{u}}^2(n_i) + \mathbf{u}(n_i)\cdot\mathbf{S}(\mathbf{r}(n_i))\cdot\mathbf{u}(n_i)\right) \\ & -\textstyle\int d\mathbf{r}\frac{w}{2}\left(Tr[\sigma(\mathbf{r}):\sigma(\mathbf{r})] - Tr(\sigma(\mathbf{r}))^2\right) + i\int d\mathbf{r}\psi(\mathbf{r}):\sigma(\mathbf{r}) \\ & +i\textstyle\sum_i \lambda_i Lk \end{aligned} \tag{27}$$

with

$$\mathbf{h}(\mathbf{r}(n_i)) = \mathbf{I}\left(\frac{3(1-\overline{\theta})}{4l} + \frac{i\lambda(1+\overline{\theta})}{2}\right) + \frac{i(1+\overline{\theta})}{2}\psi(\mathbf{r}(n_i));$$

$$\mathbf{S}(\mathbf{r}(n_i)) = \mathbf{I}\left(-\frac{3}{2l} + i\lambda_i\right) + i\psi(\mathbf{r}(n_i)) ;$$

$$A = \frac{\beta\varepsilon(1+\overline{\theta})}{4} C = \frac{\beta\varepsilon}{2} ; k = \frac{(1+\overline{\theta})}{2} \tag{28}$$

I is the identity matrix. The partition function is now given by:

$$Z = \overline{\int\int} \prod_{n_i,\mathbf{r}} D\mathbf{u}_i(n_i) D\sigma(\mathbf{r}) D\psi(\mathbf{r}) \prod_i d\lambda_i exp[L_{im}(\sigma(\mathbf{r}), \psi(\mathbf{r}), \theta(n_i), \mathbf{u}_i(n_i))] \tag{29}$$

The average of the log (Z) is performed non-pertubatively with the replica trick [59]. This trick implies that all fields are replicated and couple to the quenched sequence realization $[\delta\theta(n_i)]$. m-replicas are introduced and the free energy is averaged over all realization of the Gaussian sequence disorder distribution with $\overline{\delta\theta^2} = <(\delta\theta(n_i)\delta\theta(n'_j))> = \delta(n_i - n'_j)4f(1-f)l$.

Below, the effective disorder-averaged action functional, is expressed with the principal axis representation of the average tensor fields ψ and σ. In that representation the tensor fields ψ^{ij}, and σ^{ij} become diagonal, i.e. $\sigma^{ij} \to \sigma_\alpha$ and $\psi^{ij} \to \psi_\alpha$.

$$L_{im} = \sum_{i,\beta} \int dn_i \sum_\beta \left(-A\sum_\alpha (\dot{u}_\alpha^\beta)^2(n_i) - \sum_\alpha (u_\alpha^\beta(n_i))^2 \mathbf{h}_\alpha^\beta\right)$$

$$+ \left(\sum_i \int dn_i \frac{\overline{\delta\theta^2}}{8} \left(\sum_\beta \left(C\sum_\alpha (\dot{u}_\alpha^\beta)^2(n_i) + (u_\alpha^\beta(n_i))^2 \mathbf{S}_\alpha^\beta\right)\right)^2\right)$$

$$- \int d\mathbf{r}^\beta \frac{w}{2}[Tr[\sigma^\beta : \sigma^\beta] - (Tr(\sigma^\beta)^2)] + i\sum_\beta \int d\mathbf{r}^\beta \psi^\beta : \sigma^\beta + i\sum_{i,\beta} \lambda_i^\beta Lk \tag{30}$$

The disorder averaged Lagrangian is no longer local in replica indexes but is still local in chain index. Thus the partition function can be exactly written as:

$$Z = \overline{\int\int} D_{fields} exp(Mlog(L_{im-sc}[fields])) exp(L_{im-nci}[fields]) \tag{31}$$

where *sc* stands for single-chain-index, and *nci* for not-dependent-on-chain-index.

It is easy to see that a normal mode representation of eq. 30 ([57]) presents computational complications in doing conformational and orientational averages in L_{im-sc}. The Rouse modes couple and exact summation of the Rouse modes, i.e. the crux of the non-perturbative solution in the homopolymer nematogen problem [57], cannot be efficiently carried out for sequence-disordered LCP's.

Next we shift the partition function in eq. 25 to a Hamiltonian representation. For computational convenience, we exchange the Lagrangian with a Hamiltonian representation [60].

After some tedious algebra we derive the imaginary time Hamiltonian:

$$\begin{aligned}
H_{sc} = & -\tfrac{1}{4A}\sum_\beta(\mathbf{p}^\beta)^2+\sum_{\beta,\alpha}h_\alpha^\beta(q_\alpha^\beta)^2+\tfrac{3\overline{\delta\theta^2}}{128A^4}(\sum_\beta(\mathbf{p}^\beta)^2)^2\\
& +\tfrac{\overline{\delta\theta^2}C2}{32A^2}\left(\sum_\beta(\mathbf{p}^\beta)^2(\sum_{\alpha,\gamma}S_\alpha^\gamma(q_\alpha^\gamma)^2)+\sum_\beta(\sum_{\alpha,\gamma}S_\alpha^\gamma(q_\alpha^\gamma)^2(\mathbf{p}^\beta)^2)\right)\\
& -\tfrac{\overline{\delta\theta^2}}{8}\left(\sum_{\beta,\alpha}S_\alpha^\beta(q_\alpha^\beta)^2\right)^2+i\sum_\beta\lambda^\beta Lk\\
H_{nci} = & -\sum_\beta\int d\mathbf{r}_\beta\tfrac{w}{2}[Tr[\sigma^\beta:\sigma^\beta]-(Tr(\sigma^\beta)^2)]+\\
& i\sum_\beta\int d\mathbf{r}^\beta\psi^\beta:\sigma^\beta
\end{aligned} \tag{32}$$

with

$$\begin{aligned}
h_\alpha &= \frac{3(1-\overline{\theta})}{4l}+i\frac{(1+\overline{\theta})(\lambda+\psi_\alpha)}{2}\\
S_\alpha &= -\frac{3}{2l}+i(\lambda+\psi)
\end{aligned} \tag{33}$$

where $\mathbf{p}^\beta$, and $\mathbf{q}^\beta$ are the canonical "momentum" and "spatial coordinates" of the β copy of the system. The eq. above is the Hamiltonian rotated to a principal axis representation of the orientational tensors. Next the replica symmetric free energy is computed in the Hamiltonian reference frame:

$$\begin{aligned}
H_{ref} &= \sum_\alpha(-\frac{1}{4A}\hat{p}_\alpha^2+h_\alpha\hat{q}_\alpha^2)\\
h_\alpha &= \frac{3(1-\overline{\theta})}{4l}+\frac{(1+\overline{\theta})(\lambda+\psi_\alpha)}{2}\\
S_\alpha &= -\frac{3}{2l}+(\lambda+\psi)
\end{aligned} \tag{34}$$

in eq. 32, we made the substitutions: $i\lambda\to\lambda$ and $i\psi_\alpha\to\psi_\alpha$ to obtain h_α and S_α in eq. 34. Based on eq. 34, creation/annihilation operators, $(\mathbf{a}^+,\mathbf{a})$ and their components (a_α^+,a_α) are introduced:

$$\hat{q}_\alpha = \frac{a_\alpha+a_\alpha^+}{\sqrt{2m\omega_\alpha}}\ ;\ \hat{p}_\alpha=(a_\alpha-a_\alpha^+)\frac{\sqrt{m\omega_\alpha}}{2}$$

$$m=2A\ ;\ \omega_\alpha=(\frac{h_\alpha}{A})^{\frac{1}{2}}:[a_\alpha,a_\beta^+]=\delta_{\alpha,\beta}$$

$$;\ a_\beta|n_\alpha>=\sqrt{n_\beta}|n_\alpha-1>\delta_{\alpha,\beta}\ ;\ a_\beta^+|n_\alpha>=\sqrt{n_\beta+1}|n_\alpha+1>\delta_{\alpha,\beta}\ ;$$

$$\prod_{\alpha=1,3} |n_\alpha> = |\mathbf{n}> ; \mathbf{a} = \prod_{\alpha=1,3} a_\alpha ; \prod_{\alpha=1,3} a_\alpha |\mathbf{n}> = \sqrt{n_1}\sqrt{n_2}\sqrt{n_3}|\mathbf{n}-\mathbf{1}>$$
$$; \mathbf{a}^+|\mathbf{n}> = \sqrt{n_1+1}\sqrt{n_2+1}\sqrt{n_3+1}|\mathbf{n}+\mathbf{1}> \quad (35)$$

Using these commutation relations for long chains, i.e. $\frac{\beta\epsilon}{2}/L << 1$ we derive orientational averages in the free energy in the ground state $|0>$. In the limit of long chains the ground state calculation is exact and reproduces precisely in the limit of f→1 the many chain stiff LCPs [57]; using the commutation relations for creation-annihilation components we obtain:

$$< p_\alpha^2 > = < \frac{m\omega_\alpha}{2}(a_\alpha - a_\alpha^+)^2 > = \frac{m\omega_\alpha}{2} < -2a_\alpha a_\alpha^+ + 1 > = \frac{m\omega_\alpha}{2} \quad (36)$$

$$< \mathbf{p}^2 > = -\frac{m}{2}\sum_\alpha \omega_\alpha \quad (37)$$

In a similar way it can be shown that:

$$< \mathbf{q}^2 > = -\sum_\alpha \frac{1}{2m\omega_\alpha} \quad (38)$$

Let us now derive more complicated averages such as $\sum_\alpha S_\alpha q_\alpha^2 \sum_\beta S_\beta q_\beta^2$; first the sum may be separated by components:

$$\sum_\alpha S_\alpha q_\alpha^2 \sum_\beta S_\beta q_\beta^2 = \sum_\alpha S_\alpha^2 q_\alpha^4 + \sum_{\alpha\neq\beta} S_\alpha q_\alpha^2 S_\beta q_\beta^2 \quad (39)$$

Let us now calculate q_α^4; all terms involving product of four operators, involve averages of 16 products of creation/annihilation operators. Substituting the operator form of q_α in q_α^4 and doing the averages only 2 terms have non-zero averages:

$$q_\alpha^4 = (\frac{1}{2\omega_\alpha})^2 < a^2 a^{+2} + aa^+aa^+ > = 3(\frac{1}{2\omega_\alpha})^2 \quad (40)$$

The final result for the orientational averages in eq. 6 is:

$$\sum_\alpha S_\alpha q_\alpha^2 \sum_\beta S_\beta q_\beta^2 = \sum_\alpha 3S_\alpha^4(\frac{1}{2\omega_\alpha})^2 + \sum_{\alpha,\beta} \frac{S_\alpha^2}{2m\omega_\alpha}\frac{S_\beta^2}{2m\omega_\beta} \quad (41)$$

In the computation of $< \sum_\alpha (p_\alpha)^4 >$, five of the 16 terms have non-zero averages:

$$< \sum_\alpha (p_\alpha)^4 > = (\tfrac{m\omega_\alpha}{2})^2 < a_\alpha^2 a_\alpha^{+2} + 4a_\alpha a_\alpha^+ a_\alpha a_\alpha^+ - 2a_\alpha a_\alpha^+ - 2a_\alpha a_\alpha^+ + 1 >$$
$$= \sum_\alpha 3\left(\tfrac{m\omega_\alpha}{2}\right)^2 \quad (42)$$

for the $< \sum_\alpha p_\alpha^2 q_\alpha^2 S_\alpha^2 >$ we obtain:

$$< \sum_\alpha p_\alpha^2 q_\alpha^2 S_\alpha^2 > = \tfrac{1}{4} < a_\alpha^2 a_\alpha^{+2} - 4a_\alpha a_\alpha^+ a_\alpha a_\alpha^+ + 2a_\alpha a_\alpha^+ + 2a_\alpha a_\alpha^+ - 1 > = \tfrac{1}{4} S_\alpha^2 \tag{43}$$

while for $< \sum_\alpha q_\alpha^2 p_\alpha^2 S_\alpha^2 >$:

$$< \sum_\alpha q_\alpha^2 p_\alpha^2 S_\alpha^2 > = \tfrac{1}{4} < a_\alpha^2 a_\alpha^{+2} - 4a_\alpha a_\alpha^+ a_\alpha a_\alpha^+ + 2a_\alpha a_\alpha^+ + 2a_\alpha a_\alpha^+ - 1 > = \sum_\alpha \tfrac{1}{4} S_\alpha^2 \tag{44}$$

The power of the creation/annihilation formalism is in its simplicity which makes the method convenient to use by comparison with tedious orientational averages of high moments performed in the path integral formalism. Using equations 32-44 after some lengthy algebra the free energy per segment is obtained in a compact form:

$$\overline{F} = F_{FH} + \frac{1}{2}\sum_\alpha \left(\frac{h_\alpha}{A}\right)^{\frac{1}{2}} + \frac{\overline{\delta\theta^2}3}{64A^3}\sum_\alpha h_\alpha + \frac{\overline{\delta\theta^2}C}{32A^2}\sum_\alpha S_\alpha - \frac{\overline{\delta\theta^2}}{64A}\sum_\alpha \frac{S_\alpha^2}{h_\alpha} - \lambda k + \frac{1}{\rho}\left(\frac{w}{2}\left[\left(\sum_\alpha \sigma_\alpha\right)^2 - \sum_\alpha \sigma_\alpha^2\right] - \sum_\alpha \psi_\alpha\sigma_\alpha\right) \tag{45}$$

First we note three self-consistent relations among conjugate fields ψ_α and σ_α obtained by minimizing the free energy with respect to σ_α fields:

$$\psi_\alpha = w\sum_\alpha \sigma_\alpha - w\sigma_\alpha \,;\ \alpha = x,y,z \tag{46}$$

and the sum rule in Einstein notation $\sigma^{ii} = \rho_A$. Using eq. 46, the free energy can be expressed solely with the fields ψ_α. The free energy expressed by ψ_α has a simpler convenient form and is the one used in the present work:

$$\overline{F} = F_{FH} + \tfrac{1}{2}\sum_\alpha \left(\tfrac{h_\alpha}{A}\right)^{\frac{1}{2}} + \tfrac{\overline{\delta\theta^2}3}{64A^3}\sum_\alpha h_\alpha + \tfrac{\overline{\delta\theta^2}C}{32A^2}\sum_\alpha S_\alpha - \tfrac{\overline{\delta\theta^2}}{64A}\sum_\alpha \tfrac{S_\alpha^2}{h_\alpha} - \lambda k + \tfrac{1}{2\rho w}\left(\sum_\alpha \psi_\alpha\right)^2 - \rho w \tag{47}$$

with h_α and S_α given in eq. 34.

Now, we seek saddle point equations for the free energy; there are seven such equations; three of them are given in eq. 46, the other four in concise form are:

$$\frac{C\overline{\delta\theta^2}}{32A^2} + \frac{\psi_\alpha}{\rho w} + \frac{3\overline{\delta\theta^2}k}{64A^3} + \frac{0.25k}{A\left(\frac{h_\alpha}{A}\right)^{0.5}} - \frac{\overline{\delta\theta^2}S_\alpha}{32Ah_\alpha} + \frac{\overline{\delta\theta^2}kS_\alpha^2}{64Ah_\alpha^2} = 0; \quad \alpha = 1,2,3 \tag{48}$$

and

$$\frac{3C\overline{\delta\theta^2}}{32A^2}+\frac{9\overline{\delta\theta^2}k}{64A^3}+\sum_{\alpha=1,2,3}\left(\frac{0.25k}{A(\frac{h_\alpha}{A})^{0.5}}-\frac{\overline{\delta\theta^2}S_\alpha}{32Ah_\alpha}+\frac{\overline{\delta\theta^2}kS_\alpha^2}{64Ah_\alpha^2}\right)-k=0 \quad (49)$$

$$\frac{3C\overline{\delta\theta^2}}{32A^2}+\frac{9\overline{\delta\theta^2}k}{64A^3}+\sum_{\alpha=1,2,3}\left(\frac{0.25k}{A(\frac{h_\alpha}{A})^{0.5}}-\frac{\overline{\delta\theta^2}S_\alpha}{32Ah_\alpha}+\frac{\overline{\delta\theta^2}kS_\alpha^2}{64Ah_\alpha^2}\right)-k=0 \quad (50)$$

ρ is the segment density; The principal axis representation of σ^{ij} is:

$$\sigma^{ij}=\begin{pmatrix} a-b & 0 & 0 \\ 0 & a+b & 0 \\ 0 & 0 & 2a \end{pmatrix} \quad (51)$$

For uniaxial ordering the orientational order parameter, $< S >$, is given by $< S >= \frac{-3a}{b}$. $1 > \quad < S > \quad > 0$ signals uniaxial nematic ordering while $-0.5 < \quad < S > \quad < 0$ signals discotic ordering. In Einstein notation, in the isotropic phase, $\sigma^{ii}(\mathbf{r}) = \rho_A(\mathbf{r})$, otherwise, $\sigma^{ii}(\mathbf{r}) \neq \rho_A(\mathbf{r})$. The homopolymer limit of the free energy, i.e. f=1 in eq. 45, reproduces exactly the GE free energy of a many-chains LCPs obtained by non-perturbative summation of polymer Rouse modes, (viz. eq. 32 in [57]), and the minimal value for uniaxial ordering, $< S >$=0.25, obtained analytically in [57]), is reproduced herein.

7. Phase diagram analysis for DLCPs

Lets us turn attention to details of numerical solution. Unconstrained direct iteration [61] of principal axis of the fields ψ_α with $\alpha = 1,2,3$ and λ for the non-linear self-consistent equations obtained in eq.'s 48 and 49 is performed to compute ordering in sequence disordered LCP's. In order to ensure convergence we mix consecutive iterations of all fields. An mixing factor of 10^{-4} is normally sufficient for robust convergence. The convergence criteria for consecutive iterated fields is 10^{-9}. The direct iteration procedure is first employed for small segment densities; the low density solution for the iterated fields is then used as starting point for the direct iteration of ordering at a higher density.

This iteration approach works well in computation of order parameter values at and above continuous phase transitions; at a first order transition (which is the case in the present problem) the discontinuity in the principal axis values of the tensor fields ψ^α at the N/I transition leads to a numerical overshoot of the critical N/I threshold density ~ 0.2. Thought, the discontinuity in the order parameter precludes convergence toward the ordered phase from below the N/I threshold at all values of the iteration mixing factor, once the overshoot is spotted, direct iteration is safely carried out backward from higher to lower densities, and the N/I transition point is obtained with 10^{-9} accuracy; iteration from above the N/I density threshold is carried out and robust convergence is attained. In all calculations chemical potentials of all solutions are computed and only the stable solutions with the lowest chemical potential are displayed in figures.

The numerical study allows to inspect the effects of entropic contributions carried by the flexible segments, the entropy loss due to stiffness of the mesogenic units, and the energy gain by alignment of stiff segments on orientational order parameter $\langle S \rangle$, the N/I density threshold and the orientational strength at the transition.

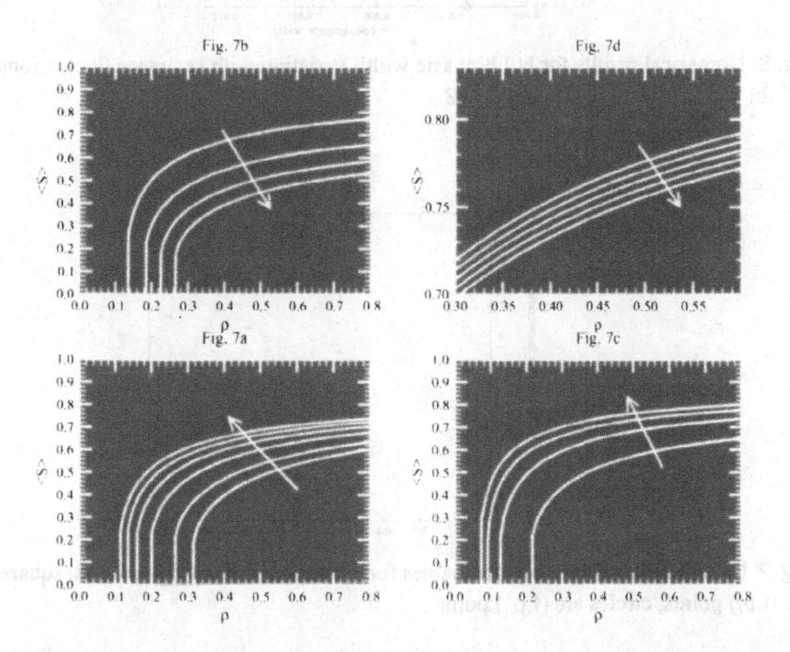

Fig. 7: Numerical results for uniaxial ordering - $\langle S \rangle$ variation with segment density, ρ and property X, 7a. X=w; f=0.95, βε =5, l=8. w values follow from upper to lower $\langle S \rangle$ curves; w=14, 12, 10, 8, 6, 5. 7b. X = βε e; f=0.95, w=10, l=8. βε values follow from upper to lower $\langle S \rangle$ curves; βε =10, 8, 6, 4. 7c. X = segment length of flexible spacers, l; f=0.95, βε e=10, w=10. l values follow from upper to lower $\langle S \rangle$ curves; l=2, 4, 6, 8, 10.

Fig. 7a depicts numerical results for the effects of mesogen/mesogen interaction alignment strength on ordering at a small fraction of flexible spac-

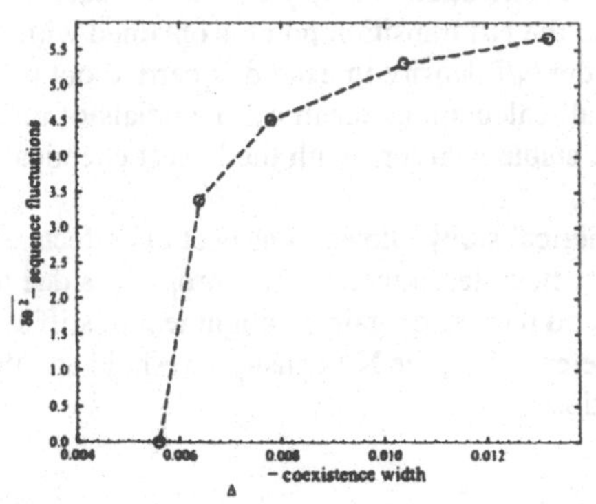

Fig. 8: Numerical results for N/I biphasic width variation with sequence fluctuations, $\overline{\delta\theta^2}$ for $f > 0.5$; u=0, w=10, βe=5, l=8.

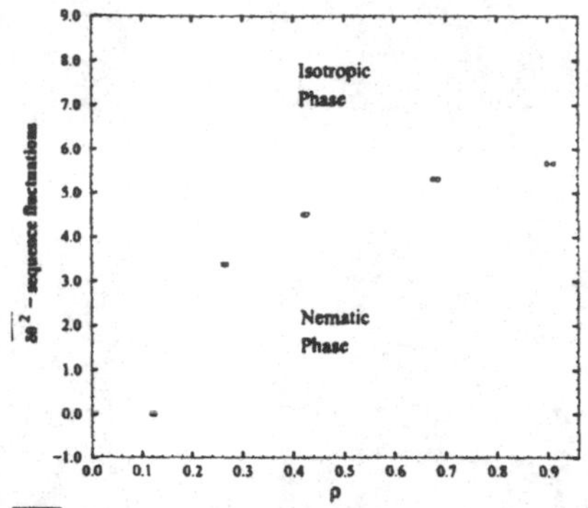

Fig. 9: N/I phase diagram in $\overline{\delta\theta^2}$ ρ variables for $f > 0.5$; u=0, w=10, βe=5, l=8; squares are (f,ρ_i) points, circles are (f,ρ_n) points.

ers. Numerical stable solutions are displayed for lyotropic orientational ordering. The arrow in fig. 7a indicates the increase in interaction alignment across curves. Fig. 7a suggests that the energetic gain from increase in alignment strength of stiff segments, increases the overall ordering and shifts the N/I transition density threshold to lower values; indeed, stronger alignment energies can induce orientational ordering of an entropy-dominated lower-density-state, and also increase the overall orientational strength $< S >$ at the N/I transition.

Fig. 7b depicts the effect of stiffness of the mesogen segments on orientational uniaxial ordering. The mesogenic effect on ordering is associated with

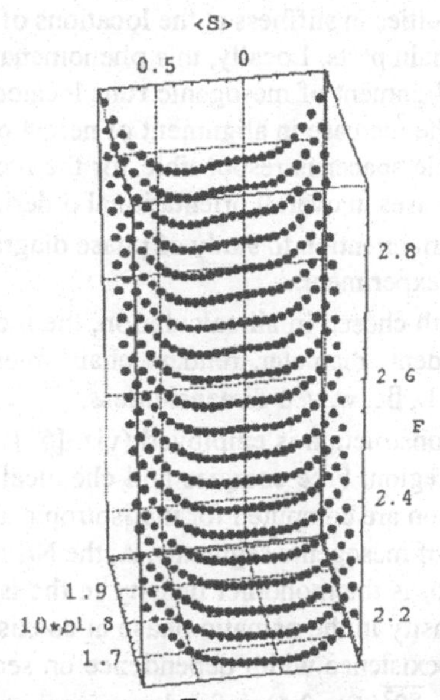

Fig. 10: Non-equilibrium free-energy F= $\frac{Fdl}{Vk_bT}$ for f= 0.938, w=10, βe=5, l=8, d - is monomer diameter

low entropy content carried by the mesogen units in comparison with flexible segments, and also with the energetic gain from and increase in alignment strength with stiffness. These contributions have an associating effect on ordering. Our numerical calculations indicate that an increment in stiffness of mesogenic segments increases the overall alignment of segments and shifts the orientational ordering threshold to lower values.

Fig. 7c depicts the effects of entropy associated with the flexible segments on local uniaxial ordering at a small sequence disorder values. Flexible segments carry a higher entropic content than stiff segments and based on purely thermodynamical arguments increase in segment flexibility should decrease orientational ordering.

Unexpectedly, our calculation shows that increase in flexibility increases marginally the effect of alignment. This result can be explained on a physical basis; the increment in entropy with increase in size of the flexible spacer, is, in the present case small since the composition is fixed at a small fraction of flexible segments. On the other hand changes in flexible spacer size has important effects on the interaction energies. For small fractions of flexible segments, the chain may be viewed as a stiff chain carrying short, highly flexible regions on the chain of contour. The presence of sequence disorder induces occasional random discontinuities in stiffness at the locations of flexible segments among two mesogenic chain parts. Locally, this phenomena promotes bending of the chain and local alignment of mesogenic runs located on adjacent sides of the flexible spacer. The increase in alignment of neighboring mesogenic runs separated by a flexible spacer is responsible for the local energetic gain and for the marginal increases in overall orientational ordering described in fig. 7d.

Let us now turn attention to study of phase diagram of DLCP solution and comparison with experiment.

The unit length chosen in all calculation, the monomer hard sphere (temperature independent) diameter, renders chain microscopic interactions and characteristics, l, L, $\beta\varepsilon$, w, v, ρ dimensionless.

A Maxwell construction is employed (viz. [57]) for the calculation of the N/I coexistence region. Free energies and chemical potentials in the vicinity of the N/I transition are computed for the isotropic and the nematic phases for several fractions of mesogenic segments. Δ, the N/I coexistence width is given by $\Delta = \rho_i - \rho_n$; ρ_i is the monomer density in the isotropic phase while ρ_n is the monomer density in the nematic phase at coexistence. In fig. 8 numerical results for the coexistence width dependence on sequence fluctuations of the mesogenic units - $\overline{\delta\theta^2}(f > 0.5)$ defined previously are depicted. $\overline{\delta\theta^2}$ is 0 for a semi-flexible homopolymer and maximal at f=0.5. Large sequence fluctuations increase the N/I coexistence width significantly, 3 folds that of a semi-flexible homopolymer. Fig. 9 depicts the effect of sequence fluctuations on transition threshold and coexistence.

The entropy carried by sequence fluctuation-induced mesophases rich in flexible segments destroys the nematic ordering and below a specific fraction of mesogen segments (which is density dependent), the nematic phase loses stability, while the isotropic phase remains stable. For the parameters values of fig. 9 at melt densities, ($\rho = 1$), the critical fraction of mesogen segments for isotropization is $f_c = 0.76$. The N/I coexistence width, displayed as the horizontal distance among an adjacent square (ρ_i) and circle (ρ_n) is small at all densities. In the equi-stiff limit defined as the scenario wherein the difference in stiffness of A and B segments is negligible, sequence heterogeneity effects vanish in the infinite chain limit in agreement with [54]; in many experiments, (i.e. $(MBPE - (methylene)_y$ with y=17, 18, 20 [62]) one segment is stiff, the other is flexible the equi-stiff homopolymer limit does not apply, and the rigorous treatment of anisotropy dissimilarity of A and B segments developed herein is crucial. For long chains ($\frac{\beta e}{2}/L << 1$) with significant fluctuations in sequence stiffness, the sequence-heterogeneity effect on the N/I biphasic region does not vanish, but in most cases is small with a $\Delta T \sim 10^o$ contribution to the N/I biphasic width.

Fig. 10 depicts the three dimensional free energy surface for the disordered liquid crystalline heteropolymer computed numerically from eq. 45 at arbitrary values of the orientational order parameter $< S >$ (away from equilibrium). Below $\rho_i = 0.169$ we find a global minima with $< S > = 0$ i.e. the isotropic phase. For $\rho > \rho_i$ a maxima with positive $< S >$ followed by a meta-stable minima i.e. a weakly discotic phase characterized by a small negative $< S >$, appears. Unlike the liquid-gas transition, the meta-stable discotic phase persists at all densities leading to unusual domain behavior and influences the nucleation during phase separation process. The excess free energy among the homogeneous and the phase separated state along the transition state pathway, $< S >_{min} (\rho)$ connecting the coexistence values of ρ, $\rho_i = 0.169$ and $\rho_n = 0.1785$ is obtained from the non-equilibrium free energy surface (viz. fig. 10) by computing $F(< S >_{min} (\rho), \rho)$, where $< S >_{min} (\rho)$ is the minimum of the non-convex $F(\rho)$. The excess free energy in the meta-stable state found is very small i.e. $\Delta F(\rho = 0.1705) = 1.9 * 10^{-5}$. The critical domain size from nucleation theory is given by $R^* = \frac{2\gamma}{\Delta F}$. Using for the nematic-isotropic inter-domain interfacial free energy γ, $\sim 4.\frac{mN}{M}$ which is typical of hydrophobic-hydrophilic polymer mixtures (viz. [64]) and the segment length $l = 1. * 10^{-9} M$ (i.e. l=8 in our calculation) yields $R^* = 1.7910^{-6} M$. The critical domain size predicted is in agreement with experimental observations of optical micrographs for meso-domain size in disordered liquid crystalline heteropolymers [53].

Ackgnowledgment

We thank NSF for finanacial support, grant no. 9972138.

References

1. T. R. C. Boyde, *J. Chromatogr.*, **124**, 219, (1976)
2. R. W. Veatch, *J. Pet. Technol.*, 677, April (1983)
3. R. A. Siegel, M. Galamarzian, B. A. Firestone, and B. C. Moxley, *J. Control Release*, **8**, 179, (1988)
4. *Biopolymer Gels*, A. L. Clark, *Curr. Opin.in Coll. and Int. Sci.*, **6**, 712, (1996)
5. *Textbook of Polymer Science*, John Wiley and Sons, F. W. Billmeyer, (1971)
6. T. Tanaka, C. Wang, V. Pande and A. Yu. Grosberg, *Faraday Discus.*, **102**, 201, (1996)
7. A. Keller, *Faraday Discuss.*, **101**, 1, (1995)
8. S. Panyukov and I. Rabin, *Phys. Rep.*, **269**, 1, (1996)
9. P. M. Goldbardt, H.E. Castillo and A. Zippelius, *Adv. Phys.*, **45**, 393, (1996)
10. L. Gutman and E. I. Shakhnovich, *J. Chem. Phys.*, **107**, 1247, (1997)
11. L. Gutman and E. I Shakhnovich, *J. Chem. Phys.*, **109**, 2947, (1998)
12. M. Annaka and T. Tanaka, *Nature*, **355**, 430, (1992)
13. P. G. deGennes, *J. Phys. Lett.*, **69**, 40, (1979)
14. R. M. Briber and B. J. Bauer, *Macromolecules*, **21**, 3296, (1988)
15. L. Chikina, M. Daoud, *J. Polym. Sci. B*, **36**, 1507, (1998)
16. M. Annaka, T. Tanaka, *Phys. A*, 40, (1994)
17. T. Shiomi, H. Ishimatsu, T. Eguchi and K. Imai, *Macromol.*, **23**, 4970, (1990)
18. E. I. Shakhnovich and A. M. Gutin, *J. Phys. (Fr)*, **50**, 1843, (1989)
19. V. S. Pande, A. Y. Grosberg, and T. Tanaka, *J. Phys. II France*, **4**, 1771, (1994)
20. F. Rindfleisch, T. P. DiNoia, and M. A. McHugh, *J. Phys. Chem.* **100**, 15581, (1996)

21. R. C. Sutton, L. Thai, J. M. Hewitt, C. L. Voycheck, and J. S. Tan, *Macromol.*, **21**, 2432, (1988)
22. D. Bratko, A. K. Chakraborty, E.I. Shakhnovich, *J. Chem. Phys.*, **106**, 1264, (1997)
23. G. H. Fredrickson and S. T. Milner, *Phys. Rev. Lett.*, **67**, 835, (1991)
24. A. M. Gutin and E. I. Shakhnovich, *J. Chem. Phys.* **100**, 5290, (1994)
25. L. Gutman and E.I. Shakhnovich, *J. Chem. Phys.*, in preparation
26. L. Gutman and E.I. Shakhnovich, M. Shibayama, M. Annaka *Phys. Rev. Lett.*, in preparation
27. F. Simoni *Liq. Cryst.*, **24**, 83, (1998)
28. T. E. Creighton, *BioEssays*, **8**, 57, (1988)
29. L. Gutman and E. I. Shakhnovich, *J. Chem. Phys.*, **107**, 1247, (1997)
30. L. Gutman and E. I. Shakhnovich, *J. Chem. Phys.*, **109**, 2947, (1998)
31. P. M. Goldbardt, H.E. Castillo and A. Zippelius, *Adv. Chem. Phys.*, **45**, 393, (1996)
32. A. M. Gutin and E. I. Shakhnovich, *J. Chem. Phys.*, **100**, 5290, (1994)
33. E. I. Shakhnovich and A. M. Gutin, *Biophys. Chem.*, **34**, 187, (1989)
34. G. Parisi, *J. Phys. A: Math. Gen.*, **13**, 1887, (1990); M. Mezard and G. Parisi, *J. Phys. I*, **1**, 809, (1991)
35. P. J. Flory, *Proc. Roy. Soc.* **234**, 73, (1956)
36. L. Onsager *Ann. N. Y. Acad. Sci.*, **51a**, 627, (1949)
37. W. Maier and A. Z. Saupe *Naturoforsch*, **12**, 882, (1959)
38. A. Yu Grosberg and A. R. Khokhlov *Soc. Sci. Rev. A. Phys.*, **8**, 147, (1987)
39. P. J. Flory and G. Ronca *Mol. Crys. Liq. Cryst.*, **54**, 289, (1979)
40. B. Y. Ha and D. Thirumalai *J. Chem. Phys.*, **106**, 4243, (1997)
41. B. Jung and B. L. Schurman *Macromol.*, **22**, 477, (1989)
42. R. D. Kamier and G. S. Grest *Phys. Rev. E*, **55**, 1197, (1997)
43. M. Dijkstra and D. Frenkel *Phys. Rev. E*, **51**, 5891, (1995)
44. Z. Y. Chen *Macromol.*, **26**, 3419, (1993)
45. A. R. Khokhlov and A. N. Semenov *Physica Amsterdam*, **112A**, 605, (1985)
46. R. Podgornick *Phys. Rev. E.*, **54**, 5268, (1996); ibid *Phys. Rev. E.*, **52**, 5170, (1995)
47. R. K. Baharadwaj and R. H. Boyd *Macromol.*, **31**, 7682, (1998)
48. G. D. Butzbach, J. H. Wendorff and H. J. Zimmermann *Makromol. Chem. Rapid Commun.*, **6**, 821, (1985)
49. G. C. Rutledge *Macromol.*, **25**, 3984, (1992)
50. V. Percec and Y. Tsuda *Macromol.*, **55**, 1197, (1997)
51. A. Blumstein and T. Oomanan *Macromol.*, **15**, 1264, (1982)
52. J. S. Moore and S. I. Stupp *Macromol.*, **20**, 273, (1987)
53. P. G. Martin and S. I. Stupp *Macromol.*, **21**, 1222, (1988); ibid **21**, 1288, (1988)
54. G. H. Fredrickson and L. Leibler *Macromol.*, **23**, 531, (1990)
55. S. I. Stupp, J. S. Moore, and P. G. Martin *Macromol.*, **21**, 1228, (1988)
56. L. Gutman and A. K. Chakraborty *J. Chem. Phys.*, **101**, 10074, (1994); **103**, 10733, (1995)
57. A. M. Gupta and S. F. Edwards *J. Chem. Phys.*, **98**, 1588, (1993)
58. K. F. Freed *Adv. Chem. Phys.*, **22**, 1, (1972)
59. K. Binder and A. P. Young *Rev. Mod. Phys.*, **58**, 801, (1986)
60. M. Swanson *Path Integrals and Quantum Processes, Academic Press, INC, Harcourt Brace Jovanovich, Publishers*, (1992)
61. B. Carnahan, H A. Luther and J. O. Wilkes *Applied Numerical Methods (Wiley, New York)*, (1969)
62. V. Percec and Y. Tsuda *Macromol.*, **23**, 3509, (1990)
63. A. Blumstein R. B. Blumstein, M. M. Gauthier, O. Thomas and J. Asrar *Mil. Cryst., Liq. Cryst. (Lett.)*, **92**, (1983), 87
64. J. E. Mark *Physical Properties of Polymers Handbook* AIP Press, Woodbury NY, (1996)

CONFORMATIONAL TRANSITIONS IN PROTEINS AND MEMBRANES

JEREMY C. SMITH, ZOE COURNIA, ANTOINE TALY, ALEXANDER L. TOURNIER, DAN MIHAILESCU AND G. MATTHIAS ULLMANN
Computational Molecular Biophysics, Interdisciplinary Center for Scientific Computing (IWR), Im Neuenheimer Feld 368, Universität Heidelberg, 69120 Heidelberg, Germany

1. Introduction

An understanding of protein and biological membrane function requires the realization that these objects are dynamic. This present survey treats simulation-based methods for investigating internal motions in soluble and membrane proteins and for probing the dynamics of membranes themselves.

2. Dynamics and Conformational Change in Proteins

In the early days of structural biology the dynamic aspect of function was not fully recognized. The resolution at atomic-detail of protein structures using X-ray crystallography dates back to the late 1950s.[1] Since then, thousands of native protein structures have been solved. The interpretation of how they function has been somewhat biased by the limitations inherent in the earlier X-ray crystallographic analyzes, which were at the time limited to purely structural results. Thus, protein function has been interpreted largely in terms of static structures. Taking enzymes (catalytic proteins) as an example, the lock-and-key model of enzyme function of Emil Fischer in 1890 was taken as the model for enzyme function. However, this model was found to be inaccurate for a number of enzymes, and in 1958, Koshland proposed the idea of an induced fit mechanism of substrate binding. In the induced fit model the binding of a substrate to an enzyme is accompanied by a conformational change that aligns the catalytic groups in their correct orientations. The induced fit model has been shown to be an accurate representation of substrate binding for many enzymes.

An essential part of the induced fit model is enzyme flexibility, and it is now widely accepted that flexibility is required for catalytic activity[2–6]. In

J. Samios and V.A. Durov (eds.), Novel Approaches to the Structure and Dynamics of Liquids: Experiments, Theories and Simulations, 485–502.

many cases crystallography has provided direct evidence for the existence of different enzyme conformations in the presence and absence of substrate [5]. Control, catalytic conversion and product release may also require flexibility in the protein. However, the relationship between enzyme flexibility and activity is not as well understood as the structure-activity relation, and the timescales and forms of the functionally-important motions remain poorly characterized. The same can be said for the non-catalytic classes of proteins, many of which also undergo conformational change accompanying their functions.

Central to present discussions of the physical basis of protein function the fascinating characteristic that the native state of a protein comprises a large number of slightly different structures that correspond to local minima in the potential energy surface of the system. The presence of these 'conformational substates' in proteins was suggested as a means to explaining the temperature dependence of ligand binding to myoglobin [2, 7]. Detailed discussions of protein conformational substates, with particular reference to myoglobin, have been given elsewhere [8–10]. Interconversion between these substates may occur if the kinetic energy is sufficient. Transitions among some of these substates may be global (collective) or local, and, in principle, either may be of biological importance [11].

A major challenge for computational and experimental biochemistry is to determine the forms and timescales of functional motions in proteins and membranes. These motions can be associated with progress along a reaction coordinate, and may have a direct effect on the reaction dynamics and/or a thermodynamic (potential of mean force) effect. Understanding large-scale transitions requires characterization of the end-states and of the pathways between them. The end states are often known from X-ray crystallographic experiments. However, the pathways between the end-states, i.e. the sequence of structures accompanying the transition from one stable state to another, are difficult to determine experimentally. Computer simulation techniques can in principle provide the required information. Using empirical energy functions, searches of protein conformational space can be used to determine plausible pathways. However, with present-day computer power the sampling obtained with standard MD is insufficient if the functional conformational change to be studied occurs on a timescale longer than about a nanosecond, as is frequently the case. Therefore, specialized techniques are required. Although the development of these techniques is still in its infancy, a number do exist and have been used to examine important conformational changes in several proteins [10, 12–15].

A related question concerns equilibrium protein fluctuations. A variety of experiments have demonstrated the existence of a dynamical transition in hydrated proteins at around 180-220 K, characterized by deviation from linearity of the temperature-dependence of the mean-square displacement $\langle u^2 \rangle$ [16–29]. The protein transition has dynamical aspects in common with the liquid-glass transition. For example, as in glass-forming liquids, proteins exhibit diffusive motions above the transition and are trapped in harmonic potential wells below. Experiments have shown that in several proteins biological function ceases below the dynamical transition [19, 26, 30].

An important physical question concerns the environmental effect on the dynamical transition. Molecular dynamics (MD) simulations [31, 32] and neutron scattering experiments [33] have shown that isolated or dehydrated proteins present dynamical transition behavior. However, a number of experiments have indicated that when a protein is solvated the dynamical transition is strongly coupled to the solvent [16, 19, 20, 29, 33–37]. Highly viscous solvents, such as trehalose, suppress dynamical transition behavior [35, 38, 39], and neutron scattering experiments on enzymes in a range of cryosolvents showed that the dynamical transition behavior of the protein solution resembles that of the pure solvent [20, 40–42].

The observed solvent coupling leads to the question of whether the dynamical transition in a solvated protein is controlled by the solvent or whether the intrinsic anharmonicity of protein dynamics also plays a role. Computer simulations can be used to investigate the physical origin of the dynamical transition [36, 43, 44]. In this context, an innovative method was used by Vitkup *et al.* to probe features of the dynamical transition due to the protein and those due to the solvent [45]. The approach consisted of using the Nosé-Hoover thermostat [46] to set and maintain the protein and its solvent at two different temperatures. Recently, the work of Vitkup *et al.*[45] was complemented with simulations of hydrated myoglobin performed under varying conditions [47, 48]. The results of these simulations confirm the strong influence of the solvent on the protein fluctuations and permit reconsideration of the extent that solvent determines high-temperature fluctuations. Two main results are seen (i) Low temperature solvent cages the protein fluctuations. (ii) Heating the solvent while keeping the protein cold drives the protein fluctuations to values intermediate between those in the fully cold and fully hot systems. The Nosé-Hoover results thus confirm, in accord with previous studies[20, 45], that solvent strongly influences the dynamical transition in proteins.

In Figure 1 are shown the side-chain fluctuations in MD simulations of hydrated myoglobin as a function of distance from the protein center of mass. The dynamical transition is seen to be most pronounced in the outer parts of the protein, i.e., those close to the solvent shell - above the transition the outer shells exhibit both stronger fluctuations and a larger change in gradient (inset to Figure 1) than the inner atoms. The solvent transition drives dynamical transition behavior primarily in the side-chain atoms of the external protein regions,

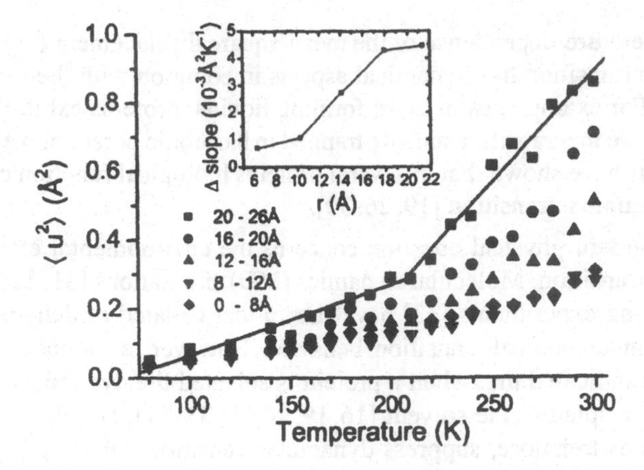

Figure 1. Mean-square fluctuations of the protein side-chain heavy atoms for 5 different shells, each 4 Å thick (except for the inner shell (8 Å) and outer shell (6 Å)). The inset shows the difference in slopes of lines fitted below and above 220 K as a function of distance from the protein center of mass. Linear fits to the data above and below 220K are also shown for the outermost shell.

i.e., those closest to the solvent. This dovetails with the recently-proposed "radially-softening" description of protein dynamics, based on MD-simulation analysis of quasielastic neutron scattering data, in which it was shown that the average dynamical properties of a protein at 300 K vary smoothly with increasing distance from the protein core, involving a gradual increase of the diffusive amplitudes and a narrowing and shift to shorter (picosecond) times of the distribution of diffusive relaxation processes [49].

3. Conformational Changes in Protein Complexes

Also protein complexes can show a dynamic behavior and need dynamics for their function. One example of such a complex is the one formed by plastocyannin and cytochrome *f*. These two proteins are involved in the electron transfer in higher plant photosynthesis. Plastocyanin accepts an electron from reduced cytochrome *f*, and donates an electron to the oxidized form of photosystem I. When plastocyanin and cytochrome *f* are non-invasively cross-linked, the resulting covalent complex cannot undergo the internal electron-transfer reaction, which is fast within the electrostatic complex [50, 51]. The unreactivity of the crosslinked complex was taken as evidence that the two proteins dock and react with each other in different configurations [51]. This prediction and this analysis were nicely corroborated by subsequent publication of the structure of cytochrome *f*[52], which showed that the positively-charged patch and the heme are relatively far apart.

The reaction of plastocyanin and cytochrome *f* involves the association and a subsequent electron-transfer reaction.

$$\mathrm{pc(II)} + \mathrm{cytf(II)} \rightleftharpoons \mathrm{pc(II)/cytf(II)} \rightarrow \mathrm{pc(I)/cytf(III)} \quad (1)$$

The Roman numerals are the oxidation states of copper and iron, and the slant represents the diprotein complex. We simulated the association of plastocyanin and cytochrome *f* by a Monte Carlo simulation[53]. The obtained complexes were clustered into six different groups. One representative out of each family was subject to a molecular dynamics simulation in order to relax side chains and to remove conformational strain. The energy and the relative electronic coupling of each of the resulting complex structures was calculated using the Poisson-Boltzmann equation and the Pathways model, respectively. The most stable structure and the structure with the highest relative electronic coupling were not identical. This finding is in agreement with proposal from Qin and Kostić[51] that a rearrangement is required prior to the electron transfer. A subsequent NMR study of the association of plastocyanin and cytochrome *f* by Marcellus Ubbink et al.[54] supported this idea further.

Another well-studied example of a dynamic protein complex is the complex formed by plastocyanin and cytochrome *c*. These two proteins from an electron transfer active complex under in vitro conditions. Since the two proteins are very well characterized, structurally and functionally, and these proteins are easy to handle, the electron transfer complex formed by these two proteins was deeply investigated. The plastocyanin-cytochrome *c* complex can serve as a model for many other electron transfer complexes. The electron transfer reaction can be described as a multi-step reaction. In the first step the two proteins encounter and form a complex. Than the actual electron transfer reaction takes place. Finally the two proteins separate.

The iron in cytochrome *c* can be substituted by zinc or tin without perturbing the protein structure nor its association with other proteins [55]. Zinc or tin cytochrome *c* is redox inactive in its ground state (Zncyt). A laser flash excites zinc cytochrome *c* into its triplet state. The triplet zinc cytochrome *c* (^{3}Zncyt) has a fairly negative redox potential ($E°$=-0.88 V vs. NHE for zinc cytochrome *c* and $E°$=-0.4 V vs. NHE for tin cytochrome *c*) and can easily reduce oxidized plastocyanin[55, 56]. This radical is rereduced by the reduced plastocyanin and recombines to the ground state of zinc cytochrome *c*. The possibility to study photoinduced electron transfer reaction enables to separate the dynamic motion that is associated with the association process, i.e., the docking dynamics (bimolecular rate constant), from the dynamics that is associated with the dynamic within the complex immediately prior to the electron transfer (unimolecular rate constant).

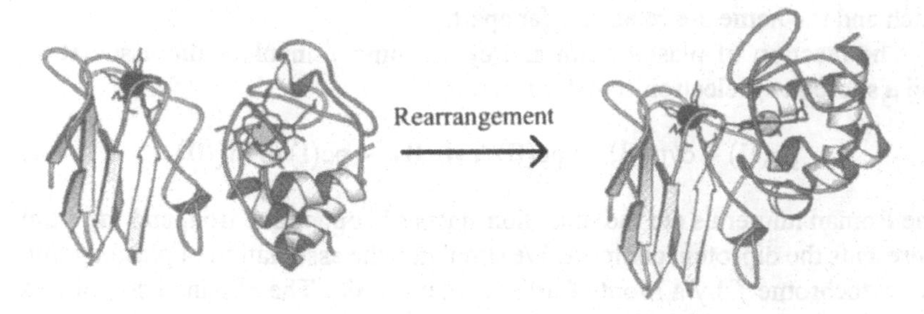

Figure 2. **Rearrangement in the plastocyanin-cytochrome *c* complex. The two proteins associate in an electron transfer inactive orientation and rearrange transiently to a orientation that is more favorable for electron transfer. The figure shows a structural model of this rearrangement process.**

Plastocyanin and cytochrome *c* can be non-invasively cross-linked by carbodiimids which introduce an amid bond between the carboxyl group of aspartates and glutamates and the amine groups of lysines. While the electron transfer is fast within the complex that results from electrostatic association (k_{et}=1300 s^{-1}), it becomes undetectably slow in the covalently-crosslinked complex ($k_{et} < 0.2\ s^{-1}$). When cytochrome *c* is replaced by zinc cytochrome *c*, the electron transfer reaction in the electrostatic complex and in the covalent complex is accelerated. Nevertheless, the electron transfer in the covalent complex (k_F=$2.2\times10^4\ s^{-1}$) is slower than in the electrostatic complex (k_F=2.5×10^5 s^{-1}). This finding led to the suggestion that the initial docking orientation, which is most likely similar to the crosslinked orientation, is not the reactive orientation [56–59] which was corroborated by a theoretical study[60]. It was suggested that a rearrangement from an initial docking orientation to a reactive orientation takes place in the plastocyanin cytochrome *c* complex. In order to probe this hypothesis, the viscosity dependence of the electron transfer reaction between zinc cytochrome *c* and plastocyanin in the electrostatic complex was investigated[56, 59]. In this reaction, a biphasic behavior is expected. The slower phase describes the association of the two proteins, while the faster phase describes the electron transfer within the formed diprotein complex. If the initial, most-stable orientation would be the same as the reactive orientation, only the slower reaction rate should depend on the viscosity of the medium. It was, however, observed that both reaction rates depend on the viscosity of the medium. The electron transfer rate in the crosslinked complex was, however, independent of the viscosity of the medium. The unimolecular electron transfer rate in the electrostatic complex approaches the electron transfer rate of the cross-linked complex in the limit of high viscosity[56, 59]. The same behavior can be observed in the reaction with Tin cytochrome *c*. Despite the different driving force for the electron transfer reaction, the reaction rate in the Tin cytochrome *c*-plastocyanin complex is virtually identical to that of the zinc cytochrome *c*-plastocyanin complex. Thus the reaction rate is independent of the driving force of the electron transfer reaction. Consequently, the rate determining step is not the electron transfer reaction. The reaction is gated by another process. These findings strongly support the idea that a rearrangement from the initial binding orientation to the electron transfer orientation takes place and that the initial binding orientation is very similar to the crosslinked orientation. Based on electrostatic[61] and electron transfer pathway calculations[60], we proposed a structural model the initial and the reactive complex which is given in Figure 2.

Theoretical studies led to experiments to probe the type of rearrangement and to investigate the rearrangement pathway. Several mutants that involve residues in the lower or in the upper acidic patch were constructed for that goal[62, 63]. It was found that residues closer to the copper binding site, i. e., in the upper acidic patch, influence the rate of rearrangement, while residues remote from the the copper binding site do not. This finding was confirmed by studies that involve the temperature dependence of the rate of rearrangement[64, 65] and activation barrier measurements[66, 67]. Also a NMR study [68] on the diprotein complex confirmed the high degree of flexibility and indicated a rearrangement from the acidic towards the hydrophobic patch as depicted in Figure 2.

4. Coupling between Conformational and Protonation State Changes in Membrane Proteins

Many membrane proteins are involved in electron and proton transfer reactions across a membrane[69]. Protonatable groups play a prominent role in these reactions, because they can either function as proton acceptor or donor in proton transfer reactions or influence the redox potential of adjacent redox-active groups. The titration behavior of protonatable groups in proteins can often considerably deviate from the behavior of isolated compounds in aqueous solution. This deviation is caused by interactions of the protonatable group with other charges in the protein and also by changes in the dielectric environment of the titratable group when the group is transferred from aqueous solution into the protein [70–73]. The situation can be even more complicated. Because the charge of protonable residues depends on pH, also their interaction is pH-dependent, which can lead to titration curves that can not be described by normal sigmoidal titration curves[74, 75].

The photosynthetic reaction center (RC) is the membrane protein complex that performs the initial steps of conversion of light energy into electrochemical energy [76, 77] by coupling electron transfer reactions to proton transfer. The bacterial RC of *Rb. sphaeroides* is composed of three subunits: L, M and H. The L and M subunits have pseudo-two-fold symmetry. Both the L and M subunits consist of five transmembrane helices. The H subunit caps the RC on the cytoplasmic side and possesses a single N-terminal transmembrane helix. The RC binds several cofactors: a bacteriochlorophyll dimer, two monomeric bacteriochlorophylls, two bacteriopheophytins, two quinones, a non-heme iron and a carotenoid. The non-heme iron lies between the two quinone molecules. The primary electron donor, a bacteriochlorophyll dimer called the special pair, is located near the periplasmic surface of the complex, and the terminal electron acceptor, a quinone called Q_B is located near the cytoplasmic side. While Q_A is a one-electron acceptor and does not protonate directly, Q_B accepts two electrons and two protons to form the reduced Q_BH_2 molecule. The first reductions of Q_A and of Q_B are accompanied by pK_a shifts of residues that interact with the semiquinone species [78]. The reductions induce substochiometric proton uptake by the protein.[79–81] The number of protons taken up by the protein upon reduction of the quinones is an observable which is directly dependent on the energetics of the system and intimately coupled to the thermodynamics of the electron transfer process between the states $Q_A^-Q_B$ and $Q_AQ_B^-$. The pH-dependence of the proton uptake associated with the formation of Q_A^- and Q_B^- in wild type RCs have been determined for *Rb. sphaeroides* and *Rb. capsulatus* [82–85]. Using X-ray structural analysis, it has been shown that a major conformational difference exists between the RC handled in the dark (the ground state) or under illumination (the charge-separated state) [86]. The main difference between the two structures concerns Q_B itself, which was found in two different positions about 4.5 Å apart. In the dark-adapted state in which Q_B is oxidized, Q_B is found mainly in the distal position and only a small percentage in the proximal position. Under illumination, i.e., when Q_B is reduced, Q_B is seen only in the proximal position. The crystal was grown at pH=8 [87]. The reaction center structures with proximal or distal Q_B are called RC^{prox} and RC^{dist}, respectively [88]. The proton uptake upon the first Q_B reduction and the pH-dependent conformational equilibrium between RC^{prox} and RC^{dist} are shown in Figures 3 and 4, respectively.

By continuum electrostatic calculations, we investigated the pH-dependence of the proton uptake associated with the reduction of Q_B.[89] The two experimentally-observed conformations of the RC were considered: with Q_B bound in the proximal or the distal binding site. Comparing the calculated and experimental pH-dependence of the proton uptake revealed that a pH-dependent conformational transition is required to reproduce the experimental proton uptake curve (Figure 4). Neither the individual conformations nor a static mixture of the two conformations with a pH-independent population are capable to reproduce the experimental proton uptake profile. We presented

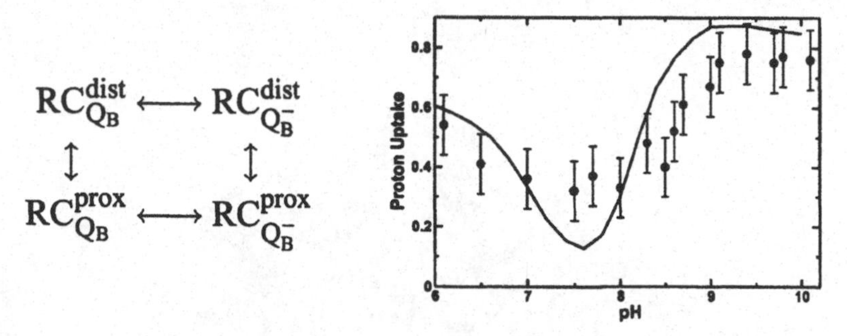

Figure 3. **pH dependence of the proton uptake upon Q_B reduction. The symbols in the diagram show the experimentally-determined proton uptake. The line shows a proton uptake calculated by electrostatic calculations using a model that takes conformational transition between the two different reaction center positions RC^{dist} and RC^{prox} into account.**

a new picture[89] in which the position of Q_B depends not only on the redox state of Q_B, but also on pH. This prediction will now be studied experimentally by X-ray crystallography at different pH values.

5. Effects of Peptides and Sterols on Membrane Dynamics

Of particular importance in biology is the effect on the membrane structure and dynamics of interacting molecules, such as peptides and sterols. In recent work gramicidin S, [cyclo-(Leu-D-Phe-Pro-Val-Orn)$_2$, (GS)] a cyclic decapeptide (Figure 5) that is particularly suitable for pursuing peptide:membrane MD

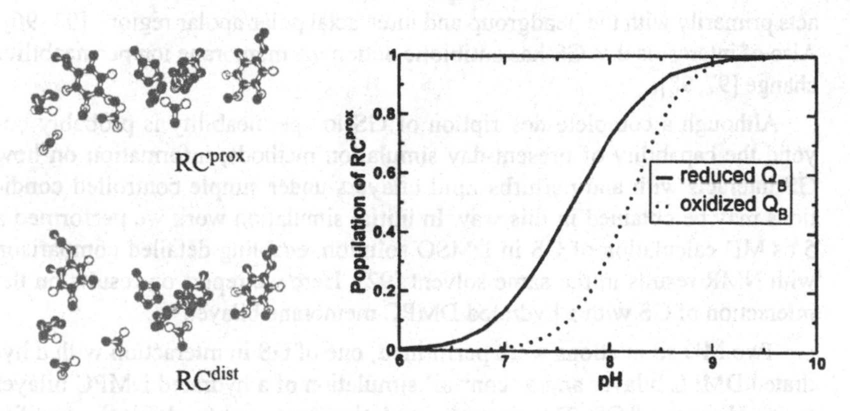

Figure 4. **Conformational Equilibrium between RC^{prox} and RC^{dist} structures. The population of RC^{prox} shown for oxidized (dashed line) and reduced (solid line) Q_B depends on pH. In the neutral pH range, both conformations are populated.**

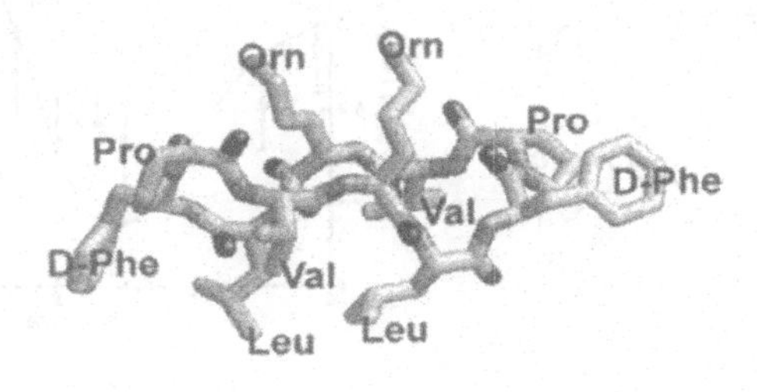

Figure 5. Gramicidin S

studies, was examined in interaction with a DMPC membrane. The sequence and structure of the peptide are relatively simple. NMR, X-ray and MD studies indicate that the backbone adopts an antiparallel β-sheet with two Type II' β-turns in various solutions of different polarity and in the crystalline form [90–92]. One consequence of this is that the GS structure is amphipathic, with the hydrophobic side chains on one side of the molecule and the hydrophilic ones on the other, and this provides a logical geometry for interaction with a lipid membrane, with the polar side of GS at the lipid:water interface and the nonpolar side interacting with the lipid tails. Considerable experimental evidence exists that this is indeed the case and a variety of biophysical studies on the interaction of GS with model lipid membranes have confirmed that it interacts primarily with the headgroup and interfacial polar/apolar regions [93–96]. Also of interest is that GS has antibiotic action via membrane ion permeability change [97, 98].

Although a complete description of GS ion permeability is probably beyond the capability of present-day simulation methods, information on how GS interacts with and perturbs lipid bilayers under simple controlled conditions may be obtained in this way. In initial simulation work we performed a 5 ns MD calculation of GS in DMSO solution, enabling detailed comparison with NMR results in the same solvent [92]. Here we report on results on the interaction of GS with a hydrated DMPC membrane bilayer.

Two MD simulations were performed, one of GS in interaction with a hydrated DMPC bilayer and a 'control' simulation of a hydrated DMPC bilayer in the absence of GS. The control simulation was used to check the validity of the simulation method with respect to spectroscopic and diffraction data, and to use for comparison of the lipid molecule structure and dynamics in the

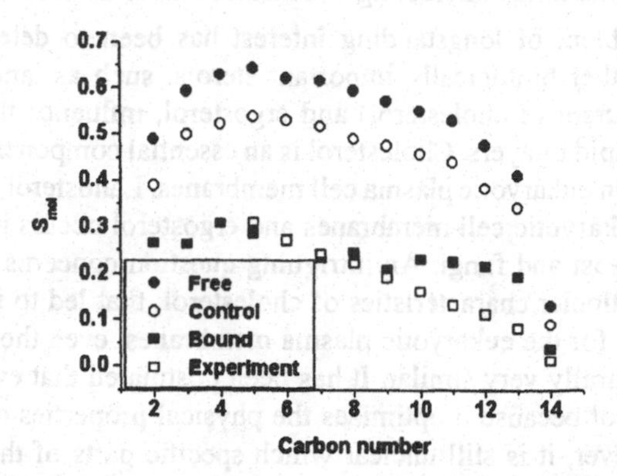

Figure 6. Molecular order parameter ($S_{\mathrm{mol}} = -2S_{\mathrm{CD}}$) averaged between the chains sn-1 and sn-2 of DMPC: "free" lipids (•); control simulation (○); "bound" lipids (■) and derived from NMR quadrupole splitting experiments on fully-deuterated multilamellar dispersions of DMPC doped with GS (□) [94]

presence and absence of GS. Detailed results will be described elsewhere.

During NVE equilibration GS diffused into the bilayer, with its center of mass 3-4 Å deeper into the hydrophobic core than at the beginning of the equilibration. After about 1 ns of NVE equilibration the GS molecule regained its initial membrane/water interfacial position with the backbone and membrane planes parallel to each other. During the 3.0 ns NPT production dynamics the average position of GS relative to the membrane remained practically unchanged.

Nine lipids from the GS-containing layer have average interaction energies greater than $k_B T$, and in what follows these are considered as "bound" to the GS. The remaining eight GS-layer lipids are "free" as are the 21 from the non-GS layer. It is of interest to examine whether the ordering of the bound lipids is different from that of the free ones. A suitable quantity for this examination is the order parameter of the carbon-deuterium bond, S_{CD}, defined by $S_{\mathrm{CD}} = \langle 1/2(3\cos^2\theta(t) - 1)\rangle$ where $\theta(t)$ is the angle between the carbon-deuterium bond vector and the bilayer normal. "$\langle\rangle$" means 'time average'. For both kinds of lipids, an ensemble average and a time average for the two lipid tails was calculated to obtain the molecular order parameter plots ($-2S_{\mathrm{CD}}$) in Figure 6. The figure also shows experimental data obtained using NMR quadrupole splitting experiments on fully-deuterated multilamellar dispersions of DMPC doped with GS at a molar ratio of 1:5.5 at 305 K[94]. The results show that the free lipids are more ordered than in the control simulation. The bound lipid order parameters are close to the experimental values of Zidovetzki et al[94]. Clearly, the lipids interacting with GS are more disordered.

A related problem of longstanding interest has been to determine how cholesterol and other biologically important sterols, such as lanosterol (the evolutionary precursor of cholesterol) and ergosterol, influence the structure and dynamics of lipid bilayers. Cholesterol is an essential component and plays an important role in eukaryotic plasma cell membranes. Lanosterol is the major constituent of prokaryotic cell membranes and ergosterol occurs in the membranes of some yeast and fungi. An intriguing question concerns elucidating which are the particular characteristics of cholesterol, that led to its selection through evolution for the eukaryotic plasma membranes, even though its precursors are structurally very similar. It has been postulated that evolution has selected cholesterol because it optimizes the physical properties of the membrane [99]. However, it is still unclear which specific parts of this molecule provide this optimization. A central question concerns the possible different characteristics of the high-frequency motions of these sterols in membranes.

Although structural aspects of sterols in membranes and especially for cholesterol have been investigated so far [100–103], our knowledge of sterol dynamics is quite limited. Sterol dynamics has up until now been examined for the thre sterols [104, 105] using quasielastic neutron scattering (QENS). Interestingly, a strong anisotropy in the cholesterol motion was observed together with discrete rotation around its long axis. It was suggested that these motions are important for the increase of membrane stiffness upon cholesterol insertion [104]. The QENS data for the three sterols [105] also suggest that slight modifications of the sterol structure have a drastic effect on the molecular dynamics of these molecules in lipid bilayers, which in turn may be related to the membrane micromechanical properties. Cholesterol showed an amplitude of its out-of-plane motion of 1.0-1.1 nm, more than a factor of three higher than measured for the other two sterols.

The combination of QENS with MD should help us investigate the dynamical effects of sterols in membranes and examine biologically-relevant structure-function relationships from a dynamical stand-point. QENS yields information on the same time-scale (1 ps - 10 ns) as the MD simulation, which makes the comparison straightforward. However, while QENS gives us information on the average dynamic behavior of the molecules, using MD simulations it is possible to decompose the simulation-derived dynamic structure factor into motional components and thus shed light on which particular parts of cholesterol provide the optimal physical properties of the membrane. The central goal would then be to establish a model of sterol dynamics in the membrane and to determine the influence of sterols on membrane lipid dynamics. Although several MD simulations of lipid/sterol systems have been published over the past decade [100–103], their analysis has been largely restricted to structural properties, and none has provided or investigated specific motional models. To model better and more accurately the physical properties of cholesterol and its effect on the structure and dynamics of the lipid bilayer, a force field for cholesterol has been derived [106] based on the CHARMM force field [107]. The new parameter set was tested using energy minimization and MD against the existing X-ray crystal structure of cholesterol, which was well reproduced. Work is in progress using these parameters in more realistic and reliable biomembrane simulations so as to gain new insights into the structural evolution of cell membranes.

6. Conclusion

The dynamics of biological molecules is crucial for their function. It is, however, often relatively hard to obtain direct insights in the dynamics of molecules from experiments. A combination of experimental and simulation studies can help in such situations. Quantities that have been measured experimentally can be calculated from simulation data and compared with each other. If both agree, the simulations can be used to obtain a microscopic interpretations of the experimentally-observed phenomena. Simulations and theoretical analysis can thus not only help to interpret experimental data but also guide the design of new experiments. The examples given here testify the richness of the biological information that can be addressed in using simulation techniques.

References

1. Kendrew, J. C., G. Bodo, H. M. Dintzis, R. G. Parrish, H. Wyckoff, and D. C. Phillips (1958) A Three-dimensional Model of the Myoglobin Molecule Obtained by X-ray Analysis. *Nature* **181**, 662.
2. Frauenfelder, H., G. A. Petsko, and D. Tsernoglou (1979) Temperature-dependent X-ray diffraction as a probe of protein structural dynamics. *Nature* **280**, 558.
3. Artymiuk, P. J., C. C. Blake, D. E. Grace, S. J. Oatley, D. C. Phillips, and M. J. Sternberg (1979) Crystallographic studies of the dynamic properties of lysozyme. *Nature* **280**, 563.
4. Karplus, M. and G. A. Petsko (1990) Molecular dynamics simulations in biology. *Nature* **347**, 631–639.
5. Gerstein, M., A. M. Lesk, and C. Chothia (1994) Structural mechanisms for domain movements in proteins. *Biochemistry* **33**, 6739.
6. Huber, R. and W. S. J. Bennett (1983) Functional significance of flexibility in proteins. *Biopolymers* **22**, 261.
7. Austin, R. H., K. W. Beeson, L. Eisenstein, H. Frauenfelder, and I. C. Gunsalus (1975) Dynamics of ligand binding to myoglobin. *Biochemistry* **14**, 5355.
8. Frauenfelder, H., S. G. Sligar, and P. G. Wolynes (1991) The energy landscapes and motions of proteins. *Science* **254**, 1598.
9. Ansari, A., J. Berendzen, S. F. Bowne, H. Frauenfelder, I. E. Iben, T. B. Sauke, E. Shyamsunder, and R. D. Young (1985) Protein states and proteinquakes. *Proc. Natl. Acad. Sci. USA* **82**, 5000.
10. Elber, R. and M. Karplus (1987) Multiple conformational states of proteins: a molecular dynamics analysis of myoglobin. *Science* **235**, 318.

11. Ostermann, A., R. Waschipky, F. G. Parak, and G. U. Nienhaus (2000) Ligand binding and conformational motions in myoglobin. *Nature* **404**, 205.
12. Ech-Cherif el-Kettani, M. A. and J. Durup (1992) Theoretical determination of conformational paths in citrate synthase. *Biopolymers* **32**, 561.
13. Fischer, S. and M. Karplus (1992) Conjugate Peak Refinement : an algorithm for finding reaction paths and accurate transition states in systems with many degrees of freedom. *Chem. Phys. Lett.* **194**, 252–261.
14. Schlitter, J., M. Engels, and P. Kruger (1994) Targeted molecular dynamics: a new approach for searching pathways of conformational transitions. *J. Mol. Graph.* **12**, 84.
15. Wroblowski, B., J. F. Diaz, J. Schlitter, and Y. Engelborghs (1997) Modelling pathways of alpha-chymotrypsin activation and deactivation. *Protein Eng.* **10**, 1163.
16. Fitter, J., R. E. Lechner, and N. A. Dencher (1997) Picosecond molecular motions in bacteriorhodopsin from neutron scattering. *Biophys. J.* **73**, 2126.
17. Doster, W., S. Cusack, and W. Petry (1989) Dynamical transition of myoglobin revealed by inelastic neutron scattering. *Nature* **337**, 754.
18. Doster, W., S. Cusack, and W. Petry (1990) Dynamic instability of liquidlike motions in a globular protein observed by inelastic neutron scattering. *Phys. Rev. Lett.* **65**, 1080.
19. M. Ferrand, M., A. J. Dianoux, W. Petry, and G. Zaccai (1993) Thermal motions and function of bacteriorhodopsin in purple membranes: effects of temperature and hydration studied by neutron scattering. *Proc. Natl. Acad. Sci. USA* **90**, 9668.
20. Reat, V., R. Dunn, M. Ferrand, J. L. Finney, R. M. Daniel, and J. C. Smith (2000) Solvent dependence of dynamic transitions in protein solutions. *Proc. Natl. Acad. Sci. USA* **97**, 9961.
21. Bicout, D. J. and G. Zaccai (2001) Protein flexibility from the dynamical transition: a force constant analysis. *Biophys. J.* **80**, 1115.
22. Parak, F., E. N. Frolov, R. L. Mossbauer, and V. I. Goldanskii (1981) Dynamics of metmyoglobin crystals investigated by nuclear gamma resonance absorption. *J. Mol. Biol.* **145**, 825.
23. Cohen, S. G., E. R. Bauminger, I. Nowik, S. Ofer, and J. Yariv (1981) Dynamics of the Iron-Containing Core in Crystals of the Iron-Storage Protein, Ferritin, through Mossbauer Spectroscopy. *Phys. Rev. Lett.* **46**, 1244.
24. Knapp, E. W., S. F. Fischer, and F. Parak (1982) Protein Dynamics from Mossbauer Spectra. The Temperature Dependence. *J. Am. Chem. Soc.* **86**, 5042.
25. Tilton, Jr., R. F., J. C. Dewan, and G. A. Petsko (1992) Effects of temperature on protein structure and dynamics: X-ray crystallographic studies of the protein ribonuclease-A at nine different temperatures from 98 to 320 K. *Biochemistry* **31**, 2469.
26. Rasmussen, B. F., A. M. Stock, D. Ringe, and G. A. Petsko (1992) Crystalline ribonuclease A loses function below the dynamical transition at 220 K. *Nature* **357**, 423.
27. Lee, A. L. and A. J. Wand (2001) Microscopic origins of entropy, heat capacity and the glass transition in proteins. *Nature* **411**, 501.
28. Green, J. L., J. Fan, and C. A. Angell (1994) The Protein-Glass Analogy: Some Insights from Homopeptide Comparisons. *J. Phys. Chem.* **98**, 13780.
29. Teeter, M. M., A. Yamano, B. Stec, and U. Mohanty (2001) On the nature of a glassy state of matter in a hydrated protein: Relation to protein function. *Proc. Natl. Acad. Sci. USA* **98**, 11242.
30. Parak, F., E. N. Frolov, A. A. Kononenko, R. L. Mossbauer, V. I. Goldanskii, and A. B. Rubin (1980) Evidence for a correlation between the photoinduced electron transfer and dynamic properties of the chromatophore membranes from Rhodospirillum rubrum. *FEBS Letters* **117**, 368.
31. Smith, J., K. Kuczera, and M. Karplus (1990) Dynamics of myoglobin: comparison of simulation results with neutron scattering spectra. *Proc. Natl. Acad. Sci. USA* **87**, 1601.
32. Hayward, J. A. and J. C. Smith (2002) Temperature dependence of protein dynamics: computer simulation analysis of neutron scattering properties. *Biophys. J.* **82**, 1216.
33. Paciaroni, A., S. .Cinelli, and G. Onori (2002) Effect of the environment on the protein

dynamical transition: a neutron scattering study. *Biophys. J.* **83**, 1157.
34. Fitter, J. (1999) The temperature dependence of internal molecular motions in hydrated and dry alpha-amylase: the role of hydration water in the dynamical transition of proteins. *Biophys. J.* **76**, 1034.
35. Cordone, L., M. Ferrand, E. Vitrano, and G. Zaccai (1999) Harmonic behavior of trehalose-coated carbon-monoxy-myoglobin at high temperature. *Biophys. J.* **76**, 1043.
36. Steinbach, P. J. and B. R. Brooks (1993) Protein hydration elucidated by molecular dynamics simulation. *Proc. Natl. Acad. Sci. USA* **90**, 9135.
37. Steinbach, P. J. and B. R. Brooks (1996) Hydrated myoglobin's anharmonic fluctuations are not primarily due to dihedral transitions. *Proc. Natl. Acad. Sci. USA* **93**, 55.
38. Hagen, S. J., K. Hofrichter, and W. A. Eaton (1995) Protein reaction kinetics in a room-temperature glass. *Science* **269**, 959.
39. Walser, R., A. E. Mark, and W. F. van Gunsteren (2000) On the temperature and pressure dependence of a range of properties of a type of water model commonly used in high-temperature protein unfolding simulations. *Biophys. J.* **78**, 2752.
40. Tarek, M. and D. J. Tobias (2000) The dynamics of protein hydration water: a quantitative comparison of molecular dynamics simulations and neutron-scattering experiments. *Biophys. J.* **79**, 3244.
41. Bizzarri, A. R., A. Paciaroni, and S. Cannistraro (2000) Glasslike dynamical behavior of the plastocyanin hydration water. *Phys. Rev. E* **62**, 3991.
42. Tarek, M. and D. J. Tobias (2002) Role of protein-water hydrogen bond dynamics in the protein dynamical transition. *Phys. Rev. Lett.* **88**, 138101.
43. Smith, J. C. (1991) Protein Dynamics: Comparison of simulations with inelastic neutron scattering experiments. *Q. Rev. Biophys.* **24**, 227.
44. Kneller, G. R. and J. C. Smith (1994) Liquid-like side-chain dynamics in myoglobin. *J. Mol. Biol.* **242**, 181.
45. Vitkup, D., D. Ringe, G. A. Petsko, and M. Karplus (2000) Solvent mobility and the protein 'glass' transition. *Nat. Struct. Biol.* **7**, 34.
46. Hoover, W. G. (1985) Canonical Dynamics: Equilibrium phase-space distributions. *Phys. Rev. A* **31**, 1695.
47. Tournier, A. L., D. Huang, S. M. Scwarzl, S. Fischer, and J. C. Smith (2002) Time-resolved computational protein biochemistry: Solvent effects on interactions, conformational transitions and equilibrium fluctuations. *Faraday Discuss.* **122**, 243.
48. Tournier, A. L., J. Xu, and J. C. Smith (2002) *submitted.*
49. Dellerue, A., A. J. Petrescu, J. C. Smith, and M. C. Bellissent-Funel (2001) Radially softening diffusive motions in a globular protein. *Biophys. J.* **81**, 1666.
50. Qin, L. and N. M. Kostić (1992) Electron Transfer Reactions of Cytochrome *f* with Flavin Semiquinones and with Plastocyanin. Importance of Protein-Protein Electrostatic Interactions and of Donor-Acceptor Coupling. *Biochemistry* **31**, 5145–5150.
51. Qin, L. and N. M. Kostić (1993) Importance of Protein Rearrangement in the Electron-Transfer Reaction between the Physiological Partners Cytochrome *f* and Plastocyanin. *Biochemistry* **32**, 6073–6080.
52. Martinez, S. E., D. Huang, A. Szczepaniak, W. A. Cramer, and J. L. Smith (1994) Crystal Structure of Chloroplast Cytochrome *f* Reveals a Novel Cytochrome Fold and Unexpected Heme Ligation. *Structure* **2**, 95–105.
53. Ullmann, G. M., E. W. Knapp, and N. M. Kostić (1997) Computational Simulation and Analysis of the Dynamic Assocition between Plastocyanin and Cytochrome *f*. Consequences for the Electron-Transfer Reaction. *J. Am. Chem. Soc.* **119**, 42–52.
54. Ubbink, M., M. Ejdebäck, B. G. Karlson, and D. S. Bendall (1998) The Structure of the Complex of Plastocyanin and Cytochrome *f*, Determined by Paramagnetic NMR and Restrained Rigid Body Molecular Dynamics. *Structure* **6**, 323–335.
55. Ye, S., C. Shen, T. M. Cotton, and N. M. Kostić (1997) Characterization of Zinc-Substituted Cytochrome c by Circular Dichroism and Resonance Raman Spectroscopic Methods. *J. Inorg. Biochem.* **65**, 219–226.

56. Zhou, J. S. and N. M. Kostić (1993) Gating of the Photoinduced Electron Transfer from Zinc Cytochrome *c* and Tin Cytochrome *c* to Plastocyanin. Effects of the Solution Viscosity on the Rearrangement of the Metalloproteins. *J. Am. Chem. Soc.* **115**, 10796–10804.
57. Peerey, L. M. and N. M. Kostić (1989) Oxidoreduction Reactions Involving the Electrostatic and the Covalent Complex of Cytochrome *c* and Plastocyanin: Importance of the Rearrangement for the Intracomplex Electron-Transfer Reaction. *Biochemistry* **28**, 1861–1868.
58. Peerey, L. M., H. M. Brothers, J. T. Hazzard, G. Tollin, and N. M. Kostić (1991) Unimolecular and Bimolecular Reactions Involving Diprotein Complexes of Cytochrome *c* and Plastocyanin. Dependence of Electron-Transfer Reactivity on Charge and Orientation of the Docked Metalloproteins. *Biochemistry* **30**, 9297–9304.
59. Zhou, J. S. and N. M. Kostić (1992) Photoinduced electron Transfer from Zinc Cytochrome *c* to plastocyanin is Gated by Surface Diffusion within the Metalloprotein Complex. *J. Am. Chem. Soc.* **114**, 3562–3563.
60. Ullmann, G. M. and N. M. Kostić (1995) Electron-Tunneling Paths in Various Electrostatic Complexes between Cytochrome *c* and Plastocyanin. Anisotropy of the Copper-Ligand Interactions and Dependence of the Iron-Copper Electronic Coupling on the Metalloprotein Orientation. *J. Am. Chem. Soc.* **117**, 4766–4774.
61. Roberts, V. A., H. C. Freeman, A. J. Olson, J. A. Tainer, and E. D. Getzoff (1991) Electrostatic Orientation of the Electron-Transfer Complex Between Plastocyanin and Cytochrome *c*. *J. Biol. Chem.* **266**, 13431–13441.
62. Crnogorac, M. M., C. Shen, S. Young, Ö. Hansson, and N. M. Kostić (1996) Effects of Mutations in Plastocyanin on the Kinetics of the Protein Rearrangement Gating the Electron-Transfer Reaction with Zinc Cytochrome *c*. Analysis of the Rearrangement Pathway. *Biochemistry* **35**, 16465–16474.
63. Crnogorac, M., G. M. Ullmann, and N. M. Kostić (2001) Effects of pH on Protein Association. Modification of the Proton Linkage Model and Experimental Verification of the Modified model in the Case of Cytochrome c and Plastocyanin. *J. Am. Chem. Soc.* **123**, 10789–10798.
64. Ivković-Jensen, M. M. and N. M. Kostić (1996) Effects of Temperature on the Kinetics of the Gated Electron-Transfer Reaction between Zinc Cytochrome *c* and Plastocyanin. Analysis of Configurational Fluctuation of the Diprotein Complex. *Biochemistry* **35**, 15095–15106.
65. Ivković-Jensen, M. M. and N. M. Kostić (1997) Effects of Viscosity and Temperature on the Kinetics of the Electron Transfer Reaction between the Triplet State of Zink Cytochrome *c* and Cupriplastocyanin. *Biochemistry* **36**, 8135–8144.
66. Ivković-Jensen, M. M., G. M. Ullmann, S. Young, Ö. Hansson, M. Crnogorac, M. Edjebäck, and N. M. Kostić (1998) Effects of Single and Double Mutations in Plastocyanin on the Rate Constant and Activation Parameters of the Gated Electron-Transfer Reaction between the Triplet State of Zinc Cytochrome *c* and Cupriplastocyanin. *Biochemistry* **37**, 9557–9569.
67. Ivković-Jensen, M. M., G. M. Ullmann, M. M. Crnogorac, M. Ejdebäck, S. Young, Ö. Hansson, , and N. M. Kostić (1999) Comparing the Rates and the Activation Parameters for the Forward Reaction between the Triplet State of Zinc Cytochrome c and Cupriplastocyanin and the Back Reaction between the Zinc Cytochrome c Cation Radical and Cuproplastocyanin. *Biochemistry* **38**, 1589–1597.
68. Ubbink, M. and D. S. Bendall (1997) Complex of Plastocyanin and Cytochrome *c* Characterized by NMR Chemical Shift Analysis. *Biochemistry* **36**, 6326–6335.
69. Ullmann, G. M. (2001) *Charge Transfer Properties of Photosynthetic and Respiratory Proteins*. pp. 525–584. In: H. S. Nalwa (Ed.): Supramolecular Photosensitive and Electroactive Matrials. Academic Press New York.
70. Ullmann, G. M. and E. W. Knapp (1999) Electrostatic Computations of Protonation and Redox Equilibria in Proteins. *Eur. Biophys. J.* **28**, 533–551.

71. Beroza, P. and D. A. Case (1998) Calculation of Proton Binding Thermodynamics in Proteins. *Methods Enzymol.* **295**, 170–189.
72. Briggs, J. M. and J. Antosiewicz (1999) Simulation of pH-dependent Properties of Proteins Using Mesoscopic Models. *Rev. Comp. Chem.* **13**, 249–311.
73. Sham, Y. Y., Z. T. Chu, and A. Warshel (1997) Consistent Calculation of pK_a's of Ionizable Residues in Proteins: Semi-mircoscopic and Mircoscopic Approaches. *J. Phys. Chem. B* **101**, 4458–4472.
74. Onufriev, A., D. A. Case, and G. M. Ullmann (2001) A Novel View on the pH Titration of Biomolecules. *Biochemistry* **40**, 3413–3419.
75. Ullmann, G. M. (2003) Relations between Protonation Constants and Titration Curves in Polyprotic Acids: A Critical View. *J. Phys. Chem. B* **107**, in press.
76. Okamura, M., M. Paddock, M. Graige, and G. Feher (2000) Proton and electron transfer in bacterial reaction centers. *Biochim. Biophys. Acta* **1458**, 148–163.
77. Sebban, P., P. Maróti, and D. K. Hanson (2001) Electron and proton transfer to the quinones in bacterial photosynthetic reaction centers: insight from combined approaches of molecular genetics and biophysics. *Biochimie* **77**, 677–694.
78. Wraight, C. A. (1979) Electron acceptors of bacterial photosynthetic reaction centers. II. H^+ binding coupled to secondary electron transfer in the quinone acceptor complex. *Biochim. Biophys. Acta* **548**, 309–327.
79. Rabenstein, B., G. M. Ullmann, and E. W. Knapp (1998) Energetics of the Electron Transfer and Protonation reactions of the Quinones in The Photosynthetic reaction center of *Rhodopseudomonas viridis*. *Biochemistry* **37**, 2488–2495.
80. Rabenstein, B., G. M. Ullmann, and E. W. Knapp (1998) Calculation of Protonation Patterns in Proteins with Conformational Relaxation - Application to the Photosynthetic Reaction Center. *Eur. Biophys. J.* **27**, 628–637.
81. Rabenstein, B., G. M. Ullmann, and E. W. Knapp (2000) Electron Transfer between the Quinones in the Photosynthetic Reaction Center and its Coupling to Conformational Changes. *Biochemistry* **39**, 10487–10496.
82. Maróti, P. and C. A. Wraight (1988) Flash-induced H^+ Binding by Bacterial Photosynthetic Reaction Centers: Influences of the Redox States of the Acceptor Quinones and Primary Donor. *Biochim. Biophys. Acta* **934**, 329–347.
83. McPherson, P. H., M. Y. Okamura, and G. Feher (1988) Light-induced Proton Uptake by Photosynthetic Reaction Centers from *Rhodobacter sphaeroides* R-26. I. Protonation of the One-Electron States $D^+Q_A^-$, DQ_A^-, $D^+Q_AQ_B^-$ and $DQ_AQ_B^-$. *Biochim. Biophys. Acta* **934**, 348–368.
84. Tandori, J., J. M. M. Valerio-Lepiniec, M. Schiffer, P. Maroti, D. Hanson, and P. Sebban (2002) Proton Uptake of *Rhodobacter sphaeroides* Reaction Center Mutants Modified in The Primary Quinone Environment. *Photochem. Photobiol.* **75**, 126–133.
85. Sebban, P., P. Maróti, M. Schiffer, and D. Hanson (1995) Electrostatic dominoes: long distance propagation of mutational effects in photosynthetic reaction centers of *Rhodobacter capsulatus*. *Biochemistry* **34**, 8390–8397.
86. Stowell, M. H. B., T. M. McPhillips, D. C. Rees, S. M. Soltis, E. Abresch, and G. Feher (1997) Light-Induced Structural Changes in Photosynthetic Reaction Center: Implications for Mechanism of Electron-Proton Transfer. *Science* **276**, 812–816.
87. Allen, J. (1994) Crystallization of the reaction center from Rhodobacter sphaeroides in a new tetragonal form. *Proteins* **20**, 283–286.
88. Lancaster, C. R. D. and H. Michel (1997) The coupling of light-induced electron transfer and proton uptake as derived from crystal structures of reaction centres from Rhodopseudomonas viridis modified at the binding site of the secondary quinone. *Structure* **5**, 1339–59.
89. Taly, A., P. Sebban, J. C. Smith, and G. M. Ullmann (2003) The structural changes in the Q_B pocket of the photosynthetic reaction center depend on pH: a theoretical analysis of the proton uptake upon Q_B reduction. *Biophys. J.* **84**, in press.
90. Jones, C. R., C. T. Sikakana, S. Hehir, M.-C. Kuo, and W. A. Gibbons (1978) The

quantitation of nuclear Overhauser effect methods for total conformational analysis of peptides in solution. Application to gramicidin S. *Biophys. J.* **24**, 815.

91. Hull, E., R. Karlsson, P. Main, M. M. Woolfson, and E. J. Dodson (1978) The crystal structure of a hydrated gramicidin S-urea complex. *Nature* **75**, 206.
92. Mihailescu, D. and J. C. Smith (1999) Molecular dynamics simulation of the cyclic decapeptide antibiotic, gramicidin S, in dimethyl sulfoxide solution. *J. Phys. Chem.* **9**, 1586.
93. Datema, K. P., K. P. Pauls, and M. Bloom (1986) Deuterium nuclear magnetic resonance investigation of the exchangeable sites on gramicidin A and gramicidin S in multilamellar vesicles of dipalmitoylphosphatidylcholine. *Biochemistry* **25**, 3796.
94. Zidovetzki, R., U. Banerjee, D. W. Harrigton, and S. I. Chan (1988) NMR study of the interaction of polymyxin B, gramicidin S, valinomycin with dimyristoyllecithin bilayers. *Biochemistry* **27**, 5686.
95. Prenner, E. J., R. N. A. H. Lewis, K. C. Neuman, S. M. Gruner, L. H. Kondejewski, R. S. Hodges, and R. N. McElhaney (1997) Nonlamellar phase induced by the interaction of gramicidin S with lipid bilayers. A possible relationship to membrane-disrupting activity. *Biochemistry* **37**, 7906.
96. Higashijima, T., T. Miyazawa, M. Kawai, and U. Nagai (1986) Gramicidin S analogues with D-Ala, Gly, or L-Ala residues in place of the D-Phe residue: molecular conformations and interactions with phospholipid membrane. *Biopolymers* **25**, 2295.
97. Katsu, T., H. Kobayahi, T. Hirota, Y. Fujita, K. Sato, and U. Nagai (1987) Structure-activity relationship of gramicidin S analogues on membrane permeability. *Biochim. Biophys. Acta* **899**, 159.
98. Portlock, S. H., M. J. Clague, and R. J. Cherry (1990) Leakage of internal markers from erythrocytes and lipid vesicles induced by melittin, gramicidin S and alamethicin: a comparative study. *Biochim. Biophys. Acta* **1030**, 1.
99. Bloch, K. (1985) *Cholesterol, evolution of structure and function.* pp. 1–24. in Biochemistry of Lipids and Membranes, Eds. J. E. Vance and D. E. Vance, Benjamin/Cummins Pub. Co. Inc., New York.
100. Smordynev, A. and M. L. Berkowitz (2001) Molecular Dynamics Simulation of the structure of DMPC bilayers with Cholesterol, Ergosterol, and Lanosterol. *Biophys. J.* **80**, 1649.
101. Pasenkiewicz-Gierula, M., T. Rog, K. Kitamura, and A. Kusumi (2000) Cholesterol Effects on the Phosphatidylcholine Bilayer polar region: A Molecular Dynamics study. *Biophys. J.* **78**, 1376.
102. Chiu, S. W., E. Jacobsson, and H. L. Scott (2001) Combined Monte Carlo and Molecular Dynamics simulation of hydrated lipid-cholesterol lipid bilayers at low Cholesterol concentration. *Biophys. J.* **80**, 1104.
103. Tu, K., M. Klein, and D. Tobias (1998) Constant-Pressure Molecular Dynamics investigation of cholesterol effects in a Dipalmitoylphosphatidylcholin bilayer. *Biophys. J.* **75**, 2147.
104. Gliss, C., O. Ranedl, H. Casalta, E. Sackmann, R. Zorn, and T. Bayerl (2001) Anisotropic motion of cholesterol in oriented DPPC bilayers studies by quasielastic neutron scattering: the liquid-ordered phase. *Biophys. J.* **77**, 331.
105. Endress, E., H. Heller, H. Casalta, M. F. Brown, and T. M. Bayerl (2002) Anisotropic motion and molecular dynamics of cholesterol, lanosterol and ergosterol in lecithin bilayers studied by quasi-elastic neutron scattering. *Biochemistry* **41**, 13078.
106. Cournia, Z., A. Vaiana, J. C. Smith, and G. M. Ullmann (2003) *submitted. Pure Appl. Chem.*
107. Brooks, B. R., R. E. Bruccoleri, B. D. Olafson, D. J. States, S. Swaminathan, and M. Karplus (1983) CHARMM: A Program for Macromolecular Energy, Minimization, and Dynamics Calculation. *J. Comp. Chem.* **4**, 187–217.

STRUCTURE, THERMODYNAMICS AND CRITICAL PROPERTIES OF IONIC FLUIDS

WOLFFRAM SCHRÖER
Institut für Anorganische und Physikalische Chemie, Universität Bremen, D-28359 Bremen, Germany

HERMANN WEINGÄRTNER
Physikalische Chemie II, Ruhr-Universität Bochum, D-44780 Bochum, Germany

1. INTRODUCTION

Ionic fluids such as molten salts and electrolyte solutions are of central interest in Chemical Physics and Physical Chemistry since the first days of those sciences. They play an important role in many applied fields such as electrochemistry, chemical engineering or the geosciences. If compared to simple neutral fluids, two properties provide major challenges for theory, namely the long-range nature of the Coulomb interactions and the high figures of the Coulomb energy at small ion separations.

The long-range Coulomb interactions cause severe technical problems in theory and simulations. For example, integrals as the second virial coefficient that exist in normal fluids diverge for Coulomb interactions [1]. This long-range nature also challenges the hypothesis that liquid-gas and liquid-liquid transitions all belong to the Ising universality class [2-6]. The Ising model assumes only next neighbor interactions, and is limited to transitions driven by r^{-p} interactions with $p > 4.97$ [4,5]. It has been speculated that r^{-1} Coulomb interactions suppress the non-classical critical fluctuations, giving rise to mean-field critical behavior [2], as suggested by a mean-field-like liquid-vapor coexistence curve of NH_4Cl [7]. NH_4Cl is a rare example for a molten salt, where the critical point can be reached. Friedman [8] then pointed out that liquid-liquid phase transitions in electrolyte solutions near room temperature might be more suited to investigate the problem of ionic criticality. In the late 1980s, Pitzer and coworkers followed this path, indeed finding evidence for an apparent mean-field behavior [9,10].

The second striking difference between ionic and nonionic fluids concerns the strength of the intermolecular forces. In normal fluids at ambient conditions, the size of the intermolecular interactions is roughly of the order of the thermal energy. At a quantitative level, this fact is expressed by introducing a reduced temperature $T^* = kT / \phi_{min}$ as the ratio between the thermal energy kT and the depth of the interaction potential $\phi(r)$ at its minimum. For simple neutral fluids T^* is of the order of unity. In contrast, in typical molten salts the interionic interactions are more than an order of

J. Samios and V.A. Durov (eds.), Novel Approaches to the Structure and Dynamics of Liquids: Experiments, Theories and Simulations, 503–537.

magnitude larger than kT and can even reach the magnitude of chemical bonds, *i.e.* $T^* << 1$. In electrolyte solutions, the strength of the Coulomb interactions depends on the dielectric constant ε of the solvent. In aqueous solutions T^* may come close to unity but in the solvents of low dielectric constant of interest here, one has $T^* << 1$ like in molten salts [11,12].

Such strong interactions have major consequences for the structure ionic liquids. While ordinary liquids can in good approximation be described in the van der Waals picture as hard core bodies in a see of an average potential, the Coulomb interactions may give rise to the formation of ion pairs and higher ion clusters as long-lived molecule-like entities. These entities separate, however, into free ions by rather mild changes of the conditions *e.g.* by dilution or change of the solvent. One way to describe this ion distribution therefore resorts to a "chemical picture" in which the fluid is regarded as a system of ions, ion pairs and, may be higher clusters, in chemical equilibrium. This is true even for the simplest model fluid, which is the model of equal-sized, charged hard-spheres in a dielectric continuum termed Restricted Primitive Model (RPM) [11].

In the "physical picture" ion-pairs are just consequences of large values of the Mayer f-functions that describe the ion distribution [1]. The technical consequence, however, is a major complication of the theory: the high-temperature approximations of the f-functions, which are applied in the so called analytical methods of statistical mechanics [13] such as the Mean Spherical Approximation (MSA) or the Percus-Yevick approximation (PY) are not appropriate in ionic systems. Even Hypernetted Chain theory (HNC) [1], which does not involve the high temperature approximation but applies the Mayer functions in full, becomes unreliable, because the neglected bridge graphs give long range contributions to the direct correlation function, and therefore affect the structure and thermodynamic properties.

A way out of these difficulties is a blend of chemical and physical descriptions in which ion configurations of high energy are described by the mass action law, while the evaluation of the Mayer functions cares for the remainder of the interactions. Note, that Debye-Hückel (DH) theory [14], which is derived from classical electrostatics and statistical thermodynamics, is also a high-temperature approximation, whose range of validity can be largely extended by a mass action law for ion pair formation [15].

Because phase transitions driven by Coulomb interactions fall into the regime, where ion pairs and higher clusters play a major role, the development of an adequate electrolyte theory for this regime is of crucial importance. For example, any renormalization (RG) group analysis will depend on the construction of a simple mean-field Hamiltonian that adequately describes the interionic forces [4]. More specifically, this calls for *analytical* expressions for the free energy, because Taylor expansions have to be performed in the analysis of critical phenomena [4]. At present, such analytical expressions are only provided by DH-type approaches, which have been developed further by including the interactions of dipolar ion pairs [16,17]. For simulations but also for non-analytical results of the popular integral equation methods of liquid-state theory this is possible only after fitting the free energy resulting from many calculations. It is fair to say that the need for a transparent analytical theory has triggered this revival of DH theory, thus giving a new impetus to electrolyte theory.

In the present article, we review recent progress in this subject area. In the first part, we address in Section 2 the problem of the electrolyte solution structure at conditions of low reduced temperature, where phase separations are known to occur.

We discuss new experimental results derived from electrical conductance and dielectric data and compare the conclusions with results from Monte-Carlo (MC) simulations, integral equation theories, and DH-type approaches. Thereby, we consider in some detail new DH-based theories of the hard sphere ionic fluid (RPM), which in recent theories of ionic phase transitions have played a major role. In Section 3, we consider experimental and theoretical results concerning the location of the two-phase regime in ionic fluids with special regard to liquid-liquid transitions in electrolyte solutions. Having discussed the location of the phase transition, we review in Section 4 experimental results on near-critical behavior of ionic fluids, and attempts to rationalize this behavior by modern theories of critical phenomena.

2. THE PARTICLE DISTRIBUTION IN SYSTEMS WITH STRONG INTERIONIC INTERACTIONS

2.1 HISTORY

Much of our knowledge on electrolyte solutions results from studies of aqueous systems with solution properties, which are fundamentally different from those considered here. It is therefore adequate to start with some comments on the history of the field treated here. This history can be cartooned as fight between two extreme views, assuming dissociating salt molecules and complete dissociation into free ions, respectively.

The ion hypothesis goes back to the famous paper of Arrhenius [19], who introduced a mass-action law for a chemical equilibrium between free ions and salt molecules, although the idea that the electrical conductance results from ion dissociation seems to go back to Clausius [18]. The mass action law implies complete dissociation in the limit of infinite dilution and a decrease of the degree of dissociation with increasing salt concentration. Arrhenius also suggested that the degree of dissociation α at any concentration is equal to the ratio of the observed equivalent conductance Λ to its limiting value Λ^{∞} at infinite dilution, *i.e.* $\alpha = \Lambda / \Lambda^{\infty}$.

Arrhenius theory applies well to solutions of weak acids and bases in water, but fails in the case of strong electrolytes such as ordinary salts. This problem was eventually solved by Debye and Hückel [14] who assumed complete dissociation, but considered the Coulomb interactions between the ions by a patchwork theory based on both statistical mechanics and macroscopic electrostatics. Their theory enabled for a satisfactory explanation of many experimental observations, above all for aqueous solutions.

Because the idea of complete dissociation was so simple, it was generally expected that deviations from DH theory, as *e.g.* observed in solvents of low dielectric constant, will find a physical explanation, and would not require a return to the idea of undissociated salt molecules. Bjerrum (Bj), however, developed a theory, which combined the Arrhenius and DH approaches by assuming a chemical equilibrium between ion-pairs and free ions [15]. This concept allows taking into account interactions of ions at short range, which are not adequately described in DH theory. Although in the second half of the 20th century new theoretical methods such as simulation techniques and the so-called analytical methods of liquid-state theory have been developed [13], it is fair to say that the DH-Bj model is a good guide to theory. It also includes a theory for the mass action constant as a function of the dielectric constant ε of the solvent. Many experimental investigations, *e.g.* reviewed by Kraus [20], have confirmed Bjerrum's concept. Note however that an ε-dependence of the association constant was recognized a long time before Bjerrum by Nernst [21], and was verified in numerous conductance studies by Walden and coworkers, the results of which are *e.g.* collected in the data compilation of Landolt-Börnstein [22].

While much of the knowledge about ion association in organic solvents was derived from electrical conductance data, we know today that another important feature of electrolyte solutions in low-ε solvents is the existence of liquid-liquid transitions [11]. Of course, this is of prime interest in studies of near-critical behavior of ionic fluids, but this critical point may also serve as an important target for testing electrolyte theories. Again, the first evidence for such transitions was already presented at the beginning of the 20th century by Walden [23], but did not receive further attention. As such transitions are normally preempted by crystallization [24], it needed the design of low-melting salts [9,10,24] to enable more extensive studies of such transitions. Thereby, pilot experiments by Pitzer [9,10] paved the way for the work reviewed in this article.

2.2 EXPERIMENTAL OBSERVATIONS

We first consider the available experimental information on the ion distribution at the typical conditions considered in this article. In the absence of X-ray or neutron scattering data, the electrical conductance provides a major source for structural information. In systems showing phase separation, a minimum in the concentration dependence of the equivalent conductance Λ is a major feature [24-26]. Because the molecular forces causing this minimum also play a crucial role in driving the phase transition, it is instructive to discuss this conductance behavior in some detail.

We consider first the low-concentration branch, where Λ decreases rapidly with increasing molar concentration C of the salt. In contrast to the situation encountered for aqueous systems, in low-ε solvents one does however not reach the limiting regime of the famous Debye-Hückel-Onsager $C^{1/2}$ law [27], which results from long-range interionic interactions between free ions. Rather, even at the lowest concentrations association to ion-pairs controls the concentration dependence of Λ. For symmetrical electrolytes with charges $q_+ = z_+e = |q_-| = |z_-|\,e$, where e is the fundamental charge and the z_i are signed charge numbers, the underlying mass action law reads

$$\frac{2(1-\alpha)}{\alpha^2} = K(T)\,\rho\,\frac{\gamma_+\gamma_-}{\gamma_P}, \tag{1}$$

where $\rho = (N_+ + N_-)/V$ is the total number density of the ions, $K(T)$ is the association constant, and the γ_i are the activity coefficients of the free ions and ion pairs, respectively. Bjerrum's expression for the association constant is [15]

$$K(T) = \int_\sigma^R \exp\left[\frac{R}{r}\right] 4\pi r^2\, dr, \tag{2}$$

where $R = \beta q^2/\varepsilon$ is the so-called Bjerrum length and σ is the collision diameter. The evaluation of this low-concentration branch of the conductance curve enables the determination of the association constant $K(T)$ for ion pair formation.

The high concentration branch, where Λ increases, is more difficult to interpret. Note that the appearance of a conductance minimum is not limited to systems showing liquid-liquid phase separation, but depends on the size of the dielectric constant of the medium and on the size and charges of the ions. For univalent cations and anions it generally occurs at low dielectric constant of the solvent, $\varepsilon < 10$ say.

A minimum of Λ implies the reappearance of charged species by redissociation of the pairs into free ions and/or by the formation of charged ion clusters. Its origin has been of long-standing and controversial debate. Fuoss and Kraus [28] attributed this minimum to the formation of charged ion triplets, allowing to describe the concentration dependence of Λ by two coupled mass action laws for pair and triplet formation. Still today, this triple ion scenario is a popular hypothesis, adopted by many workers in the field. However, other effects must come into play, because eventually, the ion pairs and ion clusters have to redissociate to obtain the fully dissociated pattern of the molten salt. The corresponding redissociation is now known to result from the interactions of the free ions with the ion pairs that stabilize the free ions [16]. Furthermore, the increase of the dielectric permittivity due to the formation of ion pairs causes a reduction of the association constant, also driving redissociation [17,29,30].

For illustration, Fig. 1 shows isotherms of the Walden product, which is the product of the molar conductivity Λ and the shear viscosity η. We consider two demixing ionic solutions slightly above their critical temperatures as a function of the molar concentration of the salt. The systems are tetra-*n*-butyl ammonium picrate (Bu_4NPic) dissolved in 1-tridecanol [24] and 1-chloroheptane [25]. The data were normalized to the values at the critical points. Both sets of data fall onto a common curve, which clearly indicates corresponding states behavior of the degree of dissociation that controls the conductance.

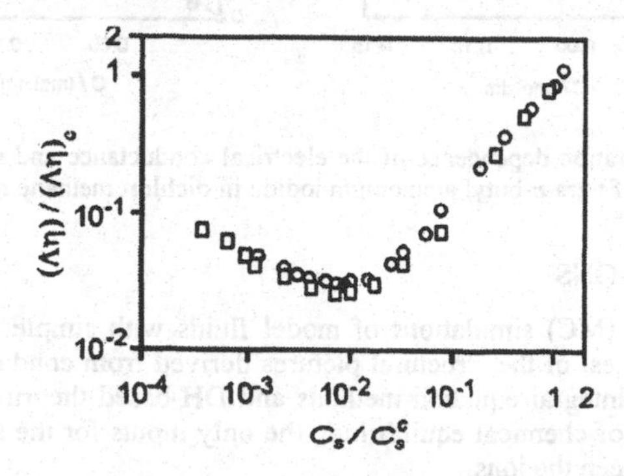

Figure 1. Conductance for tetra-*n*-butyl ammonium picrate (Bu_4NPic) dissolved in tridecanol (O) and 1-chloroheptane (□) along isotherms slightly above the liquid-liquid critical points of these systems. The molar conductance and molar salt concentrations are normalized to their values at the critical points.

Another important source of information is dielectric spectroscopy, because dipolar ion pairs result in an observable mode in the dielectric spectrum, and thus contribute to the static dielectric constant of the solution. Thus, the dielectric spectra reflect two modes caused by the reorientation of solvent dipoles and of ion-pair dipoles, allowing the separation of the static dielectric constant into ion pair and solvent contributions, respectively (in apolar solvents one solely observes ion pair reorientation [31]).

For illustration, Fig. 2 shows results for Bu_4N-iodide (Bu_4NI) in dichloromethane (CH_2Cl_2) [12] (see also [30]). The observed increase of the total dielectric constant ε may form an essential ingredient for understanding the solution properties. This increase results solely from the ion pair contribution. The slight decrease of the solvent contributions is readily explained by the decrease in the particle density of the solvent. The static dielectric constant does, however, not increase linearly with the concentration of the salt. The decreasing slope at high salt concentrations may result from the redissociation of the ion pairs, which causes the minimum of the conductivity, but at a quantitative level, redissociation alone is not sufficient to account for this effect (see Section 2.5). Rather, one has to account for correlations between the ion pairs with a tendency to antiparallel orientations, so that the dipolar pairs do not fully contribute to the dielectric constant. Then, the well-known Kirkwood g factor for orientational correlations [32] may serve as tool to interpret the data, implying Kirkwood factors $g < 1$. In the cluster picture, this result can be rationalized as the formation of quadruple ions formed by antiparallel pairs.

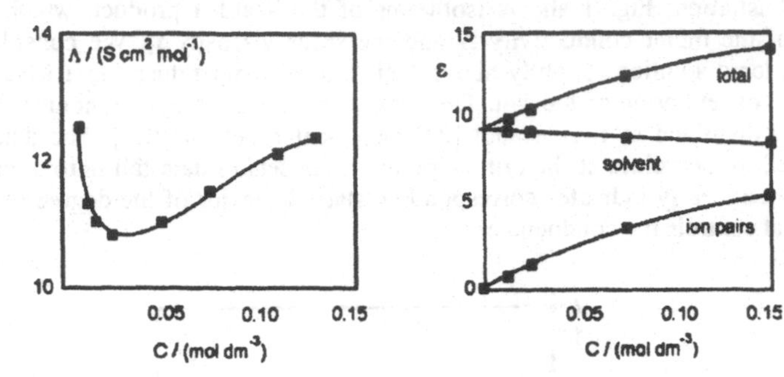

Figure 2. Concentration dependence of the electrical conductance and static dielectric constant of solutions of tetra-*n*-butyl ammonium iodide in dichloromethane near the conductance minimum.

2.3 SIMULATIONS

Monte Carlo (MC) simulations of model fluids with simple Hamiltonians provide an ultimate test of the structural pictures derived from conductance and dielectric data. While integral equation methods and DH-based theories theory start with the presumption of chemical equilibrium, the only inputs for the simulations are the interactions between the ions.

The simplest model for a generic ionic fluid is the "primitive model" of charged hard spheres in a dielectric continuum with the dielectric constant ε. Thus, the potential is just the Coulomb potential between the ions labeled i and k, which is cut-off at the collision diameter σ_{ik}, which is the mean of the diameters of the ions

$$\phi_{ik}(r) = \begin{matrix} q_i q_k / \varepsilon r & r > \sigma_{ik} \\ \infty & r \leq \sigma_{ik} . \end{matrix} \qquad (3)$$

Primitive models satisfy the corresponding states principle [11,12]. Then, the thermodynamic state is completely specified by a reduced temperature T^* and reduced ion density ρ^*. The energy scale used for defining T^* is set by the Coulomb energy at contact

$$T^* = \frac{kT\varepsilon\sigma_{+-}}{|q_+ q_-|}. \tag{4}$$

The reduced density is

$$\rho^* = \rho\sigma_{+-}^3 . \tag{5}$$

Most simulations concern the so-called restricted primitive model (RPM), which consists of equal-sized charged hard spheres with charges $q_+ = |q_-|$.

It should be noted that the scaling of the Coulomb potential by ε is only a first, albeit very successful approximation, because it assumes the dielectric permittivities ε_{in} inside the particle and ε in the medium to be equal. This is, e.g. correct for the gas of charged hard spheres. For charged hard spheres embedded in a dielectric continuum other interactions as "cavity interactions" based on continuum electrostatics should be included. The cavity term is repulsive and decays as r^{-4} [33].

The determination of the ion distribution from simulation data employs an analysis of the particle separations found in the simulated box. This implies a somewhat arbitrary definition of the relevant pairs and higher ion clusters, e.g. by counting ions separated by (1–2)σ as being paired [34,35]. The picture resulting from such simulations is quite incomplete, but some important general results can be obtained from the simulation data.

Considering first the low-density regime, some data are available at $\rho^* = 0.004$, which is one order of magnitude below the critical density of the RPM [34]. At this condition, the density of free ions decreases when lowering the reduced temperature T^*, and near $T^* = 0.1$ the concentrations of ion pairs and free ions become equal. At $T^* = 0.05$ (which is roughly the critical temperature of the RPM) the free ions practically vanish. Moreover, below $T^* = 0.1$ higher clusters begin to appear at the cost of the expense of pairs. Thus, when lowering T^*, the ion pair concentration passes a maximum. This maximum corresponds to a maximum in the specific heat and a sudden decrease of the dielectric constant. These effects occur still well above the critical temperature. The changes become less sharp with increasing ion density. The results are in good agreement with the conclusions drawn from conductance and dielectric data.

Further results are available at conditions near the liquid-vapor coexistence of the RPM [35]. In real molten salts liquid-vapor coexistence occurs between a highly conducting melt and an almost insulating vapor of neutral molecules and higher clusters with only very few individual ions. Likewise, liquid-liquid phase transitions occur between highly dilute, almost non-conducing solutions, and concentrated solutions that show a high conductance. These conclusions are more-or-less confirmed by the simulation data, which highlight the role of neutral ion pairs and probably higher neutral ion clusters. However, at higher concentrations this description in terms of clusters certainly becomes inappropriate. In view of the dominance of neutral species, it is even worthwhile to ask whether, at some stage, the vapor will become insulating, as found in two dimensions for the $d = 2$ primitive model [36]. An analysis of the wave vector-dependent dielectric constant $\varepsilon(k)$ by Caillol [34] and results from MD simulations of a more refined model for NaCl [37] both indicate that the saturated vapor is conducting.

In real systems, cations and ions are usually of different size and often also of different charge. It has been speculated that this asymmetry may be responsible for the peculiar critical behavior of ionic fluids [5]. Only a few simulations are however available for cluster configurations in primitive models with ions differing in the size or/and charges [38-40]. In general, these simulations appear to show a more rich variation of clusters than the RPM. While for the RPM at $T^* = 0.05$ and $\rho^* = 0.002$ clusters of 2,4,6 particles are found, much larger clusters involving up to 60 ions occur, if for similar conditions the sizes of the ions differ by a factor of 10. Moreover, in addition to the predominant chain-like clusters of the RPM, configurations such as rings are found. If the ions differ in the charge, even more cluster configurations are observed. Such data will provide insights into the behavior of charged colloidal suspensions and polyelectrolytes.

2.4 INTEGRAL EQUATION THEORIES

Liquid-state statistical mechanics provides a number of tools for determining the equation of state from which integral equation have received special attention [13]. Integral equation theories are theories for the pair correlation function $g_{\alpha\beta}(\mathbf{r}_\alpha,\mathbf{r}_\beta)$, which relates the density of particle β at location $\mathbf{r}_\beta$ to that of α at $\mathbf{r}_\alpha$ ($\alpha,\beta = +,-$). In isotropic, homogeneous fluids, the pair correlation function depends only on the radial distance r of the particles. Clearly, in ionic systems $g_{+-}(r) = g_{-+}(r)$, while $g_{++}(r) = g_{--}(r)$ only holds, when the ions are equal in size and in magnitude of the charge.

Integral equation theories start with the Ornstein-Zernike (OZ) equation, where the total correlation function $h_{\alpha\beta}(r) = g_{\alpha\beta}(r) - 1$ is split up into the direct correlation function $c_{\alpha\beta}(r)$ for pair interactions plus an indirect term, which reflects these α,β interactions mediated by all other particles γ. For isotropic homogenous systems the OZ equation reads [13]

$$h_{\alpha\beta}(r_{\alpha\beta}) = c_{\alpha\beta}(r_{\alpha\beta}) + \sum_{\gamma} \int c_{\alpha\gamma}(r_{\alpha\gamma})\ \rho_\gamma\ h_{\gamma\beta}(r_{\gamma\beta})\ d\vec{r}_\gamma . \qquad (6)$$

Approximations for $c_{\alpha\beta}(r)$ are then used as closures to solve the OZ relation for $h_{\alpha\beta}(r)$. Once $h_{\alpha\beta}(r)$ and $c_{\alpha\beta}(r)$ are available, the equation of state can be extracted via well-known equations for the internal energy, pressure, or compressibility [13].

The Helmholtz free energy A is readily obtained from the energy by the so-called thermodynamic integration with respect to β, carried out for a fixed distribution of particles:

$$\Phi = \beta A / V = \int_0^\beta u\ d\beta . \qquad (7)$$

Φ is the reduced free energy density. Alternatively, the Helmholtz free energy is calculable from the pressure or the compressibility by integration with respect to the density. In principle, these three routes are equivalent, but due to approximations involved in the different closures of the OZ equation, this is not the case and thermodynamic consistence can hardly be achieved from such theories. The free energy allows evaluating the activity coefficients of the various species, which, in turn, enables the determination of the degree of dissociation in the mass action law, Eqn. (1).

From the various closures applied to nonionic fluids, only the mean spherical approximation (MSA) seems to be suited for treating electrolyte solutions in media of low dielectric constant. The Percus-Yevick (PY) closure is unsatisfactory for long-range potentials [13]. The hypernetted chain approximation (HNC), widely used in electrolyte thermodynamics of highly dissociated systems, suffers from the neglect of the bridge graphs [1,13], which give long range contributions to the direct correlation function, and affect the structure and thermodynamic properties at the conditions of interest [35].

Thus, from all integral equation methods, only the MSA is attractive for gaining structural and thermodynamic informations of ionic fluids at the conditions of interest here. This is particularly true for models with hard sphere repulsion such as the RPM. As already mentioned, the approximate nature of $g_{\alpha\beta}(r)$ leads to severe thermodynamic inconsistencies, and only the energy route to the thermodynamic functions provides sensible results. Attempts were therefore made to construct thermodynamically consistent theories such as the Generalized MSA (GMSA) [41] in which an *ad hoc* term is added to $c_{\alpha\beta}(r)$ to gain consistency. Similar in spirit is the self-consistent OZ approximation (SCOZA) [42], in which a closure is constructed by imposing self-consistency on the compressibility and energy equation of state. In view of the fact that for the neutral lattice gas this theory remains quite successful in the critical region, application of the SCOZA to ionic fluids seems promising [43].

Specifically, the structural information is gained by applying the mass action law (1) and estimating the activity coefficients from the free energy of the MSA. Thereby, Bjerrum's expression (2) for the association constant has occasionally been replaced by a slightly different expression going back to Ebeling [44], which results from the virial expansion and remains correct in the high temperature limit. Based on these concepts, various versions of the MSA have been applied to systems at low T^*. All versions seem to predict the minimum in the degree of dissociation observed for such systems. With regard to critical point predictions, Stell [5] and Guillot and Guissani [45] have given interesting discussions of this work. However, as we know from experiments and simulations, interactions of the pairs with the remainder of the ionic fluid play an important role at low T^*. It has proven to be quite difficult to extend MSA theory to incorporate these interactions of the pairs, and all attempts have resorted on approximations [45,46]. We quote for example the pairing MSA (PMSA) of Stell and coworkers [46].

2.5 DEBYE-HÜCKEL-TYPE THEORIES

The need for an analytical expressions for the equation of state have led to a revival of the classical approach due to Debye, Hückel and Bjerrum. DH theory is based on macroscopic electrostatics, which becomes exact for large particles. A further advantage of DHBj theory is that interactions between dipolar ion pairs and free ions and/or other pairs can be incorporated in a natural and transparent way [16,17].

In the following, we review some important foundations of DH theory with particular regard to new developments that incorporate the interactions of the dipolar ion pairs with the free ions and with other ion pairs in solution. In terms of Eqn. (1) this means that apart from the activity coefficients of the ions, the activity coefficient γ_p of the pairs is considered. Setting $\gamma_p = 1$, and applying Bjerrum's expression for the association constant, we then obtain the lowest level of approximation considered here, which is denoted as DH-Bj theory. In pilot work by Fisher and Levin [16] denoted by FL, DH-Bj theory is extended by considering the interactions of the pairs with the free ions. Finally, Weiss and Schröer (WS) [17] have supplemented this type of approach by adding a term for dipole-dipole interactions between pairs, thereby also accounting for the ε-dependence of the association constant.

DH theory starts with the Poisson-equation of classical macroscopic electrostatics, which connects the charge density ρ_q and the electrostatic potential ψ in a medium with the dielectric permittivity ε

$$\nabla^2\Psi = -\frac{4\pi}{\varepsilon}\rho_q. \tag{8}$$

The charge density ρ_q at separation r from an ion with charge q_+ is given by the charge-charge correlation functions $g_{++}(r)$ and $g_{+-}(r)$ and number densities ρ_+ and ρ_- of the ions: by $\rho_q = \rho_+ q_+ g_{++}(r) + \rho_- q_- g_{--}(r)$. The correlation functions can be written in terms of the potential of the mean force w_{ij}. In DH theory w_{ij} is approximated by the exact result of macroscopic electrostatics $w_{ij}(r) = q_j \Psi_{ij}(r)$, where Ψ_{ij} the potential due to the ion labeled i felt by the ion j. Consequently the corresponding pair-correlation functions are

$$g_{ij}(r) = \exp\left(-\beta q_j \Psi_{ij}(r)\right). \tag{9}$$

Combining these equations leads to the famous Poisson-Boltzmann equation. With the presumption of electro neutrality, the expansion in first order of β yields the Helmholtz equation or linearized Poisson-Boltzmann equation

$$\nabla^2\Psi = \kappa^2\Psi, \tag{10}$$

which is the basis of classical DH theory [14]. For charge-symmetrical systems the reciprocal Debye length κ is given by

$$\kappa^2 = 4\pi\beta q^2(\rho_+ + \rho_-)/\varepsilon \tag{11}$$

or in terms of reduced quantities by

$$x = \kappa\sigma = \sqrt{4\pi\rho^*/T^*}. \tag{12}$$

For $\kappa = 0$, Eqn. (10) reduces to the Laplace equation, which is used to calculate the potential inside the particle that is modeled as dielectric sphere with a central multipole. For this case the solution in spherical coordinates is given by a series of Legendre polynomials $P_l(x)$ and related powers of r

$$\Psi(r,\theta) = \sum_l \left(A_l r^{-l-1} + B_l r^l\right) P_l(\cos\theta). \tag{13}$$

This series arises naturally, whenever we express the Coulomb potential of a charge separated by a distance s from the origin in terms of spherical coordinates. The positive powers result when $r < s$, while for $r > s$ the potential is described by the negative powers. Similarly the solutions of the linearized Poisson–Boltzmann equation are generated by the analogous expansion of the shielded Coulomb potential $\exp[\kappa\cdot r]/r$ of a non-centered point charge. Now the expansion for $r > s$ involves the modified spherical Bessel-functions $k_l(x)$, while for $r < s$ the functions are the same as for the unshielded Coulomb potential:

$$\Psi(r,\theta) = \sum_l \left(C_l k_l(\kappa r) + D_l r^l\right) P_l(\cos\theta). \tag{14}$$

The coefficients A_l, B_l, C_l, D_l are calculable from the boundary conditions and from the known behavior in certain limits. The potential of an ion is given by the $l = 0$ terms with, while the potential of a dipole is determined by the $l = 1$ terms.

For further evaluation, we consider a sphere with the dielectric permittivity ε_0 and a central multipole. The sphere is surrounded by a dielectric continuum of the dielectric constant ε that also contains free point charges. Matching the potential and the normal component of the field at the boundary of the sphere determine the coefficients B_l and C_l. The coefficients A_l and D_l are fixed by the known potentials of the central particle inside the sphere and at infinity. DH theory is the solution for the case $l = 0$:

$$A_0 = \frac{q}{\varepsilon_0},\ B_0 = -\frac{q}{\varepsilon}\frac{(\varepsilon - \varepsilon_0 + x)}{\varepsilon_0\ \sigma\ (1-x)},\ C_0 = \frac{q}{\varepsilon}\frac{\exp(x)}{(1+x)}\ . \tag{15}$$

B_0 is the potential inside the sphere caused by charge distribution induced the by the central charge. The potential energy of this charge is therefore qB_0. This potential energy separates into a contribution due to the ion-ion interactions that is determined by κ and a term due to the dipole-dipole interactions, which is governed by $\varepsilon - \varepsilon_0$.

In FL theory [16] and WS theory [17] theory the collision diameter σ is chosen as radius of the excluded volume, which, to some extend, takes into account the volume of the field particles. The theory becomes much more complex when the ions differ in size [47]. Then, in a certain region around an ion only one kind of ions can fill the space, two different boundaries must be considered, and electro neutrality is not possible between the two boundaries, so that the Helmholtz equation becomes inhomogeneous in this region.

For $l = 1$ the boundary conditions for an excluded sphere with radius σ_D yield

$$A_1 = \frac{\mu}{\varepsilon_0},\ B_1 = -\ \mu\left(\frac{2\ (\varepsilon-\varepsilon_0)\ (1+\kappa\ \sigma_D)\ +\ \varepsilon\ (\kappa\ \sigma_D)^2}{\varepsilon_0\ \sigma_D^3\ (2\varepsilon+\varepsilon_0)\ (1+\kappa\ \sigma_D)+\varepsilon\ (\kappa\ \sigma_D)^2}\right),$$

$$C_1 = \left(\frac{3\mu\ \exp\ (\kappa\ \sigma_D)}{(2\varepsilon+\varepsilon_0)\ (1+\kappa\ \sigma_D)+\varepsilon\ (\kappa\ \sigma_D)^2}\right). \tag{16}$$

Here, B_1 is the field inside the sphere resulting from the distribution of the external charges induced by the permanent central dipole and the polarization of the sphere. In the language of dielectric theory, B_1 defines the reaction field factor [48]

$$R = \frac{2(\varepsilon-\varepsilon_0)(1+\kappa\sigma_D)+\varepsilon(\kappa\sigma_D)^2}{\sigma_D^3(2\varepsilon+\varepsilon_0)(1+\kappa\sigma_D)+\varepsilon(\kappa\sigma_D)^2}. \tag{17}$$

The potential energy of the dipole is $\mu B_1 = -\mu^2 R$. The part depending on $(\varepsilon - \varepsilon_0)$ represents the dipole-dipole interactions while the $(\kappa\sigma_D)^2$ term concerns the dipole-ion interactions. For $\kappa = 0$, Eqn. (17) reduces to the reaction-field expression of dielectric theory derived from Laplace equation. Interestingly, even for $\varepsilon = \varepsilon_0$ a reaction field results because of the polarization of the ionic cloud.

Clearly, the sum of the ion-dipole interactions must be equal to the sum of the dipole-ion interactions. By equating the dipole-ion and the ion-dipole of B_1 and B_0 we get for $\sigma = \sigma_D$

$$y = \frac{(\varepsilon-\varepsilon_0)\left((2\varepsilon+\varepsilon_0)(1+x)+\varepsilon x^2\right)}{9\varepsilon(1+x)} = \frac{4\pi\varepsilon\rho_D^*}{9T^*}, \tag{18}$$

where $\rho_D{}^* = \rho_D\sigma^3$. Eqn. (18) is a generalization of the well-known Fröhlich-Onsager-Kirkwood equation [49] for a mixture of ions and dipoles, which for the ion-free case reduces to standard expressions known from dielectric theory. The equation becomes slightly more complicated when $\sigma \neq \sigma_D$ [17]. The ionic correction to the Fröhlich-Onsager-Kirkwood equation may formally be written as a Kirkwood g-factor [48]

$$g = \left(1 + \frac{\varepsilon x^2}{(2\varepsilon + \varepsilon_0)(1+x)}\right)^{-1}. \qquad (19)$$

The shielding by the charges makes $g < 1$ like the anti-parallel correlation in the absence of ions, which both reduce the dielectric constant. Near criticality we have $x = \kappa\sigma \cong 1$, and the reduction is of the order of 15%.

If the ion-pairs are the only dipoles in the fluid, the energy density is given by

$$u = \frac{1}{2}\left((\rho_+ + \rho_-) q B_0 + \rho_D \mu B_1\right), \qquad (20)$$

where the mass action law connects the densities of the ions and of the dipoles. In polar solvents, Eqn. (20) is supplemented by contributions of the solvent dipoles.

Again, for calculating the equation of state via the Helmholtz energy thermodynamic integration of the energy according to Eqn. (7) is required. The charging process applied by Debye and Hückel [14] is just an approximation to this thermodynamic integration for fixed dielectric permittivity. Actually, the thermodynamic integration is, however, a rather involved task, because both ε and κ are functions of β, and κ is a function of ε. Therefore, Weiss and Schröer [17] performed an expansion in $(\varepsilon-\varepsilon_0)$ up to second order, which includes the ε-dependence of the Bjerrum constant.

A great simplification occurs when the contribution of the ion pairs to the dielectric constant is neglected as in the theory of Fisher and Levin [16]. In this approximation, B_0 represents the ion-ion interactions while B_1 gives the interaction of an ion-pair with the surrounding charges only. Neglecting the difference of the diameters of the ions and the dipoles, within this approximation the free energy densities of the ions and ion pairs are

$$\Phi_I = -\frac{\rho_I^*}{4\pi}\left(\ln(1+x) - x + \frac{1}{6}x^2\right), \qquad (21)$$

$$\Phi_D = -\frac{\rho_D^*}{x^2 T^*}\left(\ln\left(1 + x + \frac{x^2}{3}\right) - x + \frac{1}{6}x^2\right). \qquad (22)$$

For comparison, the Onsager reaction field yields the free energy density of a dipolar fluid [50] as

$$\Phi_{\mathrm{Ons}} = -\frac{1}{4\pi}\left(\frac{(\varepsilon-\varepsilon_0)^2}{\varepsilon} - \ln\left(\varepsilon\left(\frac{2\varepsilon+\varepsilon_0}{3\varepsilon}\right)^3\right)\right). \qquad (23)$$

Clearly, Eqns.(20-23) concern the electrostatic interactions only, so that a suitably chosen hard-core contribution, *e.g.* of Carnahan-Starling type [51] must be added to the free energy densities. Differentiation with respect to the densities of the species finally yields the chemical potential and the activity coefficients required in the evaluation of the mass action law determining the concentrations of free ions and ion pairs.

Using these procedures, Weingärtner, Weiss and Schröer [25] have calculated the degree of dissociation α over a wide range of conditions from subcritical states at $T^* = 0.04$ up to distinctly supercritical states at $T^* = 0.15$ for different approximations of the outlined electrostatic model. They considered the WS, FL and the DHBj models, but choose the Ebeling expression for the association constant. All theories yield a conductance minimum. Detailed analysis [25] shows that DI and DD interactions considered in FL and WS theory, are essential ingredients for rationalizing the observed conductance behavior near the two-phase regime. Thus one has not to resort to the assumption of charged triple ions. Pure pairing theories such as DHEb fail at $T^* < 0.08$, which is distinctly above criticality. However, near criticality both FL theory overestimates dissociation, and WS theory deviates even more. Note that the same is true for many versions of integral equation theories such as the PMSA [46].

In WS theory, this high ionicity is a consequence of the increase of the dielectric constant induced by dipolar pairs. The direct dipole-dipole contribution of the free energy favors pair formation [17]. One can expect that an account for neutral quadruple ions, as predicted by the MC studies, will improve the performance of DH-based theories. First, the coupled mass action equilibria reduce dissociation. Second, quadrupoles yield no direct contribution to the dielectric constant so that the increase of ε and diminution of the association constant becomes less pronounced than estimated form the WS approach. Such an effect is suggested from dielectric constant data for electrolyte solutions at low T^* [31].

3. PHASE DIAGRAM AND LOCATION OF THE CRITICAL POINT

3.1 MONTE-CARLO SIMULATIONS

Perhaps the most exciting property of the RPM at low reduced temperatures is a fluid phase transition with an upper critical point. By corresponding states arguments this critical point corresponds to the liquid-vapor transition of molten salts and to some liquid-liquid transitions in electrolyte solutions in solvents of low dielectric constant [11,12]. Many attempts have been made to locate this critical point by MC simulations and analytical theories.

MC simulations of critical properties are extremely involved and require much more computer force than *e.g.* the calculation of the pair-correlation functions. Methodological developments in MC techniques were addressed in a recent review [52]. As discussed later, simulations may also reflect the nonclassical critical fluctuations, thus allowing to identify the critical exponents and universality class. Most simulations use now the Gibbs ensemble technique, by which direct simulations of phase equilibria are possible by running two simulations in physically detached, but thermodynamically connected boxes that are representative of the coexisting phases. Particle transfer and volume exchanges between the boxes lead to an establishment of phase equilibrium. A further essential step, which allows large-scale simulations, was the observation that MC simulations may be carried out on a lattice [53]. If the separation of the lattice points is chosen about one order of magnitude smaller than the diameter of the particles, results of such simulations prove to be almost identical with those of conventional MC simulations. The virtue of this technique is not only the speeding up of the simulations, but also the possibility to investigate the change from a lattice to the fluid.

The first question to be answered by the simulations concerns the existence of the phase transition. The early calculations for the RPM of Stell *et. al.* [54] based on the MSA and other integral equations have indeed indicated the existence of a two-phase regime with an upper critical point. However, the case of the hard sphere dipole fluids may caution that a prediction based on such a mean field theory may be in error. In the latter case, it was taken for granted for a long time that this system has a gas-liquid phase transition. This conclusion is intuitive, because a critical point could easily be estimated from a generalized van der Waals theory. Instead, simulations that are more recent have failed to indicate such a transition, but gave evidence that, as the temperature is lowered the dipolar spheres associate to polymer like chains [55].

Meanwhile, a liquid-vapor transition of the RPM is well established, but over the years, the figures for the critical parameters of the RPM have changed appreciably, as illustrated in some reviews [3–5]. One reason for these inconsistencies is associated with finite-size scaling effects, as the critical fluctuations exceed the dimension of the simulation box. Therefore, conventional simulations can be regarded as "mean-field simulations". To obtain accurate data, the size L of the simulation box is varied and the results are extrapolated by finite-size scaling methods to an infinite sample sizes [56]. For the RPM the finite-size scaling technique leads to a reduction of the critical temperature and shifts the critical density to higher values. There is now a series of accurate MC studies [57–60], which, based on finite-size scaling corrections, locate the critical of the RPM near

$$T_c^* = 0.049, \qquad \rho_c^* = 0.08. \tag{24}$$

Going one step further, simulations are now also conducted for investigating the influence of the relative size of the ions and a charge asymmetry on critical parameters [38-40]. It is found that the critical temperature is reduced by the asymmetry of the particle size, as is the critical density. The charge asymmetry causes a reduction of the critical temperature, but an enhancement of the critical density. Notably these findings are in agreement with DH-type calculations, but not with MSA results.

Although free ions are certainly present in ionic fluids, it is questionable if their concentration is really dominating the thermodynamics. Following an old idea of Stillinger [61], Camp and Patey [62,63] and more recently Pangiotopoulos and co-workers [64] considered a fluid of dumbbells representing the ion pairs. It turned out that the coexistence curves comes out almost the same as for the RPM. We recall that the cousin model of the hard sphere dipolar fluid has no liquid-gas transition at all [55].

3.2 ANALYTICAL THEORIES

The determination of the phase diagram and of the critical point by analytical theories usually requires the calculation of the excess part of the reduced free energy density Φ^{ex}, which is then supplemented by an appropriately chosen hard-core contribution. As already noted, in DH theory Φ^{ex} is determined from thermodynamic integration of the energy of the ions and dipoles. For any temperature the spinodal is defined by the stability condition $(\partial^2\Phi/\partial\rho^2) = 0$, and the maximum of the spinodal with respect to T defines the critical point. The coexistence curve can be obtained by expanding the free energy density about the critical point and solving for ρ at a given temperature. Alternatively, one can exploit the condition that at coexistence pressure and chemical potentials are equal in the two phases.

Interestingly, in both DH and MSA theory, the excess free energy density of the ions supplemented by the ideal gas free energy density predicts already a phase transition without accounting for pair formation. However, there are gross deviations from simulation results. In DH theory the analytical results are $x_c = 1$, $T_c^* = 1/16$ and $\rho_c^* = 1/64\pi$. While the critical temperature comes out almost correctly, the density is too small by more than an order of magnitude. This flaw is corrected by including the pairing concept and employing the mass action law with the association constant proposed by Bjerrum. Actually, it turns out [16,17] that at the critical point $x = \kappa\sigma \cong 1$ in all DH-based theories. Recalling that κ is evaluated at the free ion density, all effects that increase the free ion density will displace the critical density towards lower values and *vice versa*. Including the Bjerrum pairing the critical point is calculated in fair agreement with those simulations for the RPM, which do not use the finite size scaling technique. The coexistence curve, however, has an unnatural banana-shaped form [16].

At a better level of approximation, in FL theory, the interactions of the ion pairs with the surrounding ion cloud are included. Then, good agreement of the phase-diagram with the mean-field simulations is obtained. Nevertheless, there are some objections against this theory. It appears that cancellation of errors contributes to the impressing good performance of the FL theory. For example, when the ideal gas free energy density is replaced by the more appropriate expressions of van der Waals or Carnahan and Starling, the good agreement is lost [17,45].

It is instructive to consider some other possible versions [4,16,17,45]. Combining the DH limiting law for the ionic free energy with the ideal gas free energy does not yield a phase transition. This is also true, if the electrostatic energy is used instead of the free energy resulting from thermodynamic integration. In both cases, however, phase transitions are obtained with the van der Waals hard-core free energy or the Carnahan-Starling hard sphere contribution. However, then the critical density and the critical temperature come out more near to that of van der Waals fluids.

Finally, one can carry the approximations to the point, where dipole-dipole interactions between ion pairs are taken into account as well [17,45]. Thus, in WS theory [17] the effects of dipolar interactions between ion pairs of the resulting change of the dielectric permittivity are worked out. At least in the version adopted by them, this approximation shifts the critical density to figures, which are too small. The physical origin of this overestimate is directly related to the overestimate of the ionicity already discussed in Section 2.5. It is suggested that an account for the presence of antiparallel ion pairs, as indicated by experimental dielectric data and by the ion clusters observed in MC simulations, may largely remove the discrepancies.

3.3 EXPERIMENTAL RESULTS

We turn now to the question, whether the critical points predicted theoretically have some relation to critical points of real systems. We recall that the best RPM critical parameters are $T_c^* \cong 0.049$ and $\rho_c^* \cong 0.08$. Both figures are much lower than observed for simple nonionic fluids, for example $T_c^* = 1.31$ and $\rho_c^* = 0.32$ for the Lennard-Jones fluid [13]. Low values of T_c^* are typical signatures for phase transition driven by Coulombic interactions [11]. By corresponding states arguments this critical point should correspond to the liquid-vapor critical point of molten salts and liquid-liquid critical points of electrolyte solutions [11,12].

Let us first consider the liquid-vapor phase transition of molten salts. For simple electrolytes such as NaCl various physical properties have been recorded up to about 2000 K, but this is still far below the critical point. From the available experimental data, and by extrapolations guided by simulations and theory, one expects $T_c \cong 3300$ K for NaCl [37]. With $\sigma \cong 0.276$ nm and $\varepsilon = 1$, this maps onto $T_c^* \cong 0.05$. Moreover, the critical mass density $d_c = 0.18$ g cm^{-3} estimated in ref. [37] implies $\rho_c^* \cong 0.08$. The quantitative agreement of these extrapolated critical parameters with predictions for the RPM is of course to some degree accidental.

Phase transitions driven by Coulomb interactions also correspond to some liquid-liquid phase transitions with critical points near room temperature. From the critical parameters of the RPM one concludes that such liquid-liquid transitions should occur in solvents of low dielectric constant, for 1:1-electrolytes typically at $\varepsilon \cong 5$. The major difficulty to find such systems arises from the interference of crystallization, driven by high melting points of salts. Low-melting salts, which usually comprise large organic cations and anions, now enable the systematic design of suitable systems, and meanwhile many systems are known that satisfy these conditions. Usually, these are based on low-melting tetraalkyl ammonium salts, which enable to tune the location of the two-phase regime by changing the alkyl residues at the cation or the anion [24].

However, there are systems with gross deviations of the liquid-liquid critical points from these predictions. Liquid-liquid immiscibilities have been observed with some of these salts even in aqueous solutions [24]. In such cases, the ionic forces are not expected to drive the phase separation. Rather, solvophobic effects of salts with large ions in solvents of high cohesive energy density may be responsible for these transitions [24]. Still other mechanisms are known [6].

Instructive examples for the interplay of Coulomb and solvophobic forces are tetraalkyl ammonium picrates dissolved in homologous alcohols. Extending earlier work [24], Kleemeier *et al.* [65] observed critical points of tetra-*n*-butyl ammonium picrate (Bu_4NPic) in a homologous series of 10 alcohols with dielectric constants ranging from 16.8 (2-propanol) to 3.6 (1-tetradecanol). For mapping the critical data onto reduced variables, we estimate the separation of the ionic charges to be $\sigma = 6.6$ Å, based on the Stokes radius of the cation and the van der Waals radius of the oxygen, which is assumed to be the center of the charges in the anion. Fig. 3 shows the resulting reduced critical temperatures as a function of the dielectric constant of the solvents. For long chain lengths of the alcohols, when the solvent becomes almost non-polar, one indeed approaches this "Coulombic limit" $T_c^* \cong 0.05$ of the RPM. However, for shorter chain length, *i.e.* higher dielectric constants of the solvents, T_c^* increases, and one clearly moves from a Coulombic mechanism to a distinctly non-Coulombic mechanism for phase separation. In other words, specific interactions increasingly shift the critical temperature away from the RPM prediction.

Interestingly, there are now observations of liquid-liquid coexistence in solutions of some newly designed room-temperature molten salts, *e.g.* based on imidazolium cations. These "ionic liquids" have prospects for engineering applications. At present these systems are mostly treated by typical non-electrolyte equations of state that ignore the long-range terms associated with Coulombic interactions [66]. In seeking for molecular-based interpretations, one should however not forget that in such systems Coulomb interactions might form the major driving force for phase transitions. As an illustrative example, we also show in Fig. 3 the location of the critical point in the corresponding-states plot derived from literature data [66] for solutions of 1-butyl-3-methylimidiazolium hexafluorophosphate and own measurements of 1-butyl-3-methylimidiazolium tetrafluoroborate in the lower normal alcohols.

The effective ion diameter of those salts should be markedly smaller than that of the picrate, where the alkyl groups bury the center of charge in the cation. We can estimate the characteristic separation of the ions from bond lengths and van der Waals radii to be about 4.60 Å for the tetrafluoroborate and 4.95 Å for the hexafluorophosphate. Then, the reduced critical temperatures of the imidazolium salts practically fall onto the same line as those of the tetralkylammonium salts, impressively showing the need for including Coulombic effects in the data analysis.

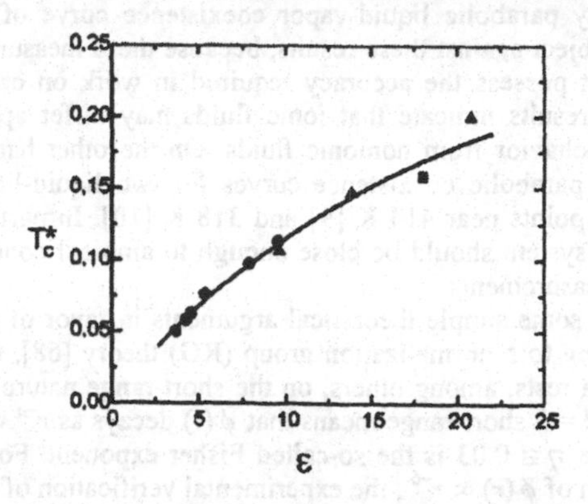

Figure 3. Reduced critical temperature vs. dielectric constant of the solvent of tetra-*n*-butyl ammonium picrate (•), 1-butyl-3-methylimidiazolium hexafluorophosphate [24] (▲) and (■) in homologous alcohols.

Refs. [3] and [6] summarize several other systems, where specific non-Coulombic interactions come into play, so that the critical point is largely displaced from what is expected for a generic ionic system. One system is molten NH_4Cl [7] which, although highly conducting, exhibits a too low critical temperature $T_c^* \cong 0.02$ and a too high critical density $\rho_c^* \cong 0.2$. This salt decomposes, however, in the gaseous phase into HCl and NH_3, which largely affects the equation of state. Another system is ethyl ammonium nitrate ($EtNH_3NO_3$) dissolved in 1-octanol, which, as summarized in ref. [3], has been used in many investigations of ionic criticality. Here the critical temperature corresponds to the expected value, but the critical density is located in the salt-rich regime. A rationale, supported by conductance data, is that hydrogen bonds between cations and anions cause extensive ion pairing, which stabilizes the pairs in excess to what is expected from electrostatic interactions. A detailed discussion of these aspects of non-Coulombic contributions is given in ref. [6]. The results caution that, if one aims at investigating the properties of generic ionic fluids, little can be learnt from data for such systems with specific interactions.

4. CRITICALITY OF IONIC FLUIDS

4.1 STATEMENT OF THE PROBLEM

Considering a liquid-gas or liquid-liquid coexistence curve, any analytical equation of state such as the van der Waals equation predicts a parabolic top, which simply follows from a series expansion of the free energy in powers of temperature and density. Already van der Waals was aware of distinct deviations from this "classical" or "mean-field" behavior, and Verschaffelt showed that the top of such coexistence curves is approximately cubic [67]. Today we know that this anomaly is but one example for the universal nonanalytic behavior of fluid properties near critical points [68]. Based on the available experimental evidence and simple theoretical arguments, in the late 1980s Pitzer [9,10] conjectured, however, that ionic fluids might behave classical.

Considering first the experimental evidence, such a classical behavior followed from an apparently parabolic liquid-vapor coexistence curve of NH_4Cl [7]. It is however easy to object against these results, because these measurements near 1150 K certainly do not possess the accuracy required in work on critical phenomena. Nevertheless, the results indicate that ionic fluids may differ appreciably in their effective critical behavior from nonionic fluids. On the other hand, Pitzer and coworkers observed parabolic coexistence curves for two liquid-liquid phase transitions with critical points near 414 K [9] and 318 K [10]. In particular, the critical point of the latter system should be close enough to ambient conditions to perform highly accurate measurements.

There are also some simple theoretical arguments in favor of a classical critical behavior. According to renormalization group (RG) theory [68], the universality of critical phenomena rests, among others, on the short-range nature of the interaction potential $\phi(r)$. In $d = 3$ short-range means that $\phi(r)$ decays as r^{-p} with $p > d + 2 - \eta = 4.97$ [69], where $\eta \cong 0.03$ is the so-called Fisher exponent. For insulating fluids with a leading term of $\phi(r) \propto r^{-6}$, the experimental verification of Ising-like criticality is unquestionable. The bare Coulombic interaction ($p = 1$) and interactions of charges with rotating dipoles ($p = 4$) do not fall into this class. Thus, intuitively, analytical (classical) critical behavior is a possible expectation. However, in real ionic fluids the screening of the electrostatic interactions to shorter range by counter ions may restore an Ising-like criticality. An analogous situation is encountered with liquid metals, where the electrons screen the interactions of the Coulombic interactions of the cores [70].

We anticipate here that later work [71] has not confirmed Pitzer's experimental observations, although there is no clear evidence for the origin of the flaws, and many aspects remain puzzling. Rather, it seems now that crossover phenomena occur from an Ising-like asymptotic behavior to mean-field behavior further away from the critical point [3,65,72,73]. Such a behavior is also expected for nonionic fluids, but occurs so far away, that conditions close to mean-field behavior are never reached [72].

Assuming that Ising-like criticality with crossover is indeed the correct interpretation, this would imply that the temperature distance of the crossover regime from the critical point, as characterized by the so-called Ginzburg temperature ΔT_{Gi}^*, is much smaller than observed for nonionic fluids. Actually, such a behavior was observed some time before Pitzer's work from coexistence curve data for the system $Na + NH_3$ [74]. There is a transition to metallic states in concentrated solutions, but in dilute solutions and near criticality ionic states prevail. Thus, an essential aspect of more recent work has been the experimental characterization of crossover behavior, and its explanation by crossover theory [73].

However, even if the critical point of an ionic fluid were mean-field-like, this does not necessarily imply that this behavior is related to the long-range nature of the interionic forces. Crossover may also be controlled by the approach toward a real or virtual tricritical point which in $d = 3$ is mean-field-like. This aspect plays an important role in recent theoretical and experimental studies (see Section 4.7).

4.2 EXPERIMENTAL RESULTS FOR BINARY SYSTEMS

4.2.1 Coexistence curves of binary systems

Non-analytical divergences at critical points result from fluctuations of the order parameter M, which is a measure of the dissimilarity of the coexisting phases. In pure fluids, one identifies M with the density difference of the coexisting phases. In solutions, M is related to some concentration variable. At a quantitative level, these divergences are described by crossover theory [72] or by asymptotic scaling laws and corrections to scaling, which are expressed in the form of a so-called Wegner series [68]. With regard to the coexistence curve, the difference ΔM in the two coexisting phases is represented by the series

$$\Delta M = B_0 \tau^{\beta} (1 + B_1 \tau^{\Delta} + B_2 \tau^{2\Delta} + \cdots). \tag{25}$$

$\tau = |T - T_c| / T_c$ characterizes the temperature distance from the critical point, β is a universal critical exponent and the B_i are a substance-specific amplitudes. For Ising-like systems $\beta \cong 0.326$, while in the mean-field case $\beta = 0.5$ exactly. The correction terms depend on substance-specific amplitudes and on the universal correction exponent $\Delta \cong 0.5$. In the asymptotic regime Eqn. (25) reduces to the power law $\Delta M = B_0 \tau^{\beta}$. For practical data evaluation, one often characterizes the deviations from asymptotic behavior by an effective exponent defined by the local slope in the double-logarithmic plot of the order parameter *vs*. the reduced temperature:

$$\beta_{eff} = \mathrm{d} \ln \Delta M / \mathrm{d} \ln \tau . \tag{26}$$

Using this procedure Singh and Pitzer [10] indeed found a parabolic coexistence curve for the low-melting salt *n*-hexyl-triethyl-ammonium *n*-hexyl-triethylborate ("Pitzer's salt") dissolved in diphenylether. If one imposes an Ising-like asymptotic criticality upon this data, one finds a very low Ginzburg crossover temperature, never observed in other cases [10, 73].

Actually, the critical temperature $T_c \cong 317$ K reported by Singh and Pitzer could not be reproduced in later work [71], which, depending on the sample, yielded values between 288 and 309 K (see also Table 2 of ref. [3]). These differences seem to indicate a considerable chemical instability of Pitzer's salt, so that decomposition products displace T_c. In view of these observations, coexistence curve measurements were repeated with a sample, which was tempered until stability of T_c was reached. These experiments, as well as determinations of other critical exponents discussed later, yielded plain Ising behavior [71].

Tetrabutylammonium picrate (Bu_4NPic) in long-chain alcohols behaves more reproducible, as the salt is chemically quite stable [24,75]. Moreover, in such cases critical points can be observed for a homologous series of systems with a gradual shift from Coulombic to solvophobic behavior of the phase transitions. Thus, systems including 2-propanol, 1-decanol, 1-dodecanol, 1-tri-decanol, and 1-tetradecanol as solvents were employed in precise studies of the coexistence curves [65].

The evaluation of the coexistence curve data for these systems indicates some subtle problems, because for two-component systems several choices of the order parameter exist, *e.g.* the mole fraction or volume fraction. In terms of the mole fraction, the phase diagrams of ref. [65] are highly skewed and located in the solvent-rich regime. Such highly skewed phase diagrams show a strong resemblance with those of polymer solutions in poor solvents, suggesting looking for theoretical analogies [4,76]. The volume fraction, however, leads to a more symmetric phase diagram, and the asymptotic range becomes larger. Symmetric coexistence curves with large asymptotic ranges are often taken as criteria for an optimum choice of the order parameter.

Fig. 4 shows the effective exponents β_{eff}, defined by Eqn. (26) and calculated on the basis of volume fractions. Asymptotic Ising behavior is obvious, but deviations occur away from T_c. Although they are quite small, they show a systematic increase with increasing chain length of the alcohol, thus suggesting an increased tendency for crossover to the mean-field case, when the Coulombic contribution becomes essential.

These results are unique in the sense that, apart from the system Na + NH_3, these are the only cases, with clear evidence for crossover phenomena quite close to T_c. However, ref. [3] summarizes work on coexistence curves for some other ionic systems with non-polar solvents as toluene and cyclohexane, which show normal Ising behavior without particular noticeable deviations.

The diameter $(M_1 + M_2)/2$ of the coexistence curve is represented by the series [68]

$$(M_1 + M_2)/2 = M_c + D_0\tau^{1-\alpha}\left(1 + D_1\tau^{\Delta} + D_2\tau^{2\Delta} + \ldots\right), \tag{27}$$

involving the Ising exponent $\alpha \cong 0.11$ of the heat capacity. In a mean-field system with $\alpha = 0$ and $\beta = 1/2$ the diameter is "rectilinear". For a long time, the diameter anomaly in nonionic systems was a matter of controversy, because the deviations from rectilinear behavior are small, and there is an additional spurious 2β contribution, when an improper order parameter is chosen in data evaluation [77]. In studies of ionic systems, there were repeated reports on diameter anomalies, while in some other cases no such evidence was observed [3]. A careful investigation of the picrate systems yielded a substantial anomaly [65] with all reasonable choices of the order parameter, including the volume fraction. The data are consistent with an $(1-\alpha)$ anomaly. Large diameter anomalies are expected, when the intermolecular interactions depend on the density. In the systems considered here, the dilute phase is essentially composed of ion pairs, while the concentrated phase is an ionic melt. However, any general conclusion is weakened by the fact that with Pitzer's system no such anomaly was observed [71].

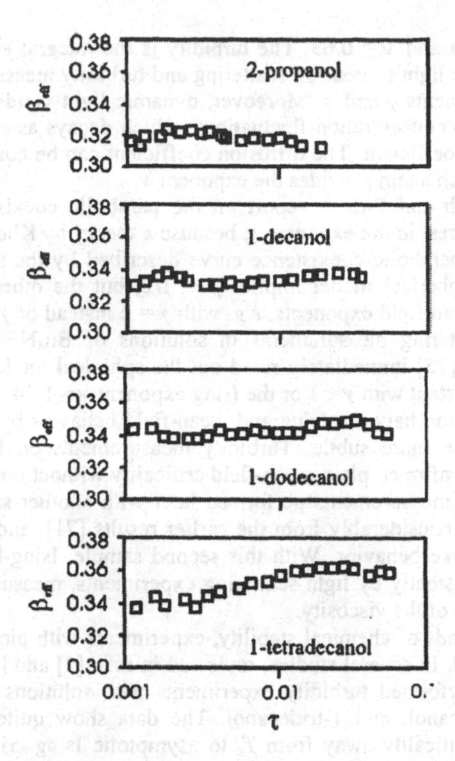

Figure 4. Effective exponents β_{eff} of tetra-*n*-butyl ammonium picrate dissolved in several alcohols as a function of the reduced temperature based on volume fractions as order parameter.

4.2.2 Scattering and Turbidity

The non-analytical divergences at critical points result from fluctuations of the order parameter, which can be observed by scattering experiments. The intensity I of single scattering in binary systems is determined by the concentration fluctuations, which in a rather good approximation are described by the Ornstein-Zernicke equation

$$I = \frac{\chi}{1+q^2\xi^2} . \tag{28}$$

The OZ equation depends on the scattering vector q and involves the osmotic compressibility χ and the correlation length ξ of the fluctuations, the temperature dependence of which is given by scaling laws of the form analogous to Eqn. (25)

$$\chi = \chi_0 \tau^{-\gamma}(1+\chi_1\tau^{\Delta}\ldots) \ , \tag{29}$$

$$\xi = \xi_0 \tau^{-\nu}(1+\xi_1\tau^{\Delta}\ldots) . \tag{30}$$

γ and ν satisfy the equality $\gamma = \nu(2-\eta) \cong 2\nu$, where $\eta \cong 0.03$ is the Fisher exponent. In the mean-field case $\gamma = 1$ and $\nu = 0.5$ and $\eta = 0$, so that $\gamma = 2\nu$ exactly. In the

Ising case, $\gamma = 1.24$ and $\nu = 0.63$. The turbidity is the integral about all scattering vectors. Thus, static light or neutron scattering and turbidity measurements enable to determine the exponents γ and ν. Moreover, dynamic light yields the time correlation function of the concentration fluctuations, which decays as $\exp(-2Dq^2)$, where D is the diffusion coefficient. The diffusion coefficient can be converted to the correlation length, which again provides the exponent ν.

Following Singh and Pitzer's report on the parabolic coexistence curve [10], there was large interest in the exponent γ, because a theory by Kholodenko and Beyerlein predicted a parabolic coexistence curve described by the so-called spherical model [78]. The spherical model implies $\beta = 1/2$, but the other exponents differ largely from the mean-field exponents, *e.g.* with $\gamma = 2$ instead of $\gamma = 1$. By static and dynamic light scattering measurements in solutions of Bu_4NPic in 1-tridecanol, Weingärtner *et al.* [75] immediately ruled out the spherical model by showing that the results are consistent with $\gamma = 1$ or the Ising exponent $\gamma = 1.24$ rather than $\gamma = 2$.

The discrimination between Ising and mean-field behavior by scattering experiments proved to be more subtle. Turbidity measurements on Pitzer's system by Zhang *et al.* [79] confirmed plain mean-field criticality without noticeable crossover. However, turbidity measurements performed later with another sample in the same laboratory differed considerably from the earlier results [71], and the data were indicative for Ising-like behavior. With this second sample, Ising-like criticality was also obtained consistently by light-scattering experiments, measurements of the coexistence curve and of the viscosity.

Again, on grounds of chemical stability, experiments with picrate solutions may be more convenient. In several studies, reviewed in refs. [2] and [3], Narayanan and Pitzer therefore performed turbidity experiments with solutions of Bu_4NPic in 1-undecanol, 1-dodecanol, and 1-tridecanol. The data show quite sharp crossovers from mean-field criticality away from T_c to asymptotic Ising criticality, which occurs almost within one decade of the reduced temperature. Crossover is closest to T_c for the highest homologue, where Coulombic interaction is expected to be strongest. Adding 1,4-butanediole to 1-dodecanol shifts the crossover region further away from T_c, and for pure 1,4-butanediol clear Ising behavior is observed. Later, Kleemeier [80] performed turbidity measurements for the picrates in conjunction with static and dynamic light scattering experiments. His results confirm the essential features of Pitzer's data, but crossover seems to be smoother than observed earlier. Clearly, these results confirm the picture developed above from coexistence curve data [65]. Attention may also be drawn to systems in nonpolar solvents as cyclohexane and toluene, where Ising-like criticality was deduced from both, measurements of the coexistence curve and light-scattering experiments [3,80].

In concluding this section, we like to draw attention to the amplitudes ξ_0 derived in the scattering experiments. As outlined in Section 4.6, ξ_0 enters into theoretical expressions for the crossover temperature. Thereby, large ξ_0-values favor a small Ising regime. In simple nonelectrolyte mixtures ξ_0 is generally found to be of the order of the molecular diameters, $\xi_0 = 0.1$ - 0.4 nm say. It is therefore notable that in the studies mentioned above considerably larger amplitudes were observed, in parts yielding $\xi_0 > 1$ nm. Such large amplitudes were however obtained in systems with small or large corrections to scaling, so that large ξ_0-values seem to be a signature of ionic systems not related to the mean-field *vs*. Ising problem in an obvious way.

4.2.3 Some other critical properties

Particularly interesting are weak divergences in Ising systems, which are absent in the mean-field case. One such case is the specific heat, which diverges with the exponent α. Kaatze and coworkers [81] have indeed shown the presence of such an a anomaly in $EtNH_3NO_3$ + *n*-octanol, but as already mentioned this system shows an anomalous location of the critical point, indicating that non-Coulombic interactions play a considerable role in driving the phase separation.

Another property related to α is the electrical conductance, which diverges as $\tau^{1-\alpha}$. Bonetti and Oleinikova [82] proved the existence of an Ising-type $(1 - \alpha)$ anomaly for some picrate systems, finding no essential difference between Coulombic and solvophobic systems.

Finally, the viscosity of Ising-like systems is known to exhibit a weak divergence of the form

$$\eta / \eta_b = (Q_0\xi)^z = (Q_0\xi_0)^z |\tau|^{-y}, \qquad (31)$$

where η_b is the background viscosity, Q_0 a system-dependent wave vector, $(Q_0\xi_0)^z$ a system-dependent critical amplitude, y the viscosity exponent and $z = \nu \cdot y$. For an Ising system, mode coupling theory and dynamic RG theory both yield $z = 0.065$, *i.e.* $y = 0.041$ in agreement with the best experimental data. For mean-field systems there seems to be no decisive answer, but probably mean-field behavior excludes an anomalous viscosity enhancement, except perhaps for logarithmic corrections.

Viscosity measurements were first reported for the Bu_4NPic + 1-tridecanol [83], and ensure that the scaling law (31) with Ising exponent is applicable. Further away from T_c, the anomaly was however found to vanish, in accordance with crossover to mean-field behavior. With ξ_0 taken from light scattering data, the wave vector, Q_0 was found to fall in the broad range of values reported for nonionic fluids. Wiegand *et al.* [84] observed a viscosity anomaly of Ising-type for Pitzer's system as well. Presuming the absence of a critical viscosity anomaly for mean-field systems, these experiments clearly prove the Ising-like character of the critical points in ionic solutions.

4.3 EXPERIMENTAL RESULTS FOR MULTICOMPONENT SYSTEMS

One option for explaining classical critical behavior and unusual crossover is the existence of a tricritical point, which in $d = 3$ is mean-field like [68, 72]. The existence of a tricritical point requires the presence of a further phase transition line, in addition to the liquid-liquid coexistence curve. This is difficult to imagine in binary systems, although the coupling of some charge ordering or of an insulator-conductor transition to the phase separation can be thought of. While polymer solutions indeed seem to be well described by a crossover scenario [76] based on the coupling of the liquid-liquid phase transition to the fluctuations of the radius of gyration, in binary electrolyte solutions no real evidence for such coupling has been found so far.

In ternary systems, such a crossover scenario is more easily accounted for. For example, tricritical behavior is obtained, if three phases have a common critical point. There is a large body of experimental work on ternary systems of the general type salt + water + organic cosolvent. Many electrolytes have a salting-out effect, which increases the miscibility gaps of the binary system, or, generates a gap not present in the binaries. In the literature, there have been repeated reports on an apparent mean-field behavior in such ternary systems. For example, measurements of the refractive index along a near-critical isotherm of the system 3-methylpyridine (3-MP) + H_2O doped with NaCl yielded an exponent $\delta = 3.05$ close to the mean-field value $\delta = 3$, while for Ising systems $\delta \cong 4.8$ [85]. Viscosity data for the same system indicated an Ising-like exponent, but a shrinking of the asymptotic range by added NaCl [86].

Recently, Jacob et al. [87] performed light scattering measurements in mixtures of 3-MP + H_2O + NaBr. The data indicate a comparatively sharp crossover regime in the range close to the critical point. When increasing the salt concentration, the tendency to approach mean-field behavior at T_c becomes more pronounced, and eventually, at 17 mass % NaBr pure mean-field behavior was obtained.

As viewed from the Coulombic *vs.* solvophobic dichotomy, the mean-field-like behavior of the ternary systems is somewhat unexpected. Thermodynamic properties of the salt-free systems clearly point towards a hydrophobic phase separation mechanism. One would then expect the ions just to enhance the forces already present in the binary mixture, which in turn, would imply that the critical point is Ising-like. In fact, further thorough examinations of the NaBr + water + 3-methylpyridine system did not confirm the reported results. Checking carefully the homogeneity of the samples in the concentration range of 10-18 mass% NaBr, viscosity, coexistence curves, turbidity and light scattering data, shown in Fig.(5), provided neither evidence for the pronounced crossover nor the observed mean-field behavior reported for 17 mass % NaBr [88-91].

4.4 MC SIMULATIONS OF NEAR-CRITICAL BEHAVIOR

In standard simulations, long-range fluctuations cannot be treated adequately because the size of the boxes is too small. In this respect, the simulated systems are small in spite of the periodic boundary conditions applied. Therefore, special techniques are required to elucidate the nature of the critical point of a particular system. The finite scaling technique provides the adequate tools [58]. Using this technique MC simulations yield reliable values for the critical point and allow conclusions about the universality class of model fluids such as of the square well fluid or the RPM [60]. Even today, this is an appreciable task.

In work, that is more recent Caillol *et al.* [58] and Lujten *et al.* [92] agree that the simulations suggest an Ising-like critical point. Lujiten *et al.* claim an accuracy of the exponents obtained from the simulations, which even allows distinguishing between Ising criticality and other models with rather similar exponents as the XY model and the self-avoiding walk model. An ultimate test would be the unambiguous proof of the weak α-anomaly of the specific heat. Whether such an anomaly can be extracted from the existing simulation data, is subject of a highly controversial debate [93–95].

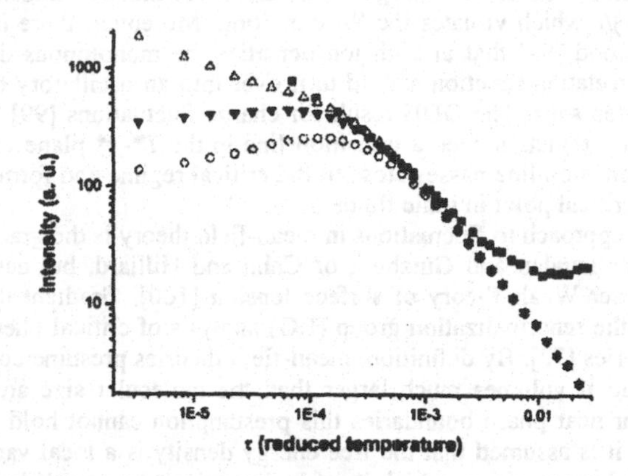

Figure 5. Temperature dependence of the light-scattering intensity (90°) at critical composition for 17 mass % NaBr in the ternary system NaBr/water/ 3-methylpyridine. In the figure the results (■) reported in [87] are compared with recent measurements of Wagner et.al. [91]. Ising values are obtained for γ and ν after correcting the experimental data (○) for the loss due to turbidity (Δ) and multiple scattering (▼).

4.5 MEAN-FIELD THEORIES OF FLUCTUATIONS

The appearance of the critical opalescence, which is caused by inhomogenities of the refractive index, spectacularly reflects the growing density or concentration fluctuations when approaching a critical point. In ionic fluids, both density and charge fluctuations are of relevance. Density-density correlations are reflected by the sum combination

$$h_q(r) = h_{\alpha\alpha}(r) + h_{\alpha\beta}(r), \tag{32}$$

while charge-charge correlations are reflected by the difference combination [13]

$$h_\rho(r) = h_{\alpha\alpha}(r) - h_{\alpha\beta}(r)\,. \tag{33}$$

In DH theory we have $h_\rho(r) = 0$ and $h_q(r)$ is short-range without relation to fluctuations near critical points.

Obviously, the correlation functions of DH theory do not say anything at all about density fluctuations that are crucial in any description of critical behavior. To treat density fluctuations, Lee and Fisher [96] have extended DH theory to non-uniform, slowly varying ion densities. This extension yields a functional of the Helmholtz free energy, which, in turn, allows deriving expressions for density correlations. This generalized DH (GDH) theory provides a universal limiting law for the correlation length of the density fluctuations, which, as $\rho \to 0$, diverge as $(T\rho)^{-1/4}$. Universal means that the expression is independent from model-specific quantities such as the diameter σ. The correctness of this universal limiting law was confirmed by explicit consideration of the cluster expansion of $h_S(r)$ [97].

The charge-charge correlation function $h_D(r)$ has to satisfy the rigorous the Stillinger-Lovett (SL) second-moment condition [13]

$$\rho \int h_q(r)\, 4\pi r^2\, dr = -24\pi / \kappa^2\,. \tag{34}$$

Simple DH theory describes charge fluctuations via the well-known screening decay as $\exp(-\kappa r)/r$ which violates the SL condition. Moreover, there is an old argument by Kirkwood [98] that at high ion densities the monotonous decay of the charge-charge correlation function should turn over into an oscillatory behavior, as observed for molten salts. The GDH result for charge fluctuations [99] satisfies the SL condition, and predicts indeed a transition line in the T^*-ρ^* plane to oscillating behavior. This transition line passes close to the critical regime and forms a possible scenario for a tricritical point in ionic fluids.

The classical approach to fluctuations in mean-field theory is the gradient theory, often termed after Landau and Ginzburg or Cahn and Hilliard, but actually going back to the van der Waals theory of surface tension [100]. Gradient theory is the starting point of the renormalization group (RG) analysis of critical phenomena and of crossover theories [72]. By definition, mean-field theories presume constant composition in a fluid if volumes much larger than the molecular size are compared. Near criticality or near phase boundaries this presumption cannot hold. In Landau-Ginzburg theory it is assumed that the free energy density is a local variable determined by the local composition, which itself is determined by a probability function constructed from the Boltzmann factor involving the local free energy density [101].

For the global density ρ_c at criticality, the local free energy density can be written as an expansion in powers of the density deviations $\tilde{\rho}(\mathrm{r}) = (\rho^* / \rho_c^* - 1)$ from this composition plus the corresponding gradients. Considering the critical composition and keeping the leading terms only, one gets

$$\Phi\left(T^*,\rho_c^*,\tilde{\rho}(r^*)\right) = \frac{1}{2}c_2\left[\tilde{\rho}(r^*)\right]^2 + \frac{1}{4!}c_4\left[\tilde{\rho}(r^*)\right]^4 + \frac{1}{2}c_{2g}\left(\nabla^*\tilde{\rho}(r^*)\right)^2 + \ldots \quad (35)$$

The star on the gradient indicates that the coordinates are scaled by the particle diameter σ, and index "g" denotes quantities related to the gradient terms. The system-dependent coefficients c_2 and c_4 are just the second and fourth derivatives of the free energy density, respectively. Quite generally, $c_2 = c_2^o \cdot \tau$. For the critical isochore, variation of Φ yields an exponential decay of density fluctuations determined by the mean field correlation length

$$\xi^* = \xi_0^* \cdot \tau^{-1/2}, \quad (36)$$

where $\xi_0^* = \sqrt{c_{2g} / c_2^o}$. Similarly, for $\tau \le 0$ the coexistence curve is given by

$$\tilde{\rho} = B^* \cdot \tau^{1/2} = \left(\frac{3!c_2^o}{c_4}\right)^{1/2} \cdot \tau^{1/2} \quad (37)$$

Finally, the temperature independent amplitude of the susceptibility is

$$\chi_0^* = 1 / c_2^o. \quad (38)$$

Whenever an appropriate mean field theory is available, it seems straightforward to work out the coefficients c_2 and c_4. By functional differentiation, it can be shown that c_{2g} is the second moment of the direct correlation function [102].

$$c_{2g} = \frac{1}{6}\int r^{*2} c(r^*) 4\pi r^{*2} dr^*. \quad (39)$$

However, there are subtle problems concerning the direct correlation function $c(r^*)$, for which several approximations are available. First, in van der Waals theory $c(r^*)$ is equal to $-\beta\phi(r^*)$ outside the excluded volume and equal to −1 inside. A straightforward generalization of the van der Waals approach is then [103]

$$c(r^*) = -\int g(r^*)\phi(r^*)d\beta\,, \tag{40}$$

which corresponds to the method by which the free energy is calculated from an ansatz for the energy by thermodynamic integration. The flaw in this approach is that $g(r^*)$ is largely density dependent and the particular variation of the density should be considered in working out the integral (40). Note that in van der Waals type approaches, which always lack thermodynamic consistency, $c(r^*)$ is usually not part of the theory.

A second approach employs a reformulation of the direct correlation function in terms of the total correlation function $h(r^*)$, which is derived from graph theory [104]. The leading terms are

$$c(r^*) = -\beta\cdot\phi(r^*) + \frac{1}{2}h(r^*)^2 + \dots \tag{41}$$

For the RPM, the first term cancels, and the second is dominant. In principle, $c(r^*)$ defined in this way may also vary locally.

Third, one may also calculate c_{2g} by working backward from the Laplace transformation of the pair correlation function, as done by Lee and Fisher [96] for the pair correlation function of their generalized Debye-Hückel theory.

4.6 CROSSOVER CRITERIA

The Landau-Ginzburg expansion (35) applies, however, only in a range where the fluctuations are sufficiently small. The so-called Ginzburg temperature ΔT^*_{Gi} approximately marks the end of the classical regime on approaching the critical point along the critical isochore. There are several ways of defining such a Ginzburg temperature [68], *e.g.* by requiring that in d dimensions [103]

$$\left(2\xi^*\right)^d \geq r_0^{*d}\,, \tag{42}$$

where $r_0^{*d} = c_4/\left(c_2\right)^2$ is the volume required to allow the density fluctuations to be described by means of Gaussian statistics. Therefore, Landau theory applies, if

$$\Delta T^{*4-d} \geq \frac{c_4^2}{2^{2d}\,c_2^{0\,4-d}\,c_{2g}^d}\,, \tag{43}$$

For $d = 3$ we get

$$\Delta T^*_{Gi} = \frac{c_4^{\,2}}{64\cdot c_2^0\cdot c_{2g}^{\,3}}\,. \tag{44}$$

The numerical factor is to some degree arbitrary; various values have been used in the literature [72]. Note that Eqn. (44) can be reformulated in terms of amplitudes appearing in the scaling laws of the critical phenomena. Also, in view of the condition defining the Ginzburg temperature one can define a short-range parameter

$$\Delta r_0^* = \left(c_4 / c_2^{02}\right)^{1/3}, \tag{45}$$

so that the criterion can be written in different equivalent ways

$$\Delta T_{Gi}^* = \frac{9}{16} \frac{\chi_0^{*2}}{B^{*4} \cdot \xi_0^{*6}} = \frac{1}{64}\left(\frac{\Delta r_0^*}{\xi_0^*}\right)^6. \tag{46}$$

It is quite tricky that the predictions for the Ginzburg temperature depend sensitively on the approach by which c_{2g} is calculated.

Considering now results from DH theory, Eqn. (40) leads to a very small Ginzburg number [103], apparently confirming rapid crossover found in some experiments. However, the GDH theory leads to a Ginzburg number, which is large, even if compared to values for ordinary non-ionic fluids [96]. By accounting for the density dependence of the pair correlation function in Eqn. (40), a large Ginzburg number is found as well [105]. Large values are also found with Eqn. (41), which, based on a cluster expansion [104], is rated to be superior if compared to Eqn.(40). A further argument in favor of a large Ginzburg number is founded in thermodynamic consistency. The susceptibility derived from the free energy density agrees with that obtained by integrating the direct correlation function given by Eqn. (41) [105]. Note, in non-ionic and non-polar fluids, Eqns. (40) and (41) become equivalent because $\beta\phi(r)$ dominates, which vanishes in the ionic and polar fluids.

The experiments yield indications of a crossover in the region of, $\Delta T_{Gi}^* \approx .01$ [43]. This figure is neither very small, of order 10^{-4} say, as resulting from the straightforward use of Eqn. (40), nor large, of the order of unity say, as estimated from the GDH theory and by using Eqn. (41).

4.7 MODELS THAT PREDICT A TRICRITICAL POINT

The Ginzburg analysis just described does not provide an explanation for the observations of crossover and mean field behavior in ionic systems. At present, the search for a solution to this problem is focused on the possibility of a tricritical point, which in three-dimensional systems is mean-field like [68]. Generally, a tricritical point is defined (a) as the critical point, where three phases become identical. Alternatively a tricritical point can be defined (b) as the point where a second order λ- line of an order transition becomes first order or (c) as the point, where a λ- line cuts a first order coexistence curve at their critical point. Regarding the inaccessible two-phase region as an area in an appropriate phase diagram, the definitions (b) and (c), which are relevant for the ionic fluids, can be regarded as special cases of (a). In fact, there are quite a few theoretical arguments, which suggest coupling of the critical fluctuations to some real or virtual tricritical point in ionic fluids. Notably, there are models closely related to the RPM that have indeed a tricritical point.

In particular, as first demonstrated by Dickmann and Stell [106], the lattice analogue of the RPM, which is the Coulomb gas with cations and anions on lattice sites interacting by Coulombic interactions, has no liquid-gas critical point but a tricritical point. This is quite unexpected, because for uncharged fluids the nature of the critical point is unchanged when considering a lattice model and the related continuum model. In fact, much of our knowledge about phase transitions of fluids comes from investigations of the Ising lattice model. Dickmann and Stell considered a lattice-based version of the RPM with a ratio $\zeta = 1$ of the ion size to the lattice unit cell dimension of unity, so that each charge just occupies one lattice site. This model yielded a first-order transition coexistence between a disordered and an antiferro-electrically ordered phase terminating at a tricritical point. Above the tricritical point, a line of second-order transitions between a disordered and antiferroelectric phase was predicted. Thus, there is a tricritical rather than critical point, and there is no liquid-vapor transition

Kobelev *et al.* [107] were able to solve the lattice version of the DH theory and its refinement due to Fisher and Levin (FL). They derived closed form expressions for the free energy for a *d*-dimensional ionic lattice system. The predicted critical temperatures for cubic lattices lie about 60% above that given by the continuum theories. However, allowing for sublattice ordering, a second order λ-line is found, which terminates at a tricritical point at a reduced temperature T^*=0.38 and ρ^*=0.36 for the simple cubic lattice. In accord with the treatment of Dickmann and Stell the theory predicts suppression of the liquid–gas transition and criticality by the order–disorder transitions at higher temperatures. However, in an anisotropic lattice obtained by stretching a cubic lattice in one direction, the low-density gas-liquid phase separation reappears and the phase diagram exhibits critical, tricritical and triple points [108].

Obviously, the criticality of the ionic fluids is quite different from that of the lattice. Therefore, the question arises, if tricriticality proven for the RPM lattice may be of any relevance for real ionic fluids and their continuum models. Panagiotopoulos and Kumar [53] performed MC simulations for several integer ratios $1 \leq \zeta \leq 5$. For $\zeta = 1$ the results of simulations and the analytical lattice calculations essentially agree. The tricritical point is located at $T^* = 0.15$ and $\rho^* = 0.48$. Considering larger figures for ζ, the tricritical point was shifted to very high density, and ordinary liquid-vapor critical points were observed. Already at $\zeta = 4$ the critical parameters of the lattice and continuum RPM agreed closely. The simulations show that the order-disorder transition is vanishing, when the grid of the lattice becomes smaller than the diameter of the particles by one order of magnitude.

Simulations, which check the hypothesis of order transitions in ionic fluids, are not available. The GDH theory of Lee and Fisher predicts an order-transition between the isotropic fluid and a phase with periodic charge density waves [96,97]. The corresponding transition line cuts the liquid-liquid line near their critical point, so that a tricritical point is a possibility. The possible presence of both gas-liquid and tricritical points in ionic fluids has been predicted by Ciach and Stell [109,110] for a model, in which additional short-range forces supplement the lattice Coulomb forces.

Furthermore, the *2d* RPM also yields a tricritical point, which, however, has a different physical basis [36,111]. Here tricriticality is founded in the insulator-conductor transition, which changes from 2nd. order to first order. Notably, in real ionic solutions the conductivity shows two points of inflection one at low densities, which corresponds to the conductor insulator transition in *2d*, and one near the criticality. Although accompanied by a maximum of the specific heat [63], those changes of the conductivity are soft transitions determined by the mass action law and not cooperative λ-transitions, required to allow for a tricritical point [25].

Concluding we state, while different theoretical arguments suggest a possibility of a tricritical point in ionic systems, experimentally so far no clear evidence was found. The observations of mean field behavior or crossover in a near-critical region remain puzzling.

5. SUMMARY

Two properties render electrolyte theories difficult, namely the long-range nature of the Coulomb interactions and the high figures of the Coulomb energy at small ion separations.

In solvents of low dielectric constant, where the Coulomb interactions are particularly strong, electrical conductance and dielectric spectra suggest that the ion distribution involves dipolar ion pairs, which then interact with the free ions and with other dipolar pairs. The ion pairs cause an increase of the dielectric constant, which in turn stabilizes the free ions, thus leading to redissociation at high salt concentrations. Extending the approach of Debye-Hückel and Bjerrum, theory accounts for ion pairing, ion – ion pair and ion pair – ion pair interactions and rationalizes the basic features of the ion distribution in accordance with experiments and MC-simulations.

It also predicts a fluid-phase transition at low reduced temperatures, which is in close agreement with simulation results and also agrees, using corresponding states considerations, with the experimentally observed liquid-liquid phase transitions. Nevertheless, a systematic variation of the reduced critical temperature with the dielectric constant of the solvent indicates some limitations of the analogy of the liquid-liquid transition in ionic solutions to the liquid-gas transition of the RPM.

The long-range nature of the Coulomb potential driving these transitions raises questions concerning their universality class. MC-simulations of the RPM involving finite size scaling techniques are at least consistent with Ising criticality. The overwhelming majority of experiments also suggest that the liquid-liquid phase transition in ionic solutions belongs to the Ising universality class. Experiments, which supported the expectation of mean field criticality, could not be reproduced in later work. However, there remains a puzzling observation of crossover to mean-field behavior rather close to the critical point, which is not understood. Different theories suggest a crossover scenario, which involves a tricritical point. However, the decisive experiments proving e.g. the hypothesis of charge density waves are not available.

Acknowledgements

We have to thank many coworkers, which, over the years, have participated in these projects. The pre- and reprints received from Profs. M.E. Fisher, J. Segers, M. Anisimov and A. Kolomeisky are much appreciated. The work was supported, in parts, by the Deutsche Forschungsgemeinschaft.

REFERENCES

1. See *e.g.*: Friedman, H. L. and Dale, W. D. T. (1977) Electrolyte solutions at equilibrium, in: *Modern Theoretical Chemistry*, **5**, Part A, 85-135.
2. Pitzer, K. S. (1995) Ionic fluids: Near-critical and related properties, *J. Phys. Chem.*, **99**, 13070-7.
3. Weingärtner, H. and Schröer, W. (2001) Criticality of ionic fluids, *Adv. Chem. Phys.*, **116**, 1-66.
4. Fisher, M. E. (1995) The story of coulombic criticality, *J. Stat. Phys.*, **75**, 1-36.
5. Stell, G. (1995) Criticality and phase transitions in ionic fluids, *J. Stat. Phys.*, **78**, 197-238.

6. Weingärtner, H., Kleemeier, M., Wiegand, S., and Schröer, W. (1995) Coulombic and non-coulombic contributions to the criticality of ionic fluids, *J. Stat. Phys.*, **78**, 169-96.
7. Buback, M. and Franck, E. U. (1972) Measurements of vapor pressure and critical data of ammonium halides, *Ber. Bunsenges. Phys Chem.*, **76**, 350-4.
8. Friedman, H. L. (1972) discussion remark, *J. Solution Chem.*, **2**, p. 354.
9. De Lima, M. C. P., Schreiber, D. R., and Pitzer, K. S. (1985) Critical point and phase separation for an ionic system, *J. Phys. Chem.*, **89**, 1854-5.
10. Singh, R.R., and Pitzer, K.S. (1990) Near-critical coexistence curve and critical exponent of an ionic fluid, *J Chem. Phys.*, **92**, 6775-8.
11. Friedman, H.L. and Larsen, B. (1979) Corresponding states for ionic fluids, *J. Chem. Phys.* **70**, 92-100.
12. Weingärtner, H. (2001) Corresponding states for electrolyte solutions, *Pure Appl. Chem.*, **73**, 1733-48.
13. Hansen, J.-P. and McDonald, I. R. (1986) *Theory of Simple Liquids*, Academic Press, New York.
14. Debye, P. and Hückel, E. (1923) The theory of electrolytes, *Physik. Z.*, **24**, 185-206.
15. Bjerrum, N. (1927) Ionic association. I. Influence of ionic association on the activity of ions at moderate degrees of association, *Kgl. Danske Videnskab. Selskab. Math.-fys. Medd.*, **7**, 1-48.
16. Levin, Y. and Fisher, M. E. (1996) Criticality in the hard-sphere ionic fluid, *Physica A*, **225**, 164-220, references cited therein.
17. Weiss, V. C. and Schröer, W. (1998) Macroscopic theory for equilibrium properties of ionic-dipolar mixtures and application to an ionic model fluid, *J. Chem. Phys.*, **108**, 7747-57.
18. Clausius, R. (1857) Über die Elektrizitätsleitung in Elektrolyten (On the electrical Conductance in Electrolytes), *Pogg. Ann.* **101**, 338-60.
19. Arrhenius, S. (1887) Über die Dissoziation der in Wasser gelösten Stoffe (On the dissociation of substances dissolved in water), *Z. Phys. Chem.*, **1**, 631-48.
20. Kraus, C. A. (1956) The ion-pair concept: its evolution and some applications, *J. Phys. Chem.*, **60**, 129-41.
21. Nernst, W. (1894), Dielektrizitätskonstante und chemisches Gleichgewicht (Dielectric constant and chemical equilibrium), *Z. Phys. Chem.*, **13**, 531-6.
22. Cruse, K. (1960) in: Landolt-Börnstein, Berlin, 6th edition, Vol. II, part 7.
23. Walden, P. and Centnerszwer, M. (1903) Über Verbindungen des Schwefeldioxyds mit Salzen, *Z. Phys. Chem.*, **42**, 432-68.
24. Weingärtner, H., Merkel, T., Maurer, U., Conzen, J. P., Glasbrenner, H., and Käshammer, S. (1991) Coulombic and solvophobic liquid-liquid phase-separation in electrolyte solutions, *Ber. Bunsenges. Phys. Chem.* **95**, 1579-86.
25. Weingärtner, H., Weiss, V. C., and Schröer, W. (2000) Ion association and electrical conductance minimum in Debye-Hückel-based theories of the hard sphere ionic fluid, *J. Chem. Phys.*, **113**, 762-70.
26. Schreiber, D. R., De Lima, M. C. P., and Pitzer, K. S. (1987) Electrical conductivity, viscosity, and density of a two-component ionic system at its critical point, *J. Phys. Chem.*, **91**, 4087-91.
27. Onsager, L. (1927) The theory of electrolytes, *Physik. Z.*, **28**, 277-98.
28. Fuoss, R. M. and Kraus, C. A. (1933) Properties of electrolytic solutions. IV. The conductance minimum and the formation of triple ions due to the action of Coulomb forces, *J. Am. Chem. Soc.*, **55**, 2387-99.
29. Cavell, E. A. S. and Knight, P. C. (1968) Effect of concentration changes on permittivity of electrolyte solutions, *Z. Phys. Chem.*, **57**, 331-4.
30. Gestblom, B. and Songstad, J. (1987) Solvent properties of dichloromethane. VI. Dielectric properties of electrolytes in dichloromethane., *Acta Chem. Scand. Ser. B*, **41**, 396-40.

31. Weingärtner, H., Nadolny, H. G., and Käshammer, S. (1999) Dielectric properties of an electrolyte solution at low reduced temperature, *J. Phys. Chem. B*, **103**, 4738-43.
32. Kirkwood, J.G. (1936) The theory of dielectric polarization, *J. Chem. Phys.*, **4**, 592-601.
33. Ramanathan, P. S. and Friedman, H. L. (1971) Refined model for aqueous 1-1-electrolytes, *J. Chem. Phys.*, **54**, 1086-99.
34. Caillol, J. M. (1995) A Monte Carlo study of the dielectric constant of the restricted primitive model of electrolytes on the vapor branch of the coexistence line, *J. Chem. Phys.*, **102**, 5471-5479.
35. Caillol, J. M. and Weis, J. J. (1995) Free energy and cluster structure in the coexistence region of the restricted primitive model, *J. Chem. Phys.*, **102**, 7610-21.
36. Kosterlitz J. M. and Thouless D. J. (1973) Ordering, metastability, and phase transitions in two-dimensional systems., *J. Phys. C.*, **6**, 1181-203.
37. Guissani, Y. and Guillot, B. (1994) Coexisting phases and criticality in NaCl by computer simulation, *J. Chem. Phys.*, **101**, 490-509.
38. Yan, Q. L. and de Pablo J. J. (2001) Phase equilibria and clustering in size-asymmetric primitive model electrolytes, *J. Chem. Phys.*, **114**, 1727-31.
39. Yan, Q. L. and de Pablo, J. J. (2002) Phase equilibria of charge-, size-, and shape-asymmetrical models of electrolytes, *Phys. Rev. Lett.*, **88**, 095504/1-4.
40. Romero-Enrique, J.M., Orkoulas, G., Panagiotopoulos, A. Z., and Fisher, M. E. (2000) Coexistence and criticality in size-asymmetric hard-core electrolytes, *Phys. Rev. Lett.*, **85**, 4558-61.
41. Høye, J. S., Lebowitz, J. L., and Stell, G. (1974) Generalized mean spherical approximations for polar and ionic fluids, *J. Chem. Phys.*, **61**, 3253-60.
42. Høye, J. S. and Stell, G. (1977) New self-consistent approximations for ionic and polar fluids, *J. Chem. Phys.*, **67**, 524-9.
43. Pini, D., Stell, G., and Wilding, N. B. (1998) A liquid-state theory that remains successful in the critical region, *Mol. Phys.*, **95**, 483-94.
44. Ebeling, W. (1972) Theory of ion-pair formation in electrolytes, *Z. Phys. Chem. (Leipzig)*, **249**, 140-2.
45. Guillot, B. and Guissani, Y. (1996) Towards a theory of coexistence and criticality in real molten salts, *Mol. Phys.*, **87**, 37-86.
46. Zhou,Y., Yeh, S., and Stell, G. (1995) Criticality of charged systems. I. The restricted primitive model, *J. Chem. Phys.*, **102**, 5785-95.
47. Zuckerman, D. M., Fisher, M. E., and Bekiranov, S. (2001) Asymetric primitive model electrolytes: Debye-Hückel theory, criticality, and energy bounds, *Phys Rev. E*, **64**, 011206/1-13.
48. Schröer, W. (2001) Generalization of the Kirkwood-Fröhlich theory of dielectric polarization for ionic fluids, *J. Mol. Liquids*, **92**, 67-76.
49. Fröhlich, H. (1958) *Theory of Dielectrics*, Oxford University Press, Oxford.
50. Sutherland, J. W. H., Nienhuis, G., and Deutch, J. M. (1974) Thermodynamics of pure and multicomponent dipolar hard-sphere fluids, *Mol. Phys.*, **27**, 721-39.
51. Mansoori G. A., Carnahan N. F., Starling, K. E. and Leland, T. W., Jr. (1971) Equilibrium thermodynamic properties of mixture of hard spheres, *J. Chem. Phys.*, **54**, 1523-5.
52. Panagiotopoulos, A. Z. (2000) Monte Carlo methods for phase equilibria of fluids, *J. Phys. Cond. Matter*, **12**, R25-R52.
53. Panagiotopoulos, A. Z. and Kumar, S. K. (1999) Large lattice discretization effects on the phase coexistence of ionic fluids, *Phys. Rev. Lett.*, **83**, 2981-84.
54. Stell, G., Wu, K. C., and Larsen, B. (1976) Critical point in a fluid of charged hard spheres, *Phys. Rev. Lett.*, **37**, 1369-72.
55. See *e.g.* Weis, J. J. and Levesque, D. (1993) Chain formation in low density dipolar hard spheres: a Monte Carlo study, *Phys. Rev. Lett.*, **71**, 2729-32.
56. Bruce, A. D. and Wilding, N. B. (1992), *Phys. Rev. Lett.*, **68**, 193-6.

57. Panagiotopoulos, A. Z.(2002) Critical parameters of the restricted primitive model, *J. Chem. Phys.*, **116**, 3007-11.
58. Caillol, J.-M., Levesque, D. and Weis, J.-J. (2002) Critical behavior of the restricted primitive model revisited., *J. Chem. Phys.*, **116**, 10794-800.
59. Orkoulas, G., Panagiotopoulos, A. Z., and Fisher, M. E. (2000) Criticality and crossover in accessible regimes, *Phys. Rev. E*, **61**, 5930-39.
60. Orkoulas, G. and Panagiotopoulos, A. Z. (1999) Phase behavior of the restricted primitive model and square-well fluids from Monte Carlo simulations in the grand canonical ensemble, *J. Chem. Phys.*, **110**, 1581-90.
61. Stillinger, F. H. and Lovett, R (1968) Ion pair theory of concentrated electrolytes. I. Basic concepts, *J. Chem. Phys.*, **48**, 3858-68.
62. Camp, P. J. and Patey, G. N. (1999) Ion association and condensation in primitive models of electrolyte solutions, *J. Chem. Phys.*, **111**, 9000–8.
63. Camp, P. J. and Patey, G. N. (1999) Ion association in model ionic fluids, *Phys. Rev. E*, **60**, 1063–66.
64. Romero-Enrique, J. M., Rull, L. F., and Panagiotopoulos, A. Z. (2002) Dipolar origin of the gas-liquid coexistence of the hard-core 1:1 electrolyte model., *Phys. Rev. E*, **66**, 041204/1-10.
65. Kleemeier, M., Wiegand, S., Schröer, W. and Weingärtner, H. (1999) The liquid-liquid phase transition in ionic solutions: Coexistence curves of tetra-n-butylammonium pricrate in alkyl alcohols, *J. Chem. Phys.*, **110**, 3085–99.
66. Marsh, K. N., Deev, A., Wu, A. C-T., Tran, E., and Klamt, A. (2002) Room temperature ionic liquids as replacements for conventional solvents - a review, *Kor. J. Chem. Eng.*, **19**, 357-62.
67. Verschaffelt, J. E. (1900) On the critical isothermal line and the densities of vapor and liquid in isopentane and carbon dioxide, *Proc. Kon. Acad.*, **2**, 588-92.
68. See *e.g.*: Pfeuty, P. and Tolouse, G. (1977) *Introduction to Renormalization Group and Critical Phenomena*, Wiley, New York.
69. Kayser, R. F. and Raveche, H. J. (1984) Asymptotic density correlations and corrections to scaling for fluids with non-finite range interactions, *Phys. Rev. A*, **29**, 1013–15.
70. Hensel, F. (1990) Critical behavior of metallic liquids, *J. Phys. Condens. Matter*, **2** (Suppl. A), SA33-SA45.
71. Wiegand S., Briggs M. E., Levelt Sengers J. M. H., Kleemeier M. and Schröer W. (1998) Turbidity, light scattering, and coexistence curve data for the ionic binary mixture triethyl n-hexyl ammonium triethyl n-hexyl borate in diphenyl ether., *J. Chem. Phys.* **109**, 9038-51.
72. Anisimov, M. A. and Sengers, J. V. (2000) The Critical Region, in *Equations of State for Fluids and Fluid mixtures*, Eds. Sengers, J. V., Kayser, R. F., Peters, C. J., and White, H.J., Elsevier, Amsterdam.
73. Gutowski, K., Anisimov, M. A., and Sengers, J.V. (2001) Crossover criticality in ionic solutions, *J. Chem. Phys.* **114**, 3133-48.
74. Chieux, P. and Sienko, M. J. (1970) Phase separation and the critical index for liquid-liquid coexistence in the sodium-ammonia system, *J. Chem. Phys.*, **53**, 566-70.
75. Weingärtner, H., Wiegand, S., and Schröer, W. (1992) Near-critical light scattering of an ionic fluid with liquid-liquid phase transition, *J. Chem. Phys.*, **96**, 848-51.
76. Anisimov, M. A., Agayan, V. A. and Gorodetskij, E. E. (2000) Scaling and crossover to tricriticality in polymer solutions *JETP Lett.*, **72**, 578-82.
77. See *e.g.*: Kumar, A., Krishnamurthy, H. R., and Gopal, E. S. R.. (1983) Equilibrium critical phenomena in binary liquid mixtures, *Phys. Reports* **98**, 57-143.
78. Kholodenko, A. L. and Beyerlein, A. L. (1990) Comment on "Near-critical coexistence curve and critical exponent of an ionic fluid", *J. Chem. Phys.*, **93**, 8405.
79. Zhang, K. C., Briggs, M. E., Gammon, R. W., and Levelt Sengers, J. M. H. (1992) The susceptibility critical exponent for a nonaqueous ionic binary mixture near a consolute point, *J. Chem. Phys.*, **97**, 8692-97.

80. Kleemeier, M. (2001), *Untersuchungen zum kritischen Verhalten des Flüssig-Flüssig Phasenübergangs in ionischen Lösungen*, Ph. D. Thesis, University of Bremen.
81. Heimburg, T., Mirzaev, S.Z., and Kaatze, U. (2000) Heat capacity behavior in the critical region of the ionic binary mixture ethylammonium nitrate – *n*-octanol, *Phys. Rev.* E, **62,** 4963-76.
82. Oleinikova, A. and Bonetti, M. (2001) Electrical conductivity of highly concentrated electrolytes near the critical solute point: A study of tetra-n-butylammonium picrate in alcohols of moderate dielectric constant, *J. Chem. Phys.*, **115**, 9871-82.
83. Kleemeier, M., Wiegand, S., Derr, T., Weiss, V., Schröer, W., and Weingärtner, H. (1996) Critical viscosity and Ising-to-mean-field crossover near the upper consolute point of an ionic solution, *Ber. Bunsenges. Phys. Chem.*, **100**, 27-32.
84. Wiegand, S., Berg, R.F. and Levelt Sengers, J. M. H. (1998) Critical viscosity of the ionic mixture triethyl *n*-hexyl ammonium triethyl n-hexyl borate in diphenyl ether, *J. Chem. Phys.*, **109**, 4533-45.
85. Bulavin, L., Oleinikova, A., and Petrovitskij, A. V. (1996) Influence of ions on the critical behavior of a binary mixture near the consolute point, *Int. J. Thermophys.*, **17**, 137-45.
86. Oleinikova, A.; Bulavin, L., and Pipich, V. (1997) Critical anomaly of shear viscosity in a mixture with an ionic impurity, *Chem. Phys. Lett.*, **278**, 121-26.
87. Jacob, J., Anisimov, M. A., Kumar, A., Agayan, V. A. and Sengers, J. V. (2000) Novel phase-transition behavior in an aqueous electrolyte solution, *Int. J. Thermophys.*, **21**, 1321-38.
88. Gutkowski K. I., Bianchi H. L. and Japas M. L. (2003) Critical Behavior of a ternaryx ionic system: A controversy, *J. Chem. Phys.*, **118**, 2808-14.
89. Wagner, M., Stanga, O., and Schröer, W. (2002) Tricriticality in the ternary system 3-methylpyridine + water + NaBr? Measurements of the viscosity, *Phys. Chem. Chem. Phys.*, **4**, 5300-06.
90. Wagner, M., Stanga, O., and Schröer, W. (2003) Tricriticality in the ternary system 3-methylpyridine + water + NaBr? The coexistence curves, *Phys. Chem. Chem. Phys.*, **5**, 1225-34.
91. Wagner, M., Stanga, O., and Schröer, W. (2003) in preparation
92. Luijten, E., Fisher, M.E., and Panagiotopoulos, A. Z. (2002) Universality Class of Criticality in the Restricted Primitive Model Electrolyte, *Phys. Rev. Lett.*, 88, 185701/1-4.
93. Luijten, E., Fisher, M.E., and Panagiotopoulos, A. Z. (2001) The heat capacity of the restricted primitive model electrolyte, *J. Chem. Phys.*, **114**, 5468-71.
94. Valleau , J. and Torrie, G. (1998) Heat capacity of the restricted primitive model near criticality, *J. Chem. Phys.*, **108**, 5169–72.
95. Valleau, J. and Torrie, G. (2002) Further remarks on the heat capacity of the restricted primitive model near criticality, *J. Chem. Phys.*, **117**, 3305–3309.
96. Lee, B. P. and Fisher, M. E. (1996) Density fluctuations in an electrolyte from generalized Debye-Hueckel theory., *Phys. Rev. Lett.*, **76**, 2906–9.
97. Bekiranov, S. and Fisher, M.E. (1999) Diverging correlation lengths in electrolytes: Exact results at low densities, *Phys. Rev. E*, **59**, 492-511
98. Kirkwood, J. G. (1934) Statistical mechanics of liquid solutions, *Chem. Rev.*, **19,** 275–307.
99. Lee, B. P. and Fisher, M. E. (1997) Charge oscillations in Debye-Hückel theory, *Europhys. Lett.*, **39**, 611–16.

100. See *e.g.*: Widom, B. and Rowlinson, J. S. (1979) Translation of: "The thermodynamic theory of capillarity under the hypothesis of a continuous variation of density" by J. D. van der Waals, *J. Stat. Phys.*, **20**, 197.
101. See *e.g.*: Landau, L. D. and Lifshitz, E. M.(1958) *Statistical Physics*, Pergamon, New York.
102. Rowlinson, J.S. and Widom B.(1982) *Molecular Theory of Capillarity* , Clarendon, Oxford.
103. Weiss, V.C. and Schröer, W (1997) On the Ginzburg temperature of ionic and dipolar fluids, *J. Chem. Phys.,* **106**, 1930-40.
104. Stell G. (1964) Cluster expansion for classical systems, in H.L. Frisch and J. L. Lebowitz (eds.), *The equilibrium theory of classical fluids* , Benjamin, New York, pp.171-267.
105. Schröer W. and Weiss V.C. (1998) Ginzburg criterium for the crossover behavior of model fluids, *J. Chem. Phys.*,**109**, 8504-8513.
106. Dickman R.(1999) , unpublished work, cited by Stell G., New results on some ionic –fluid problems in *New Approaches to Problems in Liquid State Theory*, Caccamo, C., Hansen, J.-P., and Stell, G. (eds.)., NATO ASI Series C, Kluwer, Dordrecht, pp 71-89.
107. Kobelev, V., Kolomeisky, A. B., and Fisher, M. E. (2002) Lattice models of ionic systems, *J. Chem. Phys.*, **116**, 7589–98.
108. Kobelev, V. and Kolomeisky, A.B., Anisotropic lattice models of electrolytes, *J. Chem. Phys.*, **117**, 8879–85.
109. A. B., Ciach, A. and Stell, G. (2002) Criticality and tricriticality in ionic systems, *Physica A*, **306**, 220–9.
110. Ciach, A. and Stell, G. (2001) Why the Ising and continuous-space models of ionic systems exhibit essentially different critical behavior, *J. Chem. Phys.*, **114**, 382–6.
111. Levine, Y and Fisher, M.E. (1994) Coulombic criticality in general dimensions, *Phys. Rev. Lett.*, **73**, 2716-19.

INDEX